VECTOR OPERATIONS

RECTANGULAR COORDINATES

$$\nabla u = \hat{\mathbf{x}}\frac{\partial u}{\partial x} + \hat{\mathbf{y}}\frac{\partial u}{\partial y} + \hat{\mathbf{z}}\frac{\partial u}{\partial z} \tag{1-37}$$

$$\nabla \cdot \mathbf{A} = \frac{\partial A_x}{\partial x} + \frac{\partial A_y}{\partial y} + \frac{\partial A_z}{\partial z} \tag{1-42}$$

$$\nabla \times \mathbf{A} = \hat{\mathbf{x}}\left(\frac{\partial A_z}{\partial y} - \frac{\partial A_y}{\partial z}\right) + \hat{\mathbf{y}}\left(\frac{\partial A_x}{\partial z} - \frac{\partial A_z}{\partial x}\right) + \hat{\mathbf{z}}\left(\frac{\partial A_y}{\partial x} - \frac{\partial A_x}{\partial y}\right) \tag{1-43}$$

$$\nabla^2 u = \frac{\partial^2 u}{\partial x^2} + \frac{\partial^2 u}{\partial y^2} + \frac{\partial^2 u}{\partial z^2} \tag{1-46}$$

CYLINDRICAL COORDINATES

$$\nabla u = \hat{\boldsymbol{\rho}}\frac{\partial u}{\partial \rho} + \hat{\boldsymbol{\varphi}}\frac{1}{\rho}\frac{\partial u}{\partial \varphi} + \hat{\mathbf{z}}\frac{\partial u}{\partial z} \tag{1-85}$$

$$\nabla \cdot \mathbf{A} = \frac{1}{\rho}\frac{\partial}{\partial \rho}(\rho A_\rho) + \frac{1}{\rho}\frac{\partial A_\varphi}{\partial \varphi} + \frac{\partial A_z}{\partial z} \tag{1-87}$$

$$\nabla \times \mathbf{A} = \hat{\boldsymbol{\rho}}\left(\frac{1}{\rho}\frac{\partial A_z}{\partial \varphi} - \frac{\partial A_\varphi}{\partial z}\right) + \hat{\boldsymbol{\varphi}}\left(\frac{\partial A_\rho}{\partial z} - \frac{\partial A_z}{\partial \rho}\right) + \hat{\mathbf{z}}\left[\frac{1}{\rho}\frac{\partial}{\partial \rho}(\rho A_\varphi) - \frac{1}{\rho}\frac{\partial A_\rho}{\partial \varphi}\right] \tag{1-88}$$

$$\nabla^2 u = \frac{1}{\rho}\frac{\partial}{\partial \rho}\left(\rho\frac{\partial u}{\partial \rho}\right) + \frac{1}{\rho^2}\frac{\partial^2 u}{\partial \varphi^2} + \frac{\partial^2 u}{\partial z^2} \tag{1-89}$$

SPHERICAL COORDINATES

$$\nabla u = \hat{\mathbf{r}}\frac{\partial u}{\partial r} + \hat{\boldsymbol{\theta}}\frac{1}{r}\frac{\partial u}{\partial \theta} + \hat{\boldsymbol{\varphi}}\frac{1}{r \sin \theta}\frac{\partial u}{\partial \varphi} \tag{1-101}$$

$$\nabla \cdot \mathbf{A} = \frac{1}{r^2}\frac{\partial}{\partial r}(r^2 A_r) + \frac{1}{r \sin \theta}\frac{\partial}{\partial \theta}(\sin \theta A_\theta) + \frac{1}{r \sin \theta}\frac{\partial A_\varphi}{\partial \varphi} \tag{1-103}$$

$$\nabla \times \mathbf{A} = \frac{\hat{\mathbf{r}}}{r \sin \theta}\left[\frac{\partial}{\partial \theta}(\sin \theta A_\varphi) - \frac{\partial A_\theta}{\partial \varphi}\right] + \frac{\hat{\boldsymbol{\theta}}}{r}\left[\frac{1}{\sin \theta}\frac{\partial A_r}{\partial \varphi} - \frac{\partial}{\partial r}(r A_\varphi)\right]$$
$$+ \frac{\hat{\boldsymbol{\varphi}}}{r}\left[\frac{\partial}{\partial r}(r A_\theta) - \frac{\partial A_r}{\partial \theta}\right] \tag{1-104}$$

$$\nabla^2 u = \frac{1}{r^2}\frac{\partial}{\partial r}\left(r^2\frac{\partial u}{\partial r}\right) + \frac{1}{r^2 \sin \theta}\frac{\partial}{\partial \theta}\left(\sin \theta\frac{\partial u}{\partial \theta}\right) + \frac{1}{r^2 \sin^2 \theta}\frac{\partial^2 u}{\partial \varphi^2} \tag{1-105}$$

ELECTROMAGNETIC FIELDS

2ND EDITION

ELECTROMAGNETIC FIELDS

ROALD K. WANGSNESS

PROFESSOR OF PHYSICS
UNIVERSITY OF ARIZONA

JOHN WILEY & SONS

NEW YORK   CHICHESTER   BRISBANE   TORONTO   SINGAPORE

Library of Congress Cataloging in Publication Data:

Wangsness, Roald K.
  Electromagnetic fields.

  Includes indexes.
  1. Electromagnetic fields.   I. Title.
QC665.E4W36   1986   537   85-24642
ISBN 0-471-81186-6

Printed in the United States of America

Printed and bound by the Hamilton Printing Company.

10

FOR CLEO

PREFACE

This is a textbook for use in the usual year-long course in electromagnetism at an intermediate level given for advanced undergraduates. I have done my best to make it as student oriented as possible by writing it in a systematic and straightforward way without any sleight of hand, and minimum use of "It can be shown that... ." I have also tried to make clear the motivations for each step in a derivation or each new concept as it is introduced. At appropriate points I have pointed out the source of many of the simple mistakes commonly made by students and have suggested how they may be prevented. There is extensive cross referencing in the text so that there should be no doubt as to the detailed source of any specific result or its relation to the rest of the subject; this will also make the book much more useful as a reference source well after the course has been completed.

In preparing this revision, I have concentrated on adhering to these goals while trying to improve the clarity and organization of the material, since my aim is to produce a useful text with an adequate, rather than encyclopedic, coverage at a reasonable mathematical level. Consequently, most of the changes are scattered throughout the text. For example, I have improved the discussion of electric and magnetic energies in the presence of matter in Sections 10-8, 10-9, and 20-6. In addition, the beginning treatment of radiation has been simplified somewhat in Sections 28-1 and 28-2, and the proof about the electric field in a cavity in a conductor has been corrected in Section 6-1. The principal and most noticeable change has been the addition of a completely new Chapter 27 on Circuits and Transmission Lines. I hope that this will turn out to be useful and satisfactory to those who have expressed a desire to have this material included.

The emphasis continues to be on the properties and sources of the field vectors, and I hope that I have succeeded in making clear the shifts in concepts and points of view that are involved in the change from action at a distance to fields. The overall treatment is generally that of a macroscopic and empirical description of phenomena, although the microscopic point of view is presented in the discussion of conductivity in Sections 12-5 and 24-8. However, Appendix B briefly surveys the microscopic origins of electromagnetic properties, and it is written and organized so that, if desired, it can be taken up section by section at an appropriate intermediate point. Thus Section B-1 could be discussed anytime after Section 10-7, and most of Section B-2 can follow after Section 20-5 while the last part on ferromagnetism can follow Section 20-7; finally, Section B-3 could be covered after Section 24-8 has been mastered and the student has worked Exercise 24-28. Similarly, even more flexibility is possible since separate sections of Appendix A that deals with the motion of charged particles can be studied anytime after the corresponding force term involving $\mathbf{E}$ and/or $\mathbf{B}$ has been obtained.

SI units are used throughout; in practice, this means we use MKSA units. It is virtually certain, however, that at some time a student will encounter material in Gaussian units and will need some guidance on what to do about it. This is the purpose of Chapter 23 in which other unit systems are discussed, but only after the general theory as given by Maxwell's equations has been systematically described. I have written this chapter primarily in terms of the purely practical aspects of how to recognize an equation written in other unit systems, how to put it in more familiar form if desired, and exactly what numbers one should put into an equation in Gaussian units in order to get a correct answer.

In Chapter 9, the boundary conditions satisfied by an *arbitrary* vector at a surface of discontinuity in properties are obtained in the general form involving its divergence and curl. Not only does this help the student by showing the importance of knowing these particular source equations but it simplifies later discussion because, as each new vector is defined, its boundary conditions can be found at once without having, in effect, to rederive them each time.

I have added several new clarifying examples and exercises so that now there are over 145 worked-out examples in the text. Virtually all of the standard ones are included, many of them presented in more detail than is usually found, with particular emphasis on the crucial stage of setting up the problem, since this is often what causes students so much difficulty. By now, there are a total of 587 exercises included. Some are numerical to give an idea of typical orders of magnitude, some are similar to examples of the text, many refer to completely different situations, and some involve extensions of the theory. Many of these exercises will be found to be suitable for use as additional examples for classroom analysis. Answers to odd-numbered exercises are provided except, of course, for those in which the answer is included in the statement of the problem.

I have benefited over the years from discussions with and the questions of many students and my colleagues; I am grateful for their contributions to the final character of this book. In particular, I thank my colleagues C. Y. Fan, K. C. Hsieh, E. W. Jenkins, and J. E. Treat for their useful and helpful comments. I am also grateful for the suggestions and remarks given to me by H. Asmat, R. C. Henry, A. T. Mense, K. Park, J. R. Ravella, and several reviewers.

Tucson, Arizona   ***Roald K. Wangsness***
January 1986

CONTENTS

ELECTROMAGNETIC FIELDS

INTRODUCTION

...Faraday, in his mind's eye, saw lines of force traversing
all space where the mathematicians saw centres of force
attracting at a distance: Faraday saw a medium where they
saw nothing but distance: Faraday sought the seat of the
phenomena in real actions going on in the medium, they
were satisfied that they had found it in a power of action
at a distance impressed on the electric fluids.

— J. C. Maxwell,
A Treatise on Electricity and Magnetism

It has been more than 100 years since Maxwell wrote the above in the preface to his now-famous book. His aim was to put the field concepts, which Faraday had been so instrumental in developing, into mathematical forms that would be convenient to use and would emphasize the fields as basic to a coherent description of electromagnetic effects. At that time, it had been only slightly more than 50 years since Oersted and Ampère had shown the relation between electricity and magnetism—subjects that had been studied and developed completely separately over a long period. The emphasis had been primarily on the forces exerted between electric charges and between electric currents and the idea of shifting to electric and magnetic fields as the primary features had little acceptance and was often, in fact, viewed with outright hostility.

As the title of this book indicates, times have changed, and our main interest here is the study of the nature, properties, and origins of electromagnetic fields, that is, of electric and magnetic vector quantities that are defined as functions of time and of position in space. Forces, and associated concepts such as energy, have not disappeared from the subject, of course, and it is desirable to begin with forces and to define the field vectors in terms of them. Nevertheless, our principal aim is to express our descriptions of phenomena in terms of fields in as complete a manner as we possibly can. This emphasis on fields has proved to be extremely rewarding and it is difficult to imagine how electromagnetic theory could have been developed to its present state without it.

This book contains more material than is normally covered in the usual one-year course; all of it, however, is of interest and value to a serious student of physics.

The points of view of all authors are generally not the same, and no book discusses every detail of a given subject. Here is a short list of relatively recent books on electromagnetism that are written at roughly the same level as this one.

D. M. Cook, *The Theory of the Electromagnetic Field*, Prentice-Hall, Englewood Cliffs, N.J., 1975.

P. Lorrain and D. R. Corson, *Electromagnetic Fields and Waves*, Second Edition, Freeman, San Francisco, 1970.

M. H. Nayfeh and M. K. Brussel, *Electricity and Magnetism*, Wiley, New York, 1985.

J. R. Reitz, F. J. Milford, and R. W. Christy, *Foundations of Electromagnetic Theory*, Third Edition, Addison-Wesley, Reading, Mass., 1979.

A. Shadowitz, *The Electromagnetic Field*, McGraw-Hill, New York, 1975.

The following books discuss electromagnetism at a more advanced level:

J. D. Jackson, *Classical Electrodynamics*, Second Edition, Wiley, New York, 1975.

E. J. Konopinski, *Electromagnetic Fields and Relativistic Particles*, McGraw-Hill, New York, 1981.

W. K. H. Panofsky and M. Phillips, *Classical Electricity and Magnetism*, Second Edition, Addison-Wesley, Reading, Mass., 1962.

A. M. Portis, *Electromagnetic Fields: Sources and Media*, Wiley, New York, 1978.

J. A. Stratton, *Electromagnetic Theory*, McGraw-Hill, New York, 1941.

(Finally, a note on notation: in this book, the symbols $=$, $\simeq$, $\approx$, $\sim$, $\neq$ always mean, respectively, equal to, approximately equal to, of the order of magnitude of, proportional to, and different from.)

1

VECTORS

In the study of electricity and magnetism, we are constantly dealing with quantities that need to be described in terms of their directions as well as their magnitudes. Such quantities are called vectors and it is well to consider their properties in general before we meet specific examples. Using the notation and terminology that has been developed for this purpose enables us to state our results more compactly and to understand their basic physical significance more easily.

1-1   DEFINITION OF A VECTOR

The properties of the *displacement of a point* provide us the essentials required for our definition. If we start at some point P_1 and move in some arbitrary way to another point P_2, we see from Figure 1-1 that the *net* effect of the motion is the same as if the point were moved directly along the straight line D from P_1 to P_2 as indicated by the direction of the arrow. This line D is called the displacement and is characterized by both a magnitude (its length) and a direction (from P_1 to P_2). If we now further displace our point along E from P_2 to still another point P_3, we see from Figure 1-2 that the new net effect is the same as if the point had been given the single displacement along F from P_1 to P_3. Accordingly, we can speak of F as the resultant, or sum, of the successive displacements D and E, so that Figure 1-2 shows the fundamental way in which displacements are combined or added to obtain their resultant.

A *vector* is a generalization of these considerations in that it is defined as any quantity that has the same mathematical properties as the displacement of a point. Thus we see that a vector has a magnitude; it has a direction; and the addition of two vectors of the same intrinsic nature follows the basic rule illustrated in Figure 1-2. Because of the first two properties, we can represent a vector by a directed line such as those already used for displacements. A vector is generally printed in boldface type, thus, **A**; its magnitude will be represented by $|\mathbf{A}|$ or by A.

A *scalar* is a quantity that has magnitude only. For example, the mass of a body is a scalar, whereas its weight, which is the gravitational force acting on the body, is a vector.

Because of the nature of a vector as a directed quantity, it follows that a parallel displacement of a vector does not alter it, or, in other words, two vectors are equal if they have the same magnitude and direction. This is illustrated in Figure 1-3 where we see that $\mathbf{A} = \mathbf{A}'$. Now we can investigate what mathematical operations we can perform with and on vectors.

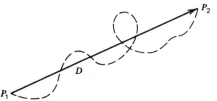

Figure 1-1.   *D* is the displacement of the point from P_1 to P_2.

3

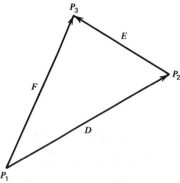

Figure 1-2. *F* is the resultant of the displacements *D* and *E*.

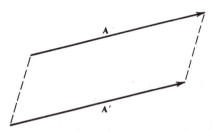

Figure 1-3. These two vectors are equal.

1-2 ADDITION

According to our basic rule we find that, if we take **A** and add **B**, we obtain the sum **C** shown as a solid line in Figure 1-4. We also see that, if we take **B** and then add **A**, we get the same vector **C**. Therefore addition of vectors has the commutative property that

$$\mathbf{C} = \mathbf{A} + \mathbf{B} = \mathbf{B} + \mathbf{A} \tag{1-1}$$

By proceeding in the same manner, one can establish the associative property of vector addition:

$$\mathbf{D} = (\mathbf{A} + \mathbf{B}) + \mathbf{C} = \mathbf{A} + (\mathbf{B} + \mathbf{C}) = (\mathbf{A} + \mathbf{C}) + \mathbf{B} \tag{1-2}$$

and so on.

If we reverse a displacement such as *D* in Figure 1-1 by retracing it in the opposite direction, the net effect is then no displacement; hence it is appropriate to define the negative of a vector as a vector of the same magnitude but reversed in direction, for then we should obtain $\mathbf{A} + (-\mathbf{A}) = 0$, as we would want. Then we can easily subtract a vector by adding its negative:

$$\mathbf{A} - \mathbf{B} = \mathbf{A} + (-\mathbf{B}) \tag{1-3}$$

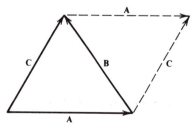

Figure 1-4. The sum of two vectors does not depend on the order in which they are added.

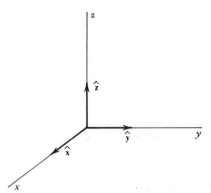

Figure 1-5. â is a unit vector in the direction of A.

Figure 1-6. Unit vectors for rectangular coordinates.

The product of a scalar s and a vector, which we write as either $s\mathbf{A}$ or $\mathbf{A}s$, is then merely the sum of s vectors $\mathbf{A}$, or is a vector with a magnitude equal to $|s|$ times the magnitude of $\mathbf{A}$, and is in the same direction as $\mathbf{A}$ if s is positive, and in the opposite direction to $\mathbf{A}$ if s is negative.

1-3 UNIT VECTORS

A *unit vector* is defined as a vector of unit magnitude and will be written with a circumflex above it, thus, $\hat{\mathbf{e}}$; since unit vectors are always taken to be dimensionless we will have $|\hat{\mathbf{e}}| = 1$. If, for example, a unit vector $\hat{\mathbf{a}}$ is chosen to have the direction of $\mathbf{A}$, then we can write

$$\mathbf{A} = A\hat{\mathbf{a}} \qquad \text{and} \qquad \hat{\mathbf{a}} = \frac{\mathbf{A}}{A} \tag{1-4}$$

This point is illustrated in Figure 1-5.

A particularly convenient set of unit vectors can be associated with a rectangular coordinate system. They are written $\hat{\mathbf{x}}, \hat{\mathbf{y}}, \hat{\mathbf{z}}$ and are defined to be in the directions of the x, y, and z axes respectively, as shown in Figure 1-6. In other words, each is in the direction of increasing value of the corresponding rectangular coordinate. We also see that any one of this set is perpendicular to each of the other two.

As we will see, it is often convenient and advantageous to define other unit vectors.

1-4 COMPONENTS

In order to proceed further, it is convenient to refer our vectors to particular coordinate systems. From Figure 1-7, we see that we can write a vector $\mathbf{A}$ as the sum of three properly chosen vectors, each of which is parallel to one of the axes of a rectangular coordinate system; that is, $\mathbf{A} = \mathbf{A}_x + \mathbf{A}_y + \mathbf{A}_z$. It is more useful, however, to write each of these terms as the product of a scalar and the unit vectors of Figure 1-6. Thus,

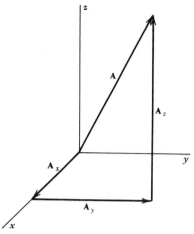

Figure 1-7. **A is the sum of the rectangular vector components.**

we write $\mathbf{A}_x = A_x\hat{\mathbf{x}}$, and so on, and the above expression becomes

$$\mathbf{A} = A_x\hat{\mathbf{x}} + A_y\hat{\mathbf{y}} + A_z\hat{\mathbf{z}} \tag{1-5}$$

The three scalars A_x, A_y, A_z are called the *components* of $\mathbf{A}$; hence we see that a vector can be specified by three numbers. The components can be positive or negative; for example, if A_x were negative, then the vector $\mathbf{A}_x$ of Figure 1-7 would have a direction in the sense of decreasing values of x.

From Figure 1-7, it is seen that the magnitude of a vector can be expressed in terms of its components as

$$A = |\mathbf{A}| = \left(A_x^2 + A_y^2 + A_z^2\right)^{1/2} \tag{1-6}$$

In Figure 1-8, we illustrate the fact that $\mathbf{A}$ makes specific angles with respect to each of the axes; these angles α, β, γ are called the *direction angles* of $\mathbf{A}$ and are measured from the positive directions of their respective axes. Figure 1-9 shows the plane containing both $\mathbf{A}$ and $\hat{\mathbf{x}}$ and we see that A_x is given by $A_x = A\cos\alpha$. Combining this with (1-6), we get

$$l_x = \cos\alpha = \frac{A_x}{A} = \frac{A_x}{\left(A_x^2 + A_y^2 + A_z^2\right)^{1/2}} \tag{1-7}$$

where l_x is called a *direction cosine*. Similar expressions hold for the other two direction angles β and γ and their associated direction cosines l_y and l_z, so we see from (1-6) and (1-7) that, if we know the rectangular components of a vector, we can calculate its magnitude and direction.

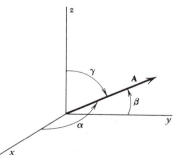

Figure 1-8. **Definition of direction angles.**

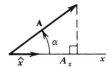

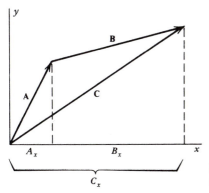

Figure 1-9. A_x is the x component of A.

Figure 1-10.   A component of a sum equals the sum of the corresponding components.

If we now combine (1-4), (1-5), and (1-7), we find that the unit vector **â** can also be written as

$$\hat{\mathbf{a}} = l_x\hat{\mathbf{x}} + l_y\hat{\mathbf{y}} + l_z\hat{\mathbf{z}} \qquad (1\text{-}8)$$

so that the components of a unit vector in a given direction are simply the direction cosines associated with the direction. If we now apply the general result (1-6) to the specific vector **â**, we get the important relation involving direction cosines

$$l_x^2 + l_y^2 + l_z^2 = 1 \qquad (1\text{-}9)$$

which can also be obtained from (1-7) and its analogues.

The addition of vectors illustrated in Figure 1-4 is easily expressed in terms of the rectangular components. From Figure 1-10, we see that a component of the sum **C = A + B** is given by the sum of the corresponding components, that is,

$$C_x = A_x + B_x \qquad C_y = A_y + B_y \qquad C_z = A_z + B_z \qquad (1\text{-}10)$$

1-5   THE POSITION VECTOR

We now consider a simple specific example of a vector. As shown in Figure 1-11, the location of a particular point P in space can be specified by the vector **r** drawn from the origin of a suitably and conveniently chosen coordinate system; this vector **r** is called the *position vector* of the point. In terms of the rectangular coordinate system of Figure 1-6, the components of **r** are just the rectangular coordinates (x, y, z) of the point; thus we have

$$\mathbf{r} = x\hat{\mathbf{x}} + y\hat{\mathbf{y}} + z\hat{\mathbf{z}} \qquad (1\text{-}11)$$

Similarly, another point P' with coordinates (x', y', z') will be located by its position vector $\mathbf{r}' = x'\hat{\mathbf{x}} + y'\hat{\mathbf{y}} + z'\hat{\mathbf{z}}$ as shown in Figure 1-12. Now that we have located the two points individually, we can describe the position of one with respect to the other by drawing a vector from P' to P; this vector **R** is called the *relative position vector* of P

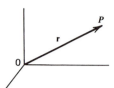

Figure 1-11.   r is the position vector of the point P.

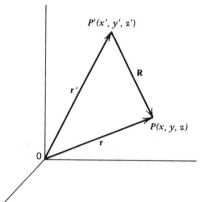

Figure 1-12.   R is the relative position vector of P with respect to P'.

with respect to P'. We see from Figure 1-12 that $\mathbf{r}' + \mathbf{R} = \mathbf{r}$ so that

$$\mathbf{R} = \mathbf{r} - \mathbf{r}' \qquad (1\text{-}12)$$

Using (1-10) and (1-11), we can write $\mathbf{R}$ in component form as

$$\mathbf{R} = (x - x')\hat{\mathbf{x}} + (y - y')\hat{\mathbf{y}} + (z - z')\hat{\mathbf{z}} \qquad (1\text{-}13)$$

and therefore

$$R = \left[(x - x')^2 + (y - y')^2 + (z - z')^2\right]^{1/2} \qquad (1\text{-}14)$$

because of (1-6). We will have frequent occasion to use these results. We note that the relative position of P' with respect to P is given by a vector $\mathbf{R}'$ drawn from P to P' and, in fact, $\mathbf{R}' = -\mathbf{R}$.

Although we have not specified our coordinate system other than to say that it was arbitrarily and conveniently chosen, once this choice has been made in a particular case, the coordinate system is said to be "fixed in space" and its unit vectors $\hat{\mathbf{x}}$, $\hat{\mathbf{y}}$, and $\hat{\mathbf{z}}$ are to be taken as constant in direction as well as magnitude. In other words, the fixed coordinate system that we use is simply one of the inertial reference frames familiar to us from classical mechanics.

We now turn to multiplication of vectors; two types are defined.

1-6   SCALAR PRODUCT

We define the *scalar product* of two vectors as the scalar equal to the product of the magnitudes of the vectors and the cosine of the angle between them, or

$$\mathbf{A} \cdot \mathbf{B} = AB \cos \Psi \qquad (1\text{-}15)$$

Because of the notation the scalar product is also called the *dot product*.

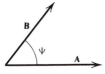

Figure 1-13. **The angle involved in the scalar product.**

We see from Figure 1-13 that we can get a simple interpretation of the scalar product: $(B \cos \Psi)A$ = component of **B** along the direction of **A** times the magnitude of **A** = $(A \cos \Psi)B$ = component of **A** along **B** times the magnitude of **B**.

It is clear from (1-15) that the order of terms does not change the scalar product, that is,

$$\mathbf{A} \cdot \mathbf{B} = \mathbf{B} \cdot \mathbf{A} \qquad (1\text{-}16)$$

and that if two vectors are perpendicular then $\mathbf{A} \cdot \mathbf{B} = 0$ and conversely. Furthermore, the square of a vector can be interpreted as the vector dotted with itself; the result is the square of its magnitude and we can write

$$\mathbf{A}^2 = \mathbf{A} \cdot \mathbf{A} = A^2 \qquad (1\text{-}17)$$

If we knew only the rectangular components of **A** and **B**, it would be inconvenient to calculate $\mathbf{A} \cdot \mathbf{B}$ from (1-15) since this would necessitate finding the angle between **A** and **B**. Fortunately, it is possible to express $\mathbf{A} \cdot \mathbf{B}$ directly in terms of the rectangular components. Since the angle between each pair of unit vectors defined in Figure 1-6 is 90°, we easily find from (1-15) that

$$\hat{\mathbf{x}} \cdot \hat{\mathbf{y}} = \hat{\mathbf{y}} \cdot \hat{\mathbf{z}} = \hat{\mathbf{z}} \cdot \hat{\mathbf{x}} = 0 \qquad (1\text{-}18)$$

and from (1-17) that

$$\hat{\mathbf{x}} \cdot \hat{\mathbf{x}} = \hat{\mathbf{y}} \cdot \hat{\mathbf{y}} = \hat{\mathbf{z}} \cdot \hat{\mathbf{z}} = 1 \qquad (1\text{-}19)$$

Writing **A** and **B** each in the form (1-5), we can multiply them together term by term to get

$$\mathbf{A} \cdot \mathbf{B} = \left(A_x \hat{\mathbf{x}} + A_y \hat{\mathbf{y}} + A_z \hat{\mathbf{z}} \right) \cdot \left(B_x \hat{\mathbf{x}} + B_y \hat{\mathbf{y}} + B_z \hat{\mathbf{z}} \right)$$
$$= A_x B_x \hat{\mathbf{x}} \cdot \hat{\mathbf{x}} + A_x B_y \hat{\mathbf{x}} \cdot \hat{\mathbf{y}} + A_x B_z \hat{\mathbf{x}} \cdot \hat{\mathbf{z}} + \ldots$$

and, after we use (1-18) and (1-19) to simplify the resulting nine terms, we find that

$$\mathbf{A} \cdot \mathbf{B} = A_x B_x + A_y B_y + A_z B_z \qquad (1\text{-}20)$$

Suppose now that $\hat{\mathbf{e}}$ is a unit vector in some specific direction. If we let A_e be the component of **A** in this direction, we see from (1-15) that it can be found from

$$A_e = \mathbf{A} \cdot \hat{\mathbf{e}} \qquad (1\text{-}21)$$

1-7 VECTOR PRODUCT

This is also called the *cross product* and is written $\mathbf{A} \times \mathbf{B}$. It is a vector perpendicular to *both* **A** and **B** and its magnitude is defined as

$$|\mathbf{A} \times \mathbf{B}| = AB \sin \Psi \qquad (1\text{-}22)$$

Its direction is given by the following right-hand rule: if the fingers of the right hand are curled in the sense necessary to rotate **A** through the smaller angle into coincidence with **B**, the thumb points in the direction of $\mathbf{A} \times \mathbf{B}$. This rule is illustrated in Figure 1-14.

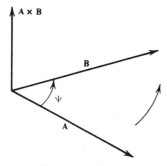

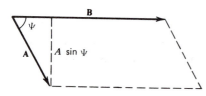

Figure 1-14. Definition of the direction of the cross product.

Figure 1-15. Interpretation of the magnitude of a cross product as an area.

If we look at the plane containing **A** and **B** shown in Figure 1-15, we can get a simple interpretation of the cross product. We see from the figure and (1-22) that the magnitude of the cross product is equal to the area of the parallelogram with **A** and **B** as sides.

From the definition of the direction of a cross product shown in Figure 1-14, it is evident that the order of terms is important, since it is seen that

$$\mathbf{B} \times \mathbf{A} = -(\mathbf{A} \times \mathbf{B}) \tag{1-23}$$

If **A** and **B** are parallel, then it follows from (1-22) that $\mathbf{A} \times \mathbf{B} = 0$, and conversely. In particular,

$$\mathbf{A} \times \mathbf{A} = 0 \tag{1-24}$$

For the unit vectors along the axes shown in Figure 1-6, if we use (1-22), the right-hand rule, the facts that they are mutually perpendicular, and that the cross product is perpendicular to both vectors, we find that

$$\hat{\mathbf{x}} \times \hat{\mathbf{y}} = \hat{\mathbf{z}} \qquad \hat{\mathbf{y}} \times \hat{\mathbf{z}} = \hat{\mathbf{x}} \qquad \hat{\mathbf{z}} \times \hat{\mathbf{x}} = \hat{\mathbf{y}} \tag{1-25}$$

and, from (1-24), that

$$\hat{\mathbf{x}} \times \hat{\mathbf{x}} = \hat{\mathbf{y}} \times \hat{\mathbf{y}} = \hat{\mathbf{z}} \times \hat{\mathbf{z}} = 0 \tag{1-26}$$

The vector product can also be conveniently written in terms of the rectangular components. Proceeding similarly to what we did in obtaining (1-20), we write **A** and **B** in the form (1-5), multiply them together term by term, and use (1-23), (1-25), and (1-26) to simplify the results. We find that

$$\mathbf{A} \times \mathbf{B} = \left(A_y B_z - A_z B_y \right)\hat{\mathbf{x}} + \left(A_z B_x - A_x B_z \right)\hat{\mathbf{y}} + \left(A_x B_y - A_y B_x \right)\hat{\mathbf{z}} \tag{1-27}$$

This can be written as an easily remembered determinant

$$\mathbf{A} \times \mathbf{B} = \begin{vmatrix} \hat{\mathbf{x}} & \hat{\mathbf{y}} & \hat{\mathbf{z}} \\ A_x & A_y & A_z \\ B_x & B_y & B_z \end{vmatrix} \tag{1-28}$$

It will be left as exercises to verify that

$$\mathbf{A} \cdot (\mathbf{B} \times \mathbf{C}) = (\mathbf{A} \times \mathbf{B}) \cdot \mathbf{C} = \begin{vmatrix} A_x & A_y & A_z \\ B_x & B_y & B_z \\ C_x & C_y & C_z \end{vmatrix} \tag{1-29}$$

and that

$$\mathbf{A} \times (\mathbf{B} \times \mathbf{C}) = \mathbf{B}(\mathbf{A} \cdot \mathbf{C}) - \mathbf{C}(\mathbf{A} \cdot \mathbf{B}) \tag{1-30}$$

In (1-29) we see that the dot and the cross can be interchanged without affecting the value of this triple scalar product; hence the parentheses are not really needed. In the triple vector product (1-30), however, the parentheses are important because $(\mathbf{A} \times \mathbf{B}) \times \mathbf{C} = -\mathbf{C} \times (\mathbf{A} \times \mathbf{B})$ from (1-23).

Division of vectors is not defined.

1-8 DIFFERENTIATION WITH RESPECT TO A SCALAR

Suppose that **A** is a continuous function of some scalar variable σ, so that we can write $\mathbf{A} = \mathbf{A}(\sigma)$. This is equivalent to the three scalar equations: $A_x = A_x(\sigma)$, $A_y = A_y(\sigma)$, $A_z = A_z(\sigma)$. If σ is changed to $\sigma + \Delta\sigma$, **A** can generally change in both magnitude and direction as shown in Figure 1-16. The change in **A** is $\Delta\mathbf{A} = \mathbf{A}(\sigma + \Delta\sigma) - \mathbf{A}(\sigma)$. We can then define the derivative of the vector **A** with respect to the scalar σ as follows:

$$\frac{d\mathbf{A}}{d\sigma} = \lim_{\Delta\sigma \to 0} \frac{\Delta\mathbf{A}}{\Delta\sigma} = \lim_{\Delta\sigma \to 0} \frac{\mathbf{A}(\sigma + \Delta\sigma) - \mathbf{A}(\sigma)}{\Delta\sigma} \tag{1-31}$$

This process has yielded another vector from a vector. Familiar examples of (1-31) are the velocity and acceleration of a particle that are successive derivatives of the position vector with respect to the time.

If **A** is written in terms of its rectangular components, then, since these unit vectors are constant, we see from (1-31) that the components of the derivative are the derivatives of the respective components, and we have

$$\frac{d\mathbf{A}}{d\sigma} = \frac{dA_x}{d\sigma}\hat{\mathbf{x}} + \frac{dA_y}{d\sigma}\hat{\mathbf{y}} + \frac{dA_z}{d\sigma}\hat{\mathbf{z}} \tag{1-32}$$

Once we have defined the derivative of **A**, we can proceed to its differential $d\mathbf{A}$, which is to represent an infinitesimal change in **A**. This is obtained, in effect, by multiplying (1-32) by $d\sigma$ to give

$$d\mathbf{A} = dA_x\hat{\mathbf{x}} + dA_y\hat{\mathbf{y}} + dA_z\hat{\mathbf{z}} \tag{1-33}$$

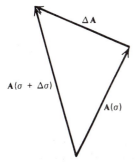

Figure 1-16. $\Delta\mathbf{A}$ is the change in the vector corresponding to the change $\Delta\sigma$ of the scalar.

Applying this to the position vector of (1-11), we get

$$d\mathbf{r} = dx\,\hat{\mathbf{x}} + dy\,\hat{\mathbf{y}} + dz\,\hat{\mathbf{z}} \tag{1-34}$$

1-9 GRADIENT OF A SCALAR

Suppose we have a scalar quantity u that is a function of position so that we can write $u = u(x, y, z)$. Such a situation is called a *scalar field*. An example would be the temperature at each point in a room. At some other point, which is displaced by $d\mathbf{s}$ from the first, the value of the scalar will have changed to $u + du$ (Figure 1-17). In fact,

$$du = \frac{\partial u}{\partial x}\,dx + \frac{\partial u}{\partial y}\,dy + \frac{\partial u}{\partial z}\,dz \tag{1-35}$$

where we should remember that the derivatives are evaluated at the original point, that is, $\partial u/\partial x = (\partial u/\partial x)_P$, and so on. Although we have written the displacement as $d\mathbf{s}$, it is clearly the change $d\mathbf{r}$ in the position vector of the point in this case, so that

$$d\mathbf{s} = dx\,\hat{\mathbf{x}} + dy\,\hat{\mathbf{y}} + dz\,\hat{\mathbf{z}} \tag{1-36}$$

according to (1-34). Comparing (1-35) and (1-36) with (1-20), we see that we can also write du as the scalar product of $d\mathbf{s}$ and the vector

$$\nabla u = \hat{\mathbf{x}}\frac{\partial u}{\partial x} + \hat{\mathbf{y}}\frac{\partial u}{\partial y} + \hat{\mathbf{z}}\frac{\partial u}{\partial z} \tag{1-37}$$

so that

$$du = d\mathbf{s} \cdot \nabla u \tag{1-38}$$

The vector that we have obtained in this way and that is written in terms of its rectangular components in (1-37) is called the *gradient* of u and is also often written grad u. We can regard (1-38) as the general definition of ∇u since it is written in a form that is independent of a particular coordinate system. In other words, the gradient is that quantity that will give the change in the scalar when it is dotted with the displacement.

In order to understand the meaning of the gradient, let us consider Figure 1-18, in which is indicated a series of surfaces each of which is made up of those points for which u has the same value; in other words, these are surfaces of constant u, with corresponding values $u_1, u_2, u_3 \ldots$. Now a displacement such as $d\mathbf{s}_1$, which takes us to

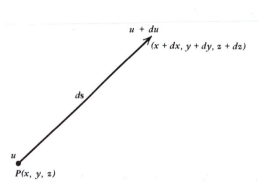

Figure 1-17.   Change in a scalar function of position u as a result of the displacement $d\mathbf{s}$.

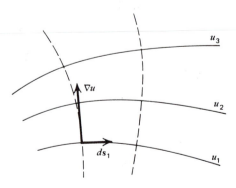

Figure 1-18.   Surfaces of constant u. The gradient is perpendicular to such a surface.

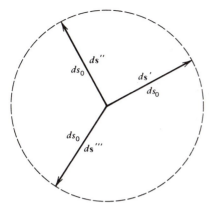

Figure 1-19. Displacements of constant magnitude but different directions.

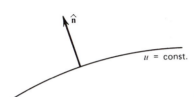

Figure 1-20. Definition of the normal vector.

a point on the same surface, does not take one to a point where u has changed. Therefore, $du_1 = d\mathbf{s}_1 \cdot \nabla u = 0$. Comparing this with (1-15), we see that ∇u and $d\mathbf{s}_1$ are perpendicular; thus ∇u is perpendicular to a surface of constant u, as is also shown in Figure 1-18.

Now let us consider displacements of constant magnitude ds_0 from a given point but of varying directions such as $d\mathbf{s}'$, $d\mathbf{s}''$, and $d\mathbf{s}'''$ of Figure 1-19. We see from (1-38) and (1-15) that the change in u resulting from any of these displacements is given by $du = ds_0 |\nabla u| \cos \Psi$ and is different for each displacement only because of the different value of the angle Ψ between it and the fixed direction of ∇u. We now see that du will be a maximum when $\cos \Psi = 1$ or $\Psi = 0$ so that ∇u and the corresponding displacement are parallel. In other words, the direction of the gradient is also the direction in which the scalar has its maximum rate of change, and its magnitude is just that maximum rate of change of u.

A unit vector $\hat{\mathbf{n}}$ that is perpendicular to a given surface at a given point is called the *normal vector* and is illustrated for a surface u = const. in Figure 1-20. But we have just seen that ∇u is also perpendicular to the surface so that $\hat{\mathbf{n}}$ and ∇u are parallel. Thus, by using (1-4), we can write

$$\hat{\mathbf{n}} = \frac{\nabla u}{|\nabla u|} \tag{1-39}$$

In this case, $\hat{\mathbf{n}}$ will also give the direction in which u is increasing.

■ **Example**

Suppose we consider a two-dimensional case where $u = y^2 - kx$ where k = const. If we write $y^2 = kx + u$ we see that the surfaces of constant u are curves in the xy plane. In fact, they are parabolas as illustrated for $k = 1$ and some specific values of u in Figure 1-21. Substituting this expression for u (with $k = 1$) into (1-37), we get

$$\nabla u = -\hat{\mathbf{x}} + 2y\hat{\mathbf{y}}$$

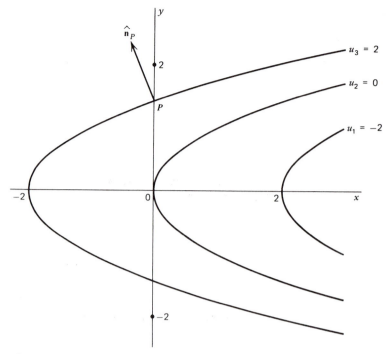

Figure 1-21. Surfaces of constant u for the example of Section 1-9.

so that, from (1-6), $|\nabla u| = (1 + 4y^2)^{1/2}$ and therefore

$$\hat{\mathbf{n}} = \frac{-\hat{\mathbf{x}} + 2y\hat{\mathbf{y}}}{\left(1 + 4y^2\right)^{1/2}} \tag{1-40}$$

according to (1-39). In using this result, we of course have to use the value of y corresponding to a point on the curve for the given u. For example, let us evaluate $\hat{\mathbf{n}}$ for the point P of Figure 1-21 where the parabola corresponding to $u_3 = 2$ crosses the positive y axis. Here $x_P = 0$, and $y_P^2 = x_P + u_3 = 2$ so that $y_P = \sqrt{2}$. Substituting this into (1-40), we get $\hat{\mathbf{n}}_P = -\frac{1}{3}\hat{\mathbf{x}} + \frac{2}{3}\sqrt{2}\,\hat{\mathbf{y}}$. This vector, with its negative x component and considerably larger positive y component is also shown in the figure. ∎

1-10 OTHER DIFFERENTIAL OPERATIONS

It is quite possible that the components of a vector can also depend on position, so that, for example, $A_x(x, y, z)$, and so on. Then $\mathbf{A}$ will generally change as we go from one point to another, in magnitude or direction or both, and we can write $\mathbf{A} = \mathbf{A}(x, y, z) = \mathbf{A}(\mathbf{r})$. In the last term we have used a compact notation that we find convenient for expressing $\mathbf{A}$ as a function of the coordinates of the point through its position vector. A vector whose value is thus given at every point in space is called a *vector field*. We now consider some specific ways in which it can change with position.

Looking back at (1-37), we see that ∇u can be interpreted as the product of u and the *del operator* given by

$$\nabla = \hat{\mathbf{x}}\frac{\partial}{\partial x} + \hat{\mathbf{y}}\frac{\partial}{\partial y} + \hat{\mathbf{z}}\frac{\partial}{\partial z} \tag{1-41}$$

It will be shown in Section 29-3 that this somewhat abstract operator has the same mathematical properties as the displacement of a point and hence can be treated as a legitimate vector. We can now perform two differential operations of interest by using our two forms of multiplication of vectors.

Using (1-20) and (1-41), we get

$$\nabla \cdot \mathbf{A} = \frac{\partial A_x}{\partial x} + \frac{\partial A_y}{\partial y} + \frac{\partial A_z}{\partial z} \tag{1-42}$$

This scalar product is called the *divergence* of $\mathbf{A}$ and is often written div $\mathbf{A}$.

Using (1-27) and (1-41), we obtain

$$\nabla \times \mathbf{A} = \hat{\mathbf{x}}\left(\frac{\partial A_z}{\partial y} - \frac{\partial A_y}{\partial z}\right) + \hat{\mathbf{y}}\left(\frac{\partial A_x}{\partial z} - \frac{\partial A_z}{\partial x}\right) + \hat{\mathbf{z}}\left(\frac{\partial A_y}{\partial x} - \frac{\partial A_x}{\partial y}\right) \tag{1-43}$$

This vector product is called the *curl* of $\mathbf{A}$ and is often written curl $\mathbf{A}$. It can also be conveniently written as a determinant as in (1-28):

$$\nabla \times \mathbf{A} = \begin{vmatrix} \hat{\mathbf{x}} & \hat{\mathbf{y}} & \hat{\mathbf{z}} \\ \dfrac{\partial}{\partial x} & \dfrac{\partial}{\partial y} & \dfrac{\partial}{\partial z} \\ A_x & A_y & A_z \end{vmatrix} \tag{1-44}$$

The significance of the names "divergence" and "curl" will become clearer as we use them in situations in which they naturally arise.

Another operator of great interest and utility is the *Laplacian*:

$$\nabla^2 = \nabla \cdot \nabla = \frac{\partial^2}{\partial x^2} + \frac{\partial^2}{\partial y^2} + \frac{\partial^2}{\partial z^2} \tag{1-45}$$

For example, if applied to a scalar,

$$\nabla^2 u = \frac{\partial^2 u}{\partial x^2} + \frac{\partial^2 u}{\partial y^2} + \frac{\partial^2 u}{\partial z^2} \tag{1-46}$$

while the expression $\nabla^2\mathbf{A}$ represents the three equations in which ∇^2 operates on each of the three rectangular components of $\mathbf{A}$, that is,

$$\nabla^2 A_x = \frac{\partial^2 A_x}{\partial x^2} + \frac{\partial^2 A_x}{\partial y^2} + \frac{\partial^2 A_x}{\partial z^2} \tag{1-47}$$

with similar expressions for the other two components.

We also note two useful results involving del. Because of (1-24), the curl of a gradient is zero:

$$\nabla \times \nabla u = 0 \tag{1-48}$$

Similarly, the divergence of a curl is zero,

$$\nabla \cdot (\nabla \times \mathbf{A}) = 0 \tag{1-49}$$

because of (1-29) and (1-24).

We now turn to some integrals involving vectors. Although many possibilities can be imagined, two particular ones are of interest to us and we take them up in turn.

1-11 THE LINE INTEGRAL

Let us imagine starting at some given initial point $P_i(x_i, y_i, z_i)$ and moving to a given final point $P_f(x_f, y_f, z_f)$ along a specific given curve C (a "line" or "path") as shown

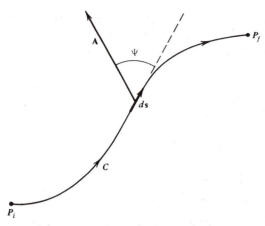

Figure 1-22.   Relations for the calculation of a line integral.

in Figure 1-22. The whole traversal of this path can be treated as the vector addition of a sequence of infinitesimal displacements $d\mathbf{s}$ along C. We assume that there is a vector field $\mathbf{A}$ so that its value can be found at all points along the way. At each intermediate step, we evaluate $\mathbf{A}$, multiply its component along $d\mathbf{s}$ with the magnitude of $d\mathbf{s}$, and add all of these quantities. The result is called the *line integral* of $\mathbf{A}$ along C and is given by

$$\int_i^f A \cos \Psi \, ds = \int_C A \cos \Psi \, ds = \int_C \mathbf{A} \cdot d\mathbf{s} \qquad (1\text{-}50)$$

Perhaps the most familiar example of a line integral is that of work done on a particle; in that case, $\mathbf{A}$ would be a force acting on the particle.

If the path of integration follows a closed curve, for example, a circle, then the initial and final points will coincide. We write the line integral for this case as

$$\oint_C \mathbf{A} \cdot d\mathbf{s}$$

This integral is sometimes called the *circulation* of $\mathbf{A}$; it may or may not be zero depending upon $\mathbf{A}$ as we will see.

If $\mathbf{r}$ is the position vector of each point on C, then $d\mathbf{s} = d\mathbf{r}$ and we can use (1-20) and (1-34) to write

$$\int_C \mathbf{A} \cdot d\mathbf{s} = \int_C \left(A_x \, dx + A_y \, dy + A_z \, dz \right) \qquad (1\text{-}51)$$

In using (1-51), we must be sure to take into account the fact that dx, dy, and dz cannot be varied independently because the coordinates x, y, z are related by the equation describing the path. Similarly, the expressions for $A_x = A_x(x, y, z)$ etc. must also take this interdependence into account. These considerations are best illustrated by looking at a specific case.

■ **Example**

Let $\mathbf{A} = x^2 \hat{\mathbf{x}} + y^2 \hat{\mathbf{y}} + z^2 \hat{\mathbf{z}}$ and let us choose as our path that part of the parabola $y^2 = x$ between the origin $(0,0,0)$ and the point $(2, \sqrt{2}, 0)$; this curve is exactly the parabola illustrated for $u_2 = 0$ in Figure 1-21. Here $z = \text{const.}$, so that $dz = 0$ and the integrand of (1-51) becomes simply

$$A_x \, dx + A_y \, dy = x^2 \, dx + y^2 \, dy$$

We can write this as a function of one variable by using the equation of the path. Since $y^2 = x$, $2y\,dy = dx$ or $dy = dx/2\sqrt{x}$ and $y^2\,dy = \frac{1}{2}\sqrt{x}\,dx$; thus we get

$$\int_C \mathbf{A} \cdot d\mathbf{s} = \int_0^2 \left(x^2 + \tfrac{1}{2}\sqrt{x} \right) dx = \left[\tfrac{1}{3}x^3 + \tfrac{1}{3}x^{3/2} \right]_0^2 = \tfrac{2}{3}\left(4 + \sqrt{2} \right) \qquad \blacksquare$$

1-12 VECTOR ELEMENT OF AREA

Before we consider integrals similar to the above but instead representing sums over given areas, we will find it helpful to look in some detail at the representation of area in vector terms. In Figure 1-23 we show an infinitesimal element of area da that has some particular orientation with respect to the coordinate axes. We also see that a direction can be associated with this area, that is, the unit vector $\hat{\mathbf{n}}$, which is normal to the surface. Thus, we can associate a vector $d\mathbf{a}$ with this element of area and write it as

$$d\mathbf{a} = da\,\hat{\mathbf{n}} \tag{1-52}$$

by following the general form of (1-4). It is clear, however, that there is some ambiguity in this definition because we could equally well have chosen $\hat{\mathbf{n}}$ to be in the opposite direction and it would still be perpendicular to the area element da. Hence we need to supplement (1-52) with a convention that will tell us what to do; there are two cases to be considered.

First, da may be part of an open surface, that is, it is bounded by a closed curve C; a page of this book is an example of such an open surface. In this case, the first step is to *choose* a sense of traversal around the bounding curve; once this is done, curl the fingers of the right hand in the sense of traversal and then, by convention, the direction in which the thumb is pointed is to be taken as the direction of $\hat{\mathbf{n}}$. This right-hand rule is illustrated in Figure 1-24; note how the direction of $\hat{\mathbf{n}}$ would be reversed if C were traversed in the opposite sense.

Second, da may be part of a closed surface. In this case, there is no bounding curve C but the surface divides a volume into an inside and an outside; the surface of a ball is such an example. Here the direction of $\hat{\mathbf{n}}$ is always chosen to point *from* the inside *to* the outside. This is illustrated in Figure 1-25 where the corresponding directions of the outward normal are shown for several points.

When (1-52) is combined with an expression for $\hat{\mathbf{n}}$ of the form (1-8), $d\mathbf{a}$ can be written in component form as

$$d\mathbf{a} = da_x\,\hat{\mathbf{x}} + da_y\,\hat{\mathbf{y}} + da_z\,\hat{\mathbf{z}} \tag{1-53}$$

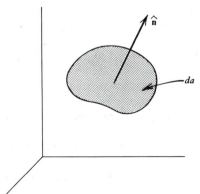

Figure 1-23.   The normal vector for an element of area.

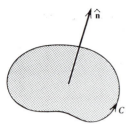

Figure 1-24. Definition of the direction of the normal vector.

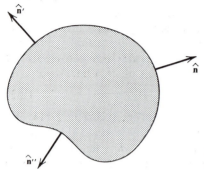

Figure 1-25. Various outward normals for a closed surface.

where

$$da_x = l_x\, da \qquad da_y = l_y\, da \qquad da_z = l_z\, da \qquad (1\text{-}54)$$

and where l_x, l_y, l_z are the components of $\hat{\mathbf{n}}$, that is, its direction cosines. Our experience with multiple integrals has made us familiar with the use of $dx\,dy$ as an element of area in the xy plane, and it is clear that this expression is somehow related to one of the components of $d\mathbf{a}$. In order to find this relationship, let us consider Figure 1-26, which shows a rectangular area element da of sides b and c so that $da = bc$. The plane of the area is parallel to the y axis so that $\hat{\mathbf{n}}$ is parallel to the xz plane and makes an angle γ with the z axis; Figure 1-27 shows a side view looking along the y axis toward the origin. The figures also show the projection of the area onto the xy and yz planes; these projections are rectangles of area $dx\,dy$ and $dy\,dz$ respectively; it is also evident from the figure that $dy = c$. The projection of da onto the xz plane is the line labeled b in Figure 1-27. The other two direction angles of $\hat{\mathbf{n}}$ are obtained by comparing these two figures with Figure 1-8 and are seen to be $\alpha = 90° - \gamma$ and $\beta = 90°$, so that the direction cosines are $l_x = \sin\gamma$, $l_y = 0$, and $l_z = \cos\gamma$. Substituting these values into (1-54) and using Figure 1-27, we find that $da_z = da\cos\gamma = (b\cos\gamma)c = dx\,dy$, which is just the projection of the area da onto the xy plane, that is, the projection perpendicular to the corresponding coordinate axis. Similarly, we find $da_x = dy\,dz$ while $da_y = 0$ for this special case.

Now suppose that n_z were negative so that $\gamma > 90°$ while everything else was kept the same; this is illustrated in Figure 1-28. Comparing with Figure 1-26, we see that the projections on the xy and yz planes will still be rectangles of respective areas $dx\,dy$ and $dy\,dz$. Now, however, $\cos\gamma$ will be negative so that $da_z = da\cos\gamma = -dx\,dy$. On the other hand, since $\alpha = \gamma - 90°$, $\cos\alpha$ will be positive and $da_x = dy\,dz$ as before; da_y is still zero.

These considerations can be generalized to a situation in which $\hat{\mathbf{n}}$ makes arbitrary angles with all axes. The magnitude of a given component of $d\mathbf{a}$ along a given axis will be equal to its projection on the coordinate plane perpendicular to the axis and will be

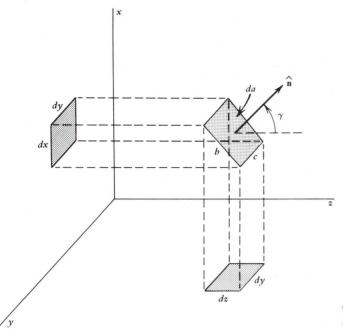

Figure 1-26. Determination of the components of a vector area element.

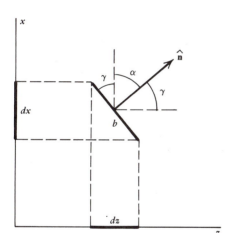

Figure 1-27. Side view of situation in Figure 1-26.

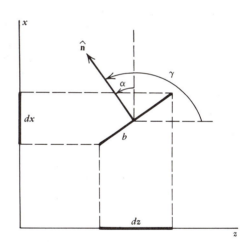

Figure 1-28. An area element with a negative z component.

given, in rectangular coordinates, by the product of the corresponding differentials. Since we always treat these differentials as positive as written, the actual component will be obtained by multiplying this product by a plus or a minus sign depending on the sign of the corresponding component of $\hat{\mathbf{n}}$. Thus we have succeeded in finding a representation of the rectangular components of an element of area and we can write

$$da_x = \pm\, dy\, dz \qquad da_y = \pm\, dz\, dx \qquad da_z = \pm\, dx\, dy \qquad (1\text{-}55)$$

where the plus sign is to be used for a given component if the direction angle of $\hat{\mathbf{n}}$ with the corresponding axis is less than 90°, while the negative sign is used when the direction angle is greater than 90°.

1-13 THE SURFACE INTEGRAL

Consider a surface S; as shown in Figure 1-29, we can divide S into vector elements of area $d\mathbf{a}$ as discussed in the previous section. We assume the existence of a vector field $\mathbf{A}$ so that its value can be found at all points of S. At each element of area, we evaluate $\mathbf{A}$, multiply its component in the direction of $d\mathbf{a}$ by the magnitude of $d\mathbf{a}$, and add all of these quantities. The result is called the *surface integral* of $\mathbf{A}$ over S and is given by

$$\int_S A \cos \Psi \, da = \int_S \mathbf{A} \cdot \hat{\mathbf{n}} \, da = \int_S \mathbf{A} \cdot d\mathbf{a} \tag{1-56}$$

This integral is also called the *flux* of the vector $\mathbf{A}$ through the surface S. We have written (1-56) with a single integral sign for convenience, but it actually represents a double integral.

If the surface is a closed surface, it is useful to indicate this explicitly by writing the integral as

$$\oint_S \mathbf{A} \cdot d\mathbf{a}$$

The value of this integral may or may not be zero, depending on $\mathbf{A}$ as we will see.

Using (1-20) and (1-53), we can write (1-56) in terms of rectangular coordinates as

$$\int_S \mathbf{A} \cdot d\mathbf{a} = \int_S (A_x \, da_x + A_y \, da_y + A_z \, da_z) \tag{1-57}$$

In using (1-57), we would need to use (1-55) and then be sure to take into account the fact that dx, dy, and dz cannot be varied independently because the coordinates x, y, z are related by the equation describing the surface S. This interdependence must also be taken into account in determining the limits of integration and in the writing of the components such as $A_x = A_x(x, y, z)$. Let us illustrate these points with a specific case.

■ **Example**

Let $\mathbf{A} = yz\hat{\mathbf{x}} + zx\hat{\mathbf{y}} + xy\hat{\mathbf{z}}$ and let us choose our area S to be that part of a circle of radius a centered at the origin and in the first quadrant of the xy plane as shown in Figure 1-30; the equation of the circle is $x^2 + y^2 = a^2$. Here the area is perpendicular to the z axis so that $\hat{\mathbf{n}}$ can be taken as either $\hat{\mathbf{z}}$ or $-\hat{\mathbf{z}}$. Let us arbitrarily choose $\hat{\mathbf{n}} = \hat{\mathbf{z}}$; then the only component of $d\mathbf{a}$ is $da_z = dx \, dy$ as obtained from (1-55) with the plus sign since the direction angle $\gamma = 0$, and (1-57) becomes

$$\int_S \mathbf{A} \cdot d\mathbf{a} = \int_S xy \, dx \, dy \tag{1-58}$$

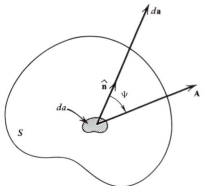

Figure 1-29. Relations for the calculation of a surface integral.

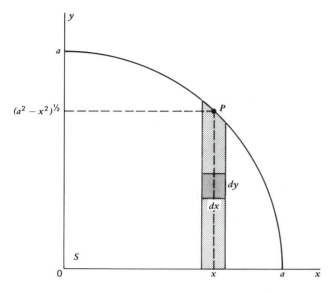

Figure 1-30. Area of integration for the example of Section 1-13.

The element of area is shown more darkly shaded in the figure. We choose to evaluate this integral by first integrating over y while keeping x constant; this will add up the contributions from the less darkly shaded strip; the upper limit of integration on y corresponds to the location of the point P and is found from the equation of the circle for this fixed value of x, that is, $y_P = (a^2 - x^2)^{1/2}$. After this is done, we integrate over all possible values of x, summing the contributions from similar strips, and thereby covering the whole area S. Thus, (1-58) becomes

$$\int_S \mathbf{A} \cdot d\mathbf{a} = \int_0^a x \, dx \int_0^{(a^2 - x^2)^{1/2}} y \, dy = \int_0^a x \, dx \left[\tfrac{1}{2} y^2 \right]_0^{(a^2 - x^2)^{1/2}}$$

$$= \tfrac{1}{2} \int_0^a x (a^2 - x^2) \, dx = \tfrac{1}{2} \left[\tfrac{1}{2} a^2 x^2 - \tfrac{1}{4} x^4 \right]_0^a = \tfrac{1}{8} a^4 \qquad \blacksquare$$

We now turn to two important theorems involving the types of integrals we have just discussed.

1-14 THE DIVERGENCE THEOREM

Consider a volume V enclosed by a surface S. *Gauss' divergence theorem* states that

$$\oint_S \mathbf{A} \cdot d\mathbf{a} = \int_V \nabla \cdot \mathbf{A} \, d\tau \qquad (1\text{-}59)$$

The integrals are taken over the total surface S and throughout the volume V whose volume element is $d\tau$. Again, for convenience, we have written the volume integral with a single integral sign although in reality it is a triple integral. Since S is a closed surface, the unit normal $\hat{\mathbf{n}}$ used for $d\mathbf{a}$ is the outward normal according to our convention of Section 1-12 as shown in Figure 1-25.

This theorem relates a surface integral of a vector to the volume integral of its divergence. The surface integral depends only on the values of $\mathbf{A}$ on the surface, whereas the volume integral requires a knowledge of $\nabla \cdot \mathbf{A}$ (but not of $\mathbf{A}$ itself) throughout the whole volume.

We will prove this theorem by direct evaluation. In rectangular coordinates, the volume element is

$$d\tau = dx \, dy \, dz \qquad (1\text{-}60)$$

and, using (1-42), we can write the volume integral as a sum:

$$\int_V \nabla \cdot \mathbf{A} \, d\tau = \int_V \frac{\partial A_x}{\partial x} \, dx \, dy \, dz + \int_V \frac{\partial A_y}{\partial y} \, dx \, dy \, dz + \int_V \frac{\partial A_z}{\partial z} \, dx \, dy \, dz \quad (1\text{-}61)$$

Consider the first integral. Our first step will be to integrate over x while keeping y and z constant at the values y_0 and z_0. Thus we will be summing up the contributions from a rod of cross section $dy \, dz$. This rod is shown in Figure 1-31 where its projection on the yz plane is also indicated. The rod intersects the surface S at the points P_1 and P_2 and thus defines two elements of area $d\mathbf{a}_1$ and $d\mathbf{a}_2$ of S whose directions are shown. (For clarity, the rest of V and S are not shown in the figure.) The coordinates of the points P_1 and P_2 will be (x_1, y_0, z_0) and (x_2, y_0, z_0) respectively where x_1 and x_2 are found as the values that satisfy the equation defining the surface S; thus x_1 and x_2 are the limits of integration for x. In this step, then, $\partial A_x / \partial x$ will be a function only of x since y and z are constant, so that $(\partial A_x / \partial x) \, dx = dA_x$.

Therefore, the first integral in (1-61) becomes

$$\int \int dy \, dz \int_{x_1}^{x_2} \frac{\partial A_x}{\partial x} \, dx = \int \int [A_x(x_2, y_0, z_0) - A_x(x_1, y_0, z_0)] \, dy \, dz \quad (1\text{-}62)$$

In the integrand of this result, the term in brackets is the difference in A_x when evaluated at the points P_2 and P_1, which we can write as $A_x(P_2) - A_x(P_1)$. From (1-55) and Figure 1-31, we see that $da_{2x} = dy \, dz$ while $da_{1x} = -dy \, dz$ because the angles made by these areas with the x axis are less than $90°$ and greater than $90°$ respectively. The integrand in (1-62) thus can be written as

$$A_x(P_2) \, da_{2x} - A_x(P_1)(-da_{1x}) = A_{x2} \, da_{2x} + A_{x1} \, da_{1x} \quad (1\text{-}63)$$

which is just equal to the total contribution to the surface integral of $A_x \, da_x$ arising from the areas $d\mathbf{a}_2$ and $d\mathbf{a}_1$ intercepted on the surface S by this rod. Thus, when we perform the integrations over y and z in (1-62), we will be summing the contributions of all rods of this type; the contribution from each rod will be $A_x \, da_x$ from its share of the area, so that the final result will be the surface integral of $A_x \, da_x$ over the whole area S. In other words, we have found that

$$\int_V \frac{\partial A_x}{\partial x} \, dx \, dy \, dz = \oint_S A_x \, da_x \quad (1\text{-}64)$$

In the same manner, the last two integrals in (1-61) will be found equal to

$$\oint_S A_y \, da_y \quad \text{and} \quad \oint_S A_z \, da_z$$

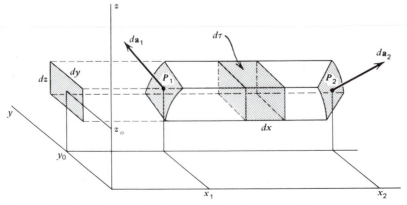

Figure 1-31. Volume used in deriving the divergence theorem.

respectively, so that, if we add these to (1-64), substitute into (1-61), and use (1-20), we find that

$$\int_V \nabla \cdot \mathbf{A} \, d\tau = \oint_S (A_x \, da_x + A_y \, da_y + A_z \, da_z) = \oint_S \mathbf{A} \cdot d\mathbf{a}$$

which is exactly (1-59) and proves the theorem.

We have proved this only for a volume bounded by a single surface, but we can easily extend the proof to a region bounded by several surfaces, such as a hollow ball. Figure 1-32 shows a volume V surrounded by two surfaces S_1 and S_2; two representative outward normals to the volume are shown as $\hat{\mathbf{n}}$ and $\hat{\mathbf{n}}'$. We now imagine a plane intersecting the volume and dividing it into two volumes V_2 and V_1; the trace of this plane is shown by the dashed lines AB and CD. The volume V_2 is now enclosed by a single surface composed of those portions of S_2 and S_1 to the left of the plane plus the plane surfaces of intersection shown by AB and CD; the outward normal to the new part of the bounding surface is shown as $\hat{\mathbf{n}}_2$. Similar remarks apply to V_1; its corresponding outward normal is $\hat{\mathbf{n}}_1$. Applying (1-59) to each of these volumes and adding, we obtain

$$\int_{V_1 + V_2} \nabla \cdot \mathbf{A} \, d\tau = \int_{S_1} \mathbf{A} \cdot d\mathbf{a} + \int_{S_2} \mathbf{A} \cdot d\mathbf{a} + \int_{ABCD} \mathbf{A} \cdot \hat{\mathbf{n}}_2 \, da + \int_{ABCD} \mathbf{A} \cdot \hat{\mathbf{n}}_1 \, da$$

In the last two integrals, the normals are oppositely directed so that at each point of $ABCD$, $\hat{\mathbf{n}}_2 = -\hat{\mathbf{n}}_1$ while the values $\mathbf{A}$ and da are the same; thus these integrals cancel each other and we are left with

$$\int_{V_1 + V_2} \nabla \cdot \mathbf{A} \, d\tau = \int_{S_1 + S_2} \mathbf{A} \cdot d\mathbf{a}$$

which is the same as (1-59) because the total volume is $V_1 + V_2$ and the total bounding surface is $S_1 + S_2$.

This proof can obviously be generalized to an arbitrary number of bounding surfaces by introducing as many intersecting surfaces like $ABCD$ as would be necessary.

If we now apply the divergence theorem to a particular simple situation, we can get a useful and illuminating result. Consider a point P at the center of a small volume ΔV. If ΔV is very small, $\nabla \cdot \mathbf{A}$ will be nearly constant throughout the volume so that it is

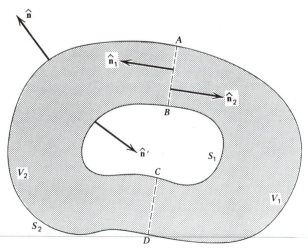

Figure 1-32. A volume bounded by two surfaces.

easily seen that the volume integral in (1-59) can be written as

$$\int_{\Delta V} \nabla \cdot \mathbf{A}\, d\tau = \langle \nabla \cdot \mathbf{A} \rangle_P\, \Delta V$$

where $\langle \nabla \cdot \mathbf{A} \rangle_P$ is the average value of $\nabla \cdot \mathbf{A}$ in the vicinity of P. Putting this into (1-59) and dividing by ΔV, we get

$$\langle \nabla \cdot \mathbf{A} \rangle_P = \frac{1}{\Delta V} \oint_S \mathbf{A} \cdot d\mathbf{a} \tag{1-65}$$

If we now let $\Delta V \to 0$, while P remains at the center, the average value of $\nabla \cdot \mathbf{A}$ in the vicinity of P becomes the value of $\nabla \cdot \mathbf{A}$ at P; writing this simply as $\nabla \cdot \mathbf{A}$, we have

$$\nabla \cdot \mathbf{A} = \lim_{\Delta V \to 0} \frac{1}{\Delta V} \oint_S \mathbf{A} \cdot d\mathbf{a} \tag{1-66}$$

This important expression for $\nabla \cdot \mathbf{A}$ is in a form independent of a particular coordinate system, in contrast to (1-42), and hence can be taken as a general definition of the divergence of a vector in the same sense that (1-38) is the general way of defining a gradient. The result (1-66) also gives us a much better understanding of the significance of the divergence since we see that it is a measure of the outward flux of the vector through a small area about the point in question.

It is possible to start with the definition (1-66) and, by evaluating the flux of $\mathbf{A}$ through the surface around a volume $\Delta x\, \Delta y\, \Delta z$, find the expression for $\nabla \cdot \mathbf{A}$ in rectangular coordinates; the result is, of course, (1-42).

1-15 STOKES' THEOREM

Consider a surface S enclosed by a curve C. *Stokes' theorem* states that

$$\oint_C \mathbf{A} \cdot d\mathbf{s} = \int_S (\nabla \times \mathbf{A}) \cdot d\mathbf{a} \tag{1-67}$$

and hence relates the line integral of a vector about a closed curve to the surface integral of its curl over the enclosed area. Since S is an open surface, the direction of $d\mathbf{a}$ in (1-67) is determined from the chosen sense of traversal around C according to (1-52) and the right-hand rule illustrated in Figure 1-24; it is *essential* that this sign convention be used in any application of (1-67).

It is interesting to note that (1-67) does not require S to have any particular shape other than that it be bounded by C; thus there are many possibilities in choosing the surface. In general, the values of the integrand $(\nabla \times \mathbf{A}) \cdot d\mathbf{a}$ will be different for the points on all these surfaces, but (1-67) tells us that the sum of all these terms will be the same, since the line integral depends only on the values of $\mathbf{A}$ along the common perimeter. This point is illustrated somewhat schematically in Figure 1-33. Suppose, for

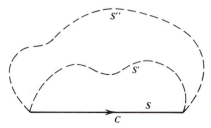

Figure 1-33. Surfaces with the same bounding curve.

simplicity, that C is a closed curve lying in a plane such as a circle. S could be taken as the plane area enclosed by the circle; when viewed from the side, C and S would both appear as the line shown. The dashed lines represent traces of other possible surfaces S', S'', and so on, all of which have C as a boundary, and the integral of $(\nabla \times \mathbf{A}) \cdot d\mathbf{a}$ over any of them will give the same result.

We prove this theorem by direct evaluation of the surface integral by using our previous expressions in rectangular coordinates. Using (1-20) and (1-43), we find that

$$\int_S (\nabla \times \mathbf{A}) \cdot d\mathbf{a} = \int_S \left(\frac{\partial A_x}{\partial z} \, da_y - \frac{\partial A_x}{\partial y} \, da_z \right) + \int_S \left(\frac{\partial A_y}{\partial x} \, da_z - \frac{\partial A_y}{\partial z} \, da_x \right)$$
$$+ \int_S \left(\frac{\partial A_z}{\partial y} \, da_x - \frac{\partial A_z}{\partial x} \, da_y \right) \tag{1-68}$$

where we have grouped the terms by components of $\mathbf{A}$. Consider the first integral that we designate by I_x. We evaluate it by first integrating over a strip of width dx, which is parallel to the yz plane, and a distance x from it. Then, by integrating over x, we sum up the contributions from all of the strips into which S can be subdivided. Initially, we assume S to be simple enough that we can choose the orientation of our axes so that both y and z increase as we go from the beginning to the end of the strip. This situation is illustrated in Figure 1-34, which also shows the projections of the strip on the xz and xy planes as an aid to understanding the orientation of the surface. P_1 and P_2 are, respectively, the initial and final points of our integration, that is, they are the points of intersection of the strip with the bounding curve C and their coordinates satisfy the equation describing C. The area element $d\mathbf{a}$ is shown at an intermediate stage of the integration; we see that the direction angles of $d\mathbf{a}$ have values such that α

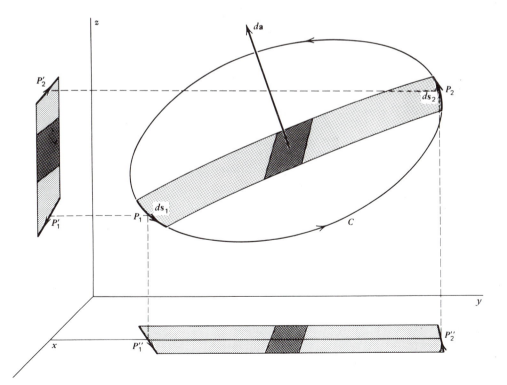

Figure 1-34.   Area used in deriving Stokes' theorem.

and γ are less than $90°$, while $\beta > 90°$. (A pencil held normal to a piece of cardboard oriented in the same sense as the shaded strip will quickly make this clear.) Thus, according to (1-55), $da_y = -dx\,dz$ and $da_z = dx\,dy$ and we can write

$$I_x = -\int_{strips} dx \int_{P_1}^{P_2} \left(\frac{\partial A_x}{\partial y}\,dy + \frac{\partial A_x}{\partial z}\,dz \right) \qquad (1\text{-}69)$$

In the term in parentheses, dy and dz are not independent because y and z are related by the equation for S and the value of x involved. Since this integrand is being evaluated on the strip for which $x = $ const., and hence $dx = 0$, we can add $(\partial A_x/\partial x)\,dx = 0$ to it to give it a quickly recognizable form:

$$\frac{\partial A_x}{\partial x}\,dx + \frac{\partial A_x}{\partial y}\,dy + \frac{\partial A_x}{\partial z}\,dz = dA_x$$

As a result, (1-69) becomes

$$I_x = -\int_{strips} dx \int_{P_1}^{P_2} dA_x = -\int_{strips} \left[A_x(P_2) - A_x(P_1) \right] dx \qquad (1\text{-}70)$$

If we turn again to Figure 1-34, and consider the displacements $d\mathbf{s}_1$ and $d\mathbf{s}_2$ along C at the corresponding limits, we see that, at P_1, $d\mathbf{s}_1$ has a positive x component so we can write $ds_{1x} = dx$, while at P_2, $d\mathbf{s}_2$ has a negative x component with the result that $ds_{2x} = -dx$. Consequently, the integrand of (1-70) can be written as

$$-A_x(P_2)(-ds_{2x}) + A_x(P_1)(ds_{1x}) = A_{x2}\,ds_{2x} + A_{x1}\,ds_{1x} \qquad (1\text{-}71)$$

which is just equal to the total contribution to the line integral of $A_x\,ds_x$ arising from the displacements $d\mathbf{s}_2$ and $d\mathbf{s}_1$ intercepted on the curve C by this strip. Thus, when we perform the final integration over x in (1-70), that is, adding up the contributions of all of the strips, the contribution from each strip will be $A_x\,ds_x$ from its share of the bounding curve; the final result will be the line integral of $A_x\,ds_x$ over the whole curve C. In other words, we have found that

$$\int_S \left(\frac{\partial A_x}{\partial z}\,da_y - \frac{\partial A_x}{\partial y}\,da_z \right) = \oint_C A_x\,ds_x \qquad (1\text{-}72)$$

Similarly, the last two integrals of (1-68) can be shown to have the respective values

$$\oint_C A_y\,ds_y \qquad \text{and} \qquad \oint_C A_z\,ds_z$$

Substituting these results, along with (1-72), into (1-68), we find that

$$\int_S (\nabla \times \mathbf{A}) \cdot d\mathbf{a} = \oint_C (A_x\,ds_x + A_y\,ds_y + A_z\,ds_z) = \oint_C \mathbf{A} \cdot d\mathbf{s}$$

which is exactly (1-67) and proves the theorem.

The theorem can be extended to the case of a surface bounded by more than one curve by a method similar to that we used for the divergence theorem. An example of such a situation is shown in Figure 1-35. Note the direction of traversal of the inner bounding curve; this sense is chosen to keep the area of interest to one's left as one moves along the curve and is seen to be equivalent to the right-hand rule illustrated in Figure 1-24. We would divide S into surfaces each bounded by a single curve by introducing as many pairs of coincident lines as we would need; two such pairs are shown dashed in the figure. Then Stokes' theorem can be applied to each of these surfaces and the results added. The contributions to the line integrals from the introduced lines would cancel since they are traversed in opposite directions and the

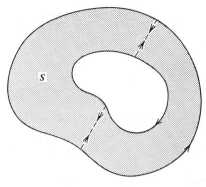

Figure 1-35. An area bounded by two curves.

final result would again be (1-67). A similar procedure will also dispel any remaining qualms relating to our assumption that the surface in Figure 1-34 was "flat enough" for us to orient our axes so that y and z always increased as we integrated along the strip. If the surface is very convoluted, we can divide it into pieces each of which is flat enough so that, if necessary, we can choose a different set of axes for each piece which will satisfy the requirements of Figure 1-34; such a division is illustrated in Figure 1-36. When we apply the theorem to each piece and add the results, the contributions from the dividing lines will cancel each other, and we will again obtain (1-67).

Consider a point P at the center of a small area $\Delta a \hat{\mathbf{n}}$. When (1-67) is applied to this case, $(\nabla \times \mathbf{A}) \cdot \hat{\mathbf{n}}$ will be nearly constant over the area, so that, much as in the last section, we can write

$$\langle (\nabla \times \mathbf{A}) \cdot \hat{\mathbf{n}} \rangle_P = \frac{1}{\Delta a} \oint_C \mathbf{A} \cdot d\mathbf{s}$$

where the left-hand side is the average value near P of the component of $\nabla \times \mathbf{A}$ in the direction of $\hat{\mathbf{n}}$ according to (1-21). Now if we let $\Delta a \to 0$, the average value near P becomes the value at P, or

$$(\nabla \times \mathbf{A}) \cdot \hat{\mathbf{n}} = \lim_{\Delta a \to 0} \frac{1}{\Delta a} \oint_C \mathbf{A} \cdot d\mathbf{s} \qquad (1\text{-}73)$$

which gives the component of $\nabla \times \mathbf{A}$ in a given direction in terms of the line integral of $\mathbf{A}$ about a small area normal to this direction. Thus we can take (1-73) as a general definition of the component of the curl in a given direction. If we do this for three

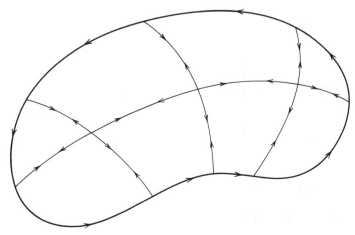

Figure 1-36. Division of an area for a general proof of Stokes' theorem.

mutually perpendicular directions (such as $\hat{\mathbf{x}}$, $\hat{\mathbf{y}}$, and $\hat{\mathbf{z}}$), we will get the components of $\nabla \times \mathbf{A}$ in each of these directions and thus obtain the whole vector $\nabla \times \mathbf{A}$. When this procedure is carried out for rectangular coordinates, the result is, of course, (1-43).

1-16 CYLINDRICAL COORDINATES

Up to now, we have used only rectangular coordinates with their constant unit vectors. However, many problems are more conveniently stated and worked in other systems of coordinates, and we want to see what happens to many of our results. We need to deal with only two of the more important ones.

The first is *cylindrical coordinates* in which the location of a point P is specified by the three quantities ρ, φ, z whose definition is illustrated in Figure 1-37; this figure also shows the position vector $\mathbf{r}$ of the point along with three new unit vectors that we will define shortly. We see that when $\mathbf{r}$ is projected onto the xy plane, ρ is the length of this projection while φ is the angle that it makes with the positive x axis; z is the same as the corresponding rectangular coordinate. The relation between the cylindrical and rectangular coordinates of P is seen from the figure to be

$$x = \rho \cos \varphi \qquad y = \rho \sin \varphi \qquad z = z \tag{1-74}$$

so that

$$\rho = \left(x^2 + y^2 \right)^{1/2} \qquad \tan \varphi = \frac{y}{x} \tag{1-75}$$

We can now define a set of three mutually perpendicular unit vectors as follows: first, $\hat{\mathbf{z}}$ is the same as the rectangular $\hat{\mathbf{z}}$; second, $\hat{\boldsymbol{\rho}}$ is chosen to be in the direction of increasing ρ and is perpendicular to $\hat{\mathbf{z}}$ so that $\hat{\boldsymbol{\rho}}$ is parallel to the xy plane; finally, $\hat{\boldsymbol{\varphi}}$ is defined to be perpendicular to both of these and in the direction shown. We see that $\hat{\boldsymbol{\varphi}}$

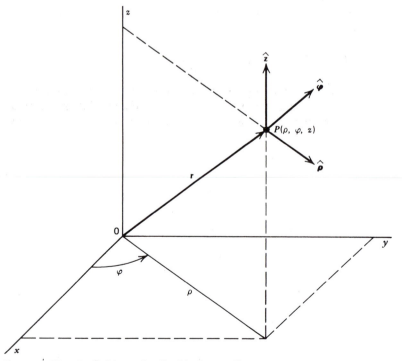

Figure 1-37. Definition of cylindrical coordinates.

is perpendicular to the semiinfinite plane $\varphi = $ const., and its direction is therefore in the sense of increasing φ. These unit vectors are shown at the location of P in order to emphasize that they are functions of P in the sense that if P is displaced, both $\hat{\boldsymbol{\rho}}$ and $\hat{\boldsymbol{\varphi}}$ change their directions, although $\hat{\mathbf{z}}$ does not change. Thus, these unit vectors are not all constants, in contrast to $\hat{\mathbf{x}}$, $\hat{\mathbf{y}}$, and $\hat{\mathbf{z}}$.

Since $\hat{\boldsymbol{\rho}}$, $\hat{\boldsymbol{\varphi}}$, and $\hat{\mathbf{z}}$ are mutually perpendicular unit vectors, they satisfy relations analogous to (1-18), (1-19), and (1-25):

$$\hat{\boldsymbol{\rho}} \cdot \hat{\boldsymbol{\rho}} = \hat{\boldsymbol{\varphi}} \cdot \hat{\boldsymbol{\varphi}} = \hat{\mathbf{z}} \cdot \hat{\mathbf{z}} = 1$$

$$\hat{\boldsymbol{\rho}} \cdot \hat{\boldsymbol{\varphi}} = \hat{\boldsymbol{\varphi}} \cdot \hat{\mathbf{z}} = \hat{\mathbf{z}} \cdot \hat{\boldsymbol{\rho}} = 0 \qquad (1\text{-}76)$$

$$\hat{\boldsymbol{\rho}} \times \hat{\boldsymbol{\varphi}} = \hat{\mathbf{z}} \qquad \hat{\boldsymbol{\varphi}} \times \hat{\mathbf{z}} = \hat{\boldsymbol{\rho}} \qquad \hat{\mathbf{z}} \times \hat{\boldsymbol{\rho}} = \hat{\boldsymbol{\varphi}}$$

The rectangular components of $\hat{\boldsymbol{\rho}}$ and $\hat{\boldsymbol{\varphi}}$ are found from inspecting Figure 1-37; it is helpful to imagine them projected onto the xy plane, and the results are

$$\hat{\boldsymbol{\rho}} = \cos\varphi\,\hat{\mathbf{x}} + \sin\varphi\,\hat{\mathbf{y}} \qquad \hat{\boldsymbol{\varphi}} = -\sin\varphi\,\hat{\mathbf{x}} + \cos\varphi\,\hat{\mathbf{y}} \qquad (1\text{-}77)$$

These equations can be solved for $\hat{\mathbf{x}}$ and $\hat{\mathbf{y}}$ to give their components in cylindrical coordinates:

$$\hat{\mathbf{x}} = \cos\varphi\,\hat{\boldsymbol{\rho}} - \sin\varphi\,\hat{\boldsymbol{\varphi}} \qquad \hat{\mathbf{y}} = \sin\varphi\,\hat{\boldsymbol{\rho}} + \cos\varphi\,\hat{\boldsymbol{\varphi}} \qquad (1\text{-}78)$$

By differentiating (1-77), we can find explicitly how $\hat{\boldsymbol{\rho}}$ and $\hat{\boldsymbol{\varphi}}$ vary as P is displaced:

$$\frac{d\hat{\boldsymbol{\rho}}}{d\varphi} = \hat{\boldsymbol{\varphi}} \quad \text{and} \quad \frac{d\hat{\boldsymbol{\varphi}}}{d\varphi} = -\hat{\boldsymbol{\rho}} \qquad (1\text{-}79)$$

Since $\hat{\boldsymbol{\rho}}$, $\hat{\boldsymbol{\varphi}}$, and $\hat{\mathbf{z}}$ are mutually perpendicular, we can express any vector $\mathbf{A}$ in terms of its components along these directions; by analogy with (1-5), we write $\mathbf{A}$ in the form

$$\mathbf{A} = A_\rho\hat{\boldsymbol{\rho}} + A_\varphi\hat{\boldsymbol{\varphi}} + A_z\hat{\mathbf{z}} \qquad (1\text{-}80)$$

For the particular case of the position vector $\mathbf{r}$, we see from Figure 1-37 that

$$\mathbf{r} = \rho\hat{\boldsymbol{\rho}} + z\hat{\mathbf{z}} \qquad (1\text{-}81)$$

which can also be obtained from (1-11) by substitution of (1-74) and the use of (1-77). We can also find the differential of $d\mathbf{r}$ from (1-81) and (1-79):

$$d\mathbf{r} = d\rho\,\hat{\boldsymbol{\rho}} + \rho\,d\hat{\boldsymbol{\rho}} + dz\,\hat{\mathbf{z}} = d\rho\,\hat{\boldsymbol{\rho}} + \rho\,d\varphi\,\hat{\boldsymbol{\varphi}} + dz\,\hat{\mathbf{z}} \qquad (1\text{-}82)$$

so that its components in the directions of increasing ρ, φ, and z are $d\rho$, $\rho\,d\varphi$, and dz, respectively. These component displacements are shown in Figure 1-38, and we see that they correspond to the distance moved by P resulting from the change of any one coordinate while the other two are held fixed. The shaded volume element has as sides just the components of $d\mathbf{r}$ given by (1-82). Therefore, the element of volume in cylindrical coordinates is given by

$$d\tau = \rho\,d\rho\,d\varphi\,dz \qquad (1\text{-}83)$$

We also see from this figure that the areas perpendicular to the unit vectors are $\rho\,d\varphi\,dz$, $d\rho\,dz$, and $\rho\,d\rho\,d\varphi$, so that the components of an element of area $d\mathbf{a}$ are given by

$$da_\rho = \pm\rho\,d\varphi\,dz \qquad da_\varphi = \pm d\rho\,dz \qquad da_z = \pm\rho\,d\rho\,d\varphi \qquad (1\text{-}84)$$

where the proper sign is to be chosen in the same manner as for (1-55).

Now we can go on to find what our differential operators become when expressed in this system. If $u = u(\rho, \varphi, z)$, then

$$du = \frac{\partial u}{\partial \rho}\,d\rho + \frac{\partial u}{\partial \varphi}\,d\varphi + \frac{\partial u}{\partial z}\,dz$$

and, on comparing this with (1-82) and the general definition of the gradient given in (1-38), and noting that $d\mathbf{s} = d\mathbf{r}$ in this case, we find that the expression for the gradient

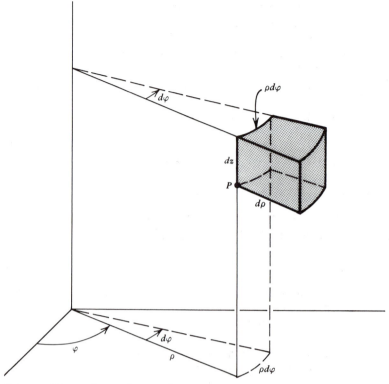

Figure 1-38. Volume element in terms of cylindrical coordinates.

in cylindrical coordinates is

$$\nabla u = \hat{\rho}\frac{\partial u}{\partial \rho} + \hat{\varphi}\frac{1}{\rho}\frac{\partial u}{\partial \varphi} + \hat{z}\frac{\partial u}{\partial z} \tag{1-85}$$

so that the del operator can be written as

$$\nabla = \hat{\rho}\frac{\partial}{\partial \rho} + \hat{\varphi}\frac{1}{\rho}\frac{\partial}{\partial \varphi} + \hat{z}\frac{\partial}{\partial z} \tag{1-86}$$

By applying (1-86) to (1-80), we can find corresponding expressions for the divergence and curl. In doing this, however, we must remember that the unit vectors are not constant and we have to take (1-79) into account; in addition, we need to use (1-76). We illustrate this process for $\nabla \cdot \mathbf{A}$:

$$\nabla \cdot \mathbf{A} = \left(\hat{\rho}\frac{\partial}{\partial \rho} + \hat{\varphi}\frac{1}{\rho}\frac{\partial}{\partial \varphi} + \hat{z}\frac{\partial}{\partial z}\right)\cdot\left(A_\rho\hat{\rho} + A_\varphi\hat{\varphi} + A_z\hat{z}\right)$$

$$= \hat{\rho}\cdot\left(\frac{\partial A_\rho}{\partial \rho}\hat{\rho} + \frac{\partial A_\varphi}{\partial \rho}\hat{\varphi} + \frac{\partial A_z}{\partial \rho}\hat{z}\right)$$

$$+ \hat{\varphi}\cdot\frac{1}{\rho}\left(\frac{\partial A_\rho}{\partial \varphi}\hat{\rho} + A_\rho\frac{\partial \hat{\rho}}{\partial \varphi} + \frac{\partial A_\varphi}{\partial \varphi}\hat{\varphi} + A_\varphi\frac{\partial \hat{\varphi}}{\partial \varphi} + \frac{\partial A_z}{\partial \varphi}\hat{z}\right)$$

$$+ \hat{z}\cdot\left(\frac{\partial A_\rho}{\partial z}\hat{\rho} + \frac{\partial A_\varphi}{\partial z}\hat{\varphi} + \frac{\partial A_z}{\partial z}\hat{z}\right)$$

$$= \frac{\partial A_\rho}{\partial \rho} + \frac{A_\rho}{\rho} + \frac{1}{\rho}\frac{\partial A_\varphi}{\partial \varphi} + \frac{\partial A_z}{\partial z}$$

which is usually written as

$$\nabla \cdot \mathbf{A} = \frac{1}{\rho} \frac{\partial}{\partial \rho} (\rho A_\rho) + \frac{1}{\rho} \frac{\partial A_\varphi}{\partial \varphi} + \frac{\partial A_z}{\partial z} \qquad (1\text{-}87)$$

In the same manner, the expression for the curl becomes

$$\nabla \times \mathbf{A} = \hat{\boldsymbol{\rho}} \left(\frac{1}{\rho} \frac{\partial A_z}{\partial \varphi} - \frac{\partial A_\varphi}{\partial z} \right) + \hat{\boldsymbol{\varphi}} \left(\frac{\partial A_\rho}{\partial z} - \frac{\partial A_z}{\partial \rho} \right) + \hat{\mathbf{z}} \left[\frac{1}{\rho} \frac{\partial}{\partial \rho} (\rho A_\varphi) - \frac{1}{\rho} \frac{\partial A_\rho}{\partial \varphi} \right] \qquad (1\text{-}88)$$

while $\nabla \cdot \nabla u$ turns out to be

$$\nabla^2 u = \frac{1}{\rho} \frac{\partial}{\partial \rho} \left(\rho \frac{\partial u}{\partial \rho} \right) + \frac{1}{\rho^2} \frac{\partial^2 u}{\partial \varphi^2} + \frac{\partial^2 u}{\partial z^2} \qquad (1\text{-}89)$$

where u is a scalar function of position.

It is very important to remember that (1-86), (1-87), (1-88), and (1-89) *cannot* be obtained from the corresponding expressions in rectangular coordinates as given by (1-41), (1-42), (1-43), and (1-46) by the simple replacement of x, y, z by ρ, φ, z. Similarly, (1-44) and (1-47) can only be used for rectangular coordinates; see (1-120) for the definition of $\nabla^2 \mathbf{A}$ for other coordinate systems. You would be surprised at how often these mistakes are made.

1-17 SPHERICAL COORDINATES

In this system, the location of a point P is specified by the three quantities r, θ, φ shown in Figure 1-39. We see that r is the distance from the origin and thus the magnitude of the position vector $\mathbf{r}$, θ is the angle made by $\mathbf{r}$ with the positive z axis, while φ is again the angle made with the positive x axis by the projection of $\mathbf{r}$ onto the xy plane. The relations between the rectangular and spherical coordinates are seen to be

$$x = r \sin \theta \cos \varphi \qquad y = r \sin \theta \sin \varphi \qquad z = r \cos \theta \qquad (1\text{-}90)$$

so that

$$r = \left(x^2 + y^2 + z^2 \right)^{1/2} \qquad \tan \theta = \frac{\left(x^2 + y^2 \right)^{1/2}}{z} \qquad \tan \varphi = \frac{y}{x} \qquad (1\text{-}91)$$

We define a set of mutually perpendicular unit vectors $\hat{\mathbf{r}}$, $\hat{\boldsymbol{\theta}}$, and $\hat{\boldsymbol{\varphi}}$ in the sense of increasing r, θ, and φ, respectively, as shown in Figure 1-39; we see that as the location of P is changed, all three of these vectors also change. They satisfy relations analogous to (1-76):

$$\hat{\mathbf{r}} \cdot \hat{\mathbf{r}} = \hat{\boldsymbol{\theta}} \cdot \hat{\boldsymbol{\theta}} = \hat{\boldsymbol{\varphi}} \cdot \hat{\boldsymbol{\varphi}} = 1$$
$$\hat{\mathbf{r}} \cdot \hat{\boldsymbol{\theta}} = \hat{\boldsymbol{\theta}} \cdot \hat{\boldsymbol{\varphi}} = \hat{\boldsymbol{\varphi}} \cdot \hat{\mathbf{r}} = 0 \qquad (1\text{-}92)$$
$$\hat{\mathbf{r}} \times \hat{\boldsymbol{\theta}} = \hat{\boldsymbol{\varphi}} \qquad \hat{\boldsymbol{\theta}} \times \hat{\boldsymbol{\varphi}} = \hat{\mathbf{r}} \qquad \hat{\boldsymbol{\varphi}} \times \hat{\mathbf{r}} = \hat{\boldsymbol{\theta}}$$

Their rectangular components are found from inspection of the figure to be

$$\hat{\mathbf{r}} = \sin \theta \cos \varphi \hat{\mathbf{x}} + \sin \theta \sin \varphi \hat{\mathbf{y}} + \cos \theta \hat{\mathbf{z}}$$
$$\hat{\boldsymbol{\theta}} = \cos \theta \cos \varphi \hat{\mathbf{x}} + \cos \theta \sin \varphi \hat{\mathbf{y}} - \sin \theta \hat{\mathbf{z}} \qquad (1\text{-}93)$$
$$\hat{\boldsymbol{\varphi}} = -\sin \varphi \hat{\mathbf{x}} + \cos \varphi \hat{\mathbf{y}}$$

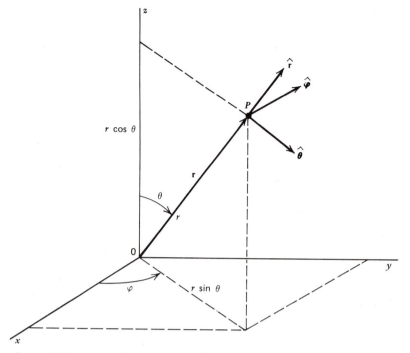

Figure 1-39.   Definition of spherical coordinates.

and therefore

$$\hat{\mathbf{x}} = \sin\theta\cos\varphi\hat{\mathbf{r}} + \cos\theta\cos\varphi\hat{\boldsymbol{\theta}} - \sin\varphi\hat{\boldsymbol{\varphi}}$$

$$\hat{\mathbf{y}} = \sin\theta\sin\varphi\hat{\mathbf{r}} + \cos\theta\sin\varphi\hat{\boldsymbol{\theta}} + \cos\varphi\hat{\boldsymbol{\varphi}} \qquad (1\text{-}94)$$

$$\hat{\mathbf{z}} = \cos\theta\hat{\mathbf{r}} - \sin\theta\hat{\boldsymbol{\theta}}$$

By differentiating (1-93), we obtain

$$\frac{\partial\hat{\mathbf{r}}}{\partial\theta} = \hat{\boldsymbol{\theta}} \qquad \frac{\partial\hat{\mathbf{r}}}{\partial\varphi} = \sin\theta\hat{\boldsymbol{\varphi}}$$

$$\frac{\partial\hat{\boldsymbol{\theta}}}{\partial\theta} = -\hat{\mathbf{r}} \qquad \frac{\partial\hat{\boldsymbol{\theta}}}{\partial\varphi} = \cos\theta\hat{\boldsymbol{\varphi}} \qquad (1\text{-}95)$$

$$\frac{\partial\hat{\boldsymbol{\varphi}}}{\partial\theta} = 0 \qquad \frac{\partial\hat{\boldsymbol{\varphi}}}{\partial\varphi} = -\sin\theta\hat{\mathbf{r}} - \cos\theta\hat{\boldsymbol{\theta}}$$

A vector **A** will be written in component form as

$$\mathbf{A} = A_r\hat{\mathbf{r}} + A_\theta\hat{\boldsymbol{\theta}} + A_\varphi\hat{\boldsymbol{\varphi}} \qquad (1\text{-}96)$$

The position vector is

$$\mathbf{r} = r\hat{\mathbf{r}} \qquad (1\text{-}97)$$

and its differential is found with the use of (1-93) and (1-95) to be

$$d\mathbf{r} = dr\,\hat{\mathbf{r}} + r\,d\theta\,\hat{\boldsymbol{\theta}} + r\sin\theta\,d\varphi\,\hat{\boldsymbol{\varphi}} \qquad (1\text{-}98)$$

These component displacements are shown in Figure 1-40. Since the shaded volume element has as sides just the components of (1-98), the volume element will be

$$d\tau = r^2\sin\theta\,dr\,d\theta\,d\varphi \qquad (1\text{-}99)$$

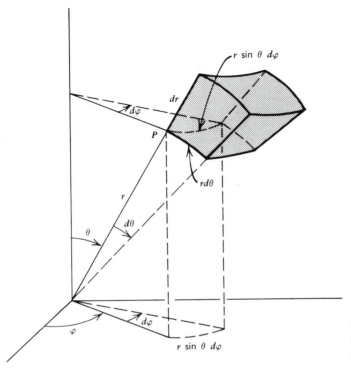

Figure 1-40. Volume element in terms of spherical coordinates.

while the components of an element of area $d\mathbf{a}$ will be given by

$$da_r = \pm r^2 \sin\theta\, d\theta\, d\varphi \qquad da_\theta = \pm r \sin\theta\, dr\, d\varphi \qquad da_\varphi = \pm r\, dr\, d\theta \quad (1\text{-}100)$$

If $u = u(r, \theta, \varphi)$, then

$$du = \frac{\partial u}{\partial r} dr + \frac{\partial u}{\partial \theta} d\theta + \frac{\partial u}{\partial \varphi} d\varphi$$

so that the gradient as obtained from (1-38) and (1-98) is

$$\nabla u = \hat{\mathbf{r}} \frac{\partial u}{\partial r} + \hat{\boldsymbol{\theta}} \frac{1}{r} \frac{\partial u}{\partial \theta} + \hat{\boldsymbol{\varphi}} \frac{1}{r \sin\theta} \frac{\partial u}{\partial \varphi} \qquad (1\text{-}101)$$

showing that the del operator is

$$\nabla = \hat{\mathbf{r}} \frac{\partial}{\partial r} + \hat{\boldsymbol{\theta}} \frac{1}{r} \frac{\partial}{\partial \theta} + \hat{\boldsymbol{\varphi}} \frac{1}{r \sin\theta} \frac{\partial}{\partial \varphi} \qquad (1\text{-}102)$$

If we now proceed to use (1-102) and (1-96) in the same manner as in the last section, taking account of (1-92) and (1-95), we find:

$$\nabla \cdot \mathbf{A} = \frac{1}{r^2} \frac{\partial}{\partial r}\left(r^2 A_r\right) + \frac{1}{r \sin\theta} \frac{\partial}{\partial \theta}\left(\sin\theta A_\theta\right) + \frac{1}{r \sin\theta} \frac{\partial A_\varphi}{\partial \varphi} \qquad (1\text{-}103)$$

$$\nabla \times \mathbf{A} = \frac{\hat{\mathbf{r}}}{r \sin\theta}\left[\frac{\partial}{\partial \theta}\left(\sin\theta A_\varphi\right) - \frac{\partial A_\theta}{\partial \varphi}\right] + \frac{\hat{\boldsymbol{\theta}}}{r}\left[\frac{1}{\sin\theta}\frac{\partial A_r}{\partial \varphi} - \frac{\partial}{\partial r}\left(r A_\varphi\right)\right]$$

$$+ \frac{\hat{\boldsymbol{\varphi}}}{r}\left[\frac{\partial}{\partial r}\left(r A_\theta\right) - \frac{\partial A_r}{\partial \theta}\right] \qquad (1\text{-}104)$$

$$\nabla^2 u = \frac{1}{r^2} \frac{\partial}{\partial r}\left(r^2 \frac{\partial u}{\partial r}\right) + \frac{1}{r^2 \sin\theta} \frac{\partial}{\partial \theta}\left(\sin\theta \frac{\partial u}{\partial \theta}\right) + \frac{1}{r^2 \sin^2\theta} \frac{\partial^2 u}{\partial \varphi^2} \qquad (1\text{-}105)$$

Remarks similar to those following (1-89) also apply here. You *cannot* get the correct spherical coordinate expressions for (1-102) through (1-105) from the corresponding ones in rectangular coordinates by simply replacing the symbols x, y, z by r, θ, φ.

1-18 SOME VECTOR RELATIONSHIPS

We list below some general results that are useful in evaluating and simplifying many of the expressions encountered when using vectors; we have already seen examples in (1-29) and (1-30). The first set of these expressions can be verified by straightforward, but sometimes tedious, calculations. Such verifications are most easily done in rectangular coordinates since the unit vectors involved are constants. For completeness, it should be noted that $\nabla^2 \mathbf{A}$ in (1-120) can be resolved as we did in (1-47) only for rectangular components; for other coordinate systems, (1-120) should be taken as a definition of $\nabla^2 \mathbf{A}$ to be evaluated from the rest of the identity.

$$(\mathbf{A} \times \mathbf{B}) \cdot (\mathbf{C} \times \mathbf{D}) = (\mathbf{A} \cdot \mathbf{C})(\mathbf{B} \cdot \mathbf{D}) - (\mathbf{A} \cdot \mathbf{D})(\mathbf{B} \cdot \mathbf{C}) \qquad (1\text{-}106)$$

$$\frac{d}{d\sigma}(u\mathbf{A}) = \frac{du}{d\sigma}\mathbf{A} + u\frac{d\mathbf{A}}{d\sigma} \qquad (1\text{-}107)$$

$$\frac{d}{d\sigma}(\mathbf{A} \cdot \mathbf{B}) = \frac{d\mathbf{A}}{d\sigma} \cdot \mathbf{B} + \mathbf{A} \cdot \frac{d\mathbf{B}}{d\sigma} \qquad (1\text{-}108)$$

$$\frac{d}{d\sigma}(\mathbf{A} \times \mathbf{B}) = \frac{d\mathbf{A}}{d\sigma} \times \mathbf{B} + \mathbf{A} \times \frac{d\mathbf{B}}{d\sigma} \qquad (1\text{-}109)$$

$$\nabla(u + v) = \nabla u + \nabla v \qquad (1\text{-}110)$$

$$\nabla(uv) = u\nabla v + v\nabla u \qquad (1\text{-}111)$$

$$\nabla(\mathbf{A} \cdot \mathbf{B}) = \mathbf{B} \times (\nabla \times \mathbf{A}) + \mathbf{A} \times (\nabla \times \mathbf{B}) + (\mathbf{B} \cdot \nabla)\mathbf{A} + (\mathbf{A} \cdot \nabla)\mathbf{B} \qquad (1\text{-}112)$$

$$\nabla(\mathbf{C} \cdot \mathbf{r}) = \mathbf{C} \quad \text{where} \quad \mathbf{C} = \text{const.} \qquad (1\text{-}113)$$

$$\nabla \cdot (\mathbf{A} + \mathbf{B}) = \nabla \cdot \mathbf{A} + \nabla \cdot \mathbf{B} \qquad (1\text{-}114)$$

$$\nabla \cdot (u\mathbf{A}) = \mathbf{A} \cdot (\nabla u) + u(\nabla \cdot \mathbf{A}) \qquad (1\text{-}115)$$

$$\nabla \cdot (\mathbf{A} \times \mathbf{B}) = \mathbf{B} \cdot (\nabla \times \mathbf{A}) - \mathbf{A} \cdot (\nabla \times \mathbf{B}) \qquad (1\text{-}116)$$

$$\nabla \times (\mathbf{A} + \mathbf{B}) = \nabla \times \mathbf{A} + \nabla \times \mathbf{B} \qquad (1\text{-}117)$$

$$\nabla \times (u\mathbf{A}) = (\nabla u) \times \mathbf{A} + u(\nabla \times \mathbf{A}) \qquad (1\text{-}118)$$

$$\nabla \times (\mathbf{A} \times \mathbf{B}) = (\nabla \cdot \mathbf{B})\mathbf{A} - (\nabla \cdot \mathbf{A})\mathbf{B} + (\mathbf{B} \cdot \nabla)\mathbf{A} - (\mathbf{A} \cdot \nabla)\mathbf{B} \qquad (1\text{-}119)$$

$$\nabla \times (\nabla \times \mathbf{A}) = \nabla(\nabla \cdot \mathbf{A}) - \nabla^2 \mathbf{A} \qquad (1\text{-}120)$$

where

$$(\mathbf{A} \cdot \nabla)\mathbf{B} = \hat{\mathbf{x}}\left(A_x \frac{\partial B_x}{\partial x} + A_y \frac{\partial B_x}{\partial y} + A_z \frac{\partial B_x}{\partial z} \right)$$

$$+ \hat{\mathbf{y}}\left(A_x \frac{\partial B_y}{\partial x} + A_y \frac{\partial B_y}{\partial y} + A_z \frac{\partial B_y}{\partial z} \right)$$

$$+ \hat{\mathbf{z}}\left(A_x \frac{\partial B_z}{\partial x} + A_y \frac{\partial B_z}{\partial y} + A_z \frac{\partial B_z}{\partial z} \right) \qquad (1\text{-}121)$$

If we take $\mathbf{A}$ to be a constant, but otherwise arbitrary, vector in (1-115) and (1-118), while $\mathbf{B}$ is taken as constant in (1-116), and use (1-59) and (1-67), we obtain three interesting and occasionally useful integral theorems:

$$\oint_S u\, d\mathbf{a} = \int_V \nabla u\, d\tau \tag{1-122}$$

$$\oint_S \mathbf{A} \times d\mathbf{a} = -\int_V (\nabla \times \mathbf{A})\, d\tau \tag{1-123}$$

$$\oint_C u\, d\mathbf{s} = -\int_S \nabla u \times d\mathbf{a} \tag{1-124}$$

We can obtain an alternative way of looking at the meanings of the gradient and curl from (1-122) and (1-123) by considering a small volume ΔV and proceeding as we did to get (1-66); we find that

$$\nabla u = \lim_{\Delta V \to 0} \frac{1}{\Delta V} \oint_S u\, d\mathbf{a} \tag{1-125}$$

$$\nabla \times \mathbf{A} = \lim_{\Delta V \to 0} \frac{1}{\Delta V} \oint_S d\mathbf{a} \times \mathbf{A} \tag{1-126}$$

Finally, if we integrate (1-115) over a volume and use (1-59), we get

$$\oint_S u\mathbf{A} \cdot d\mathbf{a} = \int_V [\mathbf{A} \cdot (\nabla u) + u(\nabla \cdot \mathbf{A})]\, d\tau \tag{1-127}$$

Consider the particular case in which $u = B_x$; then (1-127) becomes

$$\oint_S B_x(\mathbf{A} \cdot d\mathbf{a}) = \int_V [\mathbf{A} \cdot \nabla B_x + B_x(\nabla \cdot \mathbf{A})]\, d\tau \tag{1-128}$$

Since there are similar expressions involving the other two components of **B**, we recognize that (1-128) will also lead us to

$$\oint_S \mathbf{B}(\mathbf{A} \cdot d\mathbf{a}) = \int_V [(\mathbf{A} \cdot \nabla)\mathbf{B} + \mathbf{B}(\nabla \cdot \mathbf{A})]\, d\tau \tag{1-129}$$

1-19 FUNCTIONS OF THE RELATIVE COORDINATES

As we proceed, we will find that we are constantly dealing with functions that depend *only* on the differences of the coordinates, that is, they are functions solely of the combinations $x - x'$, $y - y'$, and $z - z'$. We see from (1-13) that these combinations are simply the components of the relative position vector **R**, hence the name "relative coordinates" for $x - x'$ and so on. Functions of this type have properties that will enable us to simplify much of our later work and it is convenient to consider them now.

Let f be such a function; f could be a scalar or a component of a vector. Since f depends only on the relative coordinates, we write it variously as $f(\mathbf{R}) = f(x - x', y - y', z - z') = f(X, Y, Z)$ where $X = x - x'$, $Y = y - y'$, $Z = z - z'$. Using the chain rule of differentiation, we find that

$$\frac{\partial f}{\partial x} = \frac{\partial f}{\partial X}\frac{\partial X}{\partial x} = \frac{\partial f}{\partial X} \qquad \text{and} \qquad \frac{\partial f}{\partial x'} = \frac{\partial f}{\partial X}\frac{\partial X}{\partial x'} = -\frac{\partial f}{\partial X}$$

so that

$$\frac{\partial f(\mathbf{R})}{\partial x} = -\frac{\partial f(\mathbf{R})}{\partial x'} \tag{1-130}$$

with similar expressions for the y and z derivatives. Following (1-41), we can define a del operator ∇' in terms of the primed coordinates as

$$\nabla' = \hat{\mathbf{x}}\frac{\partial}{\partial x'} + \hat{\mathbf{y}}\frac{\partial}{\partial y'} + \hat{\mathbf{z}}\frac{\partial}{\partial z'} \tag{1-131}$$

If now, we calculate the gradient of f according to (1-37) and use (1-130) and (1-131), we get

$$\nabla f(\mathbf{R}) = -\nabla' f(\mathbf{R}) \tag{1-132}$$

which shows us that when we are dealing with functions of the relative coordinates the ∇ and ∇' operators can be interchanged provided that the sign is also changed.

If the vector $\mathbf{A}$ is a function only of the relative coordinates, $\mathbf{A}(\mathbf{R})$, so that $A_x(\mathbf{R})$, $A_y(\mathbf{R})$, and $A_z(\mathbf{R})$, we can apply (1-130) and its analogues to (1-42) and (1-43) and obtain results similar to (1-132):

$$\nabla \cdot \mathbf{A}(\mathbf{R}) = -\nabla' \cdot \mathbf{A}(\mathbf{R}) \tag{1-133}$$

$$\nabla \times \mathbf{A}(\mathbf{R}) = -\nabla' \times \mathbf{A}(\mathbf{R}) \tag{1-134}$$

The Laplacian operator defined in (1-45) is unchanged, however:

$$\nabla^2 f(\mathbf{R}) = \nabla'^2 f(\mathbf{R}) \tag{1-135}$$

■ **Example**

The magnitude of the relative position vector R as given by (1-14) is an important example of the kind of function we have been discussing. By direct differentiation of (1-14) we find that

$$\frac{\partial R}{\partial x} = \frac{x - x'}{R} = -\frac{\partial R}{\partial x'} \tag{1-136}$$

with similar expressions for the y and z derivatives. If we combine these results with (1-41), (1-131), and (1-13), we get

$$\nabla R = -\nabla' R = \frac{\mathbf{R}}{R} = \hat{\mathbf{R}} \tag{1-137}$$

where $\hat{\mathbf{R}}$ is the unit vector in the direction of the relative position vector. Similar results are found for functions of R:

$$\frac{\partial g(R)}{\partial x} = \frac{dg}{dR}\frac{\partial R}{\partial x} = \frac{dg}{dR}\hat{R}_x \tag{1-138}$$

so that

$$\nabla g(R) = -\nabla' g(R) = \frac{dg(R)}{dR}\hat{\mathbf{R}} \tag{1-139}$$

and, in particular,

$$\nabla(R^n) = -\nabla'(R^n) = nR^{n-1}\hat{\mathbf{R}} \tag{1-140}$$

An especially important case of (1-140) corresponds to $n = -1$:

$$\nabla\left(\frac{1}{R}\right) = -\nabla'\left(\frac{1}{R}\right) = -\frac{\hat{\mathbf{R}}}{R^2} = -\frac{\mathbf{R}}{R^3} \tag{1-141}$$

The x component of (1-141) is

$$\frac{\partial}{\partial x}\left(\frac{1}{R}\right) = -\frac{\partial}{\partial x'}\left(\frac{1}{R}\right) = -\frac{(x - x')}{R^3} \tag{1-142}$$

If we differentiate this once more, we get

$$\frac{\partial^2}{\partial x^2}\left(\frac{1}{R}\right) = -\frac{1}{R^3} + \frac{3(x - x')^2}{R^5} \tag{1-143}$$

with the use of (1-136). There will be similar expressions for the y and z second derivatives; if we add them to (1-143), use (1-14) and (1-45), we find that

$$\nabla^2\left(\frac{1}{R}\right) = -\frac{3}{R^3} + \frac{3R^2}{R^5} = 0$$

so that, if we also recall (1-135),

$$\nabla^2\left(\frac{1}{R}\right) = \nabla'^2\left(\frac{1}{R}\right) = 0 \qquad (R \neq 0) \tag{1-144}$$

We have added the parenthetical term in (1-144) to remind us that all of our calculations from (1-141) on were made with the implicit assumption that $R \neq 0$ so that we were dealing with finite quantities. If we recall that $\nabla^2 = \nabla \cdot \nabla$, we see from (1-141) that (1-144) can also be written as

$$\nabla \cdot \left(\frac{\hat{\mathbf{R}}}{R^2}\right) = \nabla \cdot \left(\frac{\mathbf{R}}{R^3}\right) = 0 \qquad (R \neq 0) \tag{1-145}$$

$$\nabla' \cdot \left(\frac{\hat{\mathbf{R}}}{R^2}\right) = \nabla' \cdot \left(\frac{\mathbf{R}}{R^3}\right) = 0 \qquad (R \neq 0) \tag{1-146}$$

It will be helpful later to know that

$$\nabla \times \left(\frac{\hat{\mathbf{R}}}{R^2}\right) = -\nabla \times \nabla\left(\frac{1}{R}\right) = 0 \tag{1-147}$$

which follows from (1-141) and (1-48). ∎

1-20 THE HELMHOLTZ THEOREM

We will not prove this theorem at this time but simply quote it as an aid to understanding the motivations for many of the procedures we will be following. We will, in effect, prove it eventually, but piece by piece.

The theorem deals with the question of what information we need to calculate a vector field. Basically the answer is that if the divergence and curl of a vector field are known everywhere in a finite region, then the vector field can be found uniquely. We consider a vector field $\mathbf{F} = \mathbf{F}(x, y, z) = \mathbf{F}(\mathbf{r})$ and assume that the functions $\nabla \cdot \mathbf{F} = b(\mathbf{r})$ and $\nabla \times \mathbf{F} = \mathbf{c}(\mathbf{r})$ are given to us everywhere in a finite volume V, that is, they are known functions of position. Then, if we define the following two functions,

$$\Phi(\mathbf{r}) = \frac{1}{4\pi}\int_V \frac{b(\mathbf{r}')\,d\tau'}{|\mathbf{r} - \mathbf{r}'|} \tag{1-148}$$

$$\mathscr{A}(\mathbf{r}) = \frac{1}{4\pi}\int_V \frac{\mathbf{c}(\mathbf{r}')\,d\tau'}{|\mathbf{r} - \mathbf{r}'|} \tag{1-149}$$

the theorem tells us that $\mathbf{F}$ can be found from

$$\mathbf{F} = \mathbf{F}(\mathbf{r}) = -\nabla\Phi + \nabla \times \mathscr{A} \tag{1-150}$$

In these expressions, $|\mathbf{r} - \mathbf{r}'|$ is seen to be the magnitude of the relative position vector as given by (1-12) and shown in Figure 1-12.

Since $\mathbf{F}$ can be found from a knowledge of them, its divergence and curl are often called the *sources* of the field. The point $\mathbf{r}$ where we evaluate $\mathbf{F}$ is called the *field point*, while the point **r′** where the sources are evaluated for purposes of integration is called the *source point*; $d\tau'$ is then a volume element at the location of the source point. Similarly, the derivatives in (1-150) involve the components of the field point **r**. The functions Φ and $\mathscr{A}$ are called *scalar* and *vector potentials*, respectively, because **F** is obtained from them by differentiation. For reasons that will become clear later, the divergence is known as the "charge" source while the curl is called a "current" source.

If there are sources at infinity, so that they are not all confined to a finite volume, the above theorem is not absolutely correct, because one must also include certain surface integrals involving **F**. We will handle such situations by special methods when we encounter them.

Much of the content of modern electromagnetism involves various electric and magnetic field vectors. Consequently, much of our effort will be devoted to finding expressions for their divergence and curl from fundamental experimental results; this theorem shows us why we want to know them. When the complete set of these source equations are found, they are called *Maxwell's equations* and form the fundamental description of the electromagnetic field.

EXERCISES

I really understand the theory;
I just can't work the problems.

—Anonymous

1-1 Verify Equation 1-2 by graphical methods.

1-2 Given the two vectors $\mathbf{A} = 2\hat{\mathbf{x}} - 3\hat{\mathbf{y}} - 4\hat{\mathbf{z}}$ and $\mathbf{B} = 6\hat{\mathbf{x}} + 5\hat{\mathbf{y}} + \hat{\mathbf{z}}$, find the magnitudes and angles made with the x, y, and z axes for $\mathbf{A} + \mathbf{B}$ and $\mathbf{A} - \mathbf{B}$.

1-3 Find the relative position vector **R** of the point $P(2, -2, 3)$ with respect to $P'(-3, 1, 4)$. What are the direction angles of **R**?

1-4 Given the two vectors $\mathbf{A} = \hat{\mathbf{x}} + 2\hat{\mathbf{y}} + 3\hat{\mathbf{z}}$ and $\mathbf{B} = 4\hat{\mathbf{x}} - 5\hat{\mathbf{y}} + 6\hat{\mathbf{z}}$, find the angle between them. Find the component of **A** in the direction of **B**.

1-5 Given the vectors $\mathbf{A} = 2\hat{\mathbf{x}} + 3\hat{\mathbf{y}} - 4\hat{\mathbf{z}}$ and $\mathbf{B} = -6\hat{\mathbf{x}} - 4\hat{\mathbf{y}} + \hat{\mathbf{z}}$. Find the component of $\mathbf{A} \times \mathbf{B}$ along the direction of $\mathbf{C} = \hat{\mathbf{x}} - \hat{\mathbf{y}} + \hat{\mathbf{z}}$.

1-6 Verify (1-29) and (1-30).

1-7 Show that $\mathbf{A} \cdot (\mathbf{B} \times \mathbf{C})$ equals the volume of a parallelepiped if **A**, **B**, and **C** are the vectors representing the three edges with a common corner.

1-8 A family of hyperbolas in the xy plane is given by $u = xy$. Find ∇u. Given the vector $\mathbf{A} = 3\hat{\mathbf{x}} + 2\hat{\mathbf{y}} + 4\hat{\mathbf{z}}$, find the component of **A** in the direction of ∇u at the point on the curve for which $u = 3$ and for which $x = 2$.

1-9 The equation giving a family of ellipsoids is

$$u = \frac{x^2}{a^2} + \frac{y^2}{b^2} + \frac{z^2}{c^2}$$

Find the unit vector normal to each point of the surface of these ellipsoids.

1-10 Verify (1-48) and (1-49) by direct calculation.

1-11 Do the example of Section 1-11 by integrating over y rather than x and thus show that the same result is obtained.

1-12 Find the surface integral of **r** over a surface of a sphere of radius a and center at the origin. Also find the volume integral of $\nabla \cdot \mathbf{r}$ and compare your results.

1-13 Given the vector field $\mathbf{A} = xy\hat{\mathbf{x}} + yz\hat{\mathbf{y}} + zx\hat{\mathbf{z}}$. Evaluate directly the flux of **A** through the surface of a rectangular parallelepiped of sides

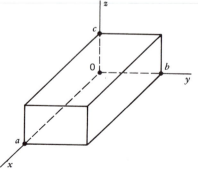

Figure 1-41. Rectangular parallelepiped with a corner at the origin.

a, b, c with origin at one corner and edges along the positive directions of the rectangular axes as shown in Figure 1-41. Evaluate $\int \nabla \cdot \mathbf{A} \, d\tau$ over the volume of this same parallelepiped and compare your results.

1-14 Calculate directly the line integral $\oint \mathbf{A} \cdot d\mathbf{s}$ of the vector $\mathbf{A} = -y\hat{\mathbf{x}} + x\hat{\mathbf{y}}$ around the closed path in the xy plane with straight sides given by: $(0, 0) \rightarrow (3, 0) \rightarrow (3, 4) \rightarrow (0, 4) \rightarrow (0, 0)$. Also calculate the surface integral of $\nabla \times \mathbf{A}$ over the enclosed area and show that (1-67) is satisfied.

1-15 Given the vector field $\mathbf{A} = x^2 y\hat{\mathbf{x}} + xy^2\hat{\mathbf{y}} + a^3 e^{-\beta y} \cos \alpha x \hat{\mathbf{z}}$ where a, α, β are constants. Evaluate directly the line integral of $\mathbf{A}$ over the closed path in the xy plane shown in Figure 1-42. The straight portions are parallel to the axes, and the curved portion is the parabola $y^2 = kx$ where $k = \text{const}$. Evaluate the surface integral of $\nabla \times \mathbf{A}$ over the area S enclosed by C and compare your results.

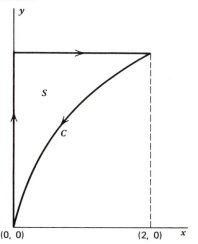

Figure 1-42. Path of integration for Exercise 1-15.

1-16 Verify (1-88) and (1-89).

1-17 Verify (1-103), (1-104), and (1-105).

1-18 Given $\mathbf{A} = a\hat{\mathbf{x}} + b\hat{\mathbf{y}} + c\hat{\mathbf{z}}$ where a, b, c are constants. Is $\mathbf{A}$ a constant vector? Find the cylindrical and spherical components of $\mathbf{A}$, expressing them in terms of ρ, φ, z and r, θ, φ respectively.

1-19 Given $\mathbf{A} = a\hat{\boldsymbol{\rho}} + b\hat{\boldsymbol{\varphi}} + c\hat{\mathbf{z}}$ where a, b, c are constants. Is $\mathbf{A}$ a constant vector? Find $\nabla \cdot \mathbf{A}$ and $\nabla \times \mathbf{A}$. Find the rectangular and spherical components of $\mathbf{A}$, expressing them in terms of x, y, z and r, θ, φ, respectively.

1-20 Given $\mathbf{A} = a\hat{\mathbf{r}} + b\hat{\boldsymbol{\theta}} + c\hat{\boldsymbol{\varphi}}$ where a, b, c are constants. Is $\mathbf{A}$ a constant vector? Find $\nabla \cdot \mathbf{A}$ and $\nabla \times \mathbf{A}$. Find the rectangular and cylindrical components of $\mathbf{A}$, expressing them in terms of x, y, z and ρ, φ, z, respectively.

1-21 Find $\nabla \cdot \mathbf{r}$ for the position vector $\mathbf{r}$ expressed in rectangular, cylindrical, and spherical coordinates, thus showing that the same result is obtained in all cases.

1-22 Given $\mathbf{A} = a\rho\hat{\boldsymbol{\rho}} + b\hat{\boldsymbol{\varphi}} + cz\hat{\mathbf{z}}$ where a, b, c are constants. Find $\oint \mathbf{A} \cdot d\mathbf{a}$ over the surface of a right circular cylinder of length L and radius ρ_0. The axis of the cylinder is along the positive z axis and the origin is at the center of the lower circular face. Also find $\int \nabla \cdot \mathbf{A} \, d\tau$ over the volume enclosed by the cylinder and compare your results.

1-23 Given the vector $\mathbf{A} = 4\hat{\mathbf{r}} + 3\hat{\boldsymbol{\theta}} - 2\hat{\boldsymbol{\varphi}}$, find its line integral around the closed path shown in Figure 1-43. The curved portion is the arc of a circle of radius r_0 centered at the origin. Also find the surface integral of $\nabla \times \mathbf{A}$ over the enclosed area and compare your results.

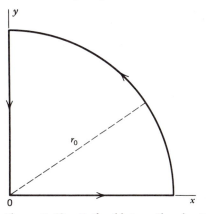

Figure 1-43. Path of integration for Exercise 1-23.

1-24 Verify (1-115) and (1-118).

1-25 By applying the divergence theorem to the special case in which $\mathbf{A}$ is a constant but otherwise arbitrary vector, show that the total vector area of a closed surface is zero, that is, $\oint d\mathbf{a} = 0$. Similarly, show that $\oint d\mathbf{s} = 0$. Do these results surprise you?

1-26 Verify (1-122), (1-123), and (1-124) by using the method suggested in the text.

2

COULOMB'S LAW

The phenomena nowadays associated with the term "electricity" go back many centuries to observations that when certain naturally occurring materials were rubbed, they acquired the ability to exert forces on other objects. A typical material of this sort is amber (*ēlektron* in Greek). The process is called "electrification by friction" or "triboelectrification" and, in order to describe the altered state of the matter, one says that it has become "charged" or has an "electric charge" on it.

After many experiments and much thought, it was finally concluded that electrification by friction did not represent a process of creation of electric charge, but rather a *separation* of two types of charge that were originally present in equal amounts in the uncharged "neutral" material. These two types of charge are arbitrarily called "positive" and "negative." Positive charge is *defined* as that which is left on a glass rod after it has been rubbed with a silk cloth; since the process is one of separation, the silk cloth will be left with a negative charge equal in magnitude to that on the glass rod. As implied by these remarks, and confirmed by all subsequent experiments, electric charge is *conserved* in the sense that net charge cannot be created or destroyed; we will put this fundamental experimental result in quantitative terms in Chapter 12.

The forces between electric charges can be forces of repulsion as well as of attraction. The first quantitative investigation of the dependence of these forces on the magnitudes of the charges and the distance between them was made by Coulomb in 1785 and the result is known as Coulomb's law.

2-1 POINT CHARGES

We use the symbol q to represent electric charge. In a general situation, the charge of an object will be distributed in some manner on or throughout it and the force between two objects will depend on these distributions as well as on the total amount of each charge. As a result, it is convenient to begin with the case of a *point charge* in which it is assumed that all of the charge is located at a geometrical point in space. This is obviously an idealization, but can be approximated very well in the laboratory by making any distances of separation that are involved very large compared to the dimensions of the charged objects.

In order to make further progress one has to be able to compare the magnitudes of two point charges q_1 and q_2. This can be done by introducing another arbitrary point charge q, putting it at a fixed distance R from q_1, and measuring the resultant force $\mathbf{F}_1$ on q; this is illustrated in Figure 2-1a. Then q_1 is removed and replaced by q_2 at the same distance R from q; the new force $\mathbf{F}_2$ on q can then be measured as indicated in Figure 2-1b. Since both q and R are the same in the two cases, the difference in the forces can only be due to the difference in the numerical values of the charges q_1 and q_2, and it is natural to ascribe the magnitudes of the forces as being directly proportional to the magnitudes of q_1 and q_2. Accordingly, we can *define* the ratio of their magnitudes as equal to the ratio of the magnitudes of the forces they produce on

Figure 2-1.   Comparing charges by comparing the forces they exert.

the arbitrary charge q; thus, we get

$$\frac{|q_1|}{|q_2|} = \frac{|\mathbf{F}_1|}{|\mathbf{F}_2|} \tag{2-1}$$

for q and R both constant.

   Once this procedure for comparing magnitudes has been established, one can proceed to the study of how the force between two point charges depends on their relative sizes. In addition, one can now assign absolute values to the charges by *choosing* a charge of unit magnitude in some arbitrary but convenient way, and by using (2-1) with the numerical value $|q_{unit}| = 1$.

2-2   COULOMB'S LAW

This basic experimental law refers to the situation illustrated in Figure 2-2 in which we have two point charges q and q' separated by a distance R; we assume the charges to be fixed in position and that there is no other matter, that is, the charges are situated in a vacuum. The force *on q due to q'* will be written as $\mathbf{F}_{q' \to q}$. Thus, q' is being treated as the origin of the force, and we can refer to q' as the "source" and its location given by $\mathbf{r}'$ as the "source point." Since q is the charge for which the force is to be found, we say it is at the "field point" located by $\mathbf{r}$. Then, according to (1-12), $\mathbf{R}$ is the relative position vector of q with respect to q' and is seen to be directed from the source point to the field point as is its corresponding unit vector $\hat{\mathbf{R}}$. Thus, using (1-12) and (1-4), we have

$$\mathbf{R} = \mathbf{r} - \mathbf{r}' \qquad R = |\mathbf{r} - \mathbf{r}'| \qquad \hat{\mathbf{R}} = \frac{\mathbf{R}}{R} \tag{2-2}$$

In terms of all of these quantities, *Coulomb's law* says that

$$\mathbf{F}_{q' \to q} = \frac{1}{4\pi\epsilon_0} \frac{qq'}{R^2} \hat{\mathbf{R}} \tag{2-3}$$

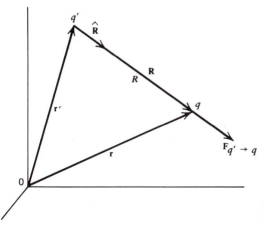

Figure 2-2.   Position vectors involved in Coulomb's law.

so that the force is proportional to the product of the charges and to the inverse square of the distance between them; it is seen to be similar to gravitation in these respects.

The factor $1/4\pi\epsilon_0$ is a constant of proportionality whose numerical value will depend on the system of units being used; it is written in this form for later convenience. We will be using exclusively the International System of units (SI, for *Systéme International* d'Unites), which is essentially the same as the MKSA system. This means that distance is measured in meters, mass in kilograms, time in seconds, force in newtons, energy in joules, and so on. Charge is defined in this system in terms of electric current which is rate of flow of charge. The unit of current is called an *ampere*, while the unit of charge is given the name *coulomb* and is defined by 1 coulomb = 1 ampere-second. We defer giving the precise definition of the ampere in terms of the magnetic forces between currents until Section 13-2; in the meantime, we can still take the coulomb as a known charge unit for our purposes. Other systems of units that are used in electromagnetism are discussed in Chapter 23.

The significance of this to us is that the units of all of the physical quantities in Coulomb's law have been already chosen so that the constant of proportionality must be found *by experiment*; this necessity is analogous to that of finding the gravitational constant appearing in the law of gravitation. The result turns out to be that

$$\epsilon_0 = 8.85 \times 10^{-12} \text{ (coulomb)}^2/\text{newton-(meter)}^2$$

$$= 8.85 \times 10^{-12} \text{ farad/meter} \tag{2-4}$$

The constant ϵ_0 is called the *permittivity of free space*, and is generally written in the last form from which we see by comparison of both forms that 1 farad = 1 (coulomb)^2/joule. It is also useful to note that

$$\frac{1}{4\pi\epsilon_0} = 9 \times 10^9 \frac{\text{meter}}{\text{farad}} \tag{2-5}$$

to an accuracy that will be sufficient for our purposes.

From (2-3), we see that if $qq' > 0$, so that both charges are of the same sign, then $\mathbf{F}_{q' \to q}$ is in the direction of $\hat{\mathbf{R}}$, that is, the force is repulsive as is seen from Figure 2-2. On the other hand, if $qq' < 0$, so that the charges have opposite signs, $\mathbf{F}_{q' \to q}$ is in the opposite direction to $\hat{\mathbf{R}}$, that is, the force on q is one of attraction toward q'. This is often summarized by the statement that "like" charges repel one another, while "unlike" charges attract.

We can also write Coulomb's law completely in terms of $\mathbf{R}$ by combining (2-2) and (2-3) to give

$$\mathbf{F}_{q' \to q} = \frac{qq'\mathbf{R}}{4\pi\epsilon_0 R^3} \tag{2-6}$$

If we wanted the force of q on q', which we write as $\mathbf{F}_{q \to q'}$, the only change necessary would be to use the relative position vector of q' with respect to q, that is, $\mathbf{R}' = \mathbf{r}' - \mathbf{r}$, so that

$$\mathbf{F}_{q \to q'} = \frac{q'q\mathbf{R}'}{4\pi\epsilon_0 R'^3} \tag{2-7}$$

We see from (2-2) that $\mathbf{R}' = -\mathbf{R}$ and since the magnitude of each is equal to $|\mathbf{r} - \mathbf{r}'|$, we see from (2-6) and (2-7) that

$$\mathbf{F}_{q \to q'} = -\mathbf{F}_{q' \to q} \tag{2-8}$$

which shows that the Coulomb forces are equal and opposite even though the individual charges may differ greatly in magnitude.

We are assuming a static situation, that is, the charges are at rest at fixed positions. This means that in order for q to be in equilibrium there must be an additional mechanical force $\mathbf{F}_{q,m}$ on it so that the net force will be zero; in other words, we must have

$$\mathbf{F}_{q' \to q} + \mathbf{F}_{q,m} = 0 \tag{2-9}$$

Similar remarks apply to q'.

2-3 SYSTEMS OF POINT CHARGES

Now suppose that, in addition to q, there are a number N of point charges distributed at fixed positions throughout otherwise empty space. We designate each charge by q_i and its position vector by $\mathbf{r}_i$ where $i = 1, 2, \ldots, N$. This situation is illustrated in Figure 2-3; for clarity, the individual position vectors are not shown, but the unit vectors $\hat{\mathbf{R}}_i$ corresponding to the relative positions of q with respect to the q_i are shown. Each of these charges can exert a force on q, $\mathbf{F}_{q_i \to q}$, which will be of the general form given by (2-3) or (2-6). The experimental facts of the superposition properties of forces are already familiar to us from mechanics; hence $\mathbf{F}_q$, the total force on q, will be given by the vector sum of the individual forces so that

$$\mathbf{F}_q = \sum_{i=1}^{N} \mathbf{F}_{q_i \to q} = \sum_{i=1}^{N} \frac{qq_i\hat{\mathbf{R}}_i}{4\pi\epsilon_0 R_i^2} = \sum_{i=1}^{N} \frac{qq_i\mathbf{R}_i}{4\pi\epsilon_0 R_i^3} \tag{2-10}$$

where

$$\mathbf{R}_i = \mathbf{r} - \mathbf{r}_i \qquad R_i = |\mathbf{r} - \mathbf{r}_i| \qquad \hat{\mathbf{R}}_i = \frac{\mathbf{R}_i}{R_i} \tag{2-11}$$

The last form in (2-10) is often more convenient to write as a starting point in solving problems, whereas we will usually use the form expressed in terms of the unit vectors for general discussions. Equation 2-10 expresses the fact that the total force can be found as the sum of the individual forces between pairs that are calculated from Coulomb's law as if the other charges were not present. Again we assume that the individual charges are at rest, and are kept at rest, by mechanical forces of some sort as may be required.

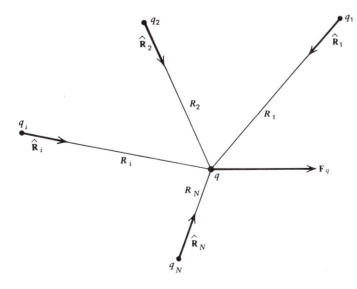

Figure 2-3. Calculation of total force due to more than one charge.

■ **Example**

If we express all positions in rectangular coordinates, we can easily write down an explicit form for (2-10). Using (1-13) and (1-14), and noting that the various charges are designated by the subscripts i rather than by primes, we find that (2-10) becomes

$$\mathbf{F}_q = \sum_{i=1}^{N} \frac{qq_i}{4\pi\epsilon_0} \frac{[(x - x_i)\hat{\mathbf{x}} + (y - y_i)\hat{\mathbf{y}} + (z - z_i)\hat{\mathbf{z}}]}{[(x - x_i)^2 + (y - y_i)^2 + (z - z_i)^2]^{3/2}} \tag{2-12}$$

In a sense, (2-12) provides a simple recipe for solving problems, since once the values of all the charges and their positions in rectangular coordinates are given, all that remains is to substitute these numbers into (2-12) and to simplify the result as much as possible.

■

2-4 CONTINUOUS DISTRIBUTIONS OF CHARGE

We often encounter situations in which the other charges are so close together compared to the other distances of interest that we can regard them as being continuously distributed, much as we can treat a glass of water, on a laboratory scale, as a continuous distribution of mass by neglecting its molecular structure. We can deal with such a case by considering a region of the charge distribution that is so small that the charge within it can be written as dq' and treated as a point charge; this is illustrated in Figure 2-4. We can still use (2-10), but now the sum will become an integral over the complete charge distribution so that

$$\mathbf{F}_q = \frac{q}{4\pi\epsilon_0} \int \frac{dq'\hat{\mathbf{R}}}{R^2} \tag{2-13}$$

where (2-2) continues to be applicable.

If the charges are distributed throughout a volume, we can introduce a *volume charge density* ρ, which is defined as the charge per unit volume and hence will be measured in coulombs/(meter)^3. (We will write this charge density as ρ_{ch} in the infrequent cases in which it might be confused with the ρ of cylindrical coordinates.) Then the charge contained in a small source volume $d\tau'$ will be given by

$$dq' = \rho(\mathbf{r}') \, d\tau' \tag{2-14}$$

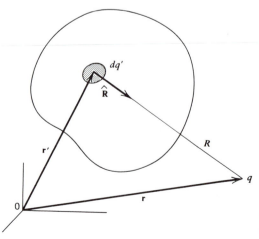

Figure 2-4. Charge element of a continuous distribution.

as shown in Figure 2-5*a*, and (2-13) will become

$$\mathbf{F}_q = \frac{q}{4\pi\epsilon_0} \int_{V'} \frac{\rho(\mathbf{r}')\hat{\mathbf{R}}\, d\tau'}{R^2} \tag{2-15}$$

We have written $\rho = \rho(\mathbf{r}')$ because, in general, the volume density can vary with the location of the source point; the integral in (2-15) is to be taken over the total volume V' containing the charge distribution.

When we say that $d\tau'$ is a "small" volume, we mean that it is small on a macroscopic, laboratory scale. On the other hand, it must be large on a microscopic, atomic scale so that it will contain many atoms and/or molecules. Only in this way can we treat ρ as a continuously varying function of position. If $d\tau'$ were made comparable to, or smaller than, atomic sizes, then at most locations $d\tau'$ would contain no charge and ρ would be practically always zero. The charge density would be different from zero only when $d\tau'$ included an electronic or nuclear charge, but this would lead to such widely fluctuating values for ρ that it would no longer be a useful concept.

Similarly, the charges can often be idealized to lie on a surface or along a line. We introduce analogous charge densities: the *surface charge density* σ defined as the charge per unit area, and the *linear charge density* λ defined as the charge per unit length; they will have units of coulomb/(meter)^2 and coulomb/meter, respectively, and can also vary with position in general. From these definitions, we get

$$dq' = \sigma(\mathbf{r}')\, da' \qquad \text{or} \qquad dq' = \lambda(\mathbf{r}')\, ds' \tag{2-16}$$

as indicated in Figure 2-5*b* and *c*. For such cases, (2-13) will become

$$\mathbf{F}_q = \frac{q}{4\pi\epsilon_0} \int_{S'} \frac{\sigma(\mathbf{r}')\hat{\mathbf{R}}\, da'}{R^2} \tag{2-17}$$

$$\mathbf{F}_q = \frac{q}{4\pi\epsilon_0} \int_{L'} \frac{\lambda(\mathbf{r}')\hat{\mathbf{R}}\, ds'}{R^2} \tag{2-18}$$

where (2-17) is to be integrated over the total surface S' on which there is a surface distribution while (2-18) covers the whole line L' occupied by a linear distribution of charge.

Finally, if all the possibilities we have discussed are simultaneously present, the total force on q would be obtained from the sum of all forces due to the various charge distributions, that is,

$$\mathbf{F}_q = \mathbf{F}_q(\text{from points}) + \mathbf{F}_q(\text{from volumes})$$
$$+ \mathbf{F}_q(\text{from surfaces}) + \mathbf{F}_q(\text{from lines}) \tag{2-19}$$

A very important word of warning: you can avoid a lot of trouble, wasted time, and wrong answers by remembering, and *following*, these two simple, although almost trivial sounding rules—(1) *always* draw the relative position vector, and hence $\hat{\mathbf{R}}$, *from* the source point *to* the field point; (2) *never* write the location of a *source* point as $\mathbf{r}$ or

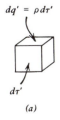

$dq' = \rho\, d\tau'$

$d\tau'$

(a)

$dq' = \sigma\, da'$

da'

(b)

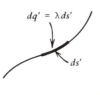

$dq' = \lambda\, ds'$

ds'

(c)

Figure 2-5.   Definition of various charge densities.

(x, y, z) and so on, but instead use $\mathbf{r}'$, or (x', y', z') or some sort of label, such as was done in (2-10) and (2-11).

2-5   POINT CHARGE OUTSIDE A UNIFORM SPHERICAL CHARGE DISTRIBUTION

As an example of the effect of a continuous charge distribution, we will evaluate (2-15) for a case in which q is located outside a sphere containing a uniform distribution of charge, that is, for which $\rho = \text{const}$. We choose the origin at the center of the sphere of radius a and let q be on the z axis so that $z > a$; the situation is shown in Figure 2-6 in which only one octant of the sphere is pictured. We use spherical coordinates to describe the source point $\mathbf{r}'$ and for doing the integration. Figure 2-7 shows the plane containing the z axis, $\mathbf{r}'$, and $\mathbf{R}$. We see that $\mathbf{r} = z\hat{\mathbf{z}}$ and $\mathbf{r}' = r'\hat{\mathbf{r}}'$ from (1-11) and (1-97), and therefore $\mathbf{R} = z\hat{\mathbf{z}} - r'\hat{\mathbf{r}}'$ according to (2-2); it then follows from (1-17), (1-19), (1-92), (1-15), and Figure 2-7 that

$$R^2 = z^2 + r'^2 - 2zr'\hat{\mathbf{z}} \cdot \hat{\mathbf{r}}' = z^2 + r'^2 - 2zr'\cos\theta'$$

We see, in fact, that this value of R^2 is exactly that given by the law of cosines as applied to Figure 2-7. If we now obtain $\hat{\mathbf{R}}$ from these results, we find that (2-15) becomes

$$\mathbf{F}_q = \frac{q\rho}{4\pi\epsilon_0} \int_{\text{sphere}} \frac{(z\hat{\mathbf{z}} - r'\hat{\mathbf{r}}')\, d\tau'}{(z^2 + r'^2 - 2zr'\cos\theta')^{3/2}} \tag{2-20}$$

where ρ has been removed from under the integral sign because it is constant.

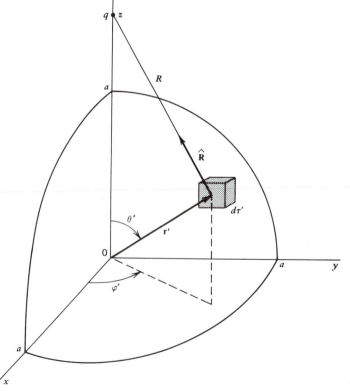

Figure 2-6.   Point charge outside a uniform spherical charge distribution.

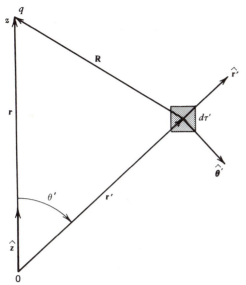

Figure 2-7. Another view of the situation in Figure 2-6.

Because $\hat{\mathbf{r}}'$ is not constant during the integration, it will be convenient to find $\mathbf{F}_q$ in terms of its rectangular components. Following (1-21), we dot each side of (2-20) with $\hat{\mathbf{z}}$, and use (1-19), (1-93), and (1-18); then, after writing $d\tau'$ in the form (1-99), we get

$$F_{qz} = \frac{q\rho}{4\pi\epsilon_0} \int_0^{2\pi} \int_0^{\pi} \int_0^a \frac{(z - r'\cos\theta')r'^2 \sin\theta' \, dr' \, d\theta' \, d\varphi'}{(z^2 + r'^2 - 2zr'\cos\theta')^{3/2}} \tag{2-21}$$

The integration over $d\varphi'$ can be performed at once and gives 2π. We do the integration over $d\theta'$ next. For this purpose, it is convenient to introduce a new variable $\mu = \cos\theta'$. Then $d\mu = -\sin\theta' \, d\theta'$, and, if $f = f(\cos\theta')$ is a function of $\cos\theta'$, we make the indicated substitutions and get the general and useful result that

$$\int_0^{\pi} f(\cos\theta') \sin\theta' \, d\theta' = \int_{-1}^{1} f(\mu) \, d\mu \tag{2-22}$$

When this is done, and the 2π from the integral over $d\varphi'$ is included, (2-21) becomes

$$F_{qz} = \frac{q\rho}{2\epsilon_0} \int_0^a r'^2 \, dr' \int_{-1}^{1} \frac{(z - r'\mu) \, d\mu}{(z^2 + r'^2 - 2zr'\mu)^{3/2}} \tag{2-23}$$

The integral over μ can be found from tables to be

$$\left. \frac{(z\mu - r')}{z^2(z^2 + r'^2 - 2zr'\mu)^{1/2}} \right|_{-1}^{1} = \frac{1}{z^2}\left(\frac{z - r'}{|z - r'|} + \frac{z + r'}{|z + r'|} \right) \tag{2-24}$$

where we have written the terms $[(z \pm r')^2]^{1/2}$, which appear, as $|z \pm r'|$ in order to emphasize that we must be sure that we get a positive value for the square root. In our case, we have assumed that q is outside the sphere so that $z > a$; since $r' \le a$, we will always have $z > r'$ so that $|z - r'| = z - r'$. Now $|z + r'| = z + r'$ since we have taken z to be positive and r' always is. When these are substituted into (2-24), the integral over μ in (2-23) is seen to be just $2/z^2$, which is constant as far as the integration over r' is concerned, so that (2-23) becomes

$$F_{qz} = \frac{q\rho}{\epsilon_0 z^2} \int_0^a r'^2 \, dr' = \frac{q\rho a^3}{3\epsilon_0 z^2} \tag{2-25}$$

Before we discuss this result, let us find the remaining components.

If we dot (2-20) with $\hat{\mathbf{x}}$, and use (1-93), we see that the resulting integrand will be proportional to $\cos \varphi'$ and therefore

$$F_{qx} = \hat{\mathbf{x}} \cdot \mathbf{F}_q \sim \int_0^{2\pi} \cos \varphi' \, d\varphi' = 0$$

Similarly,

$$F_{qy} = \hat{\mathbf{y}} \cdot \mathbf{F}_q \sim \int_0^{2\pi} \sin \varphi' \, d\varphi' = 0$$

The fact that these two components vanish is a consequence of the "symmetry" of the situation as we can see with the use of Figure 2-8. The charge element contained in $d\tau'$ will produce its contribution $d\mathbf{F}'$ to the total force and this contribution will have a horizontal component. Corresponding to $d\tau'$, however, is another volume element $d\tau''$ that is the reflection of $d\tau'$ in the line $\mathbf{r}$, and hence is at the same distance R from q. The equal charge contained in $d\tau''$ will produce a contribution $d\mathbf{F}''$ to the total force. Since $d\mathbf{F}''$ and $d\mathbf{F}'$ have the same magnitude, we see that the horizontal component of $d\mathbf{F}''$ will be equal and opposite to that of $d\mathbf{F}'$. Thus, when the contributions of this pair are added, the horizontal forces will cancel, although the vertical ones will not. Since all charge elements in the sphere can be paired off in this way, the total force will have no net horizontal components F_{qx} and F_{qy} while F_{qz} will not vanish, as we found above. As we will see, symmetry considerations such as these can often simplify our work and one should try to be aware of and look for them.

Since only the z component is different from zero, the total force will be in the z direction and

$$\mathbf{F}_q = \frac{q\rho a^3 \hat{\mathbf{z}}}{3\epsilon_0 z^2} \tag{2-26}$$

from (2-25) and (1-5). We see that if $q > 0$ and $\rho > 0$, then $\mathbf{F}_q$ is directed away from the sphere as expected since q would be repelled by all of the positive charges; similarly, if $\rho < 0$, $\mathbf{F}_q$ is directed toward the sphere, that is, the force on q is attractive. We can write (2-26) in an interesting and instructive form if we express it in terms of

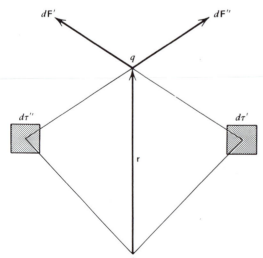

Figure 2-8. Force contributions from two symmetrically located charge elements.

the total charge Q' contained within the sphere. From (2-14), we get

$$Q' = \int dq' = \int \rho \, d\tau' = \rho \int_{\text{sphere}} d\tau' = \tfrac{4}{3}\pi a^3 \rho \qquad (2\text{-}27)$$

since ρ is constant; when this is used to eliminate ρ in (2-26), we find that

$$\mathbf{F}_q = \frac{qQ'\hat{\mathbf{z}}}{4\pi\epsilon_0 z^2} \qquad (2\text{-}28)$$

We see from Figure 2-7 that z is the distance from the center of the sphere to q, so that, on comparing (2-28) with (2-3), we find that this uniform sphere of charge acts *as if* it were a single point charge located at the center of the sphere as far as its effect on a charge located outside the sphere is concerned. As we will see later, this is not the case if the point of interest is located inside the sphere.

Actually, our result is even more general than it first appears. The location of q was taken to be along the z axis for convenience in evaluating the integral. As we previously found from Figure 2-7, the position vector of q with respect to the center of the sphere is $\mathbf{r} = z\hat{\mathbf{z}}$ so that $|\mathbf{r}| = r = z$ and $\hat{\mathbf{r}} = \hat{\mathbf{z}}$, which enables us to rewrite (2-28) for any location of q in terms of its spherical coordinates as

$$\mathbf{F}_q = \frac{qQ'\hat{\mathbf{r}}}{4\pi\epsilon_0 r^2} \qquad (2\text{-}29)$$

[This also follows from (1-90) and (1-93) since the location of q in Figure 2-7 corresponds to the special case $\theta = 0$.]

EXERCISES

2-1 Two point charges q' and $-q'$ are on the x axis with coordinates a and $-a$, respectively. Find the total force on a point charge q located at an arbitrary point in the xy plane.

2-2 Four equal point charges q' are located at the corners of a square of side a. The square lies in the yz plane with one corner at the origin and its sides parallel to the positive axes. Another point charge q is on the x axis at a distance b from the origin. Find the total force on q.

2-3 Eight equal point charges q are located at the corners of a cube of edge a, which has the location and orientation of the figure shown in Figure 1-41. Find the total force on the charge at the origin.

2-4 Repeat the calculation of Section 2-5 for the case in which q is outside the sphere but below it, that is, z is negative and $|z| > a$. Show that your result is consistent with (2-26) and (2-29).

2-5 Repeat the calculation of Section 2-5 for the case in which q is inside the sphere ($z < a$) to show that $\mathbf{F}_q = (q\rho z/3\epsilon_0)\hat{\mathbf{z}}$.

2-6 A sphere of radius a contains charge distributed with constant volume density ρ. Its center is on the z axis at a distance b from the origin with $b > a$. A point charge q is located on the y axis at a distance c from the origin where $c > b$. Find the force on q.

2-7 A line charge of length L with $\lambda = $ const. lies along the positive z axis with its ends located at $z = z_0$ and $z_0 + L$. Find the total force on this line charge due to a uniform spherical charge distribution with center at the origin and radius $a < z_0$.

2-8 The surface of a sphere of radius a is charged with a constant surface density σ. What is the total charge Q' on the sphere? Find the force produced by this charge distribution on a point charge q located on the z axis for $z > a$ and for $z < a$.

2-9 Two line charges of the same length L are parallel to each other and located in the xy plane as shown in Figure 2-9. They each have the same

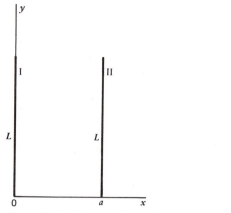

Figure 2-9.   The two line charges of Exercise 2-9.

linear charge density λ = const. Find the total force on II due to I.

2-10   The line charge I in Figure 2-9 now has a linear charge density $\lambda = Ay^2$ where A is a constant. What are the units of A? What is the total charge on I? Find the total force due to I on a point charge q placed on the x axis at $x = a$.

2-11   Charge is distributed over the surface of a circle of radius a lying in the xy plane with origin at the center. The surface density is given in cylindrical coordinates by $\sigma = A\rho^2$ where A is a constant. What are the units of A? What is the total charge on the circle? Find the force produced by this charge distribution on a point charge located on the z axis.

3

THE ELECTRIC FIELD

Coulomb's law is an example of what is known as an "action at a distance" law. It provides us with a straightforward way of calculating the force on a given charge when the relative position with respect to the source charge is known. Coulomb's law does not purport to describe how the first charge "knows" the other one is present. If, for example, the position of the source charge is changed, the force on the first charge will also be changed and again given by Coulomb's law. The implication is that this change will occur instantaneously, but again there is no suggestion as to how this altered state of affairs is brought about. As a result of these and similar considerations, it has been found convenient and useful to make a mental division of the interaction between the two charges into two aspects: first, one assumes that the source charge produces "something" at the field point, and, second, that this "something" then interacts with the charge at the field point to produce the resultant force on it. This "something," which acts as a kind of intermediary between the two charges, is called the electric field and is what we discuss now.

3-1 DEFINITION OF THE ELECTRIC FIELD

If we look again at (2-10), we see that q is a common factor of all of the terms so that $\mathbf{F}_q$ can be written as a product of q and a quantity that is independent of q but does depend on the values of all of the other charges and their locations with respect to q. This quantity is called the *electric field* $\mathbf{E}$; thus, we write (2-10) in the form

$$\mathbf{F}_q = q\mathbf{E} \tag{3-1}$$

where

$$\mathbf{E}(\mathbf{r}) = \sum_{i=1}^{N} \frac{q_i \hat{\mathbf{R}}_i}{4\pi\epsilon_0 R_i^2} \tag{3-2}$$

Equation 3-1 provides us with the definition of the electric field and we see that we can interpret it as a quantity such that, when it is multiplied by a point charge, the result is the force on the point charge. It also follows from (3-1) that $\mathbf{E}$ will be measured in newtons/coulomb. Equation 3-2 then provides us with a prescription for calculating $\mathbf{E}$ at the location $\mathbf{r}$ (the "field point") for a given distribution of point charges; we are still using (2-11) of course. We note that q is *not* included among the source charges in (3-2), that is, we do not envision a charge exerting a force on itself.

If the source charges have a continuous distribution, we can combine (3-1) with our previous results (2-15), (2-17), and (2-18) to get corresponding expressions for $\mathbf{E}$:

$$\mathbf{E}(\mathbf{r}) = \frac{1}{4\pi\epsilon_0} \int_{V'} \frac{\rho(\mathbf{r}')\hat{\mathbf{R}}\, d\tau'}{R^2} \tag{3-3}$$

$$\mathbf{E}(\mathbf{r}) = \frac{1}{4\pi\epsilon_0} \int_{S'} \frac{\sigma(\mathbf{r}')\hat{\mathbf{R}}\, da'}{R^2} \tag{3-4}$$

$$\mathbf{E}(\mathbf{r}) = \frac{1}{4\pi\epsilon_0} \int_{L'} \frac{\lambda(\mathbf{r}')\hat{\mathbf{R}}\, ds'}{R^2} \tag{3-5}$$

If all locations are given in rectangular coordinates, we get an explicit expression for $\mathbf{E}$ from (2-12):

$$\mathbf{E}(\mathbf{r}) = \sum_{i=1}^{N} \frac{q_i}{4\pi\epsilon_0} \frac{\left[(x - x_i)\hat{\mathbf{x}} + (y - y_i)\hat{\mathbf{y}} + (z - z_i)\hat{\mathbf{z}}\right]}{\left[(x - x_i)^2 + (y - y_i)^2 + (z - z_i)^2\right]^{3/2}} \tag{3-6}$$

Finally, if all of the possibilities we have discussed are simultaneously present, we see from (3-1) and (2-19) that the total $\mathbf{E}$ at a given point will be obtained as the vector sum of the contributions from all of the various charge distributions producing the field.

If the charge distribution is simple enough, $\mathbf{E}$ can be easily calculated by a straightforward integration. We look at two such examples; one for a linear distribution of charge, and the other for a surface distribution. After we have done this, we discuss in more detail the significance of what we have done.

3-2   FIELD OF A UNIFORM INFINITE LINE CHARGE

We assume that $\lambda = $ const. and choose the z axis to coincide with the charge distribution as shown in Figure 3-1. We choose the origin so that the field point P will lie in the xy plane for convenience; then we have $\mathbf{r} = \rho\hat{\boldsymbol{\rho}}$ and $\mathbf{r}' = z'\hat{\mathbf{z}}$ so that $\mathbf{R} = \rho\hat{\boldsymbol{\rho}} - z'\hat{\mathbf{z}}$ and $R^2 = \rho^2 + z'^2$. We also see from the figure that $ds' = dz'$ in this case, and therefore (3-5) becomes

$$\mathbf{E} = \frac{\lambda}{4\pi\epsilon_0} \int_{-\infty}^{\infty} \frac{(\rho\hat{\boldsymbol{\rho}} - z'\hat{\mathbf{z}})\, dz'}{(\rho^2 + z'^2)^{3/2}} = \frac{\lambda\rho\hat{\boldsymbol{\rho}}}{4\pi\epsilon_0} \int_{-\infty}^{\infty} \frac{dz'}{(\rho^2 + z'^2)^{3/2}} \tag{3-7}$$

The last form is obtained because the $\hat{\mathbf{z}}$ component of the integral vanishes since the integrand is an odd function of z' and $\hat{\boldsymbol{\rho}}$ is a constant with respect to the integration variable z'. The integral in (3-7) is found to be

$$\frac{z'}{\rho^2(\rho^2 + z'^2)^{1/2}}\Bigg|_{-\infty}^{\infty} = \frac{1}{\rho^2}[(1) - (-1)] = \frac{2}{\rho^2} \tag{3-8}$$

so that the final result is

$$\mathbf{E} = \frac{\lambda}{2\pi\epsilon_0\rho}\hat{\boldsymbol{\rho}} \tag{3-9}$$

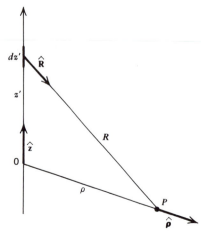

Figure 3-1.   Calculation of the field due to a uniform infinite line charge.

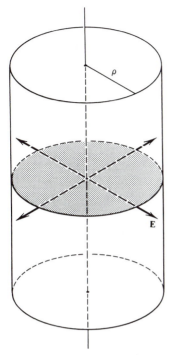

Figure 3-2.   A cylinder of constant electric field magnitude for a uniform infinite line charge.

Thus the electric field has only a radial component. It is directed away from the line charge if $\lambda > 0$, as it should be since a positive charge q would be repelled, while it is directed radially inward if λ is negative. The magnitude of **E** varies inversely with the distance ρ from the line charge.

Since (3-9) is independent of the angle φ, we see that a surface of constant magnitude of **E** will be a cylinder of radius ρ with the line charge as the axis of the cylinder. A portion of this cylinder is shown in Figure 3-2 in which are also indicated some directions of **E** for $\lambda > 0$ on a circle formed by the intersection of a plane perpendicular to the z axis and the cylinder.

3-3   FIELD OF A UNIFORM INFINITE PLANE SHEET

We assume the surface charge density σ to be constant on an infinite plane which we take to be the xy plane. It is convenient to choose the z axis to pass through the field point P. We will use rectangular coordinates for integration. We see then from Figure 3-3 that $\mathbf{r} = z\hat{\mathbf{z}}$ and $\mathbf{r}' = x'\hat{\mathbf{x}} + y'\hat{\mathbf{y}}$. Since the element of area is $da' = dx'\, dy'$, (3-4) becomes, with the use of (2-2), (1-13), and (1-14):

$$\mathbf{E} = \frac{\sigma}{4\pi\epsilon_0} \int_{-\infty}^{\infty}\int_{-\infty}^{\infty} \frac{(-x'\hat{\mathbf{x}} - y'\hat{\mathbf{y}} + z\hat{\mathbf{z}})\, dx'\, dy'}{\left(x'^2 + y'^2 + z^2\right)^{3/2}} \tag{3-10}$$

We see at once that $E_x = E_y = 0$ since the $\hat{\mathbf{x}}$ and $\hat{\mathbf{y}}$ terms in the integrand are odd functions of x' and y', respectively. Therefore, (3-10) reduces to

$$\mathbf{E} = \frac{\sigma\hat{\mathbf{z}}}{4\pi\epsilon_0} \int_{-\infty}^{\infty} z\, dx' \int_{-\infty}^{\infty} \frac{dy'}{\left(x'^2 + y'^2 + z^2\right)^{3/2}} \tag{3-11}$$

The integral over y' is identical in *form* to that in (3-7) and from (3-8) equals

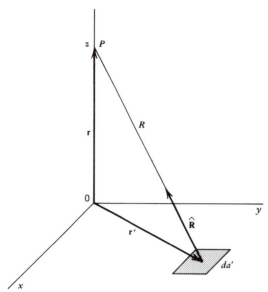

Figure 3-3. Calculation of the field due to a uniform infinite plane sheet.

$2/(x'^2 + z^2)$; (3-11) thus becomes

$$\mathbf{E} = \frac{\sigma \hat{\mathbf{z}}}{2\pi\epsilon_0} \int_{-\infty}^{\infty} \frac{z \, dx'}{x'^2 + z^2} = \pm \frac{\sigma}{2\epsilon_0} \hat{\mathbf{z}} \qquad (3\text{-}12)$$

where the plus sign is to be used for $z > 0$, while the minus sign is used when $z < 0$. It is often convenient to write (3-12) as

$$\mathbf{E} = \frac{\sigma}{2\epsilon_0} \left(\frac{z}{|z|} \right) \hat{\mathbf{z}} \qquad (3\text{-}13)$$

which automatically gives the correct signs for $\mathbf{E}$.

We see from (3-12) that $\mathbf{E}$ is always directed away from a positively charged plane ($\sigma > 0$), and is always directed toward the plane when $\sigma < 0$. These directions correspond to those of the force on a positive charge q placed at some point. It is interesting to note that the magnitude of $\mathbf{E}$ is independent of position, that is, $\mathbf{E}$ has the same value no matter how near one is to the plane or how far away one is from it; this arises essentially from the fact that no matter where the field point is located, there is always an infinite amount of charge "visible" to it. These properties of $\mathbf{E}$ are indicated in Figure 3-4 which is drawn for $\sigma > 0$, and shows an edge on view of the charged

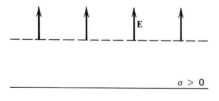

Figure 3-4. Electric field due to a uniform infinite plane sheet.

plane. The dashed lines are traces of planes above and below the charged plane and parallel to it. The figure looks the same if it is turned upside down, as it should since our original choice of direction for positive z was completely arbitrary. Similarly, the appearance is the same if one views it looking from behind the page rather than toward the page. In other words, our result (3-12) is completely consistent with the basic "symmetry" of the source charge distribution. $\mathbf{E}$ also changes discontinuously in direction as one passes through the charged plane; if one passes from below to above, for example, the total change is $E_z(\text{above}) - E_z(\text{below}) = \sigma/\epsilon_0$ as found from (3-12).

3-4 WHAT DOES ALL OF THIS MEAN?

We managed rather easily to introduce an auxiliary quantity that enabled us to divide the interaction between two charges into conceptually different parts. We did this by defining a new vector field $\mathbf{E}$ and we have ways that, in principle, allow us to find it at any point once the charges that are its sources are given. It is natural, however, to ask if there is really anything useful to be gained in doing this.

One can quite easily adopt the point of view that this is done merely for mathematical convenience, if for no other reason than one saves some writing by not having to carry the symbol q along in all of the equations, but can calculate $\mathbf{E}$ first and then insert q as a last step by means of (3-1). We can thus regard the calculation of $\mathbf{E}$ as merely providing us with a sort of contingency statement distributed throughout space in the sense that $\mathbf{E}(\mathbf{r})$, combined with (3-1), tells us what *would* happen *if* we were to put a point charge q at $\mathbf{r}$.

On the other hand, our formulas (3-2) through (3-6) enable us to calculate the electric field at $\mathbf{r}$ *whether or not* there is a charge there to be subject to a force. This fact provides us with a strong temptation to make a conceptual leap and regard $\mathbf{E}$ as an actual physical entity in its own right. Most of these ideas originated with Faraday, and he felt that the presence of charges actually changed the physical properties of space, and that $\mathbf{E}$ was a manifestation of this altered state. For him, the electric field was a very real physical quantity.

If we adopt the attitude that $\mathbf{E}$ is a physical quantity, the question then naturally arises as to how one would measure it. At first glance, this appears to be very simple: one merely puts a point charge q at rest at the point $\mathbf{r}$ of interest, measures the force $\mathbf{F}_q$ on it, and then (3-1) tells us that $\mathbf{E}(\mathbf{r}) = \mathbf{F}_q/q$. A possible problem now arises with the recognition that the presence of q now subjects the source charges q_i of (3-2) to new forces as given by (2-8) and they will no longer be in equilibrium, although eventually equilibrium will be reestablished. In the idealized case in which we can assume the q_i to be rigidly attached to fixed positions, the new electrical force can be compensated by a new mechanical force that is produced without deformation of the support; then (2-9) as applied to a given q_i will still hold, but the positions of the q_i will not have changed. The value of $\mathbf{E}$ as given by (3-2) will be exactly what it was before q was introduced and $\mathbf{E}$ will be correctly given by (3-1). In a real case, however, the new values of the mechanical forces required for equilibrium can generally only be obtained by deformation of the support, such as by the bending of a rod or by a stretching or compression of a spring; thus, when the new equilibrium configuration is attained, the positions of the q_i will have been changed, the value of (3-2) will generally be different, and the net result will be that the very act needed to measure the preexisting field has altered it. (In addition, as we will see later, if the source charges are associated with conductors, they will ordinarily need to move about on the conductor in order to come into mutual equilibrium, again leading us to the conclusion that the field can be different.) This problem of changing what you are trying to measure is not unique to electromagnetism, of course, and one usually tries to solve it in the same general way, that is, by minimizing the disturbance as much as possible, while still being able to get ·a measurable effect. In order to apply this idea to **E**, one imagines the charge on which the force is to be measured to be very small and then try to go to the limit in which it approaches zero; if we use δq to denote this "test charge," and $\delta \mathbf{F}$ the measured force on it, then we would require that **E** be determined by

$$\mathbf{E}(\mathbf{r}) = \lim_{\delta q \to 0} \frac{\delta \mathbf{F}}{\delta q} \tag{3-14}$$

Although the use of the electric field may be regarded as a convenient artifice if one is dealing solely with electrostatics, when one comes to handling other problems, particularly time dependent ones, such as that briefly alluded to in the introductory paragraph of this chapter, it has been found to be virtually impossible to do this without an extensive use of vector fields. We find it useful to define several other vector fields as we proceed, and whether we want to regard them as real physical quantities or not, we will certainly treat them *as if* they were. We want to study their properties extensively as well as applications of them; among these properties are their differential source equations, that is, their divergence and curl. We already know that the sources of the electric field are charges of any type, but we want to *restate* this in the form of explicit expressions for $\nabla \cdot \mathbf{E}$ and $\nabla \times \mathbf{E}$; we obtain them, along with other information, in the next two chapters.

EXERCISES

3-1 Two point charges q and $-q$ are located on the y axis at $y = a$ and $-a$, respectively. Find **E** for any point in the xy plane. For what points, if any, will $E_x = 0$?

3-2 Four point charges are located at the corners of a square in the xy plane. Their values and locations are as follows: $q, (0,0)$; $2q, (0, a)$; $3q, (a, 0)$; $-4q, (a, a)$. Find **E** at the center of the square.

3-3 Consider a cube of edge a with the location and orientation of the figure in Figure 1-41. There is a point charge q at each corner except for that at $(a, a, 0)$. Find **E** at the empty corner.

3-4 Repeat the calculation of Section 3-2 for a general field point $\mathbf{r} = \rho \hat{\boldsymbol{\rho}} + z \hat{\mathbf{z}}$ and thus show that the same result is obtained. Is this physically reasonable?

3-5 Repeat the calculation of Section 3-3 for a general field point (x, y, z) and thus show that the same result is obtained.

3-6 Repeat the calculation of Section 3-3 using cylindrical coordinates for the source point.

3-7 A uniform infinite line charge is parallel to the z axis and intersects the xy plane at the point

$(a, b, 0)$. Find the rectangular components of **E** produced at the point $(0, c, 0)$.

3-8 Two infinite plane sheets with equal constant surface charge density σ are parallel to the xy plane and located as shown in Figure 3-5. Find **E** for all values of z.

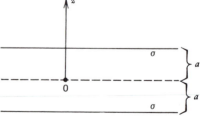

Figure 3-5. The two infinite plane sheets of Exercise 3-8.

3-9 Two infinite plane sheets have equal and opposite constant surface charge density σ. They are parallel to the xy plane and located as shown in Figure 3-6. Find **E** for all values of z.

3-10 The circular arc of radius a shown in Figure 3-7 lies in the xy plane and has a constant linear charge density λ and center of curvature at the origin. Find **E** at an arbitrary point on the z

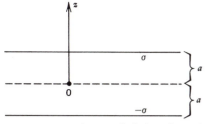

Figure 3-6. The two infinite plane sheets of Exercise 3-9.

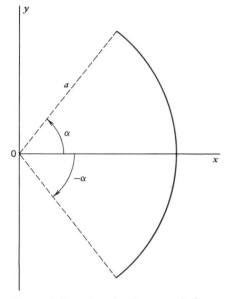

Figure 3-7. The circular arc of charge of Exercise 3-10.

axis. Show that when the curve is a complete circle your answer becomes

$$\mathbf{E} = \frac{\lambda a z \hat{\mathbf{z}}}{2\epsilon_0 (a^2 + z^2)^{3/2}}$$

3-11 Charge is distributed with constant linear charge density λ on the line of finite length shown in Figure 3-8. Find $\mathbf{E}$ at P. With the aid of the distances R_2 and R_1, express $\mathbf{E}$ in terms of the angles α_2 and α_1 shown. Find $\mathbf{E}$ for the special case for which $L_2 = L_1 = L$ and P is in the xy plane.

3-12 Charge is distributed with constant surface charge density σ on a circular disc of radius a.

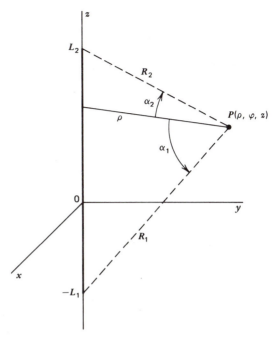

Figure 3-8. The finite line charge of Exercise 3-11.

The disc lies in the xy plane with center at the origin. Show that the electric field at a point on the z axis is given by

$$\mathbf{E} = \hat{\mathbf{z}}\frac{\sigma}{2\epsilon_0}\left(\frac{z}{|z|}\right)\left[1 - \frac{|z|}{(a^2 + z^2)^{1/2}}\right] \quad (3\text{-}15)$$

What does this become as $a \rightarrow \infty$?

3-13 An infinitely long cylinder has its axis coinciding with the z axis. It has a circular cross section of radius a and contains a charge of constant volume density ρ_{ch}. Find $\mathbf{E}$ at all points, both inside and outside the cylinder. *Hints:* use cylindrical coordinates for integration; for convenience, choose the field point on the x axis (Will this be general enough?); you will probably need this definite integral

$$\int_0^\pi \frac{(A - B\cos t)\,dt}{A^2 - 2AB\cos t + B^2}$$

$$= \begin{cases} \dfrac{\pi}{A} & \text{if } A^2 > B^2 \\[2mm] 0 & \text{if } A^2 < B^2 \end{cases} \quad (3\text{-}16)$$

4

GAUSS' LAW

We can anticipate from the general definition of the divergence given in (1-66) that the flux of $\mathbf{E}$ through a closed surface, that is, its surface integral, will be worth investigating. As we will see, the inverse square nature of Coulomb's law makes it possible to evaluate this flux over a surface of arbitrary size and shape.

4-1 DERIVATION OF GAUSS' LAW

We want to show that

$$\oint_S \mathbf{E} \cdot d\mathbf{a} = \frac{1}{\epsilon_0} \sum_{\text{inside}} q_i = \frac{Q_{\text{in}}}{\epsilon_0} \tag{4-1}$$

where Q_{in} is the net charge contained *within* the volume enclosed by the arbitrary closed surface S. Now from (3-2) we quickly obtain

$$\oint_S \mathbf{E} \cdot d\mathbf{a} = \frac{1}{4\pi\epsilon_0} \sum_i q_i \oint_S \frac{\hat{\mathbf{R}}_i \cdot d\mathbf{a}}{R_i^2} \tag{4-2}$$

There are two cases to consider.

1. q_i is inside S (Figure 4-1). If we recall our general discussion in connection with Figure 1-26, we see that

$$\frac{\hat{\mathbf{R}}_i \cdot d\mathbf{a}}{R_i^2} = \frac{da \cos \Psi}{R_i^2} = \frac{\text{area} \perp \text{to } \hat{\mathbf{R}}_i}{R_i^2} = d\Omega \tag{4-3}$$

where $d\Omega$ = element of solid angle subtended at q_i by the area da. In order to evaluate the integral in (4-2), we consider a sphere S_0 of radius R_0 with q_i as the center. This *same* solid angle $d\Omega$ will intercept the area $d\mathbf{a}_0$ on this sphere; as we see from the figure, $d\mathbf{a}_0$ is parallel to $\hat{\mathbf{R}}_i$ so that if we apply (4-3) to this case, $d\Omega$ also equals da_0/R_0^2. Therefore, the integral in (4-2) can equally well be written as

$$\oint d\Omega = \oint_{S_0} \frac{da_0}{R_0^2} = \frac{1}{R_0^2} \oint_{S_0} da_0 = \frac{4\pi R_0^2}{R_0^2} = 4\pi \tag{4-4}$$

since R_0 is constant for all points on the surface of the sphere. Thus, the total solid angle subtended by any surface about a point within it is 4π, and we can write

$$\oint_S \frac{\hat{\mathbf{R}}_i \cdot d\mathbf{a}}{R_i^2} = 4\pi \qquad \text{(if } \mathbf{r}_i \text{ is inside } S) \tag{4-5}$$

2. q_i is outside S (Figure 4-2). Here we consider the two elements of area $d\mathbf{a}_1$ and $d\mathbf{a}_2$ of S that are cut out by the same solid angle $d\Omega$ but that are on opposite sides of S. Their distances from q_i will be written as R_{i1} and R_{i2}. As before, we will have

$$\frac{\hat{\mathbf{R}}_i \cdot d\mathbf{a}_1}{R_{i1}^2} = \frac{da_1 \cos \Psi_1}{R_{i1}^2} = d\Omega$$

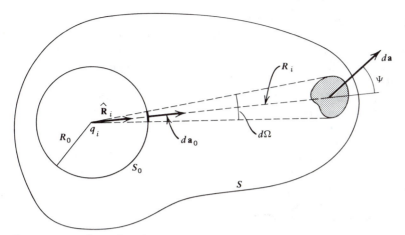

Figure 4-1.   The point charge is inside the surface S.

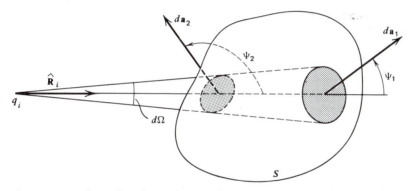

Figure 4-2.   The point charge is outside the surface S.

but now, since $\Psi_2 > \pi/2$, so that $\cos \Psi_2$ is negative,

$$\frac{\hat{\mathbf{R}}_i \cdot d\mathbf{a}_2}{R_{i2}^2} = \frac{da_2 \cos \Psi_2}{R_{i2}^2} = -d\Omega$$

and therefore

$$\frac{\hat{\mathbf{R}}_i \cdot d\mathbf{a}_1}{R_{i1}^2} + \frac{\hat{\mathbf{R}}_i \cdot d\mathbf{a}_2}{R_{i2}^2} = 0 \qquad (4\text{-}6)$$

so that the net contribution of these two elements of area to the integral in (4-2) is zero. Since all of the elements of area of S can be paired off in this way, all of their contributions to the integral will mutually cancel, and therefore

$$\oint_S \frac{\hat{\mathbf{R}}_i \cdot d\mathbf{a}}{R_i^2} = 0 \qquad (\text{if } \mathbf{r}_i \text{ is outside } S) \qquad (4\text{-}7)$$

Hence, from (4-2), (4-5), and (4-7),

$$\oint_S \mathbf{E} \cdot d\mathbf{a} = \frac{1}{4\pi\epsilon_0} \sum_{\text{inside}} q_i \oint_S \frac{\hat{\mathbf{R}}_i \cdot d\mathbf{a}}{R_i^2} = \frac{1}{\epsilon_0} \sum_{\text{inside}} q_i = \frac{Q_{\text{in}}}{\epsilon_0} \qquad (4\text{-}8)$$

and we have proved Gauss' law as stated in (4-1), and which is now seen to be a direct consequence of the inverse square law of force between point charges.

We can note here some of the interesting implications of this result. Any charges outside the surface do not affect the value of the *integral*, although their values and locations can clearly affect the particular value of **E** at each point on the surface. Similarly, the integral depends only on the total value of the charges inside the surface and thus is independent of their specific locations inside; however, if they are placed at new locations, the value of **E** at a particular point on the surface can be expected to be changed, but the value of the complete integral will again be unaffected. Since the result given in (4-1) is a simple sum, we see that each charge contributes independently to the total flux of **E** through *S*; therefore, a given point charge *q* has a total flux of **E** equal to q/ϵ_0 through any closed surface surrounding it.

If we now assume the charges within *S* to be continuously distributed with density ρ, we can use (2-14) and write

$$Q_{in} = \int_V \rho \, d\tau$$

where *V* is the total volume enclosed by *S*. If we also apply the divergence theorem (1-59), we can write (4-8) as

$$\int_V \nabla \cdot \mathbf{E} \, d\tau = \frac{1}{\epsilon_0} \int_V \rho \, d\tau \tag{4-9}$$

Since this result holds for any arbitrary volume *V*, it will be true for an infinitesimal one, and we can equate the integrands to give

$$\nabla \cdot \mathbf{E} = \frac{\rho}{\epsilon_0} \tag{4-10}$$

This important result is one of Maxwell's equations and is *equivalent* to Coulomb's law of force between point charges. (Later, we will find it important to recall that the charge density ρ includes *all* charges from any source whatsoever since the electric field is produced, by its definition, by all charges.)

4-2 SOME APPLICATIONS OF GAUSS' LAW

If the charge distribution has sufficient symmetry, Gauss' law provides a simple and convenient way of calculating the electric field. The principal problem is that of choosing a suitable closed surface of integration. The main things to look for are surfaces on which **E** has a constant magnitude, and surfaces for which **E** is either parallel to or perpendicular to the surface both for ease in integration and for avoiding difficulties with an unknown dependence of **E** on position. This process is most easily demonstrated with examples, and we will look at a sample of each of our types of continuous charge distribution.

■ **Example**

Uniform infinite line charge. We assume $\lambda = $ const. and again let the line charge coincide with the *z* axis to be specific; we will use cylindrical coordinates. For an infinite line, it doesn't make any difference where you are located along it since the charge still extends to infinity in both directions; hence we conclude that **E** cannot depend on *z*. Similarly, there is nothing to distinguish one value of φ from another since the charge distribution looks the same no matter from where you view it perpendicular to z. Therefore, $\mathbf{E}$ must also be independent of φ and, at most, can only depend on the distance ρ from the line; thus we conclude that $\mathbf{E} = \mathbf{E}(\rho)$.

Now suppose that $\mathbf{E}$ had a component parallel to the line, that is, $E_z(\rho) \neq 0$. In this example, the choice of positive or negative direction for the z axis is completely arbitrary, that is, there is no real distinction between "up" and "down." But if E_z were different from zero, this fact would distinguish up from down; hence we must conclude that $E_z = 0$. Similarly, there is no reason to prefer the sense of increasing φ over the sense of decreasing φ, so that $E_\varphi = 0$. Therefore, we conclude from the symmetry of the situation that $\mathbf{E}$ can only be radial and that we can write $\mathbf{E} = E_\rho(\rho)\hat{\rho}$. Thus, a surface of constant magnitude of $\mathbf{E}$ is an infinitely long cylinder of radius ρ whose axis coincides with the line charge; a finite portion of this cylinder chosen to be of length L is shown in Figure 4-3, and we note that its outward normal $\hat{\mathbf{n}}_c$ is exactly $\hat{\rho}$, and therefore parallel to $\mathbf{E}$. All of this suggests that we use the curved surface of this cylinder as part of our surface of integration. We can now get a closed surface of integration by choosing the rest of it to consist of the two circular cross sections of radius ρ shown, thus making our final surface of integration a right circular cylinder. The outer normals to these upper and lower surfaces are shown as $\hat{\mathbf{n}}_u$ and $\hat{\mathbf{n}}_l$ and are seen to equal $\hat{\mathbf{z}}$ and $-\hat{\mathbf{z}}$, respectively. Although E_ρ has an unknown dependence on ρ on these circular areas, $\mathbf{E}$ is perpendicular to the vector areas and hence its contribution to the flux will vanish.

After this lengthy discussion, the actual evaluation of the surface integral is almost anticlimactic. We can write it as the sum of integrals over the curved surface, and the upper and lower faces (which we designate as c, u, and l). We also remember that $E_\rho(\rho)$ is constant on the curved surface since ρ is constant. Therefore, (4-1) becomes, in

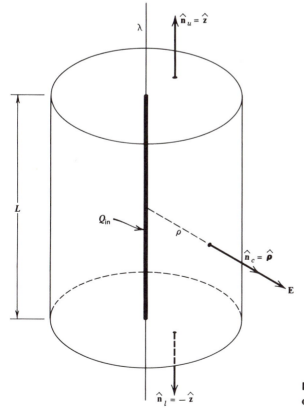

Figure 4-3. Calculation of the field of a uniform infinite line charge.

this case,

$$\oint_S \mathbf{E} \cdot d\mathbf{a} = \int_c E_\rho(\rho)\hat{\boldsymbol{\rho}} \cdot \hat{\mathbf{n}}_c \, da + \int_u E_\rho\hat{\boldsymbol{\rho}} \cdot \hat{\mathbf{n}}_u \, da + \int_l E_\rho\hat{\boldsymbol{\rho}} \cdot \hat{\mathbf{n}}_l \, da$$

$$= E_\rho(\rho)\int_c da + 0 + 0 = E_\rho(\rho)2\pi\rho L = \frac{Q_{in}}{\epsilon_0} = \frac{\lambda L}{\epsilon_0}$$

where we used (1-76) and (2-16), the latter to find the total charge inside the cylinder which is that on the line of length L. We see that when we solve the above for E_ρ, the arbitrary length L cancels and we get $E_\rho(\rho) = \lambda/2\pi\rho\epsilon_0$ so that

$$\mathbf{E} = \frac{\lambda}{2\pi\epsilon_0\rho}\hat{\boldsymbol{\rho}} \tag{4-11}$$

which is exactly the result (3-9) that we previously obtained by direct integration. ∎

■ **Example**

Uniform infinite plane sheet. Here we assume that σ = const., and, for definiteness, that the sheet coincides with the xy plane. An edge on view of the situation is shown in Figure 4-4. Since the choice of the origin as well as of the orientation of the x and y axes is completely arbitrary in this example, $\mathbf{E}$ must be independent of x and y. Similarly, there is no basic distinction here between left and right, or into or out of the paper, so that $\mathbf{E}$ cannot have any components parallel to the sheet; thus $\mathbf{E}$ has only a z component which, at most, can be a function of z, the distance from the sheet. There is also no real difference here between up and down, so that $\mathbf{E}$ must always point away

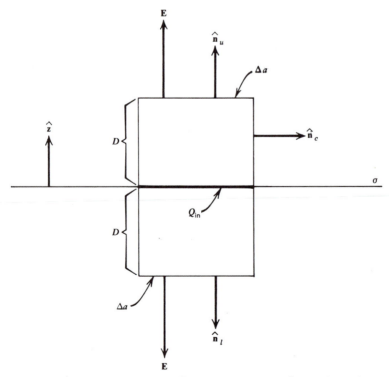

Figure 4-4. Calculation of the field of a uniform infinite plane sheet.

from the sheet or always toward it depending on the sign of σ. Therefore, **E** must have the form $\mathbf{E} = \pm E(z)\hat{\mathbf{z}}$ with the upper sign for $z > 0$ and the lower for $z < 0$; $E(z)$ can be positive or negative, the figure being drawn for the case in which $E(z)$ is positive.

These considerations suggest choosing the closed surface of integration to be that of a right cylinder extending an equal distance D above and below the sheet, with faces parallel to the sheet and of area Δa; the outward normals are shown as $\hat{\mathbf{n}}_u = \hat{\mathbf{z}}$ and $\hat{\mathbf{n}}_l = -\hat{\mathbf{z}}$ in the figure. **E** has the constant magnitude $E(D)$ on these faces. The curved surface connecting these faces will have its sides perpendicular to the sheet; one of its outward normals is shown as $\hat{\mathbf{n}}_c$ and we see that $\hat{\mathbf{n}}_c \cdot \hat{\mathbf{z}} = 0$ for all parts of the curved surface. We also see that Q_{in} is the charge intercepted on the sheet by the cross section of the cylinder and, hence, equals $\sigma \Delta a$.

Therefore, in this case, (4-1) becomes:

$$\oint_S \mathbf{E} \cdot d\mathbf{a} = \int_u E(D)\hat{\mathbf{z}} \cdot \hat{\mathbf{z}}\, da + \int_l E(D)(-\hat{\mathbf{z}}) \cdot (-\hat{\mathbf{z}})\, da + \int_c E(z)\hat{\mathbf{z}} \cdot \hat{\mathbf{n}}_c\, da$$

$$= E(D)\, \Delta a + E(D)\, \Delta a + 0 = 2E(D)\, \Delta a = \frac{Q_{\text{in}}}{\epsilon_0} = \frac{\sigma \Delta a}{\epsilon_0}$$

From this we find that $E(D) = \sigma/2\epsilon_0$ and has turned out to be actually independent of D; therefore, the electric field for this sheet is given by

$$\mathbf{E} = \pm \frac{\sigma}{2\epsilon_0}\hat{\mathbf{z}} \tag{4-12}$$

which is exactly the result (3-12) that we obtained by direct integration. ∎

■ **Example**

Spherically symmetric spherical charge distribution. The charge distribution is contained within a sphere of radius a. We do not necessarily assume the charge density to be constant, but we do assume it to be independent of angle so that at most $\rho = \rho(r)$; this is what "spherically symmetric" means. From this we can conclude that **E** itself is radial with a magnitude independent of angle and therefore can be written in the form $\mathbf{E} = E_r(r)\hat{\mathbf{r}}$.

Since the magnitude of **E** is constant on a sphere of radius r, we choose such a sphere as our surface of integration; its outward normal is also $\hat{\mathbf{r}}$. Therefore, the left side of (4-1) becomes, with the use of (1-92):

$$\oint_S \mathbf{E} \cdot d\mathbf{a} = \oint_S E_r(r)\hat{\mathbf{r}} \cdot \hat{\mathbf{r}}\, da = E_r(r)\oint_S da = 4\pi r^2 E_r(r) \tag{4-13}$$

and therefore

$$E_r(r) = \frac{Q_{\text{in}}}{4\pi\epsilon_0 r^2} \tag{4-14}$$

where

$$Q_{\text{in}} = \int_{V(r)} \rho(r')\, d\tau' \tag{4-15}$$

and where $V(r)$ is the volume of the sphere of radius r. There are two cases to consider.

1. Outside the sphere of charge, $r > a$. Here $\rho(r') = 0$ if $r' > a$, and the volume of integration in (4-15) reduces to $V(a)$, the total volume of the charge distribution, so that

$$Q_{\text{in}} = \int_{V(a)} \rho(r')\, d\tau' = Q \tag{4-16}$$

where Q is the total charge contained in the sphere. Then (4-14) becomes

$$E_r(r) = \frac{Q}{4\pi\epsilon_0 r^2} \qquad (r > a) \tag{4-17}$$

and the electric field outside is the same *as if* all of the charge were a point charge located at the center of the sphere. We recall from (3-1) and (2-29) (where the total charge was written Q') that this is what we found by direct integration for a sphere of constant charge density, but now we see that this conclusion is generally true for any sphere of charge as long as the charge distribution is spherically symmetric.

2. Inside the sphere of charge, $r < a$. In this case, we can use (1-99) to write (4-15) as

$$Q_{in} = \int_0^{2\pi}\int_0^{\pi}\int_0^r \rho(r')r'^2 \sin\theta' \, dr' \, d\theta' \, d\varphi' = 4\pi\int_0^r \rho(r')r'^2 \, dr' \tag{4-18}$$

which, when substituted in (4-14), gives the field inside as

$$E_r(r) = \frac{1}{\epsilon_0 r^2}\int_0^r \rho(r')r'^2 \, dr' \qquad (r < a) \tag{4-19}$$

We cannot proceed any further until we know the explicit form of $\rho(r')$. ∎

■ **Example**

As a special case of the previous example, we assume $\rho = $ const., that is, we have a uniformly charged sphere. Then the integral in (4-19) becomes

$$\int_0^r \rho r'^2 \, dr' = \rho\int_0^r r'^2 \, dr' = \tfrac{1}{3}\rho r^3$$

so that

$$E_r(r) = \frac{\rho r}{3\epsilon_0} \qquad (r < a) \tag{4-20}$$

We can express this in terms of the total charge Q by using (2-27); the result is

$$E_r(r) = \frac{Qr}{4\pi\epsilon_0 a^3} \qquad (r < a) \tag{4-21}$$

which shows that the field increases linearly with distance as we proceed out from the zero value at the center; outside, it varies inversely with the square of the distance from the center according to (4-17). We note that both (4-17) and (4-21) give the same value

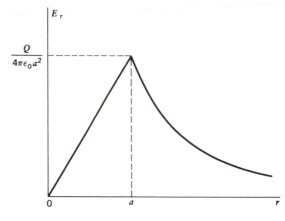

Figure 4-5.   Field of a uniform sphere of charge of radius a as a function of distance r from the center.

$Q/4\pi\epsilon_0 a^2$ at the surface of the sphere of charge where $r = a$; thus the electric field is continuous across the surface. These results for the uniformly charged sphere are shown in Figure 4-5.

Other spherically symmetric charge distributions $\rho(r)$ will give a different dependence of E_r on r *inside* the sphere, but exactly the same as shown in Figure 4-5 outside the sphere when expressed in terms of the total charge contained within the sphere $r = a$. ∎

4-3   DIRECT CALCULATION OF $\nabla \cdot \mathbf{E}$

Although we have succeeded in our aim of finding the divergence of $\mathbf{E}$ as given by (4-10), it was obtained somewhat indirectly, and it is of some interest to see how this same result can be obtained by a direct approach which starts from our basic defining equation for $\mathbf{E}$. We find it convenient to start with the expression (3-3) appropriate for a continuous charge distribution. If we dot both sides of (3-3) with ∇, we get

$$\nabla \cdot \mathbf{E} = \frac{1}{4\pi\epsilon_0} \nabla \cdot \int_{V'} \frac{\rho(\mathbf{r}')\hat{\mathbf{R}} \, d\tau'}{R^2} \tag{4-22}$$

Now the derivatives in ∇ are all taken with respect to the field point coordinates x, y, and z, which is where we want to know the value of $\nabla \cdot \mathbf{E}(\mathbf{r})$. The definite integral in (4-22) is taken over a specific source volume V'; the limits of integration will be either constants or, at worst, functions of the source point variables, that is, primed variables. (In the example of Section 1-13, we have already seen a two dimensional case in which the limits involved the variables of integration.) Therefore, the only dependence of the integral on $\mathbf{r}$ can be in the integrand, and, since the definite integral can be regarded as the limit of a sum, we can use (1-114) as generalized to a sum of more than two terms and interchange the order of differentiation and integration. If we then use (1-115) and note that $\rho(\mathbf{r}')$ and $d\tau'$ are constants as far as ∇ is concerned since they involve only primed source point coordinates, we find that (4-22) becomes

$$\nabla \cdot \mathbf{E} = \frac{1}{4\pi\epsilon_0} \int_{V'} \rho(\mathbf{r}')\nabla \cdot \left(\frac{\hat{\mathbf{R}}}{R^2} \right) d\tau' \tag{4-23}$$

But now we see from (1-145) that the integrand is zero as long as $R \neq 0$, so that any contribution to (4-23) must come from the region corresponding to $R = |\mathbf{r} - \mathbf{r}'| = 0$, that is, from the immediate neighborhood of the field point. Thus, in order to evaluate (4-23), with its problem of an infinite integrand at a point, we use a limiting process; we integrate over a small but finite volume $\Delta\tau'$ around the field point P and then we can go to the limit $\Delta\tau' \to 0$. It is easiest, but not necessary, to choose $\Delta\tau'$ to be a sphere of radius R about P as shown in Figure 4-6; then we can eventually let $R \to 0$. We note that, as usual, $\hat{\mathbf{R}}$ is drawn toward the field point, that is, toward the center of the sphere. Since $|\mathbf{r} - \mathbf{r}'| \simeq 0$ within this sphere, $\rho(\mathbf{r}')$ is nearly constant and equal to $\rho(\mathbf{r})$ so that we can take it out of the integral; this procedure will become exact as $R \to 0$. If we do this, we get

$$\nabla \cdot \mathbf{E} = \frac{\rho(\mathbf{r})}{4\pi\epsilon_0} \int_{\Delta\tau'} \nabla \cdot \left(\frac{\hat{\mathbf{R}}}{R^2} \right) d\tau' \tag{4-24}$$

We can now use (1-133) to express the derivatives in the integrand in terms of the primed coordinates; once this has been done (including the change in sign), we can use the divergence theorem (1-59) to convert (4-24) into a surface integral over the surface

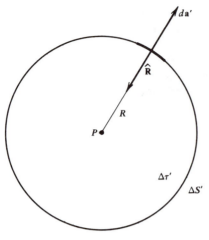

Figure 4-6. Small sphere surrounding the field point.

$\Delta S'$ enclosing $\Delta \tau'$:

$$\nabla \cdot \mathbf{E} = \frac{\rho(\mathbf{r})}{4\pi\epsilon_0} \oint_{\Delta S'} \frac{(-\hat{\mathbf{R}}) \cdot d\mathbf{a}'}{R^2} \qquad (4\text{-}25)$$

Now, if we recall (4-3), we see that the integrand is just $d\Omega$, taking into account the fact that the angle between $\hat{\mathbf{R}}$ and $d\mathbf{a}'$ is 180° as seen from Figure 4-6; therefore, since P is inside $\Delta S'$, the integral in (4-25) will be exactly 4π according to (4-5), and (4-25) becomes

$$\nabla \cdot \mathbf{E} = \frac{\rho(\mathbf{r})}{\epsilon_0} \qquad (4\text{-}26)$$

which is just what we found before. We note that this derivation reenforces our recognition that ρ must be evaluated at the point where we know $\mathbf{E}(\mathbf{r})$ and want the value of $\nabla \cdot \mathbf{E}$ at this same point.

EXERCISES

4-1 The rectangular parallelepiped of Figure 1-41 with $a > b > c$ is filled with charge of constant density ρ. A sphere of radius $2a$ is constructed with its center at the origin. Find the flux $\oint \mathbf{E} \cdot d\mathbf{a}$ through the surface of this sphere. What is the flux when the center of the sphere is at the corner (a, b, c)?

4-2 A sphere of radius a has its center at the origin and a charge density given by $\rho = Ar^2$ where $A = $ const. Another sphere of radius $2a$ is concentric with the first. Find the flux $\oint \mathbf{E} \cdot d\mathbf{a}$ through the surface of the larger sphere.

4-3 The infinite line charge of Figure 4-3 is surrounded by an infinitely long cylinder of radius ρ_0 whose axis coincides with the line charge. The surface of the cylinder carries a charge of constant surface density σ. Find $\mathbf{E}$ everywhere. What

particular value of σ will make $\mathbf{E} = 0$ for all points outside the charged cylinder? Is your answer reasonable?

4-4 A charge of constant volume density has the form of a slab of thickness a. The faces of the slab are infinite planes parallel to the xy plane. Choose the origin midway between the faces and find $\mathbf{E}$ everywhere.

4-5 A sphere of radius a has a charge density that varies with distance r from the center according to $\rho = Ar^{1/2}$ where $A = $ const. Find $\mathbf{E}$ everywhere.

4-6 Two concentric spheres have radii a and b with $b > a$. The region between them (i.e., $a \leq r \leq b$) is filled with charge of constant density. The charge density is zero everywhere else. Find $\mathbf{E}$ at all points and express it in terms of the total charge Q. Do your results reduce to the correct values as $a \rightarrow 0$?

4-7 An infinitely long cylinder has a circular cross section of radius a. It is filled with charge of constant volume density ρ_{ch}. Find **E** for all points both inside and outside the cylinder. Are your results consistent with (4-11)?

4-8 Two infinitely long coaxial cylinders have radii a and b with $b > a$, as shown in Figure 4-7. The region between them is filled with charge of volume density given in cylindrical coordinates by $\rho_{ch} = A\rho^n$ where A and n are constants. The charge density is zero everywhere else. Find **E**

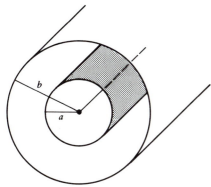

Figure 4-7. The coaxial cylinders of Exercise 4-8.

everywhere. For what values of n and a should your results reduce to those of Exercise 4-7? Do they?

4-9 The region between the infinitely long coaxial cylinders of Figure 4-7 is filled with charge whose volume density in cylindrical coordinates is $\rho_{ch} = Ae^{-\alpha\rho}$. Find **E** everywhere. Under what simple circumstances should your results here and for Exercise 4-8 reduce to the same values for **E**? Do they?

4-10 The average electrostatic field in the earth's atmosphere in fair weather has been found experimentally to be approximately given by $\mathbf{E} = -E_0(Ae^{-\alpha z} + Be^{-\beta z})\hat{\mathbf{z}}$. All of the empirical constants are positive and z is the height above the (locally flat) surface. Find the average charge density in the atmosphere as a function of height. What is its sign? (Over land, a typical value for E_0 is $\simeq 250$ newtons/coulomb.)

4-11 A certain electric field is given by $\mathbf{E} = E_0(\rho/a)^3\hat{\boldsymbol{\rho}}$ for $0 < \rho < a$ and $\mathbf{E} = 0$ otherwise. Find the volume charge density.

4-12 An electric field in the region $r > a$ is given by $E_r = 2A\cos\theta/r^3$, $E_\theta = A\sin\theta/r^3$, $E_\varphi = 0$ where $A = $ const. Find the volume charge density in this region.

5

THE SCALAR POTENTIAL

Up to now, our description of electrostatic effects has been done completely in terms of the vector field $\mathbf{E}$. By rewriting our expression for $\mathbf{E}$, we will see that we will be able to express substantially the same information in terms of a scalar field that will be much more convenient for many purposes.

5-1 DEFINITION AND PROPERTIES OF THE SCALAR POTENTIAL

Our basic defining equation for the electric field is given by (3-2). If we use (1-141) to replace $\hat{\mathbf{R}}_i/R_i^2$ by $-\nabla(1/R_i)$, and then use (1-110) to write the sum of the derivatives as the derivative of the sum, we find that (3-2) becomes

$$\mathbf{E}(\mathbf{r}) = -\sum_i \frac{q_i}{4\pi\epsilon_0} \nabla\left(\frac{1}{R_i}\right) = -\nabla \sum_i \frac{q_i}{4\pi\epsilon_0 R_i} \tag{5-1}$$

where, as usual, $R_i = |\mathbf{r} - \mathbf{r}_i|$. Thus, we see that if we define

$$\phi(\mathbf{r}) = \sum_{i=1}^{N} \frac{q_i}{4\pi\epsilon_0 R_i} \tag{5-2}$$

we can write

$$\mathbf{E}(\mathbf{r}) = -\nabla\phi(\mathbf{r}) \tag{5-3}$$

so that

$$\nabla \times \mathbf{E} = 0 \tag{5-4}$$

according to (1-48). The scalar field ϕ is called the *scalar potential* or the *electrostatic potential*. The unit in which the scalar potential is measured is called a *volt*; we see from (5-3) that the electric field can also be stated in volts/meter, which, in fact, is the most commonly used unit. Combining this with our previous unit of newton/coulomb for $\mathbf{E}$, we find that 1 volt = 1 joule/coulomb.

Since the curl of the electrostatic field is zero everywhere according to (5-4), we find from Stokes' theorem (1-67) that

$$\oint_C \mathbf{E} \cdot d\mathbf{s} = 0 \tag{5-5}$$

where C is an arbitrary closed path. This result shows us explicitly that the electrostatic field is an example of what is known as a *conservative field*.

In summary, we have found that the vector electrostatic field can be written as the negative gradient of the scalar potential, and (5-2) provides us with a method of calculating ϕ at any desired field point $\mathbf{r}$ once we are given the values and locations of the source point charge distributions. For example, if everything is given in rectangular coordinates, we can use (1-14) to write (5-2) as

$$\phi(\mathbf{r}) = \phi(x, y, z) = \sum_{i=1}^{N} \frac{q_i}{4\pi\epsilon_0\left[(x - x_i)^2 + (y - y_i)^2 + (z - z_i)^2\right]^{1/2}} \tag{5-6}$$

Since ϕ is a scalar quantity, it is generally easier to proceed indirectly by evaluating the sum (5-2) and then finding $\mathbf{E}$ by differentiation by (5-3), rather than by the direct process of evaluating the vector sum (3-2); this is one reason why ϕ is of practical interest.

If the source charges have a continuous distribution, we can proceed exactly as we did in Sections 2-4 and 3-1 by using (2-14) and (2-16) to write (5-2) in integral form; the results are

$$\phi(\mathbf{r}) = \frac{1}{4\pi\epsilon_0} \int_{V'} \frac{\rho(\mathbf{r}')\, d\tau'}{R} \tag{5-7}$$

$$\phi(\mathbf{r}) = \frac{1}{4\pi\epsilon_0} \int_{S'} \frac{\sigma(\mathbf{r}')\, da'}{R} \tag{5-8}$$

$$\phi(\mathbf{r}) = \frac{1}{4\pi\epsilon_0} \int_{L'} \frac{\lambda(\mathbf{r}')\, ds'}{R} \tag{5-9}$$

where the integrals are taken over all volumes, surfaces, or lines containing the source charges or, in fact, over all space, since in regions where there is no charge the respective charge densities will vanish. Finally, if all of the possibilities we have discussed are simultaneously present, we see from (5-1) that the total ϕ at a given point will be the scalar sum of all of the contributions from (5-2) and (5-7) through (5-9) and the total $\mathbf{E}$ at a given point can be found as the negative gradient of this total scalar potential.

There is a certain amount of ambiguity remaining in our results, however, which arose when we went from (5-1) to (5-2). Suppose that, instead of (5-2), we had chosen to define ϕ as

$$\phi(\mathbf{r}) = \sum_{i=1}^{N} \frac{q_i}{4\pi\epsilon_0 R_i} + C \tag{5-10}$$

where C is a constant but otherwise completely arbitrary. When this expression is substituted into (5-3), we get exactly the same field $\mathbf{E}$ as given by (5-1) since $\nabla C = 0$. In other words, the scalar potential in principle always includes an arbitrary additive constant and we can assign this constant any convenient value without changing any essential features of a given problem. It is inconvenient to carry this constant along in all one's equations so that the *usual convention* is to simply choose $C = 0$, thus bringing us back to (5-2) and the equations that follow from this. In order to see the significance of this choice, let us assume that all of the charges in (5-2) occupy a finite region in space; then we can always choose our field point $\mathbf{r}$ to be so far away from all of the charges that all of the $R_i \rightarrow \infty$ and $\phi(\mathbf{r}) \rightarrow 0$. In other words, with the choice of (5-2), the scalar potential will vanish "at infinity." While this is the standard convention and the one we will generally use, there are some situations where it is convenient to choose other points, such as the surface of the earth, to have a value of zero for the potential; if we do this, we will mention it explicitly.

There is another useful way of expressing the relation between $\mathbf{E}$ and ϕ. Let us consider the line integral of $\mathbf{E}$ between an initial point P_i at $\mathbf{r}_1$ and a final point P_f at $\mathbf{r}_2$ as illustrated in Figure 1-22:

$$\int_1^2 \mathbf{E} \cdot d\mathbf{s} = \int_1^2 - \nabla\phi \cdot d\mathbf{s} = - \int_1^2 d\phi = -(\phi_2 - \phi_1) = -[\phi(\mathbf{r}_2) - \phi(\mathbf{r}_1)]$$

where we have used (5-3) and (1-38). We can write this, using (5-5), as

$$\Delta\phi = \phi(\mathbf{r}_2) - \phi(\mathbf{r}_1) = -\int_1^2 \mathbf{E} \cdot d\mathbf{s} = \int_2^1 \mathbf{E} \cdot d\mathbf{s} \tag{5-11}$$

giving us a relation between the change in the scalar potential $\Delta\phi$ and the line integral of the electric field. Since the result depends only on the values of ϕ at the end points, the value of the integral is *independent of the path*. We recall that this is another way of saying that the field **E** is a conservative field. [If the path is closed, $\mathbf{r}_2 = \mathbf{r}_1$ and (5-11) gives (5-5) again.] We note that $\Delta\phi$ given by (5-11) is independent of any arbitrary additive constant that might be included in the definition of ϕ since this constant will cancel when the difference is evaluated. We can use (5-11) to find the difference in potential between two points if the field **E** has been found by other methods, such as those illustrated in the last two chapters, since the known value of **E** can be used and the integral evaluated over *any* convenient path as it is independent of the path which may be chosen. As we will see, this procedure is sometimes useful when the charges are assumed to extend to infinity so that ϕ may not go to zero at infinity, or even becomes infinite there as well as at various inconvenient points, or in other situations in which (5-7) through (5-9) are difficult to apply.

A surface on which ϕ is constant is called an *equipotential surface*. We saw in Section 1-9 that the gradient of a scalar is normal to a surface of constant value of the scalar and in the direction of the maximum rate of change of the scalar. As a result, we see from (5-3) that the direction of **E** will be perpendicular to the equipotential surface and in the sense of decreasing ϕ. This is illustrated in Figure 5-1, in which the equipotential surfaces are shown as solid curves and the dashed lines are drawn to indicate the direction of **E** at each point for the case in which $\phi_3 > \phi_2 > \phi_1$. (A line that is defined to be tangent to the direction of **E** at each of its points is called a "line of force" or a "line of **E**." It is conventional to draw them so that the density of the lines is proportional to the magnitude of **E**, that is, the numerical value of **E** is greater in regions where the lines are closer together than it is in regions where the lines are farther apart.) This mutually orthogonal relationship illustrated in the figure enables us to obtain a visualization of the situation by sketching the field lines once we have solved for ϕ so that the equipotential surfaces can be found; the converse way of sketching the shape of the equipotentials from the known lines of **E** is also useful.

■ **Example**

Single point charge. Let us illustrate some of these ideas with the simplest possible example, that of a point charge Q located at $\mathbf{r}'$. In this case, (5-2) reduces to the single

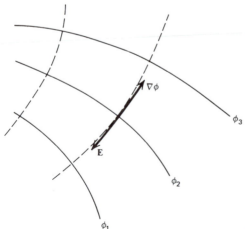

Figure 5-1. Equipotential surfaces (solid) and lines of electric field (dashed).

term

$$\phi(\mathbf{r}) = \frac{Q}{4\pi\epsilon_0 R} \qquad (5\text{-}12)$$

where $R = |\mathbf{r} - \mathbf{r}'|$. When this is combined with (5-3) and (1-141), the result is

$$\mathbf{E}(\mathbf{r}) = \frac{Q\hat{\mathbf{R}}}{4\pi\epsilon_0 R^2} \qquad (5\text{-}13)$$

which is, of course, in agreement with what (3-2) becomes for a single charge. The value of ϕ given by (5-12) as a function of the distance R from Q is shown in Figure 5-2 for both possible signs of Q. The equipotential surfaces are found by solving (5-12) for R and letting ϕ have a definite value; the resulting equation is

$$R = \frac{Q}{4\pi\epsilon_0\phi} \qquad (5\text{-}14)$$

so that these surfaces correspond to $R = $ const., that is, they are spheres centered on the charge Q. This situation is shown in Figure 5-3 in which we have assumed Q to be positive so that $\phi_3 > \phi_2 > \phi_1$. According to Figure 5-1, $\mathbf{E}$ must be perpendicular to these spheres and in the sense of decreasing ϕ. Therefore $\mathbf{E}$ must be directed radially outward from Q, in agreement with (5-13). ∎

In principle, an equation like (5-7) provides us with the complete solution to a problem in electrostatics. As we will see, however, some problems are not stated in such a way that (5-7) can be used directly, or it is too difficult to do so. For example, we may know the charge distribution only within a finite region of space, and it is unknown outside this region, but the values of ϕ on the bounding surfaces of the region may be given instead. Then (5-7) cannot generally be used directly; however, another approach

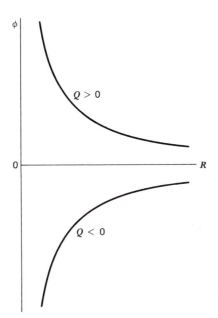

Figure 5-2. Potential as a function of distance R from a point charge Q.

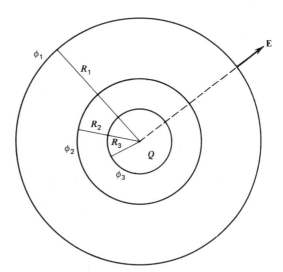

Figure 5-3. Equipotential spheres due to a point charge.

can be formulated. If we combine (5-3) with (4-10), we can eliminate **E** and we find, with the use of (1-45), that

$$\nabla^2 \phi = -\frac{\rho}{\epsilon_0} \tag{5-15}$$

In other words, the scalar potential satisfies this differential equation that is known as *Poisson's equation*. In regions where $\rho = 0$, (5-15) simplifies to *Laplace's equation*:

$$\nabla^2 \phi = 0 \tag{5-16}$$

We will defer the discussion of the detailed use of these equations and the justification of such an approach until Chapter 11.

Finally, we can point out here that our source equations for **E**, (4-10) and (5-4), along with the other results (5-3) and (5-7), agree completely with the Helmholtz theorem quoted in (1-148) through (1-150).

We consider next a few more elaborate examples of the calculation of the scalar potential for continuous charge distributions.

5-2   UNIFORM SPHERICAL CHARGE DISTRIBUTION

This is exactly the same charge distribution we considered in Section 2-5, that is, a total charge Q contained in a sphere of radius a with constant charge density $\rho = 3Q/4\pi a^3$. We again use the coordinate system shown in Figures 2-6 and 2-7; the point in these figures showing the location of q is now to be interpreted as the field point where we want to find ϕ. As before, $R^2 = z^2 + r'^2 - 2zr'\cos\theta'$, so that (5-7) becomes

$$\phi = \frac{\rho}{4\pi\epsilon_0} \int_0^{2\pi} \int_0^{\pi} \int_0^a \frac{r'^2 \sin\theta' \, dr' \, d\theta' \, d\varphi'}{(z^2 + r'^2 - 2zr'\cos\theta')^{1/2}} \tag{5-17}$$

where we have used (1-99) and taken the constant value of ρ outside of the integral. The integration over $d\varphi'$ can be performed at once and gives 2π. If we again let $\mu = \cos\theta'$, and use (2-22), we find that (5-17) becomes

$$\phi = \frac{\rho}{2\epsilon_0} \int_0^a r'^2 \, dr' \int_{-1}^1 \frac{d\mu}{(z^2 + r'^2 - 2zr'\mu)^{1/2}} \tag{5-18}$$

The integral over μ can be found from tables to be

$$-\frac{(z^2 + r'^2 - 2zr'\mu)^{1/2}}{zr'} \Bigg|_{-1}^1 = \frac{1}{zr'}(|z + r'| - |z - r'|) \tag{5-19}$$

There are now two cases to be considered.

1. Outside the sphere. Here $z > a$; since we are assuming z to be positive for simplicity, and since $r' \le a$, we have $z > r'$, so that the parentheses in (5-19) become $(z + r') - (z - r') = 2r'$ and the integral over μ equals $2/z$. Inserting this into (5-18) and integrating over r', we find the potential at a point outside the sphere, ϕ_o, to be

$$\phi_o = \frac{\rho a^3}{3\epsilon_0 z} = \frac{Q}{4\pi\epsilon_0 z} \tag{5-20}$$

2. Inside the sphere. Here $z < a$, so that r' can be both greater than or less than z. When $z < r' < a$, (5-19) becomes

$$\frac{1}{zr'}[(z + r') - (r' - z)] = \frac{2}{r'}$$

while, when $r' < z < a$, (5-19) is $2/z$ as before. Since the integral over μ has these different values for different ranges of r', we have to evaluate the integral over r' as the sum of two integrals each with an integrand appropriate for the range of r' involved; thus, our expression for the potential inside, ϕ_i, as obtained from (5-18) becomes

$$\phi_i = \frac{\rho}{2\epsilon_0}\left(\int_0^z r'^2\, dr' \cdot \frac{2}{z} + \int_z^a r'^2\, dr' \cdot \frac{2}{r'}\right)$$

$$= \frac{\rho}{6\epsilon_0}(3a^2 - z^2) = \frac{Q}{8\pi\epsilon_0 a}\left(3 - \frac{z^2}{a^2}\right) \tag{5-21}$$

We note that (5-20) and (5-21) give the same value of the potential, $Q/4\pi\epsilon_0 a$, at the surface of the sphere where $z = a$.

These results can be easily generalized to an arbitrary location. The field point was taken to be along the z axis for convenience in evaluating the integral. The position vector of the field point is thus $\mathbf{r} = z\hat{\mathbf{z}}$ so that $|\mathbf{r}| = r = z$ and $\hat{\mathbf{r}} = \hat{\mathbf{z}}$; this enables us to write our results for any field point in terms of its spherical coordinate distance r from the origin, that is, from the center of the sphere. Therefore, (5-20) and (5-21) yield these general expressions for the potential at any point outside of or inside of the sphere:

$$\phi_o(r) = \frac{\rho a^3}{3\epsilon_0 r} = \frac{Q}{4\pi\epsilon_0 r} \tag{5-22}$$

$$\phi_i(r) = \frac{\rho}{6\epsilon_0}(3a^2 - r^2) = \frac{Q}{8\pi\epsilon_0 a}\left(3 - \frac{r^2}{a^2}\right) \tag{5-23}$$

As a check on our calculation of ϕ, we can see if it gives the known electric fields. If we insert (5-22) and (5-23) into (5-3), and use (1-101), we find the electric fields outside and inside the sphere to be, respectively,

$$\mathbf{E}_o = \frac{\rho a^3 \hat{\mathbf{r}}}{3\epsilon_0 r^2} = \frac{Q\hat{\mathbf{r}}}{4\pi\epsilon_0 r^2} \tag{5-24}$$

$$\mathbf{E}_i = \frac{\rho r\hat{\mathbf{r}}}{3\epsilon_0} = \frac{Qr\hat{\mathbf{r}}}{4\pi\epsilon_0 a^3} \tag{5-25}$$

and these agree with the results (4-17), (4-20), and (4-21), which we previously obtained by using Gauss' law.

According to (5-22) and (5-23), constant values of ϕ correspond to constant values of r; in other words, the equipotential surfaces are concentric spheres with centers at the origin (i.e., at the center of the charge distribution). Figure 5-4 is a plot of the potential as a function of r; the negative slope of this curve gives the electric field E_r that we have previously shown in Figure 4-5 and it is instructive to compare these two figures with this in mind.

5-3 UNIFORM LINE CHARGE DISTRIBUTION

As an example of the use of (5-11) to find the potential, let us consider the uniform infinite line charge for which we previously found the field as given by (3-9) to be $\mathbf{E} = (\lambda/2\pi\epsilon_0\rho)\hat{\boldsymbol{\rho}}$. Since $\mathbf{E}$ has only a ρ component, we see from (5-3) and (1-85) that ϕ is independent of φ and z, that is, it is a function only of ρ. This already shows us that the equipotential surfaces will be cylinders whose common axis coincides with the line charge since $\phi = $ const. leads to $\rho = $ const. Since we can use any convenient path, let us

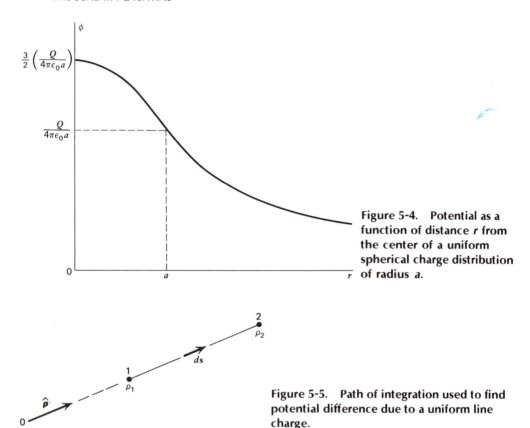

Figure 5-4. Potential as a function of distance r from the center of a uniform spherical charge distribution of radius a.

Figure 5-5. Path of integration used to find potential difference due to a uniform line charge.

choose the simple one shown in Figure 5-5; the path lies in a plane perpendicular to the charge distribution that intersects this plane at the origin. We will integrate along the straight line in the constant direction $\hat{\rho}$ from the initial point 1 to the final point 2 whose distances from the charge are ρ_1 and ρ_2, respectively. We see from the figure that $d\mathbf{s} = \hat{\rho}\, d\rho$ in this case so that the integrand in (5-11) becomes $\mathbf{E} \cdot \hat{\rho}\, d\rho = E_\rho\, d\rho$ from (1-21), and therefore (5-11) becomes

$$\phi(\rho_2) - \phi(\rho_1) = -\int_{\rho_1}^{\rho_2} \frac{\lambda\, d\rho}{2\pi\epsilon_0 \rho} = \frac{\lambda}{2\pi\epsilon_0} \ln\left(\frac{\rho_1}{\rho_2}\right) \tag{5-26}$$

This method gives us the difference in potential between the two points, and we would like to find the absolute value for ϕ. The logarithm term in (5-26) can be written as $[(-\ln \rho_2) - (-\ln \rho_1)]$ and, since any additive constant will cancel on the left side, we conclude from comparing the two sides of (5-26) that ϕ has the form

$$\phi(\rho) = -\frac{\lambda}{2\pi\epsilon_0} \ln \rho + C \tag{5-27}$$

where C is some constant. Now as $\rho \to \infty$, $\ln \rho \to \infty$ so that if we try to preserve our convention that ϕ vanish at infinity, we conclude that C must also be infinite. The source of this difficulty is that in this case the charge is not finite and extends to infinity, whereas our expectation that ϕ should vanish at infinity arose from (5-2) and the assumption that all of the charges were contained within a finite volume, as was the case in the previous example of the sphere. Even though we may be uncomfortable with it at this stage, the fact remains that the ϕ of (5-27) does give the correct field $\mathbf{E}$ by means of (5-3). It is sometimes convenient to put (5-27) into another form; if we introduce a new constant ρ_0 by writing $C = \lambda \ln \rho_0/2\pi\epsilon_0$, then (5-27) becomes

$$\phi(\rho) = \frac{\lambda}{2\pi\epsilon_0} \ln\left(\frac{\rho_0}{\rho}\right) \tag{5-28}$$

When $\rho = \rho_0$, $\phi(\rho_0) = 0$ so that the physical significance of ρ_0 is seen to be that it is the distance at which we have *chosen* ϕ to vanish, since we cannot make it do so at infinity. The derivative of ϕ as obtained from (5-28) and that gives $\mathbf{E}$ is unaffected by the choice of ρ_0, as is the potential difference that will still be given by (5-26).

Since these effects arose from our assumption that the line charge was already infinitely long, it is helpful to look at the problem afresh by starting with a uniform line charge of finite length for which we will get a finite result as we will see, and then seeing what happens as we let it get very long. We will also obtain in this way a plausible way of choosing ρ_0. Let us consider the charge distribution shown in Figure 3-8 for $\lambda = \text{const.}$, and calculate ϕ at the general point $P(\rho, \varphi, z)$. Since the charge distribution coincides with the z axis, $\mathbf{r}' = z'\hat{\mathbf{z}}$, while $\mathbf{r} = \rho\hat{\boldsymbol{\rho}} + z\hat{\mathbf{z}}$ according to (1-81). Therefore, $\mathbf{R} = \rho\hat{\boldsymbol{\rho}} + (z - z')\hat{\mathbf{z}}$ so that $R = [\rho^2 + (z - z')^2]^{1/2}$ and, since $ds' = dz'$ in this case, (5-9) becomes

$$\phi = \frac{\lambda}{4\pi\epsilon_0} \int_{-L_1}^{L_2} \frac{dz'}{[\rho^2 + (z - z')^2]^{1/2}} \tag{5-29}$$

The integral is easily evaluated with the use of tables and the result is

$$\phi = \frac{\lambda}{4\pi\epsilon_0} \ln\left\{\frac{z + L_1 + [\rho^2 + (z + L_1)^2]^{1/2}}{z - L_2 + [\rho^2 + (z - L_2)^2]^{1/2}}\right\} \tag{5-30}$$

[It will be left as an exercise to verify that when (5-30) is used in (5-3), the same $\mathbf{E}$ is obtained as was found by direct evaluation in Exercise 3-11.] Now let us assume that the line charge is very long, although not yet infinite, that is, we assume that $L_2 \gg \rho$, $L_2 \gg |z|$, $L_1 \gg \rho$, and $L_1 \gg |z|$, so that we can get an approximate expression for ϕ for situations in which our field point is not too close to the ends of the charge distribution nor too far away from it in comparison with its length. The numerator of the argument of the logarithm in (5-30) gives us no trouble and is approximately $2L_1$ since we can write $z + L_1 \simeq L_1$ and can similarly neglect ρ in comparison with L_1. We have to be more careful with the denominator since just omitting z and ρ makes it vanish; accordingly we expand it in a power series in an appropriate small quantity and keep the lowest order nonvanishing term. We use the expansion

$$(1 \pm x)^{1/2} = 1 \pm \tfrac{1}{2}x - \tfrac{1}{8}x^2 \pm \ldots \tag{5-31}$$

although here we will require only the first two terms and the upper sign. If we factor out $L_2 - z$, the denominator becomes

$$(L_2 - z)\left\{-1 + \left[1 + \frac{\rho^2}{(L_2 - z)^2}\right]^{1/2}\right\} \simeq L_2\left[-1 + \left(1 + \frac{\rho^2}{L_2^2}\right)^{1/2}\right]$$

$$\simeq L_2\left(-1 + 1 + \frac{\rho^2}{2L_2^2}\right) = \frac{\rho^2}{2L_2}$$

Substituting this into (5-30), along with $2L_1$ for the numerator, we get

$$\phi \simeq \frac{\lambda}{4\pi\epsilon_0} \ln\left(\frac{4L_2L_1}{\rho^2}\right) = \frac{\lambda}{2\pi\epsilon_0} \ln\left[\frac{(4L_2L_1)^{1/2}}{\rho}\right] \tag{5-32}$$

which is exactly of the *form* (5-27) or (5-28). Thus we are led back to the same general expression for ϕ as we approach the limit of an infinite line. We note that (5-32) is independent of z, which is reasonable since if we do not get too near the ends of this very long line charge, a displacement parallel to the line leads one to a point where the charge distribution will appear essentially unaltered.

If we now go further and let both L_2 and L_1 get infinitely large in (5-32), we again get an infinite value for ϕ. Hence this approach has not made the infinities inherent in our problem disappear, of course, but we have a better understanding of how they have arisen.

There is one interesting case in which these ambiguities in ϕ disappear because of the nature of the system.

■ **Example**

Two parallel oppositely charged lines. Let us consider a system comprised of one long uniform line charge of density λ and the other of density $-\lambda$. We assume the same values of L_2 and L_1 for them as shown in Figure 5-6. The distance ρ in (5-32) began as the cylindrical coordinate variable but is clearly the perpendicular distance from the line charge to the field point P; these distances for this case are labeled ρ_+ and ρ_- in the figure. If we assume that L_2 and L_1 are already large enough, we can use (5-32). Thus the individual potentials due to these two line charges are

$$\phi_+ = \frac{\lambda}{2\pi\epsilon_0} \ln\left[\frac{(4L_2L_1)^{1/2}}{\rho_+}\right]$$

$$\phi_- = -\frac{\lambda}{2\pi\epsilon_0} \ln\left[\frac{(4L_2L_1)^{1/2}}{\rho_-}\right]$$

so that the total potential at P is

$$\phi = \phi_+ + \phi_- = \frac{\lambda}{2\pi\epsilon_0} \ln\left(\frac{\rho_-}{\rho_+}\right) \tag{5-33}$$

and the terms depending on L_2 and L_1 have canceled. *Now* we can let L_2 and L_1 each go to infinity and we will be left with the same unambiguous result given in (5-33). In a specific case, ρ_+ and ρ_- have to be evaluated in terms of the particular coordinate system that is being used. Let us illustrate this for the case in which the line charges are

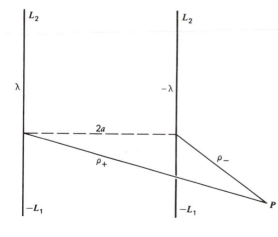

Figure 5-6. Two parallel oppositely charged lines.

taken to be parallel to the z axis, the line of length $2a$ between them lying along the x axis, the origin is chosen midway between them, and P lies in the xy plane as shown in Figure 5-7. We will use polar coordinates ρ and φ to specify the location of P. Applying the law of cosines, we find that $\rho_+^2 = a^2 + \rho^2 - 2a\rho \cos\varphi$ and $\rho_-^2 = a^2 + \rho^2 + 2a\rho \cos\varphi$, so that (5-33) can be written

$$\phi(\rho, \varphi) = \frac{\lambda}{4\pi\epsilon_0} \ln\left(\frac{\rho_-^2}{\rho_+^2}\right) = \frac{\lambda}{4\pi\epsilon_0} \ln\left(\frac{a^2 + \rho^2 + 2a\rho \cos\varphi}{a^2 + \rho^2 - 2a\rho \cos\varphi}\right) \tag{5-34}$$

and the components of $\mathbf{E}$ as found from (5-3) and (1-85) turn out to be

$$E_\rho = -\frac{\partial\phi}{\partial\rho} = \frac{\lambda}{2\pi\epsilon_0}\left[\frac{(\rho - a\cos\varphi)}{\rho_+^2} - \frac{(\rho + a\cos\varphi)}{\rho_-^2}\right] = \frac{\lambda a(\rho^2 - a^2)\cos\varphi}{\pi\epsilon_0\rho_+^2\rho_-^2} \tag{5-35}$$

$$E_\varphi = -\frac{1}{\rho}\frac{\partial\phi}{\partial\varphi} = \frac{\lambda a(\rho^2 + a^2)\sin\varphi}{\pi\epsilon_0\rho_+^2\rho_-^2} \tag{5-36}$$

while $E_z = -\partial\phi/\partial z = 0$.

The equipotential surfaces $\phi = $ const. are given by (5-34) as

$$\frac{\rho_-^2}{\rho_+^2} = e^{4\pi\epsilon_0\phi/\lambda} = \text{const.} \tag{5-37}$$

This equation is more easily interpreted in rectangular coordinates; we see by inspecting Figure 5-7 that it becomes

$$\frac{(x + a)^2 + y^2}{(x - a)^2 + y^2} = e^{4\pi\epsilon_0\phi/\lambda}$$

which, with a little algebra, can be written in the form

$$(x - a\coth\eta)^2 + y^2 = \left(\frac{a}{\sinh\eta}\right)^2 \tag{5-38}$$

where $\eta = 2\pi\epsilon_0\phi/\lambda$. This is just the equation of a circle with radius $a/\sinh\eta$ and with center displaced a distance $a\coth\eta$ along the x axis. In other words, the equipotential surfaces are cylinders with axes parallel to the z axis whose intersections with the xy plane are the circles given by (5-38). These circles are shown as the solid curves in Figure 5-8. We note that the circles whose centers lie on the positive x axis correspond to $\phi > 0$, while those with centers on the negative x axis correspond to $\phi < 0$. We also note that the yz plane ($x = 0$) is the equipotential surface for $\phi = 0$; this is easily seen

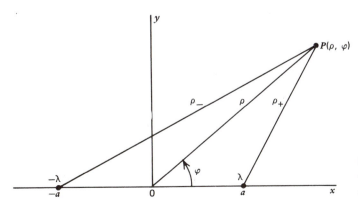

Figure 5-7.   Geometry for the potential of two line charges parallel to the z axis.

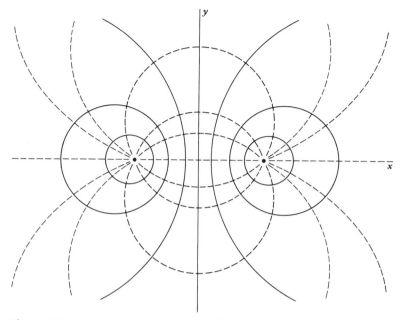

Figure 5-8. Equipotentials (solid) and lines of electric field (dashed) for two oppositely charged lines parallel to the z axis.

to be the case since we see from Figure 5-7 that each point on the y axis has $\rho_+ = \rho_-$, making $\phi = 0$ according to (5-33).

The lines of **E** are perpendicular to these equipotentials as we know and are shown as dashed lines in Figure 5-8. We can use our results for this case to illustrate how one can find an explicit equation for these curves. As we mentioned before, a line of **E** is defined in order to be tangent to the direction of **E** at the corresponding point. Therefore, if $d\mathbf{s}_{lf}$ represents a small displacement along a line of **E** (line of force), it is necessarily parallel to **E** at that point; this is shown in Figure 5-9. Therefore, we can write

$$d\mathbf{s}_{lf} = k\mathbf{E} \qquad (5\text{-}39)$$

since they have the same direction and where k is a constant of proportionality with appropriate dimensions. We can use (5-39) to obtain a differential equation for the curve describing the line of **E**.

In order to apply (5-39) to the particular case worked out above, we note that $d\mathbf{s}_{lf}$ will equal $d\mathbf{r}$ as expressed by (1-82) where $\mathbf{r}$ is the position vector of each point on the curve. Since the line of **E** lies in the ρ, φ plane, $dz = 0$ and (5-39) becomes

$$d\rho\,\hat{\boldsymbol{\rho}} + \rho\,d\varphi\,\hat{\boldsymbol{\varphi}} = k\big(E_\rho\hat{\boldsymbol{\rho}} + E_\varphi\hat{\boldsymbol{\varphi}}\big) \qquad (5\text{-}40)$$

Equating components, we get

$$d\rho = kE_\rho \qquad \text{and} \qquad \rho\,d\varphi = kE_\varphi$$

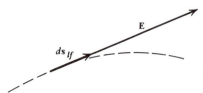

Figure 5-9. Relation between E and a displacement along a line of E.

and we can now eliminate k by dividing the first equation by the second. The result is

$$\frac{1}{\rho}\frac{d\rho}{d\varphi} = \frac{E_\rho}{E_\varphi} \tag{5-41}$$

The right-hand side of (5-41) will be a known function of ρ and φ, which, in principle, will enable us to integrate this equation to find ρ as a function of φ, that is, the equation of the line of force.

In our case, if we substitute (5-35) and (5-36) into (5-41), we find that

$$\frac{1}{\rho}\frac{d\rho}{d\varphi} = \frac{(\rho^2 - a^2)\cos\varphi}{(\rho^2 + a^2)\sin\varphi}$$

or

$$\frac{(\rho^2 + a^2)\,d\rho}{\rho(\rho^2 - a^2)} = \frac{\cos\varphi\,d\varphi}{\sin\varphi} \tag{5-42}$$

This equation can be integrated with the help of tables to give

$$\ln\left(\frac{\rho^2 - a^2}{a\rho}\right) = \ln\sin\varphi + \ln K$$

where the constant of integration is written in terms of the dimensionless constant K. From this result, we find our desired equation to be

$$\rho^2 - a^2 = Ka\rho\sin\varphi \tag{5-43}$$

For each value of K, we get a corresponding line of force. If we express (5-43) in rectangular coordinates with the use of (1-74) and (1-75), we find that it can be written

$$x^2 + \left(y - \tfrac{1}{2}Ka\right)^2 = a^2\left(1 + \tfrac{1}{4}K^2\right) \tag{5-44}$$

showing that the lines of **E** are also arcs of circles; a given circle has a radius $a(1 + \tfrac{1}{4}K^2)^{1/2}$ and its center displaced a distance $\tfrac{1}{2}Ka$ along the y axis. These are the curves that are plotted as dashed lines in Figure 5-8 and we see that, as expected, they originate on the positive charges and terminate on the negative charges since it follows from (5-43) that when $\varphi = 0$ or π, $\rho = \pm a$ or, from (5-44), that $x = \pm a$ when $y = 0$. ∎

5-4 THE SCALAR POTENTIAL AND ENERGY

In general, changes in potential are related to changes in energy. Consider a point charge q in equilibrium under the action of an electrostatic force $\mathbf{F}_q$ and a mechanical force $\mathbf{F}_{q,m}$. This condition was described by (2-9) for only two charges, but can clearly be extended to the case when more than one charge is acting on q. In this case, it is appropriate to use (3-1), so that the condition of equilibrium becomes

$$\mathbf{F}_q + \mathbf{F}_{q,m} = q\mathbf{E} + \mathbf{F}_{q,m} = 0$$

or

$$\mathbf{F}_{q,m} = -q\mathbf{E} \tag{5-45}$$

Now let us imagine moving the charge infinitely slowly from an initial point $\mathbf{r}_1$ to a final point $\mathbf{r}_2$ along some path. Under these conditions, the velocity will be essentially always zero and constant so that the acceleration will be zero. The charge will then be always in equilibrium, or nearly so, so that (5-45) applies. We assume this procedure so that we can calculate the amount of reversible work in the thermodynamic sense that is done by the external mechanical force, and by keeping the velocity zero, we can be sure that there will be no dissipative or frictional effects involved. If we let $W_{1 \to 2}$ be the work done by the external agent responsible for the mechanical force, we get

$$W_{1 \to 2} = \int_1^2 \mathbf{F}_{q, m} \cdot d\mathbf{s} = -q \int_1^2 \mathbf{E} \cdot d\mathbf{s} = q[\phi(\mathbf{r}_2) - \phi(\mathbf{r}_1)] \qquad (5\text{-}46)$$

with the use of (5-11). In other words, the work done on the charge equals the value of its charge times the change in potential. Under the circumstances we have assumed, the work done can be equated to the *change* in potential energy ΔU_e of the charge so that (5-46) becomes

$$\Delta U_e = q[\phi(\mathbf{r}_2) - \phi(\mathbf{r}_1)] = q\,\Delta\phi \qquad (5\text{-}47)$$

We note that this change ΔU_e is independent of any additive constant that may be included in ϕ. Since the right-hand side of (5-47) already has the form of a difference, it is natural to write the left-hand side in the same way, that is, $\Delta U_e = U_e(\mathbf{r}_2) - U_e(\mathbf{r}_1)$, and by comparison, we can define *the* potential energy of a charge q at $\mathbf{r}$, $U_e(\mathbf{r})$, as

$$U_e(\mathbf{r}) = q\phi(\mathbf{r}) \qquad (5\text{-}48)$$

As usual with potential energy, we can add any arbitrary constant to the right-hand side of (5-48) without changing the difference, much as we could do for ϕ, that is, as in (5-10). We will generally choose the form (5-48), however, as it has the convenient property that when ϕ vanishes at infinity, U_e does also. Since the energy U_e will be measured in joules, we see again from (5-48) that the unit of ϕ, the volt, will be equal to 1 joule/coulomb.

■   **Example**

Two point charges. Let us consider a system of two point charges q and Q separated by a distance R as shown in Figure 5-10. The potential at the location of q is given by (5-12) and when this is inserted into (5-48), we find that

$$U_e = \frac{qQ}{4\pi\epsilon_0 R} \qquad (5\text{-}49)$$

This energy can be interpreted as the work that was required to bring q from infinity to its location $\mathbf{r}$ while the charge Q was held fixed at $\mathbf{r}'$. But, because of the symmetry of

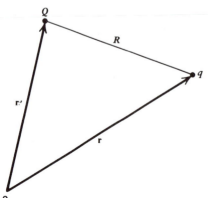

Figure 5-10. Relative positions of two point charges.

the expression, it can equally well be interpreted as the work required to bring Q from infinity to $\mathbf{r}'$ while holding q fixed at $\mathbf{r}$. In other words, it is more appropriate to regard U_e as the *mutual* potential energy of the *system* of the two charges rather than ascribing it to one charge or the other. We return to this question of the potential energy of a system of charges in detail in Chapter 7. ∎

EXERCISES

5-1 Can the vector $\mathbf{E} = (yz - 2x)\hat{\mathbf{x}} + xz\hat{\mathbf{y}} + xy\hat{\mathbf{z}}$ be a possible electrostatic field? If so, find the potential ϕ from which $\mathbf{E}$ can be obtained.

5-2 Could the vector $\mathbf{A}$ of Exercise 1-15 be interpreted as a conservative electric field? If so, find the potential ϕ from which it could be obtained by using (5-3).

5-3 Two point charges q and $-q$ are located on the z axis at $z = a$ and $-a$, respectively. Find ϕ at any point (x, y, z). Show that the xy plane is an equipotential surface and find its potential. Explain this result.

5-4 Consider the charge distribution of Exercise 3-2. Find the potential at the center of the square. Why can't you find $\mathbf{E}$ at the center of the square from your result?

5-5 Consider a cube of edge a with the location and orientation of the figure in Figure 1-41. There is a point charge q at each corner. Find ϕ at the center of the face for which $x = a$.

5-6 Within a certain region of space the electric field $\mathbf{E}$ is constant. Show that a suitable potential for this case is $\phi = -\mathbf{E} \cdot \mathbf{r} + \phi_0$ where ϕ_0 is a constant. What are the equipotential surfaces?

5-7 Obtain the results (5-22) and (5-23) by using (5-11) and the fields given by (5-24) and (5-25).

5-8 A sphere of radius a has a total charge Q distributed uniformly throughout its volume. The center of the sphere is at the point (A, B, C). Find the potential ϕ at any point (x, y, z) outside the sphere, and from this, find the rectangular components of $\mathbf{E}$ at that point.

5-9 A sphere of radius a has a charge density that varies with distance r from the center according to $\rho = Ar^n$ where $A = \text{const.}$ and $n \geq 0$. Find ϕ at all points inside and outside the sphere by using (5-7) and express your results in terms of the total charge Q of the sphere.

5-10 Find ϕ for all points inside and outside the sphere of charge of Exercise 4-5 by using (5-11). Plot ϕ as a function of r.

5-11 Find ϕ at all points for the charge distribution of Exercise 4-6. Express your answer in terms of the constant charge density ρ and plot ϕ as a function of r.

5-12 Find ϕ for all points outside and inside the cylinder described in Exercise 3-13.

5-13 Consider the charge distribution of Exercise 3-10 and Figure 3-7. Find ϕ at an arbitrary point on the z axis. Why does your result give the correct value of E_z but not of E_x?

5-14 A sphere of radius a has a constant surface charge density σ, but no volume charge density. Find ϕ for all points both outside and inside the sphere by using (5-8).

5-15 An infinite sheet of charge of constant surface density σ coincides with the xy plane. Use the expression for $\mathbf{E}$ given by (3-13) and find a potential ϕ from which $\mathbf{E}$ can be found. What are the equipotential surfaces in this case?

5-16 Find the potential due to the sheet of the previous exercise by using (5-8).

5-17 Charge is distributed with constant surface charge density σ on a circular disc of radius a lying in the xy plane with center at the origin. Show that the potential at a point on the z axis is given by

$$\phi = \frac{\sigma}{2\epsilon_0}\left[(a^2 + z^2)^{1/2} - |z|\right] \quad (5\text{-}50)$$

Verify that this gives (3-15). What does ϕ become as a becomes very large? What is the smallest value of z for which the potential due to this disc can be calculated as if it were a point charge without making an error greater than 1 percent?

5-18 Show that ϕ given by (5-30) gives the correct $\mathbf{E}$ as found in Exercise 3-11.

5-19 As a simple check on the results (5-34) through (5-36), show that they reduce to the expected results for the special cases $\varphi = 0$ and $\varphi = 90°$.

5-20 The parameter K, which characterizes a given line of force in (5-43), can be related to the magnitude E of the electric field where the line crosses the y axis, that is, when $\varphi = 90°$. Show that $K^2 = (\delta - 2)^2/(\delta - 1)$ where $\delta = \lambda/\pi\epsilon_0 aE$.

5-21 Consider the charge distribution of Figure 5-7. If one moves from a point on the x axis for which $x = b > a$ to another point on the x axis for which $x = -b$, what is the change in potential?

5-22 Express ϕ for the charge distribution of Figure 5-7 in rectangular coordinates and use this result to find E_x and E_y.

5-23 Consider the charge distribution of Exercise 5-3. How much work must be done by an external agent to change the separation of the charges from $2a$ to a? Illustrate this process on a plot of U_e versus separation R.

5-24 Consider the line charge of Figure 3-8 with $L_1 = 0$. The charge per unit length is proportional to the cube of the distance from the origin. Find the potential at a point P on the x axis for which $x = a$. From your result, find E_z at P; if you cannot do this explain why not. How much work must an external agent do to move a point charge q very slowly along the x axis from P out to infinity?

6

CONDUCTORS IN ELECTROSTATIC FIELDS

Coulomb's law assumes that the region between the charges involved is a vacuum. We are interested, of course, in what will occur when electric charges are in the presence of matter, so that at least part of the region of interest to us is not a vacuum. We will not consider this problem all at once, but in this chapter we restrict ourselves to the case in which a particular class of matter known as conductors is present. Because of the special properties of conductors, we will be able to deduce some interesting and important consequences that arise from their presence.

6-1 SOME GENERAL RESULTS

In general terms, a *conductor* can be defined as a region in which charges are free to move about under the influence of an electric field. The most common example is that of a metal in which the movable charges are "free" electrons, but it is not the only case, and we will not need to restrict ourselves here to such a specific example. We do, however, assume that we are dealing with a completely static situation on the macroscopic scale.

Now, *if* an electric field were present in a conductor, the charges would move about, and we would not have the *static* situation we are assuming. Therefore, we conclude that **E** must be zero at all points within a conductor. It then follows directly from (5-3) that ϕ is constant within the conductor, that is, the conductor is an *equipotential volume*. Therefore, we have

$$\begin{aligned} \mathbf{E}(\mathbf{r}) &= 0 \\ \phi(\mathbf{r}) &= \text{const.} \end{aligned} \qquad \text{inside a conductor} \qquad (6\text{-}1)$$

Now let us consider the situation at the surface of a conductor. The electric field in the nearby vacuum region certainly can be different from zero. Suppose that **E** made an angle with the surface as shown in Figure 6-1a. We can imagine **E** resolved into a component $\mathbf{E}_n$, which is perpendicular (normal) to the surface, and a component $\mathbf{E}_t$, which is parallel to the surface (the tangential component). Now, if $\mathbf{E}_t$ were different from zero, there would be a tangential force on the movable charges, they would move parallel to the surface, and we would no longer have the static situation we are assuming. Therefore, $\mathbf{E}_t = 0$, and the only possibility is that $\mathbf{E}_n \neq 0$; in other words, at the surface of a conductor, **E** must be normal to the surface, as shown in Figure 6-1b. But, as was illustrated in Figure 5-1, the direction of **E** is perpendicular to the equipotential surfaces, and therefore the surface of the conductor is an equipotential surface. In summary, we have

$$\begin{aligned} \mathbf{E}_n(\mathbf{r}) &\neq 0 \\ \mathbf{E}_t(\mathbf{r}) &= 0 \qquad \text{at surface of a conductor} \\ \phi(\mathbf{r}) &= \text{const.} \end{aligned} \qquad (6\text{-}2)$$

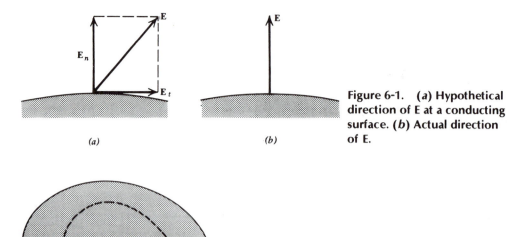

Figure 6-1. (a) Hypothetical direction of E at a conducting surface. (b) Actual direction of E.

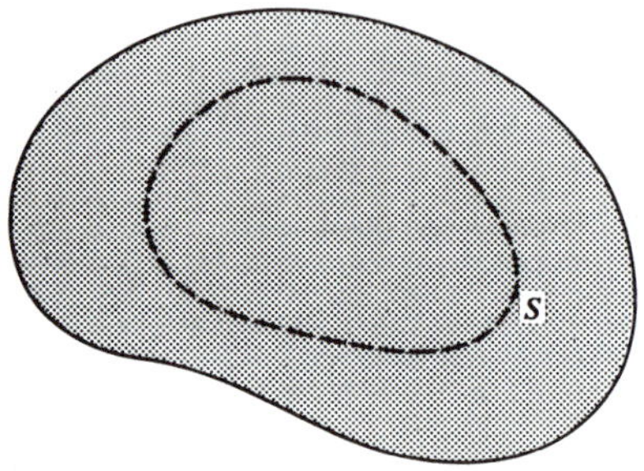

Figure 6-2. The surface S is completely inside a conductor.

(It may happen, of course, that at a particular point or points on the surface, the normal component $\mathbf{E}_n$ may also be zero, but, in any event, it is the only component that *can* be different from zero at the surface.)

Now let us apply Gauss' law (4-1) to an arbitrary closed surface that is completely in the interior of a conductor such as S shown dashed in Figure 6-2. Since $\mathbf{E} = 0$ at all points inside, it will be zero on each part of S and, in this case, (4-1) becomes

$$\oint_S \mathbf{E} \cdot d\mathbf{a} = 0 = \frac{Q_{\text{in}}}{\epsilon_0} \tag{6-3}$$

and therefore the total charge inside S, Q_{in}, is zero. Because the surface S is completely arbitrary, it can be deformed in any desired way, even to coincide with the bounding surface of the conductor and Q_{in} will still be zero. Thus, we can conclude that $Q_{\text{in}} = 0$ everywhere inside. In other words, we have found that the net charge in the interior of a conductor is always zero, so that whatever net charge is present on a conductor must reside entirely on its *surface*. This was first shown experimentally by Faraday in his famous "ice pail" experiment in which he measured the charge distribution on an ordinary metal pail. We note that this result is a direct conclusion from Gauss' law that itself is a consequence of the inverse square nature of Coulomb's law. In fact, this conclusion forms the basis for the most accurate experimental tests of the exactness of the exponent 2 in Coulomb's law, since the experiments essentially look for deviations from the prediction $Q_{\text{in}} = 0$; the results show that the exponent is indeed equal to 2 to within a few parts in 10^{16}.

In Figure 6-3, we show the situation at the surface of separation between a conducting region (1) and the vacuum region (2) where $\hat{\mathbf{n}}$ is the outer normal to the conductor. We construct a small cylinder of cross-sectional area Δa whose curved side is parallel to $\hat{\mathbf{n}}$, so that its outward normal $\hat{\mathbf{n}}_c$ is parallel to the surface of separation. The upper face of the cylinder is entirely in the vacuum region where $\mathbf{E} \neq 0$ while its lower face is entirely in the conductor where $\mathbf{E} = 0$; we choose Δa to be so small that $\mathbf{E}$ is essentially constant over the upper face. We now apply Gauss' law to this small

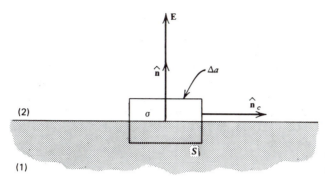

Figure 6-3. Calculation of the electric field at the surface of a conductor.

cylinder, noting that $Q_{in} = \sigma \Delta a$ in this case; therefore

$$\oint_S \mathbf{E} \cdot d\mathbf{a} = \int_{top} \mathbf{E} \cdot \hat{\mathbf{n}}\, da + \int_{side} \mathbf{E} \cdot \hat{\mathbf{n}}_c\, da + \int_{bottom} \mathbf{E} \cdot d\mathbf{a} = E \Delta a = \frac{\sigma \Delta a}{\epsilon_0}$$

since $\mathbf{E} \cdot \hat{\mathbf{n}} = E = $ const. in the first integral, $\mathbf{E} \cdot \hat{\mathbf{n}}_c = 0$ in the second, and $\mathbf{E} = 0$ in the third. Thus the value of E at the surface of a conductor is just σ/ϵ_0, and, if we combine this with (5-3) and (6-2), we can write

$$E_{at\ surface} = -\hat{\mathbf{n}} \cdot \nabla \phi = \frac{\sigma}{\epsilon_0} \tag{6-4}$$

If σ is positive, then $\mathbf{E}$ is directed away from the surface, while if σ is negative, then $\mathbf{E}$ is directed toward the conducting surface; these signs are consistent with the direction of the force that would be exerted on a positive test charge placed near the conducting surface. We note that E as given by (6-4) is exactly twice the value of E due to a plane sheet of the same σ that we found in (4-12) by a similar calculation. We can understand this difference somewhat crudely by first deciding that the total flux of $\mathbf{E}$ per unit area that can be produced by a given surface charge density is σ/ϵ_0; then, in the case of the plane sheet, this total flux could be directed equally in the two directions away from the charge, while for the conductor, since the electric field must be zero inside, only one direction is available for this total flux. As we will find in subsequent examples, σ is not necessarily a constant but can vary with position on the surface; when this is the case, $\mathbf{E}$ will also vary with location on the surface (but will always remain normal to the surface). For example, suppose that the conductor were originally neutral, that is, it had no net charge. Then, when a field is produced in its vicinity by moving other external charges up from infinity, the conductor's movable charges will not generally be already in equilibrium with respect to the new situation and will have to readjust their locations. Thus, for a short time, there will be a flow of charges around the conductor but equilibrium will eventually be established and then (6-1), (6-2), and (6-4) will all be satisfied. But since it originally had no net charge, it must still have no net charge, that is, $\oint_S \sigma\, da = 0$, thus showing that σ cannot be of the same sign everywhere and thus cannot be a constant. On the other hand, an external agent might transfer a *net* amount of charge to the conductor, that is, making it "charged." The surface charge density now could be of the same sign everywhere, and may even be constant providing there is sufficient symmetry.

■ **Example**

Isolated spherical conductor. Let us assume a net charge Q on a conducting sphere of radius a. If it is otherwise isolated, there will be no reason for preferring one direction over another so that we will have spherical symmetry. As a result, the charge Q will be distributed uniformly over the surface with a constant density $\sigma = Q/4\pi a^2$, and the electric field at the surface will be given by (6-4) as $E = Q/4\pi\epsilon_0 a^2$. Furthermore, since the field $\mathbf{E}$ at the surface is in the normal direction, it will be radial.

We can come to these same conclusions in another way. We have seen several times that, because of the spherical symmetry, the charge Q can be treated as if it were a point charge at the center of the sphere as far as effects *outside* the sphere are concerned. The potential outside will be exactly that given by (5-22), that is, $Q/4\pi\epsilon_0 r$, while the potential inside is constant and equal to the potential at the surface ($r = a$) according to (6-1) and (6-2); thus we obtain

$$\phi = \frac{Q}{4\pi\epsilon_0 r} \qquad (r \geq a)$$

$$\phi = \frac{Q}{4\pi\epsilon_0 a} \qquad (r \leq a)$$

(6-5)

The electric field outside will be given by (5-24) which becomes $\mathbf{E} = Q\hat{\mathbf{r}}/4\pi\epsilon_0 a^2$ when evaluated at the surface; this agrees exactly with the result obtained in the previous paragraph from (6-4). Inside the sphere, $\mathbf{E} = 0$, because of the constant value of the potential given by (6-5) for $r \leq a$. ∎

We have not yet exhausted the possibilities of using Gauss' law. Now let us consider a conductor with a cavity inside of it as shown in Figure 6-4. The total bounding surface of the conductor now has two parts: an outer surface S_o, and an inner surface S_i. Again let us consider an arbitrary surface S lying completely within the body of the conductor so that $\mathbf{E} = 0$ on every point of S. Then (6-3) applies in this case too so that $Q_{\text{in}} = 0$, that is, there can be no net charge contained within the surface S.

Now suppose that there is a charge inside the cavity, which we call Q_{cav}. In order that $Q_{\text{in}} = 0$ for S, there must be an equal and opposite charge somewhere within S, that is, on the conductor. But we found above that any charge on a conductor must reside entirely on its surface, and therefore this charge Q_i must be found on the inner surface S_i. Thus, we have $Q_{\text{in}} = 0 = Q_{\text{cav}} + Q_i$, or

$$Q_i = Q_{\text{inner surface}} = -Q_{\text{cav}}$$

(6-6)

which is a result that is always true.

If the conductor were neutral before Q_{cav} was inserted, it must remain neutral afterward; therefore a charge $Q_o = -Q_i = +Q_{\text{cav}}$ will appear on the rest of the surface, that is, on the outer bounding surface S_o as indicated in the figure. In this way, a charge within the cavity would make its presence known to someone outside the conductor by means of the induced charge Q_o produced on the outer surface and the electric field produced by it.

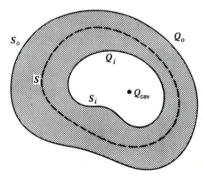

Figure 6-4. Charges on a conductor with a cavity in the interior.

Let us consider the case $Q_{cav} = 0$; then it will *always* be the case that the charge on the inner surface S_i is zero according to (6-6). (Q_o, on the other hand, would be zero only if the conductor were originally neutral and were kept that way.) Now consider a closed surface S' lying entirely within the cavity as shown in Figure 6-5. Suppose that S' is an equipotential surface whose potential ϕ' is greater than ϕ_i, the potential of S_i, which is also an equipotential surface by (6-2). Then, according to (5-11) and Figure 5-1, there will be lines of $\mathbf{E}$ generally directed from S' to S_i. Consequently, if we evaluate the surface integral of $\mathbf{E}$ over another surface S'' located between S' and S_i, we will find that

$$\oint_{S''} \mathbf{E} \cdot d\mathbf{a} \neq 0$$

and, in fact, the value of the integral will be positive since the cosine of the angle between $\mathbf{E}$ and $d\mathbf{a}$ will always be positive. But, by Gauss' law (4-1), this means that there must be a positive charge somewhere within S'', contrary to our assumption that $Q_{cav} = 0$. Thus we see that ϕ' cannot be greater than ϕ_i. In a similar way, we will find that ϕ' cannot be less than ϕ_i. Therefore, S' must be an equipotential surface with the same potential as S_i. But S' is completely arbitrary and can be made larger, made smaller, or otherwise deformed in any manner; thus, all points within the cavity can be evaluated in this way, and the net result is that the potential is constant and equal to ϕ_i within the cavity so that $\mathbf{E} = 0$ at each point in the cavity. In other words, if there is no charge within a cavity inside a conductor, the cavity will be an equipotential volume and the electric field will be zero everywhere within the cavity. In fact, the potential within the cavity will be the same as that of the conductor because of (5-11).

These general conclusions are still valid, of course, even if there is another conductor within the cavity as shown in Figure 6-6. As long as the inner conductor C' is uncharged, there will be no electric field within the cavity and C' will always have the same potential as C, the surrounding conductor. Thus, regardless of what charge may be placed on or outside of C, of whatever sign or distribution, there will be no electric field at C' and its movable charges will be unaffected. In the usual terminology, the inner conductor will be completely *shielded* or *screened*; this principle of electrostatic shielding has practical applications such as in the shielding of electronic components in circuits by enclosing them in metal containers.

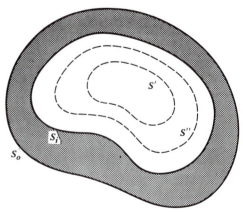

Figure 6-5. The surfaces S' and S'' are entirely within the cavity. S'' completely surrounds S'.

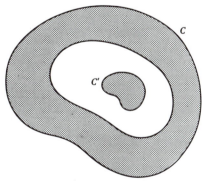

Figure 6-6. The conductor C' is in a cavity in the conductor C.

6-2 SYSTEMS OF CONDUCTORS

Let us consider a system consisting of n conductors that we number in some manner, that is, $1, 2, \ldots, j, \ldots, n$. We assume that the conductors are charged with total charges $Q_1, Q_2, \ldots, Q_j, \ldots, Q_n$. We know that each charge will be on the surface of the corresponding conductor so that they can be described by the respective surface charge densities $\sigma_1, \sigma_2, \ldots, \sigma_j, \ldots, \sigma_n$. Although we are not yet able to calculate all of these quantities, we can expect in general that the σ's will not be constant, but will usually vary with position in some manner. In any event, the potential at any point P, which we write as ϕ_P, can be found in principle from (5-8) and is

$$\phi_P = \frac{1}{4\pi\epsilon_0} \int_{S'} \frac{\sigma(\mathbf{r}') \, da'}{R} = \sum_{j=1}^{n} \frac{1}{4\pi\epsilon_0} \int_{S_j} \frac{\sigma_j(\mathbf{r}_j) \, da_j}{R_j} \tag{6-7}$$

Here we have written the total potential as the sum of the contributions from each conductor: S_j is the surface of the jth conductor, da_j the element of area of this surface at the location $\mathbf{r}_j$, and $R_j = |\mathbf{r}_P - \mathbf{r}_j|$ is the distance from da_j to the field point P at $\mathbf{r}_P$. These relations are illustrated in Figure 6-7, although the various position vectors are not shown for simplicity. Now (6-7) is completely general, and, therefore, it must hold *in particular* when P is chosen to be *any* point on the equipotential surface of the ith conductor whose potential is ϕ_i; thus, we can write

$$\phi_i = \sum_{j=1}^{n} \frac{1}{4\pi\epsilon_0} \int_{S_j} \frac{\sigma_j \, da_j}{R_{ji}} \qquad (i = 1, 2, \ldots, n) \tag{6-8}$$

where R_{ji} is the distance from the point $\mathbf{r}_j$ of the jth conductor to the particular point in question of the ith conductor. Since i could be chosen to be any of the conductors, (6-8) really represents a system of n equations, each of which has n terms on its right-hand side. [We note that the sum in (6-8) also includes the term $j = i$, that is, it includes the integral over the surface of the conductor whose total potential we are calculating.]

It will be convenient for our purposes to write (6-8) in terms of the total charges Q_j. The *average* surface charge density of the jth conductor $\langle \sigma_j \rangle$ will be just the total charge divided by the total area, that is, $\langle \sigma_j \rangle = Q_j / S_j$. The actual charge density σ_j at a

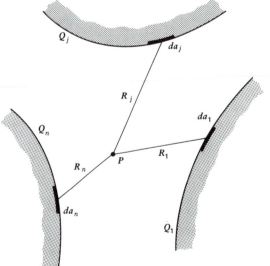

Figure 6-7. Calculation of the potential due to several charged conductors.

given point is not generally equal to the average but will be proportional to it; this allows us to write

$$\sigma_j = \langle \sigma_j \rangle f_j = \frac{Q_j}{S_j} f_j \tag{6-9}$$

where f_j is a factor that describes how the actual charge density differs from the average and itself will be a function of position on the surface of the jth conductor. Thus, $\sigma_j \sim Q_j$, and when (6-9) is substituted in (6-8) we get

$$\phi_i = \sum_{j=1}^{n} \frac{Q_j}{4\pi\epsilon_0 S_j} \int_{S_j} \frac{f_j \, da_j}{R_{ji}} \qquad (i = 1, 2, \ldots, n) \tag{6-10}$$

These equations can be written in the form

$$\phi_i = \sum_{j=1}^{n} p_{ij} Q_j \qquad (i = 1, 2, \ldots, n) \tag{6-11}$$

where

$$p_{ij} = \frac{1}{4\pi\epsilon_0 S_j} \int_{S_j} \frac{f_j \, da_j}{R_{ji}} \tag{6-12}$$

thus showing that the potential of *any* conductor depends *linearly* on the charges of *all* of the conductors, including its own. If (6-11) is written out explicitly, we get the set of equations

$$\begin{aligned}
\phi_1 &= p_{11}Q_1 + p_{12}Q_2 + \ldots + p_{1n}Q_n \\
\phi_2 &= p_{21}Q_1 + p_{22}Q_2 + \ldots + p_{2n}Q_n \\
&\cdots \qquad\qquad \cdots \qquad\qquad \cdots \\
\phi_n &= p_{n1}Q_1 + p_{n2}Q_2 + \ldots + p_{nn}Q_n
\end{aligned} \tag{6-13}$$

The set of coefficients p_{ij} defined by (6-12) are called *coefficients of potential*, and, in general, their total number is n^2. We note that (6-12) no longer contains any reference to the potentials or charges so that the p_{ij} represent purely geometric relationships. If we knew the charge distributions so that the factors f_j were known, then, in principle, we could calculate all of the coefficients p_{ij} from (6-12); we have not yet considered methods of solving electrostatic problems that would give us such information. Nevertheless, these coefficients are always *measurable* in principle, for we see from (6-11) that

$$\frac{\partial \phi_i}{\partial Q_j} = \left(\frac{\partial \phi_i}{\partial Q_j} \right)_{Q_1, \ldots Q_{j-1}, Q_{j+1}, \ldots Q_n} = p_{ij} \tag{6-14}$$

so that p_{ij} is interpretable as the ratio of the change produced in the potential of the ith conductor to the change in the charge of the jth conductor when the charges on *all of the other* conductors are kept constant.

The p_{ij} have an interesting and useful symmetry property. If we eliminate f_j in (6-12) by using (6-9), we find that

$$p_{ij} Q_j = \frac{1}{4\pi\epsilon_0} \int_{S_j} \frac{\sigma_j \, da_j}{R_{ji}} \tag{6-15}$$

which also follows from equating the corresponding terms in the sums of (6-8) and (6-11). We also have

$$Q_i = \int_{S_i} \sigma_i \, da_i \tag{6-16}$$

according to (2-16), and if we multiply these last two equations together, we get

$$p_{ij}Q_iQ_j = \frac{1}{4\pi\epsilon_0} \int_{S_i}\int_{S_j} \frac{\sigma_i\sigma_j \, da_i \, da_j}{R_{ji}} \tag{6-17}$$

If we interchange the indices i and j in (6-17), we get

$$p_{ji}Q_jQ_i = \frac{1}{4\pi\epsilon_0} \int_{S_j}\int_{S_i} \frac{\sigma_j\sigma_i \, da_j \, da_i}{R_{ij}} \tag{6-18}$$

Now the order of integration is opposite for (6-17) and (6-18) so that, for example, in (6-17) we would first integrate by keeping $\mathbf{r}_i$ fixed and letting $\mathbf{r}_j$ vary, and then finally integrate again by letting $\mathbf{r}_i$ vary, whereas these would be interchanged for (6-18). Nevertheless, each double integral will include the contributions of all pairs of area elements, and, since each specific pair always has the same distance of separation, we will have the numerical equality that $R_{ij} = R_{ji}$. Thus, we see that the integral of (6-18) is exactly that of (6-17); therefore, the left-hand sides are also equal, making $p_{ij}Q_iQ_j = p_{ji}Q_jQ_i$, or

$$p_{ji} = p_{ij} \tag{6-19}$$

The physical content of this symmetry property of the p's can be expressed according to (6-19) and (6-11) as follows: if a charge Q on conductor j brings conductor i to a potential ϕ, then the *same* charge Q placed on i would bring j to the *same* potential ϕ.

Sometimes, if the problem is simple enough to be readily solved, the p_{ij} can be more easily found from (6-11) than from (6-12).

■ **Example**

Isolated conducting sphere. This is an example we solved in the last section and the potential at the surface is given by (6-5) to be $\phi = Q/4\pi\epsilon_0 a$. For a single conductor (6-11) reduces to $\phi = p_{11}Q$ and by comparison we see that

$$p_{11} = \frac{1}{4\pi\epsilon_0 a} \tag{6-20}$$

Thus, we were able to find the single coefficient of potential for this simple case and it did turn out to depend on the geometric properties of the system. ■

6-3 CAPACITANCE

One of the earliest uses of conductors in electrostatics was for the storage of electric charge; the conductor could be charged, for example, by giving it a definite potential by means of a battery. For such an application, one would naturally be interested in the "capacity" of the conductor for charge, in much the same sense that one would refer to the capacity of a barrel in terms of how many apples it could hold. Nowadays, such systems are usually called *capacitors* and the quantitative measure of their capacity is called *capacitance*. Generally speaking, there are only two systems of conductors of basic interest in this connection: a single isolated conductor, and a system of two conductors with equal and opposite charges.

For an isolated conductor, the sum (6-11) reduces to the single term

$$\phi = p_{11}Q \tag{6-21}$$

In this case, the charge is always directly proportional to the potential and the capacitance C of a single conductor is defined as this ratio; thus

$$C = \frac{Q}{\phi} = \frac{1}{p_{11}} \tag{6-22}$$

and will be a definite property of the conductor and related to its geometry. As an example, we can consider the sphere for which we found p_{11} to be given by (6-20), so that the capacitance is

$$C_{\text{sphere}} = 4\pi\epsilon_0 a \tag{6-23}$$

and is directly proportional to the radius. Since the units of ϵ_0 were originally given as farad/meter in (2-4), we see from (6-23) that the unit of capacitance is the farad; we also see from (6-22) that 1 farad = 1 coulomb/volt, which is, of course, entirely consistent with our previous finding of 1 (coulomb)^2/joule given after (2-4).

When we now consider a system of only two conductors, the system of equations (6-13) reduces to

$$\begin{aligned} \phi_1 &= p_{11}Q_1 + p_{12}Q_2 \\ \phi_2 &= p_{21}Q_1 + p_{22}Q_2 \end{aligned} \tag{6-24}$$

along with

$$p_{12} = p_{21} \tag{6-25}$$

according to (6-19). Thus, the potential-charge relationships for this system will generally require the knowledge of three quantities: p_{11}, p_{22}, and p_{12}. When the two conductors are used as a capacitor, however, a somewhat more special situation is envisioned—one assumes the two conductors to be connected to each other by a conducting path and the process of charging up this system is *transfer of charge* from one to the other. Under these circumstances, the charge on one conductor will always be equal and opposite to the charge on the other.

Accordingly, we take for our general definition of a capacitor the following: *any* two conductors with equal and opposite charges Q and $-Q$. Even so, it is not immediately evident that one can even define a capacitance in this case. However, when we set $Q_1 = Q$ and $Q_2 = -Q$ in (6-24), we get

$$\begin{aligned} \phi_1 &= (p_{11} - p_{12})Q \\ \phi_2 &= (p_{21} - p_{22})Q \end{aligned} \tag{6-26}$$

so that the *difference in potential* between the conductors is

$$\Delta\phi = \phi_1 - \phi_2 = (p_{11} + p_{22} - p_{12} - p_{21})Q \tag{6-27}$$

and we see that for a capacitor with equal and opposite charges, the charge and the potential difference will *always* be proportional to each other so that we will be always able to characterize the system by a single parameter. We call this quantity the capacitance, C, and define it as

$$C = \frac{Q}{\Delta\phi} \tag{6-28}$$

by analogy with (6-22). Upon comparing (6-28) and (6-27), and by using (6-25), we see that the capacitance can be expressed in terms of the coefficients of potential as

$$C = \frac{1}{p_{11} + p_{22} - 2p_{12}} \tag{6-29}$$

so that in this two conductor case, the capacitance is still essentially a property reflecting the geometric relationships of the system.

■ **Example**

Spherical capacitor. Let us consider the two conductors shown in Figure 6-8; the bounding surfaces are concentric spheres of radii a, b, and c. We call the inner conductor 1, and the outer 2. It is assumed that 2 completely surrounds 1. The p's that we need in order to find C can be obtained from the general relations (6-24) by considering appropriately chosen special cases. First of all, let us suppose that $Q_1 = 0$ while $Q_2 \neq 0$, so that (6-24) becomes

$$\phi_1 = p_{12}Q_2 \quad \text{and} \quad \phi_2 = p_{22}Q_2 \tag{6-30}$$

From (6-6) it follows that Q_2 is entirely on the outer surface of radius c; therefore, if we use the appropriate radius of c rather than a in (6-5), the potential of 2 will be given by

$$\phi_2 = \frac{Q_2}{4\pi\epsilon_0 c} \tag{6-31}$$

so that (6-30) gives

$$p_{22} = \frac{1}{4\pi\epsilon_0 c} \tag{6-32}$$

Since there is no charge within the cavity, our previous discussion in connection with Figures 6-5 and 6-6 tells us that $\phi_1 = \phi_2$ so that, according to (6-30),

$$p_{12} = p_{22} \tag{6-33}$$

for this particular system.

In order to find p_{11}, let us now assume that $Q_1 \neq 0$ while $Q_2 = 0$; the latter is the net value on conductor 2 since the inner surface of radius b must have a charge $-Q_1$ according to (6-6). In this case, (6-24) yields

$$\phi_1 = p_{11}Q_1 \quad \text{and} \quad \phi_2 = p_{21}Q_1 \tag{6-34}$$

We can relate these two expressions by using (5-11); the field in the vacuum region for which $a \leq r \leq b$ can be obtained from Gauss' law, and the result will again be that, in

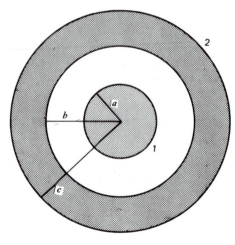

Figure 6-8. Spherical capacitor.

this region, Q_1 acts as if it were a point charge so that the field will be radial and given by (4-17) with Q replaced by Q_1. Combining all this, we get

$$\phi_1 - \phi_2 = \int_1^2 \mathbf{E} \cdot d\mathbf{s} = \int_a^b E_r \, dr$$

$$= \int_a^b \frac{Q_1 \, dr}{4\pi\epsilon_0 r^2} = \frac{Q_1}{4\pi\epsilon_0}\left(\frac{1}{a} - \frac{1}{b}\right) \tag{6-35}$$

The left-hand side can be written as $(p_{11} - p_{21})Q_1$ because of (6-34); canceling Q_1 on both sides and using (6-33) and (6-32), we finally get

$$p_{11} = p_{22} + \frac{1}{4\pi\epsilon_0}\left(\frac{1}{a} - \frac{1}{b}\right) = \frac{1}{4\pi\epsilon_0}\left(\frac{1}{a} - \frac{1}{b} + \frac{1}{c}\right) \tag{6-36}$$

Because of (6-33), the general expression for C given in (6-29) simplifies to

$$C = \frac{1}{p_{11} - p_{22}} = \frac{4\pi\epsilon_0}{\left(\dfrac{1}{a} - \dfrac{1}{b}\right)} = \frac{4\pi\epsilon_0 ab}{b - a} \tag{6-37}$$

with the use of (6-36) as well. Thus, we have been able to find the capacitance of this particular capacitor by the evaluation of the general result (6-29) as stated in terms of the coefficients of potential. ∎

Nevertheless, in most simple cases, this is not the most convenient way to proceed. What is generally done is to find the potential difference from a different and prior solution to the problem. Ordinarily, this will include a knowledge of the electric field. Then $\Delta\phi$ can be found by using (5-11) to integrate $\mathbf{E}$ over any convenient path between the two conductors; when the two conductors form a capacitor they are usually called "plates." Since we know from (6-27) that $\Delta\phi$ will be sure to turn out to be proportional to the charge Q, the ratio of the two will give the capacitance according to (6-28); thus, we have

$$\Delta\phi = \phi_+ - \phi_- = \int_{+\text{plate}}^{-\text{plate}} \mathbf{E} \cdot d\mathbf{s} = \frac{Q}{C} \tag{6-38}$$

It is convenient to write the integral in this form because $\mathbf{E}$ will be generally directed from the positively charged plate of higher potential to the negatively charged plate of lower potential so that $\mathbf{E} \cdot d\mathbf{s}$ will be positive, thus giving the correct sign to $\Delta\phi$. Let us look at two examples from this point of view.

■ **Example**

Spherical capacitor. This is the same system shown in Figure 6-8 for which we found C by the previous method. Now let us assume the inner sphere of radius a to have the positive charge Q; then the inner sphere of radius b will have the charge $-Q$ in agreement with (6-6). As before, the application of Gauss' law to a sphere of radius r such that $a < r < b$ shows us that $\mathbf{E}$ will be given by $\mathbf{E} = (Q/4\pi\epsilon_0 r^2)\hat{\mathbf{r}}$ according to (4-17). If we also integrate from the positive plate to the negative plate in the radial direction so that $d\mathbf{s} = dr\,\hat{\mathbf{r}}$, then (6-38) becomes

$$\int_+^- \mathbf{E} \cdot d\mathbf{s} = \int_a^b \frac{Q \, dr}{4\pi\epsilon_0 r^2} = \frac{Q}{4\pi\epsilon_0}\left(\frac{1}{a} - \frac{1}{b}\right) = \frac{Q}{C} \tag{6-39}$$

which gives us the same result for C as we found before in (6-37); this time the result has been obtained much more easily. ∎

Example

Parallel plate capacitor. This system consists of two conducting plates, each of area A, which are parallel to each other and separated by a distance d that is small compared to their linear dimensions. The plates need not be square, but any linear dimension should be of the order of $\sqrt{A}$ so that, in effect, we are assuming that $d \ll \sqrt{A}$. A side view of this capacitor is shown in Figure 6-9. Under these circumstances, the plates can be treated as if they were of infinite extent, and we conclude in the same manner as for the infinite sheet of Section 4-2 that $\mathbf{E}$ will be constant in magnitude and directed normal to the plates as shown by the dashed lines in the figure. Since $\mathbf{E}$ is normal to the conductor at the conductor, its magnitude is just that given by (6-4), that is, $E = \sigma/\epsilon_0$. (This value for $\mathbf{E}$ is also exactly that found for the two plane sheets of Exercise 3-9 as illustrated in Figure 3-6.) The simplest path of integration is clearly one along the direction of $\mathbf{E}$ as shown by $d\mathbf{s}$ in Figure 6-9. Then $\mathbf{E} \cdot d\mathbf{s} = E\,ds = (\sigma/\epsilon_0)\,ds$ and (6-38) becomes

$$\int_+^- \mathbf{E} \cdot d\mathbf{s} = \frac{\sigma}{\epsilon_0} \int_+^- ds = \frac{\sigma d}{\epsilon_0} = \left(\frac{d}{\epsilon_0 A}\right) Q = \frac{Q}{C} \tag{6-40}$$

where we have written $\sigma = Q/A$; it is quite accurate to treat σ as a constant in this way because the plates are effectively of infinite extent, corresponding to a uniform surface charge distribution. Solving (6-40) for C, we get the simple familiar result for the capacitance

$$C = \frac{\epsilon_0 A}{d} \tag{6-41}$$

Incidentally, this system, as well as the spherical capacitor, shows us quite graphically why capacitors are also often called "condensers." Because the field distribution is confined to the finite region between the plates, one says that the electric field has been "condensed" into this region, rather than extending throughout all space as has been the case for many of our previous examples.

Now the plates in this case are not actually infinite, of course, and we have in effect assumed that the electric field is constant and given by (σ/ϵ_0) everywhere between the plates and then drops off abruptly to zero at the edges. It will be left as an exercise to show that this is impossible because of the conservative nature of the electric field. In reality, the lines of $\mathbf{E}$ must curve outward somewhat as the edges of the plates are approached, and they must also extend outward into the region beyond the plates as indicated in Figure 6-10. However, if the plates are "large enough," we will not make any appreciable error by neglecting these "edge effects"; it is customary to do so, and we will continue to make this approximation.   ■

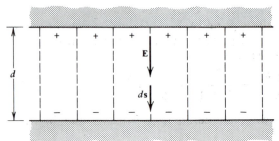

Figure 6-9.   Parallel plate capacitor.

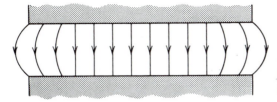

Figure 6-10. General appearance of the field of a parallel plate capacitor.

These two examples are probably enough to convince one that most of the problems involving the calculation of capacitance by means of (6-38) must have a lot of symmetry so that $\mathbf{E}$ can be easily found, usually by using Gauss' law. We will be able to handle more complicated problems later when we will have discussed other systematic ways of finding the potential ϕ as a function of position.

EXERCISES

6-1 Suppose that the two conductors shown in Figure 6-8 are originally uncharged. A charge Q is then placed on the inner conductor of radius a. Find the final static charge distribution. Find the potential ϕ for all values of r and plot your result.

6-2 An infinitely long conducting cylinder of radius a carries a total charge q_l per unit length. If ρ is the perpendicular distance from the axis of the cylinder, show that the potential outside the cylinder can be written

$$\phi(\rho) = \frac{q_l}{2\pi\epsilon_0} \ln\left(\frac{\rho_0}{\rho}\right) \quad (\rho > a) \quad (6\text{-}42)$$

where ρ_0 = const. What is the potential inside the cylinder? Verify that (6-4) is satisfied in this case. Can one define an appropriate and unique p_{11} for this case?

6-3 The farad is actually an enormous unit of capacitance. In order to illustrate this, treat the earth as a conducting sphere of radius 6.37×10^6 meters and find its capacitance.

6-4 When the set of equations (6-13) is solved for the charges, the result is another set of linear equations of the form

$$Q_i = \sum_{j=1}^{n} c_{ij}\phi_j \quad (i = 1, 2, \ldots, n) \quad (6\text{-}43)$$

where the c_{ij} are combinations of the p_{ij}. If the indices are the same, the c_{ii} are called coefficients of capacitance, and the c_{ij} with $i \neq j$ are called coefficients of induction. Find these coefficients for the system of two conductors described by (6-24) and verify that $c_{21} = c_{12}$. Show that the

capacitance of this system can be expressed as

$$C = \frac{c_{11}c_{22} - c_{12}^2}{c_{11} + c_{22} + 2c_{12}} \quad (6\text{-}44)$$

6-5 Using the results of the previous exercise, find the coefficients c_{ij} for the spherical capacitor of Figure 6-8 and verify that they give the same result (6-37) for the capacitance.

6-6 A capacitor C_1 is charged resulting in a potential difference $\Delta\phi$ between its plates. Another capacitor C_2 is uncharged. One plate of C_2 is now connected to a plate of C_1 by a conductor of negligible capacitance; the remaining plates are similarly connected. For the resultant equilibrium state, find the charge on each capacitor and the potential difference $\Delta\phi'$ between their respective plates.

6-7 The plates of two capacitors C_1 and C_2 are connected by conductors of negligible capacitance as shown in Figure 6-11a, that is, they are connected in "parallel." If a potential difference $\Delta\phi$ is now applied across the terminals T and T', show that this combination is equivalent to a single capacitor of capacitance $C_p = C_1 + C_2$. Similarly, show that the equivalent capacitance of the "series" combination shown in (b) can be found from $1/C_s = (1/C_1) + (1/C_2)$.

6-8 Consider the spherical capacitor of Figure 6-8 for the case that a and b are nearly equal. Find an approximate expression for C and write it in a form that explicitly involves the difference $\delta = b - a$ where $\delta \ll a$ or b. Interpret your result with the aid of (6-41).

6-9 The potential difference $\Delta\phi$ between the plates of a spherical capacitor is kept constant.

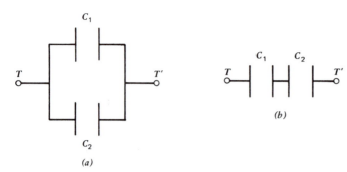

(a)

(b)

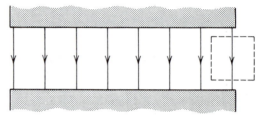

Figure 6-11. (a) Capacitors in parallel. (b) Capacitors in series.

Show that then the electric field at the surface of the inner sphere will be a minimum if $a = \frac{1}{2}b$. Find this minimum value of E.

6-10 A capacitor is made from two infinitely long conductors with coaxial cylindrical surfaces as shown in Figure 6-12. Show that the capacitance of a length L of this system is given by

$$C = \frac{2\pi\epsilon_0 L}{\ln\left(\frac{b}{a}\right)} \qquad (6\text{-}45)$$

Why doesn't the result depend on c? Does the inner conductor of radius a need to be a solid cylinder?

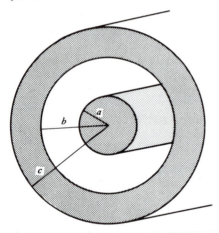

Figure 6-12. Capacitor with coaxial cylindrical plates.

6-11 Figure 6-13 illustrates the assumption we made about **E** when we neglected edge effects for the parallel plate capacitor, that is, we assumed that **E** went abruptly to zero at the edges. Actually this is impossible because $\nabla \times \mathbf{E} = 0$. Show this by first calculating $\oint \mathbf{E} \cdot d\mathbf{s}$ over the rectangular path shown, which is partly in the $\mathbf{E} \neq 0$ region and partly in the $\mathbf{E} = 0$ region. Then show that Stokes' theorem (1-67) leads to a contradic-

tion. Show qualitatively that this contradiction will be removed for the general field lines shown in Figure 6-10.

Figure 6-13. Assumed electric field for Exercise 6-11.

6-12 A sheet of conductor of thickness t and parallel faces of cross-sectional area $\geq A$ is inserted between the plates of the capacitor of Figure 6-9. The faces of the sheet are parallel to the plates of the original capacitor. Show that the capacitance is increased by

$$\Delta C = \frac{\epsilon_0 t A}{d(d-t)}$$

Why is this result independent of the distance of the sheet from either of the original plates?

6-13 Suppose that a parallel plate capacitor has rectangular plates but the plates are not exactly parallel. The separation at one edge is $d - a$ and $d + a$ at the other edge where $a \ll d$. Show that the capacitance is given approximately by

$$C \simeq \frac{\epsilon_0 A}{d}\left(1 + \frac{a^2}{3d^2}\right)$$

where A is the area of a plate. (*Hint*: recall the results of Exercise 6-7.)

6-14 Two conducting spheres have their centers a distance c apart. The radius of one sphere is a, the radius of the other is b. Show that when

$c \gg a$ and $c \gg b$, the capacitance of this system will be given approximately by

$$C \simeq 4\pi\epsilon_0 \left(\frac{1}{a} + \frac{1}{b} - \frac{2}{c} \right)^{-1}$$

[*Hints*: imagine equal and opposite charges on the spheres; how will one "look" at the location of the other?; will (6-5) still be approximately correct?]

6-15 Two infinitely long conducting cylinders have their central axes parallel and separated by a distance c. The radius of one is a and the radius of the other is b. If $c \gg a$ and $c \gg b$, find an approximate expression for the capacitance of a length L of this system.

7

ELECTROSTATIC ENERGY

We briefly touched on the question of energy as related to potential in Section 5-4 where we considered the potential energy of a single point charge and a pair of point charges. Now we want to extend these ideas to a system containing an arbitrary number of charges. We will calculate the potential energy U_e of this system, that is, the amount of reversible work that would have to be done by an external agent in order to produce the given configuration against the mutual conservative electrostatic forces of the charges. After this has been done, we can apply it to a variety of situations, and we will find it helpful to rewrite it in various ways.

7-1 ENERGY OF A SYSTEM OF CHARGES

Previously we found the expression (5-49) for the mutual potential energy of a pair of point charges. If we now call these charges q_i and q_j, rather than q and Q, the potential energy of this pair as given by (5-49) is

$$U_{eij} = \frac{q_i q_j}{4\pi\epsilon_0 R_{ij}} \tag{7-1}$$

where R_{ij} is the distance between them and, as usual, $R_{ij} = |\mathbf{r}_i - \mathbf{r}_j|$. Now let us consider a total number N of point charges in some given configuration so that the individual charges and their distances from each other are all known. We number the charges in some way so that i and j can both independently have the values $1, 2, \ldots, N$. The total potential energy of the system, U_e, will then be the sum of terms like (7-1) and where the sum is taken over all the *pairs* of charges that can be formed from the N charges:

$$U_e = \sum_{\text{all pairs}} U_{eij} \tag{7-2}$$

In this sum, we must exclude the case $i = j$ for which a given charge is paired with itself because that would imply that a charge would be exerting a force on itself, a possibility that we already excluded in connection with (3-2).

It will be convenient to express (7-2) in terms of a double sum over the indices i and j in which they independently cover their whole possible range of N values. However, if we do that we will be counting each pair twice and must correct for this. For example, consider the contribution of the pair q_3 and q_4 to (7-2) which is U_{e34}. In such a double sum, this pair will occur once when $i = 3$ and $j = 4$ and again when $i = 4$ and $j = 3$ giving a total of $U_{e34} + U_{e43} = 2U_{e34}$, according to (7-1), since $R_{43} = R_{34}$. Thus, we will have to divide the result obtained from a double sum by 2; in this way, we can write (7-2) as

$$U_e = \frac{1}{2} \sum_{\substack{i=1 \\ j\neq i}}^{N} \sum_{j=1}^{N} U_{eij} = \frac{1}{2} \sum_{\substack{i=1 \\ j\neq i}}^{N} \sum_{j=1}^{N} \frac{q_i q_j}{4\pi\epsilon_0 R_{ij}} \tag{7-3}$$

which is our desired result for the electrostatic energy.

In particular, if all of the positions of the charges are given in rectangular coordinates, we can use (1-14) to get an explicit expression for U_e as

$$U_e = \frac{1}{2} \sum_{\substack{i=1 \\ j \neq i}}^{N} \sum_{j=1}^{N} \frac{q_i q_j}{4\pi\epsilon_0 \left[(x_i - x_j)^2 + (y_i - y_j)^2 + (z_i - z_j)^2 \right]^{1/2}} \qquad (7\text{-}4)$$

We can put (7-3) into another form by regrouping:

$$U_e = \frac{1}{2} \sum_{i=1}^{N} q_i \left(\sum_{j=1}^{N} \frac{q_j}{4\pi\epsilon_0 R_{ij}} \right) \qquad (7\text{-}5)$$

If we now recall (5-2), we recognize that the term in parentheses is just $\phi_i = \phi_i(\mathbf{r}_i)$, that is, it is the scalar potential at the location of q_i due to all of the *other charges*. Hence we can write (7-5) as

$$U_e = \tfrac{1}{2} \sum_{i=1}^{N} q_i \phi_i(\mathbf{r}_i) \qquad (7\text{-}6)$$

When dealing with the potential energy of a system, it is natural to ask oneself where the energy is "stored" or "located." In mechanics, it is plausible to regard the potential energy $\frac{1}{2}kx^2$ of a stretched spring, for example, as being associated with the change in relative positions of the strained portions of the spring so that the potential energy would be "stored" at the locations of these parts of the spring. The form (7-6) that we have obtained lends itself to a similar interpretation. Since the contribution to the sum from a given charge involves only the value of the scalar potential at the location of that charge, one can, if one wishes, associate a portion of the potential energy with this charge and think of it as being "located" there.

If the charges are continuously distributed, we can rewrite the sum in (7-6) as an integral by using (2-14) and (2-16). In this way, we get the following expressions for the energies for volume, surface, and line charge distributions, respectively:

$$U_e = \tfrac{1}{2} \int_V \rho(\mathbf{r}) \phi(\mathbf{r}) \, d\tau \qquad (7\text{-}7)$$

$$U_e = \tfrac{1}{2} \int_S \sigma(\mathbf{r}) \phi(\mathbf{r}) \, da \qquad (7\text{-}8)$$

$$U_e = \tfrac{1}{2} \int_L \lambda(\mathbf{r}) \phi(\mathbf{r}) \, ds \qquad (7\text{-}9)$$

The integrals are to be taken over all regions containing charge of the particular type involved.

If the charge distribution is such that it consists of some or all of these possibilities considered above, then the total energy of the system will be a sum of quantities obtained from (7-6) and (7-7) through (7-9); for each part, the potential ϕ must be found as the resultant from all of the charges in all of their various distributions.

Actually, the integrals above can have their region of integration extended from only the volume actually occupied by the charges to all space, since in regions where there is no charge, $\rho = 0$, and these regions will contribute nothing to (7-7). Thus, we can write (7-7) equally well as

$$U_e = \tfrac{1}{2} \int_{\text{all space}} \rho(\mathbf{r}) \phi(\mathbf{r}) \, d\tau \qquad (7\text{-}10)$$

with similar expressions for (7-8) and (7-9).

■ **Example**

Uniform spherical charge distribution. As an illustration of the use of (7-7), let us consider the case of a sphere of radius a containing charge with constant volume density ρ. We considered this example in Section 5-2 and the potential ϕ is given everywhere by (5-22) and (5-23). However, we need only the potential inside the sphere, according to (7-7), and this is given by (5-23) as

$$\phi(\mathbf{r}) = \frac{\rho}{6\epsilon_0}(3a^2 - r^2) \tag{7-11}$$

If we substitute this into (7-7), use (1-99), and take all of the constants (including ρ) out of the integral, we get

$$U_e = \frac{\rho^2}{12\epsilon_0} \int_0^{2\pi} \int_0^{\pi} \int_0^a (3a^2 - r^2) r^2 \sin\theta \, dr \, d\theta \, d\varphi \tag{7-12}$$

The integral over the angles gives 4π, so that

$$U_e = \frac{\pi\rho^2}{3\epsilon_0} \int_0^a (3a^2 r^2 - r^4) \, dr = \frac{4\pi\rho^2 a^5}{15\epsilon_0} \tag{7-13}$$

If we express this in terms of the total charge of the sphere by using $Q = (4\pi a^3/3)\rho$, we get

$$U_e = \frac{3}{5}\left(\frac{Q^2}{4\pi\epsilon_0 a}\right) \tag{7-14}$$

which is smaller than the energy of two point charges Q that are separated by a distance equal to the radius of the sphere. ■

7-2 ENERGY OF A SYSTEM OF CONDUCTORS

Because of their special properties some of our results can be written in simplified form in the case of conductors. As we found in Section 6-1, the charge on a conductor is confined entirely to its surface. In addition, the potential is constant on the surface, according to (6-2), so that ϕ can be taken out of the integral in (7-8). If we now apply (7-8) to the ith conductor of our system of n which we considered in Section 6-2, we find its energy to be given by

$$U_{ei} = \tfrac{1}{2}\phi_i \int_{S_i} \sigma_i \, da_i = \tfrac{1}{2}Q_i\phi_i \tag{7-15}$$

where Q_i is its total charge as given by (6-16). The total energy of the system will be given by a sum of terms like (7-15) so that

$$U_e = \tfrac{1}{2} \sum_{i=1}^{n} Q_i\phi_i \tag{7-16}$$

The energy can also be expressed completely in terms of the charges if we use (6-11):

$$U_e = \tfrac{1}{2} \sum_{i=1}^{n} \sum_{j=1}^{n} p_{ij}Q_iQ_j \tag{7-17}$$

In the special case in which there are only two conductors, this becomes

$$U_e = \tfrac{1}{2}p_{11}Q_1^2 + p_{12}Q_1Q_2 + \tfrac{1}{2}p_{22}Q_2^2 \tag{7-18}$$

with the use of (6-25).

These results can be written in an even more compact and useful form in the cases in which the system can be described in terms of the single parameter, the capacitance C.

■ **Example**

Isolated conductor. In this case, (7-16) reduces to the single term $\frac{1}{2}Q\phi$; if we also use (6-22), we can write

$$U_e = \tfrac{1}{2}Q\phi = \frac{Q^2}{2C} = \tfrac{1}{2}C\phi^2 \tag{7-19}$$

If we apply this to a conducting sphere by using (6-23), we find its energy to be

$$U_e = \frac{1}{2}\left(\frac{Q^2}{4\pi\epsilon_0 a}\right) \tag{7-20}$$

which is less than the energy given by (7-14) for the case in which the same charge is distributed throughout the volume of the sphere rather than being confined to its surface as is the case for (7-20). ■

■ **Example**

Capacitor. Here we have equal and opposite charges on the plates. Thus, if we let $Q_1 = Q$ and $Q_2 = -Q$ in (7-18), and use (6-29) and (6-28), we find the general expression for the energy of a capacitor to be

$$U_e = \tfrac{1}{2}Q\,\Delta\phi = \frac{Q^2}{2C} = \tfrac{1}{2}C(\Delta\phi)^2 \tag{7-21}$$

which is very similar to (7-19) for a single conductor. We will find this result to be very useful in subsequent discussions. ■

7-3 ENERGY IN TERMS OF THE ELECTRIC FIELD

As we briefly remarked in the paragraph following (7-6), our energy expression easily lends itself to the interpretation that the energy of the system is directly associated with the charges and their positions. This is a point of view that is naturally consistent with the action at a distance property of Coulomb's law with its emphasis on the charges and relative positions. On the other hand, our interest is primarily in describing phenomena in terms of fields, and we would like to do the same for the energy. It will be appropriate for us to use (7-10) as our starting point.

We can write $\rho = \epsilon_0 \nabla \cdot \mathbf{E}$, because of (4-10), so that (7-10) can also be written as

$$U_e = \frac{\epsilon_0}{2} \int \phi(\nabla \cdot \mathbf{E})\, d\tau \tag{7-22}$$

We can rewrite the integrand by using (1-115), (5-3), and (1-17):

$$\phi(\nabla \cdot \mathbf{E}) = -\mathbf{E} \cdot (\nabla\phi) + \nabla \cdot (\phi\mathbf{E}) = \mathbf{E}^2 + \nabla \cdot (\phi\mathbf{E}) \tag{7-23}$$

When this is substituted into (7-22), we get

$$U_e = \frac{\epsilon_0}{2} \int \mathbf{E}^2\, d\tau + \frac{\epsilon_0}{2} \int \nabla \cdot (\phi\mathbf{E})\, d\tau \tag{7-24}$$

and if we use the divergence theorem (1-59) to write the second integral as a surface integral, we finally get

$$U_e = \frac{\epsilon_0}{2} \int_V \mathbf{E}^2 \, d\tau + \frac{\epsilon_0}{2} \oint_S (\phi\mathbf{E}) \cdot d\mathbf{a} \tag{7-25}$$

Now we have to stop and consider the fact that our starting integral in (7-10) was to be taken over all space; in order to do this, we shall find it best to start with taking V in (7-25) to be an extremely large volume, so that S is its very large bounding surface. Then we can let V become infinite; in this process, the surface S will also recede out to infinity. We assume that the charge distribution is *always* contained within a finite volume that, however, may be very large. Then V can always be chosen large enough initially to enclose all of the charges. Now as we let V get very very large, S gets so far away that the charge distribution will appear to be contained in such a small volume that at any point on S the whole charge will seem to be a point charge a distance R away. Thus, as far as the integral in (7-25) is concerned, as $R \to \infty$, we see from (5-12) and (5-13) that

$$\phi \sim \frac{1}{R} \qquad |\mathbf{E}| \sim \frac{1}{R^2} \qquad (\phi\mathbf{E}) \sim \frac{1}{R^3} \tag{7-26}$$

so that the magnitude of the integrand is decreasing as R^3. The surface of integration is increasing, however, but it will be proportional to R^2; thus, for very large R,

$$\oint_S (\phi\mathbf{E}) \cdot d\mathbf{a} \sim \frac{1}{R^3} \cdot R^2 \sim \frac{1}{R} \underset{R \to \infty}{\to} 0 \tag{7-27}$$

so that when V in (7-25) increases to include all space, the surface integral will vanish and the energy expression becomes simply

$$U_e = \int_{\text{all space}} \frac{\epsilon_0}{2} \mathbf{E}^2 \, d\tau \tag{7-28}$$

where $\mathbf{E}^2 = \mathbf{E} \cdot \mathbf{E}$. Thus, we have succeeded in expressing the energy completely in terms of the electric field. [Although the charges no longer appear explicitly in (7-28), they haven't disappeared, of course, because they are the sources of the electric field.]

The *form* of (7-28) as a volume integral, in which regions where $\mathbf{E} \neq 0$ contribute to the integral while those where $\mathbf{E} = 0$ do not, lends itself to a simple and natural *interpretation*: the electrostatic energy is distributed continuously throughout space with an *energy density* u_e given by

$$u_e = \tfrac{1}{2}\epsilon_0 \mathbf{E}^2 \tag{7-29}$$

so that the total energy can be written

$$U_e = \int_{\text{all space}} u_e \, d\tau \tag{7-30}$$

The units of u_e will be joule/(meter)^3.

We can certainly make this interpretation but we are not forced to do so, nor is (7-29) a unique possibility. For example, we could add to u_e *any* quantity χ whose volume integral over all space is zero and still be in agreement with (7-28). However, (7-29) is simple, plausible, and esthetically very appealing from the view of expressing things in terms of fields. The accepted point of view is to adopt (7-29) as correct; we will be able to justify this interpretation in more detail later, and, in fact, (7-29) turns out to be useful and accurate even when time varying fields are considered.

The energy as calculated from (7-28) should, of course, agree with that obtained from (7-6) or its variations. Let us check this in a particular case for which we have already found $\mathbf{E}$ as well as U_e.

■ **Example**

Isolated conducting sphere. We found the solution for this case in terms of the potential ϕ given by (6-5). Thus the electric field produced by this system will be

$$\mathbf{E} = \frac{Q\hat{\mathbf{r}}}{4\pi\epsilon_0 r^2} \qquad (r \geq a) \tag{7-31}$$

$$\mathbf{E} = 0 \qquad (r < a)$$

so that there is no energy contained within the volume of the sphere, according to (7-29), and the region of integration in (7-30) reduces to all space *outside* the sphere. Using (7-31) in (7-29), the energy density u_e is found to be

$$u_e = \frac{Q^2}{32\pi^2\epsilon_0 r^4} \qquad (r \geq a) \tag{7-32}$$

$$u_e = 0 \qquad (r < a)$$

so that, with the use of (1-99), (7-30) becomes

$$U_e = \frac{Q^2}{32\pi^2\epsilon_0} \int_0^{2\pi} \int_0^{\pi} \int_a^{\infty} \frac{1}{r^4} \cdot r^2 \sin\theta \, dr \, d\theta \, d\varphi = \frac{Q^2}{8\pi\epsilon_0 a} \tag{7-33}$$

in complete agreement with (7-20). ■

A useful application of (7-28) is for the calculation of capacitance. If one has found $\mathbf{E}$ by other means, then one can evaluate (7-28). We know that in such a case U_e will turn out to be proportional to Q^2 or ϕ^2 or $(\Delta\phi)^2$, depending on what is given, so that when U_e is found in this way, it can be immediately used in (7-19) or (7-21) to evaluate C. This is often a more convenient procedure than is the use of (6-38).

■ **Example**

Parallel plate capacitor. This system was illustrated in Figure 6-9 and we found the field to have the constant magnitude $E = \sigma/\epsilon_0$ between the plates and zero elsewhere. Therefore,

$$u_e = \tfrac{1}{2}\epsilon_0 E^2 = \frac{\sigma^2}{2\epsilon_0} = \frac{Q^2}{2\epsilon_0 A^2} \tag{7-34}$$

and is constant within the volume Ad between the plates and zero elsewhere. Inserting this into (7-30), we get

$$U_e = \int u_e \, d\tau = u_e(\text{volume}) = \left(\frac{Q^2}{2\epsilon_0 A^2}\right)(Ad) = \frac{Q^2 d}{2\epsilon_0 A} = \frac{Q^2}{2C} \tag{7-35}$$

which leads again to the value for C given by (6-41). ■

7-4 ELECTROSTATIC FORCES ON CONDUCTORS

In general, two charged conductors will exert forces on each other and, in principle, these forces can be calculated from Coulomb's law. It is often desirable to evaluate these forces in a different way. A similar situation is often encountered in mechanics when one finds it convenient to calculate force components as spatial rates of change of the potential energy. In addition, energy considerations provide one with another way of looking at a situation and thus can help in broadening one's overall understanding.

We want to develop a similar approach to electrostatic forces; however, in this section we restrict ourselves to conductors. Furthermore, we will not do this in complete generality, but consider only the particular case of the parallel plate capacitor. This will suffice to illustrate all of the general features of the problem in terms of an easily visualized and simple system. It will turn out, nevertheless, that our final results will be stated in such a manner that they can be seen to be actually of very general applicability.

In Figure 7-1, we show a parallel plate capacitor of capacitance C. By means of the switch S, the terminals T and T' can be connected by conducting wires to a battery B that will provide a difference of potential $\Delta\phi$ between the plates, thus charging the capacitor to the value $Q = C\Delta\phi$. Because of their opposite charges the plates will attract each other so that there will be an electrostatic force F_e on the positive plate with the direction shown. In order to have the system in equilibrium, F_e must be balanced by an equal and opposite mechanical force F_m. Our aim is to find these forces from energy considerations.

We let x be the separation of the plates. Now let us imagine that this separation is very slowly changed by the amount dx. Under these conditions, the work done on the system by the mechanical force will be given by $F_m\,dx$ and since this will be reversible work, as we saw in Section 5-4, it will be equal to the change in the total energy dU_t of the whole system, that is, $dU_t = F_m\,dx$. But since the acceleration is always nearly zero, the capacitor plate will remain in equilibrium, or be only infinitesimally different from it, so that we will continue to have $F_e = -F_m$, and therefore

$$F_e = -\frac{dU_t}{dx} \tag{7-36}$$

We have written U_t since the capacitor is not the whole system by itself; in general the battery must be included because its energy may also change as a result of the work done by F_m depending on whether the switch S is open or closed. In other words, the capacitor is not necessarily an isolated system. In this sense, our problem is similar to a situation often encountered in thermodynamics in which a thermodynamic system of interest may be in contact with heat and work reservoirs. There the problem is one of

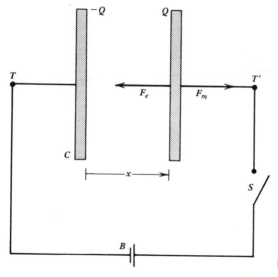

Figure 7-1.   Forces on a plate of a parallel plate capacitor.

trying to state the thermodynamic criteria of equilibrium, which refer to *isolated* systems, in terms of a function characteristic only of the nonisolated system of interest, that is, in terms of only part of the whole composite system, but that part with which we are directly concerned. In our case, we want to write (7-36) in terms of the capacitor only, if possible.

In our derivation of (7-36), we imagined a small relative displacement of the plates. However, the force F_e is a definite and measurable quantity and hence must be independent of the particular process that we visualize in order to do the calculation. In this example, there are really only two different ways in which we can carry out the infinitesimal change; that is, the switch is either open or closed. Thus, there are two possibilities to be considered and it is instructive to discuss them separately. As we will see, the details of what is going on differ in the two cases, but we should remember that we must get the same value for the force as a final result.

1. Constant charge. Here we imagine that S has been closed until the capacitor is charged, and then S is opened and kept open. This effectively removes the battery B from the total system since it will be unaffected by the displacement dx; in other words, the capacitor is isolated. In this case, $dU_t = dU_e$ where U_e is the capacitor energy and (7-36) becomes

$$F_e = -\left(\frac{dU_e}{dx}\right)_Q \qquad (Q = \text{const.}) \tag{7-37}$$

Now $U_e = Q^2/2C$ according to (7-21), but since C is a function of the plate separation, $C = C(x)$, and therefore

$$dU_e = -\frac{Q^2}{2C^2} dC \tag{7-38}$$

which makes (7-37) become

$$F_e = \frac{Q^2}{2C^2} \frac{dC}{dx} \tag{7-39}$$

[In this process, the potential difference between the plates must change in order that $Q = C\Delta\phi$ remain constant. Since this leads to $dQ = 0 = (dC)(\Delta\phi) + Cd(\Delta\phi)$, we find

$$-\frac{d(\Delta\phi)}{\Delta\phi} = \frac{dC}{C} \tag{7-40}$$

which says that the fractional change in potential difference is equal and opposite to that of the capacitance. In fact, since C will decrease as x is increased, according to (6-41) (with x replacing d), $\Delta\phi$ will increase as x is increased. The reason for this is that with $Q = $ const., σ will also be constant, as will $E = \sigma/\epsilon_0$; then, with the same E applying over a greater separation, $\Delta\phi = \int_+^- E\, ds = Ex$ will also be greater.] Before we discuss (7-39) in more detail, let us consider the other possibility.

2. Constant potential difference. This corresponds to keeping the switch S closed so that $\Delta\phi = $ const. The battery is then part of the complete system, and the capacitor is no longer isolated. In this case, the total energy change will be the sum of the energy change of the capacitor, dU_e, and that of the battery, dU_B, so that

$$dU_t = dU_e + dU_B \tag{7-41}$$

Here it is convenient to write U_e in the form $U_e = \frac{1}{2}C(\Delta\phi)^2$, which is also given by (7-21). Then, we have

$$dU_e = \frac{1}{2}(dC)(\Delta\phi)^2 \tag{7-42}$$

[If dx is positive, dC will be negative, as noted above, so that the capacitor energy U_e will decrease under these conditions, in contrast to its increase in the constant charge case as described by (7-38).]

Since $Q = C\Delta\phi$, the capacitor charge will change and this change is given by

$$dQ = (dC)(\Delta\phi) \qquad (7\text{-}43)$$

As these charges pass slowly through the battery potential difference, work will be done on them and the battery energy will change. We found in (5-46) that this work equals the charge times the potential difference, and since any work done by the battery represents a decrease in its energy, we get

$$dU_B = -dW = -(dQ)(\Delta\phi) = -(dC)(\Delta\phi)^2 = -2\,dU_e \qquad (7\text{-}44)$$

with the use of (7-43) and (7-42). Thus, the energy change in the battery is always opposite in sign to that of the capacitor and twice as great in magnitude. [If dx is positive, we saw above that dU_e will be negative, so that dU_B as given by (7-44) will be positive, that is, the battery energy will actually increase. The reason is that dQ will also be negative from (7-43) since dC is negative; as these charges return through the battery they will return to it the reversible work that the battery originally did on them in the original charging process.]

When we substitute (7-44) into (7-41), we find that $dU_t = -dU_e$ in this case so that (7-36) becomes

$$F_e = +\left(\frac{dU_e}{dx}\right)_{\Delta\phi} \qquad (\Delta\phi = \text{const.}) \qquad (7\text{-}45)$$

which when combined with (7-42) and (6-28) gives

$$F_e = \tfrac{1}{2}(\Delta\phi)^2\frac{dC}{dx} = \frac{Q^2}{2C^2}\frac{dC}{dx} \qquad (7\text{-}46)$$

which, *as it must be*, is exactly the same as (7-39).

It is perhaps well to emphasize again that the basic equation for calculating the force is (7-36) and that the apparent discrepancy in signs between (7-37) and (7-45) arises from the fact that these expressions refer to *different processes*. In the case of (7-37), the capacitor was an isolated system and it underwent the only energy change. In the case of (7-45), the capacitor was no longer an isolated system but we were *still* able to express the force in terms of the capacitor energy *alone*. This is all quite analogous to the thermodynamic situation in which one can go from the characterization of the equilibrium of a system of interest by means of its internal energy to the use of its Helmholtz or Gibbs function for the cases in which it is in contact with heat and work reservoirs. This can be done by using only generalized properties of these reservoirs as large featureless systems. We note that here too we did not need to know any of the detailed internal workings of the battery; it was sufficient to know that it is a device that is somehow capable of doing reversible work on charges.

As a matter of fact, when we review our work that led to (7-39) and (7-46), we see that, except for the parenthetical illustrative remarks we made along the way, we really did not use any detailed result that specifically required our system to be a parallel plate capacitor other than that the displacement was parallel to the mechanical force so that the element of work could be written as the simple product $F_m\,dx$. Now, however, it is necessary to introduce these details.

If we let x be the plate separation, $C = \epsilon_0 A/x$, according to (6-41). Then $dC/dx = -\epsilon_0 A/x^2 = -C/x$, and (7-39) becomes

$$F_e = -\frac{Q^2}{2Cx} = -\frac{U_e}{x} \qquad (7\text{-}47)$$

with the use of (7-21). This shows that F_e is negative, agreeing with the fact that the oppositely charged plates of Figure 7-1 will attract each other. We found in (7-35) that U_e can be written as the product of the energy density u_e and the volume Ax between the plates so that $U_e = u_e Ax$; putting this into (7-47), we find that

$$F_e = -u_e A \qquad (7\text{-}48)$$

Since the total force is proportional to the area, it is convenient to introduce the force per unit area, f_e, as equal to the magnitude of this ratio; thus, we get

$$f_e = \frac{|F_e|}{A} = u_e \qquad (7\text{-}49)$$

We see from Figure 7-1 that the direction of this force is outward from the conducting surface so that we can write the force per unit area as

$$\mathbf{f}_e = f_e \hat{\mathbf{n}} = u_e \hat{\mathbf{n}} \qquad (7\text{-}50)$$

where $\hat{\mathbf{n}}$ is the outward normal to the conductor surface as shown in Figure 7-2. Thus we have found that there is a tension or outward force per unit area on the conducting surface that is numerically equal to the value of the energy density at the surface.

Although this result was obtained by considering the specific case of a parallel plate capacitor, we can now show that it is actually of general validity. Consider a portion of the conductor surface as shown in Figure 7-3. Inside the conductor where $\mathbf{E}$ is zero, the energy density u_e is also zero. Now let us *imagine* that a small portion of the conductor surface of area Δa is given a small displacement Δx perpendicular to the surface. The volume of the region where $u_e = 0$ has been increased by $\Delta a \, \Delta x$ so that the overall energy will have been changed by the amount ΔU_e where

$$\Delta U_e = -u_e \, \Delta a \, \Delta x \qquad (7\text{-}51)$$

which is negative if Δx is positive. This energy change corresponds to a force ΔF_e on this area element given by $\Delta F_e = -\Delta U_e / \Delta x = u_e \, \Delta a$. Since ΔF_e is proportional to the area Δa, we can once again introduce a force per unit area $f_e = \Delta F_e / \Delta a$, which again turns out to be equal to u_e in agreement with (7-49). In order to determine its direction, we recall from mechanics that the states of stable equilibrium for general systems correspond to the configurations resulting in a minimum value of the potential energy, that is, the natural tendency of systems is to "attempt" to decrease their potential energy. As we saw above, a positive displacement Δx would lead to a decrease in the energy U_e; since this is what the system "wants" to do, this will determine the direction of the force to be outward from the conductor. In other words, we are led once more to (7-50) that gives the electrostatic force per unit area as always a tension, that is, in the

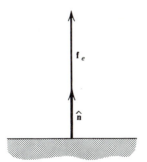

Figure 7-2. Force per unit area on the surface of a conductor.

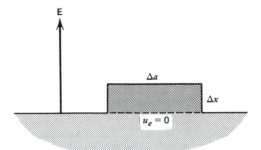

Figure 7-3. Calculation of the force on the surface of a conductor. An element of the surface is imagined displaced by Δx.

direction of the outward normal $\hat{n}$. If the internal cohesive forces of the conducting material are not large enough to counterbalance this electrostatic force, the conductor will deform. This deformation will continue until the elastic forces that are brought into being will be large enough to keep the surface in a new equilibrium situation.

If we combine (7-49) with (7-29) and (6-4), we can express f_e variously as

$$f_e = u_e = \tfrac{1}{2}\epsilon_0 E^2 = \frac{\sigma^2}{2\epsilon_0} = \tfrac{1}{2}\sigma E \qquad (7\text{-}52)$$

where E and σ must be evaluated at the particular point on the surface that is being considered.

If we multiply (7-50) by da, we will get the force on this to be $\mathbf{f}_e\, da = f_e \hat{n}\, da = f_e\, d\mathbf{a}$ so that the total force on the complete surface S of a given conductor will be given by

$$\mathbf{F}_{e,\,\text{total}} = \int\!\!\int_S f_e\, d\mathbf{a} = \frac{1}{2\epsilon_0} \int_S \sigma^2\, d\mathbf{a} \qquad (7\text{-}53)$$

and that can be used once the surface charge density has been determined as a function of position.

It is a very common error to think that the force per unit area is simply σE rather than $\tfrac{1}{2}\sigma E$ as correctly given by (7-52). Also, one should not forget that (7-53) is a *vector* equation. [For example: what is the total force on the conducting sphere whose electric field is given by (7-31)?]

EXERCISES

7-1 Consider a square of edge a. Starting at one corner and proceeding counterclockwise, one puts a point charge q at the first corner, $2q$ at the next, then $3q$, and finally $-4q$. Find U_e for this charge distribution.

7-2 A point charge q is put at each corner of a cube of edge a. Find the electrostatic energy of this system of charges.

7-3 The expression for the energy of a capacitor given by (7-21) can also be obtained in the following manner. Consider an intermediate stage when the charge is q where $0 < q < Q$. The potential difference will be q/C. Find the work required to increase the charge by dq. Then add all these work increments from the initial uncharged state to the final completely charged state and thus obtain (7-21) again.

7-4 Find the energy of the charge distribution of Exercise 5-9 by using (7-10). To what should your result reduce when $n = 0$? Does it?

7-5 Find the energy of the charge distribution of Exercise 5-17 by using (7-8).

7-6 Find the energy of a length L of the coaxial cylinders of Figure 6-12 when they are used as a capacitor with charge q_l per unit length by using (7-8). Use your result to verify again the value (6-45) for C.

7-7 Find the total gravitational energy of the earth by treating it as a homogeneous sphere of mass 5.98×10^{24} kilograms and radius 6.37×10^6 meters. [The gravitational constant is $G = 6.67 \times 10^{-11}$ newton-$(\text{meter})^2/(\text{kilogram})^2$.] If a uniform spherical charge distribution whose total charge equaled the magnitude of the electronic charge $(1.60 \times 10^{-19}$ coulombs) had the same energy, what would its radius be?

7-8 Express the energy of a system of n conductors in terms of the potentials and coefficients of capacitance and induction defined by (6-43). Show that when there are only two conductors the energy can be written as

$$U_e = \tfrac{1}{2}c_{11}\phi_1^2 + c_{12}\phi_1\phi_2 + \tfrac{1}{2}c_{22}\phi_2^2 \quad (7\text{-}54)$$

and that when these two conductors are used as a capacitor the energy is again given by (7-21) and (6-44).

7-9 Show that when (7-28) is applied to the case of the uniform spherical charge distribution the result is again (7-14). What fraction of the total energy is now regarded as being outside of the sphere?

7-10 A charge $-Q$ is on the inner sphere of Figure 6-8 and a charge Q on the outer sphere. Find the energy of this system by using (7-28) and thus show that the capacitance is (6-37).

7-11 The coaxial cylindrical conductors of Figure 6-12 are used as a capacitor with charges per unit length q_l and $-q_l$. Find the energy of a length L of this system by using (7-28) and thus show that the capacitance is (6-45).

7-12 Use (7-28) to find the energy of the charge distribution of Exercise 5-9 and verify that you get the same result as for Exercise 7-4 and that your answer reduces to the correct result when $n = 0$. What fraction of the total energy is outside the sphere?

7-13 Consider the two conductors of Figure 6-8. There is a total net charge Q_1 on the inner conductor and a total net charge Q_2 on the outer conductor where $Q_1 \neq Q_2$. Show that the total energy of this system is given by

$$U_e = \frac{1}{8\pi\epsilon_0}\left[\left(\frac{1}{a} - \frac{1}{b} + \frac{1}{c}\right)Q_1^2 \right.$$
$$\left. + \frac{2Q_1Q_2}{c} + \frac{Q_2^2}{c}\right]$$

(Show that this reduces to the expected result when $Q_2 = -Q_1$.) Do this exercise in three different ways.

7-14 A parallel plate capacitor has plates of area A. The lower plate is rigidly fastened to a table top. The upper plate is suspended from a spring of spring constant k whose upper end is rigidly fastened. The plates are originally uncharged. When the capacitor is charged to a final charge Q and $-Q$ show that the distance between the plates changes by an amount $Q^2/2k\epsilon_0 A$. Is the spring stretched or compressed?

7-15 A square metal plate of edge 20 centimeters is suspended from the arm of a balance so that it is parallel to another fixed horizontal metal plate of the same dimensions. The distance of separation between the plates is 1.5 millimeters. A difference of potential of 150 volts is now applied between the plates. What additional mass must be placed in the other arm of the balance so that the suspended plate will retain its original position?

7-16 A battery is used to charge a parallel plate capacitor to a potential difference $\Delta\phi$ and is then disconnected. The separation between the plates is now increased from d to αd where α is a constant > 1. What is the ratio of the new energy to the original energy? Is the energy increased or decreased? Where does this energy change come from or go? Verify your answer quantitatively.

7-17 There is a potential difference $\Delta\phi$ between the coaxial conductors of Figure 6-12. Find the magnitude of the force per unit area on the surface of the inner cylinder. What is its direction? What is the total force per unit length on it?

7-18 A parallel plate capacitor is formed by the metal bottom and metal movable piston of a cylinder with nonconducting walls. The cylinder is airtight and is kept at constant temperature. When the capacitor is uncharged the separation of the plates is d_0 and the pressure inside the cylinder is p_0. When a potential difference $\Delta\phi$ is applied between the plates, show that, if f is the fractional decrease in plate separation ($f > 0$), it can be found from

$$f(1 - f) = \frac{\epsilon_0}{2p_0}\left(\frac{\Delta\phi}{d_0}\right)^2$$

7-19 Suppose the two coaxial cylindrical conductors of Figure 6-12 are kept at a constant potential difference $\Delta\phi$. The cylinders are very long. Now imagine the inner conductor to be given a small displacement Δx in the direction of the common axis. Show that the inner conductor will be attracted back to its original position by a force which is given approximately by

$$F \simeq \frac{\pi\epsilon_0(\Delta\phi)^2}{\ln\left(\dfrac{b}{a}\right)}$$

8

ELECTRIC MULTIPOLES

Equation 5-2 provides us with a method by which we can, in principle, find the potential due to any given distribution of charges at any field point of interest. Let us suppose that the charges are contained in a finite volume of some reasonable size. The volume need not be of any particularly simple shape but, in fact, may be quite irregular. If we are near such a volume, we can expect that the values of the potential at different points can be quite sensitive to the details of the charge distribution. However, as we get farther and farther away, it seems clear that the finer details of the charge distribution will become less and less important and that the potential will be determined primarily by the larger scale variations in the locations and magnitudes of the charges. The extreme case would occur when we are so far away that the charge distribution would appear to act as if it were simply a point charge; we have already used this in connection with (7-26) and (7-27). What we want to do now is to consider this situation more carefully, and we will find that the effect of the charge distribution can be characterized by a set of quantities that depend on different details of the charge distribution, much as mechanical quantities such as the magnitude of the total mass and the moment of inertia of a set of mass points depend on different features of the mass distribution. These quantities are called *electric multipoles* and we will give them a specific definition. These considerations will also be helpful to us when we face the problem of describing the effects of matter in electrostatics, since, for our purposes, we will regard a piece of matter as essentially a collection of electric charges with some sort of distribution.

8-1 THE MULTIPOLE EXPANSION OF THE SCALAR POTENTIAL

The general situation is illustrated in Figure 8-1. We have a system of N point charges $q_1, q_2, \ldots, q_i, \ldots, q_N$ located in some volume V'. We choose an origin of coordinates 0 in some arbitrary way, but for convenience, either in or near V'. The position vectors of the charges are $\mathbf{r}_1, \mathbf{r}_2, \ldots, \mathbf{r}_i, \ldots, \mathbf{r}_N$. We want the potential ϕ at the field point P whose position vector with respect to this same origin is $\mathbf{r}$; thus, P is in the direction $\hat{\mathbf{r}}$ at a distance r from 0. This potential is given by (5-2) as

$$\phi(\mathbf{r}) = \sum_{i=1}^{N} \frac{q_i}{4\pi\epsilon_0 R_i} \tag{8-1}$$

where $R_i = |\mathbf{r} - \mathbf{r}_i|$. If we introduce the angle θ_i between the directions of $\mathbf{r}_i$ and $\mathbf{r}$, and use the law of cosines, we see from the figure that

$$R_i = \left(r^2 + r_i^2 - 2rr_i \cos\theta_i \right)^{1/2} \tag{8-2}$$

so that (8-1) becomes

$$\phi(\mathbf{r}) = \sum_{i=1}^{N} \frac{q_i}{4\pi\epsilon_0 \left(r^2 + r_i^2 - 2rr_i \cos\theta_i \right)^{1/2}} \tag{8-3}$$

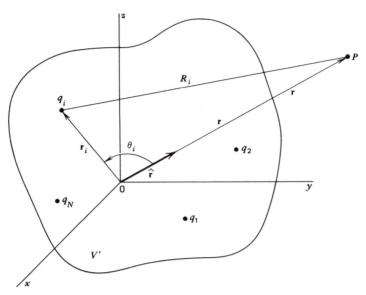

Figure 8-1.   Geometry for calculating the potential due to a system of point charges.

However, we can do nothing more specific with this without knowing the exact details of the charge distribution.

Now let us assume that P is far enough outside V' so that it is farther from the origin than any charge. Then $r > r_i$ for all i, the ratio (r_i/r) is always less than unity, and we can consider a power series expansion in this ratio. If we factor out r^2 from the square root in (8-2), we can write

$$\frac{1}{R_i} = \frac{1}{r(1+t)^{1/2}} \tag{8-4}$$

where

$$t = -2\left(\frac{r_i}{r}\right)\cos\theta_i + \left(\frac{r_i}{r}\right)^2 \tag{8-5}$$

We now use the power series

$$(1 \pm t)^{-(1/2)} = 1 \mp \tfrac{1}{2}t + \tfrac{3}{8}t^2 \mp \tfrac{5}{16}t^3 + \dots \tag{8-6}$$

with the upper sign, to expand the square root in (8-4). We will keep all terms of the order of $(r_i/r)^2$ and neglect the rest; thus we do not need to use the fourth term in (8-6), which involves t^3 and hence $(r_i/r)^3$. Inserting (8-5) into (8-6), and dropping all terms involving $(r_i/r)^3$ and $(r_i/r)^4$, we find that

$$\frac{1}{(1+t)^{1/2}} \simeq 1 - \frac{1}{2}\left[-2\left(\frac{r_i}{r}\right)\cos\theta_i + \left(\frac{r_i}{r}\right)^2\right] + \frac{3}{8}\left[-2\left(\frac{r_i}{r}\right)\cos\theta_i + \left(\frac{r_i}{r}\right)^2\right]^2$$

$$\simeq 1 + \left(\frac{r_i}{r}\right)\cos\theta_i + \frac{1}{2}\left(\frac{r_i}{r}\right)^2(3\cos^2\theta_i - 1)$$

If we divide this by r, according to (8-4), and substitute the result into (8-1), we find that

$$\phi(\mathbf{r}) = \frac{1}{4\pi\epsilon_0 r}\sum_{i=1}^{N} q_i + \frac{1}{4\pi\epsilon_0 r^2}\sum_{i=1}^{N} q_i r_i \cos\theta_i$$

$$+ \frac{1}{4\pi\epsilon_0 r^3}\sum_{i=1}^{N}\frac{q_i r_i^2}{2}(3\cos^2\theta_i - 1) + \dots \tag{8-7}$$

where we have indicated by the ... that there would be other terms of higher order, but ones which we have not calculated. The result (8-7) is called the *multipole expansion of the potential*. The individual terms in the sum are called, respectively, the monopole term, the dipole term, and the quadrupole term. We see that their dependence on the field point distance r goes successively as $1/r$, $1/r^2$, $1/r^3$, and so on, so that as we get farther away from the charge distribution, the higher-order terms in the expansion become less and less important. For convenience in later discussion, we write the sum (8-7) in the form

$$\phi(\mathbf{r}) = \phi_M(\mathbf{r}) + \phi_D(\mathbf{r}) + \phi_Q(\mathbf{r}) + \ldots \tag{8-8}$$

Although it is not necessary for our further considerations in this chapter, it is of some interest to note that the functions of the angles that have arisen in (8-7) are known functions called *Legendre polynomials*. If these functions are written $P_l(x)$, they are *defined* by the expansion

$$\frac{1}{\left(1 - 2xy + y^2\right)^{1/2}} = \sum_{l=0}^{\infty} P_l(x) y^l \qquad (|x| \le 1, \, y < 1) \tag{8-9}$$

so that they are the coefficients of y^l in the sum. A few of them are

$$P_0(x) = 1 \qquad P_1(x) = x \qquad P_2(x) = \tfrac{1}{2}(3x^2 - 1)$$

$$P_3(x) = \tfrac{1}{2}(5x^3 - 3x), \ldots \tag{8-10}$$

Once the first few are known, the others can be found by means of the *recursion relation* that they can be shown to satisfy:

$$(l + 1)P_{l+1}(x) = (2l + 1)xP_l(x) - lP_{l-1}(x) \tag{8-11}$$

We also note that $P_l(1) = 1$.

If we compare (8-9) with (8-4) and (8-5), we see that we can make the identifications $y = r_i/r$ and $x = \cos\theta_i$; both of these satisfy the parenthetical conditions given in (8-9). Then we can write

$$\frac{1}{R_i} = \frac{1}{r} \sum_{l=0}^{\infty} P_l(\cos\theta_i)\left(\frac{r_i}{r}\right)^l \tag{8-12}$$

so that (8-1) can be written in general as

$$\phi(\mathbf{r}) = \frac{1}{4\pi\epsilon_0} \sum_{l=0}^{\infty} \frac{1}{r^{l+1}} \left[\sum_{i=1}^{N} q_i r_i^l P_l(\cos\theta_i) \right] \tag{8-13}$$

whose first few terms agree with (8-7), which we obtained by a straightforward expansion. Although (8-13) gives us a complete expression for the general expansion of ϕ, we will need to consider only that part given by (8-7).

Our results are still somewhat inconvenient for our purposes because the quantities appearing in the sums still involve both the location of the field point P and the locations of the charges because of the appearance of $\cos\theta_i$. It would be nicer if we could write the terms of (8-7) and (8-8) in a form in which these two sets of variables appeared separately and explicitly, preferably in the form of a product involving something that depended on the location of the field point alone and something that involved only the charge distribution. This will be done next, but we need a different way of writing $\cos\theta_i$. We see from Figure 8-1, (1-15), (1-97), (1-8), and (1-20) that

$$\cos\theta_i = \frac{\mathbf{r} \cdot \mathbf{r}_i}{rr_i} = \hat{\mathbf{r}} \cdot \left(\frac{\mathbf{r}_i}{r_i}\right) = \frac{l_x x_i + l_y y_i + l_z z_i}{r_i} \tag{8-14}$$

where l_x, l_y, and l_z are the direction cosines of the position vector $\mathbf{r}$ of P, and x_i, y_i, and z_i are the rectangular coordinates of the location of the charge q_i. It will be convenient to look at the terms in (8-8) one by one.

1. The Monopole Term

The sum in the first term of (8-7) is easily identifiable. It is

$$\sum_{i=1}^{N} q_i = Q_{\text{total}} = Q \tag{8-15}$$

where Q is the net charge of the system. Thus, the monopole term has the form

$$\phi_M(\mathbf{r}) = \frac{Q}{4\pi\epsilon_0 r} \tag{8-16}$$

Since this is the dominant term in the potential, when we are very far away, we see that the whole distribution will act as if it were a point charge as we have already concluded.

In this context, the net charge Q is called the *monopole moment* of the charge distribution. In other words, the monopole *moment* is that feature of the charge distribution that is important for the monopole *term*.

If the charges are continuously distributed, then the sum can be replaced by an integral by means of (2-14) so that the monopole moment can be found from

$$Q = \int_{V'} \rho(\mathbf{r}') \, d\tau' \tag{8-17}$$

where the integral is taken over the volume V' of the source charge distribution. For surface and line distributions, there will be similar expressions obtained with the use of (2-16).

2. The Dipole Term

If we insert (8-14) into the second sum in (8-7), we get

$$\sum_{i=1}^{N} q_i r_i \cos\theta_i = \sum_{i=1}^{N} q_i(l_x x_i + l_y y_i + l_z z_i)$$

$$= l_x\left(\sum_i q_i x_i\right) + l_y\left(\sum_i q_i y_i\right) + l_z\left(\sum_i q_i z_i\right)$$

$$= \hat{\mathbf{r}} \cdot \left(\sum_{i=1}^{N} q_i \mathbf{r}_i\right) \tag{8-18}$$

Since the sum in parentheses in the last form involves only the properties of the charge distribution and does *not* involve the location of the field point, it is an individual property of the charge distribution *only*. It is defined as the *dipole moment* $\mathbf{p}$ of the charge distribution; that is

$$\mathbf{p} = \sum_{i=1}^{N} q_i \mathbf{r}_i \tag{8-19}$$

so that we can write

$$\sum_{i=1}^{N} q_i r_i \cos\theta_i = \hat{\mathbf{r}} \cdot \mathbf{p} = l_x p_x + l_y p_y + l_z p_z \tag{8-20}$$

When this is inserted into (8-7), we find that the dipole *term* can be written in terms of the dipole *moment* as

$$\phi_D(\mathbf{r}) = \frac{\mathbf{p} \cdot \hat{\mathbf{r}}}{4\pi\epsilon_0 r^2} = \frac{\mathbf{p} \cdot \mathbf{r}}{4\pi\epsilon_0 r^3} \tag{8-21}$$

We note that (8-21) has the form of a (scalar) product of quantities, one of which depends only on the location of the field point and one that depends only on the details of the charge distribution.

If the point P is very far away *and if* the monopole moment Q vanishes, then (8-21) will be the leading term in the expansion of ϕ and the dipole moment $\mathbf{p}$ will be the dominant feature of the charge distribution.

If the charges have a continuous distribution, then the sum (8-19) can be replaced by an integral over the volume V' so that $\mathbf{p}$ can be found from

$$\mathbf{p} = \int_{V'} \rho(\mathbf{r}')\mathbf{r}'\, d\tau' \tag{8-22}$$

and there will be similar expressions for surface and line distributions.

3. The Quadrupole Term

This term is more complicated, but it can be written in a desirable and convenient form in a reasonably straightforward manner. If we use (8-14), we find that

$$r_i^2\left(3\cos^2\theta_i - 1\right) = 3(\hat{\mathbf{r}} \cdot \mathbf{r}_i)^2 - r_i^2$$
$$= 3\left(l_x x_i + l_y y_i + l_z z_i\right)^2 - r_i^2\left(l_x^2 + l_y^2 + l_z^2\right) \tag{8-23}$$

In the last step, we have actually multiplied r_i^2 by 1 because of (1-9) and hence have not altered its value. When we multiply out the square in (8-23) and group the terms we obtain

$$r_i^2\left(3\cos^2\theta_i - 1\right) = l_x^2\left(3x_i^2 - r_i^2\right) + l_y^2\left(3y_i^2 - r_i^2\right)$$
$$+ l_z^2\left(3z_i^2 - r_i^2\right) + 6l_x l_y x_i y_i + 6l_y l_z y_i z_i$$
$$+ 6l_z l_x z_i x_i \tag{8-24}$$

We now insert (8-24) into the sum in the third term of (8-7) after factoring the $\frac{1}{2}$ out of it; we also divide the last three terms of (8-24) by noting that $6l_x l_y x_i y_i = 3l_x l_y x_i y_i + 3l_y l_x y_i x_i$. When we do all of this, we find that the sum can be written in a nice symmetrical form as follows:

$$\sum_i q_i r_i^2\left(3\cos^2\theta_i - 1\right)$$
$$= l_x^2 \sum_i q_i\left(3x_i^2 - r_i^2\right) + l_x l_y \sum_i q_i 3x_i y_i + l_x l_z \sum_i q_i 3x_i z_i$$
$$+ l_y l_x \sum_i q_i 3y_i x_i + l_y^2 \sum_i q_i\left(3y_i^2 - r_i^2\right) + l_y l_z \sum_i q_i 3y_i z_i$$
$$+ l_z l_x \sum_i q_i 3z_i x_i + l_z l_y \sum_i q_i 3z_i y_i + l_z^2 \sum_i q_i\left(3z_i^2 - r_i^2\right) \tag{8-25}$$

We note that *each* term in this expression is a product of something that depends only on the field point, that is, its direction, and a quantity that depends only on the details of the charge distribution. Accordingly, we define a set of quantities Q_{jk}, which are called the *components of the quadrupole moment tensor* as follows:

$$Q_{jk} = \sum_{i=1}^{N} q_i\left(3j_ik_i - r_i^2\delta_{jk}\right) \qquad (j, k = x, y, z) \qquad (8\text{-}26)$$

In this expression, j and k can independently be x, y, or z; the symbol δ_{jk} is the *Kronecker delta symbol* defined by

$$\delta_{jk} = \begin{cases} 1 & \text{if } j = k \\ 0 & \text{if } j \neq k \end{cases} \qquad (8\text{-}27)$$

Thus there are a total of nine Q_{jk} defined by (8-26). For example,

$$Q_{xx} = \sum_i q_i\left(3x_i^2 - r_i^2\right) \qquad Q_{xy} = \sum_i q_i 3x_i y_i \qquad (8\text{-}28)$$

Upon comparing (8-26), (8-28), and (8-25), we see that the last one can be more compactly written as

$$\sum_i q_i r_i^2\left(3\cos^2\theta_i - 1\right)$$

$$= l_x^2 Q_{xx} + l_x l_y Q_{xy} + l_x l_z Q_{xz}$$

$$+ l_y l_x Q_{yx} + l_y^2 Q_{yy} + l_y l_z Q_{yz}$$

$$+ l_z l_x Q_{zx} + l_z l_y Q_{zy} + l_z^2 Q_{zz}$$

$$= \sum_{j=x,y,z} \sum_{k=x,y,z} l_j l_k Q_{jk} \qquad (8\text{-}29)$$

Finally, when (8-29) is inserted into (8-7), we find that the quadrupole *term* can be written in terms of the quadrupole *moment* as

$$\phi_Q(\mathbf{r}) = \frac{1}{4\pi\epsilon_0 r^3} \cdot \frac{1}{2} \sum_{j=x,y,z} \sum_{k=x,y,z} l_j l_k Q_{jk} \qquad (8\text{-}30)$$

If the point P is very far away *and if both* the monopole moment Q and the dipole moment $\mathbf{p}$ are zero, then (8-30) will be the leading term in the expansion of ϕ and the quadrupole moment tensor Q_{jk} will be the dominant feature of the charge distribution.

Sometimes it is convenient to express the quadrupole term explicitly in terms of the coordinates of the field point rather than in terms of its direction cosines. We can do this by using $l_x = x/r$, $l_y = y/r$, $l_z = z/r$, according to (1-7) and (1-11), so that (8-30) becomes

$$\phi_Q(\mathbf{r}) = \frac{1}{4\pi\epsilon_0 r^5} \cdot \frac{1}{2} \sum_{j=x,y,z} \sum_{k=x,y,z} jk Q_{jk} \qquad (8\text{-}31)$$

[The forms (8-30) and (8-31) are reminiscent of the result obtained in mechanics for the kinetic energy of a rigid body when expressed in terms of the moments and products of inertia.]

If the charges have a continuous distribution, then the sum in (8-26) can be replaced by an integral, so that for a volume distribution we would have

$$Q_{jk} = \int_{V'} \rho(\mathbf{r}')\left(3j'k' - r'^2\delta_{jk}\right) d\tau' \qquad (8\text{-}32)$$

For example,

$$Q_{xx} = \int_{V'} \rho(\mathbf{r}')(3x'^2 - r'^2)\, d\tau' \qquad Q_{xy} = \int_{V'} \rho(\mathbf{r}')3x'y'\, d\tau' \qquad (8\text{-}33)$$

There will be similar expressions for surface and line charge distributions.

Although there were a total of nine quantities Q_{jk} defined by (8-26), there are actually fewer independent ones. We see from (8-26) or (8-28) that $Q_{yx} = Q_{xy}$, and so on; that is,

$$Q_{kj} = Q_{jk} \qquad (j \neq k) \tag{8-34}$$

Thus, the quadrupole moment tensor is an example of a *symmetric* tensor, and (8-34) reduces the number of independent components to six. If we now sum up the diagonal components, that is, those with $j = k$, we find that

$$Q_{xx} + Q_{yy} + Q_{zz} = \sum_i q_i \left[\left(3x_i^2 - r_i^2 \right) + \left(3y_i^2 - r_i^2 \right) + \left(3z_i^2 - r_i^2 \right) \right] = 0$$

since $r_i^2 = x_i^2 + y_i^2 + z_i^2$; thus we have

$$Q_{xx} + Q_{yy} + Q_{zz} = 0 \tag{8-35}$$

and this result, which is true regardless of the location of the origin or of the detailed nature of the charge distribution, reduces the number of independent components to five.

If the charge distribution has sufficient symmetry, the number of independent components can be reduced even further. As an extreme example, let us consider the case of axial symmetry, that is, the distribution has an axis of rotational symmetry such as is the case for a cylinder, a cone, or an egg. Let us choose the z axis to be along this axis and designate the elements in this case by Q_{jk}^a. Then, for every charge q' at (x', y', z'), say, there will be a charge of the same value at $(-x', y', z')$ and the contribution of this pair to Q_{xy}^a, for example, will be $3q'x'y' + 3q'(-x')y' = 0$. Since all of the charges can be paired in this manner, the result will be that $Q_{xy}^a = 0$. This same argument holds for the rest of the off-diagonal elements so that

$$Q_{jk}^a = 0 \qquad (j \neq k) \tag{8-36}$$

and we are down to three components. But, because of (8-35), there are really only two because

$$Q_{xx}^a + Q_{yy}^a + Q_{zz}^a = 0 \tag{8-37}$$

Furthermore, because there is no real distinction between the x and y axes in this case, for every charge at a given value x', there will be an equal charge at the same numerical value for y' for the same value of r'. Thus, the corresponding sums in (8-26) will be equal so that

$$Q_{xx}^a = Q_{yy}^a \tag{8-38}$$

When this is put into (8-37), we find that $2Q_{xx}^a + Q_{zz}^a = 0$. Therefore, $Q_{xx}^a = Q_{yy}^a = -\frac{1}{2}Q_{zz}^a$ and there is only *one* independent component of the quadrupole moment characteristic of the charge distribution. Calling $Q^a = Q_{zz}^a$ *the* quadrupole moment of this axially symmetric charge distribution, we have

$$Q_{zz}^a = Q^a$$
$$Q_{xx}^a = Q_{yy}^a = -\frac{1}{2}Q^a \tag{8-39}$$

Under these circumstances the quadrupole term simplifies considerably. If we insert (8-36) and (8-39) into (8-30), and use (1-9) to eliminate l_x and l_y, we find that ϕ_Q becomes

$$\phi_Q^a(\mathbf{r}) = \frac{Q^a}{4\pi\epsilon_0} \frac{\left(3l_z^2 - 1 \right)}{4r^3} = \frac{Q^a}{4\pi\epsilon_0} \frac{\left(3\cos^2\theta - 1 \right)}{4r^3} \tag{8-40}$$

where θ is the angle made by the position vector $\mathbf{r}$ of the field point with the axis of symmetry (the z axis). An example of such a situation as this is provided by atomic nuclei. Although their properties must be described by quantum mechanics, the treatment is quite analogous. It can be shown that if a nucleus can have a quadrupole moment at all, it must necessarily have an axis of symmetry that is the direction of its intrinsic angular momentum or "spin." Thus it is essentially the quantity Q^a that is found listed in tables of nuclear quadrupole moments; actually what is tabulated is Q^a/e where e is the electronic charge so that these nuclear quadrupole moments are given as areas.

4. Effects of the Choice of Origin

The monopole moment Q given by (8-15) is a unique property of the charge distribution. Both (8-19) and (8-26) do depend on the absolute value of the $\mathbf{r}_i$, so that the dipole moment and the quadrupole moment components are not generally unique properties of the charge distribution but depend on the choice of origin as well. Under some circumstances, however, they are independent of this choice, and we want to investigate this in more detail.

Suppose that instead of choosing the origin at 0 in Figure 8-1, we choose a new origin 0_n, which is obtained by translating our axes, without rotating them, by the displacement $\mathbf{a}$ as illustrated in Figure 8-2. The position of q_i with respect to this new origin is $\mathbf{r}_{in}$ and we see from the figure that $\mathbf{a} + \mathbf{r}_{in} = \mathbf{r}_i$ so that the old and new position vectors are related by

$$\mathbf{r}_{in} = \mathbf{r}_i - \mathbf{a} \tag{8-41}$$

If we insert these values into (8-19) in order to find the new value of the dipole moment $\mathbf{p}_n$, we get

$$\mathbf{p}_n = \sum_i q_i \mathbf{r}_{in} = \sum_i q_i \mathbf{r}_i - \mathbf{a} \sum_i q_i = \mathbf{p} - Q\mathbf{a} \tag{8-42}$$

which shows that the dipole moments as calculated with respect to these different

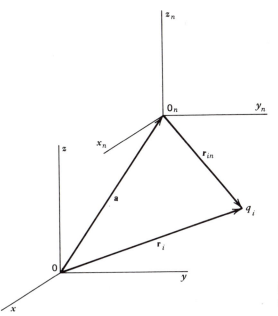

Figure 8-2. The new origin of coordinates is displaced by a from the old.

origins will generally be different. However, we now also see that the dipole moment will be independent of the choice of origin and hence a unique property of the charge distribution provided that the monopole moment vanishes, that is,

$$\mathbf{p}_n = \mathbf{p} \qquad \text{if } Q = 0 \tag{8-43}$$

In addition, the dipole term will be the leading term in the expansion (8-7). In order to make $Q = 0$, we will need at least *two* charges in our distribution; this is a reason for the name dipole.

■   **Example**

Two equal and opposite point charges. This is the simplest example for which Q vanishes. The two charges and their locations are shown in Figure 8-3. From (8-19), we find $\mathbf{p}$ to be

$$\mathbf{p} = q\mathbf{r}_+ - q\mathbf{r}_- = q(\mathbf{r}_+ - \mathbf{r}_-) = q\mathbf{l} \tag{8-44}$$

so that the dipole moment equals the product of the charge magnitude and the vector displacement $\mathbf{l}$ from the negative charge to the positive charge. Although this charge distribution is often thought of as the prototype of a dipole, and is often simply called a "dipole," its higher-order moments will generally not vanish so that (8-8) will necessarily include more than the term $\phi_D(\mathbf{r})$.   ■

Another simple example satisfying (8-43) is shown in Figure 8-4 and it is easily verified with the use of (8-19) that the dipole moment is $\mathbf{p} = 2q\mathbf{l}$.

Similar results can be obtained for the quadrupole moment. It will be sufficient for our purposes to consider only one component, Q_{xy}, say. The x and y components of (8-41) are $x_{in} = x_i - a_x$ and $y_{in} = y_i - a_y$ and when we insert these into (8-28), we find that the new component Q_{xy}^n is

$$Q_{xy}^n = 3\sum_i q_i x_{in} y_{in} = 3\sum_i q_i(x_i - a_x)(y_i - a_y)$$
$$= Q_{xy} - 3a_y p_x - 3a_x p_y + 3a_x a_y Q \tag{8-45}$$

with the use of (8-19) and (8-15). Since there will be similar expressions for the other components Q_{jk}, we can conclude that the quadrupole moment is also not generally a unique property of the charge distribution. However, we do see from (8-45) that it will be independent of the origin provided that *both* the monopole and dipole moments vanish, that is,

$$Q_{jk}^n = Q_{jk} \qquad \text{if } Q = 0 \text{ and } \mathbf{p} = 0 \tag{8-46}$$

In addition, the quadrupole term will be the leading term in the expansion (8-7).

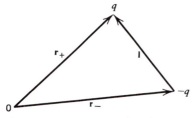

Figure 8-3.   Two equal and opposite point charges have a dipole moment in the direction of **l**.

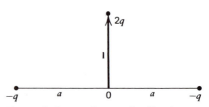

Figure 8-4.   A charge distribution that has a dipole moment.

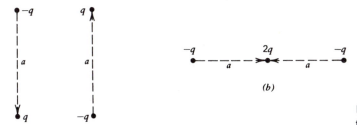

Figure 8-5.   Charge distributions that have quadrupole moments.

In general, we can satisfy the two conditions in (8-46) only with at least *four* charges in our distribution; this is a reason for the name quadrupole. The simplest examples are usually obtained by arranging two dipoles so that the total dipole moment vanishes. Two such examples are shown in Figure 8-5 and the constituent dipoles are indicated by the dashed arrows. We note that Figure 8-5*b* is a variation of Figure 8-4 with $\mathbf{l} = 0$; we also note that it corresponds to our statement about needing four charges only if we mentally divide the central charge $2q$ into two superimposed charges each of value q. Some care is required in devising such illustrative arrangements. For example, if we interchange the two charges at the top of Figure 8-5*a*, we find that the conditions of (8-46) are no longer fully satisfied so that the resultant charge configuration will have an origin-dependent quadrupole moment.

To summarize this long section, we have found that when we are sufficiently far away from an arbitrary charge distribution, the potential produced by it can be expanded in the form

$$\phi(\mathbf{r}) = \frac{1}{4\pi\epsilon_0}\left(\frac{Q}{r} + \frac{\mathbf{p}\cdot\hat{\mathbf{r}}}{r^2} + \frac{1}{2r^3}\sum_{j,k}l_jl_kQ_{jk} + \cdots\right) \qquad (8\text{-}47)$$

as obtained from (8-8), (8-16), (8-21), and (8-30). Since only the location of the field point appears in (8-47) through its distance from the origin and the direction of $\mathbf{r}$, all of the moments can themselves be thought of as being located at the origin, regardless of the actual spatial extent of the source charge distribution.

So far we have concentrated on the form of the potential. Now we want to consider some of its properties, as well as the electric field which it describes. It is convenient to look individually at each contribution arising from (8-47); the total $\mathbf{E}$ will be the vector sum of them all. The first term corresponds to a point charge at the origin, and since we are sufficiently familiar with its properties we can proceed at once to the next term.

8-2   THE ELECTRIC DIPOLE FIELD

The potential we want to investigate is given by (8-21) and will be the predominant term when $Q = 0$. Although ϕ_D was obtained by studying the potential far from the charge distribution, it is convenient to study its properties by assuming that ϕ_D holds *everywhere* in space. Such a field is commonly called *the* dipole field and is thought of as being produced by a *point dipole* $\mathbf{p}$ located at the origin. This useful fiction can be imagined to be the result of a limiting process applied to the equal and opposite charge distribution of Figure 8-3. In this process, the separation $\mathbf{l}$ is decreased to zero while the charge q is simultaneously increased so that the dipole moment, which is their product according to (8-44), remains constant and equal to $\mathbf{p}$. If we use spherical coordinates to locate the field point P, and choose the z axis to be in the direction of $\mathbf{p}$, we are led to the situation shown in Figure 8-6. Using (8-21) and (1-15), we find that we can write the

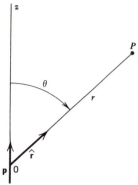

Figure 8-6. A point dipole at the origin and parallel to the *z* axis.

dipole potential as

$$\phi_D(\mathbf{r}) = \frac{p \cos \theta}{4 \pi \epsilon_0 r^2} \tag{8-48}$$

The equation giving the equipotential surfaces corresponding to ϕ_D = const. is thus

$$r^2 = \left(\frac{p}{4 \pi \epsilon_0 \phi_D} \right) \cos \theta = C_D \cos \theta \tag{8-49}$$

where the constant C_D characterizing a given surface depends on the value of ϕ_D. These equipotentials are illustrated as the solid curves in Figure 8-7; the actual surfaces would be generated by imagining this two-dimensional figure rotated about the *z* axis. Since r^2 in (8-49) must be positive, we see that C_D must be positive for $\theta < \frac{1}{2}\pi$ where $\cos \theta$ is positive; thus positive values of ϕ_D correspond to the equipotential curves in the upper half of the figure. Similarly, the curves in the lower half of the figure correspond to negative values of ϕ_D, since $\cos \theta$ is negative for $\theta > \frac{1}{2}\pi$, so that C_D must be negative also.

The components of **E** can be found from (8-48), (5-3), and (1-101); the results are

$$E_r = - \frac{\partial \phi_D}{\partial r} = \left(\frac{p}{4 \pi \epsilon_0} \right) \frac{2 \cos \theta}{r^3}$$

$$E_\theta = - \frac{1}{r} \frac{\partial \phi_D}{\partial \theta} = \left(\frac{p}{4 \pi \epsilon_0} \right) \frac{\sin \theta}{r^3} \tag{8-50}$$

and $E_\varphi \sim - \partial \phi_D / \partial \varphi = 0$. These components have a different angular dependence but both have the inverse cube variation with distance characteristic of the dipole field.

We can find the equation giving the line of **E** from our previous expression (5-39), which expresses the fact that the line is parallel to **E** at each point on it. If we write both $d\mathbf{s}_{lf}$ and **E** in spherical coordinates by using (1-98), we find that

$$dr = kE_r \quad \text{and} \quad r \, d\theta = kE_\theta \tag{8-51}$$

so that

$$\frac{dr}{r \, d\theta} = \frac{E_r}{E_\theta} = \frac{2 \cos \theta}{\sin \theta} = \frac{d \ln r}{d\theta} = \frac{2d(\ln \sin \theta)}{d\theta}$$

which integrates to

$$\ln r = \ln \sin^2 \theta + \ln K_D = \ln \left(K_D \sin^2 \theta \right)$$

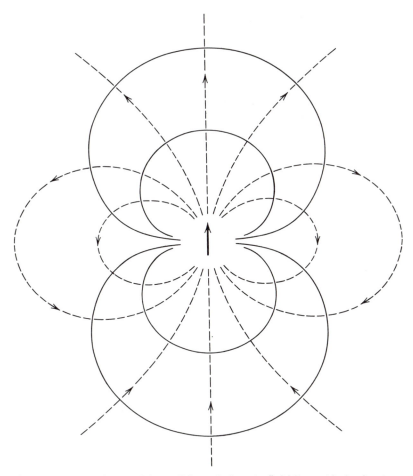

Figure 8-7.   Equipotentials (solid) and electric field lines (dashed) of a point dipole.

where K_D is a positive constant of integration. Solving for r, we find the equation for these curves to be

$$r = K_D \sin^2 \theta \qquad (8\text{-}52)$$

so that each curve is characterized by a particular value of K_D. When these curves are plotted, we obtain the lines of $\mathbf{E}$ shown dashed in Figure 8-7. They are seen to be consistent with the special cases of (8-50), which show us that $E_r = 0$ when $\theta = \frac{1}{2}\pi$, and that $E_\theta = 0$ when $\theta = 0$ and π. Be sure to verify to your own satisfaction that the directions of $\mathbf{E}$ as indicated by the arrows follow from the expressions for the components given in (8-50), and that the lines of $\mathbf{E}$ are perpendicular to the equipotential surfaces.

8-3   THE LINEAR QUADRUPOLE FIELD

The general expression for the quadrupole potential as given by (8-30) can be quite complicated depending on which components Q_{jk} are different from zero. Consequently, we investigate only the special case in which the system has an axis of symmetry so that the potential as given by (8-40) is

$$\phi_Q^a = \frac{Q^a}{4\pi\epsilon_0} \frac{(3\cos^2\theta - 1)}{4r^3} \tag{8-53}$$

Although many types of charge distributions can result in this potential, it is convenient to regard the simple charge distribution of Figure 8-5b as a prototype for this case since it certainly has the necessary symmetry; Q^a is actually negative for this situation as can be seen from (8-39) and (8-26). As a result we can refer to (8-53) as a linear (or axial) quadrupole field. The equation for the equipotential surfaces that are obtained by setting $\phi_Q^a = $ const. in (8-53) is

$$r^3 = \left(\frac{Q^a}{8\pi\epsilon_0\phi_Q^a}\right)\frac{(3\cos^2\theta - 1)}{2} = \frac{1}{2}C_Q(3\cos^2\theta - 1) \tag{8-54}$$

where the constant C_Q characterizing a given surface depends on the corresponding value of ϕ_Q^a. For simplicity, only one of these is shown as the solid curve in Figure 8-8; the others are of similar shape. Since (8-54) is independent of φ, the actual surfaces are again generated by rotating these curves around the symmetry axis (the z axis in this case.)

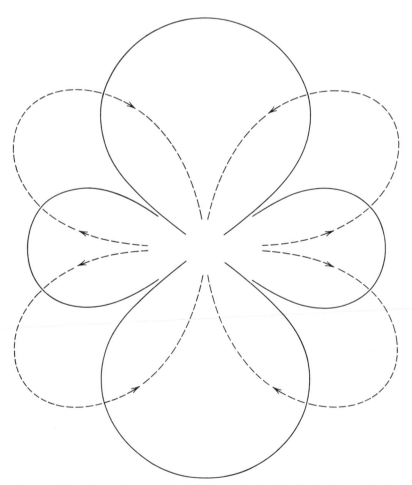

Figure 8-8. An equipotential (solid) and electric fields lines (dashed) of a linear quadrupole.

The electric field components as found from (8-53), (5-3), and (1-101) are

$$E_r = \left(\frac{3Q^a}{8\pi\epsilon_0} \right) \frac{(3\cos^2\theta - 1)}{2r^4}$$

$$E_\theta = \left(\frac{3Q^a}{8\pi\epsilon_0} \right) \frac{\cos\theta \sin\theta}{r^4} \tag{8-55}$$

$$E_\varphi = 0$$

The field components decrease as the inverse fourth power of the distance to the field point.

In order to find the equation of a field line, we again eliminate k from (8-51) by division and use (8-55). In this way, we obtain

$$\frac{dr}{r\,d\theta} = \frac{E_r}{E_\theta} = \frac{(3\cos^2\theta - 1)}{2\cos\theta \sin\theta}$$

which, when integrated, gives

$$r^2 = K_Q \sin^2\theta \cos\theta \tag{8-56}$$

where K_Q is a constant of integration. Again for simplicity, we show only one of these as the dashed curve in Figure 8-8. As was suggested for Figure 8-7, be sure to verify that the directions of **E** as indicated by the arrows are consistent with the expressions for the components given in (8-55) on the assumption that Q^a is negative, as would be the case for Figure 8-5b. We note again that, as they should be, the lines of **E** are perpendicular to this much more complicated equipotential surface.

8-4 ENERGY OF A CHARGE DISTRIBUTION IN AN EXTERNAL FIELD

In the previous chapter, we discussed the mutual electrostatic energy of an arbitrary distribution of charges that we obtained as the reversible work required to assemble these charges against their mutual conservative forces of attraction and repulsion. Under some circumstances, we are not interested in all of this energy, but only in part of it. In order to see how this can arise, let us consider a situation in which the overall charge distribution can be divided into two groups such that the charges in each group occupy their own volume, and these volumes are some distance apart so that there is no difficulty in distinguishing the groups. Such a situation is illustrated in Figure 8-9 where

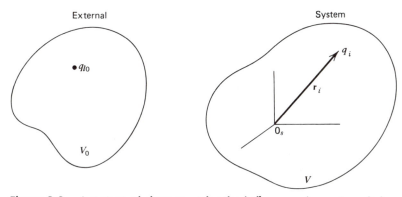

Figure 8-9.   A system of charges under the influence of an external charge distribution.

the charges in one group are labeled q_{l0} and occupy the volume V_0, while the other group is simply labeled q_i without an additional subscript, as is their volume V; we have also labeled the groups as "external" and "system" for reasons that will become clear shortly. Let us consider the single charge q_i at its location $\mathbf{r}_i$ with respect to the origin 0_s which, for later convenience, we choose to be within the system volume. According to (5-48), we can write the energy of this charge as

$$U_{ei} = q_i\phi(\mathbf{r}_i) \tag{8-57}$$

in terms of the total potential $\phi(\mathbf{r}_i)$ at its location. Now (5-2) shows that ϕ is determined by all of the charges and we can, in fact, divide the total sum into two groups, one part representing contributions from the external charges q_{l0} and the other part from the other charges in the system. Thus, we can write $\phi = \phi_s + \phi_0$ where, for example,

$$\phi_0(\mathbf{r}_i) = \sum_{l=1}^{N_0} \frac{q_{l0}}{4\pi\epsilon_0 R_{il}} \tag{8-58}$$

where N_0 is the total number of external charges and $R_{il} = |\mathbf{r}_i - \mathbf{r}_l|$ where $\mathbf{r}_l$ is the position vector of q_{l0}. The other part ϕ_s will be given by a similar expression. Inserting this into (8-57), we find that the energy of q_i can be written as the sum

$$U_{ei} = q_i\phi_s(\mathbf{r}_i) + q_i\phi_0(\mathbf{r}_i) \tag{8-59}$$

Now the first term represents the mutual energy of interaction among the charges of the system and when we sum it up over all N charges of this type we will be led to a term like (7-3). Thus this term will be the *internal energy* of the system and expresses the amount of work required to assemble it. For cases in which we are *now* interested, the system will be a definite entity in which the constituent charges will have fixed relative spatial locations and this internal electrostatic energy will be a definite constant.

The second term in (8-59) then represents an energy arising from the interactions between a system charge and the group of external charges. When this is summed over i, we get the total energy of this kind and we can regard it as being the energy of the charges of the system of interest due to the effect of the external field; if we call this energy U_{e0}, we have

$$U_{e0} = \sum_{i=1}^{N} q_i\phi_0(\mathbf{r}_i) \tag{8-60}$$

and it is this *interaction energy* which is of interest here.

We can do a similar analysis for the external charges and reach similar conclusions. However, in practical cases of interest, what is envisioned is that these external charges are set up in the laboratory by charges on capacitor plates, or on other conductors, for example, and one tries to study the system of interest by seeing how it responds to the influence of the external source charges. In such a case, we generally do not study the energy changes of the external charges so we need not concern ourselves further with that. Thus, the energy term of value to us is that *part* of the total energy represented by (8-60); consequently, in what follows, we restrict ourselves to it.

As it stands, (8-60) does not yet completely fit experimental reality. Generally, the external sources are far enough away and the system is so small in spatial extent that ϕ_0 does not vary very much over the volume V. Consequently, it is sufficiently accurate to expand $\phi_0(\mathbf{r}_i)$ in a power series about the origin and include only the first few terms of the expansion. If we use rectangular coordinates, then the series expansion of ϕ_0 can be written as

$$\phi_0(\mathbf{r}_i) = \phi_0(x_i, y_i, z_i) = \phi_0(0,0,0)$$

$$+ \left[x_i \left(\frac{\partial \phi_0}{\partial x} \right)_0 + y_i \left(\frac{\partial \phi_0}{\partial y} \right)_0 + z_i \left(\frac{\partial \phi_0}{\partial z} \right)_0 \right]$$

$$+ \frac{1}{2} \left[x_i^2 \left(\frac{\partial^2 \phi_0}{\partial x^2} \right)_0 + y_i^2 \left(\frac{\partial^2 \phi_0}{\partial y^2} \right)_0 + z_i^2 \left(\frac{\partial^2 \phi_0}{\partial z^2} \right)_0 \right.$$

$$\left. + 2x_i y_i \left(\frac{\partial^2 \phi_0}{\partial x \, \partial y} \right)_0 + 2 y_i z_i \left(\frac{\partial^2 \phi_0}{\partial y \, \partial z} \right)_0 + 2 z_i x_i \left(\frac{\partial^2 \phi_0}{\partial z \, \partial x} \right)_0 \right] + \ldots \qquad (8\text{-}61)$$

where the subscript 0 on the derivatives indicates that they are to be evaluated at the origin. It is convenient to consider the result of inserting (8-61) into (8-60) piece by piece.

The first term in (8-61) is a constant, so that when it is put into (8-60) it can be taken out of the summation and we find its contribution to U_{e0} to be

$$U_{e0M} = \phi_0(0,0,0) \sum_i q_i = Q \phi_0(0) \qquad (8\text{-}62)$$

where Q is the monopole moment as given by (8-15). Thus, this part of the interaction energy is simply that of a single point charge equal to the total charge times the external potential at the origin in harmony with (5-48).

The first bracketed term in (8-61) can be written as

$$\mathbf{r}_i \cdot (\nabla \phi_0)_0 = -\mathbf{r}_i \cdot \mathbf{E}_0 \qquad (8\text{-}63)$$

because of (5-3) where $\mathbf{E}_0$ is the external electric field. Since $\mathbf{E}_0$ is a constant also, when we put (8-63) into (8-60) we get this contribution to the energy as

$$U_{e0D} = -\mathbf{E}_0 \cdot \sum_i q_i \mathbf{r}_i = -\mathbf{p} \cdot \mathbf{E}_0 \qquad (8\text{-}64)$$

where $\mathbf{p}$ is the dipole moment as given by (8-19). Thus, this very important expression gives us the interaction energy of a dipole in an external electric field.

By now, it seems clear that the remaining terms of (8-61) will lead to an energy contribution involving the quadrupole moment components, although these terms are not yet written in the most desirable form. Since the source charges are assumed to be physically separate from the system, as in Figure 8-9, their charge density $\rho_0 = 0$ throughout the whole system volume V. Therefore, at all points within the system, ϕ_0 satisfies Laplace's equation according to (5-15) and (5-16); that is,

$$\left(\frac{\partial^2 \phi_0}{\partial x^2} \right)_0 + \left(\frac{\partial^2 \phi_0}{\partial y^2} \right)_0 + \left(\frac{\partial^2 \phi_0}{\partial z^2} \right)_0 = 0 \qquad (8\text{-}65)$$

Thus, we can add any multiple of (8-65) to the second-order terms in (8-61) without changing their value; accordingly, we multiply (8-65) by $-(r_i^2/6)$, add this to what is left of (8-61), and rearrange the terms by using the fact that mixed second partial derivatives are equal, [e.g., $(\partial^2 \phi_0/\partial x \, \partial y)_0 = (\partial^2 \phi_0/\partial y \, \partial x)_0$]; the result is that they can be also written as

$$\frac{1}{6}\left[\left(3x_i^2 - r_i^2\right)\left(\frac{\partial^2\phi_0}{\partial x^2}\right)_0 + 3x_i y_i\left(\frac{\partial^2\phi_0}{\partial x\,\partial y}\right)_0 + 3x_i z_i\left(\frac{\partial^2\phi_0}{\partial x\,\partial z}\right)_0 \right.$$

$$+ 3y_i x_i\left(\frac{\partial^2\phi_0}{\partial y\,\partial x}\right)_0 + \left(3y_i^2 - r_i^2\right)\left(\frac{\partial^2\phi_0}{\partial y^2}\right)_0 + 3y_i z_i\left(\frac{\partial^2\phi_0}{\partial y\,\partial z}\right)_0$$

$$\left. + 3z_i x_i\left(\frac{\partial^2\phi_0}{\partial z\,\partial x}\right)_0 + 3z_i y_i\left(\frac{\partial^2\phi_0}{\partial z\,\partial y}\right)_0 + \left(3z_i^2 - r_i^2\right)\left(\frac{\partial^2\phi_0}{\partial z^2}\right)_0\right] \qquad (8\text{-}66)$$

If we now put (8-66) into (8-60) and use the definition of the quadrupole moment components given in (8-26) and illustrated in (8-28), we find that the remaining contribution to the energy in the external field becomes

$$U_{e0Q} = \frac{1}{6}\left[Q_{xx}\left(\frac{\partial^2\phi_0}{\partial x^2}\right)_0 + Q_{xy}\left(\frac{\partial^2\phi_0}{\partial x\,\partial y}\right)_0 + Q_{xz}\left(\frac{\partial^2\phi_0}{\partial x\,\partial z}\right)_0 \right.$$

$$+ Q_{yx}\left(\frac{\partial^2\phi_0}{\partial y\,\partial x}\right)_0 + Q_{yy}\left(\frac{\partial^2\phi_0}{\partial y^2}\right)_0 + Q_{yz}\left(\frac{\partial^2\phi_0}{\partial y\,\partial z}\right)_0$$

$$\left. + Q_{zx}\left(\frac{\partial^2\phi_0}{\partial z\,\partial x}\right)_0 + Q_{zy}\left(\frac{\partial^2\phi_0}{\partial z\,\partial y}\right)_0 + Q_{zz}\left(\frac{\partial^2\phi_0}{\partial z^2}\right)_0\right] \qquad (8\text{-}67)$$

or more compactly as

$$U_{e0Q} = \frac{1}{6}\sum_{j=x,\,y,\,z}\sum_{k=x,\,y,\,z} Q_{jk}\left(\frac{\partial^2\phi_0}{\partial j\,\partial k}\right)_0 \qquad (8\text{-}68)$$

We can also write this in terms of the external electric field $\mathbf{E}_0$ by noting that $(\partial\phi_0/\partial j)_0 = -E_{0j}$ because of (5-3) so that

$$U_{e0Q} = -\frac{1}{6}\sum_{j=x,\,y,\,z}\sum_{k=x,\,y,\,z} Q_{jk}\left(\frac{\partial E_{0j}}{\partial k}\right)_0 \qquad (8\text{-}69)$$

which, when written out, becomes

$$U_{e0Q} = -\frac{1}{6}\left[Q_{xx}\left(\frac{\partial E_{0x}}{\partial x}\right)_0 + Q_{xy}\left(\frac{\partial E_{0x}}{\partial y}\right)_0 + Q_{xz}\left(\frac{\partial E_{0x}}{\partial z}\right)_0 \right.$$

$$+ Q_{yx}\left(\frac{\partial E_{0y}}{\partial x}\right)_0 + Q_{yy}\left(\frac{\partial E_{0y}}{\partial y}\right)_0 + Q_{yz}\left(\frac{\partial E_{0y}}{\partial z}\right)_0$$

$$\left. + Q_{zx}\left(\frac{\partial E_{0z}}{\partial x}\right)_0 + Q_{zy}\left(\frac{\partial E_{0z}}{\partial y}\right)_0 + Q_{zz}\left(\frac{\partial E_{0z}}{\partial z}\right)_0\right] \qquad (8\text{-}70)$$

showing us explicitly that the energy of a quadrupole moment in an external field depends on the spatial derivatives of the electric field components.

If we combine (8-62), (8-64), and (8-69), we get our final expression for the energy of the charge system of interest due to interactions with external source charges as

$$U_{e0} = Q\phi_0(0) - \mathbf{p}\cdot\mathbf{E}_0 - \frac{1}{6}\sum_{j,\,k} Q_{jk}\left(\frac{\partial E_{0j}}{\partial k}\right)_0 + \cdots \qquad (8\text{-}71)$$

It is possible to carry this process even further by expanding the potential ϕ_0 in terms of the multipole moments of the *external* source charges and thus getting a series for the interaction energy in terms of the energy of a dipole in the field of a dipole, and so on. We will not do this, however, but will leave the calculation of a few of the more important of such terms as exercises.

We can write (8-71) somewhat more generally. As we noted after (8-47), the moments can be regarded as being located at the origin, which in this case is the system origin 0_s of Figure 8-9. Since the system itself may be able to move, it is convenient to refer its location to a coordinate system fixed in space. Thus, if we let $\mathbf{r}$ be the position vector of 0_s with respect to this fixed system of axes, then terms in (8-71) which are all evaluated at 0_s, the "location" of the system of interest, will be evaluated at $\mathbf{r}$ so that we can finally write

$$U_{e0} = Q\phi_0(\mathbf{r}) - \mathbf{p} \cdot \mathbf{E}_0(\mathbf{r}) - \frac{1}{6}\sum_{j,k} Q_{jk}\left(\frac{\partial E_{0j}}{\partial k}\right)_{\mathbf{r}} + \dots \qquad (8\text{-}72)$$

Now that we have obtained these results, let us look at some of the consequences of the unfamiliar terms—the dipole and quadrupole energies.

1. The Dipole Energy

If we write it simply as U_D, then the energy of a dipole $\mathbf{p}$ in an external field $\mathbf{E}_0$ is

$$U_D = -\mathbf{p} \cdot \mathbf{E}_0 = -pE_0 \cos \Psi \qquad (8\text{-}73)$$

where Ψ is the angle between them as shown in Figure 8-10. This would be an appropriate form to use for a point dipole as we previously defined it. In Figure 8-11 we show this energy as a function of Ψ. We see that the range of energy variation is finite, and that the energy has a minimum at $\Psi = 0$ when the two vectors are parallel. This indicates that the tendency of the system is for its dipole moment to become aligned with the external field. We can evaluate this effect in the following way. Since the energy is a function of Ψ, that is, $U_D = U_D(\Psi)$, we know from mechanics that there will be a torque on $\mathbf{p}$, and the component of this torque in the direction of increasing Ψ, τ, will be given by

$$\tau = -\frac{\partial U_D}{\partial \Psi} = -pE_0 \sin \Psi = -|\mathbf{p} \times \mathbf{E}_0| \qquad (8\text{-}74)$$

Since τ is negative, it means that the sense of the torque on $\mathbf{p}$ is such as to rotate $\mathbf{p}$ into the direction of $\mathbf{E}_0$; thus, if we use the definition of the direction of the cross product given in Figure 1-14, we see that the vector torque $\boldsymbol{\tau}$ is in the direction of $\mathbf{p} \times \mathbf{E}_0$ as shown in Figure 8-12. Combining this with (8-74), the torque on $\mathbf{p}$ is given correctly in magnitude and direction by

$$\boldsymbol{\tau} = \mathbf{p} \times \mathbf{E}_0 \qquad (8\text{-}75)$$

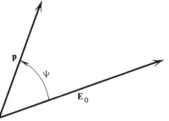

Figure 8-10.   A dipole making an angle with an external electric field.

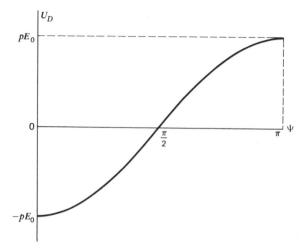

Figure 8-11.   The energy of a dipole in an external field as a function of the angle between their directions.

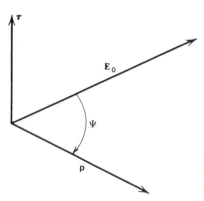

Figure 8-12.   The torque on a dipole in an external field.

We note that $\tau = 0$ when $\Psi = 0$ and π so that these values of the angle correspond to equilibrium situations; since they correspond respectively to a minimum and a maximum value of the energy U_D as shown in Figure 8-11, $\Psi = 0$ is a position of stable equilibrium, while $\Psi = \pi$ is one of unstable equilibrium. In order to have equilibrium at any intermediate angle, it will be necessary for an external agent to apply a mechanical torque τ_m equal and opposite to the value of τ given by (8-75) so that the net torque on $\mathbf{p}$ will be zero:

$$\tau + \tau_m = 0 \qquad (8\text{-}76)$$

If we look back at (8-73), we see that if $\mathbf{E}_0$ is not a constant, but a function of position $\mathbf{r}$, then it will be possible for a dipole to decrease its energy by moving to a different position; in other words, we have the possibility of a nonzero translational force $\mathbf{F}_D$ on the dipole because of the external field. Again, from mechanics, this is given by

$$\mathbf{F}_D = -\nabla U_D = \nabla(\mathbf{p} \cdot \mathbf{E}_0) \qquad (8\text{-}77)$$

The fact that $\mathbf{p}$ is a constant enables us to put this into a more useful form. If we use (1-112), we can write (8-77) as

$$\mathbf{F}_D = \mathbf{E}_0 \times (\nabla \times \mathbf{p}) + \mathbf{p} \times (\nabla \times \mathbf{E}_0) + (\mathbf{E}_0 \cdot \nabla)\mathbf{p} + (\mathbf{p} \cdot \nabla)\mathbf{E}_0 \qquad (8\text{-}78)$$

The first and third terms vanish because $\mathbf{p}$ is a constant so that all of its spatial derivatives are zero; the second term vanishes because $\mathbf{E}_0$ is conservative and $\nabla \times \mathbf{E}_0$ = 0 according to (5-4). Thus (8-78) reduces to

$$\mathbf{F}_D = (\mathbf{p} \cdot \nabla)\mathbf{E}_0 \tag{8-79}$$

For example, if we write the x component of this with the use of (1-121), we get

$$F_{Dx} = p_x \frac{\partial E_{0x}}{\partial x} + p_y \frac{\partial E_{0x}}{\partial y} + p_z \frac{\partial E_{0x}}{\partial z} \tag{8-80}$$

There are similar expressions for the other two rectangular components. The result (8-79) thus verifies our expectation that a translational force on a dipole will exist when the external field varies with position. [The torque given by (8-75) is present even in a uniform external field.]

Now all of these results for the dipole moment have been obtained in a completely general way and are applicable to the dipole moment of any type of charge distribution of interest. Nevertheless, it is of value to see that they can also be obtained quite directly and easily for the simple case of two equal and opposite charges shown in Figure 8-3 that we can take as the prototype of a point dipole by letting $\mathbf{l} \to 0$ and $q \to \infty$ so as to keep $\mathbf{p} = q\mathbf{l}$ constant. In order to emphasize this aspect, let us write the separation between them as $d\mathbf{r}$, as shown in Figure 8-13, so that we have $\mathbf{r}_+= \mathbf{r}_-+ d\mathbf{r}$. We also show the external field $\mathbf{E}_0$ and the forces on the charges which are to be calculated from (3-1).

Again, if $\phi_0(\mathbf{r})$ is the external potential, the energy will be obtained from (8-60) and is found to be

$$U_{e0} = q\phi_0(\mathbf{r}_+) - q\phi_0(\mathbf{r}_-) = q\,d\phi_0 = q\,d\mathbf{r} \cdot \nabla\phi_0 = -\mathbf{p} \cdot \mathbf{E}_0$$

with the use of (1-38), (5-3), and (8-44). Thus, this result obtained for this special situation agrees exactly with that found more generally in (8-64). The net force on the system will be

$$\mathbf{F}_{net} = \mathbf{F}_+ + \mathbf{F}_- = q\left[\mathbf{E}_0(\mathbf{r}_+) - \mathbf{E}_0(\mathbf{r}_-)\right] = q\,d\mathbf{E}_0$$

The x component of this can be found with the use of (1-38) and (8-44) to be

$$F_{net\ x} = q\,dE_{0x} = q\,d\mathbf{r} \cdot \nabla E_{0x} = (\mathbf{p} \cdot \nabla)E_{0x}$$

which is in agreement with (8-80) so that we will be led once more to (8-79). Similarly,

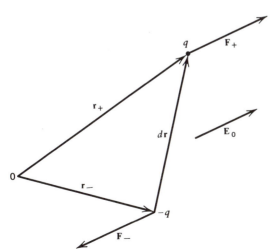

Figure 8-13. The forces on the charges of a prototype dipole.

the resultant torque in a uniform field will be given by

$$\tau = r_+ \times F_+ + r_- \times F_- = q(r_+ - r_-) \times E_0 = p \times E_0$$

and agrees exactly with (8-75).

2. The Linear Quadrupole Energy

As in the previous section, it will be sufficient for our purposes to look at the interaction energy only for the case of a charge distribution with axial symmetry. If we write it simply as U_Q^a, and substitute the values for the moment components given by (8-36) and (8-39) into the last term of (8-72), we get

$$U_Q^a = -\frac{Q^a}{6}\left[\frac{\partial E_{0z}}{\partial z} - \frac{1}{2}\left(\frac{\partial E_{0x}}{\partial x} + \frac{\partial E_{0y}}{\partial y}\right)\right] \tag{8-81}$$

and where we remember that the derivatives are to be evaluated at the system location. Again, since the external charges are all located outside the system so that $\rho_0 = 0$, we will have $\nabla \cdot E_0 = 0$ according to (4-10). We can use this, along with (1-42), to express (8-81) completely in terms of $\partial E_{0z}/\partial z$; the result is

$$U_Q^a = -\frac{Q^a}{4}\left(\frac{\partial E_{0z}}{\partial z}\right) = \frac{Q^a}{4}\left(\frac{\partial^2 \phi_0}{\partial z^2}\right) \tag{8-82}$$

where the last term follows from (5-3). Thus the energy term reduces to a very simple compact form for this special type of charge distribution. We note that it depends only on the variation along the axis of symmetry of that electric field component that is also along the symmetry axis that we chose as the z axis. If Q^a is positive, the energy will be a minimum at that location where $\partial E_{0z}/\partial z$ has its greatest positive value. Thus, there can be a translational force on this quadrupole given by

$$F_Q^a = -\nabla U_Q^a = \frac{1}{4}Q^a\nabla\left(\frac{\partial E_{0z}}{\partial z}\right) = \frac{1}{4}Q^a\frac{\partial}{\partial z}(\nabla E_{0z}) \tag{8-83}$$

EXERCISES

8-1 Carry out the expansion of $1/R_i$ to include all terms of order $(r_i/r)^3$ and thus show directly that the next term in (8-7) is correctly obtained from (8-13) with $l = 3$. This part of the expansion of ϕ is called the *octupole term*.

8-2 A *single* point charge q is located at the point (a, b, c). Find Q, p, and all components of Q_{jk} for this system. Which, if any, of these quantities are changed if a charge $-q$ located at the origin is added to the system?

8-3 Assume the charges of Figure 8-3 to be on the z axis with the origin midway between them. Find ϕ exactly for a field point on the z axis. How large must z be in order that one can approximate the exact potential on the z axis with the dipole term to an accuracy of 1 percent? How large must z be in order that one can

approximate E_z on the z axis with the dipole expression to an accuracy of 1 percent?

8-4 Evaluate (8-47) for the charge distribution of Figure 8-4 and express ϕ in rectangular coordinates. Use the origin shown and take the negative charges to be on the x axis and the positive charge on the positive y axis.

8-5 Point charges are placed at the corners of a cube of edge a. The charges and their locations are as follows: $-3q$ at $(0,0,0)$; $-2q,(a,0,0)$; $-q, (a, a, 0)$; $q, (0, a, 0)$; $2q, (0, a, a)$; $3q,(a, a, a)$; $4q,(a,0, a)$; $5q,(0,0, a)$. For this distribution find the monopole moment, the dipole moment, and all components of the quadrupole moment tensor. Show that your results satisfy (8-35). If it is possible to find a different origin of coordinates for which the dipole moment will vanish, where should this origin be located?

8-6 Show that the charge distribution of Figure 8-5b leads to (8-40) and thus evaluate Q^a for this case.

8-7 A line charge of constant charge density λ and of length L lies in the first quadrant of the xy plane with one end at the origin. It makes an angle α with the positive x axis. Find Q, $\mathbf{p}$, and all of the Q_{jk}. Express the quadrupole term of the potential due to this charge distribution in terms of the rectangular coordinates of the field point.

8-8 A sphere of radius a has a surface charge density given in spherical coordinates by $\sigma = \sigma_0 \cos \theta$ where $\sigma_0 = $ const. and the origin is at the center of the sphere. Find Q, $\mathbf{p}$, and all of the Q_{jk}. Express (8-47) for this charge distribution in terms of the spherical coordinates of a field point located outside the sphere.

8-9 Charge is distributed with constant volume density ρ throughout the figure of Figure 1-41. Find Q, $\mathbf{p}$, and all of the Q_{jk}. Interpret your result for $\mathbf{p}$. Then consider the special case in which the volume is a cube of edge a and express (8-31) in terms of the spherical coordinates of a field point located outside the cube.

8-10 If one has a charge distribution whose monopole moment is not zero, show that one can always find an origin such that the dipole moment will be zero. This point is called the *center of charge* of the distribution.

8-11 Find the analogous expression to (8-45) for Q_{xx}.

8-12 A spherically symmetric spherical charge distribution has the origin at its center. Show that $\mathbf{p}$ and all of the Q_{jk} are zero.

8-13 A point dipole $\mathbf{p}$ is located at the origin, but it has no special orientation with respect to the coordinate axes. (For example, $\mathbf{p}$ is not parallel to any of the axes.) Express its potential at a point $\mathbf{r}$ in rectangular coordinates, and find the rectangular components of $\mathbf{E}$. Show that $\mathbf{E}$ can be written in the important and general form

$$\mathbf{E}(\mathbf{r}) = \frac{1}{4\pi\epsilon_0 r^3}[3(\mathbf{p}\cdot\hat{\mathbf{r}})\hat{\mathbf{r}} - \mathbf{p}] \quad (8\text{-}84)$$

Can $\mathbf{E}$ ever be zero, other than for $r \to \infty$? If, now, $\mathbf{p}$ is located at $\mathbf{r}'$, rather than at the origin, how can $\mathbf{E}$ at $\mathbf{r}$ be obtained?

8-14 The constant K_D in (8-52) that characterizes a given line of $\mathbf{E}$ can be related to the magnitude E of the electric field for the point on the curve corresponding to $\theta = \frac{1}{2}\pi$. Find this relation.

8-15 Evaluate (8-47) for the charge distribution of Figure 8-5a and express the potential in terms of the cylindrical coordinates of the field point. Assume the figure is a square of edge a in the xy plane, origin at the center, sides parallel to the axes, and q in the first quadrant. Find the cylindrical components of $\mathbf{E}$. Find the equipotential curves in the xy plane and plot your result. Find the equation for a line of $\mathbf{E}$ in the xy plane and plot your result.

8-16 The constant K_Q in (8-56) that characterizes a given line of $\mathbf{E}$ can be related to the magnitude E of the electric field for the point on the curve for which E_r vanishes. Find this relation.

8-17 A point dipole $\mathbf{p}$ at $\mathbf{r}$ is in the field of a point charge q located at the origin. Find the energy of $\mathbf{p}$, the torque on it, and the net force on it.

8-18 The linear quadrupole of Section 8-3 is in the field of a point charge q located at the origin. Find the energy of the quadrupole and the translational force, if any, acting on it.

8-19 A point dipole $\mathbf{p}_1$ is located at $\mathbf{r}_1$ and another point dipole $\mathbf{p}_2$ is at $\mathbf{r}_2$. Show that the energy of $\mathbf{p}_2$ in the field of $\mathbf{p}_1$ is given by the *dipole-dipole interaction energy*

$$U_{DD} = \frac{1}{4\pi\epsilon_0 R^3}[(\mathbf{p}_1 \cdot \mathbf{p}_2) - 3(\mathbf{p}_1 \cdot \hat{\mathbf{R}})(\mathbf{p}_2 \cdot \hat{\mathbf{R}})]$$

$$(8\text{-}85)$$

where $\mathbf{R} = \mathbf{r}_2 - \mathbf{r}_1$. Does it make a difference if $\mathbf{R}$ is drawn from 2 to 1? Find the force $\mathbf{F}_2$ on $\mathbf{p}_2$. Evaluate $\mathbf{F}_2$ for these two special cases: (a) $\mathbf{p}_1$ and $\mathbf{p}_2$ are parallel to each other but perpendicular to $\hat{\mathbf{R}}$; (b) $\mathbf{p}_1$ and $\mathbf{p}_2$ are parallel to each other and parallel to $\hat{\mathbf{R}}$.

8-20 The linear quadrupole of Section 8-3 is in the field of a point dipole $\mathbf{p}$ located at the origin. Find the energy of the quadrupole.

8-21 A linear quadrupole Q_1^a is located at $\mathbf{r}$ and is in the field of another linear quadrupole Q_2^a located at the origin. Both are of the type of Section 8-3, that is, the axis of symmetry of each is parallel to the z axis. Find the energy of Q_1^a.

9

BOUNDARY CONDITIONS AT A SURFACE OF DISCONTINUITY

As soon as we consider the possibility of matter being present and subject to the influence of other charges, we realize that we may need to consider a situation in which we have two kinds of matter meeting at a common boundary. For example, we may have a block of wax in contact with a sheet of glass, or glass in contact with a conductor or vacuum, and so on. We can also expect that these different kinds of matter will have different electromagnetic "properties," so that these properties will change abruptly as we cross the surface of separation. As a result, it is quite possible that our various fields will be different in the two regions and it will be useful to know how they change as we go across this boundary. These changes, or lack of them as the case may be, are often called "boundary conditions," and we want to investigate them in general terms.

It turns out that these boundary conditions can be obtained for any vector in terms of its divergence and curl, so that it is convenient for us to derive them now. In this way, we will have available a set of "ready-made" expressions that we can apply to specific fields as we encounter them.

9-1 ORIGIN OF A SURFACE OF DISCONTINUITY

In Figure 9-1 we illustrate what we can call the "real situation" at a boundary between two materials or media, which we label 1 and 2. What is plotted is the variation of some as yet unspecified electromagnetic "property" as a function of position x; the "boundary" between medium 1 and 2 is indicated by the solid vertical line. As we go far enough to the left, this property becomes constant and completely characteristic of medium 1; similarly, if we go far enough to the right, the property again becomes constant, but at a different value, so that there is no doubt that we are in medium 2. Now, as illustrated, we expect that in an actual case this property will not change abruptly, but *continuously*, although possibly very rapidly, in a thin region between the two media; we call this region the "transition layer." This expectation is appropriate for our constant assumption that the physical fields we are dealing with are continuous and have continuous derivatives. Although we have indicated the extent of this transition layer by the dashed lines and even assigned it a thickness h, we really cannot hope to know anything about the specific details of what is actually occurring within it. Consequently, it is customary to replace this actual situation by an *idealized* situation, which is illustrated in Figure 9-2. We do this by imagining the transition layer to shrink to zero thickness, thus giving rise to a discontinuity in electromagnetic properties. As a result, it is possible that the fields themselves can be regarded as possibly having discontinuities in this limit as $h \to 0$, and this is what we want to consider.

Let us consider a general vector field $\mathbf{F}(\mathbf{r})$. As in Section 1-20, we write its source equations in the form

$$\nabla \cdot \mathbf{F} = b(\mathbf{r}) \qquad \text{and} \qquad \nabla \times \mathbf{F} = \mathbf{c}(\mathbf{r}) \tag{9-1}$$

132

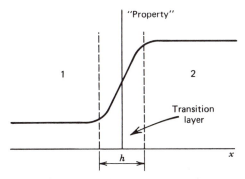

Figure 9-1. The physical origin of a transition layer between two media.

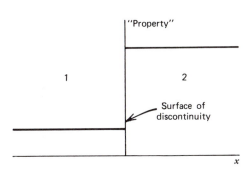

Figure 9-2. The idealized surface of discontinuity between two media.

These will provide us with our desired information but we cannot apply them directly to a discontinuity. What we will do is use these equations in the transition layer where everything is continuous and *then* go to the limit $h \to 0$.

An important definition is given by Figure 9-3, which shows the direction of the normal $\hat{\mathbf{n}}$ to the surface of discontinuity. As illustrated,

$$\hat{\mathbf{n}} = \hat{\mathbf{n}}_{\text{from 1 to 2}} \tag{9-2}$$

and we will *always* follow this sign convention and it is necessary not to forget it.

9-2 THE DIVERGENCE AND THE NORMAL COMPONENTS

The divergence theorem (1-59) combined with (9-1) yields

$$\oint_S \mathbf{F} \cdot d\mathbf{a} = \int_V \nabla \cdot \mathbf{F} \, d\tau = \int_V b(\mathbf{r}) \, d\tau \tag{9-3}$$

We apply this to a small right cylinder of height h and cross-sectional area Δa constructed in the transition layer as shown in a side view in Figure 9-4. This is done so that we will get contributions from both regions 1 and 2. The outward normals to the faces are $\hat{\mathbf{n}}_2$ and $\hat{\mathbf{n}}_1$ while $\hat{\mathbf{n}}_w$ is one of the outward normals to the curved wall. We choose the area Δa to be small enough so that it will be a good approximation to take $\mathbf{F}$ to be constant over these faces. (We contemplate that eventually we will let $\Delta a \to 0$ so that we will be obtaining relations valid at a point; in such a case, higher-order corrections obtained from, say, a power series expansion of $\mathbf{F}$ will vanish since they would be multiplied by Δa in our final results, as we will see.) The surface integral in (9-3) can then be written as

$$\oint_S \mathbf{F} \cdot d\mathbf{a} = \mathbf{F}_2 \cdot \Delta \mathbf{a}_2 + \mathbf{F}_1 \cdot \Delta \mathbf{a}_1 + W \tag{9-4}$$

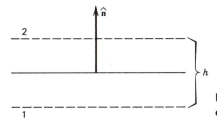

Figure 9-3. Definition of the normal to the surface of discontinuity.

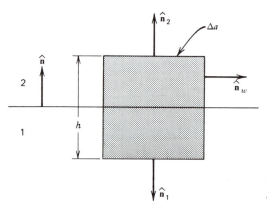

Figure 9-4. Volume used to find boundary condition from the divergence theorem.

where $\mathbf{F}_2$ and $\mathbf{F}_1$ are the values on the faces in the respective regions and W is the contribution from the curved wall. W will have some finite value and we can write $W \sim h$ by the mean value theorem. We see from the figure that $\hat{\mathbf{n}}_2 = \hat{\mathbf{n}}$ and $\hat{\mathbf{n}}_1 = -\hat{\mathbf{n}}$ so that when we use (1-52), we can write (9-4) and (9-3) as

$$\hat{\mathbf{n}} \cdot (\mathbf{F}_2 - \mathbf{F}_1)\, \Delta a + W = \int_V b\, d\tau = (hb)\, \Delta a \qquad (9\text{-}5)$$

where b should actually be written as its average value within the volume $h\, \Delta a$; again, however, we can take b to be approximately constant throughout this small volume that we will eventually let go to zero.

Now we can carry out our limiting process of letting the transition layer shrink to zero thickness by letting $h \to 0$ while we keep Δa constant. Because W is proportional to h, $W \to 0$ as $h \to 0$. On the other hand, we cannot generally be certain about the behavior of the product hb, since it may well happen that b increases in such a way in this process that $\lim_{h \to 0}(hb)$ remains finite. Upon doing this to (9-5), we see that we are able to cancel Δa from both sides of the result, and thus we get the following expression applicable to the situation at the surface of discontinuity:

$$\hat{\mathbf{n}} \cdot (\mathbf{F}_2 - \mathbf{F}_1) = \lim_{h \to 0}(hb) = \lim_{h \to 0}(h\nabla \cdot \mathbf{F}) \qquad (9\text{-}6)$$

Since $\hat{\mathbf{n}} \cdot \mathbf{F} = F_n$ is the normal component of $\mathbf{F}$, that is, that in the direction of the normal, according to (1-21), we can also write (9-6) as

$$F_{2n} - F_{1n} = \lim_{h \to 0}(hb) = \lim_{h \to 0}(h\nabla \cdot \mathbf{F}) \qquad (9\text{-}7)$$

Since this difference may be different from zero, we have the possibility of a discontinuity in the normal components of the vector $\mathbf{F}$.

9-3 THE CURL AND THE TANGENTIAL COMPONENTS

Stokes' theorem (1-67) combined with (9-1) yields

$$\oint_C \mathbf{F} \cdot d\mathbf{s} = \int_S (\nabla \times \mathbf{F}) \cdot d\mathbf{a} = \int_S \mathbf{c}(\mathbf{r}) \cdot d\mathbf{a} \qquad (9\text{-}8)$$

We apply this to a small rectangular path constructed in the transition layer and perpendicular to the surface of discontinuity as shown in Figure 9-5 so that the sides of length Δs will give us contributions from both regions. The arrows indicate the sense of integration around the path C; $\hat{\mathbf{t}}_2$ and $\hat{\mathbf{t}}_1$ are unit vectors in their respective directions

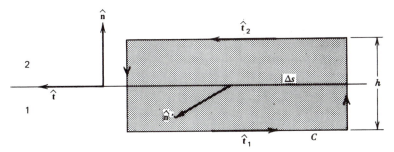

Figure 9-5.   Area used to find boundary condition from Stokes' theorem.

of integration and are parallel to the surface of discontinuity. The vector $\hat{\mathbf{n}}'$ is the normal to the area enclosed by the path and is parallel to the surface between 1 and 2; thus $\hat{\mathbf{n}}'$ is perpendicular to $\hat{\mathbf{n}}$, the normal to the surface. The figure also shows a tangential vector $\hat{\mathbf{t}}$ parallel to the plane of C defined so that $\hat{\mathbf{t}}_2 = \hat{\mathbf{t}}$ and $\hat{\mathbf{t}}_1 = -\hat{\mathbf{t}}$. Therefore, $\hat{\mathbf{n}}$, $\hat{\mathbf{t}}$, and $\hat{\mathbf{n}}'$ are a set of mutually perpendicular unit vectors and satisfy relations analogous to (1-25), that is,

$$\hat{\mathbf{n}}' = \hat{\mathbf{n}} \times \hat{\mathbf{t}} \qquad \hat{\mathbf{t}} = \hat{\mathbf{n}}' \times \hat{\mathbf{n}} \qquad \hat{\mathbf{n}} = \hat{\mathbf{t}} \times \hat{\mathbf{n}}' \tag{9-9}$$

The vector area of the rectangle as given by (1-52) is then seen to be $\hat{\mathbf{n}}' h \, \Delta s$. Applying (9-8) to this situation, we obtain

$$\oint_C \mathbf{F} \cdot d\mathbf{s} = \mathbf{F}_2 \cdot \hat{\mathbf{t}}_2 \, \Delta s + \mathbf{F}_1 \cdot \hat{\mathbf{t}}_1 \, \Delta s + \mathscr{W}$$

$$= \hat{\mathbf{t}} \cdot (\mathbf{F}_2 - \mathbf{F}_1) \, \Delta s + \mathscr{W} = \mathbf{c} \cdot \hat{\mathbf{n}}' h \, \Delta s \tag{9-10}$$

where $\mathbf{F}_2$ and $\mathbf{F}_1$ are the values of $\mathbf{F}$ in the respective regions, and $\mathscr{W}$ is the contribution to the line integral from the ends of the path. Again, strictly speaking, the values of $\mathbf{F}_2, \mathbf{F}_1$, and $\mathbf{c}$ are average values, but because Δs and $h \, \Delta s$ are so small, we can take the vectors to be nearly constant. If we replace $\hat{\mathbf{t}}$ by the middle expression in (9-9) and use (1-29), we can write (9-10) as

$$\hat{\mathbf{n}}' \cdot \left[\hat{\mathbf{n}} \times (\mathbf{F}_2 - \mathbf{F}_1) - h\mathbf{c} \right] \Delta s + \mathscr{W} = 0 \tag{9-11}$$

Now we again let the transition layer shrink to zero so that $h \to 0$ while we keep Δs constant. Similarly to before, $\mathscr{W}$ will be proportional to h and will vanish in this process. When all of this is done, we find that Δs can be canceled from both sides of what (9-11) becomes, and we are left with

$$\hat{\mathbf{n}}' \cdot \left[\hat{\mathbf{n}} \times (\mathbf{F}_2 - \mathbf{F}_1) - \lim_{h \to 0} (h\mathbf{c}) \right] = 0 \tag{9-12}$$

The orientation of our path of integration was completely arbitrary, so that $\hat{\mathbf{n}}'$ corresponds to an arbitrary direction in the surface. The only way in which (9-12) can always be true under these circumstances is for the term in brackets to be zero; thus we get

$$\hat{\mathbf{n}} \times (\mathbf{F}_2 - \mathbf{F}_1) = \lim_{h \to 0} (h\mathbf{c}) = \lim_{h \to 0} \left[h(\nabla \times \mathbf{F}) \right] \tag{9-13}$$

as our final result.

We can put (9-13) into a form that is more easily interpretable. Let us write $\mathbf{F}$ as the sum of its component normal to the surface of separation (its normal component $\mathbf{F}_n$) and its component parallel to the surface (its tangential component $\mathbf{F}_t$), that is,

$$\mathbf{F} = \mathbf{F}_n + \mathbf{F}_t = F_n \hat{\mathbf{n}} + \mathbf{F}_t \tag{9-14}$$

Consequently,

$$\hat{n} \times \mathbf{F} = F_n \hat{n} \times \hat{n} + \hat{n} \times \mathbf{F}_t = \hat{n} \times \mathbf{F}_t \qquad (9\text{-}15)$$

because of (1-24). Using this result, we see that (9-13) actually involves only the tangential components of $\mathbf{F}$:

$$\hat{n} \times (\mathbf{F}_{2t} - \mathbf{F}_{1t}) = \lim_{h \to 0} (h\mathbf{c}) \qquad (9\text{-}16)$$

We can, in fact, write our result even more explicitly in terms of the tangential components. With the use of (1-23), (1-30), (1-17), and (9-14), we find that

$$(\hat{n} \times \mathbf{F}) \times \hat{n} = \mathbf{F} - F_n \hat{n} = \mathbf{F}_t \qquad (9\text{-}17)$$

Thus, if we cross both sides of (9-13) into $\hat{n}$, and use (9-17), we get

$$\mathbf{F}_{2t} - \mathbf{F}_{1t} = \lim_{h \to 0} \left[h(\mathbf{c} \times \hat{n}) \right] = \lim_{h \to 0} \left\{ h \left[(\nabla \times \mathbf{F}) \times \hat{n} \right] \right\} \qquad (9\text{-}18)$$

and if this difference is different from zero, we have a discontinuity in the tangential components of $\mathbf{F}$.

If we combine these results with those of the last section, we see that we have obtained a way of finding how the vector $\mathbf{F}$ *at* the surface of discontinuity changes as we go *across* the bounding surface. Let us suppose that we know $\mathbf{F}_1$. Our first step would be to resolve it into its normal and tangential components F_{1n} and $\mathbf{F}_{1t}$. We can find the normal component of $\mathbf{F}_2$ from (9-7), and its tangential component from (9-18). Knowing these, we can then find the vector $\mathbf{F}_2$ by combining them according to (9-14). Since, in principle, both the normal and tangential components can change as we go across the bounding surface of discontinuity, we see that the vector $\mathbf{F}$ may have both a different magnitude and a different direction on the two sides.

9-4 BOUNDARY CONDITIONS FOR THE ELECTRIC FIELD

So far the only vector field we have considered is the electric field $\mathbf{E}$, but we have already found the information about it that we require, that is,

$$\nabla \times \mathbf{E} = 0 \qquad \text{and} \qquad \nabla \cdot \mathbf{E} = \frac{\rho}{\epsilon_0} \qquad (9\text{-}19)$$

according to (5-4) and (4-10). By comparison with (9-1), we see that in this case

$$\mathbf{c} = 0 \qquad \text{and} \qquad b = \frac{\rho}{\epsilon_0} \qquad (9\text{-}20)$$

Applying the first of these to (9-18), we see immediately that the tangential components of $\mathbf{E}$ are unchanged, that is, they are *continuous* across the surface of discontinuity:

$$\mathbf{E}_{2t} - \mathbf{E}_{1t} = 0 \qquad (9\text{-}21)$$

or, simply

$$E_{2t} = E_{1t} \qquad (9\text{-}22)$$

since the tangential components are parallel to each other. [Later, we will see that (9-21) is correct under all circumstances.]

If we use the value for b given in (9-20), we see that

$$\int_V b \, d\tau = \frac{1}{\epsilon_0} \int_V \rho \, d\tau = \frac{\rho h \, \Delta a}{\epsilon_0} = \frac{\Delta q}{\epsilon_0} \qquad (9\text{-}23)$$

where Δq is the total charge contained within the volume $h \, \Delta a$ of this portion of the transition layer. This charge Δq may occur because of the properties of the two media as we will see in the next chapter, or it may be there because we originally put it there. In any event, as we imagine the transition layer shrunk to zero thickness to produce our idealized case of Figure 9-2, this total charge Δq must be conserved. In this limit, the charge should then be described as a *surface charge* of some density σ so that $\Delta q = \sigma \Delta a$. Therefore, we get

$$\Delta q = \sigma \Delta a = \left(\lim_{h \to 0} h\rho \right) \Delta a$$

so that

$$\sigma = \lim_{h \to 0} (h\rho) \tag{9-24}$$

and therefore

$$\lim_{h \to 0} (hb) = \frac{\sigma}{\epsilon_0} \tag{9-25}$$

in this case. Inserting this into (9-6), we find the boundary condition satisfied by the normal components of the electric field to be

$$\hat{\mathbf{n}} \cdot (\mathbf{E}_2 - \mathbf{E}_1) = E_{2n} - E_{1n} = \frac{\sigma}{\epsilon_0} \tag{9-26}$$

Thus, there is a discontinuity in the normal components of $\mathbf{E}$ *only* if there is a surface charge on the surface separating the two regions. (This will also turn out to be correct under all circumstances.)

In Figure 9-6, we illustrate the application of (9-22) and (9-26) to finding the relation between $\mathbf{E}_1$ and $\mathbf{E}_2$; the figure is drawn on the assumption that $\mathbf{E}_1$ is known and that σ is positive. If α_1 and α_2 are the angles made with the normal, we see from the construction of the figure that $\alpha_2 < \alpha_1$; in other words, the vector $\mathbf{E}$ is "refracted" as we go across the boundary. In this case, its direction is changed so that it is more nearly in the direction of $\hat{\mathbf{n}}$. If σ were negative, $\mathbf{E}$ would be refracted away from the normal and $\alpha_2 > \alpha_1$.

Let us now see how well what we have obtained in this section agrees with some of our previous results for $\mathbf{E}$.

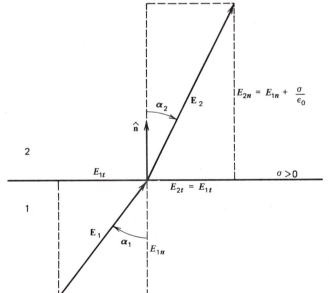

Figure 9-6. Refraction of lines of electric field at a charged surface of discontinuity.

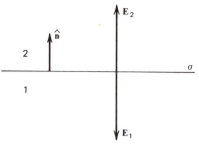

Figure 9-7. Calculation of electric field for a uniform sheet by using the boundary conditions.

- **Example**

Uniform infinite plane sheet. As given by (3-12), the electric field has the constant magnitude of $E = \sigma/2\epsilon_0$ and is directed away from the charged sheet as shown for positive σ in Figure 9-7. In this case, the surface does not separate two regions of different properties since we have assumed a vacuum on both sides, but our results are actually applicable to any surface. We see that $\mathbf{E}_2 = E\hat{\mathbf{n}}$ and $\mathbf{E}_1 = -E\hat{\mathbf{n}}$ because of (9-2) and our choice for labeling the regions. Since there are no tangential components, (9-22) is automatically satisfied. Applying (9-26), we get

$$E_{2n} - E_{1n} = E - (-E) = 2E = \frac{\sigma}{\epsilon_0}$$

so that $E = \sigma/2\epsilon_0$ in exact agreement with (3-12). (We have, in fact, already remarked on this result in the last sentence of Section 3-3.) ∎

- **Example**

Surface of a conductor. Here again there are no tangential components and the situation is that shown in Figure 6-1b. If we let the conductor region be 1, and the vacuum region outside be 2, then $\hat{\mathbf{n}}$ will be the outward normal to the surface of the conductor. In this case, $E_{1n} = 0$ because of (6-1), and $E_{2n} = E$ so that (9-26) becomes $E_{2n} - E_{1n} = E = \sigma/\epsilon_0$, agreeing exactly with (6-4), which we obtained by other means. ∎

Thus, these two examples agree with the general results we have obtained and, incidentally, show us that the use of the boundary conditions in this way provides us with a quick way of calculating the fields for simple enough situations.

9-5 BOUNDARY CONDITIONS FOR THE SCALAR POTENTIAL

Since ϕ is a scalar field rather than a vector one, we cannot use the general results of Sections 9-2 and 9-3 to discuss its behavior at a surface of discontinuity. We can, however, easily obtain what we need with the use of (5-11). If we let ϕ_2 and ϕ_1 be the values of ϕ on each side of the transition layer, then we have

$$\phi_2 - \phi_1 = -\int_1^2 \mathbf{E} \cdot d\mathbf{s} \tag{9-27}$$

where the integration is over any convenient path through the transition layer. If we choose this path to be along $\hat{\mathbf{n}}$, then we can write (9-27) as

$$\phi_2 - \phi_1 = -\int_1^2 E_n \, ds = -\langle E_n \rangle h \tag{9-28}$$

where $\langle E_n \rangle$ is the average value of the normal component of **E** in the transition layer of thickness h. Now in any real physical situation, the electric field will not become infinite as $h \to 0$, since that would imply an infinite force on a point charge. In fact, as we saw in (9-26), the most that could happen is that E_n would have a discontinuity, but would otherwise be finite. Therefore, $\langle E_n \rangle$ will always be a finite quantity so that $\lim_{h \to 0} \langle E_n \rangle h = 0$ and (9-28) will lead to

$$\phi_2 = \phi_1 \qquad (9\text{-}29)$$

Thus, the scalar potential must be continuous across a surface of discontinuity in properties.

Although ϕ is continuous, its derivatives normal to the surface need not be. Because of (1-21) and (5-3), we have

$$E_n = -\hat{\mathbf{n}} \cdot \nabla \phi \qquad (9\text{-}30)$$

so that we can write (9-26) in terms of the potential as

$$(\hat{\mathbf{n}} \cdot \nabla \phi)_2 - (\hat{\mathbf{n}} \cdot \nabla \phi)_1 = -\frac{\sigma}{\epsilon_0} \qquad (9\text{-}31)$$

[While (9-29) is certainly correct for any real physical situations, one occasionally deals with *mathematical idealizations* that do result in a discontinuity of ϕ across a surface. The most common example of this is a "dipole layer," which arises from assuming the surface of separation to contain electric dipoles of the type we discuss in the next chapter, but with an infinitesimal separation between the charges of opposite sign. However, we will not find it necessary to consider such specialized cases.]

EXERCISES

9-1 The surface of separation between regions 1 and 2 is a plane whose equation is $2x + y + z = 1$. If $\mathbf{E}_1 = 4\hat{\mathbf{x}} + \hat{\mathbf{y}} - 3\hat{\mathbf{z}}$, find the normal and tangential components of $\mathbf{E}_1$.

9-2 The actual physical relation between $\mathbf{E}_1$ and $\mathbf{E}_2$ cannot depend on our arbitrary choice of the direction for $\hat{\mathbf{n}}$. Show that if the direction of the unit normal vector were defined to be *from* region 2 *to* region 1, the relation between $\mathbf{E}_2$ and $\mathbf{E}_1$ shown in Figure 9-6 would be unaltered.

9-3 A sphere of radius a has its center at the origin. Within the sphere, the electric field is given by $\mathbf{E}_1 = \alpha\hat{\mathbf{x}} + \beta\hat{\mathbf{y}} + \gamma\hat{\mathbf{z}}$ where α, β, γ are constants. There is a surface charge density on the surface of the sphere given in spherical coordinates by $\sigma = \sigma_0 \cos\theta$ where $\sigma_0 = $ const. Find $\mathbf{E}_2$ at all points just outside the sphere and express it in rectangular coordinates.

9-4 Show that the angles in Figure 9-6 are related by

$$\tan\alpha_2 = \frac{\tan\alpha_1}{1 + \left(\dfrac{\sigma}{\epsilon_0 E_1 \cos\alpha_1} \right)}$$

How are the angles related if E_1 becomes very large? Is this reasonable?

9-5 In Figure 9-8, we show two points P_2 and P_1 located in the two media but just across the surface from each other. The point P_3 is located at the edge of the interface, wherever it may be. Use (9-27) to find the potential difference between P_2 and P_3 by following a path of integration right along the bounding surface but just inside region 2 as indicated by the curved arrow. (Why is this all right?) Similarly, find the potential difference between P_1 and P_3 by staying inside region 1 but next to the surface. From your results, show once again that the potentials at P_2 and P_1 are equal, in agreement with (9-29).

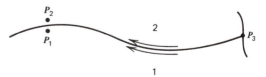

Figure 9-8. Paths of integration for Exercise 9-5.

10

ELECTROSTATICS IN THE PRESENCE OF MATTER

As we have remarked before, for our purposes, matter is to be regarded as a collection of the positive and negative charges of its constituent nuclei and electrons. We assume that an ordinary piece of matter is made up of atoms and molecules that have equal amounts of positive and negative charges, that is, that have zero monopole moments. In Chapter 6 we considered one class of matter—conductors—which we assumed to have charges that were free to move about under the influence of fields. In this chapter we assume that we are dealing with matter that is not a conductor. These latter types of material are commonly called *dielectrics* and we refer to their charges as *bound charges*. It is not always possible to make a clear-cut characterization of a given substance as a conductor or dielectric (nonconductor), but it is still useful to have these two categories available for a first rough classification scheme. Later we will be able to combine these two properties in our overall description.

10-1 POLARIZATION

Although our primary aim is to develop a macroscopic empirical description of electromagnetism, for a while we will use the microscopic picture of matter in a qualitative manner in order to get an idea of how we should proceed. There are several possibilities that come to mind when we consider the effect of an electric field on an atom or molecule.

In the absence of a field, it may well happen that the molecule has its negative electronic charge distributed symmetrically about the positively charged nuclei. In this case, the molecule will have a zero dipole moment as can be easily seen from (8-22). If, now, there is an electric field present, there will be forces exerted on these charges. The positive charges will tend to be moved in the direction of the field, and the negative charges will tend to be moved in the direction opposite to the field. Eventually, the internal forces within the molecule will produce a new state of equilibrium, but the molecular charge distribution will have been distorted from its originally spherically symmetric form. Thus, the previously "coincident" positive and negative charges will have had their overall "centers of gravity" shifted with respect to each other, and the net result will be a new charge distribution with roughly the character of that shown in Figure 8-3. But, as we know, this means that the molecule now will have a nonzero dipole moment **p**, and probably other higher-order multipoles as well. We say that the molecule now has an *induced dipole moment* and that it has become *polarized*.

Another possibility is that, because of their internal structure, some molecules already have their positive and negative charge distributions separated so that they have a dipole moment even in the absence of an electric field. Such molecules are called *polar molecules* and the dipole moment is referred to as a *permanent dipole moment*; the monopole moment is still zero, of course. An example of this type is the water molecule, H_2O. Here the negative charges tend to cluster more around the oxygen leaving the hydrogens positively charged; the net result is a charge distribution roughly like that of Figure 8-4 (with the signs of the charges reversed) and with q approximately equal to e, the magnitude of the electronic charge. In the absence of an electric field, these permanent dipole moments will generally be randomly oriented so that the *total* dipole moment of the whole piece of matter will still be zero. In the presence of a field, however, there will be a torque on the dipole that will tend to rotate the dipole into the direction of the field as we found in (8-75). This tendency can be expected to be opposed by other internal forces of the system and, especially in the cases of gases and liquids, by the temperature agitation of the molecules that will lead to collisions with their tendency to make things random. Nevertheless, when equilibrium has been finally attained in the presence of the field, we can expect that the permanent dipoles will have been rotated on the average into some degree of alignment with the direction of the field. The overall effect is that the matter will have a net dipole moment in the direction of the field. Thus we have concluded again that the material will be polarized.

A final possibility is that the permanent dipole moments of some materials are aligned to some extent even in the absence of an electric field. Such materials are called *electrets* and are said to be permanently polarized. They are comparatively rare, and not of as great commercial and technical importance as their magnetic analogues—permanent magnets.

As we noted above, we can reasonably expect the molecules, especially when distorted by the field, to have higher-order multipoles, such as quadrupole moments, associated with them in addition to their dipole moments. On the other hand, we know from Chapter 8 that the contributions of these other multipoles to the potential and field fall off more rapidly with distance from the molecule, and, in addition, their contributions generally vary with angle in a more complex manner. Since we are trying to get a macroscopic description in terms of the *average* of the microscopic behavior of the constituents of matter, we will assume that, on the average, the dominant features of matter that are of interest to us are simply those associated with the electric dipole moments. Thus, all of our considerations lead us to the following:

■ **Hypothesis**

As far as its electrical properties are concerned, neutral matter is equivalent to an assemblage of electric dipoles.   ■

Although this is obviously an hypothesis, it clearly must have worked very well in the development of this subject. Our next problem is to try to formulate this hypothesis in a quantitative manner.

For this purpose, we define the *polarization* $\mathbf{P}$ as the electric dipole moment per unit volume, so that the total dipole moment $d\mathbf{p}$ in a small volume $d\tau$ at $\mathbf{r}$ will be

$$d\mathbf{p} = \mathbf{P}(\mathbf{r})\, d\tau \qquad (10\text{-}1)$$

Thus the total dipole moment of a volume V of material will be

$$\mathbf{p}_{\text{total}} = \int_V \mathbf{P}(\mathbf{r})\, d\tau \qquad (10\text{-}2)$$

From its definition and (8-19), it follows that $\mathbf{P}$ will be measured in coulombs/(meter)^2. As implied by (10-1) and (10-2), $\mathbf{P}$ is generally expected to be a function of position within the material. As in Section 2-4 when we introduced charge density, we are assuming $d\tau$ to be small on a macroscopic scale but large on a microscopic atomic scale so that $d\tau$ actually contains many molecules. This makes $\mathbf{P}$ obtainable as an average over volumes large enough so that small fluctuations in the number of molecules within a given volume element do not prevent us from treating $\mathbf{P}$ as a smoothly varying function of position. For example, if there are n molecules per unit volume each of which has a dipole moment $\mathbf{p}$, then $\mathbf{P} = n\mathbf{p}$ provided that all of the dipoles are in the same direction.

Because of the way in which we were led to it, we expect there will be a functional relation between $\mathbf{P}$ and $\mathbf{E}$ and eventually we shall have to consider how we can determine $\mathbf{P}$. For the time being, however, we simply take it as giving us a macroscopic description of the material and we want to investigate the consequences of its existence.

10-2 BOUND CHARGE DENSITIES

Let us assume that we have a polarized object and we want to calculate the potential it will produce at a field point $\mathbf{r}$ located outside the body as shown in Figure 10-1. The dipole moment of the volume element $d\tau'$ as given by (10-1) is $d\mathbf{p}' = \mathbf{P}(\mathbf{r}')\,d\tau'$ and its contribution to the potential at $\mathbf{r}$ as obtained from (8-21) is

$$d\phi = \frac{d\mathbf{p}' \cdot \hat{\mathbf{R}}}{4\pi\epsilon_0 R^2} = \frac{\mathbf{P}(\mathbf{r}') \cdot \hat{\mathbf{R}}\,d\tau'}{4\pi\epsilon_0 R^2} \tag{10-3}$$

where, as usual, $\mathbf{R} = \mathbf{r} - \mathbf{r}'$ and corresponds to the $\mathbf{r}$ of (8-21). In order to find the total potential, we integrate this over the volume V' of the material and we get

$$\phi(\mathbf{r}) = \int_{V'} \frac{\mathbf{P}(\mathbf{r}') \cdot \hat{\mathbf{R}}\,d\tau'}{4\pi\epsilon_0 R^2} = \frac{1}{4\pi\epsilon_0} \int_{V'} \mathbf{P}(\mathbf{r}') \cdot \nabla'\left(\frac{1}{R}\right) d\tau' \tag{10-4}$$

with the use of (1-141). This can be put into a better form if we use (1-115) to write the integrand as

$$\mathbf{P} \cdot \nabla'\left(\frac{1}{R}\right) = -\frac{\nabla' \cdot \mathbf{P}}{R} + \nabla' \cdot \left(\frac{\mathbf{P}}{R}\right) \tag{10-5}$$

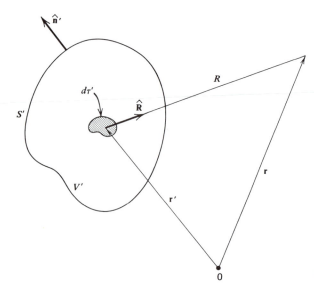

Figure 10-1. Calculation of the potential outside a polarized body.

If we insert this into (10-4) and use (1-59) and (1-52), we obtain

$$\phi(\mathbf{r}) = \frac{1}{4\pi\epsilon_0} \int_{V'} \frac{(-\nabla' \cdot \mathbf{P})\, d\tau'}{R} + \frac{1}{4\pi\epsilon_0} \int_{V'} \nabla' \cdot \left(\frac{\mathbf{P}}{R}\right) d\tau'$$

$$= \frac{1}{4\pi\epsilon_0} \int_{V'} \frac{(-\nabla' \cdot \mathbf{P})\, d\tau'}{R} + \frac{1}{4\pi\epsilon_0} \oint_{S'} \frac{\mathbf{P} \cdot \hat{\mathbf{n}}'\, da'}{R} \qquad (10\text{-}6)$$

where S' is the surface bounding V' and $\hat{\mathbf{n}}'$ is the *outer* normal to the surface as shown in the figure. Upon comparing this with (5-7) and (5-8), we see that (10-6) is *exactly* the potential ϕ that would be produced by a volume charge density ρ_b distributed throughout the volume and a surface charge density σ_b on the bounding surface where

$$\rho_b = -\nabla' \cdot \mathbf{P} \qquad (10\text{-}7)$$

$$\sigma_b = \mathbf{P} \cdot \hat{\mathbf{n}}' = P_n \qquad (10\text{-}8)$$

for then we would have

$$\phi(\mathbf{r}) = \frac{1}{4\pi\epsilon_0} \int_{V'} \frac{\rho_b\, d\tau'}{R} + \frac{1}{4\pi\epsilon_0} \oint_{S'} \frac{\sigma_b\, da'}{R} \qquad (10\text{-}9)$$

as we would expect. [In (10-8), P_n is the normal component of $\mathbf{P}$ evaluated *on* the surface.]

What we have discovered, therefore, is that, as far as its effects outside itself are concerned, the dielectric can be *replaced* by a distribution of volume and surface charge densities that are related to the polarization $\mathbf{P}$ by means of (10-7) and (10-8). The various steps that have led to our present conceptual scheme are summarized and outlined in Figure 10-2. The *total* potential at the field point will then be given by (10-9) *plus* the potential due to any other charges which might be present.

It is common practice to omit the prime in (10-7) and simply write

$$\rho_b = -\nabla \cdot \mathbf{P} \qquad (10\text{-}10)$$

with the understanding that the differentiations are made with respect to the source point coordinates.

The subscript b appearing in (10-7), (10-8), and (10-10) reflects the fact that these charge densities arise from the *bound* charges of the dielectric. Consequently, they are usually referred to as the *bound charge densities* or the *polarization charge densities*.

Although we identified these quantities by means of a formal comparison of our expression for the potential with its general form, it is possible to understand and calculate the origin of these densities in a more direct "physical" manner; we will, however, do this only qualitatively. As an extreme example, let us consider a piece of material with uniform polarization produced by dipoles of the same magnitude all pointing the same direction as shown in Figure 10-3. As in Figure 8-3, we can associate a positive charge with an arrow head, and a negative charge of the same magnitude with the tail of an arrow. Then, in a small volume within the material, as indicated by the dashed curve, we will have, on the average, as many positive as negative charges so

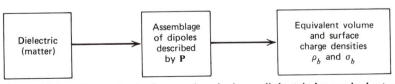

Figure 10-2.   Conceptual scheme of replacing a dielectric by equivalent charge densities.

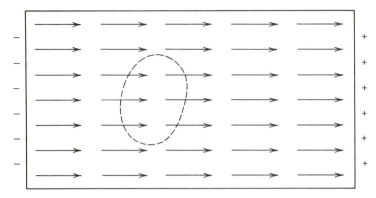

Figure 10-3.   Origin of bound surface charges for a uniformly polarized dielectric.

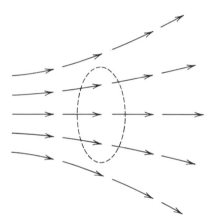

Figure 10-4.   Origin of bound volume charges in a dielectric with nonuniform polarization.

that $\rho_b = 0$ in agreement with (10-10). However, there will be no such cancellation on the faces and this will produce a surface charge density of the signs shown; these signs are in agreement with (10-8) since P_n is positive for the right side of the figure and negative for the left side. Now let us suppose that **P** is not uniform, so that the distribution of dipoles may be something like that of Figure 10-4 since the dipoles may differ in both magnitude and direction. Let us consider a fixed volume of the material as indicated by the dashed line. Before the matter was polarized, this volume will contain many neutral molecules so that $\rho_b = 0$. Now when the material is polarized and the dipoles are formed, some charges will be displaced out of the volume while others will be displaced into it. If **P** is not uniform, we see that it may well happen that the volume may end up with more charge of one sign than of the other and there will be a net bound charge density. This is exactly the effect described by (10-10), and, in fact, considerations like these can be used as an alternative way of deriving $\rho_b = -\nabla \cdot \mathbf{P}$.

Suppose we have two polarized dielectrics meeting at a common boundary as illustrated in Figure 10-5. Since each of them has a normal component at the surface, they will produce a surface charge there given by (10-8); the net surface charge density $\sigma_{b,\,\mathrm{net}}$ will be the sum of these two terms so that

$$\sigma_{b,\,\mathrm{net}} = \sigma_{b2} + \sigma_{b1} = P_{n2} + P_{n1} = \mathbf{P}_2 \cdot \hat{\mathbf{n}}'_2 + \mathbf{P}_1 \cdot \hat{\mathbf{n}}'_1 \qquad (10\text{-}11)$$

where $\hat{\mathbf{n}}'_2$ and $\hat{\mathbf{n}}'_1$ are the *outward* normals to the respective media. If we now introduce the unit vector $\hat{\mathbf{n}}$ pointing from 1 to 2 as defined in (9-2), we see that $\hat{\mathbf{n}}'_2 = -\hat{\mathbf{n}}$ and

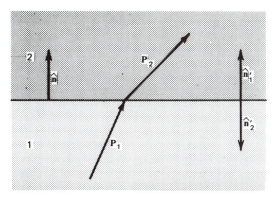

Figure 10-5.   A boundary between two regions of different polarizations.

$\hat{\mathbf{n}}'_1 = \hat{\mathbf{n}}$ so that (10-11) becomes

$$\hat{\mathbf{n}} \cdot (\mathbf{P}_2 - \mathbf{P}_1) = -\sigma_{b,\,\text{net}} \qquad (10\text{-}12)$$

This result is exactly what we get by applying our previous results (9-6) and (9-24) to this case, which is described by (10-10); thus (10-7) and (10-8) are consistent with each other.

As a further illustration of the internal consistency of our results, let us find the total bound charge Q_b of a polarized dielectric of finite extent. According to (2-14) and (2-16), this will be given by

$$Q_b = \int_{V'} \rho_b \, d\tau' + \oint_{S'} \sigma_b \, da' = -\int_{V'} \nabla' \cdot \mathbf{P} \, d\tau' + \oint_{S'} \sigma_b \, da'$$

$$= -\oint_{S'} \mathbf{P} \cdot \hat{\mathbf{n}}' \, da' + \oint_{S'} \mathbf{P} \cdot \hat{\mathbf{n}}' \, da' = 0 \qquad (10\text{-}13)$$

with the use of (10-7), (10-8), and (1-59). Thus the total bound charge is zero. But this is exactly what we would expect from our ideas of polarization as arising from the *separation* of charge in our originally neutral medium rather than from the creation of new charge. (In the case of permanent dipoles, the origin of polarization is due to reorientation of already separated charge, but this does not involve charge creation either.)

10-3 THE ELECTRIC FIELD WITHIN A DIELECTRIC

Up to this point all of our results have been obtained by considering the potential, and its corresponding electric field, at a field point in the vacuum outside the body of the dielectric. The reason is that then there is no problem about what we mean by the electric field in the sense of (3-14) since we can put a test charge there and measure the force on it without any difficulty. However, if we look at the situation inside a dielectric from this point of view, we cannot measure the force on a test charge without, say, drilling a hole in the material in order to be able to insert the charge. But if we do this, we can anticipate that we will alter the preexisting situation if for no other reason than that we will remove some bound volume charges and introduce new surface charges because we have created some new bounding surface. There are several ways in which one can proceed to answer the question of what we mean by the electric field inside the dielectric.

The simplest thing to do, and probably the best, is just to say that we calculate the potential, and its associated electric field, inside the dielectric in exactly the same way in which we do it for a point outside. In other words, we say that (10-6) or (10-9) can be used to calculate ϕ anywhere *by definition*. There is certainly nothing wrong with this and it is a perfectly reasonable thing to do; it is completely consistent with our conceptual replacement of the actual dielectric by a set of equivalent volume and surface distributions of bound charge as shown in Figure 10-2. We have discussed this approach before in Section 3-4, and we can point out again that our basic aim is to develop a macroscopic description of electromagnetism that *agrees with experiment*. We come back to this point below.

Another approach that can be used is to start with a *microscopic* picture and try to determine the macroscopic equations as appropriate averages. This is quite a complicated program to carry out accurately for all of electromagnetism, and we shall have to content ourselves with a brief outline of what is involved for electrostatics. On this scale, most of the interior region will be vacuum and the electric field will be determined by all of the nuclear charges and the atomic electrons. This electric field will have enormous variations in it as we go from points very near these charges to points which are comparatively far from them. At the same time, it will vary quite rapidly with time because of the motion of the constituent charges. Thus, when we define our macroscopic electric field, we will want to do so as an average both over the time and over a volume large enough to contain a reasonable number of molecules but not so large that we cannot consider it as infinitesimal on a laboratory scale. (This is similar to the problem of defining average charge density as we discussed in Section 2-4; there we saw that if we take too fine a subdivision, ρ can have very large fluctuations.) If we let $\mathbf{e}$ be the microscopic electric field, we want it to have the same basic vacuum properties that we have already found in (4-10) and (5-4), that is,

$$\nabla \cdot \mathbf{e} = \frac{\rho_m}{\epsilon_0} \quad \text{and} \quad \nabla \times \mathbf{e} = 0 \tag{10-14}$$

where ρ_m is the charge density defined on a microscopic scale. Our macroscopic field $\mathbf{E}$ would then be defined as

$$\mathbf{E} = \langle \mathbf{e} \rangle \tag{10-15}$$

where $\langle \mathbf{e} \rangle$ is an average over both time and space. Since the differential operators are constant as far as this averaging process is concerned, we get from (10-14) and (10-15):

$$\nabla \times \langle \mathbf{e} \rangle = \nabla \times \mathbf{E} = 0 \tag{10-16}$$

$$\nabla \cdot \langle \mathbf{e} \rangle = \nabla \cdot \mathbf{E} = \frac{1}{\epsilon_0} \langle \rho_m \rangle \tag{10-17}$$

Now ρ_m itself can be expected to have large spatial and temporal variations, but when it is averaged we expect it to reduce to the average density of the displaced bound charges; that is, $\langle \rho_m \rangle = \rho_b$, so that

$$\nabla \cdot \mathbf{E} = \frac{\rho_b}{\epsilon_0} \tag{10-18}$$

But (10-16) and (10-18) are equivalent to simply using (10-9) and (5-4) in the first place. Thus this approach leads us to exactly the same conclusions we reached in the previous paragraph.

This microscopic to macroscopic method of determining $\mathbf{E}$ in the material is somewhat different from the question of what is the actual electric field seen by the molecule, which is what will determine the actual molecular dipole moment produced. This field will not necessarily be the same as the macroscopic $\mathbf{E}$ defined above as the molecule itself contributes to the resultant field. This question is discussed in Appendix

B; there we consider the question of calculating the electromagnetic properties of matter as the resultant response of its constituent atoms and molecules, but it is not at all necessary to worry about that now.

It is helpful at this point to verify that our definitions of ϕ and $\mathbf{E}$, which we have just adopted, are in agreement with early experiments on dielectrics done by Faraday. We do this only qualitatively now, but we will be able to give a quantitative discussion in Section 10-7. We consider two simple experiments. We charge a capacitor to some charge Q, and measure the potential difference $\Delta\phi_0$ when there is a vacuum between the plates. The capacitance will be

$$C_0 = \frac{Q}{\Delta\phi_0} \tag{10-19}$$

according to (6-28). Now we take this same capacitor, completely fill the region between the plates with some "reasonable" dielectric like wax or oil, while *keeping Q constant*, and then measure the potential difference $\Delta\phi$. The capacitance is now given by

$$C = \frac{Q}{\Delta\phi} \tag{10-20}$$

The *experimental result* is that $\Delta\phi < \Delta\phi_0$, and therefore

$$C > C_0 \tag{10-21}$$

How can this result that the potential difference decreased be understood? Suppose, in order to be specific, we consider a parallel plate capacitor. Then we can assume the electric field to be directed from the positively charged plate to the negatively charged plate as in Figure 6-9, and when we calculate $\Delta\phi_0$ by means of (6-38), we get

$$\Delta\phi_0 = \int_+^- \mathbf{E}_0 \cdot d\mathbf{s} = E_0 d \tag{10-22}$$

where $\mathbf{E}_0$ is the electric field with a vacuum present and d is the plate separation. Similarly, if $\mathbf{E}$ is the electric field with the dielectric between the plates, we get

$$\Delta\phi = Ed \tag{10-23}$$

and, since $\Delta\phi < \Delta\phi_0$, while d is constant, we conclude that

$$E < E_0 \tag{10-24}$$

so that the electric field has been decreased by the presence of the dielectric. How could this have come about? In Figure 10-6, we show the capacitor with dielectric between its plates; the surfaces of the dielectric are shown slightly separated from the conductors for clarity, but we really assume that the region is completely filled. If the dielectric is uniformly polarized so that $\mathbf{P} = \text{const.}$, then $\rho_b = 0$ by (10-10). Thus the only bound charges will be surface charges given by (10-8), and they will have the indicated signs. These surface charges will produce a field E_b with the direction shown; since the charge

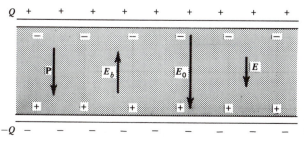

Figure 10-6.   Charges and fields for a parallel plate capacitor.

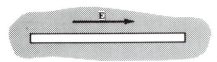

Figure 10-7. Cavity used to measure E in a dielectric.

Q will still produce the field E_0, the resultant field E is seen to be

$$E = E_0 - E_b \qquad (10\text{-}25)$$

so that $E < E_0$ in agreement with (10-24) and experiment. Thus, this method of evaluating the field inside the material by the use of equivalent bound charge agrees, in this case at least, with experiment and our general ideas about the relation between electric field and potential as expressed by (6-38).

Although we briefly alluded to the difficulty of actually measuring the electric field in the dielectric by means of a test charge, one can still ask if there is a way of doing it—obviously we will have to cut a hole in order to insert the charge. It is possible to devise such a scheme by means of a judicious use of the boundary conditions satisfied by $\mathbf{E}$; the relevant one is (9-21), which says that $\mathbf{E}_t$ is continuous. The reasoning is as follows: (1) somehow determine the direction of $\mathbf{E}$ in the dielectric (as we will see in Section 10-6, this is easily done for a large class of dielectrics); (2) make a long narrow hole in the material parallel to $\mathbf{E}$ as shown in Figure 10-7—this cavity now has a vacuum inside (such a cavity is usually called "needlelike"); (3) any bound surface charges that might alter the fields are on the *ends* of the cavity since that is the only place $\mathbf{P}$ has a normal component, and, since the ends are so far away and of such a small area, we can neglect their effects; (4) since only tangential components are involved, $\mathbf{E}_c = \mathbf{E}$ where $\mathbf{E}_c$ is the field in the cavity; (5) now insert a test charge δq into the cavity and measure the force $\delta \mathbf{F}_c$ on it. Then, by (3-14),

$$\mathbf{E}_c = \frac{\delta \mathbf{F}_c}{\delta q} = \mathbf{E} \qquad (10\text{-}26)$$

and the electric field *in the dielectric* can thus be found, in principle, from measurements made *in the cavity*. This scheme is known as the *cavity definition* of $\mathbf{E}$.

10-4 UNIFORMLY POLARIZED SPHERE

As an example of the use of bound charges, let us consider a sphere of radius a that has a constant polarization $\mathbf{P}$. We choose the z axis to be in the direction of $\mathbf{P}$ and the origin at the center of the sphere as shown in Figure 10-8; thus, $\mathbf{P} = P\hat{\mathbf{z}}$. Since $\mathbf{P}$ is constant, $\rho_b = 0$ by (10-10). We see from the figure that the outer normal to the dielectric $\hat{\mathbf{n}}' = \hat{\mathbf{r}}$, and therefore there is a bound surface charge density which is found from (10-8) and (1-94) to be

$$\sigma_b(\theta') = P \cos \theta' \qquad (10\text{-}27)$$

As we see, σ_b is not constant but varies in magnitude and sign with angle as indicated in Figure 10-9. For simplicity, we will find the potential and electric field only for points on the z axis and for positive z; in the next chapter, we solve this problem completely by a quite different method.

We see from Figure 10-10 that

$$R = (z^2 + a^2 - 2za \cos \theta')^{1/2} \qquad (10\text{-}28)$$

and, therefore, if we use (10-27), (1-100) (with $r = a$), and (2-22), we find that ϕ as

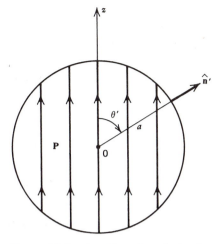

Figure 10-8.   A uniformly polarized sphere.

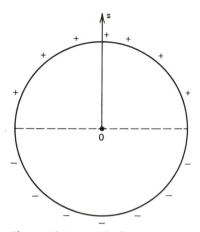

Figure 10-9.   Equivalent charge distribution of a uniformly polarized sphere.

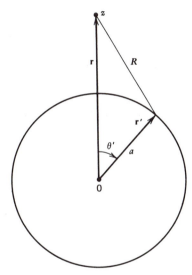

Figure 10-10.   Calculation of the potential at a point on the axis.

given by (10-9) is

$$
\phi(z) = \frac{1}{4\pi\epsilon_0} \int_0^{2\pi} \int_0^{\pi} \frac{P\cos\theta' \cdot a^2 \sin\theta'\, d\theta'\, d\varphi'}{(z^2 + a^2 - 2za\cos\theta')^{1/2}}
$$

$$
= \frac{Pa^2}{2\epsilon_0} \int_{-1}^{1} \frac{\mu\, d\mu}{(z^2 + a^2 - 2za\mu)^{1/2}} \tag{10-29}
$$

The integral can be found from tables to be

$$
-\left.\frac{(z^2 + a^2 + za\mu)(z^2 + a^2 - 2za\mu)^{1/2}}{3z^2a^2}\right|_{-1}^{1}
$$

$$
= \frac{1}{3z^2a^2}\left[(z^2 + a^2)(|z+a| - |z-a|) - za(|z+a| + |z-a|)\right] \tag{10-30}
$$

and there are two cases to be considered.

1. Outside the sphere. Here $z > a$ and $|z - a| = z - a$; also $|z + a| = z + a$ since both z and a are positive. Using these values, we find that (10-30) becomes $2a/3z^2$ and when this is put into (10-29) we find the potential outside the sphere to be

$$\phi_o(z) = \frac{Pa^3}{3\epsilon_0 z^2} \tag{10-31}$$

and therefore

$$E_{zo}(z) = -\frac{\partial \phi_o}{\partial z} = \frac{2Pa^3}{3\epsilon_0 z^3} \tag{10-32}$$

These results become more understandable if we express them in terms of the total dipole moment $\mathbf{p}$ of the sphere which, as found from (10-2), is

$$\mathbf{p} = \tfrac{4}{3}\pi a^3 P \hat{\mathbf{z}} \tag{10-33}$$

so that (10-31) and (10-32) can also be written as

$$\phi_o(z) = \frac{p}{4\pi\epsilon_0 z^2} \tag{10-34}$$

$$E_{zo}(z) = \frac{2p}{4\pi\epsilon_0 z^3} \tag{10-35}$$

Upon comparing these results with (8-48) and (8-50), and remembering that $\theta = 0$ and $r = z$ for a field point on the z axis, we see that they are exactly those of a point dipole of total moment p. This makes us suspect that the field everywhere *outside* is a dipole field corresponding to this total dipole moment; this turns out to be correct as we will see in the next chapter.

2. Inside the sphere. Here $z < a$, so that $|z - a| = a - z$; again, $|z + a| = z + a$. In this case, (10-30) becomes $2z/3a^2$, which, when put into (10-29), gives the potential inside the sphere as

$$\phi_i(z) = \frac{Pz}{3\epsilon_0} \tag{10-36}$$

and therefore

$$E_{zi}(z) = -\frac{\partial \phi_i}{\partial z} = -\frac{P}{3\epsilon_0} \tag{10-37}$$

We note that the electric field is constant; this may make us suspect that $\mathbf{E}$ is constant throughout the whole interior of the sphere, and this turns out to be the case.

It will be left as an exercise to show that these same results also hold for negative values of z, that is, E_{zo} is always in the positive direction, and E_{zi} is constant and given by (10-37) for all z. In Figure 10-11, we illustrate the directions for $\mathbf{E}$ that we have found; these directions can be understood from the source charge distribution shown in Figure 10-9.

It is a worthwhile exercise to verify that these results are in agreement with the general boundary conditions we previously obtained. We find from (10-31) and (10-36) that the potential is indeed continuous at the surface of the sphere as we found in (9-29), that is, when $z = a$, $\phi_o = \phi_i = Pa/3\epsilon_0$. Along the z axis, E_z is a normal component, and therefore we find that

$$E_{2n} - E_{1n} = E_{zo}(a) - E_{zi}(a) = \left(\frac{2P}{3\epsilon_0}\right) - \left(-\frac{P}{3\epsilon_0}\right) = \frac{P}{\epsilon_0} = \frac{\sigma_b(0)}{\epsilon_0}$$

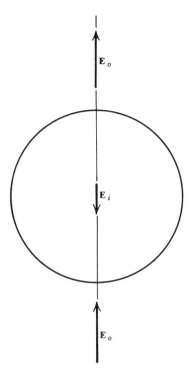

Figure 10-11. Electric fields on *z* axis due to a uniformly polarized sphere.

with the use of (10-32), (10-37), and (10-27). This value for the difference in the normal components is exactly what it should be according to (9-26).

10-5 THE D FIELD

We recall that when we defined **E** in Section 3-1, we pointed out that **E** is determined, from its definition, by all charges of whatever origin or type. In Section 10-2, we encountered a particular class of charge—the bound charge whose density is given by (10-10) as $\rho_b = -\nabla \cdot \mathbf{P}$. It is conventional and convenient to divide charge into the two broad classes of *bound charge* and *free charge* with corresponding densities ρ_b and ρ_f. As we have seen, we regard the bound charges as arising from the constituents of matter, and we generally have no control over their distribution. Free charges are essentially the rest of the charges. The name "free" is given to them because we can control their distribution to a large extent by physically moving them about, by spraying them onto or into material by electron beams, for example, and the like. This class is also generally taken to include the mobile (free) charges of conductors. This division does not always give us a clear-cut classification, but it is useful nevertheless. Thus, we can write the total charge density as the sum of these two:

$$\rho_{\text{total}} = \rho = \rho_f + \rho_b = \rho_f - \nabla \cdot \mathbf{P} \tag{10-38}$$

and when this is substituted into $\nabla \cdot \mathbf{E} = \rho/\epsilon_0$, as we found in (4-10), we obtain

$$\nabla \cdot (\epsilon_0 \mathbf{E} + \mathbf{P}) = \rho_f \tag{10-39}$$

The form of this equation, in which only the free charge density appears on the right-hand side, suggests that it may be useful to define a vector field $\mathbf{D(r)}$ as follows:

$$\mathbf{D} = \epsilon_0 \mathbf{E} + \mathbf{P} \tag{10-40}$$

for then

$$\nabla \cdot \mathbf{D} = \rho_f \qquad (10\text{-}41)$$

The vector $\mathbf{D}$ is often called the *electric displacement* or simply the *displacement*, or, even more simply, the $\mathbf{D}$ *field*. The principal characteristic of $\mathbf{D}$, and the primary reason for its definition, is its property that its divergence depends only on the free charge density. The dimensions of $\mathbf{D}$ are the same as those of $\mathbf{P}$, and thus $\mathbf{D}$ will be measured in coulombs/(meter)^2. We can think of (10-41) as an expression of Coulomb's law for the force between point charges *plus* the electrical effects of matter.

Now that we have (10-41), we can quickly find some of the properties of $\mathbf{D}$. The boundary condition satisfied by its normal component can be found from (10-41), (9-6), (9-7), and (9-24) to be

$$\hat{\mathbf{n}} \cdot (\mathbf{D}_2 - \mathbf{D}_1) = D_{2n} - D_{1n} = \sigma_f \qquad (10\text{-}42)$$

where σ_f is the surface density of free charge. Thus, the normal component of $\mathbf{D}$ will be discontinuous only if there is a *free* surface charge density; this is in contrast to $\mathbf{E}$ whose normal component is discontinuous if there is a surface density of *any* kind of charge.

Gauss' law for $\mathbf{D}$ is easily found from (10-41) and (1-59) to be

$$\oint_S \mathbf{D} \cdot d\mathbf{a} = \int_V \rho_f \, d\tau = Q_{f,\text{in}} \qquad (10\text{-}43)$$

where $Q_{f,\text{in}}$ is the net *free* charge contained within the volume V surrounded by the closed surface S. We note both the analogy to and the contrast with the corresponding result for $\mathbf{E}$ given by (4-1). Equation 10-43 can often be advantageously used for the calculation of $\mathbf{D}$ for problems of sufficient symmetry, in much the same manner as we did for $\mathbf{E}$ in Chapter 4.

Even though (10-41) contains only the free charge density, this does not mean that free charges are the only source for $\mathbf{D}$, since, according to the Helmholtz theorem of Section 1-20, we have found only one of the source equations. The remaining one is $\nabla \times \mathbf{D}$. This is easily found by taking the curl of the definition (10-40) and using (5-4); the result is

$$\nabla \times \mathbf{D} = \nabla \times \mathbf{P} \qquad (10\text{-}44)$$

so that $\mathbf{D}$ can have sources in bound charges as well as in free ones. The boundary condition satisfied by the tangential component of $\mathbf{D}$ can be found most easily from (9-21) and (10-40) to be

$$\mathbf{D}_{2t} - \mathbf{D}_{1t} = \mathbf{P}_{2t} - \mathbf{P}_{1t} \qquad (10\text{-}45)$$

It is also possible to devise a cavity definition for $\mathbf{D}$ in a manner analogous to that for $\mathbf{E}$ as illustrated in Figure 10-7. The definition is based on the continuity of the normal components of $\mathbf{D}$ in the absence of free surface charge, as stated in (10-42). For this purpose, we imagine a cavity in the form of a small right cylinder whose height is very small compared to the radius of its base. It is cut out of the dielectric so that its base is perpendicular to $\mathbf{D}$ in the dielectric as shown in Figure 10-12. If we consider a point near the center of the cavity, the edges will be too far away to affect the fields so that $\mathbf{D}_c$ (in the cavity) will be parallel to $\mathbf{D}$ (in the dielectric), and since, by construction, they have only normal components, $\mathbf{D}_c = \mathbf{D}$. Then $\mathbf{E}_c = \mathbf{D}_c/\epsilon_0 = \mathbf{D}/\epsilon_0$ since $\mathbf{P} = 0$ in the vacuum of the cavity. If we now imagine putting a small test charge δq in the cavity and measuring the force $\delta \mathbf{F}_c$ on it, we will have $\delta \mathbf{F}_c = \delta q \, \mathbf{E}_c = \delta q \, \mathbf{D}/\epsilon_0$, so that $\mathbf{D} = \epsilon_0 (\delta \mathbf{F}_c/\delta q)$. Thus, in principle, we can find $\mathbf{D}$ *in the dielectric* from measurements made *in the cavity*.

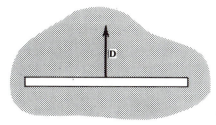

Figure 10-12. Cavity used to measure D in a dielectric.

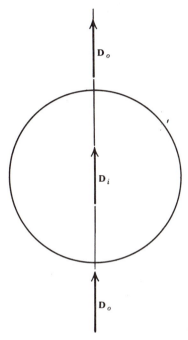

Figure 10-13. D fields on z axis due to a uniformly polarized sphere.

In spite of their apparent simplicity, (10-41) and (10-44) clearly will not be too useful until we can relate **E**, **P**, and **D** better; we will do this in the next two sections. Nevertheless, we can still briefly look at an example.

■ **Example**

Uniformly polarized sphere. This is the system we discussed in the last section for which the results are summarized in Figures 10-8, 10-9, and 10-11. We would like to obtain similar things for **D**; we will continue to consider only field points on the positive z axis. Outside the sphere where there is no matter, **P** = 0, and we find at once from (10-40) and (10-32) that

$$D_{zo}(z) = \epsilon_0 E_{zo} = \frac{2Pa^3}{3z^3} \tag{10-46}$$

The value of D_z inside the sphere as obtained from (10-40) and (10-37) is

$$D_{zi}(z) = \epsilon_0 E_{zi} + P = -\tfrac{1}{3}P + P = \tfrac{2}{3}P \tag{10-47}$$

which is seen to be independent of z and in the same direction as P. These results are illustrated in Figure 10-13. Since there is no free charge on the surface, we expect the normal components of **D** to be continuous there according to (10-42). We can verify this and check our results at the same time. We get, as expected, $D_{2n} - D_{1n} = D_{zo}(a) - D_{zi}(a) = (2P/3) - (2P/3) = 0$. ∎

10-6 CLASSIFICATION OF DIELECTRICS

As we mentioned at the end of Section 10-1, we generally expect that there will be a functional relation between the polarization and the electric field, that is, $\mathbf{P} = \mathbf{P}(\mathbf{E})$ or $P_x = P_x(E_x, E_y, E_z)$ and so on. Macroscopic descriptive electromagnetic theory does *not* predict the form of these functions, but takes them as external information. From this point of view, these relations are left to be determined from experiment or to be calculated theoretically from the microscopic properties of matter in other branches of physics such as statistical mechanics and solid state physics. This does not leave us in a hopeless position, however, because a combination of experiment and general theory shows us that most materials fall into easily classifiable groups, and this result can be used to simplify our theory and make it more useful. It is desirable to do this step by step in order to aid our understanding of the limitations of our final form.

1. Permanent Polarization

If $\mathbf{E} = 0$, there are two possibilities for the value of $\mathbf{P}(0)$. If $\mathbf{P}(0) \neq 0$, then the material is polarized even in the absence of a field and, as we noted before, it is said to have a *permanent polarization* and is called an *electret*. Although electrets do occur, we will not consider them further in this section. The situation for which $\mathbf{P}(0) = 0$ is more typical and the sort of thing we expect when we think of the polarization as being produced by the field; we shall generally use the term *dielectric* only for this latter case.

2. Nonlinear Dielectrics

Even with $\mathbf{P}(0) = 0$, it is still possible that the relation between **P** and **E** can be quite complicated. For most materials, however, this usually requires quite exceptional conditions such as very large fields, or low temperatures, or both. Thus, it is found that it is often sufficient to write **P** as a power series expansion in the components of **E**, that is, we can write

$$P_i = \sum_j \alpha_{ij} E_j + \sum_j \sum_k \beta_{ijk} E_j E_k + \ldots \tag{10-48}$$

where the indices i, j, and k take on the values x, y, and z; we note that this form satisfies our assumption that $\mathbf{P}(0) = 0$. The specific values of the coefficients $\alpha_{ij}, \beta_{ijk}, \ldots$ will depend upon the particular dielectric involved. If the second-order or higher terms in the components of **E** are required to describe the material adequately, the dielectric is called *nonlinear*. It requires experiment to determine whether (10-48) is necessary in a given case; for example, some ceramics fall into this category. We will not consider nonlinear dielectrics further, but now restrict ourselves to cases for which only the first term in (10-48) is required; such materials are called *linear dielectrics*.

3. Linear Dielectrics

In this case, the general expression relating the components of **P** to the components of **E** can be written in the form

$$P_x = \epsilon_0 \left(\chi_{xx} E_x + \chi_{xy} E_y + \chi_{xz} E_z \right)$$
$$P_y = \epsilon_0 \left(\chi_{yx} E_x + \chi_{yy} E_y + \chi_{yz} E_z \right) \tag{10-49}$$
$$P_z = \epsilon_0 \left(\chi_{zx} E_x + \chi_{zy} E_y + \chi_{zz} E_z \right)$$

where the proportionality factors χ_{ij} are called components of the *electric susceptibility tensor*. We have introduced the factor ϵ_0 so that the χ_{ij} will be dimensionless as can be seen from (10-40). In general, the χ_{ij} need not be constants but may be a function of position within the material. The χ_{ij} cannot depend on **E** because that would bring us back to the nonlinear case of (10-48). We see from the form of these relations that **P** will not be parallel to **E** even in linear dielectrics, nor will **D** generally be parallel to **E**. This situation is quite common in crystals and leads to such phenomena as double refraction. We can now proceed to our next simplifying assumption.

4. Linear Isotropic Dielectrics

We now assume in addition that *at a given point* the electrical properties of the dielectric are independent of the direction of **E**; such a condition is known as *isotropy*. Since one direction is completely equivalent to any other, **P** must necessarily be parallel to **E**, the $\chi_{ij} = 0$ if $i \neq j$, and $\chi_{xx} = \chi_{yy} = \chi_{zz}$, so that (10-49) can be written in terms of a single factor of proportionality as

$$\mathbf{P} = \chi_e \epsilon_0 \mathbf{E} \tag{10-50}$$

where χ_e is called the *electric susceptibility*. Combining (10-50) with (10-40), we find that

$$\mathbf{D} = (1 + \chi_e) \epsilon_0 \mathbf{E} = \kappa_e \epsilon_0 \mathbf{E} = \epsilon \mathbf{E} \tag{10-51}$$

where

$$\kappa_e = 1 + \chi_e = \text{dielectric constant} = \text{relative permittivity} \tag{10-52}$$

$$\epsilon = \kappa_e \epsilon_0 = \text{(absolute) permittivity} \tag{10-53}$$

The quantities χ_e, κ_e, and ϵ will characterize the electrical properties of the material and are to be found by experiment; their values can be found in many tables. For all known substances, χ_e is positive for static fields and therefore $\kappa_e > 1$. We see from (10-51) that **D** and **E** are parallel in this situation. The equation $\mathbf{D} = \epsilon \mathbf{E}$ is called a *constitutive equation*; it is not a fundamental equation of electromagnetism but is applicable only where it turns out to be applicable, so to speak.

In this case of linear isotropic dielectrics, we can also find the differential equation satisfied by the scalar potential. Using (5-3), we can write (10-51) as $\mathbf{D} = -\epsilon \nabla \phi$, and if we substitute this into (10-41), and use (1-115) and (1-45), we obtain

$$\nabla \cdot (\epsilon \nabla \phi) = \epsilon \nabla^2 \phi + \nabla \phi \cdot \nabla \epsilon = -\rho_f \tag{10-54}$$

as the equation we would have to solve for ϕ since we still must allow for the possibility that ϵ is a function of position, and we can go no further until this dependence is known.

Our next simplification leads to such an important situation that we will allot it its own section.

10-7 LINEAR ISOTROPIC HOMOGENEOUS (l.i.h.) DIELECTRICS

We now assume in addition that the electrical properties are independent of position; such materials are called electrically homogeneous. Generally gases and liquids, as well as many solids, fall into this category so that it is not such a special situation as one might think. Then the quantities χ_e, κ_e, and ϵ are *constants*; they are still characteristic of the material, however. Equations 10-50 through 10-53 are still applicable and, in addition, (10-54) simplifies to

$$\nabla^2\phi = -\frac{\rho_f}{\epsilon} \tag{10-55}$$

since now $\nabla\epsilon = 0$. By comparison with (5-15), we see that for l.i.h. dielectrics the potential ϕ again satisfies Poisson's equation but with ϵ replacing ϵ_0 and ρ_f replacing the total charge density ρ. (The bound charges of the material haven't disappeared, of course, but their effect is now summarized in the factor ϵ.) The fact that ϕ satisfies the equation (10-55) also means that for l.i.h. dielectrics we can, *with care*, bodily take over solutions we have previously found for the vacuum case and simply replace ϵ by ϵ_0; we will do little of this, however. Also, if $\rho_f = 0$, then ϕ will again satisfy Laplace's equation $\nabla^2\phi = 0$ in this region.

The boundary conditions at a surface of discontinuity can now be expressed completely in terms of **E**. If we insert (10-51) into (10-42), and remember that (9-21) is still applicable, we obtain

$$\hat{n} \cdot (\epsilon_2 \mathbf{E}_2 - \epsilon_1 \mathbf{E}_1) = \sigma_f$$
$$\mathbf{E}_{2t} - \mathbf{E}_{1t} = 0 \tag{10-56}$$

But now we see that, even if $\sigma_f = 0$, the normal components of **E** will not generally be continuous across the bounding surface separating two dielectrics, so that, as shown in Figure 10-14, the direction of **E** can change at the boundary. Thus, the lines of **E** can be refracted even in the absence of a free surface charge density and α_2 will be different from α_1.

In a l.i.h. dielectric, the bound and free charge densities are related in a simple way, as are the polarization and the displacement. If we eliminate **E** between (10-50) and (10-51), and use (10-52), we find that

$$\mathbf{P} = \frac{\chi_e}{\kappa_e}\mathbf{D} = \frac{(\kappa_e - 1)}{\kappa_e}\mathbf{D} \tag{10-57}$$

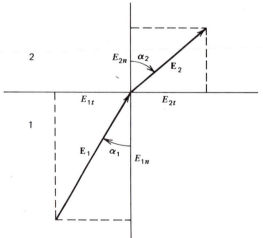

Figure 10-14. Electric fields at a boundary between two dielectrics.

which shows us that $\mathbf{P}$ and $\mathbf{D}$ are also parallel and that $|\mathbf{P}| < |\mathbf{D}|$. Now if we take the divergence of (10-57) and use (10-10) and (10-41), we get

$$\rho_b = -\frac{(\kappa_e - 1)}{\kappa_e}\rho_f \tag{10-58}$$

so that $|\rho_b| < |\rho_f|$. If we insert this result into (10-38), we find that the total charge density in a l.i.h. dielectric can always be written as

$$\rho = \frac{\rho_f}{\kappa_e} = -\frac{\rho_b}{\kappa_e - 1} \tag{10-59}$$

which shows us that the total charge density is always less than the free charge density since $\kappa_e > 1$. As a special case, we see that if $\rho_f = 0$, then $\rho_b = 0$, so that at any point in a l.i.h. dielectric where there is no free charge density, there is no bound charge density either.

From now on, we will assume that we are dealing with l.i.h. dielectrics in our examples and exercises unless we explicitly specify otherwise. At this point, we are able to discuss some examples quantitatively; we begin with the capacitor, which we first considered qualitatively in Section 10-3.

■ **Example**

Parallel plate capacitor with charges constant. In Figure 10-15, we illustrate a capacitor of total free charge Q_f with a vacuum between the plates in (a) and with a dielectric completely filling the region between the plates in (b). The directions of the various field vectors are also shown. The vacuum value of the electric field was discussed in the text immediately preceding (6-40) and was found to be $E_0 = \sigma_f/\epsilon_0$ where $\sigma_f = Q_f/A$ is the free surface charge density and A is the plate area. Since $\mathbf{P}_0 = 0$, we find from (10-40) that the displacement D_0 is

$$D_0 = \epsilon_0 E_0 = \sigma_f \tag{10-60}$$

Now Q_f and σ_f are kept constant when the dielectric is put between the plates, so that $\mathbf{D}$ will not be changed and will equal the vacuum value:

$$D = D_0 = \sigma_f \tag{10-61}$$

This result is also consistent with (10-42) since the fields are zero inside the conducting plate and therefore $D_{2n} - D_{1n} = D - 0 = \sigma_f$. The electric field has changed, however, since we find from (10-51), (10-61), and (10-60) that

$$E = \frac{D}{\epsilon} = \frac{D_0}{\kappa_e \epsilon_0} = \frac{E_0}{\kappa_e} \tag{10-62}$$

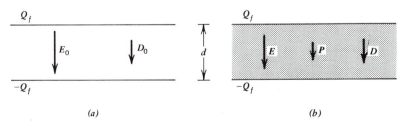

Figure 10-15. Parallel plate capacitor of constant charge. (*a*) Vacuum between the plates. (*b*) Dielectric between the plates.

and therefore $E < E_0$ in agreement with (10-24). We see now that the factor by which the electric field is reduced is exactly equal to the relative permittivity. The potential difference now is

$$\Delta\phi = \int_{+}^{-} \mathbf{E} \cdot d\mathbf{s} = Ed = \frac{E_0 d}{\kappa_e} = \frac{\Delta\phi_0}{\kappa_e} \tag{10-63}$$

with the use of (10-62) and (10-22). Thus, the potential difference is less than the vacuum value by the same factor κ_e so that $\Delta\phi < \Delta\phi_0$ in agreement with the experimental result. We can find the capacitance by expressing (10-63) in terms of the total charge:

$$\Delta\phi = \frac{E_0 d}{\kappa_e} = \frac{\sigma_f d}{\kappa_e \epsilon_0} = \left(\frac{d}{\kappa_e \epsilon_0 A} \right) Q_f$$

and therefore

$$C = \frac{\kappa_e \epsilon_0 A}{d} = \kappa_e C_0 \tag{10-64}$$

where we have used (10-20) and (6-41) to identify the vacuum value of the capacitance C_0. Thus, we see that the presence of the dielectric has increased the capacitance, in agreement with (10-21), and the ratio of the capacitance with and without the dielectric is exactly equal to the dielectric constant, that is, $C/C_0 = \kappa_e$.

The polarization as found from (10-50) and (10-62) is

$$P = \chi_e \epsilon_0 E = (\kappa_e - 1)\epsilon_0 E = \left(\frac{\kappa_e - 1}{\kappa_e} \right) \epsilon_0 E_0 \tag{10-65}$$

With the further use of (10-62), the third form can be rewritten as

$$P = \epsilon_0 (E_0 - E) \tag{10-66}$$

so that

$$E = E_0 - \frac{P}{\epsilon_0} \tag{10-67}$$

which also follows from (10-40) and (10-61), that is $D = \epsilon_0 E + P = D_0 = \epsilon_0 E_0$. This latter result has exactly the form we previously deduced in (10-25) and now we have a quantitative expression for E_b. Since $\mathbf{P}$ is constant, $\nabla \cdot \mathbf{P} = 0$ and there are no bound volume charges in agreement with (10-58) since the only free charges are on the surfaces of the conducting plates. However, there are bound surface charges on the dielectric surfaces. Their magnitude as found from (10-8), (10-65), and (10-60) is

$$|\sigma_b| = |P_n| = P = \left(\frac{\kappa_e - 1}{\kappa_e} \right) \epsilon_0 E_0 = \left(\frac{\kappa_e - 1}{\kappa_e} \right) \sigma_f \tag{10-68}$$

and the signs are exactly those already illustrated in Figure 10-6. We see from the figure that the sign of σ_b is always opposite to that of the immediately adjacent σ_f; taking this into account we can write (10-68) as

$$\sigma_b = -\left(\frac{\kappa_e - 1}{\kappa_e} \right) \sigma_f \tag{10-69}$$

which is completely analogous to the result previously found for volume charge densities in (10-58). These bound surface charges, acting as two plane infinite sheets, will produce the field E_b. We can calculate the magnitude of E_b by using our previous result (3-12) along with (10-68), and we find that

$$E_b = \frac{|\sigma_b|}{2\epsilon_0} + \frac{|\sigma_b|}{2\epsilon_0} = \frac{|\sigma_b|}{\epsilon_0} = \frac{P}{\epsilon_0} \tag{10-70}$$

This agrees exactly with (10-67), which we found by other means, and verifies our previous analysis which led to the form $E = E_0 - E_b$ given in (10-25). This field E_b is often called the *local field* and one says that the resultant field $\mathbf{E}$ is the sum of the vacuum field $\mathbf{E}_0$, produced by the free charges as if there were no matter present, plus the local field $\mathbf{E}_{loc} = \mathbf{E}_b$ produced by the bound charges arising from the polarization of the dielectric.   ∎

■ **Example**

Capacitance in general. Although we obtained (10-64) by considering the specific case of the parallel plate capacitor, the simplicity of this result, and the fact that $C = \kappa_e C_0$ no longer contains any characteristics of the parallel plate capacitor, suggests that this may actually be a general relation that holds for any capacitor. This turns out to be the case. If we combine (10-41) and (10-51), we get

$$\nabla \cdot \mathbf{E} = \frac{\rho_f}{\epsilon} = \frac{1}{\epsilon_0}\left(\frac{\rho_f}{\kappa_e}\right) \tag{10-71}$$

since ϵ is constant for a l.i.h. dielectric. We also have $\nabla \times \mathbf{E} = 0$ since $\mathbf{E}$ is still conservative. Suppose we are given a certain free charge distribution described by ρ_f. If there is no matter in the region of interest, we solve the pair of source equations to find an electric field $\mathbf{E}_0$. Now if we fill *all* of the region with a dielectric described by κ_e, and if we *keep ρ_f unchanged*, then we see from (10-71) that the problem is exactly the same as the vacuum case we solved, *except* that the source charges are smaller *everywhere* by the factor κ_e. Then, as we can see from (3-3) for instance, the field $\mathbf{E}$ we now get will also be smaller by this same factor, that is,

$$\mathbf{E} = \frac{\mathbf{E}_0}{\kappa_e}$$

is a general result. The potential difference $\Delta\phi$ between the capacitor plates as given by (6-38) will be

$$\Delta\phi = \int_+^- \mathbf{E} \cdot d\mathbf{s} = \int_+^- \frac{\mathbf{E}_0 \cdot d\mathbf{s}}{\kappa_e} = \frac{\Delta\phi_0}{\kappa_e} \tag{10-72}$$

as in the special result (10-63). We note that this result is consistent with our comments following (10-55).

Since Q_f is the same in both cases, the capacitance will be

$$C = \frac{Q_f}{\Delta\phi} = \frac{\kappa_e Q_f}{\Delta\phi_0} = \kappa_e C_0 \tag{10-73}$$

thus showing in the general case that the capacitance of any capacitor is increased by the factor κ_e when all the space between its plates is filled with a dielectric. The use of this result provides a convenient way of measuring κ_e.

If the dielectric is not homogeneous, or if not all of the space is filled with dielectric, (10-73) is not generally true. One can often treat such problems by writing the potential difference as

$$\Delta\phi = \int_+^- \mathbf{E} \cdot d\mathbf{s} = \int_+^- \frac{\mathbf{D} \cdot d\mathbf{s}}{\epsilon} \tag{10-74}$$

Then it is usually possible to express **D** in terms of the total free charge Q_f by means of (10-41), or by (10-43) if the problem has enough symmetry, so that when the integral is evaluated, the capacitance can be found from (6-38). In such a case, the boundary condition (10-42) often is of great help. Many of the exercises involve these considerations.

∎

■ **Example**

Point charge in an infinite dielectric. Suppose a point charge q is embedded in a dielectric as shown in Figure 10-16. The field of this charge will polarize the dielectric. If the dielectric has a finite size, the bound charges on the surface will contribute to the resultant field and the problem of calculating the field everywhere could be quite complicated. However, if the dielectric is of infinite extent, the effect of any bound surface charges on the outer surface can be neglected and we can assume spherical symmetry. Then we can write $\mathbf{D} = D(R)\hat{\mathbf{R}}$ and use Gauss' law for **D** as given by (10-43). If we integrate over the sphere of radius R shown dashed in the figure, we get, in the by-now-familiar way,

$$\oint_S \mathbf{D} \cdot d\mathbf{a} = \oint_S D\hat{\mathbf{R}} \cdot d\mathbf{a}\,\hat{\mathbf{R}} = 4\pi R^2 D = Q_{f,\text{in}} = q$$

so that

$$\mathbf{D} = \frac{q\hat{\mathbf{R}}}{4\pi R^2} \quad \text{and} \quad \mathbf{E} = \frac{q\hat{\mathbf{R}}}{4\pi\epsilon R^2} \tag{10-75}$$

and, as expected, the electric field is just that of a point charge reduced by the factor $\epsilon/\epsilon_0 = \kappa_e$. If we now imagine a point charge q' placed at **R**, the force on it will be

$$\mathbf{F}' = q'\mathbf{E} = \frac{qq'\hat{\mathbf{R}}}{4\pi\epsilon R^2} \tag{10-76}$$

which is simply Coulomb's inverse square law with ϵ replacing ϵ_0.

This result (10-76) is the basis for a common statement that the presence of a dielectric decreases the forces between charges by the factor $\epsilon/\epsilon_0 = \kappa_e$. We see now, however, that this is true only for a l.i.h. dielectric of infinite extent, or one so large that the bound surface charges will not affect the field and we can safely assume spherical symmetry. In fact, it is possible for the force on q' to actually increase under the right circumstances.

The polarization of the dielectric can be found from (10-57) and (10-75) to be

$$\mathbf{P} = \frac{(\kappa_e - 1)q\hat{\mathbf{R}}}{\kappa_e 4\pi R^2} \tag{10-77}$$

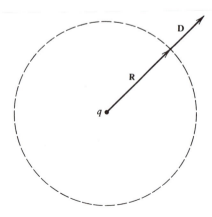

Figure 10-16.   Point charge in an infinite dielectric.

and is directed radially from q. According to (10-58), the bound charge density should vanish. It follows from (10-10), (10-77), and (1-145) that this is indeed the case. ∎

10-8 ENERGY

We recall our result (7-10) for the energy of a system of charges:

$$U_e = \tfrac{1}{2} \int_{\text{all space}} \rho \phi \, d\tau \tag{10-78}$$

When we obtained this expression, we calculated it as the amount of reversible work required to assemble a given configuration of charges, and we did not make any distinction between free and bound charges, nor was it necessary to do so. The implication was that this energy could be *retrieved* from the system as reversible work, and in that sense it is appropriate to regard it as potential energy as we did. Now when we consider charges in the presence of matter, we have to ask ourselves: what is the *useful* definition of energy? Again, this must be the work required for us to put together an assembly of charges *over which we have control* and thus energy that can, in principle, be retrieved from the system. But the only charges over which we have any sort of control are the free charges and it is only these in a broad sense that we can use to store or extract energy, as in the process of charging or discharging a capacitor in a reversible manner. Hence, the energy of interest to us is that of the free charge distribution in the presence of matter. Failure to remember the restrictive nature of this definition leads to, and has led to, a considerable amount of confusion and argument about energy relations when matter is involved.

In order to appreciate the limitations of our final result, it will be useful to begin again and (5-48), which gives the energy of a point charge in terms of the potential, provides a convenient starting point. We assume that we have some preexisting situation and increase the free charge density in a volume element $d\tau$ by the amount $\delta\rho_f$. Then $\delta\rho_f \, d\tau$ is small enough to be treated as a point charge and, by (5-48), the energy of the free charge distribution will change by

$$\delta U_e = \phi \, \delta\rho_f \, d\tau \tag{10-79}$$

The bound charges of matter are included, of course, since ϕ is the potential resulting from all charges; for example, it could be evaluated by substituting (10-38) into (5-7). As usual, we are assuming rigid mechanical supports or constraints on the charge positions so that no purely mechanical work is done. If we now integrate over all space, we get the total change in the electrostatic energy:

$$\delta U_e = \int_{\text{all space}} \phi \, \delta\rho_f \, d\tau \tag{10-80}$$

During this process, $\mathbf{D}$ will have changed and by (10-41), we initially have $\nabla \cdot \mathbf{D} = \rho_f$ and, finally, $\nabla \cdot (\mathbf{D} + \delta\mathbf{D}) = \rho_f + \delta\rho_f$, so that by using (1-114), we get $\nabla \cdot \delta\mathbf{D} = \delta\rho_f$. Then (10-80) becomes

$$\delta U_e = \int_{\text{all space}} \phi (\nabla \cdot \delta\mathbf{D}) \, d\tau \tag{10-81}$$

If we now compare this expression with (7-22), we see that they are of exactly the same form with $\delta\mathbf{D}$ replacing $\tfrac{1}{2}\epsilon_0 \mathbf{E}$. The arguments that led us from (7-22) to (7-28) can be applied in a completely analogous manner, so that (10-81) can also be written as

$$\delta U_e = \int_{\text{all space}} \mathbf{E} \cdot \delta\mathbf{D} \, d\tau \tag{10-82}$$

Finally, if we take the zero of energy to correspond to $\mathbf{D} = 0$, we can obtain the total energy U_e by integrating this from the initial state of $\mathbf{D} = 0$ to the final value:

$$U_e = \int_{\text{all space}} \int_0^{\mathbf{D}} \mathbf{E} \cdot \delta \mathbf{D} \, d\tau \tag{10-83}$$

In general, (10-83) cannot be evaluated further until the dependence of $\mathbf{E}$ upon $\mathbf{D}$ is known, and this could be quite complicated. However, for the important case of *linear isotropic* dielectrics, we can use (10-51) to write $\mathbf{E} \cdot \delta \mathbf{D} = \mathbf{D} \cdot \delta \mathbf{D}/\epsilon = \delta(\mathbf{D}^2/2\epsilon)$ so that

$$\int_0^{\mathbf{D}} \mathbf{E} \cdot \delta \mathbf{D} = \int_0^{\mathbf{D}} \delta \left(\frac{\mathbf{D}^2}{2\epsilon} \right) = \frac{\mathbf{D}^2}{2\epsilon} = \tfrac{1}{2} \mathbf{D} \cdot \mathbf{E}$$

and thus (10-83) becomes

$$U_e = \int_{\text{all space}} \tfrac{1}{2} \mathbf{D} \cdot \mathbf{E} \, d\tau \tag{10-84}$$

and gives the total reversible work done on the free charges. We can interpret (10-84) just as we did (7-28) by introducing an *energy density* u_e given by

$$u_e = \tfrac{1}{2} \mathbf{D} \cdot \mathbf{E} = \tfrac{1}{2} \epsilon E^2 = \frac{\mathbf{D}^2}{2\epsilon} \tag{10-85}$$

for linear isotropic dielectrics.

■   **Example**

Spherical capacitor. We imagine all space between the plates of the capacitor shown in Figure 10-17 to be filled with a dielectric of permittivity ϵ. The total free charge on the capacitor is Q. As usual, $\mathbf{D}$ will be radial because of the spherical symmetry, so that $|\mathbf{D}|$ will be constant on the surface of the dashed sphere of radius r. Then, if we apply (10-43) to this sphere, we get, as in the last section:

$$\oint_S \mathbf{D} \cdot d\mathbf{a} = 4\pi r^2 D = Q_{f,\text{in}} = Q$$

so that $D = Q/4\pi r^2$ and (10-85) gives

$$u_e = \frac{Q^2}{32\pi^2 \epsilon r^4} \tag{10-86}$$

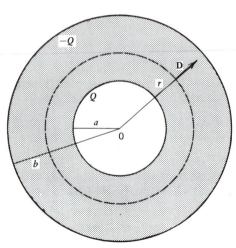

Figure 10-17.   Spherical capacitor with dielectric between the plates.

This applies to the region between the spheres; $u_e = 0$ elsewhere since the fields are different from zero only in the region occupied by the dielectric. Inserting (10-86) into (10-84), and using (1-99), we find the total energy of the capacitor to be

$$U_e = \frac{Q^2}{32\pi^2\epsilon} \int_0^{2\pi} \int_0^{\pi} \int_a^b \frac{1}{r^4} \cdot r^2 \sin\theta \, dr \, d\theta \, d\varphi = \frac{Q^2}{8\pi\epsilon}\left(\frac{1}{a} - \frac{1}{b}\right) \quad (10\text{-}87)$$

If there were a vacuum between the plates, the energy U_{e0} would be given by (10-87) with ϵ replaced by ϵ_0. Therefore, since $\epsilon/\epsilon_0 = \kappa_e$, we see that

$$U_e = \frac{U_{e0}}{\kappa_e} \quad (10\text{-}88)$$

so that $U_e < U_{e0}$ and the total energy is decreased by the presence of the dielectric in this case in which Q is kept the same.

In (7-21), we found that the energy of a capacitor is equal to $Q^2/2C$. This result was obtained by using very general properties of the charge distributions on conductors, and thus is generally true whether there are dielectrics present or not; this same result was also obtained by using only general properties of work and potential difference in Exercise 7-3. Thus, we can still apply energy methods to the calculation of capacitance, and this is sometimes more convenient. In this example, if we equate (10-87) and (7-21), we find the capacitance to be

$$C = \frac{4\pi\epsilon ab}{b - a} = \kappa_e C_0 \quad (10\text{-}89)$$

with the use of (10-53), and where C_0 is the vacuum value as given by (6-37). As it should, this specific result agrees with our general conclusion about the effect of a dielectric on capacitance as expressed in (10-73). ∎

In general, and as we just saw in the last example, the presence of a dielectric can alter the values of **D** and **E** everywhere so that the total energy as given by (10-84) can be expected to change. The exact amount of this change will normally depend on the particular manner in which the process is carried out; for example, charges kept constant, potentials kept constant, only part of space filled with the dielectric, more than one dielectric used and so on. As a result, the general problem of energy involving dielectrics can be quite complex, and it is not always possible to make an unambiguous assignment of the energy change to specific parts of the system. There is one comparatively simple case, however, in which it is possible to ascribe the energy change to the dielectric itself, and we shall consider it as an illustration.

Let us assume that initially we have vacuum everywhere and some distribution of charges which result in the fields $\mathbf{E}_0$ and $\mathbf{D}_0 = \epsilon_0 \mathbf{E}_0$ throughout all space. The initial energy U_{e0} can then be evaluated from (10-84) as

$$U_{e0} = \tfrac{1}{2} \int_{\text{all space}} \mathbf{D}_0 \cdot \mathbf{E}_0 \, d\tau \quad (10\text{-}90)$$

Now let us assume that we *keep the source charges fixed* in values and locations and introduce a dielectric of volume V into this preexisting field $\mathbf{E}_0$. (This is in contrast to the previous example where *all* of the region containing the already existing field was filled with a dielectric.) As we know, the presence of a dielectric will generally change the values of **E** and **D** everywhere and the new energy U_e can be found from (10-84) by using these new values of the fields. The *change* in the energy, $U_e - U_{e0}$, in this case, can be ascribed entirely to the presence of the dielectric, and if we call it U_{eb}, we have

$$U_{eb} = U_e - U_{e0} = \tfrac{1}{2} \int_{\text{all space}} (\mathbf{D} \cdot \mathbf{E} - \mathbf{D}_0 \cdot \mathbf{E}_0) \, d\tau \quad (10\text{-}91)$$

It can now be shown that under these conditions, (10-91) can be written as an integral over the volume V of the dielectric only. Since this is a rather lengthy calculation, we have to leave it to an exercise and simply quote the final result, which turns out to be

$$U_{eb} = -\frac{1}{2} \int_V \mathbf{P} \cdot \mathbf{E}_0 \, d\tau \tag{10-92}$$

Since this involves only the volume of the dielectric, it is reasonable to consider it to be localized *in* the dielectric and to represent the energy *of* the dielectric. Thus we can introduce an energy density for these bound charges, u_{eb}, given by

$$u_{eb} = -\frac{1}{2} \mathbf{P} \cdot \mathbf{E}_0 = -\frac{1}{2} \chi_e \epsilon_0 \mathbf{E} \cdot \mathbf{E}_0 \tag{10-93}$$

where we also used (10-50). These expressions (10-92) and (10-93) are thus appropriate for a situation in which the polarization is thought of as being produced by the field.

In all of our discussion so far, we have assumed that κ_e is constant during these processes. It turns out that many dielectrics have a κ_e that depends on the temperature, as we will see in Appendix B, so that here we are effectively assuming isothermal processes. This is consistent with our emphasis on U_e as being related to the reversible work, so that it is really more analogous to the Helmholtz function or free energy of thermodynamics, but for systems at constant temperature, there is no distinction between their changes. With this in mind, we can look on (10-92) as a contribution to the internal energy of the dielectric system.

We should also remind ourselves of the distinction between these energies and the *interaction energy* of a charge distribution and an *external field*, which we discussed in Section 8-4. In particular, we obtained (8-64) for the energy of a dipole. If we want to apply this to a polarized material, we will have to assume that the polarization is either a permanent one or that the external field is small enough so that it does not affect $\mathbf{P}$ appreciably. Then the dipole moment of a small volume will be given by (10-1) and the external interaction energy as obtained from (8-64) can be written as

$$dU_{e,\text{ext}} = -\mathbf{P} \cdot \mathbf{E}_{\text{ext}} \, d\tau \tag{10-94}$$

where we write the external field as $\mathbf{E}_{\text{ext}}$, rather than $\mathbf{E}_0$ that we have been using recently to represent vacuum values. The total interaction energy will then be obtained by integrating (10-94) over the dielectric to give

$$U_{e,\text{ext}} = -\int \mathbf{P} \cdot \mathbf{E}_{\text{ext}} \, d\tau \tag{10-95}$$

For example, if $\mathbf{E}_{\text{ext}}$ does not vary much over the volume, we can take it out of the integral and use (10-2) to get $U_{e,\text{ext}} = -\mathbf{p} \cdot \mathbf{E}_{\text{ext}}$ in agreement with (8-64).

■ **Example**

Energy of a capacitor in general. We have already seen the effect of a dielectric on a capacitor in the special case by which we obtained (10-88). Let us now briefly look at the general case. The energy is general is given by (7-21). In addition, we have obtained the general relation (10-73) for the effect on the capacitance. Therefore, if the charge Q is kept constant, the energy U_e with the dielectric between the plates will be $U_e = Q^2/2C = Q^2/2\kappa_e C_0 = U_0/\kappa_e$ where U_0 is the vacuum energy, so that the decreased energy given by

$$U_e = \frac{U_0}{\kappa_e} \qquad (Q = \text{const.}) \tag{10-96}$$

is a completely general result, in agreement with (10-88).

Now let us suppose, instead, that the potential difference $\Delta\phi$ is kept constant. Combining (7-21) with (10-73), we find that $U_e = \frac{1}{2}C(\Delta\phi)^2 = \frac{1}{2}\kappa_e C_0(\Delta\phi)^2 = \kappa_e U_0$ so that

$$U_e = \kappa_e U_0 \qquad (\Delta\phi = \text{const.}) \qquad (10\text{-}97)$$

and the energy of the capacitor has increased by the factor κ_e, in contrast to (10-96). The reason for this is that, with the increase in capacitance, the charge has also been increased and the battery had to do work on these free charges in order to separate them. In addition, it had to polarize the dielectric and the net effect of all of these changes is just that described by (10-97). ∎

10-9 FORCES

When a dielectric is polarized, the resultant bound charge densities will have forces on them due to the electric field. This general subject of forces on dielectrics, and on conductors in the presence of dielectrics, is really quite complex and it is easy to get an incorrect answer. Generally speaking, the only satisfactory way of handling these and similar problems is to use "energy methods," that is, by comparing the energy of the initial configuration of the system with that of the final configuration. Usually such problems fall into two classes: (1) the system of interest is completely isolated and thus its total energy is conserved or (2) the system is not isolated and one has to consider the possibility of energy transfer between it and external energy sources such as batteries; thus the system energy will not be conserved, although that of the system plus battery will be. Ordinarily, these two classes will correspond to our previous cases of constant free charge or constant potential difference.

Because of these considerations, we will content ourselves here with discussing only two of the simplest situations, although other common examples will be found in the exercises. Furthermore, we will not discuss the balance between electrical and mechanical forces that is required to maintain mechanical equilibrium within the system, or to produce a new one when electric fields are applied. If the dielectric is not rigid, it will generally deform under the influence of these electrical forces. This phenomenon is called *electrostriction*; it is generally a small effect, but can be of interest and importance under some circumstances.

■ **Example**

Average surface force on a dielectric. Let us consider the particular case of a capacitor. As we saw in (10-96), the energy of the capacitor for constant charge will be decreased when the dielectric is present. Since the general tendency of systems is to reduce their energy, we see that the capacitor will "want" to have the dielectric in place. In other words, there should be a force on the dielectric whose direction is such as to pull it into the region between the plates.

In order to be specific, we consider a parallel plate capacitor with square plates of side L so that $A = L^2$. We also assume we have a solid slab of dielectric of the correct size to just fit between the plates. We neglect edge effects and assume the field to be different from zero only in the region between the plates. In Figure 10-18, we show a side view in which the dielectric is partway between the plates. With the dielectric completely in place, the total change in energy can be found from (10-96) and is

$$\Delta U_e = U_e - U_0 = -\left(\frac{\kappa_e - 1}{\kappa_e}\right)U_0 \qquad (10\text{-}98)$$

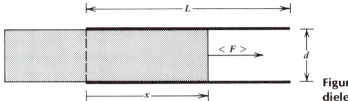

Figure 10-18. Force on a dielectric.

We let $\langle F \rangle$ be the average force on the dielectric; the total displacement of the dielectric is L, so that using (7-37) we get

$$\langle F \rangle = -\frac{\Delta U_e}{L} = \left(\frac{\kappa_e - 1}{\kappa_e} \right) \frac{U_0}{L} \qquad (10\text{-}99)$$

which is positive, showing that the dielectric will be pulled into the region between the plates. If we express this in terms of the original constant energy density u_{e0}, we find from (10-84) and (10-85) that $U_0 = u_{e0} (\text{volume}) = u_{e0} L^2 d$ so that

$$\langle F \rangle = \left(\frac{\kappa_e - 1}{\kappa_e} \right) u_{e0} (Ld)$$

Now Ld is the area of the face of the dielectric upon which $\langle F \rangle$ is acting, so that if $\langle f_a \rangle$ is the average force per unit area we get

$$\langle f_a \rangle = \left(\frac{\kappa_e - 1}{\kappa_e} \right) u_{e0} \qquad (10\text{-}100)$$

as expressed in terms of the initial vacuum situation. ■

■ **Example**

Capacitor immersed in a fluid dielectric. As an example of the force on a conductor in the presence of a dielectric, we consider a plate of a parallel plate capacitor for which all of the region where the field is different from zero is occupied by a dielectric. In Section 7-4, we evaluated this by looking at the energy changes produced by and the force required to change the plate separation by a small amount. Our treatment was general enough so that we can use the same results and need only supplement them with our knowledge of the effect of a dielectric on capacitance as given by (10-73), that is, $C = \kappa_e C_0$. If the charge is kept constant, the total force on the plates in the presence of a dielectric will still be given by (7-39), and we get

$$F_e = \frac{Q^2}{2C^2} \frac{dC}{dx} = \frac{Q^2}{2(\kappa_e C_0)^2} \frac{d(\kappa_e C_0)}{dx} = \frac{F_{e0}}{\kappa_e} \qquad (Q = \text{const.}) \qquad (10\text{-}101)$$

in terms of the force F_{e0} when there is a vacuum between the plates. Thus the force on the plates has been reduced by the factor κ_e. In using $C = \kappa_e C_0$ in this way, it is essential that the dielectric completely fill the region between the plates, as will be illustrated by some of the exercises. This means that we are in effect assuming the dielectric to be a fluid and that the capacitor is immersed in the fluid. Then, if the separation is decreased, the fluid can move out from between the plates, while if the separation of the plates is increased, the fluid can move in to fill the region and there is also fluid available for this purpose.

If the potential difference is kept constant, the middle form of (7-46) is appropriate to use, and we get

$$F_e = \tfrac{1}{2}(\Delta\phi)^2 \frac{dC}{dx} = \tfrac{1}{2}(\Delta\phi)^2 \frac{d(\kappa_e C_0)}{dx} = \kappa_e F_{e0} \qquad (\Delta\phi = \text{const.}) \qquad (10\text{-}102)$$

and, in this case, the force on the plate is increased by the factor κ_e.

Even though we realize that there are different processes involved, it still may seem somewhat paradoxical that the presence of the dielectric decreases the total force in one case while increasing it in the other. If we recall (7-37) and (7-45), we see that we should be able to understand these results somewhat more by looking at the total field energy. We found in (10-85) that the energy density is given by $u_e = \tfrac{1}{2}\mathbf{D} \cdot \mathbf{E}$. If the free charge Q is kept constant, then $\mathbf{D}$ will be constant and we find that $u_e = D^2/2\epsilon = u_{e0}/\kappa_e$. The use of (10-84) then gives $U_e = U_{e0}/\kappa_e$ and the force will decrease by the factor κ_e as in (10-101). Similarly, if the potential difference $\Delta\phi$ is kept constant, then $\mathbf{E}$ will be constant, and we are led to $u_e = \tfrac{1}{2}\epsilon E^2 = \kappa_e u_{e0}$. Thus, in this case, we will find that $U_e = \kappa_e U_{e0}$ which corresponds to the increase in the force described by (10-102). [We note that these results for the capacitor energy are exactly those we found by a somewhat different approach as given by (10-96) and (10-97).]

Further study as to exactly how these mechanical forces are affected by the presence of a dielectric shows that the pressure distribution within the fluid is altered in the presence of an electric field. The resulting pressure changes can be shown to account in detail for the changes in the force on the conductor. ∎

EXERCISES

10-1 The permanent dipole moment of the water molecule is about 6.2×10^{-30} coulomb-meter. What is the maximum polarization possible for water vapor at $100°C$ and atmospheric pressure?

10-2 Electrostatics began with the observation that charged objects attracted small pieces of matter. Describe qualitatively how a charged body can exert a net force on neutral matter and show that the force will be attractive as observed.

10-3 A slab of material has parallel faces. One coincides with the xy plane while the other is given by $z = t$. The material has a nonuniform polarization $\mathbf{P} = P(1 + \alpha z)\hat{z}$ where P and α are constants. Find the volume and surface densities of bound charge. Find the total bound charge in a cylinder of the material of cross section A and sides parallel to the z axis and thus verify directly that (10-13) holds for this case.

10-4 Consider a parallelepiped of volume $\Delta x\,\Delta y\,\Delta z$ fixed within a dielectric. If ρ_+ is the average positive bound charge density, and $\mathbf{R}_+$ is the average displacement of these charges when the material is polarized, show that the net gain of positive charge through the faces parallel to the yz plane will be

$$-\frac{\partial}{\partial x}(\rho_+ R_{+x})\,\Delta x\,\Delta y\,\Delta z$$

Similarly, find the net gain due to motion of negative charge of density $-\rho_+$ (Why?) and average displacement $\mathbf{R}_-$. Combine the results for all faces found in this way and thus show that the net gain of charge per unit volume is $\rho_b = -\nabla \cdot \mathbf{P}$.

10-5 Find the potential ϕ and E_z on the axis produced by the uniformly polarized sphere discussed in Section 10-4 for negative values of z. Show that your answers are consistent with the results found for $z > 0$ and with Figure 10-11.

10-6 Find the total positive bound charge of the uniformly polarized sphere of Figure 10-8.

10-7 A sphere of radius a has a radial polarization given by $\mathbf{P} = \alpha r^n\hat{r}$ where α and n are constants and $n \geq 0$. Find the volume and surface densities of bound charge. Find $\mathbf{E}$ outside and inside the sphere. Verify that your results for $\mathbf{E}$ satisfy the appropriate boundary conditions. Find ϕ outside and inside the sphere. Sketch your results for $\mathbf{E}$ and ϕ.

10-8 Repeat Exercise 10-7 for the case $n = -1$. Also show that $n = -2$ is not a possibility because of (10-13).

10-9 A cube of edge $2a$ has its faces perpendicular to the xyz axes and the origin at its center. It is uniformly polarized in the z direction. Find $\mathbf{E}$ at the center of the cube.

10-10 A spherical cavity of radius a is inside a very large dielectric that is uniformly polarized. Find $\mathbf{E}$ at the center of the cavity.

10-11 A cylinder of length $2L$ has its axis along the z axis and a circular cross section of radius a. The origin is at the center of the cylinder that is uniformly polarized in the direction of the axis, that is, $\mathbf{P} = P\hat{\mathbf{z}}$ where $P = $ const. (a) Find the bound charge densities ρ_b and σ_b. (b) Find the electric field for all points on the z axis for which $z \geq 0$. (c) Verify that your results in (b) satisfy the boundary condition at $z = L$. (d) From the result of (b), find $\mathbf{E}$ at the origin. (e) Sketch the result of (d) as a function of a/L. For what value of a/L does $\mathbf{E}$ at the origin have its maximum magnitude and what is its value? Is this reasonable?

10-12 An infinite cylinder of circular cross section of radius a has its axis along the z axis. It is uniformly polarized transverse to its axis, that is, $\mathbf{P} = P\hat{\mathbf{x}}$ where $P = $ const. Find $\mathbf{E}$ at any point on the axis of the cylinder. (It is easiest to choose the field point at the origin.)

10-13 Which of the results (10-55) through (10-59) are still valid if the dielectric is linear and isotropic but nonhomogeneous?

10-14 Show that the angles in Figure 10-14 satisfy the relation $\kappa_{e1} \cot \alpha_1 = \kappa_{e2} \cot \alpha_2$. Suppose region 1 is polyethylene for which $\kappa_{e1} = 2.30$, while region 2 is a glass for which $\kappa_{e2} = 4.00$. Find α_2 if $\alpha_1 = 36°$. Show that the ratio of the bound surface charge density from region 1 to the total bound surface charge density is independent of angle and evaluate it for this case.

10-15 Find the expressions analogous to (10-58) and (10-59) when the dielectric is not homogeneous and thus show that there can be a bound charge density even in the absence of free charge. What additional conditions are necessary to have $\rho_b = 0$ when $\rho_f = 0$?

10-16 Two point charges q and $-q$ are initially in a vacuum and are separated by a distance a. Then a slab of dielectric of thickness $d < a$ is inserted halfway between them, with the faces of the slab perpendicular to the line connecting the charges. Show qualitatively that the total force on q will be increased.

10-17 A point charge q is located at the center of a dielectric sphere of radius a. Find $\mathbf{D}$, $\mathbf{E}$, and $\mathbf{P}$ everywhere and plot your results. What is the total bound charge on the surface of the sphere?

10-18 An infinite dielectric with dielectric constant κ_e has a spherical cavity of radius a in it. There is a point charge q at the center of the cavity. Find ρ_b and σ_b. Find the total bound charge on the surface of the cavity. How can you reconcile your results with (10-13)?

10-19 The infinitely long coaxial conductors of Figure 6-12 have the space between them filled with a dielectric for which κ_e is given in cylindrical coordinates by $\alpha\rho^n$ where α and n are constants. There is free charge λ_f per unit length on the inner cylinder. Find $\mathbf{D}$, $\mathbf{E}$, and ρ_b everywhere between the conductors. Under what circumstances will the magnitude of $\mathbf{E}$ be constant? What will be the corresponding values of $\mathbf{D}$ and ρ_b?

10-20 An infinite slab of dielectric of thickness t and parallel faces has a constant surface density of free charge σ_f on *one* of its surfaces. Find $\mathbf{E}$ everywhere. What is the surface density of bound charge on the face which has no free charge on it?

10-21 An infinite line charge with a constant free charge λ_f per unit length coincides with the z axis. It is coaxial with a dielectric cylinder of radius a whose dielectric constant varies along the axis according to $\kappa_e = \alpha + \beta z$ where α and β are constants. Find $\mathbf{D}$, $\mathbf{E}$, $\mathbf{P}$, and ρ_b at all points within the cylinder. Is your result for ρ_b consistent with the results of Exercise 10-15?

10-22 How does the introduction of a l.i.h. dielectric into all regions of a general system affect the values of the coefficients of potential, capacitance, and induction? [These coefficients are defined in (6-11) and (6-43).]

10-23 A slab of dielectric of thickness t is inserted into a parallel plate capacitor of plate separation d and plate area A as shown in Figure 10-19. The surfaces of the slab are parallel to the plate surfaces. Find $\mathbf{D}$, $\mathbf{E}$, and $\mathbf{P}$ as functions of x, and plot your results. (Express them in terms of Q.) Find the capacitance of this system. Verify that your result for C reduces to the proper values when $t = 0$ and $t = d$.

10-24 Reconsider the example that led to (10-73) for the case in which the potential difference is kept constant while the dielectric is put between the plates. Find what happens to $\mathbf{E}$, $\mathbf{D}$, and Q and thus show that (10-73) is still correct.

10-25 The region between the plates of a parallel plate capacitor is filled with a dielectric for which κ_e varies linearly with distance from the value κ_{e1} at one plate to κ_{e2} at the other. Find

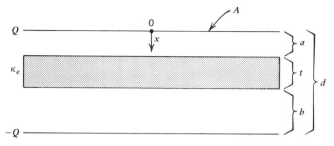

Figure 10-19. Capacitor in Exercise 10-23.

the capacitance. As a check on your result, show that C reduces to the correct expression when κ_e is constant. Also show that the result is independent of whether κ_{e1} is greater than or smaller than κ_{e2}.

10-26 The region between the plates of the spherical capacitor of Figure 10-17 is filled with a dielectric for which κ_e varies according to

$$\kappa_e = \kappa_{ea}\left[1 + \alpha\left(\frac{r-a}{b-a}\right)\right]$$

where κ_{ea} and α are constants. Find the capacitance. Does your result reduce to the correct value for $\alpha = 0$?

10-27 The coaxial cylindrical capacitor of Figure 6-12 has two different dielectrics between its plates. The value of κ_e is κ_{e1} for $a \leq \rho < \rho_0$, and κ_{e2} for $\rho_0 \leq \rho \leq b$. Find the capacitance of a length L of this system.

10-28 The region between the plates of the spherical capacitor of Figure 10-20 is filled with two l.i.h. dielectrics with permittivities shown. The total volume is divided exactly into halves by

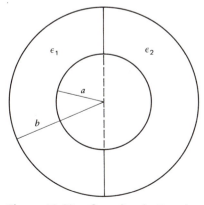

Figure 10-20. Capacitor in Exercise 10-28.

a plane that passes through the common center of the spherical conductors. Show that the capacitance is given by $C = 2\pi(\epsilon_1 + \epsilon_2)ab/(b - a)$.

10-29 The spherical capacitor of Figure 10-17 now has two different dielectrics between its plates. The value of κ_e is κ_{e1} for $a \leq r < r_0$ and κ_{e2} for $r_0 \leq r \leq b$. Find the capacitance of this system by finding the total energy of the fields between the plates.

10-30 The coaxial cylindrical capacitor of Figure 6-12 has a dielectric between its plates for which the dielectric constant varies as $\kappa_e = \kappa_0\rho^n$ where κ_0 and n are positive constants. Find the capacitance of a length L of this system by finding the energy in the fields between the plates.

10-31 A l.i.h. dielectric sphere of radius a also contains a free charge of constant density ρ_0. Find $\mathbf{D}$ and $\mathbf{E}$ for all points inside the sphere. Find the total energy of the fields inside the sphere. What is the total bound volume charge? Find the potential ϕ due to this system for all r. What is ϕ at the center of the sphere?

10-32 The parallel plate capacitor of Figure 10-18 has square plates of edge L. When the dielectric is in a distance x, show that the capacitance as a function of x is given by $C(x) = (\epsilon_0 L/d)[L + (\kappa_e - 1)x]$.

10-33 The parallel plate capacitor of Figure 10-18 has square plates of edge L. It is connected to a battery of potential difference $\Delta\phi_0$ when there is no dielectric between the plates. While it is kept connected to the battery, the dielectric is inserted halfway into the region between the plates. The battery is then disconnected. After this, the dielectric is inserted until it fills the whole region between the plates. Find the final values of $\Delta\phi$, Q, and E.

10-34 Consider the parallel plate capacitor of Figure 10-18. Find the total force on the face of the dielectric as a function of x. Does $F(x)$ increase or decrease as x increases? Show that when you average your expression for $F(x)$ over x, the result is (10-99). *Hint*: recall Exercise 10-32.

10-35 Two square conducting plates of length L on a side are placed a distance d apart to form a parallel plate capacitor. They are kept at a potential difference $\Delta\phi$ while a slab of dielectric of thickness $t < d$ that is also of length L on a side is inserted parallel to the edges of the plates. Neglect edge effects and find the average force with which the dielectric is drawn between the plates.

10-36 A parallel plate capacitor with square plates of edge L and separation d is given a charge Q and disconnected from the battery. It is then placed vertically with one edge immersed in a reservoir of liquid dielectric of mass density ρ_l. Neglect edge effects and show that the liquid will rise to a height h within the capacitor where

$$h = \frac{L}{2(\kappa_e - 1)}\left\{\left[1 + \frac{4(\kappa_e - 1)^2 Q^2}{\epsilon_0 g \rho_l L^5}\right]^{1/2} - 1\right\}$$

10-37 Two coaxial conducting cylinders like those of Figure 6-12 are lowered vertically into a reservoir of liquid dielectric of mass density ρ_l. If a potential difference $\Delta\phi$ is applied, the liquid will rise to a height h between the plates. Neglect edge effects and show that the susceptibility of the liquid is given by

$$\chi_e = \frac{\rho_l gh(b^2 - a^2)\ln(b/a)}{2\epsilon_0(\Delta\phi)^2}$$

(Do not forget the battery.)

10-38 Derive (10-92). It is best to proceed in the following way. First, write the integrand in (10-91) as $\mathbf{E}\cdot(\mathbf{D} - \mathbf{D}_0) + \mathbf{E}_0\cdot(\mathbf{D} - \mathbf{D}_0) + (\mathbf{E}\cdot\mathbf{D}_0 - \mathbf{E}_0 \cdot \mathbf{D})$. Then, with the use of (5-3), (1-115), (10-41), (1-59), and arguments analogous to those which led to (7-27), you should be able to show that the first two terms in the newly written integrand each contribute zero to (10-91). Now show that the third term is zero in the vacuum outside the dielectric. Finally, the further use of (10-40) as applied to $(\mathbf{D}_0, \mathbf{E}_0)$ and $(\mathbf{D}, \mathbf{E})$ will lead to (10-92).

10-39 We noted at the beginning of Section 10-9 that the forces on dielectrics can be ascribed to electric forces acting on the bound charges. Therefore, it may seem puzzling that we found a force on the dielectric of Figure 10-18 that is in the x direction while the electric field, which was assumed for simplicity to be like that of Figure 10-15b, is *perpendicular* to the direction of this resultant force. The origin of the force can be found in the very thing we neglected, namely the "edge effects" or curvature of the actual electric field lines as illustrated in Figure 6-10. Show qualitatively that the net force on the bound surface charges as produced by the curved field lines will really be in the x direction as we found from energy considerations.

11

SPECIAL METHODS IN ELECTROSTATICS

Up to now, we have concentrated on finding the scalar potential ϕ by integration over a given distribution of source charges by using (5-7), for example, and then obtaining the electric field from $\mathbf{E} = -\nabla\phi$. As we noted in the paragraph following (5-14), some problems are stated in such a way that this is not feasible so that it is desirable to have other methods available. Such an approach is to regard the problem of finding the potential as that of solving the partial differential equation satisfied by ϕ. This is Poisson's equation, as given by (5-15):

$$\nabla^2\phi = -\frac{\rho}{\epsilon_0} \tag{11-1}$$

where ρ is the total charge density. We also found in (10-55) that it can be written solely in terms of the free charge density for a linear isotropic homogeneous dielectric:

$$\nabla^2\phi = -\frac{\rho_f}{\epsilon} \tag{11-2}$$

If the relevant charge densities are zero, both of these reduce to Laplace's equation

$$\nabla^2\phi = 0 \tag{11-3}$$

Because of the comparative simplicity of Laplace's equation we will concentrate on solving it. We will, however, discuss one example of finding the solution of Poisson's equation in Section 11-6.

Over the years, many methods have been devised for solving these equations. Some of these methods are very general and systematic, while others are extremely specialized and of limited applicability. We will consider samples of both types. The motivation and justification for much of what we will do is provided by an important theorem that we consider next.

11-1 UNIQUENESS OF THE SOLUTION OF LAPLACE'S EQUATION

We want to show that, if we have found a solution of Laplace's equation that satisfies the given boundary conditions (i.e., it reduces to the correct preassigned values on all points of the surface surrounding the region), then this solution is unique. The term "boundary conditions" is now being used in a different sense from that of Chapter 9. There it referred to the necessary behavior of fields at a bounding surface of discontinuity between two media. Here we are assuming that we are dealing with a region that is surrounded by a surface for which the numerical value of the potential is given or known at all points; that is, we do not know the details of the source charge distribution outside this region, but we do know the potential produced by it on the surface. For example, part of the boundary may be a conductor kept at a potential of 12 volts by a battery. Then, whatever ϕ may be elsewhere, it must reduce to the value 12 volts whenever the field point is located anywhere on the conductor. Sometimes the boundary involved is very far away, that is, at infinity.

It will be sufficient for our purposes to deal with a somewhat special case. We assume that, in addition to satisfying Laplace's equation, ϕ is constant on all points of the bounding surface S:

$$\phi = \text{const.} \qquad \text{on boundary} \qquad (11\text{-}4)$$

Now if we use (1-115), (1-45), (1-17), and (11-3), we find that

$$\nabla \cdot (\phi \nabla \phi) = \nabla \phi \cdot \nabla \phi + \phi \nabla^2 \phi = (\nabla \phi)^2 \qquad (11\text{-}5)$$

which, when used in (1-59), leads to

$$\int_V (\nabla \phi)^2 \, d\tau = \int_V \nabla \cdot (\phi \nabla \phi) \, d\tau = \oint_S (\phi \nabla \phi) \cdot d\mathbf{a} = -\phi \oint_S \mathbf{E} \cdot d\mathbf{a} = 0 \quad (11\text{-}6)$$

because of (11-4) and Gauss' law (4-1) since $Q_{in} = 0$. As the integrand in the first integral of (11-6) is a sum of squares, and thereby intrinsically positive, the integral can be zero only if the integrand itself is zero everywhere; therefore

$$(\nabla \phi)^2 = \left(\frac{\partial \phi}{\partial x} \right)^2 + \left(\frac{\partial \phi}{\partial y} \right)^2 + \left(\frac{\partial \phi}{\partial z} \right)^2 = 0 \qquad (11\text{-}7)$$

according to (1-37). The expression in (11-7) is again a sum of squares, and therefore the individual terms must be zero, or

$$\frac{\partial \phi}{\partial x} = \frac{\partial \phi}{\partial y} = \frac{\partial \phi}{\partial z} = 0$$

so that $\phi = \text{const.}$ But, since $\phi = \text{const.}$ on the boundary, we see that

$$\phi = \text{const.} \qquad \text{everywhere} \qquad (11\text{-}8)$$

Now it is easy to prove our uniqueness theorem. We let $\phi_1(\mathbf{r})$ be a solution of (11-3) that satisfies the given boundary conditions. We also assume that there is another distinct solution $\phi_2(\mathbf{r})$ satisfying these same boundary conditions. We want to prove that ϕ_1 and ϕ_2 are identical. For this case, we let $\phi = \phi_1 - \phi_2$. Then $\nabla^2 \phi = \nabla^2 \phi_1 - \nabla^2 \phi_2 = 0$ because of (11-3); therefore ϕ is also a solution of Laplace's equation. On the bounding surface S, $\phi_1 = \phi_2$ so that $\phi = 0$ on the boundary. Since zero is a constant, this ϕ satisfies all of the conditions of the previous result; thus (11-8) shows that $\phi = 0$ everywhere, and therefore

$$\phi_1 = \phi_2 \qquad \text{everywhere} \qquad (11\text{-}9)$$

which is what we were to prove.

Sometimes the boundary conditions are stated in terms of the components of the field, that is, in terms of the derivatives of the potential rather than its absolute value. One can prove analogous uniqueness theorems for Laplace's equation that are applicable in such situations; an example is given in Exercise 11-1. Similarly, one can prove a uniqueness theorem for Poisson's equation, although we do not find it necessary to do so.

The significance of this result (11-9) from a practical point of view is that, once we find a solution of Laplace's equation, by any means whatsoever, which satisfies given boundary conditions, we know that it is the only solution and we *do not* need to consider the possibility that there are others. "Any means whatsoever" can include systematic methods, shrewd guesses, lucky guesses, or simply remembering past solutions and giving them a new twist.

■   **Example**

Near the end of Section 6-1, we considered conductors with cavities as illustrated in Figures 6-5 and 6-6. If there is no net charge inside, we concluded that the cavity (and any conductors therein) forms an equipotential volume with the same potential as the surrounding conductor so that the electric field is zero everywhere within the cavity. But, since the inner surface of the surrounding conductor has a constant potential, (11-4) is satisfied and thus the result (11-8) applies to these cases, showing us once again that the cavity (and any conductor within it) is an equipotential volume.   ■

The first methods of solving (11-3) that we consider are quite interesting, but also rather specialized.

11-2 METHOD OF IMAGES

We recall that Coulomb's law led us to the expression (5-2) for the potential of a system of point charges, where the contribution of each charge is proportional to $1/R$ where R is the distance from the charge to the field point. Therefore, such an expression must necessarily satisfy Laplace's equation; we also see this explicitly from (1-144). In other words, the sum of individual potentials from a set of point charges is automatically a solution of Laplace's equation. This fact is the basis for the *method of images*. Here the aim is to find a set of *fictitious* charges (image charges) that, together with any actual charges that may be present, will enable us to satisfy the boundary conditions and thus obtain the unique potential function. That is, we *try* to write the potential in the form

$$\phi = \sum_{\text{actual}} \frac{q_a}{4\pi\epsilon_0 R_a} + \sum_{\text{image}} \frac{q_i}{4\pi\epsilon_0 R_i} \tag{11-10}$$

and find the right combination as best we can. The basic idea is that the image charges will somehow *simulate* the behavior of the other source charges or of the bulk material; accordingly, the image charges are chosen to be located outside the region in which we are trying to find ϕ. This method is best illustrated by specific examples.

■   **Example**

Point charge and a semiinfinite plane grounded conductor. As shown in Figure 11-1, the point charge q is at a distance d from the yz plane that is the surface of a conductor occupying all space to the left of this plane, that is, for all negative values of x. The other half of space is a vacuum. The boundary condition is that $\phi = $ const. at $x = 0$ according to (6-2). For simplicity, we take this constant value to be zero (the conductor is "grounded"); if the actual value is a different constant, we can simply add it to our final result. Thus, our boundary condition is

$$\phi(0, y, z) = 0 \tag{11-11}$$

for all y and z. We will try to use (11-10) to satisfy (11-11) with only a single image charge q' also located on the x axis a distance d' into the conductor (and thereby replacing the conductor as far as the vacuum region is concerned). Since the coordinates of q and q' are $(d, 0, 0)$ and $(-d', 0, 0)$, respectively, we find that (11-10) together with (5-6) gives

$$\phi(x, y, z) = \frac{1}{4\pi\epsilon_0} \left\{ \frac{q}{\left[(x-d)^2 + y^2 + z^2\right]^{1/2}} + \frac{q'}{\left[(x+d')^2 + y^2 + z^2\right]^{1/2}} \right\}$$
$$\tag{11-12}$$

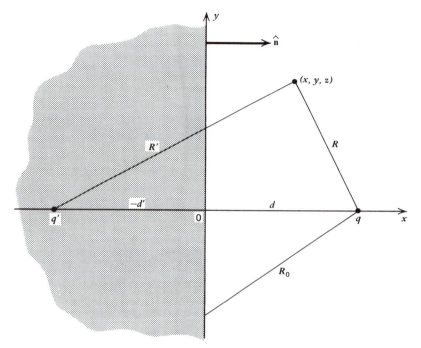

Figure 11-1. Point charge and a semiinfinite plane grounded conductor. q' is the image charge.

When this is combined with (11-11), we see that we must satisfy the condition

$$\frac{q}{(d^2 + y^2 + z^2)^{1/2}} + \frac{q'}{(d'^2 + y^2 + z^2)^{1/2}} = 0 \qquad (11\text{-}13)$$

Clearly this will always be satisfied if $d' = d$ and $q' = -q$. Hence q' is as far "behind" the boundary as q is in "front" of it, so that the term "image" charge fits very well; we note that the charge sign changed in this process—this is characteristic. If we now put these values we have just found into (11-12), we obtain the *unique* expression for the potential

$$\phi(x, y, z) = \frac{q}{4\pi\epsilon_0} \left\{ \frac{1}{\left[(x - d)^2 + y^2 + z^2\right]^{1/2}} - \frac{1}{\left[(x + d)^2 + y^2 + z^2\right]^{1/2}} \right\} \qquad (11\text{-}14)$$

and it is the complete solution to the problem. We use (11-14) only for $x \geq 0$; for $x < 0$, ϕ has the same value zero that it has at the surface of the conductor as we know from (6-1).

We can now find the electric field components from (5-3):

$$E_x = -\frac{\partial \phi}{\partial x} = \frac{q}{4\pi\epsilon_0} \left\{ \frac{(x - d)}{\left[(x - d)^2 + y^2 + z^2\right]^{3/2}} - \frac{(x + d)}{\left[(x + d)^2 + y^2 + z^2\right]^{3/2}} \right\}$$

$$E_y = -\frac{\partial \phi}{\partial y} = \frac{qy}{4\pi\epsilon_0} \left\{ \frac{1}{\left[(x - d)^2 + y^2 + z^2\right]^{3/2}} - \frac{1}{\left[(x + d)^2 + y^2 + z^2\right]^{3/2}} \right\} \qquad (11\text{-}15)$$

$$E_z = -\frac{\partial \phi}{\partial z} = \frac{qz}{4\pi\epsilon_0} \left\{ \frac{1}{\left[(x - d)^2 + y^2 + z^2\right]^{3/2}} - \frac{1}{\left[(x + d)^2 + y^2 + z^2\right]^{3/2}} \right\}$$

We can check our solution by seeing whether these have the correct properties. At the surface of the conductor, E_y and E_z are tangential components and should vanish according to (6-2); we see at once from (11-15) that this is correct, for we find that $E_y(0, y, z) = E_z(0, y, z) = 0$. At the conductor, the normal component of **E** is $E_n = \hat{\mathbf{n}} \cdot \mathbf{E} = \hat{\mathbf{x}} \cdot \mathbf{E} = E_x$ and we find that

$$E_n = E_x(0, y, z) = -\frac{qd}{2\pi\epsilon_0(d^2 + y^2 + z^2)^{3/2}} = -\frac{qd}{2\pi\epsilon_0 R_0^3} \qquad (11\text{-}16)$$

where R_0 is the distance from q to the corresponding point on the plane $x = 0$ as is also shown in the figure. But we know that a nonvanishing E_n means that there is a surface density of charge (free charge, in this case), and we find from (11-6) and (6-4) that it is given by

$$\sigma_f(y, z) = -\frac{qd}{2\pi(d^2 + y^2 + z^2)^{3/2}} \qquad (11\text{-}17)$$

This surface charge is said to have been *induced* by the point charge q. We see that σ_f is not constant over the plane; its maximum magnitude of $q/2\pi d^2$ occurs at the origin directly below q, and $\sigma_f \to 0$ as y and z go to infinity. We can find the total charge induced on the yz plane by combining (11-17) with (2-16) and (1-55):

$$q_{\text{ind}} = -\frac{qd}{2\pi}\int_{-\infty}^{\infty}\int_{-\infty}^{\infty}\frac{dy\,dz}{(d^2 + y^2 + z^2)^{3/2}} = -\frac{qd}{\pi}\int_{-\infty}^{\infty}\frac{dz}{d^2 + z^2} = -q \quad (11\text{-}18)$$

where we have used our previous results (3-7), (3-8), and (3-12) to evaluate the integrals. Thus the total induced charge turns out to be equal and opposite to the inducing charge, and hence equal to the image charge as we should have expected since the image charge simulates the overall behavior of the conductor.

In order to find the force on q, we need the value of **E** at its location. We cannot, however, use the first terms in the braces of (11-15) since they represent the contribution due to q itself, as we see from (11-12), and would mean that q is exerting a force on itself—a possibility we have constantly excluded. Putting the coordinates of q, $(d, 0, 0)$, into the remaining terms of (11-15), we find that $E_y = E_z = 0$ and $E_x = -q/16\pi\epsilon_0 d^2$, so that the force on q is

$$\mathbf{F} = q\mathbf{E} = -\frac{q^2}{16\pi\epsilon_0 d^2}\hat{\mathbf{x}} \qquad (11\text{-}19)$$

and is directed toward the conductor. This clearly represents the resultant force of attraction between q and the induced surface charge σ_f as can be verified by direct integration of (2-17). If we rewrite (11-19) as

$$\mathbf{F} = -\frac{q^2}{4\pi\epsilon_0(2d)^2}\hat{\mathbf{x}} \qquad (11\text{-}20)$$

we see that it is exactly in the form of Coulomb's law for the attraction between q and the image charge $-q$, as they are separated by the total distance $2d$.

The equation giving the equipotential surfaces in the region $x \geq 0$ is obtained by setting ϕ equal to a constant in (11-14), and the equipotential curves in the xy plane are then found by setting $z = 0$. In terms of the distances R and R' of Figure 11-1, the equation of the equipotential surface as found from (11-14) is just

$$\frac{1}{R} - \frac{1}{R'} = \frac{4\pi\epsilon_0\phi}{q} = \text{const.} \qquad (11\text{-}21)$$

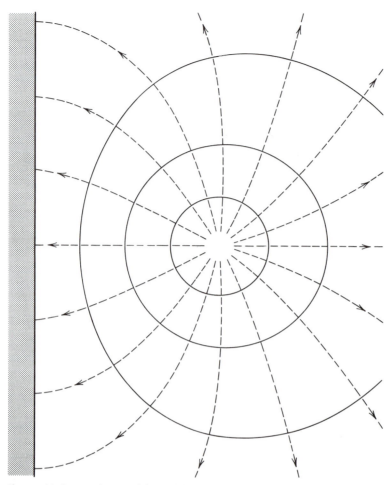

Figure 11-2. Equipotentials (solid) and electric field lines (dashed) for the system of Figure 11-1.

Some of these curves are shown as the solid lines in Figure 11-2. The dashed lines are the lines of $\mathbf{E}$. Their equation can be found from (5-39) and (11-15). ∎

■ **Example**

Point charge and a grounded conducting sphere. See Figure 11-3. We use spherical coordinates with origin at the center of the sphere of radius a and the z axis is chosen to pass through the location of q at a distance d from the center. The boundary condition is that the potential be zero on the surface of the sphere, that is,

$$\phi(a, \theta, \varphi) = 0 \tag{11-22}$$

We try to solve this problem with a single image charge q' at a distance d' from the center; we require $d' < a$ so that q' will be outside of the vacuum region. The potential at any field point P as obtained from (11-10), the law of cosines, and an

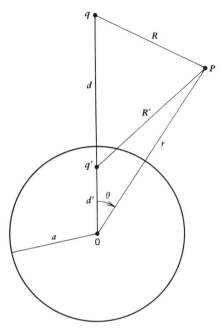

Figure 11-3. Point charge and a grounded conducting sphere.

inspection of the figure is

$$\phi(r, \theta, \varphi) = \frac{1}{4\pi\epsilon_0}\left(\frac{q}{R} + \frac{q'}{R'}\right)$$

$$= \frac{1}{4\pi\epsilon_0}\left[\frac{q}{(r^2 + d^2 - 2rd\cos\theta)^{1/2}} + \frac{q'}{(r^2 + d'^2 - 2rd'\cos\theta)^{1/2}}\right]$$

$$(11\text{-}23)$$

When this is combined with (11-22), we obtain the condition

$$\frac{q}{(a^2 + d^2 - 2ad\cos\theta)^{1/2}} + \frac{q'}{(a^2 + d'^2 - 2ad'\cos\theta)^{1/2}} = 0 \qquad (11\text{-}24)$$

from which we must find the two unknowns q' and d'. Since this will, in general, require two equations, and since (11-24) must hold for all values of θ, we can get these equations by using two particularly useful values of θ. We note that we can get rid of the square roots in the denominator by making each term in the parentheses form the square of something, and we see that this will occur for the values 0 and π for θ. When these are put into (11-24), we obtain the two equations

$$\frac{q}{d - a} + \frac{q'}{a - d'} = 0 \qquad (11\text{-}25)$$

$$\frac{q}{d + a} + \frac{q'}{a + d'} = 0 \qquad (11\text{-}26)$$

after using the fact that $d > a > d'$. Solving these, we find that

$$q' = -\frac{a}{d}q \qquad \text{and} \qquad d' = \frac{a^2}{d} \qquad (11\text{-}27)$$

In this case, the image charge again has the opposite sign to the inducing charge but is not of the same magnitude, and, in fact, $|q'| < |q|$. Substituting these results into (11-23), we find the potential that satisfies the boundary conditions, and thus gives the correct value everywhere outside the sphere, to be

$$\phi = \frac{q}{4\pi\epsilon_0} \left\{ \frac{1}{(r^2 + d^2 - 2rd\cos\theta)^{1/2}} - \frac{(a/d)}{\left[r^2 + (a^2/d)^2 - 2r(a^2/d)\cos\theta\right]^{1/2}} \right\}$$

(11-28)

The electric field components can now be found from $\mathbf{E} = -\nabla\phi$ and (1-101); the nonvanishing ones are

$$E_r = \frac{q}{4\pi\epsilon_0} \left\{ \frac{(r - d\cos\theta)}{R^3} - \frac{(a/d)\left[r - (a^2/d)\cos\theta\right]}{R'^3} \right\}$$

(11-29)

$$E_\theta = \frac{qd\sin\theta}{4\pi\epsilon_0} \left[\frac{1}{R^3} - \frac{(a/d)^3}{R'^3} \right]$$

(11-30)

At the surface of the sphere, $E_\theta(a) = 0$ as it should since it is a tangential component; however, $E_r(a) \neq 0$, and since this is a normal component at the surface, there is a surface density of charge given by

$$\sigma_f(\theta) = \epsilon_0 E_n = \epsilon_0 E_r(a, \theta) = \frac{-q(d^2 - a^2)}{4\pi a(a^2 + d^2 - 2ad\cos\theta)^{3/2}}$$

(11-31)

We can show again that the total induced charge is equal to the image charge. With the use of (2-16), (1-100), and (2-22), we get

$$q_{\text{ind}} = -\frac{q(d^2 - a^2)}{4\pi a} \int_0^{2\pi} \int_0^\pi \frac{a^2 \sin\theta \, d\theta \, d\varphi}{(a^2 + d^2 - 2ad\cos\theta)^{3/2}}$$

$$= -\frac{q(d^2 - a^2)a}{2} \int_{-1}^1 \frac{d\mu}{(a^2 + d^2 - 2ad\mu)^{3/2}}$$

(11-32)

The integral can be found from tables and is

$$\frac{1}{ad(a^2 + d^2 - 2ad\mu)^{1/2}} \bigg|_{-1}^1 = \frac{1}{ad} \left(\frac{1}{|d - a|} - \frac{1}{|d + a|} \right)$$

(11-33)

In this case, $d > a$, and (11-33) becomes $2/[d(d^2 - a^2)]$ so that

$$q_{\text{ind}} = -\frac{a}{d}q = q'$$

(11-34)

as we thought.

The charge q will be attracted toward the sphere by a force that can be found as the Coulomb force between q and its image q'. Their distance of separation is $D = d - d' = (d^2 - a^2)/d$, so that, according to (2-3),

$$\mathbf{F} = -\frac{adq^2\hat{\mathbf{z}}}{4\pi\epsilon_0(d^2 - a^2)^2}$$

(11-35)

If $d \gg a$, the force varies approximately as the inverse cube of the distance of separation.

The general appearance of the equipotentials and lines of $\mathbf{E}$ for this case are shown in Figure 11-4.   ∎

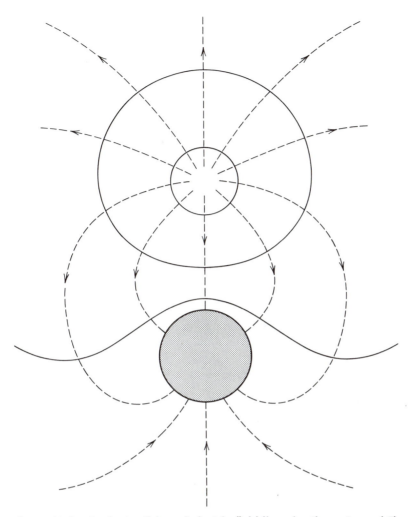

Figure 11-4. Equipotentials and electric field lines for the system of Figure 11-3.

■ **Example**

Point charge and an insulated uncharged conducting sphere. This is a variation on the previous example. The sphere is assumed to be originally neutral, and is no longer kept at a definite potential. In the presence of q, the sphere must still have zero net charge because it is no longer connected to something from which it can obtain charge as was the case in the last examples. It must also be an equipotential volume. We begin by introducing the same image charge $q' = -(a/d)q$ at the same position as the last example; this will make the surface of the sphere have constant (zero) potential. But, in order to keep the sphere neutral, we must put another charge $q'' = -q' = (a/d)q$ somewhere within the sphere. The only place we can put it *and* keep the sphere's surface an equipotential is at the center. Thus, we are led to the system of three charges shown in Figure 11-5. Since this charge distribution satisfies all of the requirements, it will lead to the correct potential and field everywhere outside the sphere. The calculation of most of the features of this example will be left as exercises, but we can easily find the final potential of the sphere. If we combine the results shown in the figure with

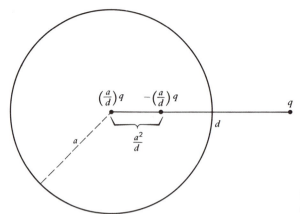

Figure 11-5. Point charge q and its images for an insulated uncharged conducting sphere.

(6-5), we find that

$$\phi_{\text{sphere}} = \phi(a) = \frac{q''}{4\pi\epsilon_0 a} = \frac{q}{4\pi\epsilon_0 d} \tag{11-36}$$

which, interestingly, is the potential that would be produced by q at the location of the center of the sphere if the sphere were not present at all. ∎

■ **Example**

Point charge and semiinfinite plane dielectric. The situation here is similar to that of Figure 11-1 except that the shaded region for all negative x is filled with a l.i.h. dielectric rather than with a conductor. As in the previous example, the potential does not have a preassigned value. This time the boundary conditions at the surface $x = 0$ are those satisfied by the components of **E** as given by (10-56) with $\sigma_f = 0$:

$$\hat{\mathbf{n}} \cdot (\epsilon_2 \mathbf{E}_2 - \epsilon_1 \mathbf{E}_1) = \epsilon_0 E_{2x} - \epsilon E_{1x} = 0$$

$$E_{2y} = E_{1y} \qquad E_{2z} = E_{1z} \tag{11-37}$$

We have taken region 1 as the dielectric ($x < 0$) and region 2 as the vacuum ($x > 0$) corresponding to the direction of $\hat{\mathbf{n}}$ shown.

In the vacuum, we try the same set of charges as in Figure 11-1, but we will assume at once that $d' = d$. Thus, as in (11-12), we get

$$\phi_2 = \frac{1}{4\pi\epsilon_0} \left\{ \frac{q}{\left[(x-d)^2 + y^2 + z^2\right]^{1/2}} + \frac{q'}{\left[(x+d)^2 + y^2 + z^2\right]^{1/2}} \right\} \tag{11-38}$$

which leads to field components

$$E_{2x} = \frac{1}{4\pi\epsilon_0} \left\{ \frac{(x-d)q}{\left[(x-d)^2 + y^2 + z^2\right]^{3/2}} + \frac{(x+d)q'}{\left[(x+d)^2 + y^2 + z^2\right]^{3/2}} \right\} \tag{11-39}$$

$$E_{2y} = \frac{y}{4\pi\epsilon_0} \left\{ \frac{q}{\left[(x-d)^2 + y^2 + z^2\right]^{3/2}} + \frac{q'}{\left[(x+d)^2 + y^2 + z^2\right]^{3/2}} \right\} \tag{11-40}$$

and where E_{2z} is obtained by replacing the y in the front of (11-40) by z.

We cannot use this same set of charges to calculate the potential in the dielectric since q' is located there, and our fictitious charges should be located outside the region

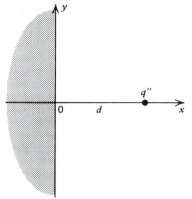

Figure 11-6. Image charge used to calculate the potential in the dielectric.

in question. Let us press our luck as much as possible and try using a *single* point charge q'' located at the position of q itself as shown in Figure 11-6. This will produce a potential in the dielectric given by

$$\phi_1 = \frac{q''}{4\pi\epsilon_0 \left[(x-d)^2 + y^2 + z^2\right]^{1/2}} \tag{11-41}$$

with the corresponding field components

$$E_{1x} = \frac{(x-d)q''}{4\pi\epsilon_0 \left[(x-d)^2 + y^2 + z^2\right]^{3/2}} \tag{11-42}$$

$$E_{1y} = \frac{yq''}{4\pi\epsilon_0 \left[(x-d)^2 + y^2 + z^2\right]^{3/2}} \tag{11-43}$$

and, again, E_{1z} is obtained by replacing y by z in the numerator of (11-43). Now we are ready to try to satisfy the boundary conditions on the field components.

The first equation of (11-37) is $\epsilon_0 E_{2x}(0, y, z) = \epsilon E_{1x}(0, y, z)$ and leads to

$$\epsilon_0(-q + q') = \epsilon(-q'') \tag{11-44}$$

while the second equation of (11-37) is $E_{2y}(0, y, z) = E_{1y}(0, y, z)$ and yields

$$q + q' = q'' \tag{11-45}$$

as does $E_{2z}(0, y, z) = E_{1z}(0, y, z)$. When we solve (11-44) and (11-45), and recall that $\epsilon/\epsilon_0 = \kappa_e$, we find that

$$q' = -\left(\frac{\kappa_e - 1}{\kappa_e + 1}\right)q \qquad q'' = \frac{2q}{\kappa_e + 1} \tag{11-46}$$

which means that we have managed to solve the problem completely.

Thus we are led to the two sets of image charges shown in Figure 11-7. The box around the region number indicates that the charges shown are to be used to calculate ϕ in that region. The complete solution to this problem is thus obtained by inserting the values of q' and q'' given by (11-46) into (11-38) through (11-43).

We can find the force on the inducing charge q at $x = d$ from (11-39) and (11-40) by using *only* the contribution from q', and it is seen to be

$$\mathbf{F} = q\mathbf{E}(d, 0, 0) = \frac{qq'\hat{\mathbf{x}}}{4\pi\epsilon_0(2d)^2} = -\left(\frac{\kappa_e - 1}{\kappa_e + 1}\right)\frac{q^2\hat{\mathbf{x}}}{16\pi\epsilon_0 d^2} \tag{11-47}$$

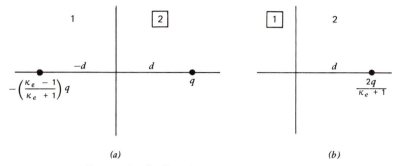

(a)   (b)

Figure 11-7.   Charges for finding the potential (a) outside the dielectric and (b) inside the dielectric.

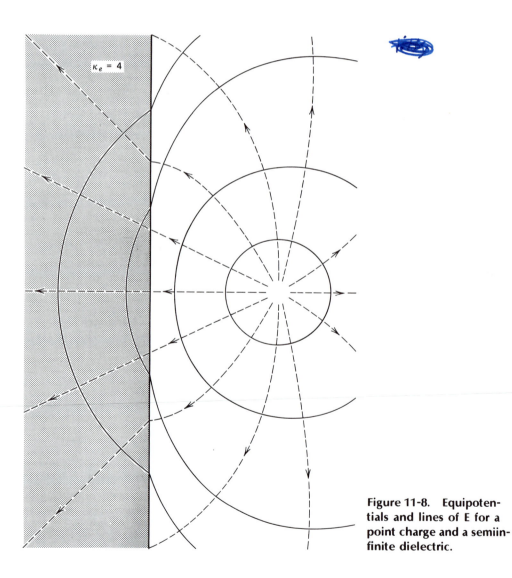

Figure 11-8.   Equipotentials and lines of E for a point charge and a semiinfinite dielectric.

and, as before, is the same as the Coulomb force of attraction between q and q', the induced image charge in the dielectric. Physically, this is just the force of attraction between q and the bound surface charges on the dielectric. We can find their density by using (10-8), (10-50), (10-52), (11-42), and (11-46):

$$\sigma_b(y, z) = P_{1n} = (\kappa_e - 1)\epsilon_0 E_{1x}(0, y, z) = -\left(\frac{\kappa_e - 1}{\kappa_e + 1}\right)\frac{qd}{2\pi(d^2 + y^2 + z^2)^{3/2}}$$

$$(11\text{-}48)$$

and is negative as we expect. By comparing (11-17) and (11-18) with (11-48), we see that the total bound charge induced on the dielectric surface will again equal the image charge q'.

The equipotential curves and lines of $\mathbf{E}$ for this system are shown in Figure 11-8. As expected, we see that the lines of $\mathbf{E}$ are refracted as they cross the bounding surface of the dielectric.   ∎

11-3 "REMEMBRANCE OF THINGS PAST"

Sometimes a solution found for a previous problem can, with appropriate interpretation, be adapted and used to solve what at first appears to be a completely new problem. We will illustrate this with two examples, both of which can be related to an example from Chapter 5.

∎ **Example**

Uniform infinite line charge and a semiinfinite plane grounded conductor. We consider an infinitely long line charge of constant charge λ per unit length that is a distance a from and parallel to the surface of a grounded conductor occupying half of all space. If we take the line to be parallel to the z axis, the surface of the conductor to be the yz plane, and choose the x axis to pass through the line charge, we get the situation in the xy plane shown in Figure 11-9. The boundary condition is that the potential be constant and equal to zero on the conducting surface, that is, $\phi(0, y, z) = 0$. Because of the similarities of Figures 11-9 and 11-1, as well as the identity of this boundary condition with (11-11), we can make a good guess that the appropriate image charge to use in this case is another infinite line charge of density $-\lambda$ located at $x = -a$. But this is exactly the charge distribution shown in Figure 5-7, for which the potential is given everywhere by (5-34), and whose equipotentials and lines of $\mathbf{E}$ are illustrated in Figure 5-8. (We

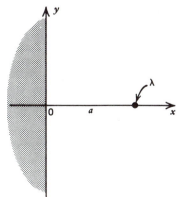

Figure 11-9. Uniform infinite line charge perpendicular to the page and parallel to a semiinfinite plane grounded conductor.

recall that the equipotential surfaces are actually cylinders whose axes are parallel to the z axis and, in fact, these axes lie in the xz plane.) As we noted in the discussion following (5-38), the yz plane ($x = 0$) is the equipotential surface for $\phi = 0$, which is exactly the boundary condition (11-11) that we have to satisfy for this problem. Therefore, the solution is given by our previous results. However, we can use (5-34) only in the unshaded vacuum region of Figure 11-9, that is, $-(\pi/2) \le \varphi \le \pi/2$. Thus, the equipotentials and lines of force for this example are given by the curves in the right half of Figure 5-8. In other words, we already have the complete solution available for this example.   ∎

We can go even further with what we found for the complete system of Figure 5-7. Suppose that one of the equipotential cylinders were replaced by a solid conductor occupying the volume enclosed by the cylinder. The surface of the conductor would be an equipotential as required and would have the potential corresponding to the surface it replaced. The electric field will be normal to the conductor as required since the lines of **E** are already normal to the equipotential cylinder. There will be a surface charge on the cylinder given by (6-4), but, as we can easily see, the *total* charge per unit length on the cylinder will still be λ (assuming a surface in the right half of Figure 5-8). Consider a Gaussian surface of integration just outside the conductor. The value of **E** will still be found from (5-35) and (5-36) so that the surface integral of Gauss' law (4-1) will be the same as if the line charge were there. But since Gauss' law equates the surface integral to the total charge inside divided by ϵ_0, regardless of its distribution, the total charge per unit length will be λ and this will be the total charge on the conductor's surface. In other words, *outside* the conductor, *nothing has changed*, and we can still use (5-34) everywhere else. [Inside, things have changed, of course: the electric field is now zero and ϕ is constant, according to (6-1).] Similar remarks apply to the left-hand part of Figure 5-8. We now are ready to consider our next example.

■   **Example**

Capacitance of two parallel cylindrical conductors. Consider two infinitely long conducting cylinders whose axes are parallel. For simplicity, we assume that they have the same radius A; their axes are separated by a distance D as shown in Figure 11-10. If we identify these cylinders with the appropriate equipotentials of Figure 5-8, they will carry charges of $-\lambda$ and $+\lambda$ per unit length. Our problem now is to relate these dimensions to our previous results and to find the potential difference. In (5-38), we found the radius of a circle of Figure 5-8 to be $a/\sinh \eta$ and the location of its center with respect to the origin is $a \coth \eta$ where $\eta = 2\pi\epsilon_0\phi/\lambda$. Therefore, if ϕ is the potential of the right-hand cylinder of charge λ, we have

$$A = \frac{a}{\sinh \eta} \tag{11-49}$$

$$\frac{D}{2} = a \coth \eta = A \cosh \eta \tag{11-50}$$

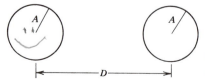

Figure 11-10.   Cross-sectional view of two parallel cylindrical conductors.

Solving the last equation for η, we find the potential of the cylinder to be

$$\phi = \frac{\lambda}{2\pi\epsilon_0} \cosh^{-1}\left(\frac{D}{2A}\right) \tag{11-51}$$

Since the potential of the other cylinder is $-\phi$, the potential difference between the two will be

$$\Delta\phi = 2\phi = \frac{\lambda}{\pi\epsilon_0} \cosh^{-1}\left(\frac{D}{2A}\right) \tag{11-52}$$

As the charge on a length L of this system will be λL, we see, from (6-28), that the capacitance for a length L will be

$$C = \frac{\lambda L}{\Delta\phi} = \frac{\pi\epsilon_0 L}{\cosh^{-1}(D/2A)} \tag{11-53}$$

If $A \ll D$, as would be appropriate for two thin wires, we find from the series expansion $\cosh^{-1} u \simeq \ln 2u - (1/4u^2) - \cdots$ that we can approximate (11-53) as

$$C \simeq \frac{\pi\epsilon_0 L}{\ln(D/A) - (A/D)^2} \simeq \frac{\pi\epsilon_0 L}{\ln(D/A)} \tag{11-54}$$

which is a commonly found expression. [This result also agrees with that of Exercise 6-15 when applied to this case.]   ∎

We now want to consider some of the more systematic methods for solving Laplace's equation.

11-4 SEPARATION OF VARIABLES IN RECTANGULAR COORDINATES

When (11-3) is expressed in rectangular coordinates with the use of (1-46), it becomes

$$\frac{\partial^2\phi}{\partial x^2} + \frac{\partial^2\phi}{\partial y^2} + \frac{\partial^2\phi}{\partial z^2} = 0 \tag{11-55}$$

We will try to solve this by assuming a solution in the form of a product of quantities each of which is a function of a single variable:

$$\phi(x, y, z) = X(x)Y(y)Z(z) \tag{11-56}$$

Upon substituting this into (11-55) and dividing by XYZ, we obtain

$$\frac{1}{X}\frac{d^2X}{dx^2} + \frac{1}{Y}\frac{d^2Y}{dy^2} + \frac{1}{Z}\frac{d^2Z}{dz^2} = 0 \tag{11-57}$$

which can also be written as

$$\frac{1}{X}\frac{d^2X}{dx^2} + \frac{1}{Y}\frac{d^2Y}{dy^2} = -\frac{1}{Z}\frac{d^2Z}{dz^2} \tag{11-58}$$

The left-hand side of (11-58) is a function only of x and y, while the right-hand side is a function only of z. But (11-58) says that both sides must always be equal for any and all values of x, y, and z—quantities that can be varied independently. This is only possible if both sides are equal to the same *constant*, which we write as $-\gamma^2$. Thus, we have

$$\frac{1}{Z}\frac{d^2Z}{dz^2} = \gamma^2 \tag{11-59}$$

Repeating this process, we find that

$$\frac{1}{X}\frac{d^2X}{dx^2} = \alpha^2 \qquad \frac{1}{Y}\frac{d^2Y}{dy^2} = \beta^2 \tag{11-60}$$

where α^2 and β^2 are also constants. These constants are not independent, for when we substitute (11-59) and (11-60) into (11-57), we find that

$$\alpha^2 + \beta^2 + \gamma^2 = 0 \tag{11-61}$$

We have seen that $d^2X/dx^2 = \alpha^2 X$ and this can be integrated at once to give

$$X(x) = a_1 e^{\alpha x} + a_2 e^{-\alpha x} \tag{11-62}$$

where a_1 and a_2 are constants of integration. Similarly,

$$Y(y) = b_1 e^{\beta y} + b_2 e^{-\beta y} \tag{11-63}$$

$$Z(z) = c_1 e^{\gamma z} + c_2 e^{-\gamma z} \tag{11-64}$$

The product of these three functions will be a solution of (11-55), provided that (11-61) is satisfied. Because of this condition, α^2, β^2, and γ^2 cannot all be positive or all negative; this means, in turn, that the constants α, β, and γ cannot all be real or all imaginary so that if one is real, at least one of the others must be imaginary. Consequently, at least one of the functions X, Y, and Z must vary exponentially with its argument, while at least one will vary sinusoidally.

Evidently, there are many combinations of α, β, γ that will satisfy (11-61), and each combination will give a solution, so that there are many possibilities. At the same time, the constants of integration a_1, a_2, b_1,... may themselves depend on the particular values of α, β, γ, so that we should write them in the form $a_1(\alpha)$, $a_2(\alpha)$, $b_1(\beta)$,.... Since Laplace's equation is a linear equation, a sum of solutions of the form (11-56) will also be a solution; if we add up all the possibilities, we see that we can write the most general solution to Laplace's equation in rectangular coordinates in the form

$$\phi(x, y, z) = \sum \left[a_1(\alpha) e^{\alpha x} + a_2(\alpha) e^{-\alpha x} \right] \left[b_1(\beta) e^{\beta y} + b_2(\beta) e^{-\beta y} \right]$$
$$\left[c_1(\gamma) e^{\gamma z} + c_2(\gamma) e^{-\gamma z} \right] \tag{11-65}$$

where the sum is to be taken over all values of α, β, γ that satisfy (11-61). Since there undoubtedly are an infinite number of combinations of α, β, γ that meet this requirement, we see that our general solution contains an infinite number of constants of integration. The basic idea in the application of (11-65) is that these constants must be determined so that ϕ will satisfy the given boundary conditions. Once this has been done, the problem is completely solved, and we know from the theorem of Section 11-1 that the solution obtained for ϕ will be unique. In using this method, then, one would start with (11-65), and satisfy the boundary conditions step by step in a systematic manner; no guesswork should be required. As usual, all of this is best illustrated with a specific example.

■ **Example**

We consider a region bounded by: (1) a semiinfinite conducting plane at $x = 0$ occupying that half of the yz plane corresponding to positive y (thus, $0 \le y \le \infty$, $-\infty \le z \le \infty$); (2) a similar plane at $x = L$; and (3) the strip between them in the xz plane (thus, $0 \le x \le L$). The region defined in this way is shown in Figure 11-11a, and its projection on the xy plane is shown shaded in Figure 11-11b.

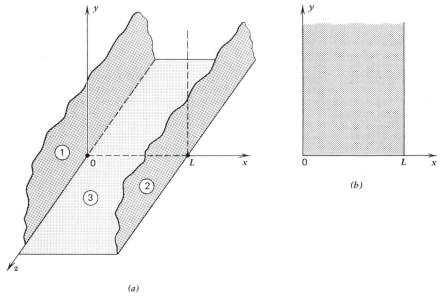

(a)

Figure 11-11. (*a*) Two semiinfinite conducting planes parallel to the *yz* plane. (*b*) Their projection on the *xy* plane.

Let us assume the following boundary conditions:

$$\text{at } x = 0 \qquad \phi(0, y, z) = 0 \tag{11-66}$$

$$\text{at } x = L \qquad \phi(L, y, z) = 0 \tag{11-67}$$

$$\text{at } y = \infty \qquad \phi(x, \infty, z) = 0 \tag{11-68}$$

$$\text{at } y = 0 \qquad \phi(x, 0, z) = f(x) \tag{11-69}$$

where $f(x)$ is some given function, that is, the potential on strip 3 varies with x *only* in some predetermined manner. This set of boundary conditions could correspond to the semiinfinite sheets being grounded conductors, while strip 3 in the xz plane is maintained in the manner described by $f(x)$ by a suitable arrangement of batteries; (11-68) corresponds to the usual requirement that the potential vanish at an infinite distance from the source charges.

Now since the region of interest extends over the complete range of z, and there is no dependence on z in the boundary conditions, this situation is actually independent of z and *is really a two-dimensional problem* so that $\phi = \phi(x, y)$. Therefore, we see from (11-65) that $\gamma = 0$ is the only allowable value, and (11-65) reduces to

$$\phi = \sum_{\alpha^2 + \beta^2 = 0} \left[A_1(\alpha)e^{\alpha x} + A_2(\alpha)e^{-\alpha x} \right] \left[b_1(\beta)e^{\beta y} + b_2(\beta)e^{-\beta y} \right] \tag{11-70}$$

where we have set $A_1(\alpha) = a_1(\alpha)[c_1(0) + c_2(0)]$, and so on. [Actually, if $\gamma = 0$, then (11-59) will integrate to $Z(z) = c_1' + c_2'' z$ but, in order to have our solution independent of z, c_2'' must be zero, which again will result in the form given by (11-70).] Since $\alpha^2 + \beta^2 = 0$, we have $\alpha^2 = -\beta^2$, so that $\alpha = i\beta$, where $i = \sqrt{-1}$, and there is only one independent quantity to be summed over and we can write

$$\phi = \sum_{\beta} \left[A_1(\beta)e^{i\beta x} + A_2(\beta)e^{-i\beta x} \right] \left[b_1(\beta)e^{\beta y} + b_2(\beta)e^{-\beta y} \right] \tag{11-71}$$

This reduction of (11-65) to (11-71) followed solely from the fact that the problem is really independent of z—which is a "boundary condition" in a generalized sense. Now we turn to the boundary conditions that give actual values of ϕ.

By comparing (11-68) and (11-71), we see that if β is positive, the term varying like $e^{\beta y}$ cannot appear because it does not vanish at infinity so that $b_1(\beta)$ must be zero. On the other hand, if β is negative, $b_2(\beta)$ must be zero. In both cases, the remaining term will vary as $e^{-|\beta|y}$, so that there is really only one possible form. For definiteness, we choose $\beta > 0$; thus we must have $b_1(\beta) = 0$, and then (11-71) becomes

$$\phi = \sum_{\beta > 0} b_2(\beta)\left[A_1(\beta)e^{i\beta x} + A_2(\beta)e^{-i\beta x}\right]e^{-\beta y} \tag{11-72}$$

[β cannot be zero, since then (11-68) will not be satisfied.] If we define $b_2(\beta)A_1(\beta) = A_\beta$, $b_2(\beta)A_2(\beta) = B_\beta$, then (11-72) can be written more simply as

$$\phi = \sum_{\beta > 0} \left(A_\beta e^{i\beta x} + B_\beta e^{-i\beta x}\right)e^{-\beta y} \tag{11-73}$$

If we now use (11-66), we get

$$\phi(0, y) = 0 = \sum_{\beta > 0} \left(A_\beta + B_\beta\right)e^{-\beta y} \tag{11-74}$$

The only way in which this can be zero for arbitrary y, since $e^{-\beta y}$ is always positive, is for each term in the sum to be zero; thus $A_\beta + B_\beta = 0$, so that $B_\beta = -A_\beta$. The term in the parentheses in (11-73) can be then written as $A_\beta(e^{i\beta x} - e^{-i\beta x}) = 2iA_\beta \sin \beta x$ and ϕ is now

$$\phi = \sum_{\beta > 0} 2iA_\beta \sin \beta x e^{-\beta y} \tag{11-75}$$

(Since ϕ must be a real quantity, we see that A_β itself must be an imaginary constant.)

Applying (11-67) to (11-75), we have

$$\phi(L, y) = 0 = \sum_{\beta > 0} 2iA_\beta \sin \beta L e^{-\beta y} \tag{11-76}$$

which shows us that $\sin \beta L = 0$ and therefore $\beta L = n\pi$ or

$$\beta = \frac{n\pi}{L} \tag{11-77}$$

where n is a *positive* integer (since β is positive). Thus our sum over β is actually a sum over n and it is convenient to write it that way; if we set $2iA_\beta = A_n$, and use (11-77), we can write (11-76) as

$$\phi(x, y) = \sum_{n=1}^{\infty} A_n \sin\left(\frac{n\pi x}{L}\right)e^{-(n\pi y)/L} \tag{11-78}$$

and all that remains to be found are the constant coefficients A_n. For this purpose, we have one boundary condition remaining.

Putting $y = 0$ into (11-78), and setting the result equal to $f(x)$, according to (11-69), we get

$$\phi(x, 0) = f(x) = \sum_{n=1}^{\infty} A_n \sin\left(\frac{n\pi x}{L}\right) \tag{11-79}$$

which shows that our remaining problem is that of expanding $f(x)$ in a Fourier series (using only the sine terms). This is done by using the first of the following results that are easily proved by direct integration

$$\int_0^L \sin\left(\frac{m\pi x}{L}\right)\sin\left(\frac{n\pi x}{L}\right) dx = \int_0^L \cos\left(\frac{m\pi x}{L}\right)\cos\left(\frac{n\pi x}{L}\right) dx = \frac{1}{2}L\delta_{mn} \tag{11-80}$$

that is, these integrals are zero when $m \neq n$, and each equals $\frac{1}{2}L$ for $m = n$, as we see from (8-27). This result (11-80) is generally known as the orthogonality and normalization property of the trigonometric functions; when this integral is zero, the functions are said to be "orthogonal." Combining this result with (11-79) enables us to "pick out" any of the coefficients we desire as follows: we multiply both sides of (11-79) by $\sin(m\pi x/L)$ and integrate over the range L of x and we get

$$\int_0^L f(x) \sin\left(\frac{m\pi x}{L}\right) dx = \sum_n A_n \int_0^L \sin\left(\frac{m\pi x}{L}\right) \sin\left(\frac{n\pi x}{L}\right) dx$$

$$= \sum_n A_n \cdot \frac{1}{2}L\delta_{mn} = \frac{1}{2}LA_m$$

since each term in the sum is zero except when $n = m$. Changing the index back to n and solving for A_n, we obtain

$$A_n = \frac{2}{L} \int_0^L f(x) \sin\left(\frac{n\pi x}{L}\right) dx \qquad (11\text{-}81)$$

Once $f(x)$ has been given, we can determine the coefficients A_n by carrying out the integration indicated in (11-81), and then the A_n can be put into (11-78). The result will be the correct unique value for the potential from which ϕ, and then $\mathbf{E}$, can be found at any point by evaluating the sum. ∎

Example

Special case. In order to have a definite example, let us now assume that $f(x) = \phi_0 = $ const. This would correspond to keeping the whole strip in the xz plane at a constant potential, as would be the case if it were a conductor connected to a single battery. (It would have to be insulated from the conducting walls by a thin strip of dielectric on each edge.) Anyhow, putting this into (11-81), we get

$$A_n = \frac{2}{L} \int_0^L \phi_0 \sin\left(\frac{n\pi x}{L}\right) dx = \phi_0 \frac{2}{n\pi}(1 - \cos n\pi) \qquad (11\text{-}82)$$

Now $(1 - \cos n\pi) = 2$ if n is an odd integer, and 0 if n is even, so that $A_n = 0$ if n is even, and $A_n = \phi_0(4/n\pi)$ if n is odd. Thus, the general solution (11-78) as applied to this special case becomes

$$\phi(x, y) = \phi_0 \frac{4}{\pi} \sum_{n \text{ odd}} \frac{1}{n} \sin\left(\frac{n\pi x}{L}\right) e^{-(n\pi y)/L} \qquad (11\text{-}83)$$

This sum cannot easily be further simplified, and in order to find ϕ at a definite point the sum will generally have to be evaluated numerically. Ordinarily, this is not a great problem for not many terms in the sum will be needed to achieve a reasonable accuracy because the successive terms get less and less important, both because of the factor $1/n$ and of the appearance of n in the exponential term.

We can now use (11-83) to calculate the electric field at any point from $\mathbf{E} = -\nabla\phi$, and we get

$$E_x = -\phi_0 \frac{4}{L} \sum_{n \text{ odd}} \cos\left(\frac{n\pi x}{L}\right) e^{-(n\pi y)/L} \qquad (11\text{-}84)$$

$$E_y = \phi_0 \frac{4}{L} \sum_{n \text{ odd}} \sin\left(\frac{n\pi x}{L}\right) e^{-(n\pi y)/L} \qquad (11\text{-}85)$$

while, of course, $E_z = -\partial\phi/\partial z = 0$. We note that these expressions have the correct units of volts/meter. We also see that $E_y = 0$ at $x = 0$ and L as it should because it is a tangential component at these conducting surfaces. E_x does not necessarily vanish there because it is a normal component and, in fact, by using (6-4), we could find the surface charge density as a function of position on these conducting surfaces.   ∎

11-5 SEPARATION OF VARIABLES IN SPHERICAL COORDINATES

If we use (1-105), we can write (11-3) as

$$\frac{1}{r^2}\frac{\partial}{\partial r}\left(r^2\frac{\partial \phi}{\partial r}\right) + \frac{1}{r^2 \sin\theta}\frac{\partial}{\partial\theta}\left(\sin\theta\frac{\partial\phi}{\partial\theta}\right) + \frac{1}{r^2 \sin^2\theta}\frac{\partial^2\phi}{\partial\varphi^2} = 0 \qquad (11\text{-}86)$$

The general solution of this equation can be found by the same systematic application of separation of variables that we have illustrated for rectangular coordinates. The results are more complicated; consequently, for our purposes, it will be sufficient to restrict ourselves only to those cases for which ϕ is independent of the angle φ, that is, to systems with axial symmetry. There are still a large number of situations which fall into this category, however. If $\phi = \phi(r, \theta)$, (11-86) simplifies somewhat to

$$\frac{\partial}{\partial r}\left(r^2\frac{\partial\phi}{\partial r}\right) + \frac{1}{\sin\theta}\frac{\partial}{\partial\theta}\left(\sin\theta\frac{\partial\phi}{\partial\theta}\right) = 0 \qquad (11\text{-}87)$$

By analogy to (11-56), we now look for a solution of the form

$$\phi(r, \theta) = R(r)T(\theta) \qquad (11\text{-}88)$$

If we substitute this into (11-87), divide the result by the product RT, and equate the resulting function of r alone to that for θ alone, we get

$$\frac{1}{R}\frac{d}{dr}\left(r^2\frac{dR}{dr}\right) = -\frac{1}{T\sin\theta}\frac{d}{d\theta}\left(\sin\theta\frac{dT}{d\theta}\right) = \text{const.} = K \qquad (11\text{-}89)$$

since each term is a function of a different independent variable so that they must be separately equal to the same constant K. If we set each in turn equal to K, we get two equations, one for R, and one for T. Doing this for the first term and carrying out the differentiation, we find that R must satisfy the equation

$$r^2\frac{d^2R}{dr^2} + 2r\frac{dR}{dr} - KR = 0 \qquad (11\text{-}90)$$

In order to make a start at solving this, we *try* a solution of the form $R = \alpha r^l$ where α and l are constants; upon substitution into (11-90), the result is that $[l(l + 1) - K]R = 0$. Since we do not want R to be zero, as this would make $\phi = 0$ everywhere, we see that we must have

$$K = l(l + 1) \qquad (11\text{-}91)$$

Equating this to the second term of (11-89), we find that the equation for T becomes

$$\frac{1}{\sin\theta}\frac{d}{d\theta}\left(\sin\theta\frac{dT_l}{d\theta}\right) + l(l + 1)T_l = 0 \qquad (11\text{-}92)$$

where we have added the subscript l to indicate the association of the solution with the constant. Now since T_l is part of the physical quantity ϕ, it must be a reasonable function. By this we mean that it should be finite, single valued, and continuous over the whole range of θ. It would take us too far afield to investigate this, but it can be shown that this is possible only if l is a positive integer, including zero; thus

$$l = 0, 1, 2, 3, \ldots \qquad (11\text{-}93)$$

As a matter of fact, the T_l can be identified with the Legendre polynomials we have already encountered in Chapter 8. If we refer back to Figure 8-1, we see that θ_i and r_i are just the corresponding spherical coordinates of the location of q_i with respect to the *fixed* direction $\hat{\mathbf{r}}$ shown; R_i is the corresponding distance from the source point to the field point and r of the figure will be a definite *constant*. Accordingly, in (8-12), we can write $1/r^{l+1}$ as a constant G_l, and *then* drop the index i on θ_i and r_i to get an expression in the same notation we are using here, that is, we can write

$$\frac{1}{R_i} = \sum_{l=0}^{\infty} G_l r^l P_l(\cos\theta) \tag{11-94}$$

We already know from (1-144) that this is *a* solution of Laplace's equation and therefore $\phi = 1/R_i$ *must* be a solution of (11-87). If we now substitute (11-94) into it, we obtain

$$\sum_{l=0}^{\infty} G_l r^l \left[l(l+1)P_l + \frac{1}{\sin\theta} \frac{d}{d\theta}\left(\sin\theta \frac{dP_l}{d\theta} \right) \right] = 0 \tag{11-95}$$

In general, this sum can be zero only if each term is zero since r is arbitrary, so that the bracketed term must be zero for each l. Upon comparing this with (11-92), we see that T_l and P_l satisfy the same differential equation and thus T_l can be taken as, at most, a constant times P_l. We will absorb any such constant into the factor $R(r)$ of (11-88) and simply let $T_l(\theta) = P_l(\cos\theta)$. Then we can write (11-88) as $R_l(r)P_l(\cos\theta)$, and since there will be a solution of this form of the linear differential equation (11-87) for each possible l, we can write the general solution of (11-87) in the form

$$\phi(r, \theta) = \sum_{l=0}^{\infty} R_l(r) P_l(\cos\theta) \tag{11-96}$$

We still have to find the general form of R_l, however.

The equation satisfied by R_l, as found from (11-90) and (11-91), now is

$$r^2 \frac{d^2 R_l}{dr^2} + 2r \frac{dR_l}{dr} - l(l+1)R_l = 0 \tag{11-97}$$

where l satisfies (11-93). We now try to solve this equation more generally than before by assuming a form $R_l = \alpha_l r^n$ where α_l is a constant and n is an integer, and, upon substitution of this into (11-97), we find that we must have $n(n+1) - l(l+1) = 0$. This equation has the *two* solutions $n = l$ and $-(l+1)$, so that the general solution of (11-97) has the form

$$R_l(r) = A_l r^l + \frac{B_l}{r^{l+1}} \tag{11-98}$$

where A_l and B_l are constants of integration. Substituting this into (11-96), we finally get the general form of the solution to Laplace's equation for an axially symmetric situation as

$$\phi(r, \theta) = \sum_{l=0}^{\infty} \left(A_l r^l + \frac{B_l}{r^{l+1}} \right) P_l(\cos\theta) \tag{11-99}$$

As we saw in (8-10), the first few Legendre polynomials are

$$P_0(\cos\theta) = 1 \qquad P_1(\cos\theta) = \cos\theta \qquad P_2(\cos\theta) = \tfrac{1}{2}(3\cos^2\theta - 1) \tag{11-100}$$

and higher-order ones can be found from the recursion relation (8-11):

$$(l+1)P_{l+1}(\cos\theta) = (2l+1)\cos\theta P_l(\cos\theta) - lP_{l-1}(\cos\theta) \tag{11-101}$$

In the last section, we found that the orthogonal properties of trigonometric functions as expressed in (11-80) were a great help in finding the expansion coefficients in rectangular coordinates. The Legendre polynomials have a similar property and it is equally helpful. It can be shown that

$$\int_0^\pi P_l(\cos\theta) P_m(\cos\theta) \sin\theta \, d\theta = \int_{-1}^1 P_l(\mu) P_m(\mu) \, d\mu = \frac{2}{2l+1}\delta_{lm} \quad (11\text{-}102)$$

where we have used (2-22). As an application of this property, we can derive a result that will be quite useful to us in what follows. Consider the sum

$$\sum_{l=0}^\infty C_l P_l(\cos\theta) = 0 \quad (11\text{-}103)$$

where the C_l are constants. Since the sum must be zero for any arbitrary value of the angle θ, it seems plausible that this can be the case only if each term in the sum is itself zero, that is, if all of the C_l are zero. We can easily show that this is the case. In (11-103), we let $\cos\theta = \mu$, multiply through by $P_m(\mu)\, d\mu$, integrate over μ from -1 to $+1$, and use (11-102); in this way we get

$$\sum_{l=0}^\infty C_l \int_{-1}^1 P_l(\mu) P_m(\mu) \, d\mu = \sum_l C_l \left(\frac{2}{2l+1}\right)\delta_{lm} = \frac{2C_m}{2m+1} = 0$$

since every term in the sum is zero except when $l = m$. Therefore, $C_m = 0$ for all m, as we suspected. Changing the index from m back to l, we have shown that

$$\text{if } \sum_{l=0}^\infty C_l P_l(\cos\theta) = 0, \qquad \text{then } C_l = 0 \quad (11\text{-}104)$$

Now let us consider some specific examples.

■ **Example**

Grounded conducting sphere in a previously uniform field. We assume that initially we have a completely uniform field $\mathbf{E}_0$ and then a conducting sphere of radius a is inserted into the region and kept at zero potential. The field $\mathbf{E}_0$ could be produced, for example, by a parallel plate capacitor whose plate separation is very large compared to the size of the sphere. We choose the z axis in the direction of $\mathbf{E}_0$, so that $\mathbf{E}_0 = E_0\hat{\mathbf{z}}$, and take the origin at the center of the sphere as shown in Figure 11-12. Even before we have found the solution for this case, we can anticipate that the lines of $\mathbf{E}$, which will finally result, will have the general form shown by the lines in the figure since they must come in normal to the conducting surface. This situation clearly has axial symmetry, making ϕ a function of r and θ only, so that (11-99) is appropriate. We already know that $\phi = 0$ for $r < a$ because of (6-1); thus we need concern ourselves only with finding ϕ for $r \geq a$. In order to do this, we need to determine the boundary conditions.

At the surface of the conducting sphere, ϕ is zero, so that one of our conditions is that

$$\phi(a, \theta) = 0 \quad (11\text{-}105)$$

At the other extreme, when we get far away from the sphere, we expect the original uniform field to be unaffected by the presence of the sphere, that is, as $r \to \infty$, $\mathbf{E}(r, \theta) \to E_0\hat{\mathbf{z}}$. Thus, in this limit, ϕ should reduce to the form of a potential appropriate to this uniform field. But since $\mathbf{E} = -\nabla\phi$, this means that for large r, $E_z = E_0 = -\partial\phi/\partial z$ which gives $\phi = -E_0 z$ as the limiting expression. Thus, our condition at the far boundary is

$$\phi(r, \theta) \underset{r \to \infty}{\to} -E_0 z = -E_0 r \cos\theta \quad (11\text{-}106)$$

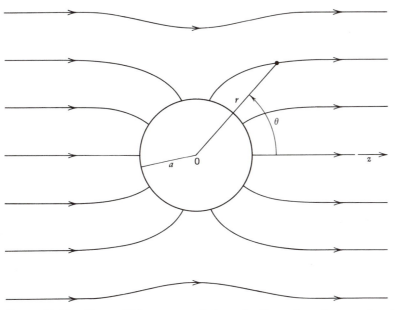

Figure 11-12. Lines of E for a grounded conducting sphere in a previously uniform field.

Putting $r = a$ in (11-99) and using (11-105) gives us

$$\sum_{l=0}^{\infty} \left(A_l a^l + \frac{B_l}{a^{l+1}} \right) P_l(\cos \theta) = 0 \qquad (11\text{-}107)$$

Now (11-104) shows that the term in parentheses is zero, so that $B_l = -a^{2l+1}A_l$, and (11-99) thus becomes

$$\phi = \sum_{l=0}^{\infty} A_l \left(r^l - \frac{a^{2l+1}}{r^{l+1}} \right) P_l(\cos \theta) \qquad (11\text{-}108)$$

Applying (11-106) to this, we get

$$\sum_{l=0}^{\infty} A_l r^l P_l(\cos \theta) = -E_0 r \cos \theta = -E_0 r P_1(\cos \theta) \qquad (11\text{-}109)$$

with the use of (11-100). If we separate out the term in the sum for $l = 1$, for convenience in writing, (11-109) takes the form

$$(A_1 + E_0) r P_1(\cos \theta) + \sum_{l \neq 1}^{\infty} A_l r^l P_l(\cos \theta) = 0 \qquad (11\text{-}110)$$

Since this must hold for all large r, it will be true in particular for a definite constant value r_0. But then (11-110) is exactly of the form (11-104), which shows us that $A_1 = -E_0$, and $A_l = 0$ $(l \neq 1)$ thus completing the evaluation of the coefficients. Consequently, the sum in (11-108) reduces to the single term for $l = 1$, and our final unique expression for ϕ is

$$\phi(r, \theta) = -E_0 r \cos \theta + \frac{a^3 E_0 \cos \theta}{r^2} \qquad (11\text{-}111)$$

Having obtained this result, we have completely solved the problem, and ϕ and $\mathbf{E}$ can be found at any desired field point.

As we see from (11-106), the first term in (11-111) is just that corresponding to the original uniform field. The second term must represent that part of the potential that arises because of the presence of the sphere. By comparing it with (8-48), we see that it is a dipole term and the dipole moment of the sphere is

$$p = 4\pi\epsilon_0 a^3 E_0 \tag{11-112}$$

Thus, the conducting sphere has acquired a dipole moment proportional to the original field and has been, in effect, polarized. The ratio of the induced dipole moment to the applied field is called the *polarizability* α and we see that

$$\alpha = \frac{p}{E_0} = 4\pi\epsilon_0 a^3 = 3\epsilon_0 V_s \tag{11-113}$$

where V_s is the volume of the sphere. This dipole moment must have its origin in the surface density of free charge on the sphere, which we can easily find. The radial component of **E** is

$$E_r = -\frac{\partial\phi}{\partial r} = \left(1 + \frac{2a^3}{r^3}\right)E_0\cos\theta \tag{11-114}$$

At the surface of the sphere, this is the normal component, and we can get the surface charge density from (6-4) as

$$\sigma_f(\theta) = \epsilon_0 E_r(a,\theta) = (3\epsilon_0 E_0)\cos\theta \tag{11-115}$$

which is seen to be proportional to $\cos\theta$. This result has the same *form* as the charge density given by (10-27) and illustrated in Figure 10-9. The surface charge has opposite signs on the two hemispheres, and it is this overall charge separation that gives rise to the dipole moment. [It is easy to verify directly that when **p** is calculated by using (11-115) in the surface integral form of (8-22), one gets (11-112); this same result was also obtained in Exercise 8-8.] The sphere was originally neutral before insertion, and we can see if it still has zero net charge by combining (11-115) and (2-16); we get

$$Q_{f,\text{total}} = \int\sigma_f\,da = \int_0^{2\pi}\int_0^\pi (3\epsilon_0 E_0\cos\theta)(a^2\sin\theta\,d\theta\,d\varphi) = 0$$

and thus the sphere is still neutral.

The remaining component of **E** is

$$E_\theta = -\frac{1}{r}\frac{\partial\phi}{\partial\theta} = -\left(1 - \frac{a^3}{r^3}\right)E_0\sin\theta \tag{11-116}$$

We see that $E_\theta(a,\theta) = 0$ as it must since it is a tangential component at the surface of the sphere.

Since ϕ as given by (11-111) has an angular dependence arising solely from the $P_1(\cos\theta)$ term of (11-99), we can conclude that the sphere has a dipole moment only. In other words, the charge distribution (11-115) is such that it produces no monopole moment, as well as zero quadrupole and higher moments; this was also verified directly for the quadrupole moment in Exercise 8-8.

The lines of **E** as found from (11-114) and (11-116) are illustrated in Figure 11-12. ∎

Example

Dielectric sphere in a previously uniform field. This example is the same as the last except that we have an uncharged dielectric sphere rather than a grounded conducting one. The boundary conditions are somewhat different because the sphere is no longer at a definite potential nor is ϕ necessarily constant within it. It will be convenient to look at the regions outside the sphere ($r > a$) and inside the sphere ($r < a$) separately; we will label the potentials and fields applicable for these regions with the subscripts o and i, respectively.

At large distances, the field again has to become uniform, so one boundary condition is exactly the same as (11-106):

$$\phi_o(r, \theta) \underset{r \to \infty}{\to} -E_0 r \cos \theta \qquad (11\text{-}117)$$

At the surface of the sphere, (11-105) no longer applies. However, the sphere's surface is a surface of discontinuity between the dielectric and the vacuum, and we know that the tangential components of **E** and the normal components of **D** are continuous there. These are written in terms of **E** only in (10-56). In this case, we have $\sigma_f = 0$, and since E_r and E_θ are the normal and tangential components, respectively, we can write (10-56) in terms of the derivatives of ϕ as

$$\left(-\epsilon_0 \frac{\partial \phi_o}{\partial r}\right)_{r=a} = \left(-\epsilon \frac{\partial \phi_i}{\partial r}\right)_{r=a} \qquad (11\text{-}118)$$

$$\left(-\frac{1}{r}\frac{\partial \phi_o}{\partial \theta}\right)_{r=a} = \left(-\frac{1}{r}\frac{\partial \phi_i}{\partial \theta}\right)_{r=a} \qquad (11\text{-}119)$$

where region 2 is taken to be the outside, while 1 is the inside. Finally, we have one more condition which we haven't met before. It is a "boundary condition" of a generalized kind, and deals with conditions at the origin ($r = 0$). By assumption, there are no free point charges within the sphere, so that there will be none at the origin. Since this is the only way in which the potential could become infinite there, we must also require that

$$\phi \text{ is finite for } r = 0 \qquad (11\text{-}120)$$

Rather than going about solving this problem in the same way as the last example, we will try to shorten our work by taking advantage of our experience. We expect the dielectric sphere to become polarized by the field and acquire a dipole moment, so it is plausible that once again only the $l = 1$ term of (11-99) will survive. Accordingly, let us assume the potential outside to be given by

$$\phi_o = \left(-A_o r + \frac{B_o}{r^2}\right) \cos \theta \qquad (11\text{-}121)$$

where A_o and B_o are constants. Since the boundary condition (11-118) will thus have a $\cos \theta$ on the left-hand side, it seems reasonable to expect the same on the right-hand side; accordingly, for the potential inside the sphere, let us choose the same general form as (11-121). Thus we also write

$$\phi_i = \left(-A_i r + \frac{B_i}{r^2}\right) \cos \theta \qquad (11\text{-}122)$$

where A_i and B_i are constants. If we can manage to find these four constants by satisfying all of the boundary conditions, we know that we will have found the *unique* solution to the problem.

Combining (11-117) and (11-121), we see that $A_o = E_0$ and we now have

$$\phi_o = \left(-E_0 r + \frac{B_o}{r^2}\right) \cos \theta \qquad (11\text{-}123)$$

We also see that if $B_i \neq 0$ in (11-122), the $1/r^2$ term will make $\phi \to \infty$ as $r \to 0$ which is not allowable by (11-120); thus this term cannot appear for the inside form of the potential, so we must really have $B_i = 0$ making (11-122) reduce to

$$\phi_i = -A_i r \cos \theta \tag{11-124}$$

Substitution of these expressions into (11-118) and (11-119) results in the following two equations:

$$E_0 + \frac{2B_o}{a^3} = \kappa_e A_i \qquad -E_0 a + \frac{B_o}{a^2} = -A_i a$$

since $\epsilon/\epsilon_0 = \kappa_e$. Solving these, we find that $A_i = 3E_0/(\kappa_e + 2)$ and $B_o = [(\kappa_e - 1)/(\kappa_e + 2)]a^3 E_0$, so that our potentials are

$$\phi_o = -E_0 r \cos \theta + \left(\frac{\kappa_e - 1}{\kappa_e + 2} \right) \frac{a^3 E_0 \cos \theta}{r^2} \tag{11-125}$$

$$\phi_i = -\left(\frac{3E_0}{\kappa_e + 2} \right) r \cos \theta \tag{11-126}$$

and the problem is completely solved.

The field inside the sphere is $\mathbf{E}_i = -\nabla \phi_i = E_i \hat{\mathbf{z}}$, and since $\mathbf{E}_0 = E_0 \hat{\mathbf{z}}$, we have

$$\mathbf{E}_i = \left(\frac{3}{\kappa_e + 2} \right) \mathbf{E}_0 \tag{11-127}$$

and we see that $|\mathbf{E}_i| < |\mathbf{E}_0|$ since $\kappa_e > 1$. Thus the electric field is constant, parallel to the original external field, but smaller than it. Consequently, the sphere will be uniformly polarized, and the polarization as obtained from (10-50), (10-52), and (11-127) is

$$\mathbf{P} = \left(\frac{\kappa_e - 1}{\kappa_e + 2} \right) 3\epsilon_0 \mathbf{E}_0 \tag{11-128}$$

and the magnitude of the total dipole moment as obtained by multiplying P by the volume of the sphere is

$$p = 4\pi \left(\frac{\kappa_e - 1}{\kappa_e + 2} \right) a^3 \epsilon_0 E_0 \tag{11-129}$$

This enables us to write (11-125) as

$$\phi_o = -E_0 r \cos \theta + \frac{p \cos \theta}{4\pi\epsilon_0 r^2} \tag{11-130}$$

where the second term agrees exactly with (8-48) as we would expect.

We can rewrite (11-127) in an interesting and instructive way by subtracting $\mathbf{E}_0$ from both sides. We find that $\mathbf{E}_i - \mathbf{E}_0 = [(1 - \kappa_e)/(\kappa_e + 2)]\mathbf{E}_0 = -\mathbf{P}/3\epsilon_0$ by using (11-128), and therefore

$$\mathbf{E}_i = \mathbf{E}_0 - \frac{\mathbf{P}}{3\epsilon_0} \tag{11-131}$$

which shows that the resultant field in the interior can be written as $\mathbf{E}_i = \mathbf{E}_0 + \mathbf{E}_{\text{loc}}$, that is, as the sum of the original external vacuum field and a local field $\mathbf{E}_{\text{loc}}$ that is proportional to the polarization and opposite to it. [We previously found the similar result (10-67) for the case in which the dielectric had the form of a slab with parallel faces.] The source of this internal field is, of course, the bound charges appearing on the surface of the dielectric where there is a discontinuity in $\mathbf{P}$, and, from (10-8), their density is given by $\sigma_b = \mathbf{P} \cdot \hat{\mathbf{n}}' = \mathbf{P} \cdot \hat{\mathbf{r}} = P \cos \theta$; this is a distribution just like (10-27) and Figure 10-9.

In writing (11-130) as we did, we were able to write the potential outside explicitly as the sum of that due to the external field and the contribution of the polarized sphere. We can get a similar result for ϕ_i by combining (11-126), (11-127), and (11-131) to give

$$\phi_i = -E_i r \cos \theta = -E_0 r \cos \theta + \left(\frac{P}{3\epsilon_0} \right) r \cos \theta \qquad (11\text{-}132)$$

There is an interesting special case of our results. If we let $\kappa_e \to \infty$, we find that $\mathbf{E}_i \to 0$, $p \to 4\pi\epsilon_0 a^3 E_0$, $\phi_i \to 0$, and $\phi_o \to$ (11-111); these are just the results corresponding to the conducting sphere of the last example as we see from (6-1), (11-105), and (11-112). Thus, we see that, as far as electrostatic effects are concerned, a conductor acts like a material of infinite dielectric constant.

The lines of $\mathbf{E}$ found for this system are shown in Figure 11-13. ∎

■ **Example**

Sphere with uniform permanent polarization. Suppose we have a sphere with uniform polarization $\mathbf{P} = P\hat{\mathbf{z}}$ where $P = $ const. but with no external field present. This is exactly the system we considered in Section 10-4 and for which we found ϕ and $\mathbf{E}$ along the z axis only. The form in which we have written (11-130) and (11-132) enables us to write down the complete solution at once, for we can set $E_0 = 0$, interpret p and P as the permanent dipole moment and polarization, respectively, and we get

$$\phi_o = \frac{p \cos \theta}{4\pi\epsilon_0 r^2} \qquad \phi_i = -E_i r \cos \theta = \left(\frac{P}{3\epsilon_0} \right) r \cos \theta \qquad (11\text{-}133)$$

The first expression shows us that the field *everywhere* outside the sphere is a dipole field as we suspected from our results for the z axis. Similarly, the second expression shows us that the electric field is uniform *everywhere* inside the sphere and is, in fact,

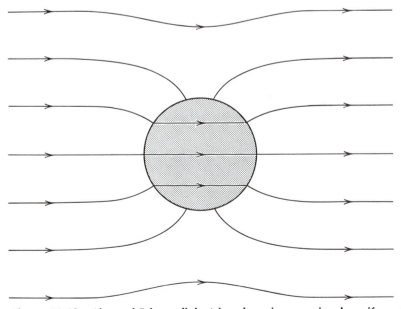

Figure 11-13. Lines of E for a dielectric sphere in a previously uniform field.

given by

$$\mathbf{E}_i = -\frac{\mathbf{P}}{3\epsilon_0} \tag{11-134}$$

in agreement with (10-37) that we previously found to hold on the z axis. ■

Since so many local fields turn out to be proportional to the polarization and opposite to it, it is common practice to write $\mathbf{E}_{loc}$ in the general form

$$\mathbf{E}_{loc} = -N\left(\frac{\mathbf{P}}{\epsilon_0}\right) \tag{11-135}$$

where N is a dimensionless constant called the *depolarizing factor*. We have already found two of them. They are $N = 1$ for an infinite slab with parallel faces and $N = \frac{1}{3}$ for a sphere, as found in (10-67) and (11-134). The result of Exercise 11-26 will show that $N = \frac{1}{2}$ for a cylinder, as could also have been surmised from Exercise 10-12.

11-6 SPHERICALLY SYMMETRIC SOLUTION OF POISSON'S EQUATION

If we combine (11-1) with (11-55) or (11-86), we will have Poisson's equation expressed in rectangular or spherical coordinates. For illustrative purposes in this section, however, we will restrict ourselves to completely spherically symmetric situations so that ϕ is a function of r only, that is, $\phi = \phi(r)$. Then ρ will necessarily be a function only of r also, and we get

$$\frac{1}{r^2}\frac{d}{dr}\left(r^2\frac{d\phi}{dr}\right) = -\frac{\rho(r)}{\epsilon_0} \tag{11-136}$$

■ **Example**

Sphere with uniform charge density. Let us consider a sphere of radius a containing charge of constant density so that $\rho = $ const. inside, while $\rho = 0$ outside. We want to find ϕ everywhere by using (11-136). Outside the sphere, this equation becomes $d(r^2 d\phi_o/dr)/dr = 0$, which can be integrated twice to give

$$\phi_o(r) = A_o + \frac{B_o}{r} \tag{11-137}$$

where A_o and B_o are constants of integration. [This is also what we get from (11-99) by dropping all the angle-dependent terms, that is, using only $l = 0$.] Inside the sphere, (11-136) becomes

$$\frac{d}{dr}\left(r^2\frac{d\phi_i}{dr}\right) = -\frac{\rho r^2}{\epsilon_0}$$

which can be easily integrated twice to give

$$\phi_i(r) = -\frac{\rho r^2}{6\epsilon_0} + A_i + \frac{B_i}{r} \tag{11-138}$$

where A_i and B_i are constants. Now all that remains to be done is to evaluate the constants of integration from the boundary conditions.

Since all of the charges are contained within a finite volume, we want ϕ to vanish at infinity, according to our discussion in connection with (5-10). Thus, as $r \to \infty$,

$\phi_o \to 0$, and we see from (11-137) that $A_o = 0$ and therefore $\phi_o = B_o/r$. Since there are no point charges at the origin, (11-120) still is applicable and shows us, that, in (11-138), $B_i = 0$. We know from (9-29) that ϕ is continuous at $r = a$, so that $\phi_o(a) = \phi_i(a)$ which gives $A_i = (B_o/a) + (\rho a^2/6\epsilon_0)$, making ϕ_i now have the form

$$\phi_i(r) = \frac{\rho}{6\epsilon_0}(a^2 - r^2) + \frac{B_o}{a} \tag{11-139}$$

Finally, since there is no surface charge, $E_n = E_r$ is continuous at $r = a$, according to (9-26), so that $-(\partial\phi_o/\partial r)_{r=a} = -(\partial\phi_i/\partial r)_{r=a}$, which leads to $(B_o/a^2) = (\rho a/3\epsilon_0)$ or $B_o = \rho a^3/3\epsilon_0$. Putting this into (11-137) (with $A_o = 0$) and (11-139), we obtain

$$\phi_o(r) = \frac{\rho a^3}{3\epsilon_0 r} \qquad \phi_i(r) = \frac{\rho}{6\epsilon_0}(3a^2 - r^2)$$

which is exactly what we obtained in (5-22) and (5-23) when we discussed this same example by other methods.   ∎

EXERCISES

11-1 Reconsider the theorem of Section 11-1, and show that if the normal component of **E** is preassigned on all points of the bounding surface, then ϕ_1 and ϕ_2 need not be equal but can differ at most by a constant.

11-2 Verify by direct integration that (11-19) is the resultant of the Coulomb force between q and the induced charge described by (11-17).

11-3 A point charge q is located in the xy plane near two grounded conducting planes intersecting at right angles as shown in Figure 11-14. The z axis lies along the line of intersection of the planes. Find and justify the image charges that, together with q, will give the potential at all points in the vacuum region $x \geq 0$, $y \geq 0$, $-\infty \leq z \leq \infty$. Find $\phi(x, y, z)$ in the vacuum region. Find $E_y(x, y, z)$. Verify that E_y vanishes on the conducting plane for which it is a tangential component. Find the surface charge density σ_f induced on the plane for which E_y is an appropriate component to use. What is the sign of σ_f? (*Hint*: recall multiple image formation in plane mirrors from geometrical optics.)

11-4 Suppose that the angle between the conducting planes of Figure 11-14 is 60° rather than the 90° shown, and that q is along the line bisecting the angle (i.e., set $a = b$ and then change the angle). Find and justify the image charges that, together with q, will give ϕ at all points in the vacuum region. What is the direction of the resultant force on q?

11-5 Suppose that q in Figure 11-1 is replaced by a point dipole $\mathbf{p} = p\hat{\mathbf{y}}$. Find the force on $\mathbf{p}$.

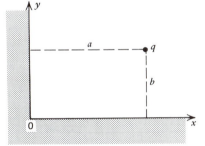

Figure 11-14.   Geometry for Exercise 11-3.

11-6 Suppose that the point where the x axis intersects the middle equipotential curve of Figure 11-2 is midway between the charge and the conductor. Find ϕ and **E** at this point.

11-7 Verify that (11-35) can also be obtained by using (11-29) and (11-30).

11-8 Using the spherical coordinate system of Figure 11-3, find ϕ at all points outside the sphere for the charge distribution of Figure 11-5. Show that your result gives (11-36). Find **E** and the force on q. Find σ_f and show that the total charge induced on the sphere is zero.

11-9 The conducting sphere of Figure 11-3 is now insulated and has a total charge Q on it. Find the potential of the sphere and the force on q.

11-10 After (11-48), we saw that the total bound surface charge on the dielectric was equal to the image charge. How do you reconcile this result with (10-13)?

11-11   For the system of Figure 11-9, find the surface density of induced charge on the conductor, the total induced charge per unit length parallel to the line charge, and the force per unit length on the line charge.

11-12   Show that the attractive force on a length L of one of the cylinders of Figure 11-10 is given by

$$F_e = - \frac{\pi \epsilon_0 L (\Delta \phi)^2}{2 \left[\cosh^{-1} (D/2A) \right]^2 (D^2 - 4A^2)^{1/2}}$$

11-13   A long wire of circular cross section of radius A is strung on poles at a height h above the ground. Neglect the sag of the wire and the curvature of the earth, and find the capacitance per unit length of this system. Also find the force attracting a unit length of the uniformly charged wire toward the earth.

11-14   Show that the potential produced by an infinite line charge of constant density λ at a distance D from the parallel axis of a circular conducting cylinder of radius A is that which would be produced by the actual line charge and an image line charge $-\lambda$ that is a distance A^2/D toward the real charge from the center of the cylinder. What is the potential of the cylinder?

11-15   For the system of Figure 11-11, find the surface charge density on the face $x = 0$.

11-16   Verify that E_x of (11-84) is zero for $y = 0$. [*Hint*: it may help to recall that $\cos u = \frac{1}{2}(e^{iu} + e^{-iu})$.]

11-17   This is a two-dimensional problem. Consider a square in the xy plane with corners at $(0,0)$, $(a,0)$, (a, a), and $(0, a)$. There is no charge nor matter inside the square. The sides perpendicular to the y axis have a potential of zero. The side at $x = a$ has the constant potential ϕ_0, while that at $x = 0$ has the constant potential $-\phi_0$. Find ϕ (x, y) for all points inside the square. Find $\mathbf{E}$ at the center of the square and evaluate the ratio of $\mathbf{E}$ to (ϕ_0/a) at this point to four significant figures.

11-18   Find the potential ϕ at all points within a cube of side L with the location and orientation of Figure 1-41. There is no charge nor matter within the cube. The potential on the face $z = L$ has the constant value ϕ_0, and the potential on all other faces is zero. Show that, to four significant figures, ϕ at the center of the cube is 0.1667 ϕ_0.

11-19   Show that the solution of Laplace's equation can be written as a sum of terms each of the form $X(x) + Y(y) + Z(z)$. Be sure to show how

these functions, or appropriate derivatives of them, are related, if in fact they are. Find the general form of $X(x)$ and interpret the corresponding electric field.

11-20   (a) Calculate directly $\int_{-1}^{1} P_l^2(\mu) \, d\mu$ for $l = 0, 1, 2$, and thus show that (11-102) is correct for these cases.
(b) Show that when the differential equation (11-92) satisfied by P_l is expressed in terms of the variable $\mu = \cos \theta$, it becomes

$$\frac{d}{d\mu} \left[(1 - \mu^2) \frac{dP_l}{d\mu} \right] + l(l + 1) P_l = 0 \quad (11\text{-}140)$$

(c) The orthogonality property expressed by (11-102) for $l \neq m$ can be shown to be a consequence of (11-140) and the fact that P_l is finite and has finite derivatives at $\mu = \pm 1$ as follows: multiply (11-140) by P_m; write down the differential equation satisfied by $P_m(\mu)$ and multiply it by P_l; subtract these two expressions and integrate the result over μ from -1 to $+1$, integrating by parts as necessary.

11-21   Find the equation for the lines of $\mathbf{E}$ corresponding to (11-111), that is, for the conducting sphere in a previously uniform field.

11-22   Suppose that instead of (11-106) we had required that at large distances $\phi \to -E_0 r \cos \theta + \phi_0$ where ϕ_0 is constant, since this will still yield a uniform field. Find ϕ under these conditions. If your solution is different from (11-111), how can you interpret the result?

11-23   A spherical cavity of radius a is within a large grounded conductor. A charge q is placed within the cavity at a distance b from the center. Find ϕ at all points within the cavity by using spherical coordinates with origin at the center and z axis passing though the location of q. Find $\mathbf{E}$ at all points within the cavity. Find $\mathbf{E}$ at the center of the cavity. Find the surface charge density induced on the wall of the cavity. What is the total induced charge on the wall?

11-24   Solve the two-dimensional form of Laplace's equation expressed in plane polar coordinates (ρ, φ) by separation of variables. Thus, show that the general solution has the form

$$\phi = A + B \ln \rho$$
$$+ \sum_{m=1}^{\infty} \left(A_m \rho^m + \frac{B_m}{\rho^m} \right)$$
$$\times (C_m \cos m\varphi + D_m \sin m\varphi) \quad (11\text{-}141)$$

where m is a positive integer and φ covers its

whole possible range. (*Hint*: ϕ must be single valued.)

11-25 An infinitely long grounded cylindrical conductor with circular cross section of radius a has its axis coinciding with the z axis. It is placed in a previously uniform electric field $\mathbf{E}_0 = E_0 \hat{\mathbf{x}}$, so that its axis is perpendicular to $\mathbf{E}_0$. Find ϕ for all points outside the cylinder. Find the surface charge density and show that the cylinder remains neutral.

11-26 The cylinder of the previous exercise is now a dielectric rather than a conductor. Find ϕ everywhere. Find $\mathbf{E}$ and $\mathbf{P}$ inside and thus verify that the depolarizing factor is $\frac{1}{2}$ in this case.

11-27 A circle of radius a lies in the xy plane with its center at the origin. The semicircular part of the boundary for $x > 0$ is kept at the constant potential ϕ_0; the other semicircle for $x < 0$ is kept at the constant potential $-\phi_0$. Find ϕ for all points within the circle. Find $\mathbf{E}$ at the center of the circle.

11-28 Although (11-99) gives the form of the general solution for a situation with axial symmetry, it is not always easy to find the coefficients A_l and B_l by setting up the problem for general values of θ. Sometimes the following approach can be used. Since (11-99) is true for all values of θ, it must be true, with the same coefficients, for the particular value $\theta = 0$, that is, for the direction along the positive z axis—the symmetry axis. Then $r = z$, and $P_l(\cos \theta) = P_l(1) = 1$, and (11-99) reduces to

$$\phi(z) = \sum_{l=0}^{\infty} \left(A_l z^l + \frac{B_l}{z^{l+1}} \right) \qquad (z > 0)$$

$$(11\text{-}142)$$

Thus, if one can solve for ϕ on the z axis, it should be possible to find A_l and B_l by comparison with (11-142), and when these coefficients are put back into (11-99), the result is an expression for ϕ which is correct for all θ. Sometimes the identification must be made by expanding $\phi(z)$ in a power series, and the coefficients found by a term by term comparison. As an example of this procedure, consider the uniform line charge of finite length of Figure 3-8, for which the potential is given by (5-30). Assume that $L_2 = L_1 = L$, and show that when $r > L$, ϕ can be written in the form

$$\phi(r, \theta) = \frac{\lambda}{2\pi\epsilon_0} \sum_{l \text{ even}} \left(\frac{L}{r} \right)^{l+1} \frac{P_l(\cos \theta)}{(l+1)}$$

where the sum is taken over only even values of l including $l = 0$.

11-29 A circular ring of radius a lies in the xy plane with origin at the center. It has a constant linear charge density λ on its circumference. Find $\phi(r, \theta)$, expressed as a series in the $P_l(\cos \theta)$, for all r. (See previous exercise.)

11-30 A system of two concentric spheres has inner radius a and outer radius b. The region between them is filled with a spherically symmetric charge distribution of volume density $\rho = \rho_0 (r/a)^n$ where $\rho_0 = $ const. and $n > 0$. The inner sphere is kept at a constant potential ϕ_1, while the outer is at a constant potential ϕ_2. Find ϕ for $a \le r \le b$ by using (11-136).

11-31 Two infinite conducting planes are parallel to the xy plane. One of them is located at $z = 0$ and is kept at a constant potential ϕ_0. The other, at constant potential ϕ_d, has $z = d$. The region between them is filled with charge with volume density $\rho = \rho_0 (z/d)^2$. Solve Poisson's equation to find ϕ for $0 \le z \le d$. Find the surface charge density on each plate.

11-32 Consider the coaxial cylinders of Figure 4-7. The inner cylinder is kept at constant potential ϕ_a and the outer at ϕ_b. There is a cylindrical sheath of constant charge density in the region between them and vacuum elsewhere. In other words, the volume charge density ρ_{ch} is: zero for $a \le \rho < \rho_1$, $A\epsilon_0$ for $\rho_1 \le \rho \le \rho_2$, and zero for $\rho_2 < \rho \le b$, where $A = $ const. Solve Poisson's equation to find ϕ for $a \le \rho \le b$.

11-33 Here are some miscellaneous results for a system of source charges occupying a *finite region*. You can easily prove them with the use of (1-111), (1-122), (1-123), (1-141), and (3-3):

$$\int_{V'} \nabla' \left(\frac{\rho}{R} \right) d\tau' = 0 \qquad \mathbf{E} = -\frac{1}{4\pi\epsilon_0} \int_{V'} \frac{(\nabla' \rho)}{R} d\tau'$$

$$\int_{V'} \nabla' \times \left(\frac{\mathbf{E}}{R} \right) d\tau' = 0$$

11-34 Occasionally, one will find in other books expressions for the image charges used for the example of the point charge and semiinfinite dielectric that are different from the ones we obtained. As shown by the ϵ_0 in the denominator of (11-41), we used vacuum properties in the region occupied by the dielectric. Another point of view is to assume in effect that the dielectric fills all space as far as calculations involving q'' are concerned. Thus, one can still use (11-41) but with an ϵ in the denominator instead. Show that, if this is done, q'' is found to be κ_e times that given in (11-46). Also show that the potential and fields in the dielectric will nevertheless be exactly the same as before, as must be the case.

12

ELECTRIC CURRENTS

Electrostatics deals with relations among electric charges that are at rest. The next principal division of electromagnetism that we will consider—magnetostatics—involves forces between moving charges. A flow of electric charge is called an *electric current*, and, in this chapter, we want to devise useful ways of describing these currents in general. We will also discuss some aspects of a particular class of currents, those in conductors.

12-1 CURRENT AND CURRENT DENSITIES

Suppose that we have someone stationed at a point P observing charges passing that point; these charges may be traveling along a metal wire or may be simply a beam of charged particles moving through space. In any event, we assume that in a time interval Δt, our observer finds that a charge Δq has passed P. Then we can define the *average current* $\langle I \rangle$ during this interval as the average rate of flow of charge

$$\langle I \rangle = \frac{\Delta q}{\Delta t} \tag{12-1}$$

Later, we will be more specific about how Δq or $\langle I \rangle$ could be measured but if, for example, the current were due to a flow of protons, each of charge e, then the measurement process could be imagined as being simply one of *counting*. Thus, if N protons passed by in this interval, $\Delta q = Ne$ and $\langle I \rangle = Ne/\Delta t$. As is implied by (12-1), the "direction" or sense of the current is defined as that of the flow of positive charge. If the moving charges had a negative charge, for example if they were electrons, the direction of $\langle I \rangle$ would be opposite to their direction of motion. The reason for this is quite simple. Suppose the region around a point originally were neutral, that is, had equal amounts of positive and negative charge. Then if negative charges were *leaving* this region, it would acquire an excess of positive charge which is the same net effect as if positive charges were *coming into* the region.

If the flow of charge is not uniform in time, we can define an *instantaneous current* I as the instantaneous rate of flow of charge:

$$I = \frac{dq}{dt} \tag{12-2}$$

For some time, we will be concerned with currents that are constant in time so that $I = $ const. and $\langle I \rangle = I$. These are called *steady currents* or *stationary currents* and describe a uniform rate of flow of charge.

As we noted in Section 2-2, the unit of charge is actually defined in terms of the unit of current which is called an *ampere*, so that, according to (12-1), 1 coulomb = 1 ampere-second. The ampere itself is defined in terms of the force between currents and we will give it a precise definition in the next chapter.

It is often convenient to think of the current as traveling along a geometric curve, as illustrated in Figure 12-1 where the arrow head indicates the direction of I and $d\mathbf{s}$ is a displacement along the line in the sense of I. This idealized situation could usefully be

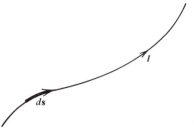

Figure 12-1.   A filamentary current.

taken to describe the flow of charge in a very thin wire or in a beam of small cross-sectional area. Such currents are called *filamentary currents*. However, we will also need to consider situations in which the flow of charge is *distributed* throughout a volume or on a surface, and we want to have suitable descriptions of them. We can do this by introducing the *current densities*.

The first of these is the *volume current density* **J**. Its direction is that of the direction of flow of charge and its magnitude J is given by the current per unit *area* through an area set perpendicular to the flow, or it is charge per unit time per unit area. We can simultaneously illustrate this definition and obtain a useful relation by considering the situation of Figure 12-2. Let us find the charge Δq, which, in a time Δt, has passed through the small area Δa on the left, which is perpendicular to **J**. By (12-1), $\Delta q = \langle I \rangle \Delta t = \langle J \rangle \Delta a \, \Delta t$ since $\langle J \rangle$ is average current per unit area. But all of the charge that has passed through Δa is contained within the volume $\Delta \tau$ of the cylinder of length Δl, so that, if we use (2-14), we also have $\Delta q = \rho \, \Delta \tau = \rho \, \Delta l \, \Delta a$ where ρ is the volume charge density. Equating these two expressions for Δq, and canceling the common factor Δa, we find that $\langle J \rangle = \rho(\Delta l / \Delta t) = \rho \langle v \rangle$ where $\langle v \rangle$ is the average speed of the charges. Now this relation clearly holds instantaneously, as well as on the average, and since the direction of **J** is defined as that of the direction of flow, which is **v**, we can write

$$\mathbf{J} = \rho \mathbf{v} \tag{12-3}$$

If the moving charges are of different types with densities ρ_i and corresponding velocities $\mathbf{v}_i$, then we see that, in Δt, the charge of type i that has passed through will be $\Delta q_i = \rho_i |\mathbf{v}_i| \, \Delta a \, \Delta t$. Then the total of all kinds will be $\Delta q = \sum_i \rho_i |\mathbf{v}_i| \, \Delta t \, \Delta a$, and gives us the natural generalization of (12-3) in the form

$$\mathbf{J} = \sum_i \rho_i \mathbf{v}_i \tag{12-4}$$

By comparing these last two results, we see that we can still use (12-3) in the general case by taking ρ as the total volume charge density and **v** as an average velocity, weighted by the densities ρ_i, much as in the calculation of the velocity of the center of mass of a collection of mass points.

Now suppose that **J** and an element of area $d\mathbf{a}$ are not parallel as is shown in Figure 12-3. We can find the charge that has passed through $d\mathbf{a}$ in a time dt in a manner similar to that we used for Figure 12-2. This time the total charge would be that

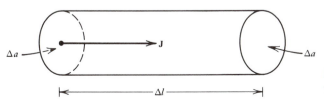

Figure 12-2.   Calculation of volume current density.

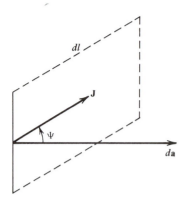

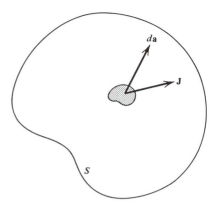

Figure 12-3. The current density and element of area are not parallel.

Figure 12-4. Calculation of total rate of flow of charge through the area *S*.

contained in the cylinder of slant height dl and volume $dl \cos \Psi\, da$ and would be given by $dq = \rho\, dl \cos \Psi\, da = \rho v \cos \Psi\, da\, dt = \rho \mathbf{v} \cdot d\mathbf{a}\, dt = \mathbf{J} \cdot d\mathbf{a}\, dt$ with the use of (12-3). Therefore, the rate of flow of charge through $d\mathbf{a}$ will be

$$\left(\frac{dq}{dt}\right)_{\text{through } d\mathbf{a}} = \mathbf{J} \cdot d\mathbf{a} \qquad (12\text{-}5)$$

If we consider an arbitrary surface *S* as shown in Figure 12-4, we can find the total rate at which charge is flowing through it by adding up the contributions of all the elements $d\mathbf{a}$ as given by (12-5). Thus we get

$$\left(\frac{dq}{dt}\right)_{\text{through } S} = \int_S \mathbf{J} \cdot d\mathbf{a} \qquad (12\text{-}6)$$

which is sometimes called the flux of charge. In (12-6), *S* can be either an open surface or a closed surface.

If, for some reason, the moving charges can be thought of as being constrained to flow on a surface, we can define a *surface current density* **K**. Its direction is that of the direction of flow of charge and its magnitude *K* is defined as equal to the current per unit *length* through a line lying in the surface and oriented perpendicular to the flow. This definition is indicated schematically in Figure 12-5a. In Figure 12-5b, we illustrate a situation where **K** is not at right angles to the line ds; the unit vector $\hat{\mathbf{t}}$ is drawn at right angles to ds and lying on the surface, as does **K**, so that it is a tangential vector. By using methods analogous to those by which we obtained (12-3) and (12-5), we can show that

$$\mathbf{K} = \sigma \mathbf{v} \qquad (12\text{-}7)$$

$$\left(\frac{dq}{dt}\right)_{\text{through } ds} = |\mathbf{K} \cdot \hat{\mathbf{t}}|\, ds \qquad (12\text{-}8)$$

where σ is the surface charge density.

Similarly, for a filamentary current, one obtains

$$I = \lambda |\mathbf{v}| \qquad (12\text{-}9)$$

where λ is the linear charge density of the flow.

In subsequent discussions, we will be constantly dealing with what is known as a *current element* and it is convenient to introduce it here. For the filamentary current of

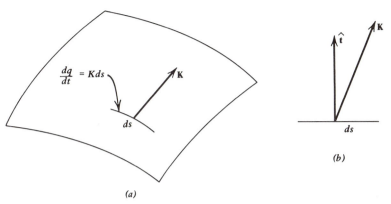

Figure 12-5. (*a*) Definition of surface current density K. (*b*) K is not perpendicular to the line *ds*.

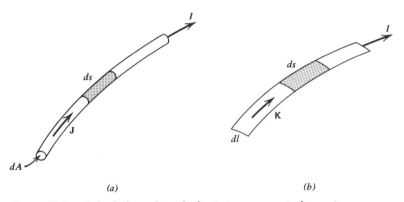

Figure 12-6. Calculation of equivalents to a current element.

Figure 12-1, it is defined simply as the product $I\,d\mathbf{s}$. In order to get an equivalent to this for distributed currents, we consider Figure 12-6a in which dA is the small cross-sectional area of the filament. Since the flow is normal to the cross section, we see from (12-5) that $I = J\,dA$ so that $I\,ds = J\,dA\,ds = J\,d\tau$ where $d\tau$ is the volume of the element shown shaded. Since $\mathbf{J}$ and $d\mathbf{s}$ are parallel in this case, we get $I\,d\mathbf{s} = \mathbf{J}\,d\tau$. Similarly, for a surface current, we get the correspondence $I\,d\mathbf{s} = \mathbf{K}\,da$ where da is the area $dl\,ds$ of the element shown shaded in Figure 12-6b. Thus we have the following *equivalent expressions* for current elements:

$$I\,d\mathbf{s} = \mathbf{J}\,d\tau = \mathbf{K}\,da \qquad (12\text{-}10)$$

These results also show the reasons for the names "volume" and "surface" current densities.

12-2 THE EQUATION OF CONTINUITY

In the introduction to Chapter 2, we mentioned that all experiments indicate that net charge is *conserved*. We can express this fundamental law of conservation of charge in a convenient quantitative manner in terms of the quantities we have just introduced. Suppose that, in Figure 12-4, the surface S is a *closed* stationary surface bounding a volume V. Now the total rate at which charge is flowing out through the surface S must equal the rate at which the total charge within V is decreasing, since the total must be constant. Therefore, if Q is the total charge within V, we find from (12-6), (2-14), and (1-59) that

$$-\frac{dQ}{dt} = \oint_S \mathbf{J} \cdot d\mathbf{a} = -\frac{d}{dt}\int_V \rho \, d\tau = -\int_V \frac{\partial \rho}{\partial t} \, d\tau = \int_V \nabla \cdot \mathbf{J} \, d\tau \qquad (12\text{-}11)$$

We were able to go from the third expression to the fourth as we did because V is a volume of constant shape and size so that any limits of integration which may be involved in the definite integral over V are independent of the time; in addition, ρ may be a function of position as well as of time. Combining the last two expressions in (12-11), we get

$$\int_V \left(\nabla \cdot \mathbf{J} + \frac{\partial \rho}{\partial t} \right) d\tau = 0 \qquad (12\text{-}12)$$

Since charge is conserved at all points, not at just part of a given volume, the integral must hold for any arbitrary volume, including an arbitrarily small one located *any-where*. Thus (12-12) can be always true only if the integrand is zero everywhere so that

$$\nabla \cdot \mathbf{J} + \frac{\partial \rho}{\partial t} = 0 \qquad (12\text{-}13)$$

This important result is called the *equation of continuity* and is a mathematical expression of the fundamental experimental result that net charge is conserved. We can note here that processes like "pair production" and "annihilation" of, say, electrons and positrons, do not violate this result because the *net* charge is constant since equal amounts of positive and negative charge are "created" or "destroyed" in these phenomena. Similar results are found to hold in more complicated reactions in nuclear and high energy physics where large numbers of particles are produced; in all cases, the net charge is conserved.

If we now combine (12-13) with (9-6) and (9-24), we get the boundary condition satisfied by the current density at a surface of discontinuity as

$$\hat{\mathbf{n}} \cdot (\mathbf{J}_2 - \mathbf{J}_1) = J_{2n} - J_{1n} = -\frac{\partial \sigma}{\partial t} \qquad (12\text{-}14)$$

Physically, this condition expresses the fact that if more charge is brought up to the surface than is taken away, charge will necessarily accumulate there, and conversely.

In the special case of steady currents, in which everything is constant in time, $\partial \rho / \partial t$ and $\partial \sigma / \partial t$ will both be zero, and these last two results simplify to

$$\nabla \cdot \mathbf{J} = 0 \qquad (12\text{-}15)$$

$$\hat{\mathbf{n}} \cdot (\mathbf{J}_2 - \mathbf{J}_1) = J_{2n} - J_{1n} = 0 \qquad (12\text{-}16)$$

Since all charge is conserved, ρ and $\mathbf{J}$ are clearly the total charge density and current density, respectively. Now let us consider their component parts. We begin with the bound charge whose density is ρ_b. Now in the *process* of polarizing a material, the bound charges will generally be moving, as we saw in Section 10-1, so that we can define a bound charge current density $\mathbf{J}_b$. Since the process of polarization involves only separation of bound charges, or reorientation of dipoles, bound charges are necessarily conserved, as shown by (10-13). Thus we must have a separate equation of continuity for bound charges, that is,

$$\nabla \cdot \mathbf{J}_b + \frac{\partial \rho_b}{\partial t} = 0 \qquad (12\text{-}17)$$

If we write $\rho_b = -\nabla \cdot \mathbf{P}$, as given by (10-10), we also have

$$\nabla \cdot \mathbf{J}_b - \frac{\partial}{\partial t} \nabla \cdot \mathbf{P} = \nabla \cdot \left(\mathbf{J}_b - \frac{\partial \mathbf{P}}{\partial t} \right) = 0$$

and, since this must be true everywhere, we can write

$$\mathbf{J}_b = \frac{\partial \mathbf{P}}{\partial t} \tag{12-18}$$

thus identifying the bound current density. This current is often called the *polarization current density* and is a consequence of the process of polarization.

Since total charge is conserved, according to (12-13), and bound charge also is, as shown by (12-17), the free charge must be conserved too and we can write

$$\nabla \cdot \mathbf{J}_f + \frac{\partial \rho_f}{\partial t} = 0 \tag{12-19}$$

which will lead to

$$\hat{\mathbf{n}} \cdot \left(\mathbf{J}_{f2} - \mathbf{J}_{f1} \right) = J_{f2n} - J_{f1n} = -\frac{\partial \sigma_f}{\partial t} \tag{12-20}$$

in exactly the way by which we obtained (12-14). In the special case of steady currents, these become

$$\nabla \cdot \mathbf{J}_f = 0 \quad \text{and} \quad \hat{\mathbf{n}} \cdot \left(\mathbf{J}_{f2} - \mathbf{J}_{f1} \right) = 0 \tag{12-21}$$

Since free charges, and thus free currents, are the ones over which we have some control, they are usually the ones in which we are most interested, and, consequently, we will be concentrating on them for some time. Free currents are often classified further into the two broad categories of *conduction currents* and *convection currents*, although the distinction between them is somewhat ill defined. Generally speaking, conduction currents include the motion of charges in conductors, that is, materials that already contain mobile charges because of their intrinsic nature. The most common example is that of currents in metals, although we can include, in this class, currents in semiconductors and solutions of electrolytes. In the last case, positive and negative ions resulting from the formation of the solution provide the ready-made carriers. On the other hand, convection currents are usually associated with the motion of charged particles in physical streams through otherwise empty space such as beams of ions, electron beams in vacuum tubes, charged particles in the solar wind, and the like. This class can also be taken to include the physical motion of macroscopically charged bodies, such as we would get by moving around a piece of glass that had been positively charged by electrification by friction. The class of most importance to us, however, is that of conduction currents and we now consider that in more detail.

12-3 CONDUCTION CURRENTS

As a convenient prototype of this case, we can think of the current in a metallic wire. In (6-1), we found that $\mathbf{E} = 0$ in the interior of a conductor for a completely *static* situation. Now, with moving charges in the wire, we no longer have a static situation, although it *may* be steady, so that it may well be that $\mathbf{E} \neq 0$ within the conductor. In fact, the very motion of the charges implies that there are forces on them that in turn implies a nonzero value of $\mathbf{E}$. Consequently, the conductor can no longer be expected to be an equipotential volume.

We find experimentally that if we apply an initial difference of potential to a conductor there will be currents in it, but if it is then left alone, the currents will eventually cease to exist, and the conductor will have attained the state of electrostatic equilibrium whose properties we have discussed in Chapter 6. It is also found that we can maintain a constant current in a conductor by means of a constant potential difference only if we also *continuously supply energy* to the system from some external source. Thus work is somewhere being done on these moving charges as they circulate in the *closed paths* of the ordinary circuit. If the total work W_q is done on a charge q as it goes around a closed path, the ratio of the two is called the *electromotive force* $\mathscr{E}$, or simply the *emf*, so that, with the use of (3-1), we have

$$\mathscr{E} = \frac{W_q}{q} = \frac{1}{q} \oint_C \mathbf{F}_q \cdot d\mathbf{s} = \oint_C \mathbf{E} \cdot d\mathbf{s} \qquad (12\text{-}22)$$

But, as we know from (5-5), the conservative electric field with which we are familiar cannot do any net work on the charge in such a case as this, so that somewhere within the circuit there *must* be a source or sources of a *nonconservative electric field* $\mathbf{E}_{nc}$; then (12-22) can be written as

$$\mathscr{E} = \oint_C \mathbf{E}_{nc} \cdot d\mathbf{s} \qquad (12\text{-}23)$$

(Since electric field can be measured in volts/meter, we see that the unit for *emf* is the volt—the same as for potential and potential difference.) Later we will discuss how some of these nonconservative electric fields can be produced, but, for now, it is sufficient to note that the most common and familiar of these sources are batteries. A battery does work on a charge that passes through it and the source of this energy is essentially due to chemical reactions of one sort or another within the battery; thus, in a sense, a battery is analogous to a pump that can do work on a fluid and raise it, for example, against the conservative gravitational field. The battery is an example of a localized source of a nonconservative field so that $\mathbf{E}_{nc}$ in (12-23) is different from zero only when the path of the charge is within the battery and $\mathbf{E}_{nc} = 0$ elsewhere on the circuit. In this case, it is possible to speak of the *emf* of the source itself as a specific quantity, and its value will be obtained from (12-23) as

$$\mathscr{E}_{\text{source}} = \int_{\text{source}} \mathbf{E}_{nc} \cdot d\mathbf{s} \qquad (12\text{-}24)$$

For simplicity, therefore, we restrict the rest of our discussion to those conducting regions that are free of nonconservative electric fields, so that we can write $\mathbf{E} = -\nabla\phi$ and hence $\nabla \times \mathbf{E} = 0$, as long as we stay *outside* of the batteries.

Since $\mathbf{E}$ will exert forces on the moving charges, there should exist some functional relation between $\mathbf{J}_f$ and $\mathbf{E}$, that is, we expect to be able to write $\mathbf{J}_f = \mathbf{J}_f(\mathbf{E})$. We will also assume for now that $\mathbf{J}_f(0) = 0$, thereby excluding superconductors from our considerations. This relation between $\mathbf{J}_f$ and $\mathbf{E}$ can be quite complex and will depend on the material. We dealt with a similar situation in Section 10-6 when we discussed the relation between $\mathbf{P}$ and $\mathbf{E}$ for dielectrics, and we could construct a similar classification scheme. However, we pass immediately to the case of a *linear isotropic* conductor, that is, we assume that we can write

$$\mathbf{J}_f = \sigma\mathbf{E} \qquad (12\text{-}25)$$

where the factor of proportionality σ is called the *conductivity*. (This is a standard notation and should not lead to any confusion; in those very few cases where we have surface charge density and conductivity in the same expression, we write the surface charge density as σ_{ch}.) Equation 12-25 is another example of a constitutive equation, as is $\mathbf{D} = \epsilon\mathbf{E}$, and only experiment can decide whether it is suitable for a given material or not. In (12-25), σ is to be independent of the field $\mathbf{E}$, although it may still depend on position, and other variables such as temperature. The unit of σ is called 1 (ohm-meter)$^{-1}$ and since $\mathbf{J}_f$ is in amperes/(meter)^2 and $\mathbf{E}$ in volts/meter, we easily find that 1 ohm = 1 volt/ampere.

When (12-25) is applicable, the boundary condition (12-21) for a steady current can be written completely in terms of $\mathbf{E}$:

$$\hat{\mathbf{n}} \cdot (\sigma_2 \mathbf{E}_2 - \sigma_1 \mathbf{E}_1) = 0 \qquad (12\text{-}26)$$

We will still have $\hat{\mathbf{n}} \times (\mathbf{E}_2 - \mathbf{E}_1) = 0$, as well, since $\nabla \times \mathbf{E} = 0$ in all of the regions we will consider. Thus, we have a situation similar to that we found for dielectrics in Figure 10-14 in which the lines of $\mathbf{E}$ are refracted as we cross a bounding surface between two media of different conductivities.

If the material is homogeneous as well, then σ = const., that is, independent of position. In other respects, σ is a characteristic of the material and is to be found from experiment or else calculated from the atomic properties of matter by means of the theories used in other branches of physics; for our purposes, we take it as a given quantity. Not all materials are linear, isotropic, and homogeneous, but the assumption of (12-25) with σ = const. holds very well for metals and solutions of electrolytes, for example.

If we have a l.i.h. conductor *and* steady currents, we can combine (12-21), (12-25), (5-3), and (1-45) to give $\nabla \cdot \mathbf{J}_f = 0 = \nabla \cdot (\sigma \mathbf{E}) = \sigma \nabla \cdot \mathbf{E} = -\sigma \nabla^2 \phi$. In other words, $\nabla^2 \phi = 0$ and the potential still satisfies Laplace's equation. This result provides one with an *experimental* method of solving Laplace's equation by setting up the required boundary values of ϕ on the boundaries of a conducting region for then, by measuring the magnitude and direction of the current density $\mathbf{J}_f$, one can find the values of $\mathbf{E}$ from $\mathbf{E} = \mathbf{J}_f/\sigma$.

The relation $\mathbf{J}_f = \sigma \mathbf{E}$ for a l.i.h. conductor is equivalent to the macroscopic empirical relation known as *Ohm's law* and, in fact, is often called the microscopic form of Ohm's law. We can see how this comes about by analyzing the situation depicted in Figure 12-7. This is a portion of a uniform conductor of length l and cross-sectional area A carrying a total current I. If $|\Delta\phi|$ is the magnitude of the potential difference between the ends, then the magnitude of $E = |\Delta\phi|/l$ according to (5-3) and (1-38) since $\nabla\phi$ is a constant in this region of constant current and dimensions. Similarly, the current density is $J_f = I/A$ by (12-6), if we assume the current to be distributed uniformly over the cross section, which turns out to be a very accurate approximation. Substituting these into (12-25), we get $I/A = \sigma|\Delta\phi|/l$, or $I = (\sigma A/l)|\Delta\phi|$. Thus, we have a relation between these macroscopic quantities and we see that the current is proportional to the potential difference, (or conversely). This is usually written in the form

$$I = \frac{|\Delta\phi|}{R} \qquad (12\text{-}27)$$

where the proportionality factor is

$$R = \frac{l}{\sigma A} \qquad (12\text{-}28)$$

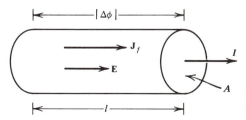

Figure 12-7. Portion of a conductor carrying a total current I.

Equation 12-27 is the empirical result discovered by Ohm and is known as Ohm's law, while the quantity R is called the *resistance*, and is measured in ohms (i.e., volts/ampere). Thus we have shown the equivalence between (12-25) and (12-27) for a l.i.h. conductor and have, simultaneously, obtained (12-28) as a method of calculating the resistance. The reciprocal of the conductivity, $1/\sigma$, is called the *resistivity* and is very often written as ρ!

As we see from the figure, $\mathbf{J}_f$ is directed longitudinally along this uniform conductor, as is $\mathbf{E}$ since they are parallel. Thus, at the surface, $\mathbf{E}$ will be tangential, and since the tangential components of $\mathbf{E}$ are continuous by (9-21), there will be a tangential field *outside* the conductor which is given by $\mathbf{E} = \mathbf{J}_f/\sigma$. This is in marked contrast to the static case where not only was $\mathbf{E} = 0$ inside the conductor, but it was necessarily normal to the surface, as we saw in (6-2) and Figure 6-1b.

We conclude this section by using some of our results to obtain an interesting and somewhat unexpected relationship.

■ **Example**

Relation between resistance and capacitance. Suppose we have two conductors of some shape. We consider two ways in which we can put them to use.

1. As a capacitor—let us fill the region between these conductors with a l.i.h. dielectric of permittivity ϵ and put equal and opposite charges on them as shown in Figure 12-8a. We want to find the capacitance. From (6-38), we know that we can find the potential difference by evaluating the integral

$$\Delta\phi = \int_+^- \mathbf{E} \cdot d\mathbf{s} \tag{12-29}$$

over any convenient path between the plates. We can write the free charge Q on the positive plate as an integral over its surface S:

$$Q = \int_S \sigma_f \, da = \int_S \epsilon \mathbf{E} \cdot d\mathbf{a} = \epsilon \int_S \mathbf{E} \cdot d\mathbf{a} \tag{12-30}$$

with the use of (2-16), (10-56), and (6-1). Putting these into (6-38), we find that the capacitance can be written as

$$C = \frac{\epsilon \int_S \mathbf{E} \cdot d\mathbf{a}}{\int_+^- \mathbf{E} \cdot d\mathbf{s}} \tag{12-31}$$

2. As a resistance—now, instead of the dielectric, let us fill the region between the plates with a l.i.h. conductor of conductivity σ as shown in Figure 12-8b. We also maintain the same potential difference $\Delta\phi$ between the plates by keeping each plate at

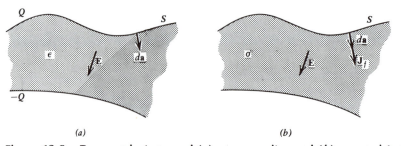

(a)   (b)

Figure 12-8. Two conductors used (a) as a capacitor and (b) as a resistance.

the *same* potential as was used before. As we saw above, ϕ satisfies Laplace's equation under these circumstances, and since the boundary conditions are exactly the same in the two cases, we know from the uniqueness theorem of Section 11-1 that $\phi(\mathbf{r})$ will be identical for each. In other words, the potential difference will again be given by (12-29) with exactly the same values of $\mathbf{E}$ at each point on the path of integration. The total current I passing between the plates can be expressed as a surface integral over the same upper plate with the use of (12-6) and (12-25); the result is that

$$I = \int_S \mathbf{J}_f \cdot d\mathbf{a} = \int_S \sigma \mathbf{E} \cdot d\mathbf{a} = \sigma \int_S \mathbf{E} \cdot d\mathbf{a} \qquad (12\text{-}32)$$

When we put (12-29) and (12-32) into (12-27), we find the resistance of this system to be given by

$$\frac{1}{R} = \frac{\sigma \int_S \mathbf{E} \cdot d\mathbf{a}}{\int_+^- \mathbf{E} \cdot d\mathbf{s}} \qquad (12\text{-}33)$$

Comparing (12-31) and (12-33), we see that $1/R\sigma = C/\epsilon$ or

$$RC = \frac{\epsilon}{\sigma} \qquad (12\text{-}34)$$

showing that these two properties of the system are not independent but are in fact related in this simple way. [This relation (12-34) is reminiscent of a typical thermodynamic result as it expresses a relation between macroscopic properties of a system without giving an indication of the absolute value of either.] This very general result also can be used as a means of measuring C indirectly, since resistance measurements are comparatively easy to perform with an ammeter and a voltmeter, while a direct electrostatic measurement of C is generally much harder to do. ∎

12-4 ENERGY RELATIONS

In the previous section, we mentioned the experimental fact that the maintenance of a steady current in a conductor is possible only if energy is constantly being supplied to the system. But in a steady situation, where things remain constant in time, there cannot be an accumulation of electrical energy either. Consequently, what is supplied as electrical energy must be converted into another form of energy, and, in fact, it is observed that the electrical energy expended appears as heat produced within the body of the conductor. We can put this into quantitative form in the following way. We saw in (5-45) and (5-46) that the work done *by the electric field* on a charge Δq is $\Delta W = -\Delta q \, \Delta\phi$ where $\Delta\phi$ is the change in potential. Therefore the rate at which electrical work is being done on the system is $\Delta W/\Delta t = -(\Delta q/\Delta t) \Delta\phi = -I \Delta\phi$ by (12-1) since we have a steady current. In the steady state, this must also be the rate of conversion of energy into heat. If we let w be the rate of production of heat per unit volume, then we find by using the dimensions used in Figure 12-7, where the volume is Al, that

$$w = \frac{(\Delta W/\Delta t)}{Al} = \left(\frac{I}{A}\right)\left(-\frac{\Delta\phi}{l}\right) = J_f E$$

Since $\mathbf{J}_f$ and $\mathbf{E}$ are parallel, we can also write this as

$$w = \mathbf{J}_f \cdot \mathbf{E} = \sigma E^2 = \frac{J_f^2}{\sigma} \qquad (12\text{-}35)$$

with the use of (12-25). The quantity w is also referred to as the power "dissipation" per unit volume. Since our unit of power is 1 watt = 1 joule/second, w will be measured in watts/(meter)^3.

12-5 A MICROSCOPIC POINT OF VIEW

This result that we have just obtained is perfectly general since it follows directly from a combination of the macroscopic laws of conservation of energy and the meaning of current and potential difference. Nevertheless, its origin may seem obscure and one may feel the need of an "explanation" in terms of the microscopic average behavior of the moving charges in the conductor. Related to this question of "why" is there a production of heat is the problem of "why" does a conductor have a resistance in the first place. Although we leave a more comprehensive discussion of these questions to Appendix B, we can easily enough get a semiquantitative answer to both of these questions from a microscopic picture.

First of all, if the electric field produced the only force on the mobile free charges, they would have a constant acceleration given by $\mathbf{a} = \mathbf{F}_q/m = q\mathbf{E}/m$ where m is the mass of the charge carrier. But a constant acceleration will result in an indefinitely increasing velocity, which is not observed. A steady current means, according to (12-3), a constant velocity and hence *zero* acceleration, that is, a zero net force. Therefore, the electrical force, which is in the direction of motion of the charges, must be balanced, at least on the average, by another force directed opposite to the motion. In order to get an idea of the origin of this force, let us consider the specific case of a metal where the free charge carriers are electrons each of charge $-e$. They move about, not in completely empty space, but among the ions of the metal that are arranged in the regular array of the crystal. The electrons can certainly collide with these ions (as well as with each other), and, when they do, their velocity will be changed. Between collisions, they will be accelerated by the electric field, but on collision, the result of this process will be abruptly changed. Thus, it is the collisions that are the origin of the other force, and it is the *average* effect in which we are interested. By analogy with similar effects involving "friction" in mechanics, we shall try to describe the overall effect of collisions as their giving rise to a force proportional to the velocity and opposite to it; thus we write the mechanical force as $\mathbf{F}_{q,m} = -\xi\mathbf{v}$ where ξ is an appropriate proportionality factor. The net force will then be the sum of the electrical and mechanical forces, or

$$\mathbf{F}_{\text{net}} = m\mathbf{a} = -e\mathbf{E} - \xi\mathbf{v} \tag{12-36}$$

We see that this equation of motion allows a situation in which the acceleration $\mathbf{a} = 0$; then the velocity will be a constant, $\mathbf{v}_d$, which is found from (12-36) to be

$$\mathbf{v}_d = -\frac{e\mathbf{E}}{\xi} \tag{12-37}$$

This velocity is commonly called the *drift velocity*; in mechanics it is usually known as the *terminal velocity*, as, for example, in the case of an object falling near the earth and also subject to the viscous (resistive) drag of the air. If n is the number of electrons per unit volume, the free charge density will be $\rho_f = n(-e)$, and if we substitute this, along with (12-37), into (12-3), we find the value of the steady free current to be

$$\mathbf{J}_f = \rho_f\mathbf{v}_d = \left(\frac{ne^2}{\xi}\right)\mathbf{E} \tag{12-38}$$

But this is exactly of the form of Ohm's law given by (12-25) and, by comparison, we see that not only have we accounted for its general form from this simplified microscopic view, but we also have obtained an expression for the conductivity:

$$\sigma = \frac{ne^2}{\xi} \tag{12-39}$$

Qualitatively, this result has reasonable properties. It is proportional to the number density of carriers, and inversely proportional to the "frictional" term ξ, so that the smaller the effect of collisions, the larger will be the current for a given **E** and *vice versa*. We also note that it is proportional to the square of the charge so that it is actually independent of the sign of the charge of the carriers. We cannot go any further in trying to calculate σ until we can calculate ξ. This, however, involves the detailed theories of solid state physics and the analysis of collisions that we cannot pursue further here. On the other hand, one can use *measured* values of σ to evaluate the factor ξ.

We have just seen how the overall effect of collisions can account for the observed *finite* value of the current. We can also see how the existence of collisions can account qualitatively for the conversion of electrical energy into heat described by (12-35). Not only does a collision change the electron velocity in direction, but by changing its magnitude, also changes the kinetic energy of both the electron and the struck ion. Thus there is a transfer of energy from the *ordered* motion of the electron produced by the field to the *disordered* random vibrations of the ions constituting the crystal. As we know from thermodynamics, this type of process is an irreversible one, so that the net effect of the collisions will be to increase the disordered energy of the metal crystal. But it is just this increase in disordered motion that is associated with the production of *heat*. As we mentioned, this is exactly what is observed, and we can see now that it too is a necessary consequence of the same microscopic processes that result in a finite conductivity.

12-6 THE ATTAINMENT OF ELECTROSTATIC EQUILIBRIUM

As we know from Chapter 6, if we put some free charge on or within a conductor, the system will not generally be in equilibrium and it will have to readjust itself by means of currents until the final electrostatic equilibrium state is attained for which all the charge is on the surface and the conductor is an equipotential volume. We do not, however, know any of the details of this process, nor do we know how long it will take in a typical case. We can use some of the results we have just obtained to give us an idea of the nature of this process; it will involve a nonsteady situation since the currents involved will eventually become zero.

We consider a l.i.h. conducting dielectric for which we can write $\mathbf{J}_f = \sigma\mathbf{E}$ and $\mathbf{D} = \epsilon\mathbf{E}$. Since the free charge density can be changing in time, we use (12-19) and the above relations, along with (10-41), to get an equation involving ρ_f only:

$$-\left(\frac{\partial\rho_f}{\partial t}\right) = \nabla\cdot\mathbf{J}_f = \nabla\cdot(\sigma\mathbf{E}) = \nabla\cdot\left(\frac{\sigma\mathbf{D}}{\epsilon}\right) = \frac{\sigma}{\epsilon}\nabla\cdot\mathbf{D} = \frac{\sigma}{\epsilon}\rho_f$$

so that ρ_f satisfies the differential equation

$$\frac{\partial\rho_f}{\partial t} = -\frac{\sigma}{\epsilon}\rho_f \tag{12-40}$$

which has the solution

$$\rho_f(t) = \rho_f(0)e^{-\sigma t/\epsilon} = \rho_f(0)e^{-t/\tau} \tag{12-41}$$

where $\tau = \epsilon/\sigma$ and $\rho_f(0)$ is the initial value. This result says that the free charge density in a l.i.h. material can only decrease! [One can make it increase, of course, but only by means other than conduction processes described by $\mathbf{J}_f = \sigma\mathbf{E}$ such as, for example, injecting a pulse of electrons that come to rest within the body of the material. However, after the termination of the pulse, the charge density that was produced will then decay exponentially as described by (12-41).]

Thus, as long as conduction processes are the only mechanism by which equilibrium ($\rho_f = 0$) is produced, we see that the approach to equilibrium will be one of exponential decay. The charge density will decrease by a factor of $1/e$ in time $\tau = \epsilon/\sigma$. This particular behavior is often given the name of *relaxation*; hence τ is called the *relaxation time*, and its value gives us an estimate of the time required for attainment of equilibrium.

Actually, the simple result we have obtained is only applicable to a poorly conducting dielectric. For a very good conductor such as a metal the situation turns out to be much more complicated.[1] In this case, the fields change so rapidly in time that one has to include magnetic effects as well and some of the phenomena that will not be discussed until Chapters 17, 21, and 26. One finds that the shape and size of the conductor affect the time taken to attain final equilibrium and that this time is generally much longer than simply τ. We can get a rough intimation as to why this can be the case by estimating the value of τ for a typical metal. For most metals, $\epsilon \simeq \epsilon_0$, and for copper $\sigma \simeq 5.8 \times 10^7$/ohm-meter, so that $\tau = \epsilon/\sigma \simeq (8.85 \times 10^{-12})/(5.8 \times 10^7) \approx 10^{-19}$ second. This is a very short time, but it is also much smaller than the average time between collisions, which is about 10^{-14} second, as can be deduced from experiment and a more detailed form of (12-39). Since we have assumed collisions to be the physical origin of a finite conductivity, it is unreasonable that effects depending on conductivity can occur on such a much shorter time scale, and hence that a simple expression like (12-41) can really describe the situation in metals.

If the bound charges in a dielectric do not have the equilibrium distribution described by all of the results of Chapter 10, they too will relax in some manner toward equilibrium. Here, however, conduction is not generally available as a mechanism, or is negligible, and the primary processes of relaxation usually involve thermal agitation in one way or another. This is a separate and large field of study all by itself and we are in no position here to consider its details, and must simply content ourselves with the observation that these phenomena exist.

EXERCISES

12-1 At a given instant, a certain system has a current density given by $\mathbf{J} = A(x^3\hat{\mathbf{x}} + y^3\hat{\mathbf{y}} + z^3\hat{\mathbf{z}})$ where A is a positive constant. (a) In what units will A be measured? (b) At this instant, what is the rate of change of the charge density at the point $(2, -1, 4)$ meter? (c) Consider the total charge Q contained within a sphere of radius a centered at the origin. At this instant, what is the rate at which Q is changing in time? Is Q increasing or decreasing?

12-2 Verify (12-7), (12-8), and (12-9).

12-3 A total charge Q is distributed uniformly throughout a sphere of radius a. The sphere is

then rotated with constant angular speed ω about a diameter. Assume that the charge distribution is unaffected by the rotation, and find $\mathbf{J}$ everywhere within the sphere. (Express it in spherical coordinates with the polar axis coinciding with the axis of rotation.) Find the total current passing through a semicircle of radius a fixed in space with its base on the axis of rotation.

12-4 A dielectric sphere of radius a is uniformly polarized. It is rotated with constant angular speed ω about the diameter parallel to $\mathbf{P}$. Assume that $\mathbf{P}$ is unaffected by the rotation and find the currents. Plot your result as a function of the ap-

[1] H. C. Ohanian, *Am. J. Phys.*, *51*, 1020–1022 (1983).

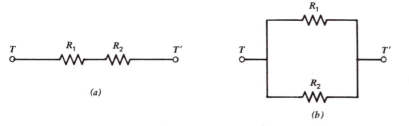

Figure 12-9. (*a*) **Resistors in series.** (*b*) **Resistors in parallel.**

propriate coordinate(s). Find the total current passing through a semicircle of radius a fixed in space with its base on the axis of rotation.

12-5 A sphere of radius a centered at the origin is made of a l.i.h. conducting material. The potential on the surface is maintained at the values given in spherical coordinates by $\phi_0 \cos\theta$ where $\phi_0 = $ const. Find the free current density J_f everywhere inside.

12-6 Two "resistors" of resistances R_1 and R_2 are connected by conductors of negligible resistance to each other and to terminals T and T' as shown in Figure 12-9*a*, that is, they are connected in "series." If a potential difference $\Delta\phi$ is now applied across the terminals, show that this combination is equivalent to a single resistor of resistance $R_s = R_1 + R_2$. Similarly, show that the equivalent resistance of the "parallel" combination shown in (b) can be found from $1/R_p = (1/R_1) + (1/R_2)$.

12-7 A long straight wire carries a steady current I. The current is distributed uniformly over the circular cross section of radius a. Consider a length l of the wire that has a resistance R. Find the electric field in the vacuum region just outside the wire and express it in terms of the quantities given. (Overall, this portion of the wire is neutral.)

12-8 Consider a material with a nonzero conductivity. Assume the material to be linear and isotropic in both its dielectric and conducting properties. Assume steady free currents. (a) If the material is inhomogeneous, show that there will be a volume density of free charge in the material given by

$$\rho_f = \mathbf{J}_f \cdot \nabla\left(\frac{\epsilon}{\sigma}\right) \qquad (12\text{-}42)$$

(b) Show that, at a surface of discontinuity between two materials, there will be a free surface charge density given by

$$\sigma_{fch} = J_n\left(\frac{\epsilon_2}{\sigma_2} - \frac{\epsilon_1}{\sigma_1}\right) \qquad (12\text{-}43)$$

(Since all complete circuits will have interfaces

between different materials, such as connections between wires and battery terminals, there will be accumulations of charge within the system that will be constant in time. Thus, the above results show us that it is these distributions of electric charge which are the sources of the electric field in systems carrying steady currents.)

12-9 Two large parallel plane conducting plates are a distance d apart. The region between them is filled with two l.i.h. materials whose surface of separation is a plane parallel to the plates. The first material (with properties σ_1 and ϵ_1) is of thickness x, while the second material (σ_2, ϵ_2) has thickness $d - x$. There is a steady current between the plates that are kept at constant potentials of ϕ_1 and ϕ_2. Find the potential at the surface of separation of the two media, and the free surface charge density there.

12-10 The region between the coaxial cylinders of Figure 6-12 is filled with a l.i.h. conducting material. If a potential difference $\Delta\phi$ is applied between the cylinders, find the current between the cylinders for a length L of the system.

12-11 Consider an infinite slab of conducting dielectric of thickness d and with plane parallel faces. Two metal cylinders, each of radius A, have their axes parallel and a distance D apart. These cylinders pass through the dielectric with their axes normal to the plane surfaces. If a potential difference $\Delta\phi$ is applied between these metal "electrodes," find the current I that will pass from one to the other.

12-12 A very long wire of radius a is suspended near the bottom of a very deep lake. Assume the lake to have a plane bottom that is a very good conductor. The wire is parallel to the bottom and at a height h above it. If the conductivity of the water is σ, find the resistance between the wire and the bottom for a length L of the system.

12-13 Show that when (12-35) is integrated over the total volume of a uniform conductor, the total rate of production of heat can be written as I^2R.

12-14 Show that, when a complete circuit includes a localized source of emf $\mathscr{E}$, the steady state current I will be given by $I = \mathscr{E}/R$ where R is the total resistance of the whole circuit. (See previous exercise.)

12-15 Show by direct integration that when a capacitor is charged by a battery of emf $\mathscr{E}$, the amount of heat developed in the circuit is equal to the final electrostatic energy of the capacitor.

12-16 Two square metal plates each 0.1 meter on a side are parallel to each other and separated by a distance of 10^{-2} meters. The region between the plates is filled with water of conductivity 10^{-3} (ohm-meter)$^{-1}$. If a potential difference of 150 volts is maintained between the plates, find the rate of change of temperature of the water. (Neglect end effects and heat losses to regions outside the water.)

12-17 A copper wire 2.5 millimeters in diameter is carrying a current of 10 amperes. Assume one free electron per copper atom and find the magnitude of the drift velocity. [Copper has a density of 8.92 grams/(centimeter)^3, an atomic weight of 63.5 grams/mole, and Avogadro's number is 6.02×10^{23}/mole.]

12-18 In Section 12-5, we discussed a microscopic picture that associated the existence of resistance in a metal with collisions between the mobile electrons and the metal ions. Based on this viewpoint, what do you think will be the effect on the resistance if the temperature of the conductor is increased?

12-19 Verify that ϵ/σ has the dimensions of time.

12-20 Find the relaxation time for a glass with a resistivity of 10^{12} ohm-meter and a relative permittivity of 4.0.

12-21 The region between the plates of a capacitor is filled with a material of finite conductivity with resultant total resistance R. Show that the charge *on the plates* will decrease to $1/e$ of its initial value in a time RC where C is the capacitance. [Note the similarity between this result and (12-34) and (12-41). But, of course, different processes are being described—or are they the same?]

13

AMPÈRE'S LAW

Interestingly, the general subject that we nowadays call *magnetism* also began with the observation that certain naturally occurring minerals could attract other materials. The name presumably derives from the association of these objects with the ancient cities of Magnesia in Asia Minor since they could be found nearby. For centuries, the subject was thought to be independent of electricity. During this time, however, people did learn about the magnetic properties of the earth and invented the magnetic compass. The first indication of the connection between electricity and magnetism occurred in 1819 when Oersted somewhat accidentally discovered that an *electric current* could exert forces on a *magnetic compass needle*. Ampère heard of Oersted's result and quickly found that an electric current could also exert a force on another electric current. He began a systematic study of these forces, and, by means of a series of ingenious and elegant experiments during the period 1820–1825, he was able to deduce the form of the basic law of force between electric currents. About 50 years later, Maxwell described Ampère's work as being "one of the most brilliant achievements in science."

These forces between steady currents are what we will take as our starting point, and we leave to a later chapter the description of magnetic materials as well as the effect of matter in general. Since currents are involved, the charges are not at rest but moving. However, as we saw in the last chapter, the charges are moving with constant average speed in *steady currents*. Under these circumstances, the forces involved will also be constant in time; thus we are justified in referring to this subject as *magnetostatics*.

13-1 THE FORCE BETWEEN TWO COMPLETE CIRCUITS

In Section 2-2 where we introduced Coulomb's law, we dealt with individual point charges. One might expect, by analogy, that here we would similarly consider "little pieces" of the current and study the forces between them. The natural thing to think of in this context is the current element of (12-10) since it involves both the magnitude of the current and its local direction. In the laboratory, however, one must, of necessity, deal with *complete closed* circuits for steady currents, so what one has to do is to take, as a fundamental experimental law, that which describes the *total force* between two complete circuits. Once this has been done, one can try to deduce from this, as best one can, a formula that gives the force between the hypothesized current elements. This is what we will do.

We consider two idealized complete circuits C and C', carrying steady filamentary currents I and I', respectively. The situation we have in mind is shown in Figure 13-1. What we want is the total force on C due to C', which we write as $\mathbf{F}_{C' \to C}$. The circuits are idealized in the sense that the batteries are not included in the diagram. It is assumed that the necessary batteries are some distance away and the conducting leads from them are twisted closely together, since one of Ampère's first experiments showed that two oppositely directed currents close together produced no effect on another current. In spite of what was said above, the result for the total force is written in terms of the current elements $I\,d\mathbf{s}$ and $I'\,d\mathbf{s}'$ shown; their location with respect to an arbitrary origin is given by their position vectors **r** and **r'** and the relative position vector is defined as usual by $\mathbf{R} = \mathbf{r} - \mathbf{r'}$, that is, it is drawn from the "source point" to the "field point." (A comparison of Figure 13-1 with Figure 2-2 is instructive.) We assume a vacuum throughout the region not occupied by currents.

The basic experimental law that gives the total force on C due to C' can be written as

$$\mathbf{F}_{C' \to C} = \frac{\mu_0}{4\pi} \oint_C \oint_{C'} \frac{I\, d\mathbf{s} \times (I'\, d\mathbf{s'} \times \hat{\mathbf{R}})}{R^2} \tag{13-1}$$

and is what we will take as *Ampère's law*. We note that it has the form of a double line integral, each of which is taken over the corresponding circuit. It strains one's credulity, of course, to have it asserted that a formula of the generality implied by (13-1) could have been deduced from a very few experiments on circuits of special and simple shapes as was done by Ampère. Equation 13-1 evidently represents a generalization from results that have been found for very many special cases.

The factor $\mu_0/4\pi$ that appears in (13-1) is a constant of proportionality whose numerical value depends on the system of units that is being used; the dimensions of μ_0 are seen to be those of force/(current)^2. In the SI unit system which we are using (in effect, MKSA units), μ_0 is *defined* to have precisely the value

$$\mu_0 = 4\pi \times 10^{-7} \text{ newton/(ampere)}^2$$

$$= 4\pi \times 10^{-7} \text{ henry/meter} \tag{13-2}$$

This constant μ_0 is called the *permeability of free space*, and is generally written in the last form from which we see, by comparison of both forms, that 1 henry = 1 joule/(ampere)^2. Since the newton and meter are already determined by other independent definitions, we see that this definition of μ_0 fixes the unit of current that is all that remains undefined in (13-1). Thus, (13-2) essentially gives us the definition of the ampere and thence that of the coulomb.

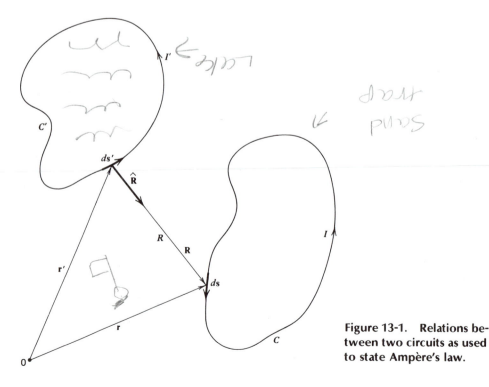

Figure 13-1. Relations between two circuits as used to state Ampère's law.

If there are several circuits that can interact with C, the force of each will be given by an expression like (13-1), and the total force on C will be the vector sum of them:

$$\mathbf{F}_C = \mathbf{F}_{\text{total on } C} = \sum_{C'} \mathbf{F}_{C' \to C} \tag{13-3}$$

Although we can imagine generalizing (13-1) and (13-3) to include distributed currents with the use of (12-10), it will be convenient to continue with filamentary currents only.

The integrand in (13-1) is more complicated from a directional point of view than is Coulomb's law as expressed by, say, (2-15), since the integrand depends on the relative orientation of the three quantities $I\,d\mathbf{s}$, $I'\,d\mathbf{s}'$, and $\hat{\mathbf{R}}$. We also note that the quantities relating to the two circuits appear unsymmetrically in the integrand. This observation may be somewhat disquieting because it may make us wonder that, if the roles of C and C' were interchanged, the force on C' due to C might not be equal and opposite to that on C due to C', as we would expect from Newton's third law as applied to these *macroscopic* complete circuits. However, this asymmetry is only apparent as we can see by rewriting our result. Using (1-30), (1-141), and (1-38), we find that we can write

$$\frac{I\,d\mathbf{s} \times (I'\,d\mathbf{s}' \times \hat{\mathbf{R}})}{R^2} = II'\,d\mathbf{s}'\left(d\mathbf{s} \cdot \frac{\hat{\mathbf{R}}}{R^2}\right) - \frac{II'\hat{\mathbf{R}}(d\mathbf{s} \cdot d\mathbf{s}')}{R^2}$$

$$= -II'\,d\mathbf{s}'\left[d\mathbf{s} \cdot \nabla\left(\frac{1}{R}\right)\right] - \frac{II'\hat{\mathbf{R}}(d\mathbf{s} \cdot d\mathbf{s}')}{R^2}$$

$$= -II'\,d\mathbf{s}'\,d_C\left(\frac{1}{R}\right) - \frac{II'\hat{\mathbf{R}}(d\mathbf{s} \cdot d\mathbf{s}')}{R^2} \tag{13-4}$$

where we have written $d_C(1/R)$ to indicate the differential change in $(1/R)$ resulting from a displacement along C. Inserting (13-4) into (13-1), we find that the total force can also be written as

$$\mathbf{F}_{C' \to C} = -\frac{\mu_0 II'}{4\pi}\oint_{C'} d\mathbf{s}' \oint_C d_C\left(\frac{1}{R}\right) - \frac{\mu_0 II'}{4\pi}\int_C\oint_{C'} \frac{(d\mathbf{s} \cdot d\mathbf{s}')\hat{\mathbf{R}}}{R^2} \tag{13-5}$$

where, in the first term, we integrate over C first and then over C'. The integral over C has the form of an integral over a closed path of the differential of a scalar, and since the final and initial points coincide, we have

$$\oint_C d_C\left(\frac{1}{R}\right) = \left(\frac{1}{R}\right)_f - \left(\frac{1}{R}\right)_i = \left(\frac{1}{R}\right)_i - \left(\frac{1}{R}\right)_i = 0$$

and the first term of (13-5) thereby vanishes, leaving us with

$$\mathbf{F}_{C' \to C} = -\frac{\mu_0 II'}{4\pi}\oint_C\oint_{C'} \frac{(d\mathbf{s} \cdot d\mathbf{s}')\hat{\mathbf{R}}}{R^2} \tag{13-6}$$

as another expression that will give us the *same* total force. We can, in fact, regard (13-6) as another version of Ampère's law in vacuum. In this form, the two circuits appear more symmetrically except for $\hat{\mathbf{R}}$, which has a definite sense. If we were to use (13-6) to calculate the force on C' due to C, we would get

$$\mathbf{F}_{C \to C'} = -\frac{\mu_0 I'I}{4\pi}\oint_{C'}\oint_C \frac{(d\mathbf{s}' \cdot d\mathbf{s})\hat{\mathbf{R}}'}{R'^2} \tag{13-7}$$

where $\mathbf{R}' = \mathbf{r}' - \mathbf{r}$. Since $R' = |\mathbf{r}' - \mathbf{r}| = R$, we have $\hat{\mathbf{R}}' = -\hat{\mathbf{R}}$; also $d\mathbf{s}' \cdot d\mathbf{s} = d\mathbf{s} \cdot d\mathbf{s}'$ by (1-16). Thus, on comparing (13-6) and (13-7), we see that

$$\mathbf{F}_{C \to C'} = -\mathbf{F}_{C' \to C} \tag{13-8}$$

as is to be expected for the total forces.

This calculation shows us that there is a lot of ambiguity in the expression for the total force on a complete circuit that, after all, is all one can really measure in the laboratory. In fact, we could write down an innumerable set of "versions" of Ampère's law by simply adding to the integrand of (13-1) any function that vanishes when integrated over a complete circuit. Long experience has shown, however, that nothing is to be gained from this and the formulation given in (13-1) is the most useful and is to be preferred for many reasons as we will see.

We are assuming a static situation, that is, the circuits are at rest at fixed positions. This means that in order for C to be in equilibrium there must be an additional mechanical force $\mathbf{F}_{C,m}$ on it so that the net force will be zero; in other words, we must have

$$\mathbf{F}_{C' \to C} + \mathbf{F}_{C,m} = 0 \tag{13-9}$$

Similar remarks apply to C'.

As we said before, Ampère's law (13-1) is a generalization of results from many special cases. As usual, in order to get any use out of this law one has to evaluate it for special cases in order to obtain results that can be checked relatively easily in the laboratory. As an illustration of the use of (13-1), we will apply it to a particularly simple and important case.

13-2 TWO INFINITELY LONG PARALLEL CURRENTS

We consider two infinitely long straight circuits carrying currents I and I'. They are parallel to each other and a distance ρ apart. In order to be specific, let us use cylindrical coordinates and choose the z axis to coincide with the source current I' as shown in Figure 13-2. These currents must each be part of a closed circuit, of course; we assume that the closing portions lie along curves at infinity (such as large semicircles) and hence are so far away that their contribution can be neglected because of the R^2 in the denominator of (13-1). We see from the figure that $\mathbf{r} = \rho\hat{\boldsymbol{\rho}} + z\hat{\mathbf{z}}$ and $\mathbf{r}' = z'\hat{\mathbf{z}}$ so that $\mathbf{R} = \mathbf{r} - \mathbf{r}' = \rho\hat{\boldsymbol{\rho}} + (z - z')\hat{\mathbf{z}}$ and $R^2 = \rho^2 + (z - z')^2$. We also see that $d\mathbf{s} = d\mathbf{r} = dz\,\hat{\mathbf{z}}$ and $d\mathbf{s}' = d\mathbf{r}' = dz'\,\hat{\mathbf{z}}$ since both ρ and $\hat{\boldsymbol{\rho}}$ are constants in this

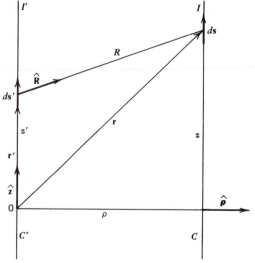

Figure 13-2. Calculation of force between two infinitely long parallel currents.

case. Using these results and the fact that $\hat{\mathbf{R}} = \mathbf{R}/R$, we find that

$$\frac{d\mathbf{s} \times (d\mathbf{s}' \times \hat{\mathbf{R}})}{R^2} = \frac{dz\,dz'\,\hat{\mathbf{z}} \times \{\hat{\mathbf{z}} \times [\rho\hat{\boldsymbol{\rho}} + (z - z')\hat{\mathbf{z}}]\}}{R^3} = -\frac{\rho\,dz\,dz'\,\hat{\boldsymbol{\rho}}}{\left[\rho^2 + (z - z')^2\right]^{3/2}}$$

(13-10)

with the use of (1-76) and (1-26). Substituting this into (13-1), we obtain

$$\mathbf{F}_{C' \to C} = -\frac{\mu_0 II' \rho\hat{\boldsymbol{\rho}}}{4\pi} \int_{-\infty}^{\infty} dz \int_{-\infty}^{\infty} \frac{dz'}{\left[\rho^2 + (z - z')^2\right]^{3/2}}$$

(13-11)

where the limits of integration will cover the circuits C and C'. If we make the substitution $t = z' - z$ so that $dt = dz'$ since z is kept constant during the integration over z', we find that the integral over z' becomes

$$\int_{-\infty}^{\infty} \frac{dt}{\left(\rho^2 + t^2\right)^{3/2}} = \frac{2}{\rho^2}$$

according to (3-7) and (3-8). Therefore, (13-11) reduces to

$$\mathbf{F}_{C' \to C} = -\frac{\mu_0 II' \hat{\boldsymbol{\rho}}}{2\pi\rho} \int_{-\infty}^{\infty} dz$$

(13-12)

If we were now to carry out the integration over z, that is, over C, we would get an infinite force. We can, nevertheless, still get something useful from our result by noting that the integrand is independent of z, so that the force on a length dz will be given by $d\mathbf{F} = -(\mu_0 II' \hat{\boldsymbol{\rho}}/2\pi\rho)\,dz$. Thus, if we introduce a *force per unit length*, $\mathbf{f}_C$, we can write

$$\mathbf{f}_C = \frac{d\mathbf{F}}{dz} = -\frac{\mu_0 II' \hat{\boldsymbol{\rho}}}{2\pi\rho} \qquad \left(\begin{array}{c}\text{parallel} \\ \text{currents}\end{array}\right)$$

(13-13)

We see that $\mathbf{f}_C$ has constant magnitude, is proportional to the product of the currents, is inversely proportional to the separation ρ, and is directed perpendicular to the currents. In fact, since the coefficient of $\hat{\boldsymbol{\rho}}$ in (13-13) is negative, we see from Figure 13-2 that this is a force of attraction of C toward C'.

In Figure 13-2, we assumed that I and I' were in the same direction. If, now, I and I' were oppositely directed, then either $d\mathbf{s}$ or $d\mathbf{s}'$ would be of opposite sign to that it previously had since the line element is defined to be in the same sense as that of the current. Then we see from (13-10) that the double cross product would have its sign changed and (13-13) would become

$$\mathbf{f}_C = \frac{\mu_0 II' \hat{\boldsymbol{\rho}}}{2\pi\rho} \qquad \left(\begin{array}{c}\text{antiparallel} \\ \text{currents}\end{array}\right)$$

(13-14)

which is one of repulsion on C.

We can sum up the qualitative nature of these results by saying that parallel ("like") currents *attract*, while antiparallel ("unlike") currents *repel*; this is sort of "opposite" to the behavior that we recall for the electrostatic forces between point charges.

A particular case of (13-14) that is of interest is that for which the currents are equal; when $I = I'$, it becomes

$$\mathbf{f}_C = \frac{\mu_0 I^2 \hat{\boldsymbol{\rho}}}{2\pi\rho}$$

(13-15)

This arrangement can be attained in practice by making the very long straight parallel lines C and C' in Figure 13-2 portions of the *same* circuit by connecting the ends at infinity to close the circuit. In this case one is certain to have equal and opposite currents so that (13-15) is applicable. But now, with μ_0 defined by (13-2), there is nothing arbitrary left in (13-15) so that the *absolute* value of I in *amperes* can be found from a determination of the *mechanical* quantities $\mathbf{f}_C$ and ρ. In other words, (13-15) can be used to *measure* I; this is essentially the method that is actually used to determine the value of the ampere. Once any current I has been found in this manner, the value I' of any other can be found in principle by using it in this arrangement so that either (13-13) or (13-14) describes the measurable force per unit length.

13-3 THE FORCE BETWEEN CURRENT ELEMENTS

Although it is often unwise to draw conclusions about an integrand from the nature and value of a definite integral, we have done it before, for example in getting (7-29), and we will be doing it again. The *form* of (13-1) is such that it can be written as

$$\mathbf{F}_{C' \to C} = \oint_C \oint_{C'} d\mathbf{F}_{e' \to e} \tag{13-16}$$

and the integrand

$$d\mathbf{F}_{e' \to e} = \frac{\mu_0}{4\pi} \frac{I\,d\mathbf{s} \times (I'\,d\mathbf{s}' \times \hat{\mathbf{R}})}{R^2} \tag{13-17}$$

given a natural interpretation as the force exerted on the current element $I\,d\mathbf{s}$ by the current element $I'\,d\mathbf{s}'$. We could call this still another version of Ampère's law and, in fact, this is very often what is done. As we pointed out, it cannot be checked directly with experiments on current elements but it does allow a verifiable generalization to moving point charges and it is consistent with all that we shall do later. In this form, (13-17) is more nearly the analogue of Coulomb's law that we would expect to find, including the inverse square dependence on the distance R between the elements.

In contrast to Coulomb's law, however, $d\mathbf{F}_{e' \to e}$ is not generally along the line connecting the elements, the direction of which is given by $\hat{\mathbf{R}}$. As a consequence, Newton's third law is not satisfied by (13-17), although we saw it to hold for the overall circuits in (13-8). In order to show this, let us calculate the force exerted on $I'\,d\mathbf{s}'$ by $I\,d\mathbf{s}$; as we saw before, we can get it by interchanging the primed and unprimed quantities in (13-17) *and* by replacing $\hat{\mathbf{R}}$ by $-\hat{\mathbf{R}}$. In this way, we find that

$$d\mathbf{F}_{e \to e'} = -\frac{\mu_0}{4\pi} \frac{I'\,d\mathbf{s}' \times (I\,d\mathbf{s} \times \hat{\mathbf{R}})}{R^2} \tag{13-18}$$

where $\mathbf{R} = \mathbf{r} - \mathbf{r}'$ still. If we add (13-17) and (13-18), we find, with the use of (1-30), that

$$d\mathbf{F}_{e' \to e} + d\mathbf{F}_{e \to e'} = \frac{\mu_0 II'}{4\pi R^2} [\hat{\mathbf{R}} \times (d\mathbf{s}' \times d\mathbf{s})] \tag{13-19}$$

which is generally different from zero, in contrast to (13-8). This sum will be zero only if $d\mathbf{s}'$ and $d\mathbf{s}$ are parallel as in Figure 13-2, or if $\hat{\mathbf{R}}$ is perpendicular to the plane formed by $d\mathbf{s}'$ and $d\mathbf{s}$.

If we do the same thing with our more symmetrical form (13-6), we would get still another possible form for the force between current elements, namely

$$d\mathbf{F}'_{e' \to e} = -\frac{\mu_0}{4\pi} \frac{[(I\,d\mathbf{s}) \cdot (I'\,d\mathbf{s}')]\hat{\mathbf{R}}}{R^2} \tag{13-20}$$

which is directed along the line connecting the elements. It is also easy to verify that

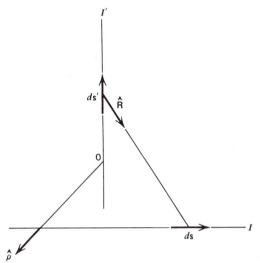

Figure 13-3. "Crossed" current elements.

this form satisfies the third law, that is

$$d\mathbf{F}'_{e' \to e} + d\mathbf{F}'_{e \to e'} = 0 \tag{13-21}$$

Although (13-17) and (13-20) will lead to the same value for the total force between complete circuits, they can give different results when applied to elements. As an example, suppose the current I of Figure 13-2 were directed into the page while I' is kept the same. Then we would get a picture for these "crossed" currents like that of Figure 13-3. In this case, $d\mathbf{s}$ and $d\mathbf{s}'$ are perpendicular so that (13-20) gives $d\mathbf{F}'_{e' \to e} = 0$, while from (13-17) we find that $d\mathbf{F}'_{e' \to e} \neq 0$. The reason for this discrepancy is that in going from (13-1) to (13-5) we, in effect, lost a term that is zero for a complete circuit but is not zero for the *integrands*; this was shown explicitly in (13-4).

In spite of the more symmetrical appearance of (13-20), its direction along the line connecting the elements, its desirable property expressed in (13-21), in contrast to (13-19), why is it that (13-17) is the accepted form for the force between current elements rather than (13-20)? Putting aside future comparison with experiment for the moment, we can see that (13-20) has an inherent problem that makes it unsuitable for our purposes. Because of its proportionality to $d\mathbf{s} \cdot d\mathbf{s}'$, (13-20) involves the cosine of the angle between the elements. Thus it cannot be written in the form of a product of the element $I d\mathbf{s}$ and something that is independent of the element, that is, a *field*. Since this product property is essential to the development of a theory based on fields, we abandon (13-20). On the other hand, (13-17) does possess the form needed to enable us to introduce a new field, as we will immediately proceed to do, and hence we adopt the form (13-1) along with (13-17) as our fundamental law of magnetostatics.

EXERCISES

13-1 Apply (13-6) to the system of Figure 13-2 and thus show directly that (13-12) is again obtained.

13-2 In order to get an idea of the magnitude of magnetic forces, find the force per unit length between very long equal and opposite 10 ampere currents separated by 1 centimeter. Find the ratio of this to the weight per unit length of the copper wire described in Exercise 12-17.

13-3 Consider the two infinitely long straight currents shown in Figure 13-4. I' coincides with the y axis. I is parallel to the yz plane, is at a

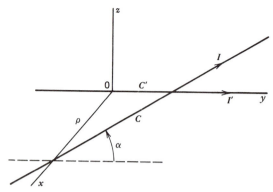

Figure 13-4. The two currents of Exercise 13-3.

distance ρ from it, crosses the x axis at $y = z = 0$, and makes the angle α with the xy plane as shown. Show that the force on I of C due to I' of C' is $-\frac{1}{2}\mu_0 II' \cot \alpha \hat{x}$.

13-4 Consider the two circuits shown in Figure 13-5. All currents lie in the same plane. C' is infinitely long. The sides of the rectangle of length b are parallel to C'. Find the total force on C. Is it attractive or repulsive?

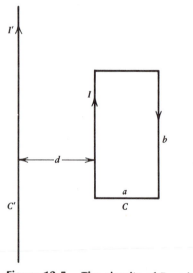

Figure 13-5. The circuits of Exercise 13-4.

13-5 An infinite plane current sheet coincides with the xy plane. The surface current density is $\mathbf{K}' = K'\hat{y}$ where $K' = $ const. A very long wire carrying a current I is parallel to the y axis and intersects the positive z axis at a distance d from the origin. Find the force per unit length on the wire.

13-6 A current I' is distributed uniformly over a very long cylinder of circular cross section of radius a. The axis of the cylinder coincides with the z axis. I' is in the direction of positive z. A very long current I is along a line parallel to the z axis; it is also in the direction of positive z. I crosses the positive x axis at a distance d from the origin. Find the force per unit length on the current I. [You will probably need to use (3-16).]

13-7 A circular loop of radius a lies in the xy plane with its center at the origin. It carries a current I' that circulates counterclockwise as seen when looking back at the origin from positive values of z. A very long current I is parallel to the x axis, is going in the positive x direction, and intersects the positive z axis at a point a distance d from the origin. Find the total force on the circuit C carrying I.

13-8 Two circles, each of radius a, carry currents circulating in the same sense. One, with current I', lies in the xy plane with center at the origin. The other, with current I, is parallel to the xy plane with its center on the positive z axis at a distance d from the origin. Find the force on the circuit with current I. Express your answer in terms of the angles φ and φ' of cylindrical coordinates, but leave your answer in integral form.

13-9 Four very long straight wires each carry current of the same value I. They are all parallel to the z axis and intersect the xy plane at the points $(0,0)$, $(a,0)$, (a,a), and $(0,a)$. The first and third have their currents in the positive z direction; the other two have currents in the negative z direction. Find the total force per unit length on the current corresponding to the point (a,a).

14

THE MAGNETIC INDUCTION

Ampère's law is another example of an "action at a distance" law. At the beginning of Chapter 3, we discussed the similar property of Coulomb's law and we concluded that it would be useful to divide it into two parts by introducing the electric field as a kind of intermediary to describe the interaction between the charges. We now want to do the same thing for the force between currents. For historical reasons, the field that we will use for this purpose is called the "magnetic induction"; the term "magnetic field" is generally used for a different vector field that we define later when we include the effects of matter.

14-1 DEFINITION OF THE MAGNETIC INDUCTION

We recall that we were able to go from Coulomb's law to the electric field because we saw in Section 3-1 that the force on a charge could be written as the product of the charge and another quantity that we called the electric field. We now see that (13-1) can be rewritten in a somewhat similar fashion as

$$\mathbf{F}_{C' \to C} = \oint_C I \, d\mathbf{s} \times \left(\frac{\mu_0}{4\pi} \oint_{C'} \frac{I' \, d\mathbf{s}' \times \hat{\mathbf{R}}}{R^2} \right) \tag{14-1}$$

The factor in parentheses is independent of the current element $I \, d\mathbf{s}$ at the position $\mathbf{r}$, but it does depend on the distribution of the other current elements with respect to $I \, d\mathbf{s}$. If we represent the quantity in parentheses by $\mathbf{B}(\mathbf{r})$, and use (1-137), we have

$$\mathbf{B}(\mathbf{r}) = \frac{\mu_0}{4\pi} \oint_{C'} \frac{I' \, d\mathbf{s}' \times \hat{\mathbf{R}}}{R^2} = \frac{\mu_0}{4\pi} \oint_{C'} \frac{I' \, d\mathbf{s}' \times \mathbf{R}}{R^3} \tag{14-2}$$

and

$$\mathbf{F}_{C' \to C} = \oint_C I \, d\mathbf{s} \times \mathbf{B}(\mathbf{r}) \tag{14-3}$$

The vector field $\mathbf{B}$, which we have defined in this way, is called the *magnetic induction*; sometimes it is also known as the *magnetic flux density* or simply as the $\mathbf{B}$ *field*. In addition, (14-2) is generally known as the *Biot-Savart law*.

We see from (14-3) that the unit of $\mathbf{B}$ will be 1 newton/(ampere-meter). This combination is not normally used, however, and the standard unit for $\mathbf{B}$ is generally written in one or another of two forms: 1 tesla = 1 weber/(meter)^2. By comparing this with the above we see that 1 weber = 1 joule/ampere = 1 volt-second.

By means of this procedure, then, we have introduced another vector field $\mathbf{B}$ that we can calculate at any field point $\mathbf{r}$ by means of (14-2) *even if* there is no current element there to have a force on it. Again, as for $\mathbf{E}$, one can regard this as all simply a mathematical convenience in that the calculation of $\mathbf{B}$ provides one with a kind of distributed contingency statement that tells us what the force on a circuit C would be *if* it were put there. On the other hand, one can, as we discussed in Section 3-4, regard $\mathbf{B}$ as a real physical entity in itself.

If there is more than one source circuit, we can find **B** for a given one, C_i, by means of (14-2), and then the resultant induction at **r** will be given by the vector sum of the individual contributions:

$$\mathbf{B}_{\text{total}} = \mathbf{B}(\mathbf{r}) = \sum_i \frac{\mu_0}{4\pi} \oint_{C_i} \frac{I_i \, d\mathbf{s}_i \times \hat{\mathbf{R}}_i}{R_i^2} \tag{14-4}$$

where $\mathbf{R}_i = \mathbf{r} - \mathbf{r}_i$ in terms of the position vector $\mathbf{r}_i$ of the current element $I_i \, d\mathbf{s}_i$ of the ith circuit. We note that $I \, d\mathbf{s}$ of C is not included in (14-4), that is, we do not envisage a current element exerting a magnetic force upon itself.

Although (14-3) was written in terms of the total force on the complete circuit C, it is natural to interpret the integrand as giving the force $d\mathbf{F}$ on the current element $I \, d\mathbf{s}$ located at **r**:

$$d\mathbf{F} = I \, d\mathbf{s} \times \mathbf{B}(\mathbf{r}) \tag{14-5}$$

This force is perpendicular to both the current element and the magnetic induction, is zero when they are parallel, and has a maximum magnitude when they are perpendicular, as we see from (1-22). This directional property of $d\mathbf{F}$ is illustrated in Figure 14-1, which corresponds to the definition of the cross product shown in Figure 1-14.

In a similar manner, we can interpret the integrand of (14-2) as the contribution $d\mathbf{B}(\mathbf{r})$ to the total induction produced by the current element $I' \, d\mathbf{s}'$ located at the source point $\mathbf{r}'$:

$$d\mathbf{B}(\mathbf{r}) = \frac{\mu_0}{4\pi} \frac{I' \, d\mathbf{s}' \times \hat{\mathbf{R}}}{R^2} = \frac{\mu_0}{4\pi} \frac{I' \, d\mathbf{s}' \times \mathbf{R}}{R^3} \tag{14-6}$$

This expression is another version of the Biot-Savart law. The directional relations given by (14-6) are shown in Figure 14-2. We see that $d\mathbf{B}$ is perpendicular to the plane formed by $d\mathbf{s}'$ and $\hat{\mathbf{R}}$, has its maximum magnitude at a field point located on a line perpendicular to $d\mathbf{s}'$, and is zero at a field point located directly forward of or behind $d\mathbf{s}'$. There is a convenient right-hand rule that *qualitatively* describes the situation of Figure 14-2: point the thumb of the right hand in the direction of $I' \, d\mathbf{s}'$ and then the fingers will curl in the correct *general sense* of the direction of $d\mathbf{B}$.

Everything we have done up to now has been stated in terms of filamentary currents. However, as we know, there are many situations that arise in which it is convenient to describe them in terms of currents distributed throughout a volume or over a surface. We can easily adapt our results to such cases by using the equivalents for current elements that we found in (12-10). For example, if the source currents are described by

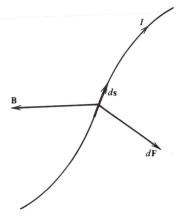

Figure 14-1. Force on a current element due to an induction B.

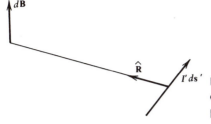

Figure 14-2. Geometrical relations between a current element and the magnetic induction it produces.

the volume current density $\mathbf{J}'(\mathbf{r}')$, the results analogous to (14-2) and (14-6) will be

$$\mathbf{B}(\mathbf{r}) = \frac{\mu_0}{4\pi} \int_{V'} \frac{\mathbf{J}'(\mathbf{r}') \times \hat{\mathbf{R}}}{R^2}\, d\tau' \tag{14-7}$$

$$d\mathbf{B} = \frac{\mu_0}{4\pi} \frac{\mathbf{J}' \times \hat{\mathbf{R}}\, d\tau'}{R^2} \tag{14-8}$$

where the integral is to be taken over the whole volume V' containing currents. Similarly, (14-3) and (14-5) become

$$\mathbf{F} = \int_V \mathbf{J}(\mathbf{r}) \times \mathbf{B}(\mathbf{r})\, d\tau \tag{14-9}$$

$$d\mathbf{F} = \mathbf{J} \times \mathbf{B}\, d\tau \tag{14-10}$$

where $\mathbf{F}$ is the total force on all of the currents contained within the volume V whose distribution is described by $\mathbf{J}(\mathbf{r})$.

Using (12-10) again, we get corresponding expressions when surface currents are involved:

$$\mathbf{B}(\mathbf{r}) = \frac{\mu_0}{4\pi} \int_{S'} \frac{\mathbf{K}'(\mathbf{r}') \times \hat{\mathbf{R}}}{R^2}\, da' \tag{14-11}$$

$$d\mathbf{B} = \frac{\mu_0}{4\pi} \frac{\mathbf{K}' \times \hat{\mathbf{R}}\, da'}{R^2} \tag{14-12}$$

$$\mathbf{F} = \int_S \mathbf{K}(\mathbf{r}) \times \mathbf{B}(\mathbf{r})\, da \tag{14-13}$$

$$d\mathbf{F} = \mathbf{K} \times \mathbf{B}\, da \tag{14-14}$$

Finally, if all of the possibilities we have mentioned are simultaneously present, the total $\mathbf{B}$ at a given point will be the sum of all of the various contributions from (14-4), (14-7), and (14-11), and it will be this resultant $\mathbf{B}$ that would be used in (14-3), (14-5), (14-9), (14-10), (14-13), or (14-14) as would be appropriate.

We have not made any use of any of our previous classifications of currents that we discussed in Section 12-2 and indeed it is a basic hypothesis of our subject that the magnetic induction $\mathbf{B}$ is produced by *all* currents from any source whatsoever, and later we will find it necessary to introduce currents of types different from any we have yet encountered.

As in the case of the electric field, an equation like (14-2) can be regarded as a "recipe" for calculating $\mathbf{B}$ once the current distribution is given, and now we want to consider a few examples of such direct calculations.

14-2 STRAIGHT CURRENT OF FINITE LENGTH

Let us consider a steady filamentary current I' and find the $\mathbf{B}$ produced by a finite length of it. If we choose the origin where the perpendicular from the field point P

intersects the current and use cylindrical coordinates, we get the situation shown in Figure 14-3a. Evidently this is not a complete circuit but a portion of one. Cases like this are useful, however, because more complicated complete circuits can frequently be thought of as being composed of parts such as this; then, once we can find **B** for each part, the resultant can be found by vector addition, as implied, for example, in (14-4).

We see from the figure that $\mathbf{r} = \rho\hat{\rho}$ and $\mathbf{r}' = z'\hat{z}$ so that $\mathbf{R} = \mathbf{r} - \mathbf{r}' = \rho\hat{\rho} - z'\hat{z}$, $R^2 = \rho^2 + z'^2$, and $d\mathbf{s}' = dz'\hat{z}$. Therefore, we find that $d\mathbf{s}' \times \mathbf{R} = \rho\,dz'\hat{z} \times \hat{\rho} = \rho\,dz'\hat{\varphi}$ and then, since $\hat{\varphi}$ is constant, (14-2) becomes

$$\mathbf{B} = \frac{\mu_0 I'\rho\hat{\varphi}}{4\pi} \int_{-L_1}^{L_2} \frac{dz'}{\left(\rho^2 + z'^2\right)^{3/2}} = \frac{\mu_0 I'\rho\hat{\varphi}}{4\pi} \left[\frac{z'}{\rho^2\left(\rho^2 + z'^2\right)^{1/2}}\right]_{-L_1}^{L_2}$$

$$= \hat{\varphi}\frac{\mu_0 I'}{4\pi\rho} \left[\frac{L_2}{\left(\rho^2 + L_2^2\right)^{1/2}} + \frac{L_1}{\left(\rho^2 + L_1^2\right)^{1/2}}\right] \tag{14-15}$$

This result can also be written in terms of the angles α_2 and α_1 whose positive sense is defined in Figure 14-3b for we see from the figure that

$$\mathbf{B} = \hat{\varphi}\frac{\mu_0 I'}{4\pi\rho}\left(\sin\alpha_2 + \sin\alpha_1\right) \tag{14-16}$$

Thus, the induction is always perpendicular to the plane formed by the current and the position vector of the field point.

We can obtain **B** for an infinitely long straight current by letting $L_2 \to \infty$ and $L_1 \to \infty$ (or, correspondingly, $\alpha_2 \to \pi/2$ and $\alpha_1 \to \pi/2$) and the result is that

$$\mathbf{B} = \frac{\mu_0 I'}{2\pi\rho}\hat{\varphi} \tag{14-17}$$

We can use this expression to compare this present approach with that of the last chapter. If we have an infinitely long current parallel to the z axis, which passes through P, as is the case in Figure 13-2, then the force on a length $d\mathbf{s}$ of it can be found by combining (14-17), (14-5), and (1-76) to give

$$d\mathbf{F} = I\,dz\,\hat{z} \times \frac{\mu_0 I'\hat{\varphi}}{2\pi\rho} = -\frac{\mu_0 II'\,dz\,\hat{\rho}}{2\pi\rho}$$

which is exactly what we found by the direct calculation that led to (13-13).

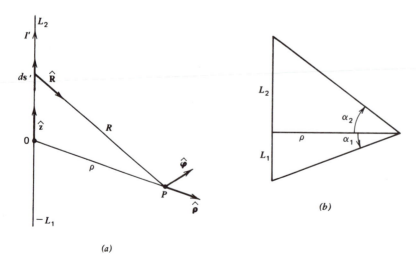

Figure 14-3.   Calculation of B due to a straight current of finite length.

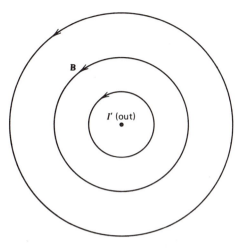

Figure 14-4. Lines of B due to an infinitely long straight current directed out of the page.

The magnitude of **B** as given by (14-17) varies inversely with the distance from the current. The surfaces of constant B are thus cylinders whose axes coincide with the current. The radius of a given cylinder is determined from $\rho = \mu_0 I'/2\pi B$. Since **B** is in the $\hat{\varphi}$ direction, the lines of **B** in a plane perpendicular to these cylinders, and hence to I', will be circles. Thus, we get a picture for **B** due to an infinitely long straight current like that shown in Figure 14-4; the figure is drawn so that I' is perpendicular to the page and coming out of it.

14-3 AXIAL INDUCTION OF A CIRCULAR CURRENT

As another example of a filamentary current, let us consider a current I' circulating along the circumference of a circle of radius a as shown in Figure 14-5. In order to be specific, we assume the circle to lie in the xy plane with the origin at its center. We

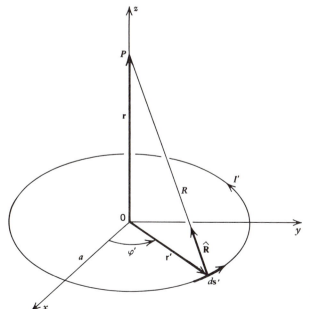

Figure 14-5. Calculation of axial induction of a circular current.

choose our field point along the z axis so that it is on the line normal to the area enclosed by the current.

We see that $\mathbf{r} = z\hat{\mathbf{z}}$ while $\mathbf{r}' = x'\hat{\mathbf{x}} + y'\hat{\mathbf{y}} = a(\cos \varphi'\hat{\mathbf{x}} + \sin \varphi'\hat{\mathbf{y}})$; we are using the polar angle φ' to locate the source point, but expressing its position vector in terms of the *constant* rectangular unit vectors. Hence, $\mathbf{R} = -a\cos \varphi'\hat{\mathbf{x}} - a\sin \varphi'\hat{\mathbf{y}} + z\hat{\mathbf{z}}$ so that $R^2 = a^2 + z^2$. In addition, $d\mathbf{s}' = d\mathbf{r}' = a\,d\varphi'(-\sin \varphi'\hat{\mathbf{x}} + \cos \varphi'\hat{\mathbf{y}})$ and therefore we get

$$d\mathbf{s}' \times \mathbf{R} = a\,d\varphi'\left[z(\cos \varphi'\hat{\mathbf{x}} + \sin \varphi'\hat{\mathbf{y}}) + a\hat{\mathbf{z}}\right]$$

with the use of (1-28). Inserting these results into (14-2), we find that

$$\mathbf{B}(z) = \frac{\mu_0 I'a}{4\pi}\int_0^{2\pi}\frac{\left[z(\cos \varphi'\hat{\mathbf{x}} + \sin \varphi'\hat{\mathbf{y}}) + a\hat{\mathbf{z}}\right]d\varphi'}{(a^2 + z^2)^{3/2}} = \frac{\mu_0 I'a^2}{2(a^2 + z^2)^{3/2}}\hat{\mathbf{z}} \quad (14\text{-}18)$$

where the $\hat{\mathbf{x}}$ and $\hat{\mathbf{y}}$ components have integrated out to zero. Thus, the induction is completely along the z axis as seems evident from the symmetry of the situation depicted in Figure 14-5. At the center of the circle ($z = 0$), the induction becomes simply

$$\mathbf{B}_{\text{center}} = \frac{\mu_0 I'}{2a}\hat{\mathbf{z}} \quad (14\text{-}19)$$

At large distances from the circle, that is, for $z \gg a$, (14-18) approximates to

$$\mathbf{B}(z) \simeq \frac{\mu_0 I'a^2}{2z^3}\hat{\mathbf{z}} \quad (14\text{-}20)$$

and varies inversely as the cube of the distance z. In this respect, it is similar to the electric field of a dipole as we saw in (10-35), for example. This similarity is not accidental, as we will learn in Chapter 19. Now let us consider one of the applications of our result.

■ **Example**

Axial induction of an ideal solenoid. Suppose we have a cylinder of length L and circular cross section of radius a. If we wind a wire uniformly around it to give us a total of N "turns," the resultant device is called a *solenoid* and is shown in Figure 14-6 (only a

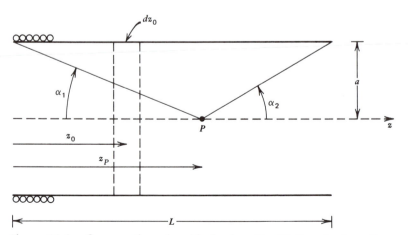

Figure 14-6.   Cross section of an ideal solenoid with its axis along the z axis.

few of the turns are shown in cross section.) If the wires are very thin and wound right next to each other, we can, to a first approximation, neglect the pitch of the winding, and we can consider it as equivalent to a set of N circular filamentary currents of radius a. (This is often referred to as an "ideal" solenoid.) The value of $\mathbf{B}$ at a point P on the axis can then be obtained by summing up the contribution of each circular turn as is given by (14-18). Consider a small portion dz_0 of the length located a distance z_0 from one end. If $n = N/L$ is the number of turns per unit length, then there will be $dN = n\, dz_0$ circular rings in this small portion, each approximately the same distance $z = z_P - z_0$ from the field point P. Hence, from (14-18), their contribution to the magnitude of $\mathbf{B}$ at P will be

$$dB = \frac{\mu_0 I' a^2 n\, dz_0}{2\left[a^2 + (z_P - z_0)^2\right]^{3/2}} \tag{14-21}$$

according to (14-4). The total value of B at P will then be

$$B = \int_0^L \frac{\mu_0 I' a^2 n\, dz_0}{2\left[a^2 + (z_P - z_0)^2\right]^{3/2}} = \frac{\mu_0 I' n a^2}{2} \int_{-z_P}^{L-z_P} \frac{dz'}{(a^2 + z'^2)^{3/2}}$$

$$= \frac{\mu_0 n I'}{2}\left\{ \frac{(L - z_P)}{\left[a^2 + (L - z_P)^2\right]^{1/2}} + \frac{z_P}{(a^2 + z_P^2)^{1/2}} \right\} \tag{14-22}$$

where we have let $z' = z_0 - z_P$ and used the value given for the integral in (14-15). This result can also be expressed very simply in terms of the angles α_1 and α_2 defined in the figure; we see that we have

$$B = \tfrac{1}{2}\mu_0 n I'(\cos \alpha_2 + \cos \alpha_1) \tag{14-23}$$

If the solenoid is infinitely long, then both α_1 and α_2 approach zero and (14-23) simplifies to

$$B = \mu_0 n I' \tag{14-24}$$

and is independent of the location of P. ∎

14-4 INFINITE PLANE UNIFORM CURRENT SHEET

As an example of a continuous current distribution, let us consider an infinite plane sheet on which there is a constant surface current density $\mathbf{K}'$. We will find the value of $\mathbf{B}$ at an arbitrary field point with the use of (14-11). Let us assume the sheet to coincide with the xy plane and that $\mathbf{K}'$ is in the y direction so that $\mathbf{K}' = K'\hat{\mathbf{y}}$ where $K' = $ const. As we see from Figure 14-7, $\mathbf{r} = x\hat{\mathbf{x}} + y\hat{\mathbf{y}} + z\hat{\mathbf{z}}$, $\mathbf{r}' = x'\hat{\mathbf{x}} + y'\hat{\mathbf{y}}$, $\mathbf{R} = (x - x')\hat{\mathbf{x}} + (y - y')\hat{\mathbf{y}} + z\hat{\mathbf{z}}$, $R^2 = (x - x')^2 + (y - y')^2 + z^2$, and $da' = dx'\, dy'$. Therefore, $\mathbf{K}' \times \mathbf{R} = K'\hat{\mathbf{y}} \times \mathbf{R} = K'[z\hat{\mathbf{x}} + (x' - x)\hat{\mathbf{z}}]$ and (14-11) becomes

$$\mathbf{B} = \frac{\mu_0 K'}{4\pi} \int_{-\infty}^{\infty}\int_{-\infty}^{\infty} \frac{[z\hat{\mathbf{x}} + (x' - x)\hat{\mathbf{z}}]\, dx'\, dy'}{\left[(x - x')^2 + (y - y')^2 + z^2\right]^{3/2}}$$

$$= \frac{\mu_0 K'}{4\pi} \int_{-\infty}^{\infty}\int_{-\infty}^{\infty} \frac{(z\hat{\mathbf{x}} + X'\hat{\mathbf{z}})\, dX'\, dY'}{(X'^2 + Y'^2 + z^2)^{3/2}} \tag{14-25}$$

where $X' = x' - x$ and $Y' = y' - y$. We see at once that the $\hat{\mathbf{z}}$ component will vanish because the integrand is an odd function of X'. Then the integral reduces to exactly the

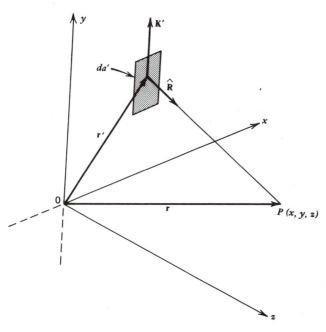

Figure 14-7. Calculation of B due to an infinite plane uniform current sheet.

same form as that of (3-11), and if we use the results given in (3-12) and (3-13) by simply replacing σ/ϵ_0 by $\mu_0 K'$ and $\hat{\mathbf{z}}$ by $\hat{\mathbf{x}}$, we obtain

$$\mathbf{B} = \pm \tfrac{1}{2}\mu_0 K'\hat{\mathbf{x}} = \tfrac{1}{2}\mu_0 K'\left(\frac{z}{|z|}\right)\hat{\mathbf{x}} \tag{14-26}$$

Thus, we see that **B** has a magnitude independent of the location of the field point, is parallel to the current sheet and at right angles to the direction of the current, and has opposite signs on the two sides of the sheet. This is summed up in Figure 14-8 showing the lines of **B** as drawn in the xz plane so that **K**′ is perpendicular to the plane of the page and coming out of it.

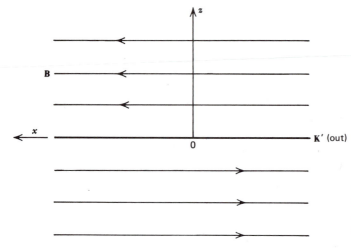

Figure 14-8. Lines of B due to an infinite plane uniform current sheet perpendicular to the page.

14-5 MOVING POINT CHARGES

If we write the volume current density as the product $\rho'\mathbf{v}'$ given by (12-3), then (14-7) becomes

$$\mathbf{B}(\mathbf{r}) = \frac{\mu_0}{4\pi} \int_{V'} \frac{\rho'\mathbf{v}' \times \hat{\mathbf{R}} \, d\tau'}{R^2} \qquad (14\text{-}27)$$

Now let us assume that the charges described by ρ' are contained within a very small volume. Then $\mathbf{r}'$ will be practically the same for all volume elements $d\tau'$ and we can take $\mathbf{r}' = $ const., thus making $\hat{\mathbf{R}}$ and R^2 constants also. If we assume in addition that all of the charges have the same velocity $\mathbf{v}'$, we can remove all of these constant factors from out of the integral and (14-27) becomes

$$\mathbf{B}(\mathbf{r}) = \frac{\mu_0}{4\pi} \frac{\mathbf{v}' \times \hat{\mathbf{R}}}{R^2} \int_{V'} \rho' \, d\tau' = \frac{\mu_0}{4\pi} \frac{q'\mathbf{v}' \times \hat{\mathbf{R}}}{R^2} \qquad (14\text{-}28)$$

where q' is the total charge. But under these conditions, q' can be regarded as a point charge, so that (14-28) gives the magnetic induction produced by a moving point charge.

By comparing (14-28) with (14-6), we see that this value of $\mathbf{B}$ is just the same as that produced by a current element

$$I' \, d\mathbf{s}' = q'\mathbf{v}' \qquad (14\text{-}29)$$

In other words, we have found that a moving point charge is the equivalent of a current element and (14-29) gives the quantitative connection. Having obtained this result, we can easily adapt some of our previous results. For example, if we combine this equivalence with (14-5) and write the force as $\mathbf{F}$ rather than $d\mathbf{F}$, we find that there will be a magnetic force on a moving point charge given by

$$\mathbf{F}_{\text{mag}} = q\mathbf{v} \times \mathbf{B} \qquad (14\text{-}30)$$

where $\mathbf{B}$ is the value of the induction at the location of the charge. Similarly, we can use (13-17) to express the magnetic force on one point charge q moving with velocity $\mathbf{v}$ due to another point charge q' of velocity $\mathbf{v}'$ and we get

$$\mathbf{F}_{q' \to q} = \frac{\mu_0 q q'}{4\pi} \frac{\mathbf{v} \times (\mathbf{v}' \times \hat{\mathbf{R}})}{R^2} \qquad (14\text{-}31)$$

Finally, if we add (14-30) to (3-1), which gives the electric force on q, we find the total electromagnetic force to be

$$\mathbf{F} = q(\mathbf{E} + \mathbf{v} \times \mathbf{B}) \qquad (14\text{-}32)$$

This important result is usually called the *Lorentz force*.

Since all of our discussion of magnetism up to now has been based on steady currents, we can expect that (14-28) and (14-31) are applicable for constant velocities, that is, zero or negligible acceleration. It does turn out that the fields produced by accelerated charges are different from these and lead to phenomena usually described as "radiation." In addition, even if the charges are moving with constant velocity, we can use (14-28) and (14-31) only if the velocities involved are "small." Although it is not at all evident at this point that this should be the case, nor even what is to be considered "small," we can anticipate some of the results of Chapter 29 and simply assert that we require that $|\mathbf{v}| \ll c$ where c is the speed of light in vacuum. [Although we have been dealing only with inductions that are constant in time, it is worth noting that the value of $\mathbf{B}$ at the fixed point $\mathbf{r}$ as given by (14-28) is actually time dependent since both R and $\hat{\mathbf{R}}$ will change as q' moves.]

We have found that the sources of **B** are currents of any type. Just as we did for **E** in Chapters 4 and 5, we would also like to express this information in terms of the differential source equations of the induction. In other words, we want to evaluate $\nabla \times \mathbf{B}$ and $\nabla \cdot \mathbf{B}$, and this is what we will obtain, along with other information, in the next two chapters.

EXERCISES

14-1 Find the magnetic induction produced by the currents of Figure 13-4 at the point on the x axis that is midway between them.

14-2 Suppose the field point P of Figure 14-3 is located at an arbitrary value of z rather than at $z = 0$ as shown. Show that the value of **B** there can still be written in the form (14-16) where α_1 and α_2 are measured positive below and above, respectively, from the perpendicular dropped from P to the line of I'.

14-3 Two infinitely long straight currents are parallel to the z axis. One of them, carrying a current I_1, intersects the xy plane at the point (x_1, y_1); the other, with current I_2, intersects the xy plane at (x_2, y_2). Find the resultant **B** produced by them at any field point (x, y, z).

14-4 A square of edge a lies in the xy plane with the origin at its center. Find the value of the magnetic induction at any point on the z axis when a current I' circulates around the square. Show that your result gives the value $2\sqrt{2}\,\mu_0 I'/\pi a$ for the induction at the center.

14-5 An ideal solenoid of length L and N turns is wound with a square cross section, that is, each turn is a square of edge a. Find the induction produced at the center of the solenoid when there is a current I' in the windings. What does this become for an infinitely long solenoid?

14-6 Two circular loops, each of radius a, have planes parallel to the xy plane and centers on the z axis a distance d apart. Each carries the same current I' circulating in the same sense. Choose the origin midway between them and find the axial field $B_z(z)$. What is B_z at the origin? Show that dB_z/dz vanishes when evaluated at the midway point. Show that the second derivative of B_z will also vanish there provided that $d = a$. What is $B_z(0)$ under these conditions? Show that $d^3 B_z/dz^3$ also vanishes at the origin. An arrangement like this with $d = a$ is called a *Helmholtz coil* and is used to produce an approximately constant induction over a small region.

14-7 The current shown in Figure 14-9 follows along the arc of a circle lying in the xy plane with

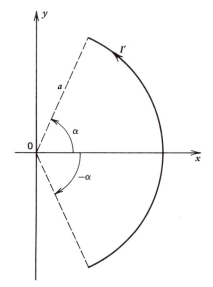

Figure 14-9. The current of Exercise 14-7.

center of curvature at the origin. Find **B** at any point on the z axis. As a check on your result, show that it reduces to (14-18) under appropriate conditions.

14-8 A wire is wound in a helix of pitch angle α on the surface of a cylinder of radius a so that N complete turns are formed. If the wire carries a current I', show that the axial component of the induction produced at the center of the helix is

$$\tfrac{1}{2}(\mu_0 NI'/a)(1 + \pi^2 N^2 \tan^2 \alpha)^{-(1/2)}.$$

14-9 An infinite plane current sheet coincides with the xy plane. Its surface current density is $\mathbf{K}' = K'\hat{\mathbf{y}}$ where $K' = $ const. Another infinite plane current sheet is parallel to the xy plane and intersects the positive z axis at $z = d$. The second current density is $\mathbf{K}' = -K'\hat{\mathbf{y}}$. Find **B** everywhere.

14-10 Repeat the preceding exercise with the surface current of the second sheet now given by $\mathbf{K}' = K'\hat{\mathbf{y}}$.

14-11 A circular dielectric disc of radius a has a uniform surface charge density σ on it. It is rotated with constant angular speed ω about an axis that is normal to the surface of the disc and passes through its center. Assume that the charge distribution is not altered by the rotation and find **B** at an arbitrary point on the axis of rotation. What is **B** at the center of the disc?

14-12 A sphere of radius a contains a total charge Q distributed uniformly throughout its volume. It is set into rotation with constant angular speed ω about a diameter. Assume that the charge distribution is not altered by the rotation and find **B** at the center of the sphere.

14-13 A dielectric sphere of radius a is uniformly polarized. It is rotated with constant angular speed ω about the diameter parallel to the polarization. Assume that the polarization is not affected by the rotation and find **B** at the point where the axis of rotation intersects the surface of the sphere, that is, at the "north pole" of the sphere. What is **B** at the center of the sphere?

14-14 An infinitely long cylinder of circular cross section of radius a carries a current I' that is uniformly distributed over the cross-sectional area. The axis of the cylinder coincides with the z axis and I' is in the positive z direction. Choose a field point on the x axis and find **B** for all values of x, both inside and outside of the cylinder.

14-15 In the circuit shown in Figure 14-10, the curved lines are semicircles with common center C. The straight portions are horizontal. At this instant a point charge q located at C has a velocity **v** directed vertically downward. Find the magnetic force on q.

14-16 Figure 14-11 shows a short thick solenoid. The N turns of the winding uniformly occupy the region between the two coaxial cylinders of radii a and b. Neglect the pitch of the windings and show that the axial component of the induction

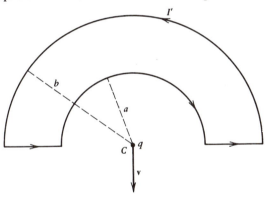

Figure 14-10. The circuit of Exercise 14-15.

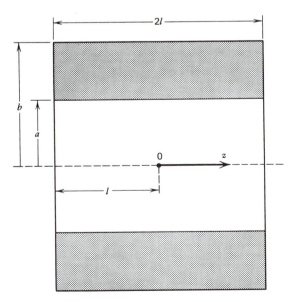

Figure 14-11. The solenoid of Exercise 14-16.

produced by a current I' is given by

$$B(z) = \frac{\mu_0 N I'}{4l(b-a)}$$

$$\times \left((l+z) \ln \left\{ \frac{b + \left[b^2 + (l+z)^2 \right]^{1/2}}{a + \left[a^2 + (l+z)^2 \right]^{1/2}} \right\} \right.$$

$$\left. + (l-z) \ln \left\{ \frac{b + \left[b^2 + (l-z)^2 \right]^{1/2}}{a + \left[a^2 + (l-z)^2 \right]^{1/2}} \right\} \right)$$

14-17 Show that if all of the distributed source currents occupy a finite volume V', the expression (14-7) for **B** can be transformed into the form

$$\mathbf{B}(\mathbf{r}) = \frac{\mu_0}{4\pi} \int_{V'} \frac{\nabla' \times \mathbf{J}'(\mathbf{r}')}{R} \, d\tau'$$

This result emphasizes the role of the nonuniformities of the source currents in producing the induction. We will not generally find it convenient to use this form.

14-18 A closed filamentary circuit C carrying a steady current I is in a region where **B** is constant. Show that the total force on C is zero. Now suppose that C is a plane circuit and the plane surface is parallel to the direction of **B**. Show that in this case there will be a nonzero torque on C that will be parallel to the plane of C.

15

THE INTEGRAL FORM
OF AMPÈRE'S LAW

The first differential source equation that we want to consider is $\nabla \times \mathbf{B}$. The general definition of the curl of a vector given by (1-73) suggests that we consider the line integral of $\mathbf{B}$ about some closed path.

15-1 DERIVATION OF THE INTEGRAL FORM

We will show that

$$\oint_C \mathbf{B} \cdot d\mathbf{s} = \mu_0 I_{\text{enc}} \tag{15-1}$$

where the integral is taken about an arbitrary closed path C and I_{enc} is the total current passing through the area *enclosed* by the curve C. The path C can be any closed curve and need *not* coincide with any real circuit. The expression (15-1) is known as *the integral form of Ampère's law*; it is also very commonly referred to as *Ampère's circuital law*.

For simplicity, we initially assume that $\mathbf{B}$ is produced by a single filamentary circuit C' carrying a current I' so that $\mathbf{B}$ is given by (14-2). Inserting this expression into the left-hand side of (15-1), we find that

$$\oint_C \mathbf{B} \cdot d\mathbf{s} = \frac{\mu_0 I'}{4\pi} \oint_C \oint_{C'} \frac{d\mathbf{s} \cdot (d\mathbf{s}' \times \hat{\mathbf{R}})}{R^2} = -\frac{\mu_0 I'}{4\pi} \oint_C \oint_{C'} \frac{(-d\mathbf{s} \times d\mathbf{s}') \cdot \hat{\mathbf{R}}}{R^2} \tag{15-2}$$

where we used (1-29) to interchange the dot and the cross in the integrand. If we now recall from Section 1-7 that the magnitude of the cross product equals the area of the parallelogram with the two vectors as sides and then, in addition, recall (4-3), we see that we should analyze (15-2) in terms of solid angles.

The general situation is illustrated in Figure 15-1. Suppose we consider a given point P on the path of integration C. The source circuit C' will subtend some total solid angle Ω at that point. Now in carrying out the integration over C, we give the point P a series of successive displacements, one of which is $d\mathbf{s}$. After P has been displaced by $d\mathbf{s}$, the source circuit C' will generally have a different aspect as viewed from P, so that the solid angle which C' subtends at the new position of P will have changed to a new value $\Omega' = \Omega + d\Omega$. Thus, $d\Omega$ = change in solid angle subtended by C' *at* P as a result of the displacement of P by $d\mathbf{s}$. But we can also obtain the same *relative* change by imagining P held fixed and giving an equal and opposite displacement $(-d\mathbf{s})$ to every point of C'. (You can easily convince yourself of this by holding this book at a reasonable distance from your eyes and then comparing the visual result of: (1) keeping the book held fixed and moving your head back, and (2) keeping your head fixed and moving the book an equal amount away from your head.) Thus we can also say that $d\Omega$ = change in solid angle produced by keeping P fixed and displacing every point of C' by $-d\mathbf{s}$; the new position and orientation of C' are also shown in the figure. Now we see that $-d\mathbf{s} \times d\mathbf{s}' = d\mathbf{a}$ = shaded area with sides $-d\mathbf{s}$ and $d\mathbf{s}'$, so that, according

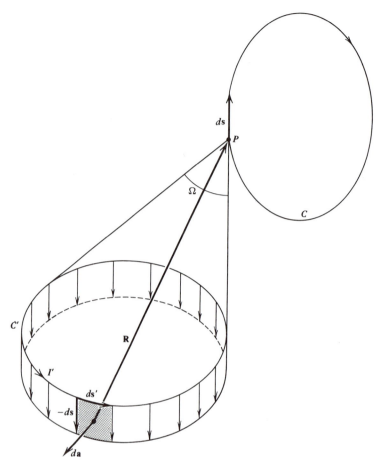

Figure 15-1. Calculation of change in solid angle subtended at *P* due to a displacement of the circuit *C'*.

to (4-3), the term under the integrals of (15-2) is just $d\mathbf{a} \cdot \hat{\mathbf{R}}/R^2 =$ solid angle subtended *at P* by $d\mathbf{a} = change$ in solid angle subtended *at P* resulting from the displacement of $d\mathbf{s}'$ by $-d\mathbf{s}$. Therefore, when we carry out the integration over C' in (15-2), we are summing up all of the contributions from all of the $d\mathbf{s}'$ of C', so that we have

$$\oint_{C'} \frac{(-d\mathbf{s} \times d\mathbf{s}') \cdot \hat{\mathbf{R}}}{R^2} = d\Omega \tag{15-3}$$

where, again, $d\Omega$ is the change in solid angle observed at P due to a displacement of P by $d\mathbf{s}$, *but* it is calculated in (15-3) by an equivalent summation over C' based on holding P fixed and displacing C'. Inserting (15-3) into (15-2) and carrying out the integration over C, we get

$$\oint_{C} \mathbf{B} \cdot d\mathbf{s} = -\frac{\mu_0 I'}{4\pi} \oint_{C} d\Omega = -\frac{\mu_0 I'}{4\pi} \Delta\Omega \tag{15-4}$$

where $\Delta\Omega$ is the total *change* in the solid angle subtended by C' at the various points on C when summed over the closed path C. It turns out that there are two cases to be considered.

1. The path C does not link the circuit C'. In this case, the relative orientations are like those shown in Figure 15-2. Here, if we start at P, then when we come back to P after completing the loop C, the final solid angle has the same value that it had initially, so that $\Delta\Omega = 0$ and (15-4) becomes

$$\oint_C \mathbf{B} \cdot d\mathbf{s} = 0 \tag{15-5}$$

2. The path C does link the circuit C'. In this case, the path of integration encloses the source current and we have a situation like that shown in Figure 15-3. It is easier to see what the change will be if we pick our initial point A just above the surface S' enclosed by C' and the final point A' just below the surface S' and then go to the limit as A' and A become coincident. First, we note that the direction of the normal $\hat{\mathbf{n}}'$ to the surface enclosed by C' is determined from the direction of I' by the standard right-hand rule shown in Figure 1-24.

   (a) At the initial point A. We see from Figure 15-4a that the angle between $d\mathbf{a}'$ and $\hat{\mathbf{R}}$ is $\Psi_A = 90° - \delta$ where δ is very small and positive. Then the solid angle subtended by $d\mathbf{a}'$ at A as given by (4-3) will be $da' \cos \Psi_A / R^2 = da' \cos(90° - \delta)/R^2$ and is *positive*. Now as A is taken closer to the surface S', $\delta \to 0$, all of the contributions to the solid angle subtended by C' at A are positive, and, since A then "sees" one half of all space, we get

$$\Omega_{\text{for } A} = \Omega_{\text{initial}} = +2\pi \tag{15-6}$$

(In order to convince yourself of this, imagine your eye located at A just above this page, and then let your eye approach near coincidence with the page.)

   (b) At the final point A'. As we see from Figure 15-4b, the angle between $d\mathbf{a}'$ and $\hat{\mathbf{R}}$ is $\Psi_{A'} = 90° + \epsilon$ where ϵ is positive and very small. Then the solid angle subtended by

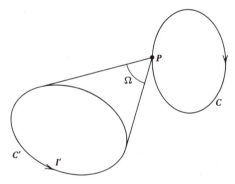

Figure 15-2. The path C does not link the circuit C'.

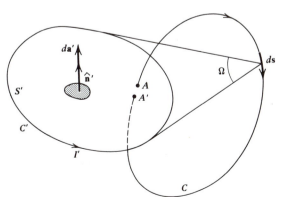

Figure 15-3. The path C links the circuit C'.

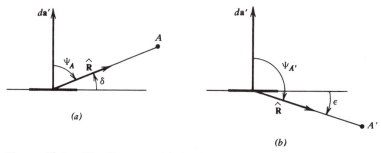

Figure 15-4. Situation near (a) the initial point A and (b) the final point A'.

$d\mathbf{a}'$ at A' will be $da' \cos(90° + \epsilon)/R^2$ and will be *negative*. Then, as A' nears the surface, $\epsilon \to 0$, all of the contributions to the solid angle will be negative, and since A' will "see" one half of all space, we get

$$\Omega_{\text{for } A'} = \Omega_{\text{final}} = -2\pi \tag{15-7}$$

Combining (15-7) and (15-6), we find that the total change in solid angle for this case will be

$$\Delta\Omega = \Omega_{\text{final}} - \Omega_{\text{initial}} = (-2\pi) - (2\pi) = -4\pi \tag{15-8}$$

so that (15-4) becomes

$$\oint_C \mathbf{B} \cdot d\mathbf{s} = \mu_0 I' \tag{15-9}$$

and equals μ_0 times the current passing through the surface enclosed by the curve C. Comparing (15-5) and (15-9), we see that the value of this line integral of $\mathbf{B}$ is different from zero only if the path of integration encloses a current.

We also see that if the sense of integration around C were reversed, then A and A' would be interchanged as initial and final points, $\Delta\Omega$ would equal $+4\pi$, and the value of the integral would be $-\mu_0 I'$, that is, the current I' would be counted as negative. We can summarize these sign results by means of Figure 15-5. Once the sense of integration about C has been chosen, this determines the positive sense of its normal $\hat{\mathbf{n}}$, as in Figure 1-24. Then, if a current I' passes through the surface enclosed by C in this same sense, as in (a) of the figure, it will make a positive contribution $\mu_0 I'$ to the integral, while if it passes through C in the opposite sense, as in (b), it contributes $-\mu_0 I'$ to the integral. (Note that Figure 15-5a corresponds to Figure 15-3.)

If there is more than one source current so that $\mathbf{B}$ will be given by an expression like (14-4), then each current I_i will contribute either 0 or $\pm\mu_0 I_i$ depending on whether I_i

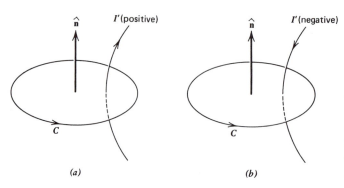

Figure 15-5. Sign conventions for current as related to sense of integration about C.

passes through C or not, as well as upon its sense of doing so, and we will get

$$\oint_C \mathbf{B} \cdot d\mathbf{s} = \sum_{\substack{i \\ \text{enclosed}}} \mu_0 I_i = \mu_0 I_{\text{enc}} \tag{15-10}$$

where I_{enc} is the *net* current enclosed by the path of integration. Thus we have obtained the integral form of Ampère's law that we initially stated in (15-1). [We note that this result has a certain "similarity" to Gauss' law (4-1).] There are some general implications of this result that we can point out. Any currents not enclosed by the path of integration do not contribute to the value of the *integral*, although they certainly can affect the value of $\mathbf{B}$ at a particular point. Also, the location of the enclosed currents does not affect the value of the *integral*, although if they are moved about within C they can affect the value of $\mathbf{B}$ at particular points on the path of integration.

We can now put (15-10) into another useful form by expressing I_{enc} in terms of the current density $\mathbf{J}$ by means of (12-6). (We will drop the prime on the source current density now for simplicity of notation.) If we do this, and use Stokes' theorem (1-67), we obtain

$$\oint_C \mathbf{B} \cdot d\mathbf{s} = \mu_0 \int_S \mathbf{J} \cdot d\mathbf{a} = \int_S (\nabla \times \mathbf{B}) \cdot d\mathbf{a} \tag{15-11}$$

where S is the open surface enclosed by C. Since C is arbitrary, (15-11) also holds for any very small area, and we can equate the integrands to get our desired source equation

$$\nabla \times \mathbf{B} = \mu_0 \mathbf{J} \tag{15-12}$$

This fundamental result is *equivalent* to Ampère's law of force between complete circuits. We see at once that, in contrast to the electrostatic field, the magnetic induction is not a conservative field since $\nabla \times \mathbf{B}$ is not always zero.

Furthermore, we can use (15-12) to find the boundary conditions satisfied by the tangential components of $\mathbf{B}$ at a surface of discontinuity. If we insert (15-12) into (9-13), we obtain

$$\hat{\mathbf{n}} \times (\mathbf{B}_2 - \mathbf{B}_1) = \lim_{h \to 0} (\mu_0 h \mathbf{J}) \tag{15-13}$$

In order to interpret this result, we consider the small area element in the transition layer that is perpendicular to the flow as shown in Figure 15-6 (Compare to Figure 9-5.) The total current ΔI through this area as given by (12-6) is $\Delta I = Jh\,\Delta s$. As the transition layer is shrunk to zero thickness, so that $h \to 0$, this total current will be squeezed into a *surface current* of density K and the constant total can be written as $\Delta I = K\,\Delta s$, as seen from Figure 12-5a. Thus, we see by comparing the two expressions for ΔI that $hJ \to K$, and since the two vectors are in the same direction, we get

$$\mathbf{K} = \lim_{h \to 0} (h\mathbf{J}) \tag{15-14}$$

and (15-13) becomes

$$\hat{\mathbf{n}} \times (\mathbf{B}_2 - \mathbf{B}_1) = \mu_0 \mathbf{K} \tag{15-15}$$

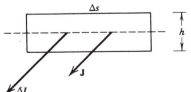

Figure 15-6. Small area element in the transition layer.

which is the required boundary condition. As in (9-18), we can use the alternative form written directly in terms of the tangential components:

$$\mathbf{B}_{2t} - \mathbf{B}_{1t} = \mu_0 \mathbf{K} \times \hat{\mathbf{n}} \qquad (15\text{-}16)$$

where the relation between the normal and tangential vectors is that illustrated in Figure 9-5.

15-2 SOME APPLICATIONS OF THE INTEGRAL FORM

Much as we were able to do with Gauss' law in Section 4-2, we can use (15-1) to calculate **B** fields if the problem has sufficient symmetry. The principal problem is that of choosing a suitable closed path for integration. The main things to look for are curves on which **B** has a constant magnitude, and curves for which **B** is either parallel to or perpendicular to the sense of traversal of the path, both for ease in integration and for avoiding difficulties with an unknown dependence of **B** on position. We illustrate this process by considering a few standard examples.

■ **Example**

Infinitely long straight current. Let us assume the current I to be uniformly distributed over the circular cross section of radius a of an infinitely long cylinder as shown in Figure 15-7. If we review the general dependence of the direction of **B** due to a current as shown in Figure 14-2 and consider the general "symmetry" of this problem, we conclude that **B** lies in the plane perpendicular to the direction of I, is everywhere tangent to the dashed circle of radius ρ shown, and its magnitude can depend on ρ at most and must be independent of z and φ. In other words, **B** has the general form $\mathbf{B} = B_\varphi(\rho)\hat{\boldsymbol{\varphi}}$. These considerations are just as valid whether the field point is inside the cylinder or is outside of it as happens to be shown in the figure. Thus, for any value of

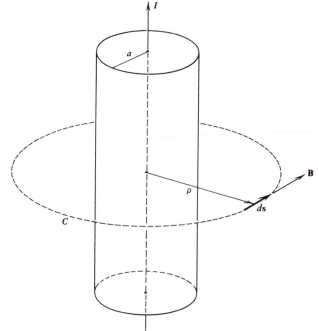

Figure 15-7. Portion of a long straight current.

ρ, let us choose as our path of integration a circle of radius ρ and go around it in the same sense as the assumed direction of **B**. Then we will have $d\mathbf{s} = \rho\, d\varphi\, \hat{\boldsymbol{\varphi}}$ according to (1-82), and (15-1) becomes

$$\oint_C \mathbf{B} \cdot d\mathbf{s} = \int_0^{2\pi} B_\varphi \hat{\boldsymbol{\varphi}} \cdot \rho\, d\varphi\, \hat{\boldsymbol{\varphi}} = 2\pi\rho B_\varphi = \mu_0 I_{\text{enc}} \tag{15-17}$$

since both ρ and $B_\varphi(\rho)$ are constant on the circle. Therefore,

$$B_\varphi(\rho) = \frac{\mu_0 I_{\text{enc}}}{2\pi\rho} \tag{15-18}$$

1. Outside the cylinder. Here $\rho > a$, and C clearly encloses the total current I; thus $I_{\text{enc}} = I$ and (15-18) becomes

$$B_\varphi(\rho) = \frac{\mu_0 I}{2\pi\rho} \qquad (\rho > a) \tag{15-19}$$

By comparing this result with (14-17), and recalling that we have dropped the prime on the source current, we see that the induction outside an infinitely long straight current is the same as if the current were a filamentary one going along the axis of the cylinder.

2. Inside the cylinder. Here $\rho < a$, and now C does not enclose all of the current but only that fraction equal to the circular area enclosed by C divided by the total cross-sectional area of the cylinder. Thus $I_{\text{enc}}/I = \pi\rho^2/\pi a^2$ so that $I_{\text{enc}} = I(\rho^2/a^2)$ and (15-18) becomes

$$B_\varphi(\rho) = \frac{\mu_0 I\rho}{2\pi a^2} \qquad (\rho < a) \tag{15-20}$$

[Both of these results (15-19) and (15-20) agree with those found for Exercise 14-14, but have been obtained much more easily here.]

We also note that both of our expressions give the same value $B_\varphi(a) = \mu_0 I/2\pi a$ at the surface of the cylinder so that B_φ is continuous across it. This is in agreement with (15-16) since **B** has only tangential components in this case and there is no surface current **K** on the cylinder. The dependence of B_φ on ρ is illustrated in Figure 15-8.   ∎

■ **Example**

Infinite uniform plane current sheet. In Figure 15-9 we show an edge-on view of the infinite plane sheet with **K** out of the page. We assume that $|\mathbf{K}| = K = \text{const.}$; this current distribution could be approximated by many closely packed wires all parallel

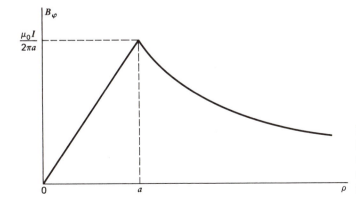

Figure 15-8. B_φ due to an infinitely long straight current as a function of distance ρ from the current axis.

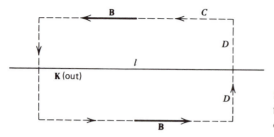

Figure 15-9. Path of integration for finding B due to an infinite uniform plane current sheet.

and all carrying the same current. Since there is no preference for angles above or below the horizontal, nor in or out of the page, we conclude, after looking at Figure 14-2 again, that **B** is perpendicular to **K**, is parallel to the plane of the sheet, and is oppositely directed on the two sides of the sheet as shown. **B** may still depend on the normal distance D from the sheet, however. Accordingly, we choose the rectangular path of integration shown dashed, with two horizontal sides each of length l and each the same distance D from the sheet; these are connected by two vertical sides of length $2D$. On the horizontal sides, **B** is parallel to $d\mathbf{s}$, so that $\mathbf{B} \cdot d\mathbf{s} = B(D)\,ds$ and $B(D)$ is constant. On the vertical sides, **B** is perpendicular to $d\mathbf{s}$ by construction so that $\mathbf{B} \cdot d\mathbf{s} = 0$ and there is no contribution to the integral from these sides; this is an advantage since the form of $B(D)$ is unknown. Since $|\mathbf{K}|$ is current per unit length, we have $I_{\text{enc}} = Kl$ for this path and (15-1) becomes

$$\oint_C \mathbf{B} \cdot d\mathbf{s} = 2Bl = \mu_0 I_{\text{enc}} = \mu_0 Kl$$

so that

$$B = \tfrac{1}{2}\mu_0 K \tag{15-21}$$

which agrees exactly with what we found in (14-26) by direct integration. Thus we have found once again that the magnitude of **B** is independent of distance from the current sheet.

This result is also in agreement with the boundary condition (15-16) for we see from Figure 15-10 that it becomes $\mathbf{B}_{2t} - \mathbf{B}_{1t} = B\hat{\mathbf{t}} - (-B\hat{\mathbf{t}}) = 2B\hat{\mathbf{t}} = \mu_0 \mathbf{K} \times \hat{\mathbf{n}} = \mu_0 K\hat{\mathbf{t}}$ so that $B = \tfrac{1}{2}\mu_0 K$ as above. ∎

■ **Example**

Infinitely long ideal solenoid. We assume that the turns of the cylindrical coil are so closely wound and that the wires are so thin that the pitch of the windings can be neglected. Then, in effect, we have a current sheet of surface density K circulating around the cylinder as indicated in Figure 15-11. If there are n turns per unit length and I is the current in the windings, the current per unit length will be nI so that

$$K = nI \tag{15-22}$$

If, in addition, we assume that the solenoid is infinitely long, we can conclude that **B**

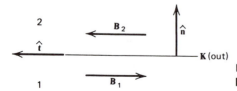

Figure 15-10. Geometry used to verify the boundary conditions on B.

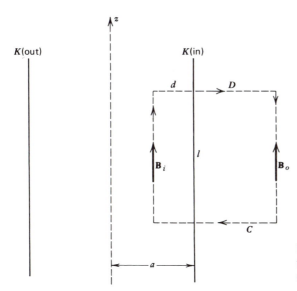

Figure 15-11. Path of integration for finding B due to an infinitely long ideal solenoid.

will be parallel to the axis of the cylinder and with the sense shown. In the figure, $\mathbf{B}_i$ is the induction inside the solenoid, while $\mathbf{B}_o$ is the value outside; both must be independent of z since the solenoid is infinitely long, making one value of z as good as another. Again we choose a rectangular path of integration C shown dashed; the vertical sides are of length l and the sense of integration is parallel to $\mathbf{B}_i$ and antiparallel to $\mathbf{B}_o$. The vertical side inside is a distance d from the current sheet and the outside vertical side is a distance D from it. The horizontal sides are of length $d + D$; since $d\mathbf{s}$ is perpendicular to $\mathbf{B}$ on them, $\mathbf{B} \cdot d\mathbf{s} = 0$ and they will not contribute to the integral. Furthermore, $I_{enc} = Kl = nIl$; then (15-1) becomes

$$\oint_C \mathbf{B} \cdot d\mathbf{s} = B_i l - B_o l = \mu_0 Kl = \mu_0 nIl$$

so that

$$B_i - B_o = \mu_0 K = \mu_0 nI \tag{15-23}$$

We see that this result is independent of both d and D; therefore, (15-23) is also in agreement with (15-16) as well if we choose region 1 to be outside of and region 2 to be inside of the solenoid.

Since (15-23) is independent of D, it will also be true as $D \to \infty$. But then, if we imagine looking back at the solenoid from a great distance away, we will see the oppositely directed currents essentially superimposed and Ampère's experimental result tells us that they will produce a negligible effect. Thus, for $D \to \infty$, $B_o = 0$ and (15-23) gives for this case:

$$B_i = \mu_0 K = \mu_0 nI \tag{15-24}$$

If we substitute this back into (15-23), which is possible since it holds for any D, we now find that

$$\mathbf{B}_o = 0 \quad \text{(everywhere)} \tag{15-25}$$

Furthermore, since the results (15-23) and (15-24) are also independent of d, we can then conclude that

$$\mathbf{B}_i = \mu_0 K\hat{z} = \mu_0 nI\hat{z} \quad \text{(everywhere)} \tag{15-26}$$

In other words, for an *infinitely long ideal* solenoid, the induction is confined entirely to the interior and is *uniform* across the cross section; it is also independent of a, the radius of the cylinder. [Our result (15-24) agrees with our previous one given in (14-24); in the latter case, however, we found this value only on the axis, but now we see that it is true for all interior points.] ∎

■ **Example**

Toroidal coil. Here we have a current I in a conductor that is wound uniformly, that is, with constant pitch, around a torus as in Figure 15-12 where only a few turns are shown. The cross section could be circular or rectangular, for example. (Such an arrangement can also be imagined to be obtained from taking a solenoid of finite length, and bending it into the shape of a doughnut, thus joining its ends.) We cannot learn everything about **B** from (15-1) for this case, so we will consider only circular paths of radius ρ which are centered at the center of the torus and are in the plane which includes the central axis of the torus. For *any* such circular paths, $d\mathbf{s} = \rho \, d\varphi \, \hat{\boldsymbol{\varphi}}$ and (15-1) becomes

$$\oint_C \mathbf{B} \cdot d\mathbf{s} = \oint_C \mathbf{B} \cdot \hat{\boldsymbol{\varphi}} \rho \, d\varphi = B_\varphi \cdot 2\pi\rho = \mu_0 I_{enc}$$

so that

$$B_\varphi = \frac{\mu_0 I_{enc}}{2\pi\rho} \tag{15-27}$$

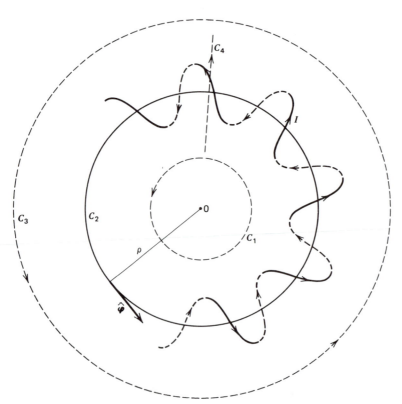

Figure 15-12. Toroidal coil and paths of integration used to calculate the induction it produces.

since, from the symmetry, B_φ will depend only on ρ and hence will be constant on such a path.

Now for a path inside the circle defined by the inner boundary of the torus, such as C_1, $I_{enc} = 0$ and $B_\varphi = 0$. For a path within the torus, such as C_2, $I_{enc} = NI$ where N is the total number of turns and (15-27) gives

$$B_\varphi = \frac{\mu_0 NI}{2\pi\rho} \tag{15-28}$$

Finally, for a path completely outside the torus, such as C_3, we find that $I_{enc} = NI - NI = 0$, according to Figure 15-5, since for every current I out of the page there is necessarily an equal one going into the page. Therefore, $B_\varphi \neq 0$ only within the torus itself.

If the dimensions of the cross section of the torus are small compared to its central radius, then $\rho \simeq$ const. for all paths C_2 and B_φ is approximately constant across the cross section.

These calculations do not tell us anything about B_ρ and B_z; they have to be found by a direct calculation from the Biot-Savart law (14-2). We will not do this, but content ourselves with the observation that they turn out to be of the order of magnitude of B_φ/N; thus they can be made negligible by using very many turns, that is, by using closely packed windings. We can, however, at least make these comments plausible by considering the path of integration C_4 shown in Figure 15-13. This is a path completely surrounding the torus and lying in the plane perpendicular to its axis, that is, in the plane $\varphi =$ const.; the trace of C_4 is also shown in Figure 15-12. For any finite pitch angle, corresponding to a finite number of turns N, we see that exactly one turn will pass through the plane enclosed by C_4; thus $I_{enc} = I$. The value of $d\mathbf{s}$ as given by (1-82) with $d\varphi = 0$ is $d\mathbf{s} = d\rho\,\hat{\boldsymbol{\rho}} + dz\,\hat{\mathbf{z}}$, so that (15-1) becomes

$$\oint_{C_4} \mathbf{B} \cdot d\mathbf{s} = \oint_{C_4} (B_\rho\,d\rho + B_z\,dz) = \mu_0 I \tag{15-29}$$

Since this is different from zero, we see that B_ρ or B_z or both must be different from zero on at least some portions of C_4. Furthermore, the value of this integral is N times smaller than that for B_φ, as we see from (15-28), so that for comparable dimensions of the paths, the magnitude of B_ρ and/or B_z will be roughly of the order of B_φ/N as we stated above. ■

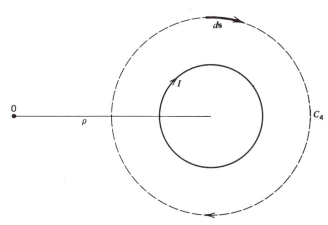

Figure 15-13.   Path of integration in a plane perpendicular to the axis of the torus.

15-3 DIRECT CALCULATION OF $\nabla \times \mathbf{B}$

Although we have already found the curl of $\mathbf{B}$ to be given by (15-12), we obtained it somewhat indirectly. It is of interest to see that we can also get this same result by operating directly on our basic defining equation for $\mathbf{B}$. It is convenient to use the expression (14-7) in terms of a continuous distribution of current. Taking the curl of this expression, we get

$$\nabla \times \mathbf{B}(\mathbf{r}) = \nabla \times \frac{\mu_0}{4\pi} \int_{V'} \frac{\mathbf{J}'(\mathbf{r}') \times \hat{\mathbf{R}}\, d\tau'}{R^2} = \frac{\mu_0}{4\pi} \int_{V'} \nabla \times \left[\frac{\mathbf{J}'(\mathbf{r}) \times \hat{\mathbf{R}}}{R^2} \right] d\tau' \quad (15\text{-}30)$$

where we were able to go from the second to the third term because the limits of integration depend on source point coordinates (x', y', z') while ∇ involves derivatives with respect to the field point coordinates (x, y, z). If we now use (1-119), we find that the quantity under the integral becomes

$$\mathbf{J}'(\mathbf{r}') \left[\nabla \cdot \left(\frac{\hat{\mathbf{R}}}{R^2} \right) \right] - \left(\frac{\hat{\mathbf{R}}}{R^2} \right) [\nabla \cdot \mathbf{J}'(\mathbf{r}')] + \left(\frac{\hat{\mathbf{R}}}{R^2} \cdot \nabla \right) \mathbf{J}'(\mathbf{r}') - [\mathbf{J}'(\mathbf{r}') \cdot \nabla] \left(\frac{\hat{\mathbf{R}}}{R^2} \right)$$

$$= \mathbf{J}'(\mathbf{r}') \left[\nabla \cdot \left(\frac{\hat{\mathbf{R}}}{R^2} \right) \right] - [\mathbf{J}'(\mathbf{r}') \cdot \nabla] \left(\frac{\hat{\mathbf{R}}}{R^2} \right) \quad (15\text{-}31)$$

where the second and third terms are zero because any component of ∇ operating on $\mathbf{J}'$ gives zero since $\mathbf{J}'$ depends only on the components of $\mathbf{r}'$. Now that we have (15-31), we can use (1-132) to change from ∇ to ∇' in the *second* term of the right-hand side in order to get an integrand that involves the primed coordinates in such a way that we can perform the integration; we leave the first term unchanged. Doing this, and inserting the result in (15-30), we find that

$$\nabla \times \mathbf{B} = \frac{\mu_0}{4\pi} \int_{V'} \mathbf{J}'(\mathbf{r}') \left[\nabla \cdot \left(\frac{\hat{\mathbf{R}}}{R^2} \right) \right] d\tau' + \frac{\mu_0}{4\pi} \int_{V'} [\mathbf{J}'(\mathbf{r}') \cdot \nabla'] \left(\frac{\hat{\mathbf{R}}}{R^2} \right) d\tau' \quad (15\text{-}32)$$

We can show that the second integral is zero by rewriting it with the use of (1-129):

$$\int_{V'} [\mathbf{J}' \cdot \nabla'] \left(\frac{\hat{\mathbf{R}}}{R^2} \right) d\tau' = -\int_{V'} \left(\frac{\hat{\mathbf{R}}}{R^2} \right) (\nabla' \cdot \mathbf{J}')\, d\tau' + \oint_{S'} \left(\frac{\hat{\mathbf{R}}}{R^2} \right) (\mathbf{J}' \cdot d\mathbf{a}') \quad (15\text{-}33)$$

Since we are considering *only* steady currents, $\nabla' \cdot \mathbf{J}' = 0$ by (12-15) and the volume integral vanishes. Now the source current distribution occupies a finite volume; consequently, the bounding surface S' can always be made large enough so that $\mathbf{J}'(\mathbf{r}') = 0$ on all points of S' and the surface integral also vanishes. Thus, as we said, the integral in (15-33) is zero, and (15-32) reduces to

$$\nabla \times \mathbf{B} = \frac{\mu_0}{4\pi} \int_{V'} \mathbf{J}'(\mathbf{r}') \left[\nabla \cdot \left(\frac{\hat{\mathbf{R}}}{R^2} \right) \right] d\tau' \quad (15\text{-}34)$$

The x component of the right-hand side of this equation has *exactly* the same form as (4-23), with $\mu_0 J_x'$ replacing ρ/ϵ_0; the result of integrating (4-23) was found to be (4-26) so that we have

$$\frac{\mu_0}{4\pi} \int_{V'} J_x'(\mathbf{r}') \left[\nabla \cdot \left(\frac{\hat{\mathbf{R}}}{R^2} \right) \right] d\tau' = \mu_0 J_x'(\mathbf{r}) \quad (15\text{-}35)$$

with similar expressions for the y and z components of (15-34). Adding these results, and finally dropping the prime on $\mathbf{J}'$ in order to agree with the standard notational convention, we get

$$\nabla \times \mathbf{B} = \mu_0 \left[J_x(\mathbf{r})\hat{\mathbf{x}} + J_y(\mathbf{r})\hat{\mathbf{y}} + J_z(\mathbf{r})\hat{\mathbf{z}} \right] = \mu_0 \mathbf{J}(\mathbf{r}) = (\nabla \times \mathbf{B})_{\text{at } \mathbf{r}}$$

which is exactly what we found before as given by (15-12).

EXERCISES

15-1 A circle of radius a lies in the xy plane with its center at the origin. Find the solid angle Ω subtended by this circle at a point on the positive z axis.

15-2 Consider the induction **B** produced by an infinitely long straight current I. Choose a convenient closed path C in the plane perpendicular to the current and which does not enclose the current. Show by direct integration that (15-5) is correct in this case.

15-3 Apply (15-1) to a circle of radius $\rho < a$ whose plane is perpendicular to the z axis of the solenoid of Figure 15-11 and thus show that $B_\varphi = 0$ in the interior.

15-4 Apply (15-1) to a circle of radius $\rho > a$ whose plane is perpendicular to the z axis of the solenoid of Figure 15-11 and show that, if the windings have a finite pitch, the average value of B_φ on the circle equals $\mu_0 I / 2\pi\rho$.

15-5 Find another current distribution that will produce the same B_φ in a torus as that which we found in (15-28).

15-6 A toroidal coil of N turns is wound on a circular cross section of radius b. If the radius of the central axis is a, that is, this is the distance from the center 0 to the center of the cross section, find the ratio b/a necessary in order that the total deviation in B_φ across the cross section be not more than 2 percent of the value at the center.

15-7 Consider the infinitely long coaxial cylindrical conductors shown in Figure 6-12. The inner conductor carries a total current I in the $\hat{\mathbf{z}}$ direction, while the outer conductor carries a current I in the $-\hat{\mathbf{z}}$ direction. Assume the currents to be uniformly distributed over their respective cross sections. Find **B** everywhere and plot your results as a function of ρ.

15-8 A certain **B** field is given in cylindrical coordinates by **B** $= 0$ for $0 < \rho < a$, **B** $= (\mu_0 I / 2\pi\rho)[(\rho^2 - a^2)/(b^2 - a^2)]\hat{\boldsymbol{\varphi}}$ for $a < \rho < b$, and **B** $= (\mu_0 I / 2\pi\rho)\hat{\boldsymbol{\varphi}}$ for $b < \rho$. Find the current density **J** everywhere. How could you produce such a **B**?

15-9 Consider a very long cylindrical beam of charged particles. The beam has a circular cross section of radius a, a uniform charge density ρ_{ch}, and the particles have the same constant velocity **v**. Find **B** inside and outside of the beam and express your result in terms of these given quantities.

15-10 Two infinitely long coaxial cylindrical surfaces have the z axis as their common axis. The inner surface of radius a carries a surface current $\mathbf{K}_1 = K_1\hat{\boldsymbol{\varphi}}$, and the outer surface of radius b carries a surface current $\mathbf{K}_2 = K_2\hat{\boldsymbol{\varphi}}$. Both K_1 and K_2 are constant. Find **B** everywhere.

15-11 In Figure 15-14, we show a portion of a current-free region in which **B** is uniform. Show that such an induction cannot drop abruptly to zero, as is indicated schematically by the location of the arrows, by applying (15-1) to the dashed rectangular path shown. Show qualitatively that a change in **B** of the general nature shown in Figure 6-10 will be compatible with (15-1).

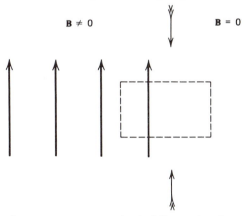

Figure 15-14. Hypothetical lines of B for Exercise 15-11.

15-12 Show that (15-12) can be obtained by a direct calculation of the curl of the expression for **B** given in Exercise 14-17.

16

THE VECTOR POTENTIAL

The remaining differential source equation of the induction that we need is $\nabla \cdot \mathbf{B}$. This is much more easily obtained than was $\nabla \times \mathbf{B}$. After we have found it, we will find that we can express much the same information as before in terms of a new vector field that we will be able to introduce.

16-1 THE DIVERGENCE OF B

The direct approach is quite easy in this case. If we use our definition of $\mathbf{B}$ as stated in terms of a filamentary current in (14-2), we find that

$$\nabla \cdot \mathbf{B} = \nabla \cdot \left[\frac{\mu_0}{4\pi} \oint_{C'} \frac{I' \, d\mathbf{s}' \times \hat{\mathbf{R}}}{R^2} \right] = \frac{\mu_0 I'}{4\pi} \oint_{C'} \nabla \cdot \left[\frac{d\mathbf{s}' \times \hat{\mathbf{R}}}{R^2} \right] \qquad (16\text{-}1)$$

since once again the limits of integration depend at most on $\mathbf{r}'$ while ∇ operates on the components of $\mathbf{r}$. We can use (1-116) to rewrite the integrand as

$$\nabla \cdot \left[d\mathbf{s}' \times \left(\frac{\hat{\mathbf{R}}}{R^2} \right) \right] = \left(\frac{\hat{\mathbf{R}}}{R^2} \right) \cdot (\nabla \times d\mathbf{s}') - d\mathbf{s}' \cdot \left[\nabla \times \left(\frac{\hat{\mathbf{R}}}{R^2} \right) \right] \qquad (16\text{-}2)$$

Now $d\mathbf{s}'$ is a function only of the components of $\mathbf{r}'$, and hence is a constant as far as ∇ is concerned; thus $\nabla \times d\mathbf{s}' = 0$. Also, (1-147) gives $\nabla \times (\hat{\mathbf{R}}/R^2) = 0$; therefore, (16-2) is identically zero and (16-1) becomes exactly

$$\nabla \cdot \mathbf{B} = 0 \qquad (16\text{-}3)$$

It is clear that this is true in the case of more than one filamentary source when $\mathbf{B}$ would be given by (14-4); the case of distributed currents will be left as an exercise.

Now that we have (16-3), we can immediately obtain the boundary condition satisfied by the normal components of $\mathbf{B}$ at a surface of discontinuity and we find from (9-6) and (9-7) that

$$\hat{\mathbf{n}} \cdot (\mathbf{B}_2 - \mathbf{B}_1) = B_{2n} - B_{1n} = 0 \qquad (16\text{-}4)$$

so that the normal components of $\mathbf{B}$ are always continuous.

This important result $\nabla \cdot \mathbf{B} = 0$ is one of Maxwell's equations. We will find that we will have no occasion to change it, even when we consider nonstatic fields, so that (16-4) as well will be generally correct. We can get an interpretation of (16-3) by comparing it with the analogous equation for the electric field as given by (4-10), namely $\nabla \cdot \mathbf{E} = \rho/\epsilon_0$. Here ρ is the net charge density and arises because electric charge exists in individual units of opposite sign so that it is possible to have $\rho \neq 0$ by having an excess of one sign or another. But we see from (16-3) that such a situation cannot arise in the magnetic case and therefore there cannot be individual units of magnetic charge analogous to electric charge. Such magnetic charges are called *magnetic monopoles*. Although their existence has been shown to be compatible with the requirements of quantum mechanics, every experimental search that has been made for them has been unsuccessful. Consequently, until any such time that magnetic monopoles have been unambiguously found to actually exist, we must continue to write $\nabla \cdot \mathbf{B} = 0$ and remember that the sources of $\mathbf{B}$ are only and always *currents*.

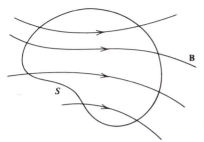

Figure 16-1. The net flux of B through the closed surface *S* is zero.

If we combine (16-3) with the divergence theorem (1-59), we find that

$$\oint_S \mathbf{B} \cdot d\mathbf{a} = 0 \tag{16-5}$$

which says that the flux of **B** through *any closed* surface is always zero. If we express this in terms of the lines of **B**, (16-5) says that there are no net lines of **B** through any closed surface, or, that there are always as many lines leaving the surface as entering it; hence the general situation will always be like that shown in Figure 16-1. This is in contrast to the electric case, where lines of **E** can begin or end on charges, as shown, for example, in Figure 8-7.

Nevertheless, we will find it very useful to consider the surface integral of **B** over a surface that is *not* closed. If we let Φ be the *magnetic flux* through a surface S, then by definition, we shall have

$$\Phi = \int_S \mathbf{B} \cdot d\mathbf{a} \tag{16-6}$$

and this can be different from zero. Since **B** is measured in webers/(meter)^2, the unit of flux will be a weber.

16-2 DEFINITION AND PROPERTIES OF THE VECTOR POTENTIAL

If we compare (16-3) with the general vector theorem (1-49) that says that the divergence of a curl is always zero, we are led to suspect that we should be able to write

$$\mathbf{B}(\mathbf{r}) = \nabla \times \mathbf{A}(\mathbf{r}) \tag{16-7}$$

The vector field $\mathbf{A}(\mathbf{r})$ introduced in this way is called the *vector potential*. We will see that in many ways it is a kind of analogue to the scalar potential ϕ. We find from (16-7) that the unit of **A** will be 1 weber/meter = 1 volt-second/meter.

If (16-7) is to be useful, it should be possible to find an explicit way of calculating **A** in terms of a given distribution of source currents. We do this by the direct approach of showing that our defining equation for **B** can actually be written in this form. If we use (1-141) and (1-118), we find that

$$\frac{d\mathbf{s}' \times \hat{\mathbf{R}}}{R^2} = -d\mathbf{s}' \times \nabla\left(\frac{1}{R}\right) = \nabla \times \left(\frac{d\mathbf{s}'}{R}\right) - \frac{(\nabla \times d\mathbf{s}')}{R} = \nabla \times \left(\frac{d\mathbf{s}'}{R}\right) \tag{16-8}$$

since $\nabla \times d\mathbf{s}' = 0$ as before. Inserting this into (14-2), we get

$$\mathbf{B} = \frac{\mu_0 I'}{4\pi} \oint_{C'} \nabla \times \left(\frac{d\mathbf{s}'}{R}\right) = \nabla \times \left(\frac{\mu_0 I'}{4\pi} \oint_{C'} \frac{d\mathbf{s}'}{R}\right) \tag{16-9}$$

when we use the fact that ∇ operates only on the components of **r**. We see that (16-9) does have the form of (16-7) and by comparison we can define **A** for this filamentary current as

$$\mathbf{A}(\mathbf{r}) = \frac{\mu_0}{4\pi} \oint_{C'} \frac{I'\,d\mathbf{s}'}{R} \tag{16-10}$$

If there is more than one filamentary source current, then we can apply (16-8) to each term of (14-4) and the resultant **A** will be given by

$$\mathbf{A}(\mathbf{r}) = \sum_i \frac{\mu_0}{4\pi} \oint_{C_i} \frac{I_i\,d\mathbf{s}_i}{R_i} \tag{16-11}$$

where $\mathbf{R}_i = \mathbf{r} - \mathbf{r}_i$. Thus, we have succeeded in showing that **B** can be written in the form (16-7) and, in addition, we have simultaneously found how to calculate **A**. We note that the contribution of each current element $I_i\,d\mathbf{s}_i$ to **A** is in the direction of the element itself.

If the currents are distributed, rather than filamentary, we can use our equivalents (12-10) to adapt (16-10), and we find that the vector potential produced by volume and surface currents will be given, respectively, by

$$\mathbf{A}(\mathbf{r}) = \frac{\mu_0}{4\pi} \int_{V'} \frac{\mathbf{J}(\mathbf{r}')\,d\tau'}{R} \tag{16-12}$$

$$\mathbf{A}(\mathbf{r}) = \frac{\mu_0}{4\pi} \int_{S'} \frac{\mathbf{K}(\mathbf{r}')\,da'}{R} \tag{16-13}$$

while that produced by a moving point charge would be obtained by our combining (14-29) and the integrand of (16-10) to give

$$\mathbf{A}(\mathbf{r}) = \frac{\mu_0}{4\pi} \frac{q\mathbf{v}}{R} \tag{16-14}$$

[As in Section 14-5, we would use (16-14) only for the case $|\mathbf{v}| \ll c$.]

The general idea behind all of this is that we would use the above expressions to find **A** and then find **B** by differentiation according to (16-7), so that the procedure would be, in this sense, analogous to that we have used in electrostatics.

It is of interest to find the divergence of **A**. Dotting ∇ into (16-10), we obtain

$$\nabla \cdot \mathbf{A} = \nabla \cdot \left(\frac{\mu_0}{4\pi} \oint_{C'} \frac{I'\,d\mathbf{s}'}{R} \right) = \frac{\mu_0 I'}{4\pi} \oint_{C'} \nabla \cdot \left(\frac{d\mathbf{s}'}{R} \right) \tag{16-15}$$

and, if we use (1-115), (1-141), and (1-16), we find that

$$\nabla \cdot \left(\frac{d\mathbf{s}'}{R} \right) = d\mathbf{s}' \cdot \nabla \left(\frac{1}{R} \right) + \frac{1}{R} (\nabla \cdot d\mathbf{s}') = -\nabla' \left(\frac{1}{R} \right) \cdot d\mathbf{s}' \tag{16-16}$$

since $\nabla \cdot d\mathbf{s}' = 0$ for the usual reason. If we insert (16-16) into (16-15), and use (1-67), we get

$$\nabla \cdot \mathbf{A} = -\frac{\mu_0 I'}{4\pi} \oint_{C'} \nabla' \left(\frac{1}{R} \right) \cdot d\mathbf{s}' = -\frac{\mu_0 I'}{4\pi} \int_{S'} \left[\nabla' \times \nabla' \left(\frac{1}{R} \right) \right] \cdot d\mathbf{a}'$$

But the integrand is always zero by (1-48) and therefore

$$\nabla \cdot \mathbf{A} = 0 \tag{16-17}$$

While (16-17) was obtained for only a single filamentary current, we see that this will still hold for more than one source current when **A** would be given by (16-11).

Although we have an integral form (16-12) that enables us to calculate **A** once the current distribution is known, it is useful to obtain the differential equation satisfied by

A in case the information is given in other forms, as we found to occur so often for ϕ, as for example in Chapter 11. If we take the curl of both sides of (16-7), use (1-120), (16-17), and (15-12), we get

$$\nabla \times \mathbf{B} = \nabla \times (\nabla \times \mathbf{A}) = \nabla(\nabla \cdot \mathbf{A}) - \nabla^2\mathbf{A} = -\nabla^2\mathbf{A} = \mu_0\mathbf{J}$$

so that

$$\nabla^2\mathbf{A} = -\mu_0\mathbf{J} \qquad (16\text{-}18)$$

which, in rectangular coordinates, is

$$\nabla^2 A_x = -\mu_0 J_x \qquad \nabla^2 A_y = -\mu_0 J_y \qquad \nabla^2 A_z = -\mu_0 J_z \qquad (16\text{-}19)$$

By comparing this with (11-1), we see that each rectangular component of **A** satisfies Poisson's equation.

If we now change our point of view somewhat, we can regard (16-7) and (16-17) as being the two differential source equations of **A** in the sense of the Helmholtz theorem of Section 1-20. But then we can immediately find the boundary conditions satisfied by the components of **A** at a surface of discontinuity in properties. We find from (16-17) and (9-6) that

$$\hat{\mathbf{n}} \cdot (\mathbf{A}_2 - \mathbf{A}_1) = A_{2n} - A_{1n} = 0 \qquad (16\text{-}20)$$

so that the normal components are continuous. If we now insert (16-7) into (9-13), we obtain

$$\hat{\mathbf{n}} \times (\mathbf{A}_2 - \mathbf{A}_1) = \lim_{h \to 0}(h\mathbf{B}) = 0 \qquad (16\text{-}21)$$

since, whatever else **B** may do in the transition layer, we certainly expect **B** to remain *finite* as the thickness of the transition layer is reduced to zero, thus making $h\mathbf{B} \to 0$ as $h \to 0$. Thus, the tangential components of **A** are also continuous. But when all of the components of a vector are continuous, the vector itself is continuous across the surface, and therefore we conclude that

$$\mathbf{A}_2 = \mathbf{A}_1 \qquad (16\text{-}22)$$

in complete analogy to the continuity of the scalar potential ϕ as expressed in (9-29).

The magnetic flux Φ can also be expressed in terms of the vector potential. If we put (16-7) into the definition (16-6), and use Stokes' theorem (1-67), we find that

$$\Phi = \int_S (\nabla \times \mathbf{A}) \cdot d\mathbf{a} = \oint_C \mathbf{A} \cdot d\mathbf{s} \qquad (16\text{-}23)$$

showing that the flux can be written as the line integral of **A** about the curve bounding the surface for which Φ is desired. Sometimes (16-23) offers an alternative way of calculating **A** if Φ can be easily found from a previous solution and if the problem has enough symmetry so that a convenient path C can be devised.

Quantities that are called "potentials" in physics generally have an ambiguity associated with their absolute value because they are ordinarily introduced as functions from which other functions of interest can be obtained by differentiation. We saw this in the case of the scalar potential that was undetermined up to a scalar additive constant as we discussed in connection with (5-10). A similar, but more complex, situation also holds for the vector potential. If we review how we went from (16-9) to (16-10) with the use of (16-7), and then remember the general vector result (1-48) that the curl of a gradient of a scalar is always zero, we see that we could add the gradient of an arbitrary scalar to (16-10) and still get the same induction. In order to make this explicit, let us assume that **A** is a "suitable" vector potential, that is, it gives the correct value of **B** when we use it in $\mathbf{B} = \nabla \times \mathbf{A}$. Now suppose that $\mathbf{A}^\dagger$ is another function given by

$$\mathbf{A}^{\dagger}(\mathbf{r}) = \mathbf{A}(\mathbf{r}) + \nabla \chi(\mathbf{r}) \tag{16-24}$$

where $\chi(\mathbf{r})$ is a scalar field, but is otherwise arbitrary. The induction $\mathbf{B}^{\dagger}$ corresponding to $\mathbf{A}^{\dagger}$ as obtained from (16-7) and (1-48) will be

$$\mathbf{B}^{\dagger} = \nabla \times \mathbf{A}^{\dagger} = \nabla \times \mathbf{A} + \nabla \times \nabla \chi = \nabla \times \mathbf{A} = \mathbf{B} \tag{16-25}$$

Thus the two inductions will be identical. However, the property that $\nabla \cdot \mathbf{A} = 0$ followed directly from our original straightforward definition (16-10), and led us to the very desirable continuity condition (16-22) as well as the simple differential equation we obtained in (16-18). Consequently, it would seem very reasonable that any vector potential we wish to use should satisfy this same condition, so that, in addition to asking that $\mathbf{A}^{\dagger}$ give the correct $\mathbf{B}$, we *require* that $\mathbf{A}^{\dagger}$ also satisfy (16-17):

$$\nabla \cdot \mathbf{A}^{\dagger} = 0 \tag{16-26}$$

This will clearly lead to some restriction on our choice of χ. We can see what this will be by substituting (16-24) into (16-26) and using (16-17) and (1-45): $\nabla \cdot \mathbf{A}^{\dagger} = 0 = \nabla \cdot \mathbf{A} + \nabla \cdot \nabla \chi = \nabla^2 \chi$, and therefore,

$$\nabla^2 \chi = 0 \tag{16-27}$$

In other words, we do not want to use just any scalar function in (16-24), but we require *in addition* that χ be a solution of Laplace's equation. [The expression (16-24) is an example of what is known as a *gauge transformation*, and we consider them again in more detail in Chapter 22 after we have completed the development of the general theory. The requirement (16-26), and hence (16-27), leads to the so-called *Coulomb gauge*.]

The vector potential is not as useful in magnetostatics as one might expect by analogy with the scalar potential in electrostatics. One of the problems is that, even in many apparently simple cases, $\mathbf{A}$ cannot be expressed in a closed form, or at least one that is reasonably familiar. It is in the discussion of time dependent problems that the vector potential is most useful. Nevertheless, we consider some specific examples to illustrate the general features that we have just considered.

16-3 UNIFORM INDUCTION

If $\mathbf{B}$ has been found by other means, then it is often possible to use (16-7) in conjunction with (16-17) to obtain $\mathbf{A}$. As an extreme example, let us consider a uniform induction given by $\mathbf{B} = B\hat{\mathbf{z}}$ where $B = $ const. Writing (16-7) in rectangular coordinates with the use of (1-43), we get the three equations

$$\frac{\partial A_z}{\partial y} - \frac{\partial A_y}{\partial z} = 0 \qquad \frac{\partial A_x}{\partial z} - \frac{\partial A_z}{\partial x} = 0 \qquad \frac{\partial A_y}{\partial x} - \frac{\partial A_x}{\partial y} = B \tag{16-28}$$

which we propose to solve by inspection. We can satisfy the first two by taking A_x and A_y to be both independent of z, while A_z is at most a function of z; however, if we keep (16-17) in mind it is probably simpler just to take $A_z = $ const. Then we see that the remaining equation in (16-28) will be satisfied if (a) $A_x = 0$, $A_y = Bx$, or, (b) $A_x = -By$, $A_y = 0$, or, (c) $A_x = -\frac{1}{2}By$, $A_y = \frac{1}{2}Bx$, and so on. We note that (c) is just one half the sum of (a) and (b). We also see that these three solutions all satisfy the condition $\nabla \cdot \mathbf{A} = 0$. Clearly, there are many other possibilities, but these are enough to illustrate the point that there is a lot of possible arbitrariness in $\mathbf{A}$. In order to emphasize that these are different functions, yet all give the same $\mathbf{B}$, the projections

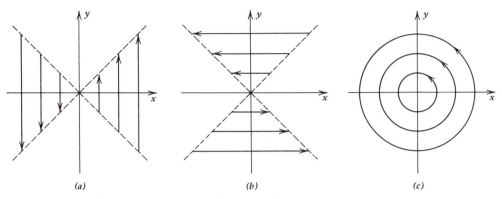

Figure 16-2. Various vector potentials that describe the same uniform induction.

of **A** on the xy plane for an arbitrary origin are illustrated as the corresponding parts (a), (b), and (c) of Figure 16-2; part (c) is obtained most easily by noting that $A^2 = A_x^2 + A_y^2 = \frac{1}{4}B^2(x^2 + y^2)$ which is the equation of a circle.

Physically, the ambiguity in **A** in this case reflects the fact that a uniform **B** can be produced within a given region in a variety of ways. For example, it could be produced in the interior of a solenoid as in (15-26), or on one side of an infinite current sheet as in (15-21), or in the region between two current sheets as in Exercise 14-9, and so on. Thus the symmetry properties of **A** will generally reflect the corresponding symmetry of the source distribution as would be found by direct calculation from (16-12), for example.

A very common and useful expression for the vector potential describing a uniform induction is based upon (c), which can be written in vector form as

$$\mathbf{A} = \tfrac{1}{2}\mathbf{B} \times \mathbf{r} \qquad (\mathbf{B} = \text{const.}) \qquad (16\text{-}29)$$

In this form, **A** is independent of the original choice of direction for **B**.

16-4 STRAIGHT CURRENTS

In Section 14-2, we found the value of **B** produced by the straight current of finite length shown in Figure 14-3. Now let us consider the same current distribution in terms of the vector potential. As before, $d\mathbf{s}' = dz'\hat{\mathbf{z}}$, $R^2 = \rho^2 + z'^2$, and if we now call the current I, (16-10) becomes

$$\mathbf{A} = \hat{\mathbf{z}}\frac{\mu_0 I}{4\pi} \int_{-L_1}^{L_2} \frac{dz'}{\left(\rho^2 + z'^2\right)^{1/2}} = \hat{\mathbf{z}}\frac{\mu_0 I}{4\pi} \ln\left[\frac{\left(\rho^2 + L_1^2\right)^{1/2} + L_1}{\left(\rho^2 + L_2^2\right)^{1/2} - L_2}\right]$$

$$= \hat{\mathbf{z}}\frac{\mu_0 I}{4\pi} \ln\left[\frac{\left(\rho^2 + L_2^2\right)^{1/2} + L_2}{\left(\rho^2 + L_1^2\right)^{1/2} - L_1}\right] \qquad (16\text{-}30)$$

where we have used (5-29) and (5-30) with $z = 0$. Before we go on to discuss the properties of this **A**, let us check our result by showing that it gives the correct **B**. If we combine (16-30) with (16-7) and (1-88), noting that $A_z(\rho)$ is the only nonzero component of **A**, we find that $B_\rho = B_z = 0$, while

$$B_\varphi = -\frac{\partial A_z}{\partial \rho}$$

$$= \frac{\mu_0 I \rho}{4\pi} \left\{ \frac{1}{\left(\rho^2 + L_2^2\right)^{1/2}\left[\left(\rho^2 + L_2^2\right)^{1/2} - L_2\right]} - \frac{1}{\left(\rho^2 + L_1^2\right)^{1/2}\left[\left(\rho^2 + L_1^2\right)^{1/2} + L_1\right]} \right\}$$

$$= \frac{\mu_0 I}{4\pi\rho}\left[\frac{L_2}{\left(\rho^2 + L_2^2\right)^{1/2}} + \frac{L_1}{\left(\rho^2 + L_1^2\right)^{1/2}} \right] \tag{16-31}$$

which is just what we found before in (14-15).

The dependence of A_z on ρ as given by the logarithm in (16-30) is exactly that which we found for the scalar potential of a line charge in (5-30), and therefore will have some of the same difficulties. For example, if we try to go to the limit of an infinitely long current by letting L_2 and L_1 become infinite, we will find according to (5-32) that

$$\mathbf{A} \simeq \hat{\mathbf{z}}\frac{\mu_0 I}{2\pi} \ln\left[\frac{\left(4L_2 L_1\right)^{1/2}}{\rho} \right] \tag{16-32}$$

This shows the dependence of **A** on ρ for a very long straight current but will go to infinity for an infinitely long one. As we did for ϕ in (5-28), we can choose another zero for our vector potential by writing

$$\mathbf{A} = \hat{\mathbf{z}}\frac{\mu_0 I}{2\pi} \ln\left(\frac{\rho_0}{\rho} \right) \tag{16-33}$$

so that ρ_0 is the distance from the infinitely long current at which we *choose* **A** to be zero, since we cannot make it do so at infinity. However, there is also a special case in which the ambiguities in **A** disappear because of the nature of the system.

Example

Two antiparallel long currents. Let us consider a system comprised of a current I in the positive z direction and a parallel current $-I$, that is, in the negative z direction; the two currents are a distance $2a$ apart as shown in Figure 16-3. (Compare with Figure 5-6.) Assuming L_2 and L_1 to be already large enough so that we can use (16-32), we

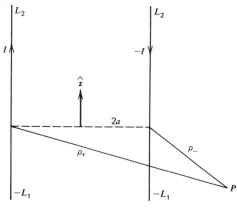

Figure 16-3. Two antiparallel long straight currents.

find the total potential at P to be

$$\mathbf{A} = \hat{\mathbf{z}}\frac{\mu_0 I}{2\pi}\left\{\ln\left[\frac{(4L_2L_1)^{1/2}}{\rho_+}\right] - \ln\left[\frac{(4L_2L_1)^{1/2}}{\rho_-}\right]\right\} = \hat{\mathbf{z}}\frac{\mu_0 I}{4\pi}\ln\left(\frac{\rho_-^2}{\rho_+^2}\right) \quad (16\text{-}34)$$

where ρ_+ and ρ_- are the distances from the respective currents. Since L_2 and L_1 no longer appear in this expression, we can let them go to infinity and we get the unambiguous *finite* result (16-34) for these oppositely directed currents; we note that we would have gotten this same result if we had started with (16-33) instead.

In a specific case, ρ_+ and ρ_- have to be evaluated in terms of the particular coordinate system that is being used. In order to illustrate this, we assume our currents that are parallel to the z axis, also lie in the xz plane, with I intersecting the x axis at a, and $-I$ at $-a$. Then the situation will correspond exactly to that shown in Figure 5-7 with the symbol λ replaced by I, and we can use (5-34) to write down the vector potential for this case as

$$\mathbf{A}(\rho, \varphi) = \hat{\mathbf{z}}\frac{\mu_0 I}{4\pi}\ln\left(\frac{a^2 + \rho^2 + 2a\rho\cos\varphi}{a^2 + \rho^2 - 2a\rho\cos\varphi}\right) \quad (16\text{-}35)$$

The surfaces of constant A_z will then be the same as those given by (5-37) and (5-38) with $\mu_0 I$ replacing λ/ϵ_0, that is, $A_z = $ const. corresponds to

$$\frac{\rho_-^2}{\rho_+^2} = e^{4\pi A_z/\mu_0 I} = \text{const.} \quad (16\text{-}36)$$

or, in rectangular coordinates,

$$(x - a\coth\eta)^2 + y^2 = \left(\frac{a}{\sinh\eta}\right)^2 \quad (16\text{-}37)$$

where, now, $\eta = 2\pi A_z/\mu_0 I$. Thus, the surfaces of constant A_z are cylinders with axes parallel to the z axis whose intersections with the xy plane are the circles described by (16-37), and, in fact, are exactly the circles shown as solid curves in Figure 5-8. As before, the circles whose centers lie on the positive x axis correspond to $A_z > 0$ while those with centers on the negative x axis correspond to $A_z < 0$; the yz plane ($x = 0$ or $\varphi = 90°$) is the surface on which $A_z = 0$.

The induction $\mathbf{B}$ can now be calculated from (16-35) and (16-7) with the use of (1-88); the results are that

$$B_\rho = \frac{1}{\rho}\frac{\partial A_z}{\partial\varphi} = -\frac{\mu_0 Ia(a^2 + \rho^2)\sin\varphi}{\pi\rho_+^2\rho_-^2} \quad (16\text{-}38)$$

$$B_\varphi = -\frac{\partial A_z}{\partial\rho} = \frac{\mu_0 Ia(\rho^2 - a^2)\cos\varphi}{\pi\rho_+^2\rho_-^2} \quad (16\text{-}39)$$

$$B_z = 0 \quad (16\text{-}40)$$

so that $\mathbf{B}$ lies completely in the xy plane. We note that the dependence on position of B_ρ is like that of $-E_\varphi$ of (5-36), while that of B_φ is like that of E_ρ in (5-35); in fact, we can connect these two fields in a formal sense by $\mathbf{B} = (\mu_0\epsilon_0 I/\lambda)\hat{\mathbf{z}} \times \mathbf{E}$ as is easily verified with the aid of (1-76). This means, however, that the lines of $\mathbf{B}$ for this case are perpendicular to the lines of $\mathbf{E}$ shown as the dashed lines of Figure 5-8; but the curves that are perpendicular to the lines of $\mathbf{E}$ in this figure are exactly the solid equipotential circles of the same figure. Thus, the lines of $\mathbf{B}$ are themselves circles of this same type described by (16-36) and (16-37) and are as shown in Figure 16-4

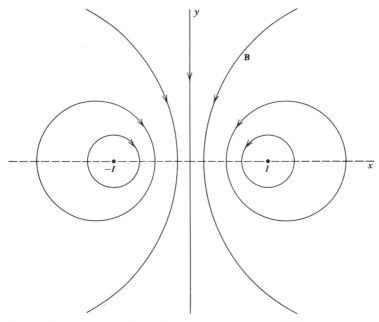

Figure 16-4. Lines of B produced by two antiparallel long straight currents.

We can also come to this same conclusion by finding the lines of **B** directly. The direction of a line of induction will be given by an equation of the same kind as (5-39), that is,

$$d\mathbf{s}_{li} = k\mathbf{B} \tag{16-41}$$

where k is a constant. When this is written in terms of the coordinates ρ and φ in the xy plane, we get the analogue of (5-41):

$$\frac{1}{\rho} \frac{d\rho}{d\varphi} = \frac{B_\rho}{B_\varphi} \tag{16-42}$$

In this specific case, if we substitute (16-38) and (16-39) into this equation, we find that

$$\frac{1}{\rho} \frac{d\rho}{d\varphi} = -\frac{\left(a^2 + \rho^2\right)\sin\varphi}{\left(\rho^2 - a^2\right)\cos\varphi}$$

or

$$\frac{\left(\rho^2 - a^2\right) d\rho}{\rho\left(a^2 + \rho^2\right)} = -\frac{\sin\varphi \, d\varphi}{\cos\varphi} \tag{16-43}$$

This equation can be integrated with the help of tables to give

$$\rho^2 + a^2 = \mathscr{K} a\rho \cos\varphi \tag{16-44}$$

where $\mathscr{K}$ is a dimensionless constant whose value characterizes a given line of induction. If we express (16-44) in rectangular coordinates, it becomes

$$\left(x - \tfrac{1}{2}\mathscr{K}a\right)^2 + y^2 = a^2\left(\tfrac{1}{4}\mathscr{K}^2 - 1\right) \tag{16-45}$$

showing that the lines of **B** are circles with centers at $\tfrac{1}{2}\mathscr{K}a$ on the x axis and with a radius $a(\tfrac{1}{4}\mathscr{K}^2 - 1)^{1/2}$, which again leads us to Figure 16-4. ∎

16-5 INFINITELY LONG IDEAL SOLENOID

Since we have found the values of **B** produced by this system as given in (15-25) and (15-26), we will use them as an example of the calculation of **A** by means of (16-23). In Figure 16-5 we show an end-on view of the solenoid with the equivalent surface current K of (15-22) circulating around it; the axis of the solenoid (z axis) is out of the page as is **B**. According to (16-13), each current element **K** da' gives a contribution to **A** in its own direction, and if we consider the respective contributions $d\mathbf{A}_1$ and $d\mathbf{A}_2$ of two symmetrically located ones $\mathbf{K}_1\, da_1'$ and $\mathbf{K}_2\, da_2'$ as shown in Figure 16-6, we see that the $\hat{\rho}$ components will cancel and we will be left with only a $\hat{\varphi}$ component. Hence we conclude that **A** has only a $\hat{\varphi}$ component, that is, the lines of **A** are circles in this plane, so that we can write

$$\mathbf{A} = A_\varphi(\rho)\hat{\varphi} \qquad (16\text{-}46)$$

since A_φ must be independent of z for this infinitely long solenoid and is independent of φ by symmetry. Thus, a suitable path of integration C to be used in (16-23) is a circle of radius ρ for which $d\mathbf{s} = \rho\, d\varphi\hat{\varphi}$; in this way, we get

$$\oint_C \mathbf{A} \cdot d\mathbf{s} = \oint_C \mathbf{A} \cdot \hat{\varphi}\rho\, d\varphi = 2\pi\rho A_\varphi = \Phi$$

so that

$$A_\varphi(\rho) = \frac{\Phi}{2\pi\rho} \qquad (16\text{-}47)$$

1. Inside the solenoid. Here $\rho < a$, and since $\mathbf{B}_i$ is uniform, according to (15-26), the flux as obtained from (16-6), (1-52), and (1-53) is

$$\Phi = \int_S B_i\hat{\mathbf{z}} \cdot d\mathbf{a}\,\hat{\mathbf{z}} = \mu_0 nI \int_S da = \mu_0 nI\pi\rho^2 \qquad (16\text{-}48)$$

which, when substituted into (16-47) gives

$$A_\varphi(\rho) = \tfrac{1}{2}\mu_0 nI\rho \qquad (\rho < a) \qquad (16\text{-}49)$$

2. Outside the solenoid. Here $\rho > a$, and since $\mathbf{B}_o = 0$, the flux enclosed is the *constant* value obtained from (16-48) for $\rho = a$, that is, $\Phi = \mu_0 nI\pi a^2$. When this is inserted into (16-47), we get

$$A_\varphi(\rho) = \tfrac{1}{2}\mu_0 nI\left(\frac{a^2}{\rho}\right) \qquad (\rho > a) \qquad (16\text{-}50)$$

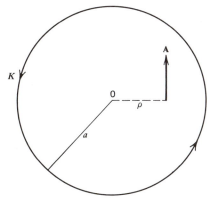

Figure 16-5.   End-on view of a long ideal solenoid.

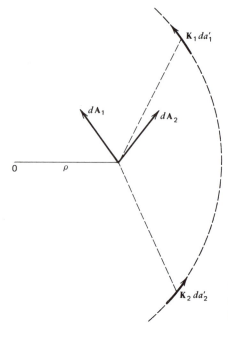

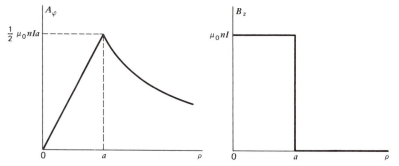

Figure 16-6.   **The resultant vector potential of the two symmetrically located current elements has only a $\hat{\varphi}$ component.**

Figure 16-7.   **Vector potential and magnetic induction of a long ideal solenoid as functions of the distance ρ from the axis.**

We note that both (16-49) and (16-50) give the same value $\frac{1}{2}\mu_0 nIa$ when $\rho = a$; this is in agreement with the continuity property of **A** as expressed by (16-22).

In Figure 16-7, we show both A_φ and B_z plotted as a function of ρ in order to contrast their behavior. This result is interesting because it shows us that *outside* the solenoid where the induction is zero, the vector potential is *different from zero*. This illustrates quite clearly that what really counts from this point of view is not the absolute value of **A** but the way in which it changes as given by $\mathbf{B} = \nabla \times \mathbf{A}$. In order to see what is happening here, we find from (1-88) that with **A** of the form (16-46), the only possible nonzero component of **B** is $B_z = \partial(\rho A_\varphi)/\rho\, \partial\rho$. Inside the solenoid, $\rho A_\varphi = \frac{1}{2}\mu_0 nI\rho^2$ from (16-49), so that $B_{zi} = \mu_0 nI$ in agreement with (15-26). Outside, however, (16-50) gives $\rho A_\varphi = \frac{1}{2}\mu_0 nIa^2 = $ const., so that $B_z = 0$ as we know.

It is also of interest to note that the circular lines of **A**, which we have found to hold for this case, correspond qualitatively to Figure 16-2c, which we deduced as a possible description for a uniform induction.

EXERCISES

16-1 We obtained the results (16-3), (16-7), and (16-17) by assuming filamentary currents. Show that these same results follow if one assumes distributed steady currents.

16-2 Apply (16-5) to a small cylinder in the interior of an infinitely long ideal solenoid. Assume the axis of the cylinder to coincide with the solenoid axis, and thus show that $B_\rho = 0$.

16-3 A certain induction has the form $\mathbf{B} = (\alpha x/y^2)\hat{\mathbf{x}} + (\beta y/x^2)\hat{\mathbf{y}} + f(x, y, z)\hat{\mathbf{z}}$ where α and β are constants. Find the most general possible form for the function $f(x, y, z)$. Find the current density $\mathbf{J}$ and verify that it corresponds to a steady current distribution.

16-4 Show that (16-29) is a suitable vector potential for any constant induction, not just one in the z direction. Also show that (16-29) satisfies (16-17).

16-5 Find the χ that transforms each of the vector potentials of Section 16-3 into the others and verify that each such χ satisfies (16-27).

16-6 Repeat the calculation that led to (16-30) for a field point with coordinates (z, ρ) rather than simply $(0, \rho)$. Show that the $\mathbf{A}$ that you obtain gives the same $\mathbf{B}$ as found for Exercise 14-2.

16-7 We always have $\nabla \cdot \mathbf{B} = 0$; also, $\nabla \times \mathbf{B} = 0$ at a point where there is no current. These equations tell us how certain derivatives of $\mathbf{B}$ are related. These relations can be used to obtain approximate expressions for $\mathbf{B}$ in cases where the general case is too hard to solve exactly while special cases are easy to do. As an example, consider the circle carrying a current. We found the induction along the axis quite easily in (14-18). If one sets up (14-2) for a general field point, one finds an integral which must be expressed in terms of elliptic functions. One can approximate this integral for points near the z axis by making a power series expansion of the integrand for small ρ, but we consider another approach. Beginning with (14-18), we write a Taylor series expansion for B_ρ as $B_\rho(\rho, z) \simeq B_\rho(0, z) + (\partial B_\rho/\partial \rho)_0 \rho$. Use (14-18) and $\nabla \cdot \mathbf{B} = 0$ to evaluate this approximation for B_ρ at a point off the axis but near the axis. Similarly, find an approximate expression for $B_z(\rho, z)$.

16-8 A circle of radius a lies in the xy plane with the origin at its center. A current I traverses this circle in the sense of increasing polar angle φ'. Find an expression for $\mathbf{A}$ produced at an arbitrary field point (x, y, z). Write it in terms of its rectangular components and express it as an integral over φ'. Do not evaluate the integral. Now assume that the field point is on the z axis and evaluate the integral to find $\mathbf{A}$ at any point on the axis. Now go back to the general expression for $\mathbf{A}$, find integral expressions for the components of $\mathbf{B}$, and then show that if the field point is on the axis, (14-18) is obtained.

16-9 A square of edge $2a$ lies in the xy plane with the origin at its center. The sides of the square are parallel to the axes, and a current I goes around it in a counterclockwise sense as seen from a positive value of z. Find $\mathbf{A}$ at all points within the square. What is $\mathbf{A}$ at the center?

16-10 Find $\mathbf{A}$ produced at any point on the z axis by the current on the circular arc shown in Figure 14-9. Why won't your result give the correct value of $\mathbf{B}$ as found in Exercise 14-7?

16-11 An infinitely long cylinder has a circular cross section of radius a and its axis along the z axis. A steady current I is distributed uniformly over the cross section and is in the positive z direction. Use (16-23) to find $\mathbf{A}$ everywhere. If $\mathbf{A}$ outside the cylinder is written in the form (16-33), what is $\mathbf{A}$ on the z axis?

16-12 A wire is wound in a helix of pitch angle α on the surface of a cylinder of radius a so that N complete turns are formed. If the wire carries a current I, show that the axial component of the vector potential produced at the center of the helix is

$$(\mu_0 I/2\pi) \ln \left\{ N\pi \tan\alpha + \left[1 + (N\pi \tan\alpha)^2 \right]^{1/2} \right\}$$

Show that this is the same as that which would be produced by a current I in a wire of length equal to that of the cylinder and going parallel to the axis along the outer surface of the cylinder. Show why this should be the case.

16-13 Find $\mathbf{A}$ for the coaxial cylinders carrying equal and opposite currents as described in Exercise 15-7. Express your answer in terms of $\mathbf{A}_0$, the value on the axis. If it is possible to make $\mathbf{A} = 0$ outside of the outer cylinder, do so, and then find the corresponding value of $\mathbf{A}_0$.

16-14 An infinite plane current sheet coincides with the xy plane. The current density has constant magnitude K and is in the positive y direction. Find the vector potential $\mathbf{A}$ everywhere. If you cannot make it vanish at infinity, express it in terms of its value at the current sheet.

16-15 A sphere of radius a contains a total charge Q distributed uniformly throughout its volume. It is set into rotation about a diameter with constant angular speed ω. Assume that the charge distribution is not altered by the rotation and find $\mathbf{A}$ at any point on the axis of rotation.

16-16 Express the constant $\mathscr{K}$ in (16-44) in terms of the magnitude of $\mathbf{B}$ when it crosses the positive x axis at its greatest distance from the origin. What would be the value of A_z whose constant magnitude corresponds to this same circle?

16-17 The results that we obtained for the two parallel infinitely long lines carrying oppositely directed currents are not as accidental as one may think. Consider any current distribution for which the currents are only in the z direction. Show that in this case, one can always write $\mathbf{B} = \nabla A_z \times \hat{\mathbf{z}}$, and thus show that the lines of $\mathbf{B}$ will always be parallel to the surfaces of constant A_z. Apply this result to the two antiparallel infinitely long currents and show that it is consistent with the directions of $\mathbf{B}$ shown in Figure 16-4.

16-18 Physically, one would expect the flux Φ to be independent of gauge. Why? Show that this is indeed the case.

17

FARADAY'S LAW OF INDUCTION

The overall general results that we have obtained so far can be summarized by the four differential source equations for the vector fields

$$\nabla \cdot \mathbf{E} = \frac{\rho}{\epsilon_0} \qquad \nabla \cdot \mathbf{B} = 0$$

$$\nabla \times \mathbf{E} = 0 \qquad \nabla \times \mathbf{B} = \mu_0 \mathbf{J} \tag{17-1}$$

as given by (4-10), (5-4), (15-12), and (16-3). In addition, we have (12-13), which expresses conservation of charge, and (14-32), which gives the force on a point charge in terms of the electric field and the magnetic induction:

$$\nabla \cdot \mathbf{J} + \frac{\partial \rho}{\partial t} = 0 \qquad \mathbf{F} = q(\mathbf{E} + \mathbf{v} \times \mathbf{B}) \tag{17-2}$$

The equations (17-1) form two completely independent sets, one for $\mathbf{E}$ and one $\mathbf{B}$, thereby implying no connection between the two field vectors. Faraday felt, or suspected, that there really was a relation between these fields and tried many experiments to verify this. About 1831, he finally succeeded in doing so, but only for the case in which things were changing in time, that is, for a *nonstatic* situation. This effect was also found independently by Henry, but it is usually referred to as *Faraday's law of induction* or simply as *Faraday's law*.

Now all of the equations in (17-1) were obtained from a study of *static* fields. Only in our derivation of the equation of continuity did we explicitly consider, for a short while, cases that were time dependent. In order to make progress when things are varying in time, we will do the only thing which is at all reasonable and *assume* that the equations (17-1) are still valid for nonstatic cases unless we are forced by experiment to modify them. In this chapter, we will see that one of them must be changed to encompass new circumstances.

17-1 FARADAY'S LAW

We will consider substantially the same experimental situation as did Faraday, although it will be somewhat simplified. Suppose that we have a closed circuit C formed by a conducting wire as shown in Figure 17-1. We also assume the presence of an induction $\mathbf{B}$ so that there will be a flux Φ through the surface S enclosed by C. If we arbitrarily choose a sense of traversal about C as indicated by the arrow, this will define the direction of the element of area $d\mathbf{a}$ by Figure 1-24 and we can then find Φ from (16-6); Φ will be positive for the choice shown in the figure. We assume that there are no batteries or other sources of emf in the circuit.

If the flux through C is constant so that $d\Phi/dt = 0$, then it is found that there is no current in the circuit. Faraday found, however, that if the flux through C is *not* constant, so that $d\Phi/dt \neq 0$, then there is a current produced in C. This current is said to be "induced" by the changing flux and is accordingly called an *induced current*. It

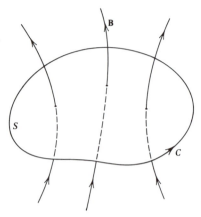

Figure 17-1. The flux through this circuit is positive.

turns out that the numerical value of the current depends on the resistance of the circuit so that it is more convenient to express the quantitative result of experiment in terms of the *induced emf* $\mathscr{E}_{ind}$, that is, the work done per unit charge as we defined it in (12-22). It is found that

$$\mathscr{E}_{ind} = -\frac{d\Phi}{dt} \tag{17-3}$$

which is *Faraday's law*. Since $\mathscr{E}_{ind}$ will be measured in volts, it is also often called the *induced voltage* or just the *voltage*; another name given to it is induced *electromotance*.

The flux Φ can change for a variety of reasons: the induction **B** may be changing in time, the circuit may be altering in shape or size so that the area enclosed by it may be changing, the circuit may be moving by translation or rotation so that it may enclose different values of **B**, or a combination of these effects may be occurring. It has been found by means of many experiments that (17-3) is valid regardless of the origins of $d\Phi/dt$. We can also note that **B** does not have to be different from zero at all parts of the surface S and it could well be zero on portions of S; the result (17-3) still turns out to be applicable. Furthermore, Faraday's law is still found to hold in the form (17-3) even in the presence of matter. Since we will not systematically consider the magnetic effects of matter until Chapter 20, we will assume materials we consider until then to be "nonmagnetic," that is, to have the same magnetic properties as a vacuum.

The negative sign is included in (17-3) to represent the "direction" or "sense" of the induced emf as *compared to* the original arbitrarily chosen sense of traversal about C. This sense is often more conveniently described by *Lenz' law*: the induced emf (and hence the induced current) has such a sense as to *oppose the change* that is producing it. The key words to remember are *oppose* and *change*; in other words, it is not the absolute value of Φ, nor its sign, which is of importance here but the way in which it is changing. (When stated in this way, Lenz' law is seen to be an example of Le Châtelier's principle, which describes the response of a system in a state of stable equilibrium to external effects which tend to alter that state.) In order to illustrate the application of Lenz' law let us consider for simplicity a circuit that is fixed in shape, size, and position so that Φ can change only if **B** changes. This is shown in Figure 17-2 and, to be specific, let us choose our indicated sense of traversal about C so that Φ is positive. Now we first suppose that $|\mathbf{B}|$ is increasing; then $d\Phi/dt$ will be positive, and $\mathscr{E}_{ind}$ will be negative by (17-3). This means that $\mathscr{E}_{ind}$ will be opposite to our original choice of traversal about C and hence will be in the "sense" of the double arrowheads of (a) of the figure; this will also be the direction of the induced current I_{ind} as shown. [$\mathscr{E}_{ind}$ is not necessarily to be thought of as localized, as may be implied by the position

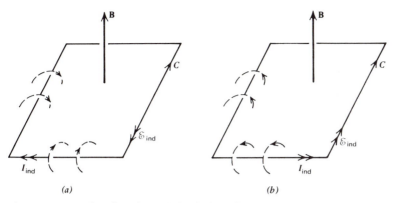

Figure 17-2. The direction of the induced current when the magnitude of B is (*a*) increasing and (*b*) decreasing.

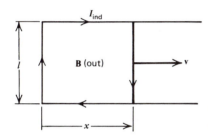

Figure 17-3. The conducting rod moves to the right with constant velocity.

of its symbol, since (17-3) describes the overall effect on the macroscopic circuit.] Let us see how our conclusions fit Lenz' law. Since Φ is increasing, the induced current will "want" to oppose this by trying to decrease the flux by producing lines of **B** opposite to the positive sense of the normal to the area, that is, "into" the circuit. The general directions of the lines of **B** produced by I_{ind} as obtained from the right-hand rule following (14-6) are shown by the dashed curves and we see that they really do tend to decrease the flux through C. Thus, the two statements lead to the same qualitative conclusion. In part (*b*) of the figure we illustrate the results that we obtain when $|\mathbf{B}|$ is assumed to be decreasing. Then, Φ is still positive but decreasing so that $d\Phi/dt$ is negative and $\mathscr{E}_{ind}$ is positive, that is, in the same sense as our originally chosen sense of traversal about C. The direction of I_{ind} now is such as to produce lines of **B** that will tend to increase the flux through the enclosed area, that is, to *oppose the change*, in accord with Lenz' law.

Before we go on to restate Faraday's law in terms of fields, let us look at an example that we can easily analyze quantitatively.

■ **Example**

In Figure 17-3, we show a conducting rod that rests on another conductor bent to form a U-shaped figure. An external agent moves the rod to the right with constant velocity **v**. We assume the presence of a constant induction **B** that is normal to and out of the page. The circuit C will then be the rectangle of sides l and x. Let us choose the sense of traversal about C to be counterclockwise, so that the positive sense of the area is out of the paper. Then the flux through C as found from (16-6) is

$$\Phi = \int_S \mathbf{B} \cdot d\mathbf{a} = Blx \tag{17-4}$$

The induced emf as obtained from (17-3) is

$$\mathscr{E}_{ind} = -\frac{d\Phi}{dt} = -Bl\frac{dx}{dt} = -Blv \tag{17-5}$$

and is negative, that is, in the clockwise sense. This will also be the direction of I_{ind} as shown. To check this against Lenz' law, we see that Φ increases as x increases. Thus I_{ind} will "want" to decrease Φ by producing lines of $\mathbf{B}$ that, within the circuit, will be directed *into* the page, in agreement with the sense of I_{ind} found above and shown in the figure. The magnitude of the induced emf is constant since v is constant. But, as x increases, more conductor is included in the circuit so that the resistance will increase. Thus, the induced current will decrease as the area enclosed by the circuit increases. ■

From (12-23), we see that we can interpret the existence of the induced emf as indicating the presence of a nonconservative *induced electric field* $\mathbf{E}_{ind}$ along the wire so that we can also write (17-3) in the form

$$\oint_C \mathbf{E}_{ind} \cdot d\mathbf{s} = -\frac{d\Phi}{dt} \tag{17-6}$$

In (9-21) we saw that the tangential components of the electric field are continuous, and we will see in the next sections that this is still correct. Thus, the electric field just outside the wire will be the same as that inside so that (17-6) also holds for a path immediately adjacent to and *outside of the circuit*. In fact, (17-6) no longer contains any specific characteristics of the wire, and, in light of the last sentence, it is natural to assume that (17-6) represents a *general physical law* relating the induced electric field to the changing flux and thus will apply to *any closed curve* whether or not there is a circuit there to show an induced current. Furthermore, at any point in space, we can write the total electric field $\mathbf{E}$ as the sum of a conservative part $\mathbf{E}_c$ and a nonconservative part $\mathbf{E}_{ind}$, that is, $\mathbf{E} = \mathbf{E}_c + \mathbf{E}_{ind}$ and then we have

$$\oint_C \mathbf{E} \cdot d\mathbf{s} = \oint_C \mathbf{E}_c \cdot d\mathbf{s} + \oint_C \mathbf{E}_{ind} \cdot d\mathbf{s} \tag{17-7}$$

But the first integral on the right is zero for the conservative field, as we know from (5-5), so that when (17-7) and (17-6) are combined we get an expression relating the total electric field $\mathbf{E}$ and the changing flux:

$$\oint_C \mathbf{E} \cdot d\mathbf{s} = -\frac{d\Phi}{dt} = -\frac{d}{dt}\int_S \mathbf{B} \cdot d\mathbf{a} \tag{17-8}$$

We take (17-8) to be our final statement of Faraday's law of induction as stated in terms of the field vectors and it holds for any closed path C and for any manner by which Φ can change. Because of the various possibilities for changing flux, it is desirable to divide our further discussion of (17-8) into two broad categories depending on whether the medium is at rest or is moving.

17-2 STATIONARY MEDIA

If the various parts of the region of interest to us are not moving, then the bounding curve C will not be instantaneously changing its shape or size, so that Φ can be changing in time only because $\mathbf{B}$ is changing in time, that is, $\mathbf{B} = \mathbf{B}(\mathbf{r}, t)$. Then if we also use (1-67), (17-8) becomes

$$\oint_C \mathbf{E} \cdot d\mathbf{s} = -\frac{d}{dt}\int_S \mathbf{B} \cdot d\mathbf{a} = -\int_S \frac{\partial \mathbf{B}}{\partial t} \cdot d\mathbf{a} = \int_S (\nabla \times \mathbf{E}) \cdot d\mathbf{a}$$

and therefore

$$\int_S \left(\nabla \times \mathbf{E} + \frac{\partial \mathbf{B}}{\partial t} \right) \cdot d\mathbf{a} = 0 \tag{17-9}$$

But since (17-8), and hence (17-9), holds for any arbitrary path of integration with its corresponding surface S, including an infinitesimal one, the integrand of (17-9) must be zero everywhere, and we get

$$\nabla \times \mathbf{E} = -\frac{\partial \mathbf{B}}{\partial t} \tag{17-10}$$

as a differential statement of Faraday's law for stationary media. With (17-10), we finally have a relation between electricity and magnetism in terms of the field vectors.

We can state (17-10) in still another way. We still have $\nabla \cdot \mathbf{B} = 0$ and therefore $\mathbf{B} = \nabla \times \mathbf{A}$ as in (16-7). Inserting this into (17-10), we get $\nabla \times \mathbf{E} = -\partial(\nabla \times \mathbf{A})/\partial t = -\nabla \times (\partial \mathbf{A}/\partial t)$, or

$$\nabla \times \left(\mathbf{E} + \frac{\partial \mathbf{A}}{\partial t} \right) = 0 \tag{17-11}$$

But we also know from (1-48) that a quantity whose curl is zero can be written as the gradient of a scalar so that we can conclude from the above that $\mathbf{E} + (\partial \mathbf{A}/\partial t) = -\nabla \phi$, or

$$\mathbf{E} = -\nabla \phi - \frac{\partial \mathbf{A}}{\partial t} \tag{17-12}$$

showing that, in general, $\mathbf{E}$ depends on both a scalar and a vector potential. In a static case, where $\partial \mathbf{A}/\partial t = 0$, (17-12) reduces to $\mathbf{E} = -\nabla \phi$ and we are back to a conservative electric field; in general, however, we can anticipate that ϕ will not always be exactly the same as the scalar potential of electrostatics.

Since $\nabla \times \mathbf{E}$ can now be different from zero, we have to reinvestigate the behavior of the tangential components of $\mathbf{E}$ at a surface of discontinuity, since our previous result (9-21) depended on $\nabla \times \mathbf{E} = 0$. If we put (17-10) into (9-18), we get

$$\mathbf{E}_{2t} - \mathbf{E}_{1t} = \lim_{h \to 0} \left\{ h \left[\left(-\frac{\partial \mathbf{B}}{\partial t} \right) \times \hat{\mathbf{n}} \right] \right\}$$

As the transition layer is shrunk to zero thickness, we certainly expect that $\partial \mathbf{B}/\partial t$ will remain *finite* so that as $h \to 0$, $h(\partial \mathbf{B}/\partial t) \to 0$; thus the right hand side of the above equation is zero and we find that

$$\mathbf{E}_{2t} = \mathbf{E}_{1t} \tag{17-13}$$

which shows that the tangential components of the electric field are still continuous, in agreement with the parenthetical remark following (9-22). We will have no reason in what follows to alter this conclusion.

■ Example

Fixed loop in an alternating induction. As an example of a stationary system involving a real circuit, let us consider the rectangular loop of sides a and b shown in Figure 17-4. We choose the z axis to lie in the plane of the loop and parallel to the side a; the origin is chosen at the center. We let the plane of the loop make an angle φ with the yz plane so that the normal $\hat{\mathbf{n}}$ to the plane lies in the xy plane and makes the same angle φ with the x axis. Furthermore, let us assume the presence of an induction $\mathbf{B}$ directed along the x axis as given by $\mathbf{B} = \hat{\mathbf{x}} B_0 \cos(\omega t + \alpha)$, that is, it is spatially constant over the

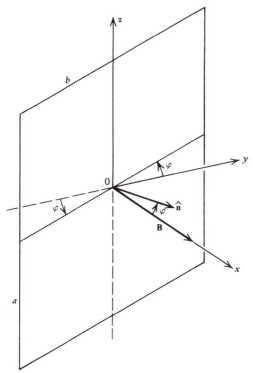

Figure 17-4. A fixed rectangular loop in an alternating induction.

area of the loop but is oscillating harmonically in time with a phase angle α that depends on the choice of the zero for t. Here $\omega = 2\pi\nu$ is the *circular* (or *angular*) *frequency* and is measured in (second)$^{-1}$ or radians/second; ν, on the other hand is the ordinary *frequency*, that is, number of oscillations per unit time and will be measured in hertz where 1 hertz = 1 (second)$^{-1}$. Then the flux through the loop as found from (16-6) is

$$\Phi = \int_S \mathbf{B} \cdot \hat{\mathbf{n}} \, da = B_0 \cos\varphi \cos(\omega t + \alpha) \int_S da = B_0 ab \cos\varphi \cos(\omega t + \alpha) \quad (17\text{-}14)$$

where ab is the area of the loop. The induced emf as given by (17-8) is then

$$\mathscr{E}_{\text{ind}} = \oint_C \mathbf{E} \cdot d\mathbf{s} = \omega B_0 ab \cos\varphi \sin(\omega t + \alpha) \quad (17\text{-}15)$$

which is seen to be proportional to the frequency of oscillation and, since it varies as $\sin(\omega t + \alpha)$, is $90°$ out of phase with $\mathbf{B}$, which varies as $\cos(\omega t + \alpha)$. For example, $\mathscr{E}_{\text{ind}}$ is a maximum when the flux is zero but changing at its maximum rate. If the loop were formed from a single wire wound into N turns, each turn would have an induced emf given by (17-15) so that the total emf for the coil would be N times as great, since it corresponds to total work per unit charge when taken around the whole circuit and is therefore additive. Hence it would be given by $N\omega B_0 ab \cos\varphi \sin(\omega t + \alpha)$. ■

■ **Example**

Let us consider an infinitely long cylindrical region containing a $\mathbf{B}$ field given in cylindrical coordinates by

$$\mathbf{B} = \begin{cases} B_0 \cos(\omega t + \alpha)\hat{\mathbf{z}} & (\rho \leq a) \\ 0 & (\rho > a) \end{cases} \quad (17\text{-}16)$$

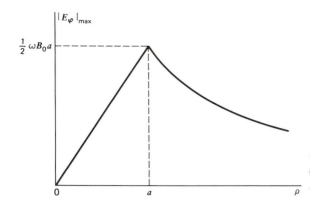

Figure 17-5. The induced electric field as a function of distance from the axis of a cylinder of radius a containing an alternating induction.

where B_0 = const. In other words, **B** is spatially constant over the area of the circle but harmonically oscillating in time; we would visualize this as being produced by an infinite ideal solenoid with an alternating current in the windings. Applying (17-10), we get $\nabla \times \mathbf{E} = \omega B_0 \sin(\omega t + \alpha)\hat{\mathbf{z}}$. The cylindrical symmetry of this problem and our previous experience leads us to expect that **E** will lie in the xy plane and have the form $\mathbf{E} = E_\varphi(\rho)\hat{\boldsymbol{\varphi}}$, that is, be tangent to circles of radius ρ. Accordingly, we choose such a circle as a path of integration, and then, for any ρ,

$$\oint_C \mathbf{E} \cdot d\mathbf{s} = \oint_C E_\varphi \hat{\boldsymbol{\varphi}} \cdot \rho\, d\varphi\, \hat{\boldsymbol{\varphi}} = 2\pi\rho E_\varphi$$

$$= \int_S (\nabla \times \mathbf{E}) \cdot d\mathbf{a} = \omega B_0 \sin(\omega t + \alpha) \int da_z \qquad (17\text{-}17)$$

with the use of (1-67) and (1-53). The area integral equals $\pi\rho^2$ if $\rho \le a$, and has the constant value πa^2 if $\rho > a$ because of (17-16). Substituting these into (17-17), we find that

$$E_\varphi = \frac{1}{2}\omega B_0 \rho \sin(\omega t + \alpha) \qquad (\rho \le a) \qquad (17\text{-}18)$$

$$E_\varphi = \frac{1}{2}\omega B_0 \left(\frac{a^2}{\rho}\right) \sin(\omega t + \alpha) \qquad (\rho > a) \qquad (17\text{-}19)$$

In Figure 17-5, we show the maximum value of $|E_\varphi|$, that is, its amplitude, as a function of ρ. As in the last example, E_φ is 90° out of phase with **B**, depending as it does on the rate of change of **B**, rather than on its absolute value. Induced electric fields produced in this general way are the basis of operation of the charged particle accelerator known as the *betatron*. ∎

17-3 MOVING MEDIA

Many of the more interesting and practical applications of Faraday's law arise when a part or all of the circuit or "medium" is moving. If we imagine our path of integration C to pass through specific points of the medium, then, as these points move, C will be carried along and Φ can change for that reason; simultaneously, **B** may be changing in time. Therefore, in order to evaluate $d\Phi/dt$, we have to compare the flux passing through the final surface enclosed by the final shape of C to that enclosed by the initial shape of C; that is, we need to evaluate

$$\frac{d\Phi}{dt} = \lim_{\Delta t \to 0} \frac{\Delta\Phi}{\Delta t} = \lim_{\Delta t \to 0} \frac{1}{\Delta t}\left[\int_{S(t+\Delta t)} \mathbf{B}(t+\Delta t) \cdot d\mathbf{a}(t+\Delta t) - \int_{S(t)} \mathbf{B}(t) \cdot d\mathbf{a}(t)\right]$$

(17-20)

In Figure 17-6, we show the initial and final locations of the bounding curve C. The element $d\mathbf{s}$ of $C(t)$ is displaced an amount $\mathbf{v}\Delta t$ as a result of the motion. In the process, it sweeps out the element of area

$$d\mathbf{a}_s = d\mathbf{s} \times \mathbf{v}\Delta t$$

(17-21)

shown shaded. We also see that $|d\mathbf{a}_s|$ is a portion of the area S_s of the "side" connecting $C(t)$ and $C(t+\Delta t)$ so that the total area

$$S_{\text{tot}} = S(t) + S_s + S(t+\Delta t)$$

(17-22)

is the bounding area of a *closed* volume through which the total flux of $\mathbf{B}$ is zero, according to (16-5). In order to evaluate the first term in the brackets of (17-20), we apply (16-5) to the volume of Figure 17-6 *at the time* $t + \Delta t$. Therefore

$$\oint_{S_{\text{tot}}} \mathbf{B}(t+\Delta t) \cdot d\mathbf{a} = 0 = -\int_{S(t)} \mathbf{B}(t+\Delta t) \cdot d\mathbf{a}(t)$$

$$+ \int_{S_s} \mathbf{B}(t+\Delta t) \cdot d\mathbf{a}_s + \int_{S(t+\Delta t)} \mathbf{B}(t+\Delta t) \cdot d\mathbf{a}(t+\Delta t) \quad (17\text{-}23)$$

where the minus sign in the first term arises because $d\mathbf{a}(t)$ points *into* the volume while the surface integral was defined in (1-56) in terms of the element of area being positive outward. In addition, the values of $\mathbf{B}$ are all evaluated at $t + \Delta t$, while the areas are written in terms of their values which define the shape of the figure in Figure 17-6. Since we are going to the limit $\Delta t \to 0$ eventually, it is appropriate to think of expanding $\mathbf{B}$ in the power series

$$\mathbf{B}(t+\Delta t) = \mathbf{B}(t) + \frac{\partial \mathbf{B}}{\partial t}\Delta t + \dots$$

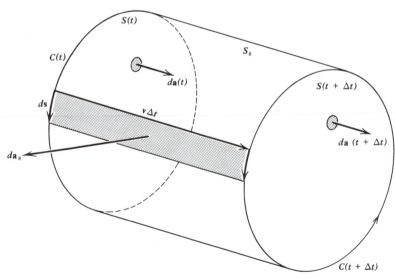

Figure 17-6. The bounding curve C at a time t and a time $t + \Delta t$ later. The line element $d\mathbf{s}$ sweeps out the area $d\mathbf{a}_s$.

and we explicitly keep only terms of first order in Δt. If we now substitute this expression into only the first and second integrals of (17-23), use (17-21), (1-23), and (1-29), we obtain

$$\int_{S(t+\Delta t)} \mathbf{B}(t+\Delta t) \cdot d\mathbf{a}(t+\Delta t) - \int_{S(t)} \mathbf{B}(t) \cdot d\mathbf{a}(t)$$

$$= \Delta t \left\{ \int_{S(t)} \frac{\partial \mathbf{B}}{\partial t} \cdot d\mathbf{a}(t) + \oint_{C(t)} [\mathbf{B}(t) \times \mathbf{v}(t)] \cdot d\mathbf{s} \right\}$$

$$+ \text{terms of the order of } (\Delta t)^2 \qquad (17\text{-}24)$$

where we were able, after substitution of (17-21), to write the integral over S_s in (17-23) as an integral over C because all of the terms involved in it are evaluated on $C(t)$, the initial bounding curve. If we now substitute (17-24) into (17-20) and let $\Delta t \to 0$, the terms originally of the order of $(\Delta t)^2$ and higher will vanish and we are left with

$$\frac{d\Phi}{dt} = \int_S \frac{\partial \mathbf{B}}{\partial t} \cdot d\mathbf{a} + \oint_C (\mathbf{B} \times \mathbf{v}) \cdot d\mathbf{s} \qquad (17\text{-}25)$$

The first term is familiar to us by now as arising from the time variation of $\mathbf{B}$, so that the second term is that arising because of the motion. If we now substitute (17-25) into (17-3) and (17-8), we will get the induced emf and the line integral of the electric field *in the moving system*. If we label these quantities with a prime, and use (1-23) again, we find that

$$\mathscr{E}' = \oint_C \mathbf{E}' \cdot d\mathbf{s} = -\int_S \frac{\partial \mathbf{B}}{\partial t} \cdot d\mathbf{a} + \oint_C (\mathbf{v} \times \mathbf{B}) \cdot d\mathbf{s} \qquad (17\text{-}26)$$

which can be written with the use of (1-67) as

$$\int_S \nabla \times (\mathbf{E}' - \mathbf{v} \times \mathbf{B}) \cdot d\mathbf{a} = -\int_S \frac{\partial \mathbf{B}}{\partial t} \cdot d\mathbf{a} \qquad (17\text{-}27)$$

so that

$$\nabla \times (\mathbf{E}' - \mathbf{v} \times \mathbf{B}) = -\frac{\partial \mathbf{B}}{\partial t} \qquad (17\text{-}28)$$

since (17-27) holds for any arbitrary bounding curve.

It is important to remember that these last three numbered equations contain quantities that are referred to, and would be measured in, *different* systems. The primed quantities are those that would be observed by someone in the moving system and hence at rest with respect to it. On the other hand, the quantities $\mathbf{v}$, $\mathbf{B}$, and $\partial \mathbf{B}/\partial t$ are those that would be measured by an observer at rest in the unprimed system of coordinates (which we will often call the laboratory system); this follows since our derivation based essentially on Figure 17-6 was all written from the point of view of one watching the moving curve C.

Keeping these remarks in mind, we can now obtain an interpretation of the argument of the curl in (17-28) by considering the force on a point charge q as described by these two hypothesized observers. From the point of view of the stationary observer, what one has is a charge q moving with velocity $\mathbf{v}$ in an induction $\mathbf{B}$ and the force on it will be given by (17-2) as $\mathbf{F} = q(\mathbf{E} + \mathbf{v} \times \mathbf{B})$ where $\mathbf{E}$ is the electric field in this system. For the observer moving along with the primed system, the charge q is at rest, and therefore the only possible electromagnetic force that could be written according to (17-2) is $\mathbf{F}' = q\mathbf{E}'$ where $\mathbf{E}'$ is the electric field in the moving system. If we restrict ourselves to cases of zero or negligible relative acceleration, as we have been consistently doing, then both of these systems can be regarded as inertial systems, and we know from mechanics that the two forces will be the same, that is, $\mathbf{F}' = \mathbf{F}$. Equating these expressions for the forces, we find the fields in the two systems to be related by

$$\mathbf{E}' = \mathbf{E} + \mathbf{v} \times \mathbf{B} \tag{17-29}$$

Now we see that the argument of the curl in (17-28) is really the electric field in the unprimed system so that the equation can be written as

$$\nabla \times \mathbf{E} = -\frac{\partial \mathbf{B}}{\partial t} \tag{17-30}$$

which is exactly the same as (17-10) that we found for the case of stationary media. In other words, Faraday's law of induction, when written in this way, has a *form* that is *independent of the motion of the medium*. [Again, when we consider relative velocities $\mathbf{v}$ that are comparable to the speed of light in a vacuum in Chapter 29, we will find that (17-29) will need to be modified somewhat; our conclusions with respect to (17-30), however, will *not* be altered.]

The terms depending on $\mathbf{v}$ in (17-26) and (17-29) are usually known as "motional" terms, while the others that depend on the time variation of $\mathbf{B}$ are often known as "transformer" terms. Some problems are most easily solved and understood by looking at things from the point of view of an observer in the moving system, so that one uses expressions for the *motional emf* and *motional electric field* given by

$$\mathscr{E}'_m = \oint_C \mathbf{E}'_m \cdot d\mathbf{s} = \oint_C (\mathbf{v} \times \mathbf{B}) \cdot d\mathbf{s} \tag{17-31}$$

$$\mathbf{E}'_m = \mathbf{v} \times \mathbf{B} \tag{17-32}$$

Now let us look at a few examples of this type.

◼   **Example**

We begin by looking again at the system illustrated in Figure 17-3. We have already analyzed this by using Faraday's law in the macroscopic form (17-3) and found the overall induced emf to be given by (17-5). Since $\mathbf{B}$ is constant in time in this example, the emf must be given completely by the motional term. As the sliding rod is the only part of the system that is moving, we see from (17-32) that it is the only place where $\mathbf{E}' \neq 0$. We also see that $\mathbf{E}'$ is directed along the moving rod as shown in Figure 17-7 and has the constant magnitude $E' = Bv$ since $\mathbf{B}$ and $\mathbf{v}$ are perpendicular. We also show the direction of $d\mathbf{s}$ that corresponds to the original choice of positive sense of traversal around the circuit on which the calculation of Φ given in (17-4) was based. Then $\mathbf{E}'$ and $d\mathbf{s}$ are oppositely directed and $\mathbf{E}' \cdot d\mathbf{s} = -E'ds = -Bv\,ds$ and (17-31) gives

$$\mathscr{E}'_m = -\int_{\text{rod}} Bv\,ds = -Blv \tag{17-33}$$

in agreement with (17-5). Since the direction of the induced current will be that of $\mathbf{E}'$, we see from Figures 17-7 and 17-3, that I_{ind} will have a clockwise sense that we found before to be consistent with Lenz' law. Although we got the correct value of the induced emf in (17-5), we were not able to "localize" it, but now we see its origin can be ascribed completely to the situation within the moving conducting rod. Furthermore, *in this example*, we can interpret the origin of the induced current as due to the magnetic force produced on a charge moving in an induction as seen by the stationary observer.   ◼

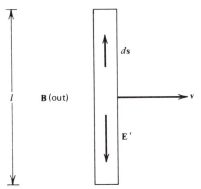

Figure 17-7.   The electric field seen by an observer moving with the rod of Figure 17-3.

■   **Example**

As a variation on the above example, we consider a conducting rod of length l moving with constant velocity perpendicular to the rod and also perpendicular to a constant **B** as shown in Figure 17-8. In this case, we do not have a complete circuit and there cannot be a circulating induced current. In fact, in the *final steady state* of the system, there can be no current at all in the rod and therefore $\mathbf{E}' = 0$ according to (12-25). But then (17-29) gives

$$\mathbf{E} = -\mathbf{v} \times \mathbf{B} \qquad (17\text{-}34)$$

so that the stationary observer sees an electric field of magnitude $E = Bv$ directed upward along the rod as shown in Figure 17-9 and that is drawn from the observer's point of view. Now this electric field must come from somewhere, and since we are dealing with a homogeneous material, it can only come from surface charges on the ends, and they will have the signs shown in the figure. Furthermore, the unprimed observer will conclude that there is a potential difference $\Delta\phi$ between the ends of the rod that can be found from (5-11) and (17-34) to be

$$\Delta\phi = \int_{+}^{-} \mathbf{E} \cdot d\mathbf{s} = El = Blv \qquad (17\text{-}35)$$

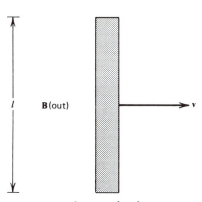

Figure 17-8.   A moving conducting rod not part of a circuit.

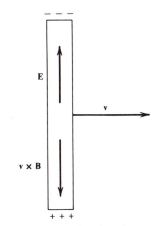

Figure 17-9.   The situation as seen by a stationary observer.

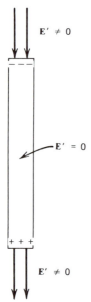

$\mathbf{E'} \neq 0$

$\mathbf{E'} = 0$

$\mathbf{E'} \neq 0$

Figure 17-10.   The situation as seen by the observer moving with the rod.

[We note that this has the same *numerical* value as the motional emf given by (17-33) for the previous example, but now we are talking about a completely different concept.]

What is the origin of the surface charges? During the original stages of the motion, the movable charges of the conductor were subject to the magnetic force $q\mathbf{v} \times \mathbf{B}$ that is directed downward along the rod. This will lead to a *separation* of the charges, the positive ones moving toward the bottom of the rod and the negative ones to the top. But these separated charges will produce an electric field pointing upward that will tend to decrease the total force on a given charge in the interior. Finally, enough charges will be separated so that the electric field $\mathbf{E}$ produced by them will lead to an upward force that just balances the downward magnetic force. This is the final equilibrium state described by (17-34).

Now the *number* of charges at the ends will be the same for both observers, since it involves only counting that they can do equally well. Therefore, the moving observer sees the *same* distribution of charges on the ends that we found above. But since $\mathbf{E'} = 0$ in this system, the rod will be seen as an *equipotential volume* and therefore the potential difference between the ends will be $\Delta\phi' = 0$, in contrast to (17-35), and is consistent with the requirement that there is no current in this system. However, these surface charges will produce an electric field $\mathbf{E'} \neq 0$ in the whole region *outside* the rod, so that the final view of things will be qualitatively like that shown in Figure 17-10; the values of $\mathbf{E'}$ *at* the surface can be found in terms of the surface charge density from (6-4). Thus, this example has been valuable in showing us that different observers will not necessarily describe a given physical situation in exactly the same way when stated in terms of fields, but their relative descriptions will depend on their relative motion.

If the rod is a dielectric, rather than a conductor, there will be no movable charges which can be separated to lead to a situation like Figure 17-9 and it is quite possible for $\mathbf{E'}$ to be different from zero. In this case, the moving dielectric can become polarized; several examples of such situations will be considered as exercises.   ∎

■ **Example**

Rotating loop. Let us consider again the loop with the dimensions and orientation shown in Figure 17-4. However, we now assume that $\mathbf{B} = B_0\hat{\mathbf{x}}$ is constant in time, while the loop is rotating as a rigid body about the z axis with constant angular speed ω so that φ is a function of the time with the form $\varphi = \omega t + \varphi_0$ where φ_0 is the value of the angle at $t = 0$.

Let us first find the induced emf from the overall view given by (17-3). We find the flux in the same manner as we used to get (17-14):

$$\Phi = \int_S \mathbf{B} \cdot \hat{\mathbf{n}}\, da = B_0 ab \cos \varphi = B_0 ab \cos (\omega t + \varphi_0) \qquad (17\text{-}36)$$

Substituting this into (17-3), we find the induced emf to be

$$\mathscr{E}_{\text{ind}} = \omega B_0 ab \sin (\omega t + \varphi_0) \qquad (17\text{-}37)$$

Again, if the loop were formed from a single wire wound into N turns, each turn would have an emf given by (17-37) so that the total emf of the coil would be N times as great, that is,

$$\mathscr{E}_{\text{ind}} = N\omega B_0 ab \sin (\omega t + \varphi_0) \qquad (17\text{-}38)$$

What we have discussed here is a rudimentary form of an electric *generator*, and it is Faraday's law applied in this manner that accounts for most of the electrical energy produced in the world today.

Although a comparison of the induced emfs given by (17-15) and (17-37) shows obvious similarities in *form*, the physical mechanisms involved are quite different, one being produced by a changing induction in a stationary coil, while the other arises in a coil moving with respect to a constant induction.

It is worthwhile looking at this same example in terms of the motional electric field. In Figure 17-11, we show the plane of the loop at some instant. The angular velocity of rotation is given by $\boldsymbol{\omega} = \omega\hat{\mathbf{z}}$ while $\mathbf{r}$ is the position vector of a point on the loop with respect to the origin at its center. The arrows indicate the positive sense of traversal about the loop that we have been using so that the normal $\hat{\mathbf{n}}$ of Figure 17-4 is here out of the paper. For rigid body rotation, the velocity of any point is given by $\mathbf{v} = \boldsymbol{\omega} \times \mathbf{r}$, and with $\mathbf{B}$ given by $\mathbf{B} = B_0\hat{\mathbf{x}}$, we find the motional electric field $\mathbf{E}'_m$ from (17-32), (1-30), and (1-20) to be

$$\mathbf{E}'_m = \mathbf{v} \times \mathbf{B} = -(\mathbf{B} \cdot \mathbf{r})\boldsymbol{\omega} = -B_0 x \omega \hat{\mathbf{z}} \qquad (17\text{-}39)$$

We now must insert this expression into (17-31) and integrate around the loop. We see

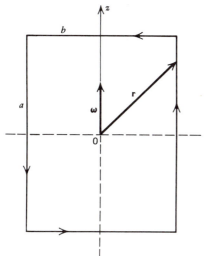

Figure 17-11.   The rectangular loop is rotating in a constant induction.

from Figure 17-11 that the ds for the horizontal portions of length b are all perpendicular to $\hat{z}$ so that, on them, $\mathbf{E}'_m \cdot d\mathbf{s} \sim \hat{z} \cdot d\mathbf{s} = 0$ and they will not contribute to the integral. On the two vertical sides of length a, we have $d\mathbf{s} = dz\,\hat{z}$, and if we let x_l and x_r be the x coordinates of the left- and right-hand sides, respectively, we find that when (17-39) is substituted into (17-31), we get

$$\mathcal{E}'_m = \int_{a/2}^{-(a/2)} (-B_0 x_l \omega\, dz) + \int_{-(a/2)}^{a/2} (-B_0 x_r \omega\, dz) = \omega B_0 a (x_l - x_r) \quad (17\text{-}40)$$

Upon referring back to Figure 17-4, we see that $x_r = -\frac{1}{2}b \sin \varphi$ while $x_l = \frac{1}{2}b \sin \varphi$; thus $x_l - x_r = b \sin \varphi$ and (17-40) becomes

$$\mathcal{E}'_m = \omega B_0 ab \sin \varphi = \omega B_0 ab \sin (\omega t + \varphi_0) \quad (17\text{-}41)$$

exactly as we found in (17-37). We now see, however, that while there is a motional electric field associated with every point on the loop as given by (17-39), only those portions that are parallel to the angular velocity contribute to the induced emf. On the other portions, the induced electric field is always perpendicular to the displacement and hence can do no work on a charge as it passes around the loop. ∎

■ **Example**

Homopolar generator. This interesting device was invented by Faraday. Consider a plane circular conducting disc of radius a rotating with constant angular speed ω in a uniform $\mathbf{B}$ that is perpendicular to the plane of the disc. This is shown in Figure 17-12 where both $\mathbf{B}$ and $\boldsymbol{\omega}$ have been chosen to be into the paper. There are sliding contacts at the central axle at 0 and at a point P on the rim of the disc. The circuit is completed with conducting wires and a resistance R. At a point with position vector $\mathbf{r}$ with respect to the origin at the center, the velocity will be $\mathbf{v} = \boldsymbol{\omega} \times \mathbf{r}$ and will be tangential as shown. Therefore, there will be a motional electric field found from (17-32) to be

$$\mathbf{E}'_m = (\boldsymbol{\omega} \times \mathbf{r}) \times \mathbf{B} = \omega B \mathbf{r} \quad (17\text{-}42)$$

and is thus seen to be radially outward. When we put this into (17-31), we see that the only contribution to the induced emf will come from that part of the disc between 0

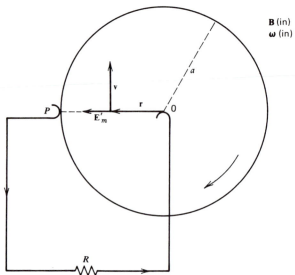

Figure 17-12. Homopolar generator.

and P and since $d\mathbf{s} = d\mathbf{r}$, we will get

$$\mathscr{E}'_m = \int_0^P \mathbf{E}'_m \cdot d\mathbf{s} = \int_0^a \omega Br \, dr = \tfrac{1}{2}\omega Ba^2 \tag{17-43}$$

The resultant current in the external circuit will have the direction shown since it corresponds to the general sense of $\mathbf{E}'_m$. Although devices like this can be built and they do operate, they are generally impractical because of the large sizes and high rotational speeds necessary to get a reasonable value of the emf. If, instead of this arrangement, an external emf, such as a battery, is used to produce a current through the system by means of the contacts, then the disc will rotate and one has a *homopolar motor*. ∎

17-4 INDUCTANCE

Although we began with Faraday's law in the form (17-3) appropriate to the system as an overall unit, we rather quickly converted it into an expression involving fields. For many applications, it is convenient to deal with properties of the system as a whole, much as capacitance proved to be a useful concept, and we now want to revert to that attitude. We begin by looking again at magnetic flux Φ; in this way, we will be led to another important geometrical property of a system known as *inductance*.

In Figure 17-13, we show two filamentary circuits C_j and C_k with their respective currents I_j and I_k. The line elements $d\mathbf{s}_j$ and $d\mathbf{s}_k$ are located by their position vectors $\mathbf{r}_j$ and $\mathbf{r}_k$ with respect to an arbitrary origin 0 and $R_{jk} = |\mathbf{r}_j - \mathbf{r}_k|$. Now the current I_k will produce an induction $\mathbf{B}_k$ at each point of the surface S_j enclosed by C_j and thence a flux $\Phi_{k \to j}$ in C_j that can be found from (16-6). For our purposes, however, it will be more useful to express the flux in terms of the vector potential by means of (16-23). Therefore, if $\mathbf{A}_k(\mathbf{r}_j)$ is the vector potential produced by circuit C_k at the point $\mathbf{r}_j$ of the

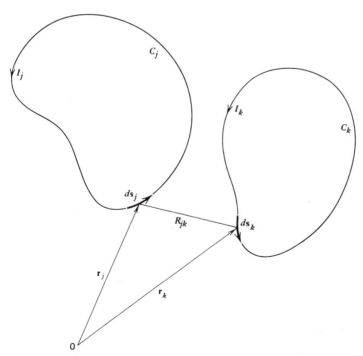

Figure 17-13. Positions of current elements of two filamentary circuits.

circuit C_j, then the flux through C_j due to C_k will be given by (16-23) and (16-10) as

$$\Phi_{k \to j} = \oint_{C_j} \mathbf{A}_k(\mathbf{r}_j) \cdot d\mathbf{s}_j = \frac{\mu_0}{4\pi} \oint_{C_j} \oint_{C_k} \frac{I_k \, d\mathbf{s}_j \cdot d\mathbf{s}_k}{R_{jk}} \tag{17-44}$$

We see that the flux through C_j is proportional to the current I_k in C_k. If we give this factor of proportionality the symbol M_{jk} and call it the *mutual inductance* of circuits j and k, we can write

$$\Phi_{k \to j} = M_{jk} I_k \tag{17-45}$$

$$M_{jk} = \frac{\mu_0}{4\pi} \oint_{C_j} \oint_{C_k} \frac{d\mathbf{s}_j \cdot d\mathbf{s}_k}{R_{jk}} \tag{17-46}$$

It can be seen from (17-46) that the mutual inductance is a purely geometrical factor relating the sizes and relative orientation of the two circuits and can be found in principle by evaluating the double integral over the two circuits. Since μ_0 was defined to have the units of henrys/meter in (13-2), we see that the unit of mutual inductance will be 1 henry.

Similarly, if we calculate the flux in C_k due to the current I_j in C_j, we will find that

$$\Phi_{j \to k} = M_{kj} I_j \tag{17-47}$$

$$M_{kj} = \frac{\mu_0}{4\pi} \oint_{C_k} \oint_{C_j} \frac{d\mathbf{s}_k \cdot d\mathbf{s}_j}{R_{kj}} \tag{17-48}$$

But since $d\mathbf{s}_k \cdot d\mathbf{s}_j = d\mathbf{s}_j \cdot d\mathbf{s}_k$, $R_{kj} = R_{jk} = |\mathbf{r}_j - \mathbf{r}_k|$, and, by using arguments similar to those used in connection with (6-19), we see that the order of integration does not affect the value of the double integral, so that

$$M_{kj} = M_{jk} \tag{17-49}$$

This tells that the flux through C_j by a current I_0 in C_k equals the flux through C_k by the same current I_0 in C_j.

It can be seen from (17-46) that the mutual inductance can be either positive or negative depending on the choices made for the senses of traversal about C_j and C_k since they can be made arbitrarily and independently and this will of course be reflected in the signs of the fluxes.

The overall induced emf arising in one circuit due to a changing current in another can be conveniently written in terms of mutual inductance; if we substitute (17-45) into (17-3), we get

$$\mathscr{E}_{k \to j} = -\frac{d\Phi_{k \to j}}{dt} = -M_{jk}\frac{dI_k}{dt} \tag{17-50}$$

since the circuits are assumed to be at rest so that M_{jk} is a constant. The emf induced in C_k can be obtained from (17-50) by interchanging the indices j and k.

If there is more than one circuit producing the resultant $\mathbf{B}$ throughout C_j, then it follows from (14-4) or (16-11) that the total flux will be the sum of terms like (17-45):

$$\Phi_j = \sum_k \Phi_{k \to j} = \sum_k M_{jk} I_k \tag{17-51}$$

Similarly, the total induced emf in C_j produced by other circuits can be obtained from (17-51) and (17-3) and is

$$\mathscr{E}_{j,\text{mutual}} = -\frac{d\Phi_j}{dt} = -\sum_k M_{jk}\frac{dI_k}{dt} \tag{17-52}$$

Although (17-46) provides a recipe by which the mutual inductance can be calculated in principle, a direct application of it often leads to complicated integrals.

■ **Example**

Two coaxial parallel rings. In Figure 17-14, we have a circle C_j of radius a lying in the xy plane with origin at the center. Another circle C_k of radius b is parallel to the xy plane and its center is on the z axis at a distance c from the origin. The relevant position vectors are seen to be $\mathbf{r}_j = a\cos\varphi_j\hat{\mathbf{x}} + a\sin\varphi_j\hat{\mathbf{y}}, \mathbf{r}_k = b\cos\varphi_k\hat{\mathbf{x}} + b\sin\varphi_k\hat{\mathbf{y}} + c\hat{\mathbf{z}}$ and therefore

$$d\mathbf{s}_j = d\mathbf{r}_j = a\,d\varphi_j\left(-\sin\varphi_j\hat{\mathbf{x}} + \cos\varphi_j\hat{\mathbf{y}}\right)$$

$$d\mathbf{s}_k = d\mathbf{r}_k = b\,d\varphi_k\left(-\sin\varphi_k\hat{\mathbf{x}} + \cos\varphi_k\hat{\mathbf{y}}\right)$$

$$R_{jk} = |\mathbf{r}_j - \mathbf{r}_k| = \left[c^2 + a^2 + b^2 - 2ab\cos\left(\varphi_j - \varphi_k\right)\right]^{1/2}$$

$$d\mathbf{s}_j \cdot d\mathbf{s}_k = ab\cos\left(\varphi_j - \varphi_k\right)d\varphi_j\,d\varphi_k$$

When these are substituted into (17-46), the result is that

$$M_{jk} = \frac{\mu_0 ab}{4\pi}\int_0^{2\pi}\int_0^{2\pi}\frac{\cos\left(\varphi_j - \varphi_k\right)d\varphi_j\,d\varphi_k}{\left[c^2 + a^2 + b^2 - 2ab\cos\left(\varphi_j - \varphi_k\right)\right]^{1/2}} \tag{17-53}$$

Although it was easy enough to write (17-53) down, it cannot be expressed in terms of elementary functions but requires the use of elliptic functions, so we leave (17-53) as it is. ■

In many cases the problem of finding the mutual inductance can be handled more easily by using the definition of it as flux per unit current as expressed by (17-45). This method thus generally requires the calculation of Φ from a previous knowledge of $\mathbf{B}$ and the use of the defining equation (16-6).

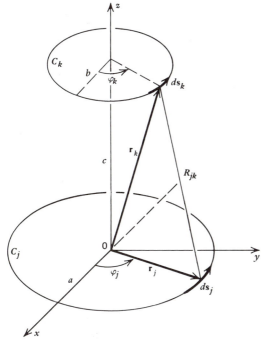

Figure 17-14. Two coaxial parallel rings.

■   **Example**

Short coil wound over a long solenoid. Suppose we have an infinitely long solenoid with n_s turns per unit length. We now suppose that another coil is wound tightly about the solenoid so that there are a total of N_c turns occupying a total length l_c as shown in Figure 17-15. The number of turns per unit length of the coil will then be $n_c = N_c/l_c$. In this case, we know how to find **B** if there is a current I_s in the windings of the solenoid; the uniform value inside as given by (15-26) is $\mathbf{B}_s = \mu_0 n_s I_s \hat{\mathbf{z}}$. If the coil is tightly wound on the solenoid surface, we can take the cross-sectional area of it to be approximately the same as that of the solenoid, S. Then since $\mathbf{B}_s$ is normal to the plane of S, the flux per turn of the coil will be $\Phi_{\text{turn}} = B_s S = \mu_0 n_s S I_s$ so that the total flux through the N_c turns will be $\Phi_c = N_c \Phi_{\text{turn}} = \mu_0 n_s N_c S I_s$. Substituting this into (17-45), we find the mutual inductance of this system to be

$$M_{cs} = \frac{\Phi_c}{I_s} = \mu_0 n_s N_c S = \mu_0 n_s n_c l_c S \qquad (17\text{-}54)$$

Since $M_{sc} = M_{cs}$, according to (17-49), we can use this result to find the flux Φ_s produced in the solenoid when there is a current I_c in the coil. It will be given by $\Phi_s = M_{sc} I_c = \mu_0 n_s n_c l_c S I_c$.   ■

It is possible, of course, for the circuit C_j to produce a flux $\Phi_{j \to j}$ that passes through itself. The coefficient of proportionality that arises in this case is called the *self-inductance* or, often, simply the *inductance*. If we use Figure 17-16 and proceed in the same manner by which we obtained (17-44) through (17-46), we find that

$$\Phi_{j \to j} = L_{jj} I_j \qquad (17\text{-}55)$$

$$L_{jj} = L_j = L = \frac{\mu_0}{4\pi} \oint_{C_j} \oint_{C_j} \frac{d\mathbf{s}_j \cdot d\mathbf{s}'_j}{R_{jj}} \qquad (17\text{-}56)$$

Here now the double integral is taken twice over the same circuit and $d\mathbf{s}_j$ and $d\mathbf{s}'_j$ are separate line elements of C_j. As indicated in (17-56), we will sometimes suppress one or more of the indices on the self-inductance when it will not lead to any confusion. By comparing Figures 17-16 and 14-2, we see that because of the sign conventions we are consistently using, $\Phi_{j \to j}$ will always be positive so that the self-inductance L_{jj} as given by (17-55) will always be a positive quantity.

If the current I_j is changing, then by (17-3), there will be an emf induced in the circuit because of its own changing flux. This is usually called the *self-induced emf* or the *back emf* and will be given by

$$\mathscr{E}_{j,\text{self}} = -L_{jj}\frac{dI_j}{dt} = -L\frac{dI_j}{dt} \qquad (17\text{-}57)$$

Finally, if there are other circuits about, the total flux in C_j will be the sum of (17-51) and (17-55), and the total emf induced in it will be the sum of its self-induced

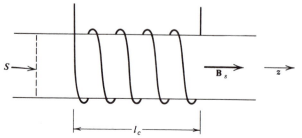

Figure 17-15. Short coil wound over a long solenoid.

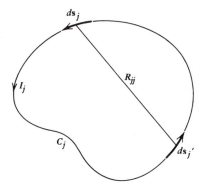

Figure 17-16. Calculation of self-inductance.

and mutually induced emfs. Thus, we can write

$$\Phi_j = \sum_k M_{jk} I_k \tag{17-58}$$

$$\mathscr{E}_j = -\sum_k M_{jk} \frac{dI_k}{dt} \tag{17-59}$$

where the index k takes on the value j as well as the values used to label the other circuits and with the understanding that when the indices are the same we use the notation

$$M_{jj} = L_{jj} = L_j \tag{17-60}$$

Although (17-56) has an appealing appearance and is useful in formal calculations, it will *not* work for true filamentary currents of zero thickness in the sense that it will always give an infinite value for the self-inductance. This occurs because in the double integration, as soon as $d\mathbf{s}_j$ and $d\mathbf{s}'_j$ coincide, $R_{jj} = 0$ and the integral diverges. The physical reason is that near such a wire $|\mathbf{B}| \sim 1/\rho$ and goes to infinity as $\rho \to 0$, leading to an infinite flux. For more realistic circuits, for which one is really dealing with distributed currents, rather than trying to develop a generalization of (17-56), one ordinarily goes back to (17-55) to find the self-inductance as the ratio of flux to current, or uses the energy methods that we will discuss in the next chapter. Consequently, we consider only one simple example of the application of (17-55).

■ **Example**

Ideal solenoid. As in the previous example, the induction is given by $B = \mu_0 n I$ and is uniform over the cross section of area S. Therefore, the flux per turn will be $BS = \mu_0 n S I$. If we consider a length l of the solenoid, there will be nl turns and the total flux will be $\mu_0 n^2 SlI$. Substituting this into (17-55), we get $L = \Phi/I$ or

$$L = \mu_0 n^2 l S \tag{17-61}$$

We note that $L \sim lS$, which is the volume of this portion of the solenoid. Also $L \sim n^2$; this arises because $B \sim n$ and the number of turns is also $\sim n$. ■

EXERCISES

17-1 Choose the positive sense of traversal about C to be the opposite of that shown in Figure 17-2 and then show that (17-3) leads to the same directions for the induced emf for the two cases discussed in the text.

17-2 In a certain region, the induction as a function of time is given by $\mathbf{B} = B_0(t/\tau)\hat{\mathbf{z}}$ where B_0 and τ are constants. Find $\mathbf{A}$. Assuming the scalar potential to be zero, find $\mathbf{E}$ from the above result for $\mathbf{A}$. Use this $\mathbf{E}$ to evaluate the left-hand side of (17-8), and thus show directly that it equals the right-hand side.

17-3 Suppose the current in the infinitely long straight circuit C' of Figure 13-5 is given by $I = I_0 e^{-\lambda t}$ where I_0 and λ are constants. Find the induced emf that will be produced in the rectangular circuit of this same figure. What is the direction of the induced current?

17-4 An infinitely long straight wire carrying a constant current I coincides with the z axis. A circular loop of radius a lies in the xz plane with its center on the positive x axis at a distance b from the origin. Find the flux through the loop. If the loop is now moved with constant speed v parallel to the x axis and away from I, find the emf induced in it. What is the direction of the induced current?

17-5 Find the emf induced in the loop of Figure 17-4 when it is rotating so that $\varphi = \omega t + \varphi_0$ while simultaneously **B** is oscillating at the same frequency, that is, $\mathbf{B} = \hat{x} B_0 \cos(\omega t + \alpha)$. Is it possible to choose the constants φ_0 and α in such a way that the induced emf is always zero?

17-6 A certain closed circuit has a total resistance R. If the total flux through it changes from Φ_i to Φ_f, show that the magnitude of the total charge that will flow through the circuit is given by $Q = (\Phi_f - \Phi_i)/R$. Note that this result is independent of the total time required or of the particular manner in which Φ varies with time.

17-7 A uniform induction **B** is parallel to the axis of a nonmagnetic cylinder of radius a and dielectric constant κ_e. If the cylinder is rotated about its axis with a constant angular velocity **ω** parallel to **B**, find the polarization produced within the cylinder and the surface charge on a length l of it.

17-8 A spherical shell of radius a rotates about the z axis with constant angular velocity **ω**. It is in a uniform induction which is in the xz plane and at an angle α with the axis of rotation. Find the induced electric field at each point on the sphere.

17-9 Find the value of $\mathscr{E}'_m$ produced by a homopolar generator of 1 meter diameter, rotating with an angular speed of 3600 revolutions/minute in an induction of 0.1 tesla.

17-10 A metal rod of length l is rotated about an axis passing through one end with constant angular speed ω. If the circle swept out by the rod is perpendicular to a uniform **B**, find the

induced emf between the ends of the rod when the final steady state has been attained.

17-11 A conducting fluid is flowing with speed v in a horizontal channel of depth w and width l in a region where the vertical component of the induction due to the earth is B_d. Two metal electrodes are placed opposite each other on the vertical walls. The electrodes are rectangular, of dimensions a and b, and are at the same distance d from the bottom of the channel with the long edges horizontal. Find (a) the resistance of the fluid column contained in the rectangular parallelepiped between the electrodes, (b) the induced emf between the electrodes, (c) the resultant current if the electrodes are connected by a wire of negligible resistance, and (d) the numerical values of (a), (b), and (c) for the case of sea water ($\sigma = 4$/ohm-meter) if $B_d = 5.5 \times 10^{-5}$ tesla, $w = 2$ meter, $l = 5$ meter, $a = b = 0.5$ meter, $v = 3$ meter/second, $d = 0.25$ meter.

17-12 Suppose that the circular loop of Exercise 17-4 is now rotated so that its center traces out a circle of radius b with center at the origin while the plane of the loop remains parallel to the z axis. In other words, the loop has an angular velocity $\boldsymbol{\omega} = \omega \hat{z}$. Find $\mathbf{E}'$ in the moving loop. Find $\oint \mathbf{E}' \cdot d\mathbf{s}$ around the loop. Explain your results from the point of view of someone in the fixed laboratory system.

17-13 Two small nonmagnetic dielectric needle shaped dust particles are moving with velocity **v** at right angles to a uniform **B**. The volume of each particle is V and the distance R between them is much larger than their linear dimensions. The particles are oriented with their axes on the same line that is perpendicular to both **v** and **B**. Find the force that the particles will exert on each other, in addition to gravitation. Is it attractive or repulsive?

17-14 An electromagnetic "eddy current" brake consists of a disc of conductivity σ and thickness d rotating about an axis passing through its center and normal to the surface of the disc. A uniform **B** is applied perpendicular to the plane of the disc over a small area a^2 located a distance ρ from the axis. Show that the torque tending to slow down the disc at the instant its angular speed is ω is given approximately by $\sigma \omega B^2 \rho^2 a^2 d$.

17-15 A very thin conducting disc of radius a and conductivity σ lies in the xy plane with the origin at its center. A spatially uniform induction is also present and given by $\mathbf{B} = B_0 \cos(\omega t + \alpha)\hat{z}$.

Find the induced current density $\mathbf{J}_f$ produced in the disc.

17-16 The previous two exercises involved currents that are produced in conductors as a result of Faraday's law and they are given the name "eddy currents." If they arise from a time varying induction, it is possible to obtain a differential equation describing their properties. Begin by combining (17-10) and (12-25), and then by applying (1-120) show that, for steady currents, the eddy currents satisfy the equation $\nabla^2 \mathbf{J}_f = \sigma\mu_0(\partial\mathbf{J}_f/\partial t)$.

17-17 The henry is a fairly large unit of inductance. In order to get an idea of its size, apply (17-61) to a solenoid 1 meter long and 2 centimeters in diameter that has a total of 600 turns.

17-18 Use (17-46) to find the mutual inductance of the circuits shown in Figure 13-5. Is your answer consistent with the result of Exercise 17-3? (*Hint*: assume the straight portion goes from $-L$ to L where L is very large and go to the limit $L \to \infty$ as late in your calculation as possible.)

17-19 The two infinitely long antiparallel currents and the rectangle of Figure 17-17 all lie in the same plane. The sides of length b are parallel to the directions of the currents. Find the mutual inductance between the circuit of the oppositely directed currents and the rectangle. Verify that in the appropriate limit your result reduces to that of the previous exercise.

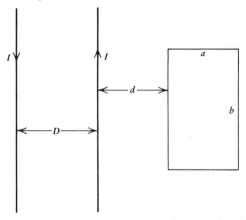

Figure 17-17. The currents and rectangle of Exercise 17-19.

17-20 A toroidal coil of N turns has the central radius of the torus equal to b and the radius of its circular cross section is a. Show that its self-inductance is $\mu_0 N^2[b - (b^2 - a^2)^{1/2}]$.

17-21 A toroidal coil of N turns has a central radius b and a square cross section of side a. Find its self-inductance.

17-22 Show that, as might be expected, when the dimensions of the cross section of a toroidal coil are small compared to the central radius, the results of the two previous exercises give the same approximate self-inductance when expressed in terms of the cross-sectional area S.

17-23 Two infinitely long hollow cylindrical conductors are parallel to each other. The radius of each is A and the parallel axes are separated by a distance D so that in cross section they are as shown in Figure 11-10. They carry equal and opposite currents I. Use the results of Exercise 16-17 and following (16-40) to show that the self-inductance of a length l of this system is simply related to the capacitance of the same length l and then apply this relation to find L.

17-24 Two infinitely long coaxial cylindrical conducting shells carry oppositely directed currents of equal magnitude. If the radius of the inner shell is a and that of the outer is b, find the self-inductance of a length l of this system.

17-25 In Figure 17-14, assume that $c \gg a$ and $c \gg b$ and show that (17-53) becomes approximately $\pi\mu_0 a^2 b^2/2c^3$.

17-26 Consider a situation where one can assume that all of the flux that passes through a coil 1 also passes through a second coil 2, and conversely. For example, we could have a very long solenoid of n_1 turns per unit length and cross-sectional area S_1 over which is wrapped another very long solenoid of n_2 turns per unit length and area S_2 such that $S_2 \simeq S_1$ as would be the case for very thin wires. Similarly, we could have one set of toroidal windings wrapped tightly over another set on the same torus. Show that under these conditions, the mutual inductance M_{12} is related to the individual self-inductances L_1 and L_2 by $|M_{12}| = \sqrt{L_1 L_2}$. This relation gives an upper limit to the mutual inductance of the two coils. In a real case, however, not all of the flux produced within a given coil will pass through the other, as would certainly be the case for the circular loops of Figure 17-14. Then the mutual inductance would be less than the value given above. It is customary in this case to write the relation as $|M_{12}| = k\sqrt{L_1 L_2}$ where k is some number less than unity and is called the *coefficient of coupling*.

18

MAGNETIC ENERGY

In Chapter 7, we evaluated the electrostatic energy of a system in terms of the reversible work required to establish a given configuration of charges. It also takes work to produce a given set of currents in circuits and our aim here is to find it and thus be able to assign a magnetic energy to the system. In the magnetic case, however, we do not have the analogue of the point charge of electrostatics so we have to proceed in a different way. After we have obtained a result in terms of the overall parameters of the system, we will find that we will be able to express the energy in terms of the magnetic induction and to think of it as being distributed throughout space.

18-1 ENERGY OF A SYSTEM OF FREE CURRENTS

We begin by considering a system of currents in a group of circuits, so that what we will actually be evaluating is the energy of *free currents*, specifically conduction currents. We want to calculate the reversible work required to start from an initial situation in which all of the currents are zero and to attain the situation in which the current in the jth circuit has its final value I_j where $j = 1, 2, \ldots, N$ with N being the total number of circuits. As we know from Chapter 13, there will be forces of attraction and repulsion between these circuits that we will assume to be all perfectly rigid and in fixed positions so that we need not be concerned with the mechanical energy of possible deformation or motion of the circuits. Furthermore, we know from (12-35) that the very existence of the currents implies a conversion of electrical energy into heat. Although this energy must be supplied by the external sources, such as batteries, it is not part of the *reversible* work required to establish the currents and hence we can exclude it from consideration here.

Suppose that we are at an intermediate stage of the process where the current in each C_j is given by its i_j so that the currents have not attained their final values I_j. Any change in any of the currents in a time interval dt will produce a change $d\Phi_j$ in the flux through C_j and, from (17-3), an induced emf in it. The external source must supply an emf equal to this value but of opposite sense in order to maintain the system. During this time, a charge $dq_j = i_j \, dt$ will have passed through it, so that the work done by the external source will be $dW_{\text{ext}, j} = \mathscr{E}_{\text{ext}} \, dq_j = -\mathscr{E}_{j, \text{ind}} i_j \, dt = i_j \, d\Phi_j$. Thus we have, as a *general* expression,

$$dW_{\text{ext}, j} = i_j \, d\Phi_j \tag{18-1}$$

as the reversible work done by the external agent and hence the change in magnetic energy. Summing up over all of the currents, we find the total work to be

$$dW_{\text{ext}} = dU_m = \sum_{j=1}^{N} i_j \, d\Phi_j \tag{18-2}$$

which we now write as the increment dU_m of the magnetic energy.

Since the rigid circuits are constant in shape and relative configuration, we see from (17-46) that the inductances will be constant, and then, according to (17-58), the only way the fluxes can change is by way of current changes so that

$$d\Phi_j = \sum_{k=1}^{N} M_{jk} \, di_k \tag{18-3}$$

If we combine (18-2) and (18-3), we obtain

$$dU_m = \sum_{j=1}^{N} \sum_{k=1}^{N} M_{jk} i_j \, di_k \tag{18-4}$$

which must now be summed up as the currents are increased from their initial values of zero to their final values I_j. There are a variety of ways in which one can imagine this *process* being carried out. For example, we could bring the currents to their final values one at a time, we could increase them all simultaneously, we could increase some first and the rest later, and so on. In any event, the final result for the energy must depend only on the *final state* and be *independent of the process* by which it was attained as is familiar to us from thermodynamic discussions of state functions. Accordingly, we will use a simple scheme for increasing all the currents simultaneously, and leave as an exercise the task of showing that the same result is obtained by changing only one current at a time.

Let us assume that at any instant t, each current is the same *fraction* f of its final value, so that $i_j(t) = f(t)I_j$ where $f(t)$ is independent of j and ranges from its initial value of zero to its final value of unity. Then $di_k = I_k \, df$ and (18-4) becomes

$$dU_m = \sum_j \sum_k M_{jk} I_j I_k f \, df \tag{18-5}$$

Summing these changes from the initial to the final state, and choosing our zero of energy such that $U_m = 0$ for the initial state of no currents, we find the magnetic energy to be given by

$$U_m = \sum_j \sum_k M_{jk} I_j I_k \int_0^1 f \, df = \frac{1}{2} \sum_{j=1}^{N} \sum_{k=1}^{N} M_{jk} I_j I_k \tag{18-6}$$

We see that (18-6) is independent of the precise manner in which f changed in time and, in that sense, is independent of the process; this is, of course, always subject to the conditions that these changes are slow enough so that the process can be considered to be reversible.

We can rewrite (18-6) in terms of the currents and fluxes by using (17-58) again to give

$$U_m = \frac{1}{2} \sum_j I_j \Phi_j \tag{18-7}$$

where Φ_j is the total flux through C_j from all sources including itself.

■ **Example**

Two circuits. With $N = 2$, we can write (18-6) with the use of (17-60) and (17-49) as

$$U_m = \frac{1}{2} \left(M_{11} I_1^2 + M_{12} I_1 I_2 + M_{21} I_2 I_1 + M_{22} I_2^2 \right)$$
$$= \frac{1}{2} L_1 I_1^2 + M_{12} I_1 I_2 + \frac{1}{2} L_2 I_2^2 \tag{18-8}$$

The first term depends only on circuit 1 and represents the work needed to establish the current I_1 against its own self-inductance and can properly be described as the "self-energy" of L_1; similar remarks apply to the third term. Thus, for a single isolated inductance L with a current I, we can write its energy as

$$U_m = \tfrac{1}{2}LI^2 \tag{18-9}$$

The middle term of (18-8) involves both currents and arises from the fact that a changing flux in one circuit produces an induced emf in the other; this term can be taken to represent the "interaction energy" of the two circuits. [It is interesting to compare (18-6) through (18-8) with the results (7-16) through (7-18) which we found for the case of conductors in an electrostatic field.]   ∎

Our results so far have been appropriate for filamentary currents and now we would like to rewrite them so as to cover the important cases where we have distributed currents. If we use (16-23) to rewrite the flux Φ_j in (18-7), we obtain

$$U_m = \frac{1}{2}\sum_j \oint_{C_j} \mathbf{A}(\mathbf{r}_j) \cdot I_j\, d\mathbf{s}_j \tag{18-10}$$

where $\mathbf{A}(\mathbf{r}_j)$ is the total vector potential at the location $\mathbf{r}_j$ of the current element $I_j\, d\mathbf{s}_j$ of C_j. Now the integral over C_j gives the contribution to U_m from all of the current elements of C_j, and then the sum over j finally produces the sum of the contributions from all of the current elements of the whole system under consideration. Therefore, if we use the first of the equivalents in (12-10), remember that we are dealing with free currents, and integrate over all regions containing currents, we can rewrite (18-10) as

$$U_m = \frac{1}{2}\int_V \mathbf{J}_f(\mathbf{r}) \cdot \mathbf{A}(\mathbf{r})\, d\tau \tag{18-11}$$

Actually, we can extend the region of integration to cover all space, since in regions in which there are no currents, $\mathbf{J}_f = 0$, and they will contribute nothing to (18-11); thus we can finally write the magnetic energy in the form

$$U_m = \frac{1}{2}\int_{\text{all space}} \mathbf{J}_f(\mathbf{r}) \cdot \mathbf{A}(\mathbf{r})\, d\tau \tag{18-12}$$

Upon comparing this result with (7-10), we see that the vector potential plays much the same role in determining the energy of a system of currents as the scalar potential does for charges. The result (18-12), as does (18-7), has a form appropriate to the attitude that the energy is associated with the currents and is "located" at their positions.

If there are surface currents, we can use the other equivalent of (12-10) to write their contribution to the energy as

$$U_m = \frac{1}{2}\int_{\text{all space}} \mathbf{K}_f(\mathbf{r}) \cdot \mathbf{A}(\mathbf{r})\, da \tag{18-13}$$

■ **Example**

Infinitely long ideal solenoid. In (15-22), we found that this system can be described in terms of a surface current $K = nI$ circulating about its outer surface so that, in accord with Figure 16-5, we can write $\mathbf{K}_f = nI\hat{\boldsymbol{\varphi}}$. Similarly, we found in (16-46) and after (16-50) that the vector potential on the surface (where $\mathbf{K}_f \neq 0$) is $\mathbf{A} = \tfrac{1}{2}\mu_0 nI\bar{a}\hat{\boldsymbol{\varphi}}$ where we write the radius as $\bar{a}$. If we insert these into (18-13) in order to obtain the energy of a length l of the solenoid, so that the area of integration is $2\pi\bar{a}l$, we find that the integrand is constant and therefore

$$U_m = \frac{1}{2}\int\left(\frac{1}{2}\mu_0 n^2 I^2\bar{a}\right)da = \frac{1}{2}\mu_0 n^2(\pi\bar{a}^2)lI^2 \tag{18-14}$$

If we compare this with (18-9), and note that the cross-sectional area is $S = \pi \bar{a}^2$, we see that the inductance is $L = \mu_0 n^2 S l$, exactly as we found in (17-61) by calculating the flux.   ∎

18-2 ENERGY IN TERMS OF THE MAGNETIC INDUCTION

As we briefly noted after (18-12), its form is appropriate to interpreting the energy as associated with and located at the currents. This is a point of view that is consistent with the action at a distance property of Ampère's law with its emphasis on the current elements and their relative orientations. Our primary interest, however, is in the field description of phenomena and we now want to express the energy in these terms.

We can use (15-12) to write $\mathbf{J}_f = (\nabla \times \mathbf{B})/\mu_0$ since we are only considering free currents at present; then (18-12) becomes

$$U_m = \frac{1}{2\mu_0} \int (\nabla \times \mathbf{B}) \cdot \mathbf{A} \, d\tau \tag{18-15}$$

We can rewrite the integrand by using (1-116), (16-7), and (1-17):

$$\mathbf{A} \cdot (\nabla \times \mathbf{B}) = \mathbf{B} \cdot (\nabla \times \mathbf{A}) - \nabla \cdot (\mathbf{A} \times \mathbf{B}) = \mathbf{B}^2 - \nabla \cdot (\mathbf{A} \times \mathbf{B}) \tag{18-16}$$

If we substitute this into (18-15) and use the divergence theorem (1-59) to write one of the resulting volume integrals as a surface integral, we find that

$$U_m = \frac{1}{2\mu_0} \int_V \mathbf{B}^2 \, d\tau - \frac{1}{2\mu_0} \oint_S (\mathbf{A} \times \mathbf{B}) \cdot d\mathbf{a} \tag{18-17}$$

Now our starting integral (18-12) was to be taken over all space so that in (18-17) we should regard V as a very large volume and S its extremely large bounding surface. Then as we let V become infinite, S will recede out to infinity. We assume that the current distribution will always be contained in a finite volume. Thus, as S gets very far away, the whole current distribution will appear to be contained within a very small volume a distance R away. If we look back at (16-12) and (14-7), we see that, as $R \to \infty$, the worst that we can expect to occur is that

$$\mathbf{A} \sim \frac{1}{R} \qquad \mathbf{B} \sim \frac{1}{R^2} \qquad \mathbf{A} \times \mathbf{B} \sim \frac{1}{R^3} \tag{18-18}$$

so that the magnitude of the integrand is decreasing as R^3. As we will see in the next chapter, however, this estimate is overly pessimistic and, in fact, at large distances the current distribution will appear as a closed loop for which we will find that

$$\mathbf{A} \sim \frac{1}{R^2} \qquad \mathbf{B} \sim \frac{1}{R^3} \qquad \mathbf{A} \times \mathbf{B} \sim \frac{1}{R^5} \tag{18-19}$$

The surface of integration is increasing as R^2, so that, for very large R

$$\oint_S (\mathbf{A} \times \mathbf{B}) \cdot d\mathbf{a} \sim \frac{1}{R^5} \cdot R^2 \sim \frac{1}{R^3} \underset{R \to \infty}{\to} 0 \tag{18-20}$$

Thus, when the volume V in (18-17) is increased to include all space, the surface integral will vanish by (18-20) and the expression for the energy becomes simply

$$U_m = \int_{\text{all space}} \frac{\mathbf{B}^2}{2\mu_0} \, d\tau \tag{18-21}$$

[We note that even if (18-18) were applicable, the surface integral would still vanish as it did in the corresponding electric case (7-27).]

The result (18-21) has the same general form as the electric result (7-28) and we can give it the same sort of *interpretation*, that is, the magnetic energy is distributed continuously throughout space with an *energy density* u_m given by

$$u_m = \frac{\mathbf{B}^2}{2\mu_0} \tag{18-22}$$

so that the total magnetic energy can be written as

$$U_m = \int_{\text{all space}} u_m \, d\tau \tag{18-23}$$

The units of u_m will be joule/(meter)^3.

As discussed after (7-30), we are not forced to interpret our expression (18-21) in this way, but it is the natural thing to do and turns out to be very useful as well as consistent with all that we will do later.

We also recall that (7-28) could be usefully applied to the calculation of capacitance. Similarly, we can use (18-21) for the calculation of inductance by combining it with (18-9). Thus, if we have found $\mathbf{B}$ by other means, we can use it to calculate the energy; then we know that U_m will turn out to be proportional to I^2 so that L can be found as $L = 2U_m/I^2$. As usual, the method is best illustrated with examples.

■   **Example**

Infinitely long ideal solenoid. We have already used this to illustrate (18-13). Using (15-26) and (15-25) in (18-22), we find the energy density to be $u_m = \frac{1}{2}\mu_0 n^2 I^2 = \text{const.}$ inside the solenoid and $u_m = 0$ outside. If we consider a length l of the solenoid of cross-sectional area S, the volume will be Sl and (18-23) becomes

$$U_m = \int \frac{1}{2}\mu_0 n^2 I^2 \, d\tau = \frac{1}{2}\mu_0 n^2 I^2 Sl$$

in complete agreement with (18-14); thus we get the same self-inductance $L = \mu_0 n^2 Sl$ as we have found twice before.   ■

■   **Example**

Coaxial line. In Figure 18-1, we show the same system previously shown in Figure 6-12, that is, two coaxial cylindrical conductors; the inner conductor has radius a while the outer conductor has an inner radius b and an outer radius c. We now assume that the conductors carry equal and opposite total currents of magnitude I uniformly distributed over the cross sections, while there is a vacuum in the region between them. We choose the z axis in the direction of the current in the inner conductor. The required values of $\mathbf{B}$ can be found by using the integral form of Ampère's law (15-1). By now we recognize from the symmetry of the situation that $\mathbf{B}$ has the form $\mathbf{B} = B_\varphi(\rho)\hat{\boldsymbol{\varphi}}$ so that it is appropriate to choose as a path of integration a circle of radius ρ. Then the evaluation of (15-1) is exactly like that of (15-17) so that, for any value of ρ, we can use (15-18) and say that

$$B_\varphi(\rho) = \frac{\mu_0 I_{\text{enc}}}{2\pi\rho} \tag{18-24}$$

and all that remains is the evaluation of I_{enc} for a given path. It will be convenient to treat the four regions labeled 1, 2, 3, 4 on the diagram separately; they correspond

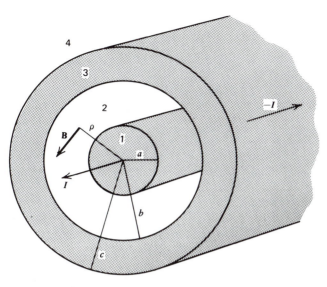

Figure 18-1.   A coaxial line carrying equal currents in opposite directions.

respectively to the inner conductor, the space between conductors, the outer conductor, and all space outside of the system. In this way, we will get a clearer idea of how each region contributes to the total self-inductance of the system.

1. $(0 \le \rho \le a)$ Here, $I_{enc}/I = (\pi \rho^2/\pi a^2) = \rho^2/a^2$ and (18-24) yields

$$B_{\varphi 1} = \frac{\mu_0 I \rho}{2\pi a^2} \tag{18-25}$$

as in (15-20). Since $\mathbf{B}^2 = B_\varphi^2$ in this case, the corresponding energy density found from (18-22) is

$$u_{m1} = \frac{\mu_0 I^2 \rho^2}{8\pi^2 a^4} \tag{18-26}$$

The volume element is $d\tau = \rho \, d\rho \, d\varphi \, dz$ by (1-83), and, if we integrate over a length l of the system, we find the energy associated with this part to be

$$U_{m1} = \int_0^l \int_0^{2\pi} \int_0^a \frac{\mu_0 I^2 \rho^2}{8\pi^2 a^4} \rho \, d\rho \, d\varphi \, dz = \frac{\mu_0 l I^2}{16\pi} \tag{18-27}$$

We note that this result is finite because the value of B_φ goes to zero as $\rho \to 0$ rather than becoming infinite there.

2. $(a \le \rho \le b)$ Here $I_{enc} = I$, and proceeding exactly as above, we obtain

$$B_{\varphi 2} = \frac{\mu_0 I}{2\pi \rho} \tag{18-28}$$

$$u_{m2} = \frac{\mu_0 I^2}{8\pi^2 \rho^2} \tag{18-29}$$

$$U_{m2} = \frac{\mu_0 l I^2}{4\pi} \ln\left(\frac{b}{a}\right) \tag{18-30}$$

3. $(b \le \rho \le c)$ Here, $I_{enc} = I - I[\pi(\rho^2 - b^2)/\pi(c^2 - b^2)] = I(c^2 - \rho^2)/(c^2 - b^2)$ and therefore

$$B_{\varphi 3} = \frac{\mu_0 I}{2\pi(c^2 - b^2)}\left(\frac{c^2}{\rho} - \rho\right) \tag{18-31}$$

$$u_{m3} = \frac{\mu_0 I^2}{8\pi^2(c^2 - b^2)^2}\left(\frac{c^4}{\rho^2} - 2c^2 + \rho^2\right) \tag{18-32}$$

$$U_{m3} = \frac{\mu_0 l I^2}{4\pi(c^2 - b^2)^2}\left[c^4 \ln\left(\frac{c}{b}\right) - \frac{1}{4}(c^2 - b^2)(3c^2 - b^2)\right] \tag{18-33}$$

4. $(c < \rho)$ Here, $I_{enc} = I - I = 0$. Therefore, $B_{\varphi 4}$, u_{m4}, and U_{m4} are all zero.

We can thus obtain the total energy by adding (18-27), (18-30), and (18-33), and when we combine this with (18-9), we find the total self-inductance of the length l of this system to be

$$L = \mu_0 l\left\{\frac{1}{8\pi} + \frac{1}{2\pi}\ln\left(\frac{b}{a}\right) + \frac{1}{2\pi(c^2 - b^2)^2}\left[c^4 \ln\left(\frac{c}{b}\right) - \frac{1}{4}(c^2 - b^2)(3c^2 - b^2)\right]\right\}$$

$$\tag{18-34}$$

which has a form clearly indicating the contribution of each region to the total. In most practical situations, the middle term in (18-34) is the principal source of the inductance. ∎

18-3 MAGNETIC FORCES ON CIRCUITS

As we know, two current carrying circuits will in general exert forces on each other and, in principle, these forces can be calculated from Ampère's law. However, as we saw for electrostatic forces in Section 7-4, it is often desirable to express forces in terms of energy changes. This is what we now want to do for magnetic forces.

For simplicity, we consider only two circuits, since this will be sufficient to illustrate all of the general features involved. The general situation will then be just that shown in Figure 13-1, which was our starting point in magnetism. As we saw in (13-9), in order to have this system in equilibrium, the magnetic force $\mathbf{F}_m$ on C must be balanced by an equal and opposite mechanical force $\mathbf{F}_{mech}$ due to an external agent, which could include springs, braces, and the like.

Now let us imagine that the position vector $\mathbf{r}$ of each point of C is slowly changed by the *same* amount $d\mathbf{r}$; thus the whole circuit is being translated by this amount but is not rotated. The other circuit C' is held fixed, as are the batteries responsible for the maintenance of the currents. (Consequently, we will never be able to treat these circuits as being completely isolated systems.) Under these conditions, the work done by the mechanical force will be reversible work and will be equal to the change dU_t in the total energy of the complete system, that is, $dU_t = \mathbf{F}_{mech} \cdot d\mathbf{r}$. But with zero acceleration, or nearly so, the circuit C will remain in equilibrium, or be only infinitesimally different from it, so that we will continue to have $\mathbf{F}_m = -\mathbf{F}_{mech}$ and then we can also write

$$dU_t = -\mathbf{F}_m \cdot d\mathbf{r} \tag{18-35}$$

Comparing this with (1-38), we see that

$$\mathbf{F}_m = -\nabla U_t \tag{18-36}$$

which is a result analogous to the one-dimensional one of (7-36). Although these are fundamental results, we would like, if possible, to relate the magnetic force to changes in the magnetic energy only. We begin by noting that the total energy change will be the sum of the magnetic energy change, dU_m, and that of the batteries, dU_B, so that

$$dU_t = dU_m + dU_B \tag{18-37}$$

As we saw for the analogous electrostatic case in Section 7-4, it is instructive and useful to visualize two possible conditions under which this displacement is carried out.

1. Constant currents. When one circuit is moved relative to the other, the fluxes through them will generally change; these will produce induced emfs and in order to keep the currents constant, the batteries will have to do work against these emfs or will have work done on them depending upon the signs of the flux changes. We can use (18-7) to write the energy of the currents I and I' as $U_m = \frac{1}{2}(I\Phi + I'\Phi')$. Then, if I and I' are constant, we have $dU_m = \frac{1}{2}(I\,d\Phi + I'\,d\Phi')$. We previously found the work required from the external sources (the batteries) in (18-2); however, we must remember that in this case the currents are already at their final values so that the dU_m of (18-2) is *not* the same as the dU_m we are discussing here. Since any work done by the batteries represents a decrease in their energy, we get

$$dU_B = -dW_{\text{ext}} = -(I\,d\Phi + I'\,d\Phi') = -2\,dU_m \tag{18-38}$$

which is opposite in sign to that of the circuits and twice as great in magnitude. Substituting this into (18-37), we get $dU_t = -dU_m$ in this case so that (18-36) becomes

$$\mathbf{F}_m = (\nabla U_m)_I \qquad \text{(constant currents)} \tag{18-39}$$

where the subscript I on the gradient reminds us that all currents are to be kept constant while evaluating the derivatives.

Since the force described by (18-39) is in the same direction as the gradient of U_m, its tendency will be to *increase* the magnetic energy of the system. Thus, the equilibrium state with constant current ($\mathbf{F}_m = 0$) will correspond to a maximum of the magnetic energy. Then we see from (18-7) that the constant current circuits will tend to adjust themselves by translation so as to enclose as much flux as possible. In the next chapter, we will see that a similar conclusion can be drawn with respect to possible rotations.

In terms of the present notation, the energy expression (18-8) can be written $U_m = \frac{1}{2}LI^2 + MII' + \frac{1}{2}L'I'^2$ where M is the mutual inductance. Since we are considering only displacements that are rigid translations, the shapes of the circuits will not change so that the self-inductances will also be constant according to (17-56); therefore, M is the only quantity affected by the displacement and (18-39) gives

$$\mathbf{F}_m = II'\,\nabla M \tag{18-40}$$

For example, the x component would be given by

$$F_{mx} = II'\frac{\partial M}{\partial x} \tag{18-41}$$

Before we discuss this in more detail, let us consider the other possibility.

2. Constant flux. As we mentioned above, when the circuits are displaced, the fluxes could be expected to change if nothing were done to prevent this. However, if the currents are appropriately adjusted during this process, one could carry it out while keeping Φ and Φ' constant. Then, the corresponding magnetic energy change obtained from (18-7) will be

$$dU_m = \frac{1}{2}(\Phi\,dI + \Phi'\,dI') \tag{18-42}$$

We also see from (18-1) that this will require no energy change in the battery so that $dU_B = 0$; the batteries' energy will nevertheless decrease because of the constant irreversible conversion of energy into heat as described by (12-35), but we are not interested in that here. Thus, for this case, (18-37) becomes $dU_t = dU_m$ so that (18-36) takes the form

$$\mathbf{F}_m = -(\nabla U_m)_\Phi \qquad \text{(fluxes constant)} \qquad (18\text{-}43)$$

[We recall the similar results (7-45) and (7-37) that we previously obtained.]

Now (18-43) must lead to the same expression (18-41) since the overall physical situation is the same, only the calculational schemes differ. Unfortunately, (18-8) is expressed in terms of the currents, while in order to use (18-43) effectively, we want an expression for U_m in terms of the fluxes. Applying (17-58) and (17-60) to this case, we get $\Phi = LI + MI'$ and $\Phi' = MI + L'I'$. Solving these for the currents, we find that

$$I = \frac{L'\Phi - M\Phi'}{LL' - M^2} \qquad I' = \frac{-M\Phi + L\Phi'}{LL' - M^2} \qquad (18\text{-}44)$$

Substitution of these into $U_m = \frac{1}{2}(I\Phi + I'\Phi')$, as obtained from (18-7), yields

$$U_m = \frac{1}{LL' - M^2}\left(\frac{1}{2}L'\Phi^2 - M\Phi\Phi' + \frac{1}{2}L\Phi'^2\right) \qquad (18\text{-}45)$$

Now, in consonance with (18-43), we can differentiate this expression while realizing that everything in it except M is constant; we find that

$$\left(\frac{\partial U_m}{\partial x}\right)_\Phi = \frac{1}{(LL' - M^2)^2}\left[ML'\Phi^2 - (LL' + M^2)\Phi\Phi' + ML\Phi'^2\right]\frac{\partial M}{\partial x}$$

$$= -II'\frac{\partial M}{\partial x}$$

with the use of (18-44). Therefore, the x component of (18-43) will be

$$F_{mx} = -\left(\frac{\partial U_m}{\partial x}\right)_\Phi = II'\frac{\partial M}{\partial x}$$

exactly as we found in (18-41) as an application of (18-39); since there will be similar expressions for the y and z components, we will again be led to (18-40) as a general result for the force on the circuit C.

- **Example**

Ampère's law. Although (18-40) has a different appearance, it must express the same thing as did Ampère's law. We can easily see that this is the case. Substituting the form of (17-48) appropriate to the present case into (18-40), we obtain

$$\mathbf{F}_m = \frac{\mu_0 II'}{4\pi}\nabla\oint_C\oint_{C'}\frac{d\mathbf{s}\cdot d\mathbf{s}'}{R} \qquad (18\text{-}46)$$

Now the rigid translation of C that we have envisaged will not affect the line elements $d\mathbf{s}$ and $d\mathbf{s}'$, nor the limits of the double integrals; in fact, the only variable as far as the ∇ operator is concerned is R. Thus we can interchange the differentiation and integration and use (1-141) to obtain

$$\mathbf{F}_m = \frac{\mu_0 II'}{4\pi}\oint_C\oint_{C'}(d\mathbf{s}\cdot d\mathbf{s}')\nabla\left(\frac{1}{R}\right) = -\frac{\mu_0 II'}{4\pi}\oint_C\oint_{C'}\frac{(d\mathbf{s}\cdot d\mathbf{s}')\hat{\mathbf{R}}}{R^2}$$

which is exactly the version of Ampère's law expressed by (13-6) and, as we know, is equivalent to (13-1) for calculating the total force on C. ∎

■ **Example**

Two interpenetrating long solenoids. Suppose we have two long ideal solenoids, one of which extends into the other a distance x as illustrated in Figure 18-2. We will assume the windings are thin enough so that their cross-sectional areas are approximately equal to the same value S. In this case, the solenoids are not infinitely long and we cannot assume that the values of **B** produced by them in the overlapping region of interest here are the same as those found in the interior far from the ends. In fact, as we will see later, the lines of **B** near the end tend to diverge quite rapidly from the axial direction as indicated schematically in Figure 18-3, and thus will not all pass through the distant turns of the other solenoid. Nevertheless, we can neglect these "end effects," provided that x is already large enough, and find the mutual inductance by considering only the flux contained within the overlapping region; we can therefore use a previous result based on Figure 17-15. Let the inner solenoid have n turns per unit length and carry a current I; the values for the outer are n' and I'. The mutual inductance can then be obtained from (17-54) and is $M = \mu_0 nn'Sx$. Using this in (18-41), we find the force to be

$$F_{mx} = \mu_0 nn'SII' = \frac{SBB'}{\mu_0} \tag{18-47}$$

where the last form, expressing it in terms of the inductions produced by each solenoid, was obtained by using (15-24). What will be the direction of this force? If I and I' both circulate about their respective solenoids in the same sense, the flux produced in one by the other will be positive and M will be positive by (17-45). Then $\partial M/\partial x$ will be positive, F_{mx} will be positive, and the inner one will be attracted into the outer one. This is consistent with our qualitative statement following (13-14) that parallel currents attract each other. If I and I' circulate in opposite senses, then the flux produced by one in the other will be negative, according to our sign conventions, making M negative, and then $\partial M/\partial x$ and F_{mx} will both be negative. Therefore, the inner solenoid will be repelled by the outer, again consistent with our qualitative statement that "unlike" currents repel each other. But all of this is also contained in (18-47) if the currents are assigned relative signs according to their relative sense of circulation. Thus, if they are in the same sense, II' will be positive, as will F_{mx}, while if they are circulating in opposite senses, $II' < 0$ and $F_{mx} < 0$ in agreement with all of the above. We can also express this same result quite nicely in terms of the inductions by writing (18-47) as

$$F_{mx} = \frac{S}{\mu_0} \mathbf{B} \cdot \mathbf{B'} \tag{18-48}$$

which will automatically give the correct sign to F_{mx}. ■

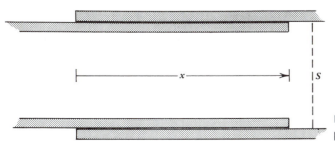

Figure 18-2. Two interpenetrating long solenoids.

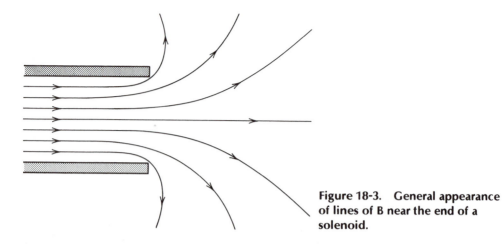

Figure 18-3.   General appearance of lines of B near the end of a solenoid.

Previously, by considering the force on a plate of a charged parallel plate capacitor, we found in (7-50) that there was a force per unit area on the conductor that was directed outward from the surface (a tension) and was numerically equal to the energy density at the surface. We can find a similar result here by looking at the specific example of what is, perhaps, the nearest analogue we have to the oppositely charged plates of a parallel plate capacitor.

■   **Example**

Two oppositely directed current sheets. In Figure 18-4, we show an edge-on view of two long parallel planes carrying surface currents of constant density K but in opposite senses, one directed out of the page and one into it. The width of the sheets is w and x is the distance between them as measured in the sense shown. If the sheets are very long and if $w \gg x$, then we can neglect the edge effects and treat them as if they were infinite plane sheets. In this case, we know from (14-26) that the **B** produced by each sheet has magnitude $\frac{1}{2}\mu_0 K$, is perpendicular to the direction of K and hence is in the

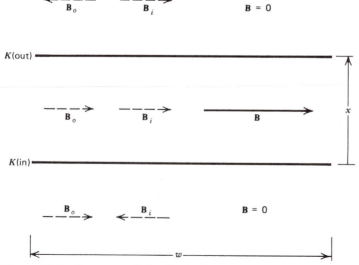

Figure 18-4.   Two oppositely directed current sheets.

plane of the page, and has opposite directions on the two sides of the sheet. These inductions are shown as the dashed arrows in the figure and are labeled by their source current ("out" or "in"). The resultant induction $\mathbf{B} = \mathbf{B}_o + \mathbf{B}_i$ is shown as the solid arrow and we see that $\mathbf{B} \neq 0$ only in the region between the sheets where it has the constant magnitude $\mu_0 K$. Putting this into (18-22), we find the energy density to be $u_m = \frac{1}{2}\mu_0 K^2 = $ const. between the plates and zero everywhere else. Therefore, the magnetic energy of a length l of this system, and corresponding volume wlx, as obtained from (18-23) is

$$U_m = \int \frac{1}{2}\mu_0 K^2 \, d\tau = \frac{1}{2}\mu_0 K^2 wlx \qquad (18\text{-}49)$$

The magnetic force on the upper sheet, which corresponds to an increase in x, as obtained from (18-39) is then found to be

$$\mathbf{F}_m = \frac{\partial U_m}{\partial x}\hat{\mathbf{x}} = \frac{1}{2}\mu_0 K^2 (wl)\hat{\mathbf{x}} \qquad (18\text{-}50)$$

We see that this force is in the positive $\hat{\mathbf{x}}$ direction and thus is repulsive, as is to be expected for these oppositely directed currents. We also note that $\mathbf{F}_m$ is proportional to wl, the area of the sheet, so that if we introduce a force per unit area, f_m, we can write

$$f_m = \frac{|\mathbf{F}_m|}{wl} = \frac{1}{2}\mu_0 K^2 = u_m \qquad (18\text{-}51)$$

and we see that its magnitude is exactly equal to the magnetic energy density. In this case, $\hat{\mathbf{x}}$ is the normal to the surface directed *from* the $\mathbf{B} \neq 0$ region *to* the $\mathbf{B} = 0$ region which enables us to write the force per unit area in vector form as

$$\mathbf{f}_m = f_m \hat{\mathbf{x}} = u_m \hat{\mathbf{x}} \qquad (18\text{-}52)$$

The direction of the force is such as to tend to move the sheet into the $\mathbf{B} = 0$ region and hence may be properly described as a *pressure*.

Combining all of the relations involved, we can write f_m variously as

$$f_m = u_m = \frac{B^2}{2\mu_0} = \frac{1}{2}\mu_0 K^2 = \frac{1}{2}KB \qquad (18\text{-}53)$$

where K and B are evaluated at the position where we want f_m. These results are all very similar to what we found in the electrostatic case as described by (7-49), (7-50), and (7-52). ∎

We have assumed throughout that the circuits are completely rigid. If the internal forces of the material are not large enough to counterbalance these magnetic forces, the conductors comprising the circuits will deform. This deformation will generally continue until the new internal elastic forces produced will be large enough to produce a new equilibrium configuration of the systems.

EXERCISES

18-1 The magnetic force on a current element $i\,d\mathbf{s}$ is $d\mathbf{F} = i\,d\mathbf{s} \times \mathbf{B}$ by (14-5). Suppose that every element of $d\mathbf{s}$ is given the same displacement $d\mathbf{r}$ while i is kept constant. Find the work done by the external agent and thus show directly that it is $i\,d\Phi$ where $d\Phi$ is the change in flux.

18-2 Consider a single self-inductance L at an intermediate stage where its current is i with $0 < i < I$. Find the work required to increase the current by di. Then add all these work increments from the initial state of zero current to the final state of current I and thus obtain (18-9) again.

18-3 Instead of the method we used to sum up (18-4), evaluate U_m in the following way. Beginning with all currents equal to zero, increase i_1 from 0 to I_1 while keeping all the others zero. Then, with I_1 at its final value, increase i_2 from 0 to I_2 and so on for each in turn. In this way, show that (18-6) is again obtained.

18-4 Use the fact that the energy of two circuits as given by (18-8) must be positive to show that $|M_{12}| \le \sqrt{L_1 L_2}$ as was discussed in another way in Exercise 17-26.

18-5 A self-inductance L, a resistance R, and a battery of emf $\mathscr{E}_b$ are all connected in series. Use energy considerations to show that the current i satisfies the differential equation $L(di/dt) + Ri = \mathscr{E}_b$. Now assume that $i \ne 0$ and the battery is switched out of the circuit. Solve the resulting equation and find the relaxation time of this system.

18-6 Consider a vacuum situation where **E** and **B** each have the same numerical values in their appropriate set of units. In other words, $E = x$ volt/meter and $B = x$ tesla. Find the ratio u_m/u_e of their respective energy densities and evaluate its numerical value.

18-7 A long cylindrical nonmagnetic conductor of radius b has a coaxial cylindrical hole of radius a drilled along it, that is, it is like Figure 18-1 with conductor in region 2 and everything else a vacuum. It carries a current I distributed uniformly over the cross section. Find the magnetic energy associated with the induction in a length l of the conductor.

18-8 A closely wound toroidal coil of N turns has the central radius of the torus equal to b and the radius of its circular cross section is a. Find its magnetic energy when there is a current I in the windings and show that this leads to the same self-inductance as found in Exercise 17-20. If $a \ll b$, show that L becomes approximately the same as that of a long ideal solenoid of length $2\pi b$. Is this reasonable?

18-9 Under what conditions, if any, will the contribution of region 1 of Figure 18-1 to the self-inductance of the coaxial line be greater than that of region 2?

18-10 A common method of making coaxial lines is to use a very thin outer conductor, that is, the radii c and b of Figure 18-1 are nearly equal. Show that under these conditions the contribution of the outer conductor to the self-inductance becomes approximately $(\mu_0 l/8\pi b)(c - b)$ and

thus show that (18-34) is consistent with the result found for Exercise 17-24.

18-11 Generalize (18-39) and (18-40) to the case in which there are more than two circuits in the system.

18-12 Use (18-41) to find the force on C of Figure 13-5 and verify that the result is the same as obtained for Exercise 13-4.

18-13 The circular loop of Exercise 17-4 now has a current I' in it that is in the sense required to make M positive. Find the force on the loop.

18-14 Find the force on the circle of radius b shown in Figure 17-14. Leave your answer in integral form but verify that it reduces to the result of Exercise 13-8 under appropriate conditions.

18-15 Use the result of Exercise 17-25 to find the force on the circle of radius b of Figure 17-14 when the two circles are far away from each other.

18-16 A long flexible spring of length l hangs vertically with its upper end rigidly fastened. It has n turns per unit length and the radius of its circular cross section is a. A mass m is fastened to the lower end of the spring. A current I passed through the spring will enable it to support the weight without its being stretched or contracted. If the mass of the spring is neglected show that $I = (1/na)(2mg/\pi\mu_0)^{1/2}$. Do this in two ways by using both (18-39) and (18-43).

18-17 A long thin coil of length l, cross-sectional area S, and n turns per unit length carries a current I. It is placed along the axis of a large circular ring of radius a which is carrying a current I'. If δ is the displacement of the center of the coil from the center of the ring along the coil axis, find the force on the coil as a function of δ.

18-18 A circular loop of wire of self-inductance L and radius r carries a current I. Show that the force tending to increase its radius is $\frac{1}{2}I^2(\partial L/\partial r)$. If T is the tension in the wire for which it will break, show that the wire will break unless T is greater than $(I^2/4\pi)(\partial L/\partial r)$.

18-19 What vacuum value of the induction will result in a magnetic pressure of one atmosphere?

18-20 A very long thin cylindrical shell of radius a carries a current I in the direction of its axis. Find the force per unit area on the shell. Does this force tend to explode the shell or to collapse it? What is the total force on a length l of the shell?

19

MAGNETIC MULTIPOLES

In Chapter 8, we saw how the scalar potential at a point outside of a finite charge distribution could be described by the various multipole moments of the system. Each multipole moment depends on a particular aspect of the detailed charge distribution. In this chapter, we want to do a similar thing for an arbitrary current distribution.

19-1 THE MULTIPOLE EXPANSION OF THE VECTOR POTENTIAL

The general situation is illustrated in Figure 19-1; compare with Figure 8-1. We have a current distribution $\mathbf{J}(\mathbf{r}')$ contained in some volume V'. We choose an origin 0 in some arbitrary way, but near or within V'. We want to find the vector potential $\mathbf{A}$ at the field point P with position vector $\mathbf{r}$, that is, in the direction $\hat{\mathbf{r}}$ and at a distance r from 0. This potential is given by (16-12) as

$$\mathbf{A}(\mathbf{r}) = \frac{\mu_0}{4\pi} \int_{V'} \frac{\mathbf{J}(\mathbf{r}')\, d\tau'}{R} \tag{19-1}$$

where

$$R = |\mathbf{r} - \mathbf{r}'| = \left(r^2 + r'^2 - 2rr'\cos\theta'\right)^{1/2} \tag{19-2}$$

follows from the law of cosines applied to the figure.

As before, we assume that P is far enough outside V' so that $r > r'$ for any portion of V'. Then we can use the expansion given by (8-12) to write (19-1) in the form

$$\mathbf{A}(\mathbf{r}) = \frac{\mu_0}{4\pi} \sum_{l=0}^{\infty} \frac{1}{r^{l+1}} \int_{V'} \mathbf{J}(\mathbf{r}')r'^l P_l(\cos\theta')\, d\tau' \tag{19-3}$$

which is the general *multipole expansion of the vector potential*. If we write out the first few terms with the use of (8-10) and (8-14), we get

$$\mathbf{A}(\mathbf{r}) = \mathbf{A}_M(\mathbf{r}) + \mathbf{A}_D(\mathbf{r}) + \mathbf{A}_Q(\mathbf{r}) + \dots$$

$$= \frac{\mu_0}{4\pi r} \int_{V'} \mathbf{J}(\mathbf{r}')\, d\tau' + \frac{\mu_0}{4\pi r^2} \int_{V'} \mathbf{J}(\mathbf{r}')(\hat{\mathbf{r}} \cdot \mathbf{r}')\, d\tau'$$

$$+ \frac{\mu_0}{4\pi r^3} \int_{V'} \frac{1}{2}\mathbf{J}(\mathbf{r}')\left[3(\hat{\mathbf{r}} \cdot \mathbf{r}')^2 - r'^2\right] d\tau' + \dots \tag{19-4}$$

where the terms in the sum are, respectively, the monopole term, the dipole term, and the quadrupole term. As in the electrostatic case, their dependence on the field point distance r goes successively as $1/r$, $1/r^2$, $1/r^3$, and so on, so that as we get farther away from the current distribution, the higher-order terms in the expansion become less and less important. These terms still involve both the field point P and the source points because of the term $\hat{\mathbf{r}} \cdot \mathbf{r}' \sim \cos\theta'$, and we want to write them as a product of something that depends only on the location of the field point and something characteristic only of the current distribution. It will be convenient to discuss these terms separately.

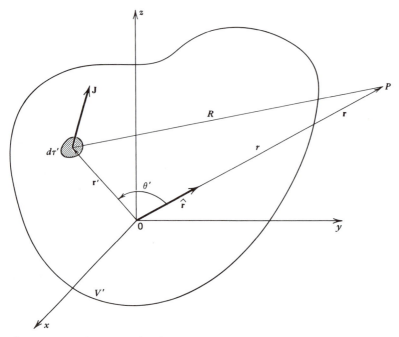

Figure 19-1. Geometry for the calculation of the vector potential at *P*.

1. The Monopole Term

In the electrostatic case, the monopole term turned out to be proportional to the monopole moment of the charge distribution, that is, its net charge. In the magnetic case, however, there are no magnetic charges, as we discussed following (16-4), and this leads us to suspect that the monopole term may actually be zero. Let us consider the x component of the integral in $\mathbf{A}_M$. If we let $\mathbf{A}$ in (1-115) be replaced by $\mathbf{J}$, while u is replaced by x', we find that

$$\nabla' \cdot (x'\mathbf{J}) = \mathbf{J} \cdot \hat{\mathbf{x}} + x'\nabla' \cdot \mathbf{J} = \mathbf{J} \cdot \hat{\mathbf{x}} = J_x$$

since $\nabla' \cdot \mathbf{J} = 0$ for stationary currents. Then, if we also use the divergence theorem (1-59), we obtain

$$\int_{V'} J_x \, d\tau' = \int_{V'} \nabla' \cdot (x'\mathbf{J}) \, d\tau' = \oint_{S'} x'\mathbf{J} \cdot d\mathbf{a}'$$

Since the currents are localized, $\mathbf{J} = 0$ on each point of S' so that the last integral vanishes and

$$\int_{V'} J_x \, d\tau' = 0$$

There will be similar results for the y and z components, and when these are combined, we get

$$\int_{V'} \mathbf{J}(\mathbf{r}') \, d\tau' = 0 \tag{19-5}$$

for localized stationary currents. Since this integral is always zero

$$\mathbf{A}_M(\mathbf{r}) = 0 \tag{19-6}$$

thus verifying our conjecture and showing that the leading term in the expansion of the vector potential is *always* the dipole term. [This also justifies our previous use of (18-20) rather than (18-18).]

While the proof of (19-5) is formally correct, it is also helpful to obtain this same result in a different way. In a steady current distribution, the currents follow closed paths, and we can think of the charges moving in filamentary tubes of currents much as shown in Figure 12-6a. Then for the jth such loop of corresponding current I_j, we see from the equivalent in (12-10) that its contribution to the first integral of (19-4) will be the sum of all its current elements $I_j \, d\mathbf{s}_j$, that is, $\oint I_j \, d\mathbf{s}_j$. Summing up over all of these closed filamentary currents into which we have divided the whole current distribution, we again get

$$\int_{V'} \mathbf{J}(\mathbf{r}') \, d\tau' = \sum_j I_j \oint_{C_j} d\mathbf{s}_j = 0$$

since the integral over $d\mathbf{s}_j$ is just the sum of the successive displacements of a point and hence is the net displacement, and when a point is taken around a closed path, its net displacement is zero so that $\oint d\mathbf{s}_j = 0$; this was also proved more formally in Exercise 1-25.

2. The Dipole Term

For convenience, let us give the symbol $\mathscr{D}$ to the integral in the dipole term:

$$\mathscr{D} = \int_{V'} \mathbf{J}(\mathbf{r}')(\hat{\mathbf{r}} \cdot \mathbf{r}') \, d\tau' \tag{19-7}$$

The process of writing this as a product in which the field point is somehow separated out is somewhat involved and it is easier to deal with a scalar quantity. If we let $\mathbf{C}$ be an *arbitrary constant* vector, then we can form the scalar product $\mathbf{C} \cdot \mathscr{D}$. After doing this, we divide the integrand into two equal parts, add and subtract the quantity $\frac{1}{2}(\mathbf{J} \cdot \hat{\mathbf{r}})(\mathbf{C} \cdot \mathbf{r}')$ under the integral, and we find that we can write the result as the sum of two integrals:

$$\mathbf{C} \cdot \mathscr{D} = \int_{V'} (\mathbf{C} \cdot \mathbf{J})(\hat{\mathbf{r}} \cdot \mathbf{r}') \, d\tau' = (\mathbf{C} \cdot \mathscr{D})_+ + (\mathbf{C} \cdot \mathscr{D})_- \tag{19-8}$$

where

$$(\mathbf{C} \cdot \mathscr{D})_+ = \frac{1}{2} \int_{V'} [(\mathbf{C} \cdot \mathbf{J})(\hat{\mathbf{r}} \cdot \mathbf{r}') + (\mathbf{J} \cdot \hat{\mathbf{r}})(\mathbf{C} \cdot \mathbf{r}')] \, d\tau' \tag{19-9}$$

$$(\mathbf{C} \cdot \mathscr{D})_- = \frac{1}{2} \int_{V'} [(\mathbf{C} \cdot \mathbf{J})(\hat{\mathbf{r}} \cdot \mathbf{r}') - (\mathbf{J} \cdot \hat{\mathbf{r}})(\mathbf{C} \cdot \mathbf{r}')] \, d\tau' \tag{19-10}$$

and we consider these separately.

Now $\mathbf{C}$ is a constant by definition, and $\hat{\mathbf{r}}$ is a constant with respect to source point derivatives, that is, with respect to ∇'. Therefore, it follows from (1-113) that

$$\nabla'(\mathbf{C} \cdot \mathbf{r}') = \mathbf{C} \quad \text{and} \quad \nabla'(\hat{\mathbf{r}} \cdot \mathbf{r}') = \hat{\mathbf{r}} \tag{19-11}$$

Using these last results, along with (1-111), we find that the bracketed part of the integrand of (19-9) can be written as

$$\mathbf{J} \cdot [\mathbf{C}(\hat{\mathbf{r}} \cdot \mathbf{r}') + \hat{\mathbf{r}}(\mathbf{C} \cdot \mathbf{r}')] = \mathbf{J} \cdot [(\hat{\mathbf{r}} \cdot \mathbf{r}')\nabla'(\mathbf{C} \cdot \mathbf{r}') + (\mathbf{C} \cdot \mathbf{r}')\nabla'(\hat{\mathbf{r}} \cdot \mathbf{r}')]$$

$$= \mathbf{J} \cdot \nabla'[(\mathbf{C} \cdot \mathbf{r}')(\hat{\mathbf{r}} \cdot \mathbf{r}')] \tag{19-12}$$

Now the last term has the form $\mathbf{J} \cdot \nabla'\mathscr{S}$ where $\mathscr{S}$ is the scalar $(\mathbf{C} \cdot \mathbf{r}')(\hat{\mathbf{r}} \cdot \mathbf{r}')$; thus, by using (1-115) and (12-15) for steady currents, we can write (19-12) further as

$$\mathbf{J} \cdot \nabla'\mathscr{S} = \nabla' \cdot (\mathscr{S}\mathbf{J}) - \mathscr{S}(\nabla' \cdot \mathbf{J}) = \nabla' \cdot [(\mathbf{C} \cdot \mathbf{r}')(\hat{\mathbf{r}} \cdot \mathbf{r}')\mathbf{J}] \qquad (19\text{-}13)$$

which is our final form for the integrand of (19-9). If we now substitute (19-13) into (19-9) and use (1-59), we finally get

$$(\mathbf{C} \cdot \mathscr{D})_+ = \tfrac{1}{2}\oint_{S'} (\mathbf{C} \cdot \mathbf{r}')(\hat{\mathbf{r}} \cdot \mathbf{r}')(\mathbf{J} \cdot d\mathbf{a}') \qquad (19\text{-}14)$$

where S' is the surface enclosing V'. However, V' encloses *all* of the currents so that $\mathbf{J} = 0$ at all area elements $d\mathbf{a}'$ of S', and therefore

$$(\mathbf{C} \cdot \mathscr{D})_+ = 0 \qquad (19\text{-}15)$$

The terms in the brackets in (19-10) can be written as

$$\mathbf{C} \cdot [\mathbf{J}(\hat{\mathbf{r}} \cdot \mathbf{r}') - \mathbf{r}'(\mathbf{J} \cdot \hat{\mathbf{r}})] = \mathbf{C} \cdot [\hat{\mathbf{r}} \times (\mathbf{J} \times \mathbf{r}')] \qquad (19\text{-}16)$$

with the use of (1-30) and (1-16). If we put this into (19-10), and use (19-15), we find that (19-8) becomes

$$\mathbf{C} \cdot \mathscr{D} = \tfrac{1}{2}\int_{V'} \mathbf{C} \cdot [\hat{\mathbf{r}} \times (\mathbf{J} \times \mathbf{r}')] \, d\tau' = \mathbf{C} \cdot \left\{ \tfrac{1}{2}\int_{V'} [\hat{\mathbf{r}} \times (\mathbf{J} \times \mathbf{r}')] \, d\tau' \right\} \qquad (19\text{-}17)$$

Since $\mathbf{C}$ is a completely arbitrary vector, (19-17) can be always true only if $\mathscr{D}$ equals the quantity in braces. Furthermore, $\hat{\mathbf{r}}$ is constant as far as the integration over the primed variables is concerned and can be taken out of the integral. Doing this, we find that

$$\mathscr{D} = \hat{\mathbf{r}} \times \tfrac{1}{2}\int_{V'} \mathbf{J} \times \mathbf{r}' \, d\tau' = \left(\tfrac{1}{2}\int_{V'} \mathbf{r}' \times \mathbf{J} \, d\tau' \right) \times \hat{\mathbf{r}} \qquad (19\text{-}18)$$

Finally, if we combine (19-18), (19-7) and (19-4), we find that the dipole term can be written as

$$\mathbf{A}_D(\mathbf{r}) = \frac{\mu_0}{4\pi r^2} \left[\tfrac{1}{2}\int_{V'} \mathbf{r}' \times \mathbf{J}(\mathbf{r}') \, d\tau' \right] \times \hat{\mathbf{r}} \qquad (19\text{-}19)$$

which has the desired form of a product of a quantity referring only to the location of the field point and something depending only on the properties of the current distribution.

The quantity in brackets is given the symbol

$$\mathbf{m} = \tfrac{1}{2}\int_{V'} \mathbf{r}' \times \mathbf{J}(\mathbf{r}') \, d\tau' \qquad (19\text{-}20)$$

and is called the *magnetic dipole moment* of the current distribution. [In some books, $\mathbf{m}$ is defined as μ_0 times the right-hand side of (19-20); this is comparatively rare nowadays, but it is a point to keep in mind.] This definition enables us to write the dipole term as

$$\mathbf{A}_D(\mathbf{r}) = \frac{\mu_0}{4\pi} \frac{\mathbf{m} \times \hat{\mathbf{r}}}{r^2} = \frac{\mu_0}{4\pi} \frac{\mathbf{m} \times \mathbf{r}}{r^3} \qquad (19\text{-}21)$$

As shown in Figure 19-2, $\mathbf{A}_D$ is perpendicular to the plane formed by $\mathbf{m}$ and $\mathbf{r}$ and its magnitude will be $A_D = \mu_0 m \sin\Psi/4\pi r^2$. The appearance of $\sin\Psi$ may be a little surprising since, by analogy with the electrostatic case as given by (8-48), one might have expected to find $\cos\Psi$ in a potential called a "dipole" potential. However, when we come to calculate the induction in the next section we will see that the analogy is really quite exact when *fields* are compared.

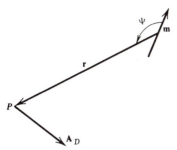

Figure 19-2.   Relation between a magnetic dipole and the vector potential it produces.

Fortunately, it will be sufficient for our purposes to consider only the dipole term, so we will not discuss the quadrupole term of (19-4) any further.

3. Effect of the Choice of Origin

The magnetic moment **m** defined in (19-20) depends on the absolute value of **r'** and, if we choose a different coordinate system, **r'** will certainly change and perhaps **m** will too. We recall that in the electrostatic case, we found in (8-43) that the electric dipole moment was independent of the choice of origin and hence was a unique property of the charge distribution provided that the electric monopole moment vanished. We saw in (19-6) that the monopole term always is zero in the magnetic case so we suspect that **m** may be independent of the choice of origin as well. Suppose that instead of choosing our origin at 0 of Figure 19-1, we choose a new origin 0_n, which is obtained by translating our axes without rotating them, by the displacement **a** as illustrated in Figure 19-3. Then we see from the figure that the old and new position vectors of **J** are related by $\mathbf{a} + \mathbf{r}'_n = \mathbf{r}'$ or $\mathbf{r}'_n = \mathbf{r}' - \mathbf{a}$. If we insert this into (19-20) to find the value of the magnetic dipole moment referred to the new axes, we get

$$\mathbf{m}_n = \tfrac{1}{2}\int_{V'} (\mathbf{r}' - \mathbf{a}) \times \mathbf{J}\, d\tau' = \tfrac{1}{2}\int_{V'} \mathbf{r}' \times \mathbf{J}\, d\tau' - \tfrac{1}{2}\mathbf{a} \times \int_{V'} \mathbf{J}\, d\tau' = \mathbf{m}$$

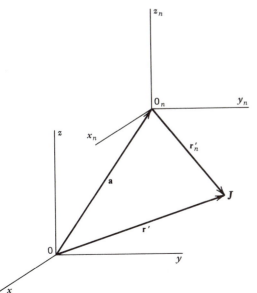

Figure 19-3.   The new origin is displaced by a from the old.

since the integral multiplying **a** is zero by (19-5). Thus, $\mathbf{m}_n = \mathbf{m}$, and the magnetic dipole moment is always a unique property characteristic of the current distribution and any convenient choice of coordinates can be used for calculating it.

19-2 THE MAGNETIC DIPOLE FIELD

The expression $\mathbf{A}_D$ given by (19-21) will be the predominant term in the vector potential when the field point is sufficiently far away from the current distribution. In order to study its properties it is convenient, as it was in Section 8-2, to assume that (19-21) holds *everywhere* in space. Then we can call it *the* magnetic dipole field and think of it as being produced by a fictional *point dipole* **m** located at the origin. Later we will consider specific current distributions that can be thought of in this way, but for now we concentrate on finding the induction **B** produced. We use spherical coordinates to locate the field point P and choose the z axis in the direction of **m**; this gives the situation shown in Figure 19-4. Then $\mathbf{m} = m\hat{\mathbf{z}}$ and (19-21) becomes

$$\mathbf{A}_D(\mathbf{r}) = \frac{\mu_0 m}{4\pi r^2}\hat{\mathbf{z}} \times \hat{\mathbf{r}} = \frac{\mu_0 m \sin\theta}{4\pi r^2}\hat{\boldsymbol{\varphi}} \tag{19-22}$$

so that the only nonzero component of $\mathbf{A}_D$ is $A_{D\varphi}$. The curves of constant values of $A_{D\varphi}$ are given by

$$r^2 = \left(\frac{\mu_0 m}{4\pi A_{D\varphi}}\right)\sin\theta = C_D \sin\theta \tag{19-23}$$

where the constant C_D characterizing a given curve depends on the value of $A_{D\varphi}$. These curves are shown as the solid lines in Figure 19-5; however, the *direction* of $\mathbf{A}_D$ is into the page on the right half of the figure and out of the page for the left half.

The induction is found from $\mathbf{B} = \nabla \times \mathbf{A}_D$. Using (19-22) and (1-104), we find the components of **B** to be

$$B_r = \frac{1}{r\sin\theta}\frac{\partial}{\partial\theta}(\sin\theta A_{D\varphi}) = \left(\frac{\mu_0 m}{4\pi}\right)\frac{2\cos\theta}{r^3}$$
$$B_\theta = -\frac{1}{r}\frac{\partial}{\partial r}(rA_{D\varphi}) = \left(\frac{\mu_0 m}{4\pi}\right)\frac{\sin\theta}{r^3} \tag{19-24}$$

while $B_\varphi = 0$. Thus **B** lies in the same plane as **m** and the field point, while $\mathbf{A}_D$ is perpendicular to this plane.

Comparing (19-24) with the corresponding results (8-50) for the electric dipole field, we see that they are proportional to the magnitude of their respective dipole moments

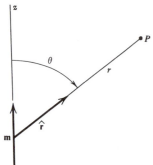

Figure 19-4. Calculation of the field produced by a magnetic dipole.

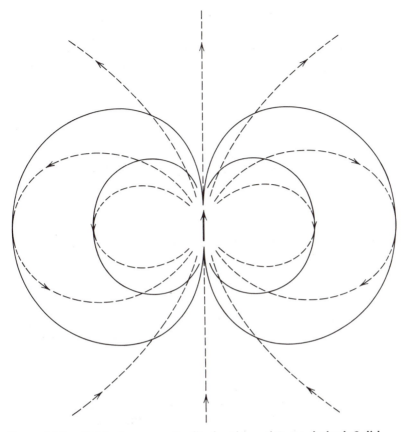

Figure 19-5.   Field of a magnetic dipole. Lines of B are dashed. Solid lines are those of constant magnitude of the vector potential that is directed into the page for the right half and out of the page for the left half.

and they have the *same* dependence on angle and distance. In fact, we can easily see that they are simply related by $\mathbf{B} = \mu_0 \epsilon_0 (m/p) \mathbf{E}$. Consequently, the lines of $\mathbf{B}$, which are shown as the dashed lines in Figure 19-5, will be given by an equation of exactly the same form as (8-52) and will be just like the dashed lines of $\mathbf{E}$ shown in Figure 8-7.

19-3 FILAMENTARY CURRENTS

So far we have concentrated on distributed currents. We can easily adapt our results to filamentary currents by using (12-10) in order to replace $\mathbf{J} \, d\tau'$ by $I \, d\mathbf{s}'$. Thus, we see from (19-20) that the magnetic dipole moment of a single filamentary circuit C' will be given by

$$\mathbf{m} = \frac{I}{2} \oint_{C'} \mathbf{r}' \times d\mathbf{s}' \tag{19-25}$$

and any convenient choice of coordinates can be used to evaluate the integral.

■   **Example**

Plane filamentary current. If the circuit lies in one plane, the integral can be given a simple and useful interpretation. Choosing the origin in this same plane, we get what is shown in Figure 19-6. Now $\mathbf{r}' \times d\mathbf{s}'$ is perpendicular to the plane of the loop, and the magnitude $|\mathbf{r}' \times d\mathbf{s}'|$ equals the area of the parallelogram with $\mathbf{r}'$ and $d\mathbf{s}'$ as sides as we found from Figure 1-15. Comparing these two figures, we see that the shaded triangular area in Figure 19-6 formed by lines from the origin to the extremities of $d\mathbf{s}'$ is just half the area of the parallelogram, that is, if $d\mathbf{a}'$ is the area of the shaded region, then $\frac{1}{2}\mathbf{r}' \times d\mathbf{s}' = d\mathbf{a}'$ and the integral in (19-25) becomes

$$\frac{1}{2}\oint_{C'} \mathbf{r}' \times d\mathbf{s}' = \int d\mathbf{a}' = \mathbf{S} = S\hat{\mathbf{n}} \tag{19-26}$$

where $\mathbf{S}$ is the total vector area enclosed by the current and $\hat{\mathbf{n}}$ is the normal to its surface given by the usual convention defined in Figure 1-24. Therefore (19-25) becomes simply

$$\mathbf{m} = I\mathbf{S} = IS\hat{\mathbf{n}} \tag{19-27}$$

so that the magnitude of the dipole moment of a plane filamentary current is just the product of the encircling current and the area enclosed by it. We note that the result is independent of the shape of the circuit. ∎

Example

Plane circular ring. If the ring is a circle of radius a, then $m = I\pi a^2$. Then at large distances, its axial induction B_z can be found from (19-24) with $\theta = 0$ and $r = z$. The result is that $B_z = \mu_0 I a^2 / 2z^3$ exactly as we found in (14-20). ∎

Plane circuits like these are often called *current loops* or *current whirls* and will produce the dipole vector potential and induction when one is far away from them. [Nearby, of course, the details of the distribution become more important and one has to go back to the exact expression (19-1).] Thus, as a prototype for a point magnetic dipole, we can take a very small current whirl with its magnetic moment perpendicular to its area as shown in Figure 19-7.

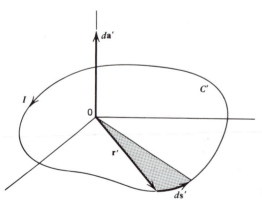

Figure 19-6. Plane filamentary current.

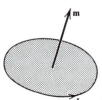

Figure 19-7. The magnetic dipole moment of a current loop is perpendicular to the area of the loop.

If the system consists of a number of filamentary circuits, then a dipole moment $\mathbf{m}_j$ can be found for the jth one by means of (19-25) and the dipole moment of the whole system will be the vector sum of these terms:

$$\mathbf{m} = \sum_j \mathbf{m}_j = \sum_j \tfrac{1}{2} I_j \oint_{C_j} \mathbf{r}_j \times d\mathbf{s}_j \tag{19-28}$$

In particular, if they are all plane circuits, then each $\mathbf{m}_j$ will have the form (19-27) and the total will be

$$\mathbf{m} = \sum_j \mathbf{m}_j = \sum_j I_j \mathbf{S}_j \tag{19-29}$$

■ **Example**

Ideal solenoid. In the example of Section 14-3, we saw that an ideal solenoid of N turns, for which the pitch of the windings could be neglected, could be treated as a set of N parallel circular current loops each carrying the current I in the same sense. Thus, if S is the cross-sectional area, the dipole moment of each will be $IS\hat{\mathbf{z}}$ by (19-27) where $\hat{\mathbf{z}}$ is in the direction of the axis. Then all of the $\mathbf{m}_j$ of (19-29) will be parallel and the total dipole moment of the solenoid will be N times that of a single loop, that is,

$$\mathbf{m} = NSI\hat{\mathbf{z}} = nlSI\hat{\mathbf{z}} \tag{19-30}$$

Consequently, when we are *far away* from the solenoid, the induction it produces will be given by (19-24) with $m = NSI$. (This is essentially the justification for Figure 18-3.)

■

19-4 ENERGY OF A CURRENT DISTRIBUTION IN AN EXTERNAL INDUCTION

In Section 8-4, we gave a fairly elaborate discussion of the energy relations involving two distinct groups of charges ("system" and "external") and concluded that, for many purposes, only part of the energy was of practical interest, namely the interaction energy of (8-60). We now want to consider a similar situation for the magnetic case.

We assume that the currents involved can be divided into two easily distinguishable groups. One of these current distributions forms a distinct physical entity (the "system" of interest) that is subject to the influences of the other current distribution (the "external" sources) that produce an induction $\mathbf{B}_0(\mathbf{r})$ at each point of the system. Again disregarding the internal energies of the two groups, we are interested in that part of the energy that we can call the magnetic *interaction energy* U_{m0}. Our immediate problem then is that of finding the appropriate expression for U_{m0}. Now in the case of two circuits, we saw that the interaction energy was given by the middle term of (18-8) and was $M_{12}I_1 I_2 = I_1 \Phi_{2 \to 1}$ where, by (17-45), $\Phi_{2 \to 1}$ is the flux through 1 produced by 2. Therefore, if we adapt this to the present case by letting I be the system current and Φ_0 be the flux produced by the external sources, we can write the interaction energy as

$$U_{m0} = I\Phi_0 \tag{19-31}$$

We can also express this in terms of the external induction $\mathbf{B}_0$ with the use of (16-6):

$$U_{m0} = I \int_S \mathbf{B}_0(\mathbf{r}) \cdot d\mathbf{a} \tag{19-32}$$

where S is the area enclosed by the system current.

Now that we have (19-31) and (19-32), we can easily generalize them. If we let I_j be the current in the jth circuit of our system of interest and Φ_{0j} is the flux through it produced by the external sources, the interaction energy for this part will be $I_j \Phi_{0j}$ so that the total interaction energy will be

$$U_{m0} = \sum_j I_j \Phi_{0j} = \sum_j I_j \int_{S_j} \mathbf{B}_0(\mathbf{r}_j) \cdot d\mathbf{a}_j \qquad (19\text{-}33)$$

Although we shall not need to use it here, it is worth transforming U_{m0} into a form suitable for distributed currents. In order to do this, we must introduce current elements and we can do this by using (16-23) to write the fluxes in terms of the externally produced vector potential $\mathbf{A}_0$, and we find that

$$U_{m0} = \sum_j \oint_{C_j} \mathbf{A}_0(\mathbf{r}_j) \cdot I_j \, d\mathbf{s}_j \qquad (19\text{-}34)$$

We can now go from the filamentary current description to the one in terms of distributed currents in exactly the same way we went from (18-10) to (18-11). In this way, we obtain a commonly found expression for the magnetic interaction energy

$$U_{m0} = \int_V \mathbf{J}(\mathbf{r}) \cdot \mathbf{A}_0(\mathbf{r}) \, d\tau \qquad (19\text{-}35)$$

where the integral is taken over the whole volume containing the currents $\mathbf{J}$ of the system of interest.

As in Section 8-4, we will now restrict ourselves to the important case in which the external sources are so far away and the system is so small in spatial extent that $\mathbf{B}_0$ does not vary much over the current distribution of the system. Then, as a first approximation, we can take $\mathbf{B}_0$ to be a constant and take it out from under the integral. Thus, if we set every $\mathbf{r}_j$ in (19-33) equal to the same value $\mathbf{r}$ that is "the" position vector of the system, (19-33) becomes

$$U_{m0D} = \mathbf{B}_0(\mathbf{r}) \cdot \left(\sum_j I_j \int_{S_j} d\mathbf{a}_j \right) = \mathbf{B}_0(\mathbf{r}) \cdot \left(\sum_j I_j \mathbf{S}_j \right) = \mathbf{m} \cdot \mathbf{B}_0(\mathbf{r}) = \mathbf{m} \cdot \mathbf{B}_0 \quad (19\text{-}36)$$

where $\mathbf{m}$ is the magnetic dipole moment of the whole system by (19-29). Thus this dipole interaction energy has the form of a product of the dipole moment characteristic of the system and the external induction.

Although we will not do it because of its complexity, it is evident that we can get further approximations to U_{m0} by expanding the components of $\mathbf{B}_0$ in a power series, much as we did for the scalar potential in (8-61). By analogy with (8-70), we can expect the next term in the expansion of the interaction energy to involve products of the spatial derivatives of the external induction with appropriately defined components of the magnetic quadrupole moment.

If $\mathbf{B}_0$ depends on position, then it is possible for a dipole $\mathbf{m}$ to change its energy by moving to another point; in other words, there can be a nonzero translational force $\mathbf{F}_D$ on the dipole. Since the system is being treated as a definite physical entity, we should assume it to be characterized by a constant current distribution. Then the appropriate expression to use for finding the force is the constant current expression of (18-39), and when we put (19-36) into it, we obtain

$$\mathbf{F}_D = \nabla U_{m0D} = \nabla(\mathbf{m} \cdot \mathbf{B}_0) \qquad (19\text{-}37)$$

where the derivatives are taken with respect to the components of $\mathbf{r}$, the position vector of the dipole. The fact that $\mathbf{m}$ is constant enables us to put this into a more useful form. If we use (1-112), we can write (19-37) as

$$\mathbf{F}_D = \mathbf{B}_0 \times (\nabla \times \mathbf{m}) + \mathbf{m} \times (\nabla \times \mathbf{B}_0) + (\mathbf{B}_0 \cdot \nabla)\mathbf{m} + (\mathbf{m} \cdot \nabla)\mathbf{B}_0 \quad (19\text{-}38)$$

The first and third terms vanish because $\mathbf{m}$ is constant. By assumption, all of the external sources are located elsewhere, so that there are no source currents $\mathbf{J}_0$ at the location of $\mathbf{m}$. Then $\nabla \times \mathbf{B}_0 = 0$ by (15-12), and (19-38) reduces to

$$\mathbf{F}_D = (\mathbf{m} \cdot \nabla)\mathbf{B}_0 \qquad (19\text{-}39)$$

showing us, as we suspected, that the translational force depends on the external induction being a function of position. We note that (19-37) and (19-39) are completely analogous to the corresponding results (8-77) and (8-79) that we found for the electric dipole; in addition, (19-39) is consistent with the result of Exercise 14-18 in which it was shown that there is no net force on a filamentary circuit in a uniform external induction.

There is one place, however, in which the analogy between the electric and magnetic cases apparently breaks down and that is in the difference in sign of the interaction energies that are given by (19-36) and (8-73) as $\mathbf{m} \cdot \mathbf{B}_0$ and $-\mathbf{p} \cdot \mathbf{E}_0$ respectively. In fact it is commonplace to find a quantity called *the* energy of a magnetic dipole in an external field which is given by

$$U'_D = -\mathbf{m} \cdot \mathbf{B}_0 = -mB_0 \cos \Psi \qquad (19\text{-}40)$$

where Ψ is the angle between $\mathbf{m}$ and $\mathbf{B}_0$. Why is this? In essence, the use of (19-40) represents the result of a complete change in point in view. What one is trying to do *in this case* is to write the expression for the force in complete *analogy* to that used in mechanics, namely, as the negative gradient of a *potential energy*. In other words, the aim is to write (19-37) in the form

$$\mathbf{F}_D = -\nabla U'_D \qquad (19\text{-}41)$$

and then *interpret* U'_D as a potential energy of the usual kind. We can easily see that the choice given in (19-40) works since (19-41) becomes $\mathbf{F}_D = -\nabla(-\mathbf{m} \cdot \mathbf{B}_0) = \nabla(\mathbf{m} \cdot \mathbf{B}_0)$, which leads us right back to the correct expression for the force. Hence it is appropriate to think of U'_D as the energy of the dipole in this sense; U'_D is often called the *orientational energy* in order to stress its role as a potential energy. Similarly, we can change the signs of (19-31) through (19-35) and reinterpret them in this manner. In this way, for example, the tendency of a circuit to readjust itself to enclose as much flux as possible, which we found after (18-39), can be seen from the negative of (19-31), $-I\Phi_0$, as corresponding to the general tendency of systems to try to attain equilibrium by minimizing their potential energy. In Figure 19-8, we show U'_D as given by (19-40) as a

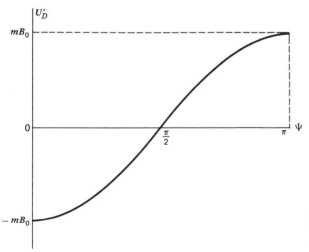

Figure 19-8. Orientational energy of a dipole in an external induction as a function of the angle between their directions.

function of Ψ. The range of energy variation is finite with the minimum at $\Psi = 0$ corresponding to the stable equilibrium state in which $\mathbf{m}$ and $\mathbf{B}_0$ are parallel and a maximum at $\Psi = \pi$ that describes the unstable equilibrium state for which $\mathbf{m}$ and $\mathbf{B}_0$ are oppositely directed.

A potential energy like U_D', which depends on the angle, implies the existence of a torque $\boldsymbol{\tau}$ on the system. Because of the similarity of (19-40) and (8-73), we can simply change the symbols in (8-75) to obtain the important expression for the torque on a magnetic dipole in an external induction:

$$\boldsymbol{\tau} = \mathbf{m} \times \mathbf{B}_0 \qquad (19\text{-}42)$$

This torque will be present even in a uniform $\mathbf{B}_0$ in contrast to $\mathbf{F}_D$; again this is consistent with the result of Exercise 14-18. We note as before that in order to have equilibrium at an intermediate angle Ψ, this magnetic torque must be balanced by an equal and opposite mechanical torque $\boldsymbol{\tau}_{\text{mech}}$ so that

$$\boldsymbol{\tau} + \boldsymbol{\tau}_{\text{mech}} = 0 \qquad (19\text{-}43)$$

Now all of the results in this section have been obtained in a completely general way and are applicable to the dipole moment of any type of current distribution of interest. Nevertheless, it is of value to see how they can be obtained directly by considering the forces on our prototype of a point magnetic dipole, that is, a small current loop.

■ **Example**

Small current loop in an external induction. We consider a filamentary current I in a circle of radius a lying in the xy plane with origin at the center as shown in Figure 19-9. Its dipole moment as obtained from (19-27) is

$$\mathbf{m} = I \pi a^2 \hat{\mathbf{z}} \qquad (19\text{-}44)$$

The force on the loop in the external induction $\mathbf{B}_0$ is given by (14-3) as

$$\mathbf{F} = I \oint_C d\mathbf{s} \times \mathbf{B}_0(\mathbf{r}) \qquad (19\text{-}45)$$

It will be convenient to work with rectangular coordinates, because of the constant unit vectors, but we will express the components in terms of the azimuthal angle φ for ease in integration over C. Then we have $\mathbf{r} = x\hat{\mathbf{x}} + y\hat{\mathbf{y}} = a\cos\varphi\,\hat{\mathbf{x}} + a\sin\varphi\,\hat{\mathbf{y}}$ so that $d\mathbf{s} = d\mathbf{r} = a\,d\varphi(-\sin\varphi\,\hat{\mathbf{x}} + \cos\varphi\,\hat{\mathbf{y}})$ and

$$d\mathbf{s} \times \mathbf{B}_0 = a\,d\varphi\left[\hat{\mathbf{x}}B_{0z}\cos\varphi + \hat{\mathbf{y}}B_{0z}\sin\varphi - \hat{\mathbf{z}}(B_{0x}\cos\varphi + B_{0y}\sin\varphi)\right] \qquad (19\text{-}46)$$

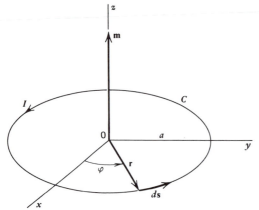

Figure 19-9.   Circular filamentary current loop.

if (1-28) is used. Since the loop is very small and $\mathbf{B}_0$ is slowly varying over it, it is appropriate to expand the components of $\mathbf{B}_0$ in a power series about the origin and keep only the terms that will give us a nonvanishing first approximation. Thus we can write

$$B_{0z}(\mathbf{r}) = B_{0z}(0) + x\left(\frac{\partial B_{0z}}{\partial x}\right) + y\left(\frac{\partial B_{0z}}{\partial y}\right) = B_{0z}(0) + a\left[\cos\varphi\left(\frac{\partial B_{0z}}{\partial x}\right) + \sin\varphi\left(\frac{\partial B_{0z}}{\partial y}\right)\right]$$

$$(19\text{-}47)$$

where we will remember that the derivatives are evaluated at the origin—the location of the dipole—and hence are constant as is $B_{0z}(0)$. There is no term $z(\partial B_{0z}/\partial z)$ in (19-47) because $z = 0$ since the loop lies in the xy plane. Substituting (19-47) into (19-46), we find the x component to be

$$(d\mathbf{s} \times \mathbf{B}_0)_x = a\,d\varphi\left\{ B_{0z}(0)\cos\varphi + a\left[\left(\frac{\partial B_{0z}}{\partial x}\right)\cos^2\varphi + \left(\frac{\partial B_{0z}}{\partial y}\right)\cos\varphi\sin\varphi\right]\right\}$$

$$(19\text{-}48)$$

The integration around the complete circuit C required by (19-45) is attained by integrating over φ from 0 to 2π. If we substitute (19-48) into (19-45), and use the appropriate integrals from the set

$$\int_0^{2\pi}\cos\varphi\,d\varphi = \int_0^{2\pi}\sin\varphi\,d\varphi = \int_0^{2\pi}\cos\varphi\sin\varphi\,d\varphi = 0$$

$$\int_0^{2\pi}\cos^2\varphi\,d\varphi = \int_0^{2\pi}\sin^2\varphi\,d\varphi = \pi$$

$$(19\text{-}49)$$

we find the x component of the force to be

$$F_x = I\oint_C (d\mathbf{s} \times \mathbf{B}_0)_x = I\pi a^2\left(\frac{\partial B_{0z}}{\partial x}\right) = m_z\left(\frac{\partial B_{0z}}{\partial x}\right) \qquad (19\text{-}50)$$

where we also used (19-44). Proceeding in exactly the same way, we find the remaining two components of the force to be

$$F_y = m_z\left(\frac{\partial B_{0z}}{\partial y}\right) \qquad (19\text{-}51)$$

$$F_z = -m_z\left[\left(\frac{\partial B_{0x}}{\partial x}\right) + \left(\frac{\partial B_{0y}}{\partial y}\right)\right] \qquad (19\text{-}52)$$

These do not yet look like the components of (19-39) but we have not used all of the information available to us. Since the external source currents are some distance from the loop, $\mathbf{J}_0 = 0$ at the loop and $\nabla \times \mathbf{B}_0 = 0$ by (15-12), and we find from (1-43) that $(\partial B_{0z}/\partial x) = (\partial B_{0x}/\partial z)$ and $(\partial B_{0z}/\partial y) = (\partial B_{0y}/\partial z)$. In addition, $\nabla \cdot \mathbf{B}_0 = 0$ by (16-3) so that, according to (1-42), $(\partial B_{0x}/\partial x) + (\partial B_{0y}/\partial y) = -(\partial B_{0z}/\partial z)$. Substituting these relations into (19-50)–(19-52), and recalling that $\mathbf{m}$ has only a z component by (19-44), we find that the force components can be written as

$$F_x = m_z\left(\frac{\partial B_{0x}}{\partial z}\right) = (\mathbf{m} \cdot \nabla)B_{0x}$$

$$F_y = m_z\left(\frac{\partial B_{0y}}{\partial z}\right) = (\mathbf{m} \cdot \nabla)B_{0y}$$

$$F_z = m_z\left(\frac{\partial B_{0z}}{\partial z}\right) = (\mathbf{m} \cdot \nabla)B_{0z}$$

with the use of (1-41) and (1-20). But these expressions are just the rectangular components of $\mathbf{F} = (\mathbf{m} \cdot \nabla)\mathbf{B}_0$, which is exactly (19-39).

A direct evaluation of the torque can be carried out in a similar manner. The force on a current element is given by (14-5) as $d\mathbf{F} = I\,d\mathbf{s} \times \mathbf{B}_0$ so that the torque on it will be $d\boldsymbol{\tau} = \mathbf{r} \times d\mathbf{F} = \mathbf{r} \times (I\,d\mathbf{s} \times \mathbf{B}_0)$. Summing this up over the whole circuit, we get

$$\boldsymbol{\tau} = \int d\boldsymbol{\tau} = I\oint_C [\mathbf{r} \times (d\mathbf{s} \times \mathbf{B}_0)] \tag{19-53}$$

From (1-30), we find that the term in brackets can be written as $d\mathbf{s}(\mathbf{r} \cdot \mathbf{B}_0) - \mathbf{B}_0(\mathbf{r} \cdot d\mathbf{s})$. Since $d\mathbf{s} = d\mathbf{r}$, we have $\mathbf{r} \cdot d\mathbf{s} = d(\frac{1}{2}\mathbf{r} \cdot \mathbf{r}) = d(\frac{1}{2}r^2)$; if we use the first-order approximation that $\mathbf{B}_0$ is constant, the contribution of the second term to (19-53) will be $-I\oint \mathbf{B}_0 d(\frac{1}{2}r^2) = -I\mathbf{B}_0\oint d(\frac{1}{2}r^2) = 0$ since the integral of the differential of a scalar over a closed path vanishes as we saw, for example, after (13-4). If we substitute the remainder of the bracketed term into (19-53), and use (1-124), we obtain

$$\boldsymbol{\tau} = I\oint_C (\mathbf{r} \cdot \mathbf{B}_0)\,d\mathbf{s} = I\int_S d\mathbf{a} \times \nabla(\mathbf{r} \cdot \mathbf{B}_0) \tag{19-54}$$

Since we are taking $\mathbf{B}_0$ to be a constant, it is just like $\mathbf{C}$ in the first part of (19-11) so that $\nabla(\mathbf{r} \cdot \mathbf{B}_0) = \mathbf{B}_0$. Inserting this into (19-54), taking $\mathbf{B}_0$ out of the integral, and using (19-27), we get

$$\boldsymbol{\tau} = \left(I\int_S d\mathbf{a}\right) \times \mathbf{B}_0 = I\mathbf{S} \times \mathbf{B}_0 = \mathbf{m} \times \mathbf{B}_0$$

exactly what we found in (19-42) from very general considerations. ∎

EXERCISES

19-1 A constant current I follows a closed path wrapped around a cylinder of radius a. In cylindrical coordinates, the position vector of a point on this circuit is given by $\mathbf{r} = a\hat{\boldsymbol{\rho}} + b\sin n\varphi\,\hat{\mathbf{z}}$ where b is a constant and n is a positive integer ≥ 2. Find the magnetic dipole moment $\mathbf{m}$ of this current distribution.

19-2 A plane circuit carrying a current I is constructed in the xy plane as follows. Cylindrical coordinates are used. Starting at the origin for $\varphi = 0$, we have $\rho = \rho_0\varphi^n$ where ρ_0 is a constant and $n > 1$. Thus a spiral is formed. This is continued until a value φ_0 of the angle is attained. Then the current follows a straight line back to the origin. Find the magnetic dipole moment of this current distribution.

19-3 A circular cylinder of radius a and length l has a total charge Q distributed uniformly throughout its volume. It is rotated about its axis with constant angular velocity $\boldsymbol{\omega}$. Assume that the charge distribution is not affected by the rotation and find the magnetic dipole moment of this system.

19-4 A dielectric sphere of radius a has a constant surface charge density σ on all parts of its surface. It is rotated about a diameter with constant angular velocity $\boldsymbol{\omega}$. Assume that the charge distribution is not affected by the rotation and find the magnetic dipole moment of this system.

19-5 A point dipole $\mathbf{m}$ is located at the origin, but it has no special orientation with respect to the coordinate axes. (For example, $\mathbf{m}$ is not parallel to any of the axes.) Express its potential $\mathbf{A}$ at a point $\mathbf{r}$ in rectangular coordinates, and find the rectangular components of $\mathbf{B}$. Show that $\mathbf{B}$ can be written in the form

$$\mathbf{B}(\mathbf{r}) = \frac{\mu_0}{4\pi r^3}[3(\mathbf{m} \cdot \hat{\mathbf{r}})\hat{\mathbf{r}} - \mathbf{m}] \tag{19-55}$$

and compare with (8-84).

19-6 Suppose the coaxial circles of Figure 17-14 are so far apart that they can be treated as dipoles. Find their mutual inductance and compare the result with that of Exercise 17-25.

19-7 Two circular loops of radii a and b lie in the same plane. Assume that the distance c between their centers is so large that the dipole approximation is appropriate and find their mutual inductance.

19-8 (a) A small rectangular loop of sides a and b, carrying a current I, lies in the xy plane with center at the origin. Find $\mathbf{A}$ at a point $\mathbf{r}$ where $r \gg a$ and $r \gg b$. (b) Using the result of (a), find $\mathbf{B}$ at this same point. (c) A point dipole $\mathbf{M} = M\hat{\mathbf{x}}$ is on the positive y axis a distance c from the origin where $c \gg a$ and $c \gg b$. Find the force and torque on $\mathbf{M}$ due to the loop. Express all of your answers in rectangular coordinates.

19-9 By using (19-35) show that there is no monopole contribution to the interaction energy analogous to that found in (8-62).

19-10 A point dipole $\mathbf{m}_1$ is located at $\mathbf{r}_1$ and another point dipole $\mathbf{m}_2$ is at $\mathbf{r}_2$. Either by direct calculation or by transcription of the corresponding results for the electric case, show that the potential energy of $\mathbf{m}_2$ in the induction of $\mathbf{m}_1$ is given by the *dipole-dipole interaction energy*

$$U'_{DD} = \frac{\mu_0}{4\pi R^3}\left[(\mathbf{m}_1 \cdot \mathbf{m}_2) - 3(\mathbf{m}_1 \cdot \hat{\mathbf{R}})(\mathbf{m}_2 \cdot \hat{\mathbf{R}})\right]$$

$$(19\text{-}56)$$

where $\mathbf{R} = \mathbf{r}_2 - \mathbf{r}_1$. Similarly, find the force $\mathbf{F}_2$ on $\mathbf{m}_2$.

19-11 Two dipoles $\mathbf{m}_1$ and $\mathbf{m}_2$ are in the same plane. The direction of $\mathbf{m}_1$ is held fixed, while $\mathbf{m}_2$ is free to rotate in the plane. If $\mathbf{m}_1$ and $\mathbf{m}_2$ make respective angles α_1 and α_2 with $\mathbf{R}$, find the relation satisfied by the angles when equilibrium is attained.

19-12 A circular ring of radius a lies in the xy plane with the origin at its center. It carries a current I circulating clockwise when viewed from a point on the positive z axis. A point dipole $\mathbf{m} = m\hat{\mathbf{z}}$ is on the z axis. Find the z component of the force on $\mathbf{m}$.

19-13 Consider the plane rectangular loop shown in Figure 17-4. Assume that the external $\mathbf{B}$ shown is uniform and that there is a constant current I about the loop in the positive sense determined by the direction of $\hat{\mathbf{n}}$. Find the torque that would have to be exerted by an external agent in order to prevent the loop from rotating.

19-14 How much work would an external agent have to do on the loop of the previous exercise in order to reverse the direction of $\hat{\mathbf{n}}$, that is, increase φ by 180°? Evaluate this in two different ways.

19-15 We could have begun the multipole expansion of the vector potential with the expression (16-11) that expresses $\mathbf{A}$ in terms of filamentary currents. Show that, if this is done, one again obtains (19-21) with $\mathbf{m}$ given by (19-28).

19-16 By analogy with Figure 8-5b, one can devise a simple model for the prototype of a magnetic quadrupole that consists of two small parallel loops with currents circulating in opposite senses and that are separated by a small distance. In particular, consider two magnetic dipoles of equal dipole moments $\pm m\hat{\mathbf{z}}$ located at $z = \pm a$; the total dipole moment of this system is zero. Show that, at large distances, the vector potential is given approximately by $\mathbf{A} = \hat{\boldsymbol{\varphi}}6\mu_0 ma \sin\theta \cos\theta/4\pi r^3$. Calculate the components of $\mathbf{B}$ and show that they are completely analogous to the linear electric quadrupole expressions given by (8-55).

20

MAGNETISM IN THE PRESENCE OF MATTER

Up to this point, we have assumed that when we have considered magnetic forces involving currents the regions involved either were vacuum or contained "nonmagnetic" conductors. Now we want to generalize our considerations to include matter of any type. As it was in Chapter 10, it will be helpful to consider the microscopic picture of matter as being a collection of atoms and molecules, that is, charged particles, in order to get an idea of how to proceed.

20-1 MAGNETIZATION

Previously, we assumed that atoms and molecules were composed of positive and negative charges and were, on the whole, electrically neutral. We now extend this picture by assuming that at least some of these charges are not at rest but are in continuous motion. Presumably these charges are moving in closed paths, the nature of which will be determined by the resultant structure of the atomic and molecular systems. From a great distance, these moving charges will appear as current whirls or magnetic dipoles. Such permanently circulating currents are often called *Ampèrian currents* since Ampère first postulated their existence in order to account for the magnetic properties of matter. We can now envisage several possibilities.

If $\mathbf{B} = 0$, it is possible that the charges are circulating in such a manner that the *net* current whirl vanishes, that is, the dipole moments of the individual charges, when added vectorially by (19-28), combine to give zero. Now let us suppose that things are arranged so that $\mathbf{B} \neq 0$, say by using external currents. We know from (19-42) that this will result in torques being exerted on the dipoles tending to align them in the direction of the induction. Under the influence of these torques, it may well happen that the paths of the moving charges will be changed enough so that in the resultant new configuration the net dipole moment will now be different from zero. Such a moment can be appropriately described as having been *induced* by the $\mathbf{B}$ field and we say that the matter has been *magnetized*.

In the absence of an external induction, it may be that the atom or molecule already has such a structure that the dipole moment associated with it is different from zero so that it has a *permanent magnetic dipole moment*. (Although it is not essential for our considerations now, it is of interest to note that such permanent dipoles *cannot* be accounted for on the basis of classical mechanics and electromagnetism alone, but are associated with the intrinsic angular momentum or "spin" of these systems that is itself essentially a consequence of quantum mechanics. However, it is not part of our task here to account for such permanent dipoles, but merely to accept their existence and to describe them in a macroscopic manner so that they can be included in our basic equations.)

Even if the material contains permanent dipoles, if $\mathbf{B} = 0$, it is still possible that they will be randomly oriented so that the *total* dipole moment of the piece of matter involved is zero, that is, it is *unmagnetized*. If, now $\mathbf{B} \neq 0$, there will be a torque on these dipoles that will tend to rotate them into alignment with the resultant induction. Generally, this aligning tendency will be offset by the randomizing processes associated with thermal agitation, such as collisions, but we can still expect that the net effect will be to produce a net dipole moment in the direction of the field and the material will again have become magnetized.

Some materials have the property that, even in the absence of a **B** field, the permanent dipoles are at least partially aligned and the material is said to be *permanently magnetized* or to be a *permanent magnet*. Such cases are generally of more technological importance than the corresponding electric cases.

As we saw in the last chapter, we can reasonably expect the molecules, especially when distorted by the presence of **B**, to have higher-order multipole moments whose effects would be described by (19-3). However, since the contributions of these higher-order terms to **A** and **B** drop off much more rapidly with distance and depend on angles in a more complicated way than those of dipoles, we assume that, for the purposes of describing the *average* properties of matter, the dominant features of interest to us are those associated with the dipole moments. Thus, all of our considerations lead us to the following:

■ **Hypothesis**

As far as its magnetic properties are concerned, neutral matter is equivalent to an assemblage of magnetic dipoles.   ■

We now have to put this hypothesis into quantitative terms. For this purpose, we define the *magnetization* **M** as the magnetic dipole moment per unit volume, so that the dipole moment $d\mathbf{m}$ in a small volume $d\tau$ at **r** will be

$$d\mathbf{m} = \mathbf{M}(\mathbf{r}) \, d\tau \tag{20-1}$$

Thus the total dipole moment of a volume V of material will be

$$\mathbf{m}_{\text{total}} = \int_V \mathbf{M}(\mathbf{r}) \, d\tau \tag{20-2}$$

From its definition and (19-20), we see that the unit of **M** will be 1 ampere/meter.

As usual, the definition (20-1) implies that $d\tau$ is large enough to include enough material so that **M** can be regarded as a smoothly varying function of position; at the same time, $d\tau$ must be small on a macroscopic scale. Because of the way in which we were led to introduce **M**, we expect that there will be some functional relation between **M** and **B**. We will consider this in more detail later, but for now, we take **M** as part of our macroscopic description of the material and we want to investigate the consequences of its existence.

20-2 MAGNETIZATION CURRENT DENSITIES

Let us assume that we have a magnetized object and we want to find the vector potential produced by it at a field point **r** outside of the body as shown in Figure 20-1. The dipole moment of the volume $d\tau'$ as given by (20-1) is $d\mathbf{m}' = \mathbf{M}(\mathbf{r}') \, d\tau'$ and its contribution to the vector potential at **r** as obtained from (19-21) is

$$d\mathbf{A} = \frac{\mu_0}{4\pi} \frac{d\mathbf{m}' \times \hat{\mathbf{R}}}{R^2} = \frac{\mu_0}{4\pi} \frac{\mathbf{M}(\mathbf{r}') \times \hat{\mathbf{R}} \, d\tau'}{R^2} \tag{20-3}$$

where, as usual, $\mathbf{R} = \mathbf{r} - \mathbf{r}'$ and corresponds to the **r** of (19-21) since **m** there was

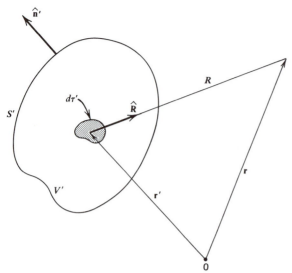

Figure 20-1. Calculation of the vector potential at a field point outside a magnetized object.

assumed to be at the origin. The total potential is obtained by integrating (20-3) over the volume V' of the material and is

$$\mathbf{A}(\mathbf{r}) = \int_{V'} \frac{\mu_0}{4\pi} \frac{\mathbf{M}(\mathbf{r}') \times \hat{\mathbf{R}} \, d\tau'}{R^2} = \frac{\mu_0}{4\pi} \int_{V'} \mathbf{M}(\mathbf{r}') \times \nabla'\left(\frac{1}{R}\right) d\tau' \qquad (20\text{-}4)$$

with the use of (1-141). We can now use (1-118) to rewrite the integrand as

$$\mathbf{M} \times \nabla'\left(\frac{1}{R}\right) = \frac{\nabla' \times \mathbf{M}}{R} - \nabla' \times \left(\frac{\mathbf{M}}{R}\right) \qquad (20\text{-}5)$$

If we now substitute this into (20-4) and use (1-123) and (1-52), we get

$$\mathbf{A}(\mathbf{r}) = \frac{\mu_0}{4\pi} \int_{V'} \frac{(\nabla' \times \mathbf{M}) \, d\tau'}{R} + \frac{\mu_0}{4\pi} \int_{V'} \left[-\nabla' \times \left(\frac{\mathbf{M}}{R}\right) \right] d\tau'$$

$$= \frac{\mu_0}{4\pi} \int_{V'} \frac{(\nabla' \times \mathbf{M}) \, d\tau'}{R} + \frac{\mu_0}{4\pi} \oint_{S'} \frac{\mathbf{M} \times \hat{\mathbf{n}}' \, da'}{R} \qquad (20\text{-}6)$$

where S' is the surface bounding the volume V' of the material and $\hat{\mathbf{n}}'$ is the outward normal as shown in Figure 20-1. Upon comparing this with (16-12) and (16-13), we see that this is exactly the vector potential that would be produced by a volume current density $\mathbf{J}_m$ distributed throughout the volume and a surface current density $\mathbf{K}_m$ on the bounding surface where

$$\mathbf{J}_m = \nabla' \times \mathbf{M} \qquad (20\text{-}7)$$

$$\mathbf{K}_m = \mathbf{M} \times \hat{\mathbf{n}}' \qquad (20\text{-}8)$$

for then we would have

$$\mathbf{A}(\mathbf{r}) = \frac{\mu_0}{4\pi} \int_{V'} \frac{\mathbf{J}_m(\mathbf{r}') \, d\tau'}{R} + \frac{\mu_0}{4\pi} \oint_{S'} \frac{\mathbf{K}_m(\mathbf{r}') \, da'}{R} \qquad (20\text{-}9)$$

as we would expect. (Compare with Equations 10-7 through 10-9.)

What we have found, therefore, is that, as far as its effects outside itself are concerned, the material can be *replaced* by a distribution of volume and surface current densities that are related to the magnetization $\mathbf{M}$ by means of (20-7) and (20-8).

The various steps which have led to our present conceptual scheme are summarized and outlined in Figure 20-2. (Compare to Figure 10-2.) The *total* vector potential at the field point will then be given by (20-6) *plus* that due to any other currents which might be present.

It is common practice to omit the primes on (20-7) and (20-8) and to simply write

$$\mathbf{J}_m = \nabla \times \mathbf{M} \quad \text{and} \quad \mathbf{K}_m = \mathbf{M} \times \hat{\mathbf{n}} \qquad (20\text{-}10)$$

with the understanding that the differentiation is with respect to source point coordinates and $\hat{\mathbf{n}}$ is the *outward* normal. We note that $\mathbf{M} \times \hat{\mathbf{n}}$ *is* tangent to the surface since it is perpendicular to $\hat{\mathbf{n}}$, as it should be to represent a surface current; this also shows that it is the tangential component of $\mathbf{M}$ that leads to the surface current. (It is essential not to forget that $\mathbf{K}_m$ is determined by the value of $\mathbf{M} \times \hat{\mathbf{n}}$ *at* the surface.)

The subscript m on these quantities reflects the fact that they are usually referred to as *magnetization current densities*. They are also often called *Ampèrian current densities* or *bound current densities*; we will not use the latter term, however, since in (12-18) we have already associated it with the displacement of bound charges as reflected in the polarization current density.

We have obtained these results by making a formal comparison of the expression (20-6) with its general form (20-9). It is also possible to understand and calculate these currents in a direct "physical" manner; we will, however, do this only qualitatively. As an extreme example, let us consider a piece of material with uniform magnetization as produced by dipoles of the same magnitude all aligned, that is, circulating in the same sense as shown in an end-on view in Figure 20-3. If we consider the immediate

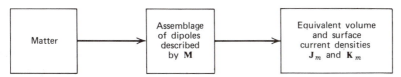

Figure 20-2. Conceptual scheme for replacing matter by equivalent current densities.

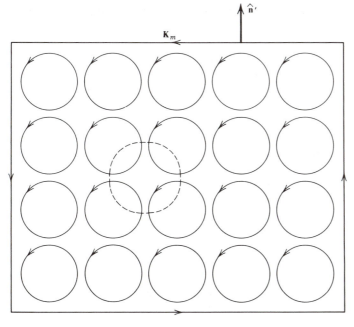

Figure 20-3. Origin of magnetization surface currents for a uniform magnetization.

neighborhood of an interior point, as indicated by the dashed line, we see that the current due to one whirl in one direction is canceled by the oppositely directed currents in adjacent whirls. Thus, in the interior of a uniformly magnetized material, the magnetization current is zero in agreement with (20-7). However, at the surface there are no adjacent currents to produce a cancellation, and, since the currents in the whirls are all circulating in the same sense, the result in effect is that of a current $\mathbf{K}_m$ circulating on the surface as indicated by the arrow. Since $\mathbf{M}$ is directed out of the page in this case, the direction of $\mathbf{K}_m$ as given by (20-8) is exactly the same as that just deduced by consideration of the current whirls.

Now let us suppose that $\mathbf{M}$ is not uniform throughout the material; for example, the current of each whirl could be different. Then, if we reconsider the region enclosed by the dashed line, we see that there will no longer generally be a complete cancellation by the oppositely directed adjacent currents and there will be a resultant current in the interior. This is exactly the effect described by (20-7) and, in fact, this picture can be used in a quantitative manner as an alternative way of deriving $\mathbf{J}_m = \nabla' \times \mathbf{M}$. There is still a surface current $\mathbf{K}_m$ in this case of nonuniform $\mathbf{M}$ of course.

Suppose that we have two magnetized materials meeting at a common boundary as illustrated in Figure 20-4. Since each of them has a tangential component at the surface, they will each produce a surface current given by (20-8). The net surface current density will be the sum of these two terms so that

$$\mathbf{K}_{m\,\text{net}} = \mathbf{K}_{m1} + \mathbf{K}_{m2} = \mathbf{M}_1 \times \hat{\mathbf{n}}'_1 + \mathbf{M}_2 \times \hat{\mathbf{n}}'_2 \tag{20-11}$$

where $\hat{\mathbf{n}}'_1$ and $\hat{\mathbf{n}}'_2$ are the *outward* normals to the respective media. If we now introduce the normal $\hat{\mathbf{n}}$ drawn from 1 to 2 by our usual convention, we see from the figure that $\hat{\mathbf{n}}'_1 = \hat{\mathbf{n}}$ and $\hat{\mathbf{n}}'_2 = -\hat{\mathbf{n}}$ so that (20-11) becomes

$$\mathbf{K}_{m\,\text{net}} = (\mathbf{M}_1 - \mathbf{M}_2) \times \hat{\mathbf{n}} = \hat{\mathbf{n}} \times (\mathbf{M}_2 - \mathbf{M}_1) \tag{20-12}$$

As an illustration of the internal consistency of our results, let us consider a piece of magnetized material of finite volume and find the total rate at which bound charge is being transferred across a plane passing through the material as shown in Figure 20-5a. Let S be the area of the plane intercepted by the material, $\hat{\mathbf{n}}$ the normal to it, and C its bounding curve as shown in (b) of the figure. The total rate at which charge is being transferred across the plane, that is, from the shaded part of the volume on the right to the unshaded volume on the left, is given by (12-6), (12-8), and Figure 12-5 as

$$\frac{dq}{dt} = \int_S \mathbf{J}_m \cdot d\mathbf{a} + \oint_C \mathbf{K}_m \cdot \hat{\mathbf{t}}\,ds \tag{20-13}$$

where $\hat{\mathbf{t}}$ is drawn perpendicular to ds as required. Substituting (20-10) into this, and

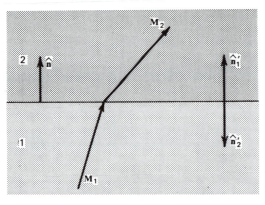

Figure 20-4. Boundary between two magnetized materials.

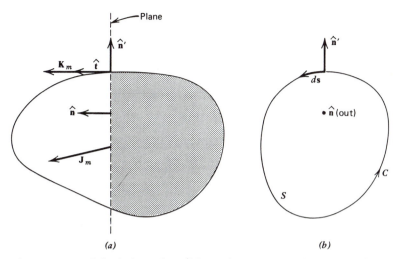

Figure 20-5. Calculation of total bound current passing across the plane.

using (1-67) and (1-29), we find that

$$\frac{dq}{dt} = \oint_C \mathbf{M} \cdot d\mathbf{s} + \oint_C (\mathbf{M} \times \hat{\mathbf{n}}') \cdot \hat{\mathbf{t}} \, ds = \oint_C \mathbf{M} \cdot d\mathbf{s} + \oint_C \mathbf{M} \cdot (\hat{\mathbf{n}}' \times \hat{\mathbf{t}}) \, ds \quad (20\text{-}14)$$

where $\hat{\mathbf{n}}'$ is the normal to the surface of the material at the point where $\mathbf{K}_m$ is to be evaluated. But we see from the figure that $(\hat{\mathbf{n}}' \times \hat{\mathbf{t}}) \, ds = -d\mathbf{s}$ so that the two integrals in (20-14) cancel and $dq/dt = 0$. Thus the magnetization currents do not transfer net charge. This is just what would be expected, for otherwise a piece of magnetized material would "spontaneously" have its charges separated, which is incompatible with our picture of magnetization as arising from a *reorientation* of existing current whirls.

The induction produced outside the material can be found from (20-6) by using $\mathbf{B} = \nabla \times \mathbf{A}$. Alternatively, we can use the current densities as found from (20-7) and (20-8) in (14-7) and (14-11) to give

$$\mathbf{B}(\mathbf{r}) = \frac{\mu_0}{4\pi} \int_{V'} \frac{\mathbf{J}_m(\mathbf{r}') \times \hat{\mathbf{R}} \, d\tau'}{R^2} + \frac{\mu_0}{4\pi} \oint_{S'} \frac{\mathbf{K}_m(\mathbf{r}') \times \hat{\mathbf{R}} \, da'}{R^2} \quad (20\text{-}15)$$

So far all of our results have been obtained by considering the vector potential, and its corresponding induction, at a field point in the vacuum region outside the matter. Here there is no difficulty in knowing what $\mathbf{B}$ is and we could imagine measuring it by finding the force on a moving point charge or the torque on a small current loop. What about the situation *inside* the material? We cannot easily measure the torque on a current loop without first drilling a hole into the material in order to insert the loop and we can anticipate that this may alter the situation since in the process we will certainly remove some volume currents and introduce some new surface currents. We faced a similar problem in the electric case and discussed it at length in Section 10-3. We could go through the same arguments that we used there, with appropriate changes of terms and symbols, and we will be led to exactly the same conclusion: the only practical and reasonable thing to do is to be consistent with our conceptual replacement of the matter by the set of equivalent magnetization current densities and assume *by definition* that (20-9) and (20-15) can be used to calculate $\mathbf{A}$ and $\mathbf{B}$ anywhere.

Nevertheless, we can still ask if it is possible to give a *cavity definition* of $\mathbf{B}$ in the same sense that we did for $\mathbf{E}$ in (10-26) and $\mathbf{D}$ after (10-45). In other words, we want to be able to cut a hole of appropriate shape within the material so that we can insert, say,

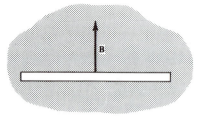

Figure 20-6.   Cavity used to measure B in the material.

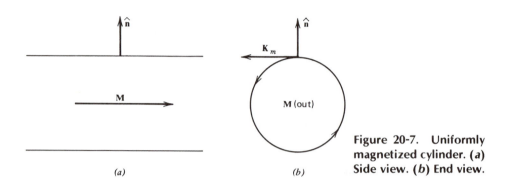

(a)

(b)

Figure 20-7.   Uniformly magnetized cylinder. (*a*) Side view. (*b*) End view.

a small current loop into the cavity and find the value of **B** in the material from a measurement of the torque on the loop in the vacuum inside the cavity. As before, we can use our knowledge of the boundary conditions satisfied by **B**; the relevant one is that the normal components of **B** are continuous as given by (16-4). But this is precisely the same boundary condition that we used to obtain the cavity definition for **D** after (10-45) so that we see exactly what is to be done. The appropriate cavity is thus a short right cylinder cut in the material with its base perpendicular to the direction of **B** there as shown in Figure 20-6. Since, by construction, only normal components are involved, we see that the value of **B** in the cavity equals that in the material, that is, $\mathbf{B}_c = \mathbf{B}$.

Finally, let us consider a simple example illustrating our results so far. We discuss a more complicated one in the next section.

■ **Example**

Infinitely long uniformly magnetized cylinder. Suppose we have an infinitely long cylinder of circular cross section that has a uniform magnetization **M** parallel to the axis of the cylinder as shown in Figure 20-7*a*. We can find the magnetization currents from (20-10). Since **M** = const., we find that $\mathbf{J}_m = \nabla \times \mathbf{M} = 0$ and there are no volume currents. Figure 20-7 shows an outward normal **n̂**, which is perpendicular to **M**, and we see that the surface current has a constant magnitude $K_m = M$ and circulates about the cylinder in the sense shown in (*b*) of the figure. But a surface current like this is exactly the equivalent of an infinitely long ideal solenoid as illustrated in Figure 15-11 and for which we have already calculated the induction. The magnitude of **B** inside is given by (15-24) and therefore in this case we have $B_i = \mu_0 K_m = \mu_0 M$ = const., while $\mathbf{B}_o = 0$ outside. Since $\mathbf{B}_i$ is in the axial direction, as is **M**, we can finally write

$$\mathbf{B}_i = \mu_0 \mathbf{M} \qquad (20\text{-}16)$$

as the complete solution to the problem.   ■

20-3 UNIFORMLY MAGNETIZED SPHERE

Let us now consider a sphere of radius a that has a constant magnetization **M**. We choose the z axis in the direction of **M** and the origin at the center of the sphere as shown in Figure 20-8; thus $\mathbf{M} = M\hat{z}$. Since **M** is constant, $\mathbf{J}_m = 0$ by (20-7). We see from the figure that the outer normal $\hat{n}' = \hat{r}'$ and therefore the surface current density as found from (20-8) is

$$\mathbf{K}_m = M\hat{z} \times \hat{r}' = M \sin \theta' \hat{\varphi}' \tag{20-17}$$

with the use of (1-94) and (1-92). Thus, in this case, the surface currents are along the lines of "latitude" on the sphere with a magnitude that is zero at the "poles" and a maximum at the "equator" as shown in Figure 20-9. If we insert (20-17) into (20-15), and use (1-100) to write $da' = a^2 \sin \theta' \, d\theta' \, d\varphi'$, while $\hat{\mathbf{R}} = \mathbf{R}/R$, we find **B** to be given in general by

$$\mathbf{B}(\mathbf{r}) = \frac{\mu_0 M a^2}{4\pi} \int_0^{2\pi} \int_0^{\pi} \frac{(\hat{\varphi}' \times \mathbf{R}) \sin^2 \theta' \, d\theta' \, d\varphi'}{R^3} \tag{20-18}$$

For simplicity in this example, we evaluate (20-18) only for a field point on the positive z axis; later, we will solve this problem completely by a quite different method. We see from Figure 20-10 that $\mathbf{r} = z\hat{z}$, $\mathbf{r}' = a\hat{r}'$, $\mathbf{R} = z\hat{z} - a\hat{r}'$, $R = (z^2 + a^2 - 2za \cos \theta')^{1/2}$ and $\hat{\varphi}' \times \mathbf{R} = z \sin \theta' \hat{r}' + (z \cos \theta' - a)\hat{\theta}' = a \sin \theta' \hat{z} + (z - a \cos \theta')(\hat{x} \cos \varphi' + \hat{y} \sin \varphi')$ with the use of (1-94), (1-92), and (1-93). If we insert these into (20-18), we see that the $\hat{x}$ and $\hat{y}$ components will vanish on integrating over φ' because of (19-49) so that **B** has only a z component, as is evident from the symmetry of the current distribution shown in Figure 20-9. Therefore, the only nonvanishing component is

$$B_z(z) = \frac{\mu_0 M a^3}{4\pi} \int_0^{2\pi} \int_0^{\pi} \frac{\sin^3 \theta' \, d\theta' \, d\varphi'}{(z^2 + a^2 - 2za \cos \theta')^{3/2}} \tag{20-19}$$

The integration over φ' gives a value of 2π; if we use $\sin^2 \theta' = 1 - \cos^2 \theta'$, and change

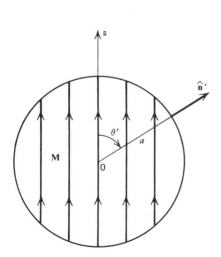

Figure 20-8. Uniformly magnetized sphere.

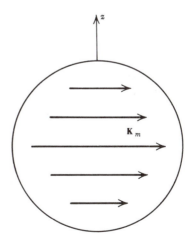

Figure 20-9. Equivalent surface currents of uniformly magnetized sphere.

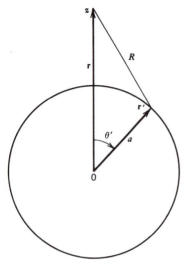

Figure 20-10.   Calculation of the induction at a point on the axis.

the variable of integration by means of (2-22), we find that this becomes

$$B_z(z) = \frac{\mu_0 M a^3}{2} \int_{-1}^{1} \frac{(1 - \mu^2) \, d\mu}{(z^2 + a^2 - 2za\mu)^{3/2}} \tag{20-20}$$

The integral can be found with the use of tables to be

$$-\frac{2(z^2 + a^2 - 2za\mu)^{1/2}}{3z^3 a^3} \left[z^2 + a^2 + za\mu + \frac{3z^2 a^2(\mu^2 - 1)}{2(z^2 + a^2 - 2za\mu)} \right] \Bigg|_{-1}^{1}$$

$$= \frac{2}{3z^3 a^3} \left\{ (z^2 + a^2)[|z + a| - |z - a|] - za[|z + a| + |z - a|] \right\} \tag{20-21}$$

and, as usual, there are two cases to be considered.

1. Outside the sphere. Here $z > a$, so that $|z - a| = z - a$ and (20-21) becomes $4/3z^3$ which, when put into (20-20), gives the induction outside the sphere to be

$$B_{zo}(z) = \frac{2\mu_0 M a^3}{3z^3} \tag{20-22}$$

This result becomes more understandable if we express it in terms of the total dipole moment of the sphere that from (20-2) is

$$\mathbf{m} = \tfrac{4}{3}\pi a^3 \mathbf{M} \tag{20-23}$$

so that (20-22) can also be written as

$$B_{zo}(z) = \frac{\mu_0}{4\pi} \frac{2m}{z^3} \tag{20-24}$$

Upon comparing this result with (19-24), and remembering that $\theta = 0$ and $r = z$ for a field point on the z axis, we see that (20-24) is exactly that of a point dipole of total moment m.

2. Inside the sphere. Here $z < a$, so that $|z - a| = a - z$ and (20-21) becomes $4/3a^3$ which, when put into (20-20), gives the induction inside the sphere to be

$$B_{zi}(z) = \tfrac{2}{3}\mu_0 M \tag{20-25}$$

which is constant and parallel to the magnetization. The numerical factor of $\tfrac{2}{3}$ is

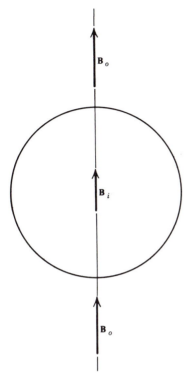

Figure 20-11. Magnetic induction on axis due to a uniformly magnetized sphere.

different from unity as found for the cylinder in (20-16) even though both of these systems have uniform magnetization; the difference in numerical factors is evidently due to the difference in geometry.

It will be left as an exercise to show that these same results also hold for negative values of z, that is, B_{z_o} is always in the positive direction and B_{z_i} is constant and given by (20-25) for all z. In Figure 20-11 we illustrate these directions for **B** that we have found. We recall that these results are similar to those we found for **E** for the uniformly polarized sphere in Section 10-4. This leads us to suspect that similar things can be said about this problem, that is, that the induction *everywhere outside* is a dipole field corresponding to the total dipole moment given by (20-23), while everywhere inside, $\mathbf{B}_i$ is constant and given by $\frac{2}{3}\mu_0\mathbf{M}$. This actually turns out to be the case as we will show in a quite different manner later in this chapter.

At the surface of the sphere ($z = a$), (20-22) gives $B_{z_o}(a) = \frac{2}{3}\mu_0 M = B_{z_i}(a)$; this is as it should be since both of these are normal components and should be equal by (16-4).

20-4 THE H FIELD

We recall that when we defined the magnetic induction in Section 14-1, we pointed out that **B**, by its definition, is determined by currents of *all* kinds. Thus, in $\nabla \times \mathbf{B} = \mu_0\mathbf{J}$, given by (15-12), **J** represents the *total* current density. In (20-10), we found a current density $\mathbf{J}_m = \nabla \times \mathbf{M}$, which is associated with the presence of matter and, as we found it useful to do for charges in Section 10-5, it is convenient to divide currents arising from moving charges into the two broad classes of *magnetization currents* and *free currents* described by the respective densities $\mathbf{J}_m$ and $\mathbf{J}_f$. The magnetization currents are to be associated with the constituents of matter and generally speaking we do not have any real control over them. As we discussed near the end of Section 12-2, free currents are those over which we can exert some control, for example by sending currents through wires with the use of batteries or by using convection currents in the form of streams of charged particles. Thus we can write the total current density as the sum of these two:

$$\mathbf{J}_{total} = \mathbf{J} = \mathbf{J}_f + \mathbf{J}_m \tag{20-26}$$

Inserting this into (15-12) and using (20-10), we find that $\nabla \times \mathbf{B} = \mu_0 \mathbf{J} = \mu_0 (\mathbf{J}_f + \nabla \times \mathbf{M})$ or

$$\nabla \times \left(\frac{\mathbf{B}}{\mu_0} - \mathbf{M} \right) = \mathbf{J}_f \tag{20-27}$$

The form of this equation, in which only the free current density appears on the right-hand side, suggests that it will be useful to define a new vector field $\mathbf{H(r)}$ by

$$\mathbf{H} = \frac{\mathbf{B}}{\mu_0} - \mathbf{M} \tag{20-28}$$

for then (20-27) becomes

$$\nabla \times \mathbf{H} = \mathbf{J}_f \tag{20-29}$$

The vector $\mathbf{H}$ is called the *magnetic field* or sometimes the $\mathbf{H}$ *field*. The principal characteristic of $\mathbf{H}$ and the primary reason for its introduction is that its curl depends only on the free current density. The dimensions of $\mathbf{H}$ are the same as those of $\mathbf{M}$ and thus $\mathbf{H}$ will be measured in ampere/meter. We can think of (20-29) as expressing Ampère's law of force between current elements *plus* the magnetic effects of matter.

Now that we have defined $\mathbf{H}$ we can easily determine some of its properties. The behavior of its tangential components at a surface of discontinuity in properties can be found by inserting the source equation (20-29) into (9-13) and (9-18) and using the analogue of (15-14) for free currents; the result is that we can express the boundary conditions in the two equivalent ways

$$\hat{\mathbf{n}} \times (\mathbf{H}_2 - \mathbf{H}_1) = \mathbf{K}_f \tag{20-30}$$

$$\mathbf{H}_{2t} - \mathbf{H}_{1t} = \mathbf{K}_f \times \hat{\mathbf{n}} \tag{20-31}$$

where $\mathbf{K}_f$ is the free surface current density. Thus, there will be a discontinuity in the tangential components of $\mathbf{H}$ only if there is a free (e.g., conduction) surface current density. This is in contrast to $\mathbf{B}$ whose tangential components will be discontinuous if there is a surface current density of any kind as we saw in (15-15) and (15-16).

The integral form of Ampère's law for $\mathbf{H}$ can be found by using (20-29), (1-67), and (12-6); the result is that

$$\oint_C \mathbf{H} \cdot d\mathbf{s} = \int_S \mathbf{J}_f \cdot d\mathbf{a} = I_{f,enc} \tag{20-32}$$

where $I_{f,enc}$ is the net free current passing through the surface S enclosed by the arbitrary path of integration C.

Equation 20-29 can be used to devise a cavity definition of $\mathbf{H}$ so that one can contemplate finding the value of $\mathbf{H}$ in the material from measurements done in the vacuum inside an appropriately shaped cavity. We assume that there are no free currents within the material. Then $\nabla \times \mathbf{H} = 0$ and its tangential components are continuous by (20-31). This is exactly the type of boundary condition that was the basis of the cavity definition of $\mathbf{E}$ given by (10-26) and we see at once what we should do. As shown in Figure 20-12, we imagine a long needlelike cavity cut into the material with its

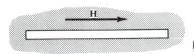

Figure 20-12. Cavity used to measure H in the material.

axis parallel to the direction of **H** there. Thus, by construction, only tangential components are involved, and since they are continuous, the magnetic field in the cavity $\mathbf{H}_c$ equals that in the material, $\mathbf{H}_c = \mathbf{H}$. Now the induction $\mathbf{B}_c$ in the cavity can be measured by determining the force on a moving charge or the torque on a small current loop. Since there is no matter in the cavity, $\mathbf{M}_c = 0$, and we see from (20-28) that $\mathbf{H}_c = \mathbf{B}_c/\mu_0$ and therefore, the field in the material is given by

$$\mathbf{H} = \mathbf{H}_c = \frac{\mathbf{B}_c}{\mu_0} \tag{20-33}$$

As the Helmholtz theorem of Section 1-20 tells us, we need one more differential source equation for the magnetic field, that is, its divergence. This is easily found by combining (20-28) with (16-3) and we see that $\nabla \cdot \mathbf{B} = 0 = \nabla \cdot [\mu_0(\mathbf{H} + \mathbf{M})] = \mu_0(\nabla \cdot \mathbf{H} + \nabla \cdot \mathbf{M})$ and therefore

$$\nabla \cdot \mathbf{H} = -\nabla \cdot \mathbf{M} \tag{20-34}$$

showing us that **H** *can* have sources associated with the Ampèrian currents of matter as well as free currents. We will return to this point later.

The boundary conditions satisfied by the normal components of **H** can be most easily obtained from the fact that the normal components of **B** are continuous, and when we substitute (20-28) into (16-4) we find that

$$\hat{\mathbf{n}} \cdot (\mathbf{H}_2 - \mathbf{H}_1) = -\hat{\mathbf{n}} \cdot (\mathbf{M}_2 - \mathbf{M}_1) \tag{20-35}$$

or

$$H_{2n} - H_{1n} = -(M_{2n} - M_{1n}) \tag{20-36}$$

In spite of the apparent simplicity of our results, particularly (20-29), they are not too useful at this point since we have to be able to relate the three vectors **B**, **M**, and **H**. Nevertheless, we can look again at some of our previous examples.

■ **Example**

Infinitely long ideal solenoid. Here we have free currents producing the fields. Since there is a vacuum everywhere else, $\mathbf{M} = 0$ and $\mathbf{H} = \mathbf{B}/\mu_0$. Thus we can use our previous results (15-25) and (15-26) and we see that $\mathbf{H}_o = 0$ while inside the solenoid

$$\mathbf{H}_i = \frac{\mathbf{B}_i}{\mu_0} = nI\hat{\mathbf{z}} \tag{20-37}$$

where n is the number of turns per meter and $\hat{\mathbf{z}}$ is along the solenoid axis. (This result is basically the reason that **H** is often given in the units of ampere-*turns*/meter.) Here we have a discontinuity in the tangential components of **H**; if we take $\hat{\mathbf{n}}$ to point from the inside (1) to the outside (2), and write $\mathbf{K}_f = nI\hat{\boldsymbol{\varphi}}$ from (15-22) and Figure 15-11, (20-31) becomes $\mathbf{H}_{2t} - \mathbf{H}_{1t} = 0 - \mathbf{H}_i = nI\hat{\boldsymbol{\varphi}} \times \hat{\mathbf{n}} = -nI\hat{\mathbf{z}}$ and therefore $\mathbf{H}_i = nI\hat{\mathbf{z}}$ in agreement with (20-37); we note that the use of the boundary conditions in this way provides us with a quick way of finding $\mathbf{H}_i$ in this case. ■

■ **Example**

Infinitely long uniformly magnetized cylinder. Here there are *no* free currents. We solved this problem for **B** above with the result that $\mathbf{B}_o = 0$ while $\mathbf{B}_i = \mu_0\mathbf{M}$. Outside the cylinder, $\mathbf{M}_o = 0$ and therefore $\mathbf{H}_o = 0$. Inside the cylinder, we find from (20-28) that $\mathbf{H}_i = (\mathbf{B}_i/\mu_0) - \mathbf{M}_i = \mathbf{M} - \mathbf{M} = 0$. Thus $\mathbf{H} = 0$ *everywhere* in spite of the resemblance of this case to the ideal solenoid as we found before. The real difference between these two examples is that the first one has free currents while the second one does not. Since $\mathbf{M}$ is uniform, $\nabla \cdot \mathbf{M} = 0$ and $\nabla \cdot \mathbf{H} = 0$ by (20-34); $\mathbf{M}$ does not have any normal components either so that our results are also consistent with (20-36). [If the cylinder were not infinitely long, there would be a discontinuity in the normal components of $\mathbf{M}$ at the ends where the surface separates the matter from the vacuum. Then, from (20-36), we can expect there would be sources of $\mathbf{H}$ and that $\mathbf{H} \neq 0$ for a cylinder of finite length.] ∎

■ **Example**

Uniformly magnetized sphere. There are also no free currents in this example. In the previous section, we found the values of $\mathbf{B}$ along the axis by a direct calculation from the magnetization currents. We can use these results to find $\mathbf{H}$ by means of (20-28). Outside the sphere, $\mathbf{M}_o = 0$ and (20-24) and (20-22) give us

$$H_{zo}(z) = \frac{B_{zo}}{\mu_0} - 0 = \frac{1}{4\pi} \frac{2m}{z^3} = \frac{2Ma^3}{3z^3} \tag{20-38}$$

Inside the sphere, $\mathbf{M}_i = M\hat{\mathbf{z}}$ and (20-25) yields

$$H_{zi}(z) = \frac{B_{zi}}{\mu_0} - M = -\frac{1}{3}M \tag{20-39}$$

Thus, in this case, $\mathbf{H} \neq 0$ both inside of and outside of the sphere and, in fact, is oppositely directed to both $\mathbf{B}$ and $\mathbf{M}$ inside the sphere as illustrated in Figure 20-13. At the surface of the sphere ($z = a$), (20-38) gives $H_{zo}(a) = \frac{2}{3}M$ so that there is a

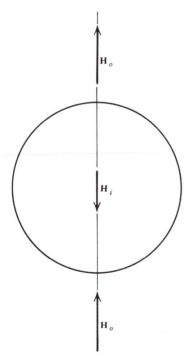

Figure 20-13. Magnetic field on axis due to a uniformly magnetized sphere.

discontinuity in the normal components of **H**. But if we take $\hat{n} = \hat{z}$ in this case, (20-36) becomes $H_{2n} - H_{1n} = H_{zo}(a) - H_{zi}(a) = \frac{2}{3}M - (-\frac{1}{3}M) = M = -(0 - M) = -(M_{2n} - M_{1n})$ exactly as it should. ∎

Both of these last two examples were characterized by the *absence* of free currents and the *presence* of a *uniform* magnetization, yet in one case **H** = 0 while **H** ≠ 0 for the other. What is the real difference between them? Although $\nabla \cdot \mathbf{M} = 0$ in the interior for both of these examples, they do differ in that there are clearly discontinuities in the normal components of **M** for the sphere, while, as we remarked above, this is not the case for the infinitely long cylinder, and the fact that **H** ≠ 0 for the sphere must somehow be related to this difference; we would like to be able to describe this somewhat more systematically and conveniently than by (20-36) alone. In addition, there is the nagging resemblance between these examples and those we discussed for the electrostatic case in Chapter 10, particularly for the uniformly magnetized and polarized spheres, and one feels that these problems are somehow related. This is, in fact, the case.

In the absence of free currents, our results (20-29), (20-31), (20-34), and (20-35) become

$$\nabla \times \mathbf{H} = 0 \qquad \nabla \cdot \mathbf{H} = -\nabla \cdot \mathbf{M}$$
$$\mathbf{H}_{2t} - \mathbf{H}_{1t} = 0 \qquad \hat{n} \cdot (\mathbf{H}_2 - \mathbf{H}_1) = \hat{n} \cdot \mathbf{M}_1 \tag{20-40}$$

where, for simplicity, we take region 2 to be a vacuum so that $\mathbf{M}_2 = 0$. Similarly, *in the absence of free charge,* our electrostatic results (5-4), (9-21), (10-39), (10-10), (9-26), and (10-12) can be written in the form

$$\nabla \times (\epsilon_0 \mathbf{E}) = 0 \qquad \nabla \cdot (\epsilon_0 \mathbf{E}) = -\nabla \cdot \mathbf{P} = \rho_b$$
$$(\epsilon_0 \mathbf{E}_{2t}) - (\epsilon_0 \mathbf{E}_{1t}) = 0 \qquad \hat{n} \cdot (\epsilon_0 \mathbf{E}_2 - \epsilon_0 \mathbf{E}_1) = \hat{n} \cdot \mathbf{P}_1 = \sigma_b \tag{20-41}$$

We also recall that **P** and **M** each represent their corresponding dipole moment per unit volume. Upon comparing these two sets of equations, we see that *by analogy* we can define magnetic analogues of charge densities by

$$\rho_m = -\nabla \cdot \mathbf{M} \qquad \sigma_m = \hat{n} \cdot \mathbf{M} \tag{20-42}$$

so that two of the equations of the set (20-40) become

$$\nabla \cdot \mathbf{H} = \rho_m \qquad \hat{n} \cdot (\mathbf{H}_2 - \mathbf{H}_1) = \sigma_m \tag{20-43}$$

and now look just like the corresponding members of (20-41). Thus we can call ρ_m the *volume density of magnetic charges* ("poles") and σ_m the *surface density of magnetic charges* ("poles") and regard them as the source of the magnetic field **H** and they will play the same role as did the corresponding electric charges in the calculation of $\epsilon_0 \mathbf{E}$. As we saw in Section 16-1, there are actually no magnetic charges so that the densities defined by (20-42) represent *fictitious charges*; however, this does not prevent them from being useful in discussing this restricted type of problem.

This analogy can be pushed even further. We recall that the fact that $\nabla \times \mathbf{E} = 0$ followed from our being able to write **E** in terms of the electrostatic scalar potential in (5-3), that is, $\mathbf{E} = -\nabla \phi$. Similarly, the fact that $\nabla \times \mathbf{H} = 0$ here enables us by using (1-48) to introduce a *magnetic scalar potential* ϕ_m so that

$$\mathbf{H} = -\nabla \phi_m \tag{20-44}$$

If we now substitute this into (20-43) and use (1-45), we find that ϕ_m satisfies Poisson's equation

$$\nabla^2 \phi_m = -\rho_m \tag{20-45}$$

while, in regions where ρ_m is zero, it satisfies Laplace's equation

$$\nabla^2 \phi_m = 0 \qquad (20\text{-}46)$$

in analogy with (11-1) and (11-3), if the former is written as $\nabla^2(\epsilon_0 \phi) = -\rho = -\rho_b$ since we are only considering cases in which $\rho_f = 0$. The uniqueness theorem of Section 11-1 that we found for ϕ will also hold for ϕ_m, since the only requisite was that the function involved satisfy Laplace's equation.

Thus what we have found is that many magnetostatic problems can be solved exactly as we did some electrostatic problems, and therefore many of the methods we developed in Chapter 11 can also be applied here. Furthermore, if we have already solved the analogous electrostatic problem, we can simply take over the solution by making the replacements $\epsilon_0 \mathbf{E} \rightarrow \mathbf{H}$, $\epsilon_0 \phi \rightarrow \phi_m$, $\mathbf{P} \rightarrow \mathbf{M}$, and so on. For example, we can write down the integral from which ϕ_m can be found from a given distribution of magnetization by changing (5-7) and (5-8) in this way:

$$\phi_m(\mathbf{r}) = \frac{1}{4\pi} \int_{V'} \frac{\rho_m(\mathbf{r}') \, d\tau'}{R} + \frac{1}{4\pi} \int_{S'} \frac{\sigma_m(\mathbf{r}') \, da'}{R}$$

$$= -\frac{1}{4\pi} \int_{V'} \frac{\nabla' \cdot \mathbf{M} \, d\tau'}{R} + \frac{1}{4\pi} \int_{S'} \frac{(\hat{\mathbf{n}}' \cdot \mathbf{M}) \, da'}{R} \qquad (20\text{-}47)$$

■   **Example**

Uniformly magnetized sphere. Let us reconsider this example from this new point of view. Since $\mathbf{M} = M\hat{\mathbf{z}} = $ const., $\rho_m = -\nabla \cdot \mathbf{M} = 0$ and there is no volume pole density. However, since $\hat{\mathbf{n}}$ in (20-42) is the $\hat{\mathbf{n}}' = \hat{\mathbf{r}}'$ of Figure 20-8, we see that there is a generally nonzero surface pole density given by

$$\sigma_m = M\hat{\mathbf{z}} \cdot \hat{\mathbf{r}}' = M \cos \theta' \qquad (20\text{-}48)$$

and we see that this is exactly of the form found in (10-27) for the bound surface charge density of the uniformly polarized sphere, and which is illustrated in Figure 10-9. Now we can see that the sources of $\mathbf{H}$ in this example can be ascribed to the discontinuities in the normal components of $\mathbf{M}$ at the surface of the sphere and, on comparing Figures 20-13 and 10-11, we can see why $\mathbf{H}_i$ is oppositely directed to $\mathbf{M}$ in the interior. In the last example of Section 11-5, we found the complete solution (11-133) for the uniformly polarized sphere, and we found that the electric field everywhere outside was a dipole field corresponding to the total dipole moment of the sphere, while it was uniform everywhere inside and given by (11-134) as $\mathbf{E}_i = -(\mathbf{P}/3\epsilon_0)$. Now we see that exactly the same things can be said about the uniformly magnetized sphere: outside, the field is a dipole one corresponding to the moment of (20-23) and the magnetic field vectors everywhere inside will be constant and equal to

$$\mathbf{B}_i = \tfrac{2}{3}\mu_0 \mathbf{M} \qquad \mathbf{H}_i = -\tfrac{1}{3}\mathbf{M} \qquad (20\text{-}49)$$

in agreement with the results (20-25) and (20-39) found for points on the axis.   ■

As we saw in (11-135), many internal or local electric fields turn out to be proportional to and opposite to the polarization. It is clear that similar results will be found in magnetostatic problems of similar type, and, accordingly, it is common practice to write

$$\mathbf{H}_{\text{loc}} = -N_m \mathbf{M} \qquad (20\text{-}50)$$

where N_m is a dimensionless constant called the *demagnetizing factor*. We see from (20-49) that $N_m = \tfrac{1}{3}$ for a sphere. It will be left as exercises to show that $N_m = 1$ for an infinite slab with parallel faces and $N_m = \frac{1}{2}$ for a cylinder, exactly as in the electrostatic case.

Historically, the approach to magnetostatic problems that we have just outlined and illustrated was quite extensively applied so that poles were used in many calculations; this is essentially the way in which **H** got its early emphasis and acquired the name magnetic *field*, which one might reasonably have expected to have been given to **B**. However, it is important to remember that the use of poles works consistently *only* in the absence of free currents and, of course, there is no experimental evidence for the existence of magnetic poles. From our present-day point of view, *all* magnetic effects are ultimately due to *currents*, and the use of poles in the way we have illustrated is a *sometimes* useful fiction, and applicable only in limited cases, generally in connection with calculations involving permanent magnets. In principle then, any problem that can be worked by using poles can be done by using magnetization currents, as we saw in several examples; it may seem harder, but it can be done. (In spite of the remark above about the necessity of the absence of free currents, it is possible to extend the use of the magnetic scalar potential to cases involving filamentary free currents. It turns out, however, that ϕ_m then depends on the solid angle subtended by the current at the field point. Since solid angles can be multiple-valued functions, this causes problems in the use of ϕ_m, and it quickly loses much of the simplicity implied above. Consequently, we will not pursue its use any further.)

Let us now consider one more simple example that will lead us naturally into our next major topic.

■ **Example**

Infinitely long straight constant current. We let the free current I coincide with the z axis and be in the positive z direction. We assume initially that there is no matter present. By the familiar symmetry arguments, **H** will have the form $\mathbf{H} = H_\varphi(\rho)\hat{\varphi}$ and we can use Ampère's integral form (20-32) by integrating over a circle of radius ρ in the xy plane. Thus, $\oint \mathbf{H} \cdot d\mathbf{s} = 2\pi\rho H_\varphi = I_{f,\text{enc}} = I$ so that $H_\varphi = I/2\pi\rho$ and therefore

$$\mathbf{H} = \frac{I}{2\pi\rho}\hat{\varphi} \quad \text{and} \quad \mathbf{B} = \frac{\mu_0 I}{2\pi\rho}\hat{\varphi} \tag{20-51}$$

since $\mathbf{M} = 0$. Now, as an extreme example, let us suppose that *all* space is filled with an obviously magnetic material such as iron. (We need to assume the presence of a thin insulating material around the wire since iron is a conductor.) What will happen? If we keep the free current I unchanged, and if the material is homogeneous enough so that the axial symmetry is still present, then we can apply (20-32) in exactly the same way and we will get precisely the same value for **H** as given by (20-51). In other words, *in this case*, **H** is *unaffected* by the presence of the matter. We can, however, expect that **M** will now be different from zero and then (20-28) shows us that the value of **B** will be altered, considerably as a matter of fact. It is also clear that we cannot evaluate **B** exactly until we know how the magnetization depends on the fields, and it is this that we now need to consider. ■

20-5 LINEAR ISOTROPIC HOMOGENEOUS MAGNETIC MATERIALS

If we recall the manner in which we were led to introduce the magnetization in Section 20-1, we should assume the existence of a functional relation between **M** and **B**, that is, we would write $\mathbf{M} = \mathbf{M}(\mathbf{B})$ and expect that the exact form of this relation would depend on the material and will have to be found by experiment. As logical as this might seem, it is not what is usually done; instead, one begins by writing a relation between **M** and **H**, that is, **M** = **M(H)**. There are essentially two reasons for this. One is primarily historical, in that at first greater importance was attached to **H** than to **B** and this seemed to be the reasonable thing to do. A much more practical reason is illustrated by the previous example where we saw that **H** was not altered in the presence of matter provided that we kept the free current constant *and* that we had an appropriate geometry. Since this is a desirable situation to have in the laboratory where one can alter the free current at will, it is convenient to consider **M** to be a measurable function of **H**.

There still remains the problem that the functional form of **M(H)** must be found for each material, either by experiment or by calculations done in other branches of physics. Fortunately, as was the case for dielectrics as discussed in Section 10-6, it is possible to classify most magnetic materials into groups in a way that can be used to simplify our theory and to make it more useful.

If **H** = 0, and **M**(0) ≠ 0, then the material is magnetized even in the absence of an external field. It is said to have a *permanent magnetization* and is called a *permanent magnet*. Many materials for which **M**(0) = 0 have a dependence of **M** on **H** that is very nonlinear and sometimes the relation is not even single valued. Many such materials are of very important practical interest, and, in fact, very often the materials that can fall into the first class also fall into the second. We do not consider these cases further in this section, but will come back to them in Section 20-7.

Rather than repeating the various simplifying classification steps outlined in Section 10-6, we proceed immediately to the simplest possible case, that of a *linear isotropic homogeneous magnetic material* for which the magnetization is proportional to and parallel to the magnetic field so that we can write

$$\mathbf{M} = \chi_m \mathbf{H} \tag{20-52}$$

Here χ_m is called the *magnetic susceptibility* and is a constant characteristic of the material. For virtually all materials that fall into this category, $|\chi_m| \ll 1$, and, in contrast to the electric case, χ_m can be of either sign. If $\chi_m > 0$, the material is called *paramagnetic*, while if $\chi_m < 0$, it is said to be *diamagnetic*. As is discussed in detail in Appendix B, *all* materials have a diamagnetic contribution to their susceptibility arising from the altered orbital motion of their constituent electrons that is produced by an applied field. In paramagnetic materials, however, this is overwhelmed by the much larger paramagnetic susceptibility due to permanent magnetic dipole moments.

Combining (20-52) with (20-28), we obtain

$$\mathbf{B} = \mu_0(1 + \chi_m)\mathbf{H} = \kappa_m \mu_0 \mathbf{H} = \mu \mathbf{H} \tag{20-53}$$

where

$$\kappa_m = 1 + \chi_m = \text{relative permeability} \tag{20-54}$$

$$\mu = \kappa_m \mu_0 = \text{(absolute) permeability} \tag{20-55}$$

Thus, in this common, but not universal, case, **B** and **H** are parallel. The relation **B** = μ**H**, or **H** = **B**/μ, is another example of a *constitutive equation* and is not a fundamental equation of electromagnetism, in contrast to (20-28) which is.

In this case, we can easily relate **M** and **B** by eliminating **H** between (20-52) and (20-53) and we find that

$$\mathbf{M} = \frac{\chi_m}{\kappa_m \mu_0}\mathbf{B} = \frac{\chi_m}{(1 + \chi_m)\mu_0}\mathbf{B} \tag{20-56}$$

so that **M** is also a linear function of **B**.

Since μ is a constant, we see that $\nabla \cdot \mathbf{B} = 0 = \nabla \cdot (\mu \mathbf{H}) = \mu \nabla \cdot \mathbf{H}$ so that

$$\nabla \cdot \mathbf{H} = 0 \tag{20-57}$$

and, therefore, from (20-52):

$$\nabla \cdot \mathbf{M} = 0 \tag{20-58}$$

which is, of course, consistent with (20-34) and, according to (20-42), shows that there can be no volume density of magnetic charges within a l.i.h. magnetic material.

The free and magnetization current densities are also simply related in a material such as this. If we take the curl of (20-52) and use (20-10), (20-29), and (20-54), we find that

$$\mathbf{J}_m = \chi_m \mathbf{J}_f = (\kappa_m - 1)\mathbf{J}_f \tag{20-59}$$

and the total current density as given by (20-26) becomes

$$\mathbf{J} = (1 + \chi_m)\mathbf{J}_f = \kappa_m \mathbf{J}_f \tag{20-60}$$

Since χ_m can be of either sign, we see that the magnetization current can be oppositely directed to the free current; the total current, however, will always be in the same direction as the free current since κ_m is positive. We also see that when $\mathbf{J}_f = 0$, both $\mathbf{J}_m$ and $\mathbf{J}$ are zero, so that at any point in a l.i.h. magnetic material where there is no free current, there is no magnetization current either. If $\mathbf{J}_f = 0$, then $\nabla \times \mathbf{H} = 0$ which, together with (20-57), shows us that there are no sources of $\mathbf{H}$ *within* the material in the absence of free currents.

When (20-53) is applicable, the boundary conditions at a surface of discontinuity in magnetic properties can be written in terms of a single vector, either $\mathbf{B}$ or $\mathbf{H}$ as one chooses. Using (16-4) and (20-30), we can write them as

$$\hat{\mathbf{n}} \cdot (\mathbf{B}_2 - \mathbf{B}_1) = 0 \qquad \hat{\mathbf{n}} \times \left(\frac{\mathbf{B}_2}{\mu_2} - \frac{\mathbf{B}_1}{\mu_1} \right) = \mathbf{K}_f \tag{20-61}$$

or, equally well, as

$$\hat{\mathbf{n}} \cdot (\mu_2 \mathbf{H}_2 - \mu_1 \mathbf{H}_1) = 0 \qquad \hat{\mathbf{n}} \times (\mathbf{H}_2 - \mathbf{H}_1) = \mathbf{K}_f \tag{20-62}$$

We see from these relations that, even if $\mathbf{K}_f = 0$, the lines of $\mathbf{B}$ (and $\mathbf{H}$) will generally have different directions on the two sides of the bounding surface, that is, they will be refracted at the boundary.

■ **Example**

Infinitely long ideal solenoid. As usual, we assume the solenoid has n turns per unit length, a cross section of area S, and a (free) current I in its winding. We now assume, in addition, that the interior is filled with a l.i.h. magnetic material of relative permeability κ_m. In this case, it would be appropriate to use the integral form of Ampère's law for $\mathbf{H}$ given by (20-32). We could proceed with this equation exactly as we did in the example of Section 15-2, and we will again obtain (20-37), that is, the magnetic field is uniform across the cross section and is given by

$$\mathbf{H} = nI\hat{\mathbf{z}} \tag{20-63}$$

while it is zero outside. Thus, $\mathbf{H}$ is unaffected by the presence of the matter. The induction $\mathbf{B}$ is now given by (20-53) and is

$$\mathbf{B} = \kappa_m \mu_0 \mathbf{H} = \kappa_m \mu_0 nI\hat{\mathbf{z}} \tag{20-64}$$

and has been increased by the factor κ_m as we see by comparison with (15-26).

Consequently, the flux per turn given by BS has been increased by this same factor and will equal $\kappa_m \mu_0 nSI$. If we now consider a length l, with its nl turns, the total flux will be $\kappa_m \mu_0 n^2 SlI$, so that the self-inductance $L = \Phi/I$, as given by (17-55), will be

$$L = \kappa_m \mu_0 n^2 lS = \kappa_m L_0 \qquad (20\text{-}65)$$

where we have used (17-61) to identify the self-inductance L_0 when the interior has a vacuum in it. Thus, we see that the presence of the matter has increased the self-inductance and the ratio $L/L_0 = \kappa_m$ is just equal to the relative permeability of the material.

∎

Example

Self-inductance in general. The simplicity of the final result $L = \kappa_m L_0$, which we have just obtained, and its independence of the details of the specific system we considered, suggest that this may actually be a general relation that holds for any self-inductance. We can see that this is actually the case. Suppose we are given a certain distribution of free currents described by $\mathbf{J}_f$, and we assume a vacuum everywhere. Then the source equations for $\mathbf{H}_0$ will be $\nabla \times \mathbf{H}_0 = \mathbf{J}_f$ and $\nabla \cdot \mathbf{H}_0 = 0$ from (20-29) and (20-57) and we can imagine solving them for $\mathbf{H}_0$. The corresponding induction will be $\mathbf{B}_0 = \mu_0 \mathbf{H}_0$ and the flux through the circuit will be $\Phi_0 = \int \mathbf{B}_0 \cdot d\mathbf{a} = \mu_0 \int \mathbf{H}_0 \cdot d\mathbf{a}$. Now let us suppose that all space occupied by fields is filled with a l.i.h. magnetic material of relative permeability κ_m; this requirement is necessary in order that there be no surfaces of discontinuity on which the normal component of $\mathbf{M}$ is different from zero, as this would introduce sources of $\mathbf{H}$ other than the free currents. Now if we keep $\mathbf{J}_f$ *unchanged*, the equations to be solved will be exactly the same as before, and therefore $\mathbf{H}$ will be unchanged, that is, $\mathbf{H} = \mathbf{H}_0$. But we see from (20-53) that, under these conditions, $\mathbf{B}$ will be increased by the factor κ_m, so that $\mathbf{B} = \kappa_m \mu_0 \mathbf{H} = \kappa_m \mu_0 \mathbf{H}_0 = \kappa_m \mathbf{B}_0$. Then when we integrate over the same surface the new flux will be $\Phi = \kappa_m \Phi_0$ and the inductance $L = \Phi/I = \kappa_m \Phi_0/I$; thus

$$L = \kappa_m L_0 \qquad (20\text{-}66)$$

will again be obtained, but now as a general result.

∎

Example

Coaxial line. In Figure 20-14, we assume the infinitely long cylindrical conductor of radius a (region 1) to be nonmagnetic and to have a total current I uniformly distributed over its cross section. The coaxial outer conductor of radius b is, for simplicity, assumed to have an infinitesimal thickness and to carry a current $-I$ uniformly distributed over its surface. Region 2 between the conductors is filled with a l.i.h. nonconducting magnetic material of permeability $\mu = \kappa_m \mu_0$. Region 3 includes all space (vacuum) outside the system. We want to find all of the field vectors everywhere.

Because of the cylindrical symmetry, $\mathbf{H}$ will, as usual, have the form $\mathbf{H} = H_\varphi(\rho)\hat{\boldsymbol{\varphi}}$ and we can use the integral form of Ampère's law (20-32) to calculate it. But this system is just like that of Figure 18-1 with $c = b$ and the present region 3 corresponding to region 4 of the previous figure. We previously worked this out in the vacuum case and we can simply take over our results by dividing them by μ_0 in order to get $\mathbf{H}$, or one can simply repeat the same calculations. Therefore, we find from (18-25), (18-28), and following (18-33) that

$$H_{\varphi 1} = \frac{I\rho}{2\pi a^2} \qquad H_{\varphi 2} = \frac{I}{2\pi\rho} \qquad H_{\varphi 3} = 0 \qquad (20\text{-}67)$$

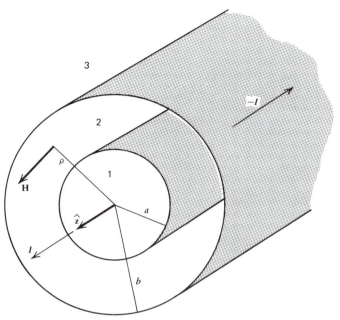

Figure 20-14.   Coaxial line carrying equal oppositely directed currents.

These are shown as a function of ρ in Figure 20-15. As a check, we will see if the necessary boundary conditions are satisfied. Since **H** has only tangential components at any of the surfaces, the relevant one here is (20-31). We see from the figure that H_φ is continuous at $\rho = a$ as it should be since there is no free surface current density there. It is also seen that H_φ is discontinuous at $\rho = b$ and this must be due to the fact that $\mathbf{K}_f$ is different from zero there. We find from Figures 20-14 and 12-5 that

$$\mathbf{K}_f = -\frac{I}{2\pi b}\hat{\mathbf{z}} \tag{20-68}$$

If we go from region 2 to 3, then $\hat{\mathbf{n}} = \hat{\rho}$ and (20-31) becomes $\mathbf{H}_{3t}(b) - \mathbf{H}_{2t}(b) = -H_{\varphi 2}(b)\hat{\boldsymbol{\varphi}} = -(I/2\pi b)\hat{\boldsymbol{\varphi}} = \mathbf{K}_f \times \hat{\mathbf{n}} = -(I/2\pi b)\hat{\mathbf{z}} \times \hat{\rho} = -(I/2\pi b)\hat{\boldsymbol{\varphi}}$ and the boundary condition at $\rho = b$ is satisfied and the origin of the discontinuity in H_φ is clear.

We can now find **B** from (20-53) and (20-67) by using $\mu_1 = \mu_3 = \mu_0$ and $\mu_2 = \mu$ and we get

$$B_{\varphi 1} = \frac{\mu_0 I\rho}{2\pi a^2} \qquad B_{\varphi 2} = \frac{\mu I}{2\pi\rho} \qquad B_{\varphi 3} = 0 \tag{20-69}$$

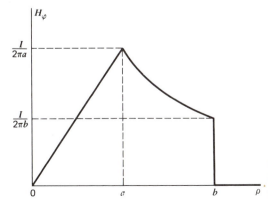

Figure 20-15.   Magnetic field produced by coaxial line as a function of distance from the axis.

These are shown as a function of ρ in Figure 20-16. We see that B_φ also has a discontinuity at $\rho = b$, but it is different from the value $\mu_0 I/2\pi b$, which it would be if there were a vacuum in region 2. In addition, B_φ has a discontinuity at $\rho = a$, while H_φ did not. These differences *must* be due to the existence of magnetization surface currents, but we cannot verify this quantitatively until we have found **M**. (Actually, we know that B_φ must satisfy the boundary conditions since H_φ does, as we saw, but it will nevertheless be instructive to work out the details.)

We can find **M** from (20-67) and (20-52) and by using $\chi_{m1} = \chi_{m3} = 0$ and $\chi_{m2} = \chi_m$, and we get

$$M_{\varphi 1} = 0 \qquad M_{\varphi 2} = \frac{\chi_m I}{2\pi\rho} \qquad M_{\varphi 3} = 0 \qquad (20\text{-}70)$$

These results are shown as a function of ρ in Figure 20-17. We see that **M** has discontinuities at both a and b, and therefore this must result in net magnetization surface currents $\mathbf{K}_m$ according to (20-12). Before we calculate them, however, we can also check our results in other ways. From (20-70) and (1-87) we find that $\nabla \cdot \mathbf{M}_2 = \rho^{-1}(\partial M_{\varphi 2}/\partial\varphi) = 0$ as it must by (20-58). Since there is no free current in the nonconducting region 2, we know from (20-59) that $\mathbf{J}_{m2} = \nabla \times \mathbf{M}_2$ given by (20-10) must be zero; we can verify this from (20-70) and (1-88) and we find that $J_{m2\rho} = -\partial M_{\varphi 2}/\partial z = 0$, $J_{m2\varphi} = 0$, and $J_{m2z} = \rho^{-1}[\partial(\rho M_{2\varphi})/\partial\rho] = \rho^{-1}[\partial(\chi_m I/2\pi)/\partial\rho] = 0$.

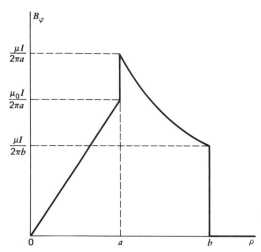

Figure 20-16. Magnetic induction produced by coaxial line as a function of distance from the axis.

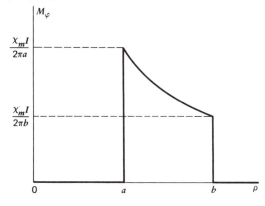

Figure 20-17. Magnetization from coaxial line as a function of distance from the axis.

In order to calculate the surface currents, we must remember that $\hat{\mathbf{n}}$ in (20-10) is the *outer normal* to the region containing the matter. Therefore, at $\rho = a$, $\hat{\mathbf{n}} = -\hat{\rho}$ and we get

$$\mathbf{K}_{m2}(a) = M_{\varphi 2}(a)\hat{\boldsymbol{\varphi}} \times (-\hat{\rho}) = \frac{\chi_m I}{2\pi a}\hat{\mathbf{z}} \qquad (20\text{-}71)$$

At $\rho = b$, $\hat{\mathbf{n}} = \hat{\rho}$ and we find that

$$\mathbf{K}_{m2}(b) = M_{\varphi 2}(b)\hat{\boldsymbol{\varphi}} \times \hat{\rho} = -\frac{\chi_m I}{2\pi b}\hat{\mathbf{z}} \qquad (20\text{-}72)$$

On the other hand, in order to apply the boundary condition on $\mathbf{B}$, which is given by (15-16) and the analogue of (20-26) as

$$\mathbf{B}_{2t} - \mathbf{B}_{1t} = \mu_0 \mathbf{K} \times \hat{\mathbf{n}} = \mu_0 \mathbf{K}_f \times \hat{\mathbf{n}} + \mu_0 \mathbf{K}_m \times \hat{\mathbf{n}} \qquad (20\text{-}73)$$

we must remember that *in this equation* $\hat{\mathbf{n}}$ is the unit normal drawn from region 1 to 2, and so on. Therefore, if we always go from the lower-numbered region to the higher-numbered region, we see from Figure 20-14 that we can always use $\hat{\mathbf{n}} = \hat{\rho}$ in (20-73).

At $\rho = a$, $\mathbf{K}_f = 0$ and we find that (20-73), (20-71), (20-54), and (20-55) give $\mathbf{B}_{2t} - \mathbf{B}_{1t} = \mu_0 \mathbf{K}_{m2}(a) \times \hat{\mathbf{n}} = (\mu_0 \chi_m I/2\pi a)\hat{\mathbf{z}} \times \hat{\rho} = [\mu_0(\kappa_m - 1)I/2\pi a]\hat{\boldsymbol{\varphi}} = [(\mu - \mu_0)I/2\pi a]\hat{\boldsymbol{\varphi}}$ in exact agreement with (20-69) and Figure 20-16. At $\rho = b$, however, $\mathbf{K}_f$ is not zero but is given by (20-68), so that (20-73) applied here becomes $\mathbf{B}_{3t} - \mathbf{B}_{2t} = (\mu_0 I/2\pi b)(-\hat{\mathbf{z}} - \chi_m \hat{\mathbf{z}}) \times \hat{\rho} = -[\mu_0(1 + \chi_m)I/2\pi b]\hat{\boldsymbol{\varphi}} = -(\mu I/2\pi b)\hat{\boldsymbol{\varphi}}$, which also agrees exactly with the values given in (20-69) and shown in Figure 20-16.

At this point, we have managed to squeeze almost everything out of this long example that we can, and we have seen that all of the relevant quantities were calculable and satisfied, in this specific case, all of the applicable general results that we have previously obtained. ∎

20-6 ENERGY

We recall our result (18-12) for the magnetic energy of a system of free currents

$$U_m = \frac{1}{2} \int_{\text{all space}} \mathbf{J}_f \cdot \mathbf{A} \, d\tau \qquad (20\text{-}74)$$

In the electric case, we concluded after a lengthy discussion following (10-78) that the *useful* definition of energy was that associated with the free charges since it represented the reversible work that could be put into and retrieved from the system by the manipulation of charges over which we have control. In the same manner, we can conclude that the energy of the free current distribution is the suitable quantity to call the magnetic energy of the system since it represents the reversible work associated with currents *that we can control*, and we can still use this definition in the presence of matter.

In order to appreciate the limitations of our final results, it will be useful to begin again and (18-1), which gives the work required from an external source to produce a small change in flux, provides a convenient starting point because it was based only on a general application of Faraday's law. If we let the changes in magnetic energy and flux be written as δU_m and $\delta \Phi_j$ for convenience, we find from (18-2) and (16-23) that

$$\delta U_m = \sum_j i_j \, \delta \Phi_j = \sum_j i_j \oint_{C_j} \delta \mathbf{A}_j \cdot d\mathbf{s}_j \qquad (20\text{-}75)$$

where $\delta \mathbf{A}_j$ is the corresponding change in the vector potential on the jth part of the system. When matter is present, the magnetization currents contribute to the flux because of the induction they produce, as described for example by (20-15). As usual, we are assuming rigid mechanical supports or constraints on the circuits so that no purely mechanical work is done.

We can now rewrite the sum in terms of the current density in exactly the same way that we went from (18-10) to (18-11) and we get

$$\delta U_m = \int_V \mathbf{J}_f \cdot \delta \mathbf{A} \, d\tau = \int_V (\nabla \times \mathbf{H}) \cdot \delta \mathbf{A} \, d\tau \tag{20-76}$$

with the use of (20-29). In this expression, V is the total volume occupied by free currents. We can, however, extend this to include all space since regions where $\mathbf{J}_f = 0$ will not contribute to the integral. Thus, we get

$$\delta U_m = \int_{\text{all space}} (\nabla \times \mathbf{H}) \cdot \delta \mathbf{A} \, d\tau \tag{20-77}$$

But $\nabla \times \mathbf{A} = \mathbf{B}$ by (16-7), so that $\nabla \times (\mathbf{A} + \delta \mathbf{A}) = \mathbf{B} + \delta \mathbf{B}$ and therefore, by (1-117), $\nabla \times \delta \mathbf{A} = \delta \mathbf{B}$. We can now rewrite the integrand by using (1-116) and (16-7):

$$(\nabla \times \mathbf{H}) \cdot \delta \mathbf{A} = \mathbf{H} \cdot (\nabla \times \delta \mathbf{A}) - \nabla \cdot (\delta \mathbf{A} \times \mathbf{H}) = \mathbf{H} \cdot \delta \mathbf{B} - \nabla \cdot (\delta \mathbf{A} \times \mathbf{H}) \tag{20-78}$$

If we compare this with (18-16) and recall how we went from (18-15) to (18-21), we see that we could proceed in exactly the same way and we will obtain

$$\delta U_m = \int_{\text{all space}} \mathbf{H} \cdot \delta \mathbf{B} \, d\tau \tag{20-79}$$

Finally, if we take the zero of energy to correspond to $\mathbf{B} = 0$, we can obtain the total energy U_m by integrating this from the initial state of $\mathbf{B} = 0$ to the final value:

$$U_m = \int_{\text{all space}} \int_0^{\mathbf{B}} \mathbf{H} \cdot \delta \mathbf{B} \, d\tau \tag{20-80}$$

In general, (20-80) cannot be further evaluated until the dependence of $\mathbf{H}$ on $\mathbf{B}$ is known, and this could be quite complicated; we will briefly look at such a situation in the next section. However, for the important case of *linear isotropic* magnetic materials, we can use (20-53) to write $\mathbf{H} \cdot \delta \mathbf{B} = \mathbf{B} \cdot \delta \mathbf{B}/\mu = \delta(\mathbf{B}^2/2\mu)$ so that

$$\int_0^{\mathbf{B}} \mathbf{H} \cdot \delta \mathbf{B} = \int_0^{\mathbf{B}} \delta \left(\frac{\mathbf{B}^2}{2\mu} \right) = \frac{\mathbf{B}^2}{2\mu} = \frac{1}{2} \mathbf{H} \cdot \mathbf{B}$$

and thus (20-80), which gives the total reversible work on the free currents, becomes

$$U_m = \int_{\text{all space}} \tfrac{1}{2} \mathbf{H} \cdot \mathbf{B} \, d\tau \tag{20-81}$$

which we can interpret in the usual way by introducing the *magnetic energy density*

$$u_m = \tfrac{1}{2} \mathbf{H} \cdot \mathbf{B} \tag{20-82}$$

so that we can write

$$U_m = \int_{\text{all space}} u_m \, d\tau \tag{20-83}$$

[Compare with (10-84) and (10-85).] If we use (20-53), we can also write this as

$$u_m = \frac{\mathbf{B}^2}{2\mu} = \frac{1}{2} \mu \mathbf{H}^2 \tag{20-84}$$

which, of course, reduces to (18-22) for a vacuum when $\mu = \mu_0$.

In Section 18-2 we saw that one of the useful applications of the energy expression was the calculation of self-inductance. If we have found the fields by other means, we can evaluate (20-83) and by equating it to (18-9), L can be found from $L = 2U_m/I^2$.

■ **Example**

Coaxial line. Let us consider the same system as shown in Figure 20-14 for which we have already found **H** and **B**. For simplicity, we shall calculate only the contribution of the region containing the matter (region 2) to L. The energy density in this region as found from (20-82), (20-67), and (20-69) is

$$u_{m2} = \frac{1}{2}H_{\varphi 2}B_{\varphi 2} = \frac{\mu I^2}{8\pi^2\rho^2} \tag{20-85}$$

Since $d\tau = \rho \, d\rho \, d\varphi \, dz$ by (1-83), the total energy contained in a length l of this region will be

$$U_{m2} = \int_0^l \int_0^{2\pi} \int_a^b \frac{\mu I^2}{8\pi^2\rho^2}\rho \, d\rho \, d\varphi \, dz = \frac{\mu I^2 l}{4\pi}\ln\left(\frac{b}{a}\right) \tag{20-86}$$

so that its contribution to the self-inductance will be

$$L_2 = \frac{\mu l}{2\pi}\ln\left(\frac{b}{a}\right) \tag{20-87}$$

Upon comparing this with the middle term of (18-34), we see that the ratio of this to the vacuum value is $L_2/L_{20} = \mu/\mu_0 = \kappa_m$ in complete agreement with (20-66) for this case in which the free currents have been kept constant. ■

The presence of matter will generally change the values of **B** and **H** everywhere and thus the magnetic energy given by (20-81) can also be expected to change. As we discussed for the electric case after (10-89), this change will generally depend on the process by which the matter is introduced into the field, and a general discussion can be quite complex. There is, however, one special case in which the energy change can be directly identified with the matter, and we consider it as an illustration.

We assume that initially we have vacuum everywhere and some distribution of free source currents that produce the fields $\mathbf{H}_0$ and $\mathbf{B}_0 = \mu_0\mathbf{H}_0$ throughout space. Now let us assume that we *keep the free source currents fixed* in value and location and introduce a material of volume V into this preexisting field $\mathbf{B}_0$. If we let **B** and **H** be the new values of the fields that arise because of the presence of the matter, the *change* in the energy of the system as obtained from (20-81) will be given by

$$U_{mm} = U_m - U_{m0} = \frac{1}{2}\int_{\text{all space}} (\mathbf{H}\cdot\mathbf{B} - \mathbf{H}_0\cdot\mathbf{B}_0)\,d\tau \tag{20-88}$$

and can be ascribed entirely to the presence of the matter. As was done in Exercise 10-38 for the analogous electric formula (10-91), it can be shown that (20-88) can be written as an integral over the volume V of the matter only. The calculation is quite long, so we shall simply quote the final result, which turns out to be

$$U_{mm} = \frac{1}{2}\int_V \mathbf{M}\cdot\mathbf{B}_0\,d\tau \tag{20-89}$$

Since this involves only the volume of the matter, it is reasonable to consider the energy to be localized *in* the matter and to represent the energy *of* the matter. Thus we can introduce an energy density u_{mm} associated with the matter given by

$$u_{mm} = \tfrac{1}{2}\mathbf{M} \cdot \mathbf{B}_0 = \tfrac{1}{2}\chi_m\mu_0\mathbf{H} \cdot \mathbf{H}_0 \qquad (20\text{-}90)$$

We note the similarity of these results to the electrostatic results (10-92) and (10-93), except for the difference in sign. These expressions are appropriate for situations in which the magnetization is produced by the field, as is implied by (20-52).

On the other hand, we have to remember the difference between these energies and the (potential) *interaction energy* of a permanent dipole in an external induction given by (19-40). If we consider a small volume $d\tau$, its dipole moment will be given by (20-1), and the interaction energy will be

$$du'_D = du'_{m,\,\text{ext}} = -\mathbf{M} \cdot \mathbf{B}_{\text{ext}}\, d\tau \qquad (20\text{-}91)$$

if we write the external induction as $\mathbf{B}_{\text{ext}}$ rather than the $\mathbf{B}_0$ of (19-40). The total interaction energy can then be obtained by integrating (20-91) over the volume V of the material to give

$$U'_{m,\,\text{ext}} = -\int_V \mathbf{M} \cdot \mathbf{B}_{\text{ext}}\, d\tau \qquad (20\text{-}92)$$

For example, if $\mathbf{B}_{\text{ext}}$ does not vary much over the material, we can take it out of the integral and use (20-2) to get $U'_{m,\,\text{ext}} = -\mathbf{m} \cdot \mathbf{B}_{\text{ext}}$ in agreement with (19-40).

When a material is magnetized, there will generally be forces on the magnetization currents due to the induction. If the material is not rigid, it can deform under the influence of these forces. This effect is called *magnetostriction*; it is generally a small effect and we will not consider it further and assume completely rigid materials.

As we saw in Section 18-3, energy considerations are often useful in discussing magnetic forces. Because of the complexity of this general subject, we illustrate these ideas with the following single example.

■ Example

Permeable rod in a long solenoid. Suppose we have a long solenoid of n turns per unit length, total length l, circular cross section of area S, and carrying a current I circulating as shown in Figure 20-18. We also suppose that a cylindrical rod of the same cross section and of permeability μ is inserted into the solenoid a distance z. We assume that the current in the solenoid is kept constant throughout and we would like to find the force on the rod. As we saw in connection with the example of the interpenetrating solenoids of Section 18-3, we can get a reasonably simple solution only by neglecting the "end effects" that will be associated with the diverging field lines near the ends of the solenoid and rod as was indicated in Figure 18-3. Since $|\chi_m| \ll 1$, the

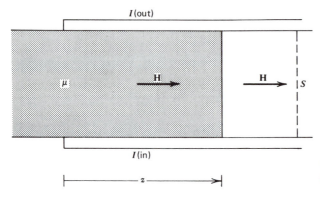

Figure 20-18. Permeable rod in a long solenoid.

magnetic field will be practically unaffected by the presence of the rod and to a first approximation can be taken from (20-63) as $\mathbf{H} = nI\hat{\mathbf{z}}$. The energy density u_{m0} in the unoccupied vacuum region of the solenoid of length $l - z$ and volume $(l - z)S$ as obtained from (20-84) is $u_{m0} = \frac{1}{2}\mu_0 n^2 I^2$. Similarly, the energy density u_{mM} in the volume zS occupied by the matter will be $u_{mM} = \frac{1}{2}\mu n^2 I^2$. Inserting these constant values into (20-83), we find that the total energy of this configuration is

$$U_m(z) = \frac{1}{2}n^2 I^2 S\left[\mu z + \mu_0(l - z)\right] \tag{20-93}$$

Since the currents are kept constant, we use (18-39) to calculate the force:

$$\mathbf{F}_m = (\nabla U_m)_I = \frac{1}{2}(\mu - \mu_0)n^2 I^2 S\hat{\mathbf{z}}$$
$$= \frac{1}{2}\chi_m \mu_0 n^2 I^2 S\hat{\mathbf{z}} \tag{20-94}$$

There are many interesting aspects to this result.

Since everything else in (20-94) is positive, the sign of $\mathbf{F}_m$ depends on the sign of χ_m. If the material is paramagnetic so that $\chi_m > 0$, the rod will be *attracted* into the solenoid, while if it were diamagnetic with a negative susceptibility, it would be *repelled*. [This is in contrast to the analogous electrostatic result (10-99), which showed that a dielectric slab would always be attracted in between the capacitor plates; this simply expresses the fact that all electric susceptibilities are positive for static fields.] We can understand this effect of the sign difference qualitatively. If the material is paramagnetic, $\mathbf{M}$ will be parallel to $\mathbf{H}$ by (20-52) and the associated surface currents $\mathbf{K}_m$ will be circulating in the same sense as I, as we saw in Figure 20-7b. Thus we have two sets of parallel currents and these will attract each other, as noted following (13-14), in agreement with the sign found from (20-94). On the other hand, in a diamagnetic material, $\mathbf{M}$ will be opposite in direction to $\mathbf{H}$, its corresponding surface currents will circulate oppositely to the sense of I, and these "unlike" currents will repel each other.

While the force $\mathbf{F}_m$ is independent of both l and z, it is proportional to the cross-sectional area S; thus we can introduce a force per unit area $\mathbf{f}_m$ as

$$\mathbf{f}_m = \frac{\mathbf{F}_m}{S} = \frac{1}{2}(\mu - \mu_0)n^2 I^2 \hat{\mathbf{z}} \tag{20-95}$$

This can be put into a more interpretable form by expressing it in terms of the energy densities in the two regions and we get

$$\mathbf{f}_m = (u_{mM} - u_{m0})\hat{\mathbf{z}} \tag{20-96}$$

In other words, the magnitude of the force per unit area is just equal to the difference in energy densities of the two regions, and its direction is such as to tend to move the material so as to *increase* the total magnetic energy of the system. This is consistent with all of our previous results relating to the general tendencies of systems to "try" to increase their magnetic energy.

Furthermore, (20-96) is in agreement with the idea expressed after (18-52) (and found from a somewhat different type of example) that magnetic forces per unit area are properly described as pressures. In effect, we found in (18-52) that we can associate a force per unit area at the surface of a given region as the product of the energy density *within* that region and the *outward normal* to the region. In this case, we would then write a pressure term at the interface for the vacuum region as $\mathbf{f}_{m0} = u_{m0}(-\hat{\mathbf{z}})$ while that for the matter would be $\mathbf{f}_{mM} = u_{mM}(+\hat{\mathbf{z}})$. Then the resultant force per unit area on the bounding surface between the two would be $\mathbf{f}_m = \mathbf{f}_{mM} + \mathbf{f}_{m0} = (u_{mM} - u_{m0})\hat{\mathbf{z}}$, which is just (20-96) again. ∎

20-7 FERROMAGNETIC MATERIALS

Up to now, we have considered in detail only l.i.h. media for which we can write $\mathbf{B} = \mu \mathbf{H}$ where μ is a constant characteristic of the material. Although many materials can be described quite well by this relation, it turns out that quite a few of the technologically important ones have a markedly different behavior, and in this section we consider briefly some of their principal characteristics.

These materials are usually called *ferromagnetic materials* since they are typified by the neighboring metals iron, cobalt, and nickel. There are, however, many alloys and nonmetals that also fall into this category. Although specific properties vary from material to material, and must be found by experiment, the general features are very similar and we shall concentrate on these. Essentially, there is no simple relation between $\mathbf{B}$ and $\mathbf{H}$ for these cases.

It is also commonly found that the magnetic properties depend on the previous history of the particular sample. Nevertheless, one can get a sample into a state where it possesses a certain regularity of behavior. This is usually accomplished by the frequent reversal of an applied field of decreasing magnitude, and we assume that this has been done. (The reason for this particular prescription will be apparent shortly.)

What one generally wants to do is to apply an external $\mathbf{H}$ field, produced by external free currents, and then to measure $\mathbf{B}$ as a function of $\mathbf{H}$. In order to accomplish this, it is clear that the sample has to be of such a shape that $\mathbf{B}$ and $\mathbf{H}$ can be measured and/or calculated conveniently and unambiguously. In particular, it is desirable that $\mathbf{H}$ can be found from a knowledge of the applied external currents *only*. If, for example, one were to use a long solenoid with an iron core as the sample, there would be discontinuities at the ends and one would have to concern oneself with the effect of the normal components of $\mathbf{M}$, that is, of the surface density of "poles." But, as we saw in (20-42), these depend on the value of $\mathbf{M}$, which is connected to the relation between $\mathbf{B}$ and $\mathbf{H}$ by $\mathbf{B} = \mu_0(\mathbf{H} + \mathbf{M})$. Since this would complicate the problem enormously, we want to choose a sample without any ends, and this immediately leads us to think in terms of a ring or torus. If the torus is tightly wound about the material, then, as we saw in the last example of Section 15-2, we can to a good approximation assume $\mathbf{H}$ to be confined completely to the interior and it can be found from the free current I in the N turns of the toroidal coil. Such a device is called a *Rowland ring*.

We can use the integral form of Ampère's law given in (20-32) to find $\mathbf{H}$. The calculation is substantially identical to that which led to (15-28) and was illustrated in Figure 15-12. For a circular path of radius ρ, $d\mathbf{s} = \rho \, d\varphi \, \hat{\boldsymbol{\varphi}}$ and (20-32) becomes

$$\oint_C \mathbf{H} \cdot d\mathbf{s} = \int H_\varphi(\rho)\rho \, d\varphi = 2\pi\rho H_\varphi(\rho) = I_{f,\text{enc}} = NI \tag{20-97}$$

so that

$$H_\varphi(\rho) = \frac{NI}{2\pi\rho} \tag{20-98}$$

We let a be the radius of the circular cross section of the ring and b be the central radius of the torus, and we shall assume that $b \gg a$. In this case, $H_\varphi(\rho)$ will be approximately constant and equal to

$$H = \frac{NI}{2\pi b} \tag{20-99}$$

Then both M and B will also be approximately constant across the cross section of area $S = \pi a^2$, and the flux through S will be

$$\Phi = SB \tag{20-100}$$

The measurement can then be carried out in a typical manner. The controllable external quantity I is changed by a small amount, and thus, by (20-99) the magnetic field will change by $\Delta H = N\Delta I/2\pi b$ and can be calculated from the known properties of the toroid. There will be corresponding changes in the induction, ΔB, and the flux $\Delta\Phi = S\Delta B$. The latter quantity can be measured by wrapping another coil about the torus and finding the total charge ΔQ_c that passes through it as a result of the induced current. If R_c is the resistance of this circuit, we find from (12-2) and (17-3) that the magnitude of ΔQ_c is

$$\Delta Q_c = \int i_{c\,\text{induced}}\, dt = \int \frac{|\mathcal{E}_{\text{ind}}|\, dt}{R_c} = \frac{1}{R_c}\int \frac{d\Phi}{dt}\, dt = \frac{\Delta\Phi}{R_c}$$

which agrees with the result of Exercise 17-6 and gives us $\Delta B = R_c \Delta Q_c/S$ thus enabling us to evaluate ΔB. In this way, the curve describing the induction B as a function of H can be obtained as the result of a series of small steps.

When a procedure like this is carried out with a monotonically increasing H, the result typically is a B versus H curve with the general appearance shown in Figure 20-19. A curve such as this is called a *magnetization curve*; this name is also often given to an M versus H curve, which will contain substantially the same information. We see at once that the relationship shown is very *nonlinear*. Eventually, as $H \to \infty$, M approaches a constant value M_s called the *saturation magnetization* since the dipoles become fully aligned. Then the relation $B = \mu_0(H + M) \to \mu_0 H + \mu_0 M_s = \mu_0 H +$ const. and the $B - H$ curve becomes linear with constant slope μ_0. For many materials, the values of H required to reach this point are impractically large.

Although when we defined $\mu = B/H$ in (20-53), we had linear systems in mind, we can stretch our definition of μ to this case by continuing to use this same equation. When μ is found from a curve like Figure 20-19, the result is like that shown in Figure 20-20. The maximum is generally of the magnitude of several thousand, although it can vary widely from material to material; the ratio μ/μ_0 also approaches the limiting value of unity as H becomes very large. Thus, μ as defined in this way is far from being a constant, but it can still be a useful and convenient way of describing the material. (We can also note here that many ferromagnetic materials are crystals, and it is found that the magnetization curves are generally different for the various directions along which H is applied.)

Now suppose that, instead of increasing H indefinitely as we assumed for Figure 20-19, we go only to the maximum value H_1 (and corresponding B_1) shown in Figure

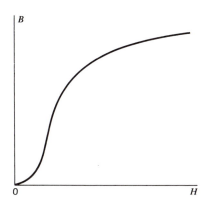

Figure 20-19. Typical B versus H curve for a ferromagnetic material.

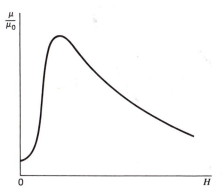

Figure 20-20. Relative permeability versus H for a ferromagnetic material.

20-21, and then start to decrease H. We find, as indicated by the directions of the arrowheads, that the curve does *not* retrace its path, and B does not decrease as rapidly as it originally increased. A general behavior like this is known as *hysteresis* (from the Greek "to be late" or "to fall short"). Thus, we see that the relation between B and H is not only nonlinear, but it is *not single valued* either. If we continue decreasing H until $H = 0$, we find that $B \neq 0$ as shown. This value B_r that remains is called the *remanent induction* or *remanence* or *retentivity*. In fact, in order to reduce B to zero, the field H must now be applied in the reverse direction so that $B = 0$ when $H = -H_c$. This value H_c is called the *coercive force* or *coercivity*. If we now continue this process of decreasing H to $-H_1$, and then reverse direction and increase it back to $+H_1$, the B *versus* H curve becomes the closed curve known as a *hysteresis loop*.

If we had gone out to $H_2 > H_1$ before starting the reversal, we would find as shown in Figure 20-22 that we would get a different hysteresis loop with different remanent induction and coercive force; similarly for the value H_3 of the maximum. In other words, the whole $B - H$ space can be filled with hysteresis loops and a given numerical value of μ does not have a great significance unless the corresponding conditions are more closely specified. We see that the tips of these hysteresis loops trace out the monotonic magnetization curve of Figure 20-19; we also see why we prescribed the application of alternating fields of gradually decreasing magnitude in order to get the sample to an initial state where $B = 0$ when $H = 0$. Finally, we can also note that if we are at a value such as H_4 and then go through a *small* cycle in H starting there, we will trace out a little hysteresis loop like that illustrated.

In systems in which hysteresis occurs, there is also an irreversible conversion of energy into heat when the system is taken around a complete cycle. In the present case, it must correspond to a conversion of magnetic energy into heat; this is in addition to the heat produced because of the existence of a conductivity and that is described by (12-35). We can use our previous general result (20-79) to demonstrate this effect.

If we let δw_m be the energy required from the external source *per unit volume* of the material, we can write the integrand of (20-79) as $\delta w_m \, d\tau$ and we have

$$\delta w_m = \mathbf{H} \cdot \delta \mathbf{B} \qquad (20\text{-}101)$$

This is equal to the shaded area shown in Figure 20-23 since we are dealing only with

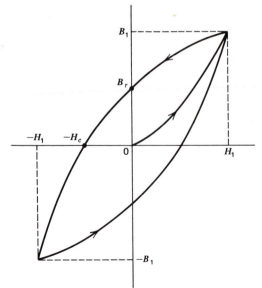

Figure 20-21. Hysteresis loop.

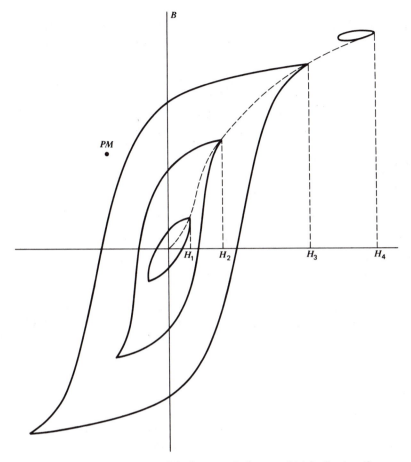

Figure 20-22.   Various possible hysteresis loops. _PM_ indicates the general state of a permanent magnet.

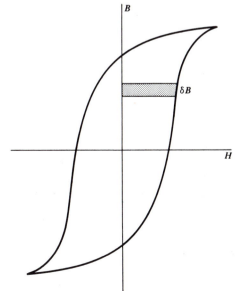

Figure 20-23.   Interpretation of _H δ B_ as an area.

isotropic materials where **B** and **H** are parallel. If we now imagine taking the system around one cycle of the hysteresis loop, the work done per unit volume will be the sum of terms like (20-101):

$$w_m = \oint_{cycle} \mathbf{H} \cdot \delta \mathbf{B} = \oint_{cycle} H \, \delta B \qquad (20\text{-}102)$$

This is numerically equal to the *area enclosed* by the hysteresis loop so that w_m is different from zero and positive. Since the initial and final states of the system are the same, the magnetic energy, which is a function only of the state, will not have changed, and therefore the work per unit volume given by (20-102) must represent an irreversible conversion of work into heat. This result can also be written in a form involving only the material. Since, by (20-28), **B** always equals $\mu_0(\mathbf{H} + \mathbf{M})$, we have

$$\delta w_m = \mathbf{H} \cdot \delta \mathbf{B} = \mu_0 \delta \left(\tfrac{1}{2} \mathbf{H}^2 \right) + \mu_0 \mathbf{H} \cdot \delta \mathbf{M} \qquad (20\text{-}103)$$

When this is inserted into (20-102), the first term will give zero, as we saw, for example, after (13-4), so that finally

$$w_m = \oint_{cycle} \mu_0 \mathbf{H} \cdot \delta \mathbf{M} \qquad (20\text{-}104)$$

The last term of (20-103), $\mu_0 \mathbf{H} \cdot \delta \mathbf{M}$, represents the work supplied to a unit volume of the material and provides the starting point for the description of magnetic systems in thermodynamics.

20-8 MAGNETIC CIRCUITS

By rewriting some of our results from the last section, we can obtain something that is both suggestive and useful. We will continue to write $B = \mu H$ with whatever value of μ is appropriate; then from (20-97), (20-99), and (20-100), we obtain

$$\Phi = \frac{\mu S N I}{2\pi b} = \frac{NI}{(l/\mu S)} = \frac{NI}{\mathscr{R}} = \frac{1}{\mathscr{R}} \oint_C \mathbf{H} \cdot d\mathbf{s} \qquad (20\text{-}105)$$

where $l = 2\pi b$ is the length of the path C around the toroid and

$$\mathscr{R} = \frac{l}{\mu S} \qquad (20\text{-}106)$$

We now recall the result of Exercise 12-14 in which, by using energy considerations, one could show that in a complete circuit of total resistance R containing an emf $\mathscr{E}$, the steady-state value of the current is

$$i = \frac{\mathscr{E}}{R} = \frac{1}{R} \oint_C \mathbf{E} \cdot d\mathbf{s} \qquad (20\text{-}107)$$

where $R = l/\sigma A$ by (12-28) where A is the cross-sectional area of the conductor of conductivity σ. Comparing these expressions, we see that there is a resemblance, and, by *analogy*, these systems are called *magnetic circuits*. Again, by analogy, there is a corresponding terminology; one can define a *magnetomotive force* or *mmf* $\mathscr{M}$ as

$$\mathscr{M} = \oint_C \mathbf{H} \cdot d\mathbf{s} = NI \qquad (20\text{-}108)$$

and $\mathscr{R}$ is called the *reluctance*. Then (20-105) can be written as

$$\Phi = \frac{\mathscr{M}}{\mathscr{R}} \qquad (20\text{-}109)$$

so that it looks like (20-107). Also from (20-106) we see that the permeability μ plays the same role in determining the reluctance as the conductivity σ does for the resistance.

The correspondence between the two cases is not exact. In the electrical circuit, the moving charges comprising the current do not leave the conductor, while in a general magnetic circuit, it is quite possible for the lines of **B** to be out in the space surrounding the material. This flux "leakage" can result in large quantitative errors in the application of (20-109). Even so, this method of discussing magnetic circuits does prove useful, particularly in aiding us to understand the overall features of the situation.

The main use of these ideas is in the design and construction of magnets. These are of two principal general classes: *electromagnets* in which the magnetic field is produced by a current in the windings, and *permanent magnets* in which one uses a ferromagnetic material that will have an induction present even in the absence of free currents. We will look at each of these, but, for simplicity, we consider only those of a toroidal shape.

Since these magnets are generally built to produce a magnetic field for experimental purposes, one needs a place to put the sample to be studied. The natural thing to do is to cut out a piece of the magnet material, thus making a "gap" of length l_0 as shown in Figure 20-24. We assume a small gap length; this is done in order that we can neglect the "fringing" properties of **B** discussed in Exercise 15-11 and illustrated for the electric field in Figure 6-10. In this way, we can assume that all of the lines of **B** present in the material are in the gap as well so that, by (20-100), we can take Φ to be constant and need not be concerned with a possible change in the effective cross-sectional area S.

■   **Example**

Electromagnet. If we let H_i be the magnetic field in the material of length $l - l_0$ where $l = 2\pi b$ as before, and H_0 the value in the gap of length l_0, we find that the mmf of (20-108) becomes

$$\oint_C \mathbf{H} \cdot d\mathbf{s} = H_i(l - l_0) + H_0 l_0 = NI \tag{20-110}$$

We assume that the surfaces forming the gap are cut so that they are perpendicular to **B**. Since the normal components of **B** are continuous, we have $B_i = B_0 = \Phi/S$;

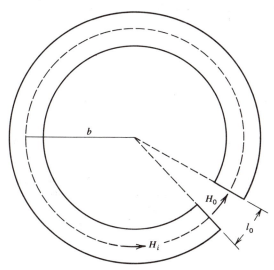

Figure 20-24. Toroidal magnet with a gap of length l_0.

furthermore, $H_i = B_i/\mu$, while $H_0 = B_0/\mu_0$. Thus we find that (20-110) can be written in terms of Φ and NI as

$$\Phi\left[\frac{(l - l_0)}{\mu S} + \frac{l_0}{\mu_0 S}\right] = NI \tag{20-111}$$

which again has the form (20-105) with the total reluctance of the circuit now given by

$$\mathscr{R} = \frac{(l - l_0)}{\mu S} + \frac{l_0}{\mu_0 S} = \mathscr{R}_i + \mathscr{R}_0 \tag{20-112}$$

In other words, these reluctances add, just as do resistances in series. We can also see that we can generalize the idea of reluctance to a circuit of varying properties so that we will have

$$\mathscr{R} = \oint_C \frac{ds}{\mu S} \tag{20-113}$$

If the gap is small ($l_0 \ll l$), then (20-112) can be approximated as

$$\mathscr{R} \simeq \frac{l}{\mu S} + \frac{l_0}{\mu_0 S} \tag{20-114}$$

Furthermore, if the material is something like iron with $\mu \approx 5000\mu_0$, it is often possible to neglect the first term and approximate $\mathscr{R}$ once again to get

$$\mathscr{R} \simeq \frac{l_0}{\mu_0 S} \tag{20-115}$$

so that the corresponding approximation to (20-111) would be

$$\Phi \simeq NI\frac{S\mu_0}{l_0} \tag{20-116}$$

This last approximation is not always valid since it depends both on the value of μ and on the relative values of the lengths l and l_0; it should really be checked before being used. In any event, (20-115) shows us that most of the reluctance is due to the vacuum (or air) in the gap, that is, the iron is more "permeable." Also since $B_i = B_0$, $H_i/H_0 = \mu_0/\mu$ so that $H_i \ll H_0$ and (20-110) becomes $\mathscr{M} = NI \simeq H_0 l_0$ showing that essentially all of the magnetomotive force producing the flux is applied "across" the gap, even though the gap is physically very small. ∎

■ **Example**

Permanent magnet. In this case, $I = 0$. The expression (20-110) is still valid, however, and we find that

$$H_i = -\frac{l_0 H_0}{(l - l_0)} \simeq -H_0\frac{l_0}{l} \tag{20-117}$$

showing that the magnetic fields in the two regions are oppositely directed. In the gap, $\mathbf{M} = 0$ and $H_0 = B_0/\mu_0$ so that $\mathbf{H}$ and $\mathbf{B}$ are in the same direction in the gap, but since $B_i = B_0$, $\mathbf{H}_i$ and $\mathbf{B}_i$ are oppositely directed within the magnet. Thus, a permanent magnet can be represented by a point in the second quadrant of a $B - H$ curve as is the point labeled *PM* in Figure 20-22.

In addition, $B_i = B_0 = \mu_0 H_0 = \mu_0(H_i + M_i)$ and we find that

$$M_i = \frac{B_i l}{\mu_0(l - l_0)} \simeq \frac{B_i}{\mu_0}\left(1 + \frac{l_0}{l}\right) \tag{20-118}$$

and is in the same direction as B_i, as is to be expected. The change in sign of **H**, which occurs at the surface separating the material from the vacuum, results from the discontinuity of **M** at this surface, that is, it arises from the surface density of "poles" as described by (20-42) and (20-43). Since M_i and H_i are oppositely directed in the material, H_i is often described as a "demagnetizing" field.   ■

EXERCISES

20-1 As we will see in Appendix B, molecular permanent magnetic dipole moments are typically of the magnitude of the Bohr magneton $\mu_e = eh/4\pi m_e = 9.27 \times 10^{-24}$ ampere-(meter)^2 where h is Planck's constant and m_e is the electron mass. Suppose each molecule of an ideal gas had a permanent moment of μ_e. Find the maximum possible magnetization for this gas at $100°$ C and one atmosphere.

20-2 A large sheet of material of thickness d has parallel faces that are perpendicular to the z axis. It is magnetized in such a way that $M = M(1 + \alpha z)\hat{z}$ where M and α are positive constants. Sketch the lines of **M**. Find J_m and K_m. Repeat this for the case where **M** is given by $M(1 + \alpha x)\hat{z}$.

20-3 A cube of edge a has the location and orientation shown for the figure in Figure 1-41. The material within the cube has a magnetization given by $M = -(M/a)y\hat{x} + (M/a)x\hat{y}$ where $M = $ const. Find the current densities J_m and K_m and sketch their directions.

20-4 A cylinder of length l and circular cross section of radius a has its axis along the z axis. It is magnetized such that $M = M\hat{x}$ where $M = $ const. Find J_m and K_m and sketch their directions.

20-5 A sphere of radius a has its center at the origin. Its magnetization is nonuniform and given by $M = (\alpha z^2 + \beta)\hat{z}$ where α and β are constants. What are the units of α and β? Find the magnetization current densities J_m and K_m as expressed in spherical coordinates.

20-6 Verify that the magnitudes and directions given for B_z by (20-22) and (20-25) also hold for negative values of z and thus agree with Figure 20-11.

20-7 A cylinder of length l and circular cross section of radius a has its axis along the z axis

and the origin at the center of the cylinder. It has a uniform magnetization $M = M\hat{z}$. Find **B** and **H** for all points on the z axis. Then use your results to discuss the limiting case of a very thin disc, that is, $l \ll a$. Find **B** outside at a point $z \gg l$. Find **B** and **H** inside the disc.

20-8 An infinitely long cylinder of radius a has its axis along the z axis. Its magnetization is given in cylindrical coordinates by $M = M_0(\rho/a)^2\hat{\varphi}$ where $M_0 = $ const. Find J_m and K_m. Verify that the total charge transferred is zero. Find the values of **B** and **H** everywhere, both inside and outside the cylinder.

20-9 A sphere of radius a has the origin at its center. It is magnetized in such a way that, in spherical coordinates, $M = M(r)\hat{r}$ and $M(0)$ is finite. Find J_m and K_m. If there are no free currents anywhere, find **B** inside the sphere. Find **H** inside the sphere. Find **H** at the surface of the sphere and just outside of it.

20-10 In order to use (20-33), one must be able to measure **B**. If one wants to do this by using a moving point charge and (14-30), one measurement of the force is not enough to determine **B**. Show that, if two measurements are made using mutually perpendicular velocities v_1 and v_2, and the forces are found to be F_1 and F_2, respectively, then **B** can be found from

$$B = \frac{1}{2qv_1^2v_2^2}\{ v_2^2 F_1 \times v_1 + v_1^2 F_2 \times v_2$$

$$+ [v_1 \cdot (F_2 \times v_2)]v_1$$

$$+ [v_2 \cdot (F_1 \times v_1)]v_2 \}$$

20-11 Since there are no magnetic poles, one should expect that the total pole strength (net magnetic charge) of a finite piece of magnetized material would be zero. Show that this is the case.

20-12 Show that the total magnetic dipole moment of a system as given by (20-2) can be written in terms of poles as

$$\mathbf{m} = \int_V \rho_m \mathbf{r} \, d\tau + \oint_S \sigma_m \mathbf{r} \, da$$

and compare with (8-22), for example. [*Hint*: recall (19-8) and (19-11).]

20-13 A very large sheet of material of thickness d has parallel faces that are perpendicular to the z axis. It is uniformly magnetized with $\mathbf{M} = M\hat{\mathbf{z}}$. Find the corresponding pole distribution, and using this find $\mathbf{H}$ everywhere. What is the demagnetizing factor in this case? Is your result compatible with those corresponding to the last sentence of Exercise 20-7?

20-14 Find the pole distribution for the cylinder of Exercise 20-4. Find ϕ_m everywhere. Show that the demagnetizing factor is $\frac{1}{2}$ for this case. (Assume that l is infinite.)

20-15 A spherical shell of radii a and b is uniformly magnetized so that $\mathbf{M} = M\hat{\mathbf{z}}$ for $a \leq r \leq b$. Choose the origin at the center of the sphere and find ϕ_m at all points on the positive z axis. Find H_z for positive values of z and sketch its behavior as a function of z. Verify that H_z satisfies the appropriate boundary conditions.

20-16 Consider a cylinder of length l and circular cross section of radius a with a permanent magnetization $\mathbf{M}$. A uniform external induction $\mathbf{B}$ is also present. Assume that $\mathbf{M}$ is unaffected by the presence of $\mathbf{B}$ and find the total torque on the cylinder. Use the result of Exercise 20-12 to express the torque in terms of the pole distribution and show that the result can be interpreted as saying that there is a force on a "point pole" q_m given by $q_m \mathbf{B}$. If there is vacuum everywhere else, show that $\mathbf{B}$ due to another point pole q'_m is given by $\mathbf{B} = \mu_0 q'_m \hat{\mathbf{R}}/4\pi R^2$, and thus deduce Coulomb's law for poles in the form $\mathbf{F}_{q'_m \rightarrow q_m} = \mu_0 q_m q'_m \hat{\mathbf{R}}/4\pi R^2$.

20-17 When you look up susceptibilities in tables, you will not generally find numerical values of χ_m. Instead, one usually finds either the *mass susceptibility* $\chi_{m,\text{mass}}$ or the *molar susceptibility* $\chi_{m,\text{molar}}$. These are defined so that $\chi_{m,\text{mass}} H$ and $\chi_{m,\text{molar}} H$ give the magnetic moment per unit mass and per mole, respectively. Find how χ_m is related to each of these; you will need to use the symbols for the mass density d and molecular weight A as well.

20-18 Suppose that we have a static situation involving a l.i.h. material with conductivity σ.

Show that then each component of $\mathbf{H}$ satisfies Laplace's equation, for example, $\nabla^2 H_x = 0$.

20-19 Lines of $\mathbf{B}$ can be "refracted" at a surface of discontinuity. Assume that there are no free surface currents on the boundary and that the angles giving the directions of $\mathbf{B}$ are measured from the normal as in the analogous electric case shown in Figure 10-14. (a) Show that the law of refraction is $\kappa_{m1} \cot \alpha_1 = \kappa_{m2} \cot \alpha_2$. (b) If $|\chi_m| \ll 1$ for both media, then $\mu_1 \simeq \mu_2 \simeq \mu_0$ and the angle of deviation $\delta = \alpha_2 - \alpha_1$ will be very small. Find an approximate expression for δ. Evaluate this result for $\alpha_1 = 45°$, region 1 a vacuum and region 2 a paramagnet for which $\chi_{m2} = 2.2 \times 10^{-5}$ (e.g., aluminum). (c) If $|\chi_m|$ is not very small, the result of (a) must be used. As extreme examples, evaluate δ for $\alpha_1 = 45°$, region 1 a vacuum, and: (i) $\chi_{m2} = -0.95$, a nearly "ideal diamagnet" (Note: if $\chi_m = -1$, then $\mathbf{B} = 0$); (ii) $\chi_{m2} = 500$, a moderate ferromagnet.

20-20 Find the expressions analogous to (20-59) and (20-60) when the material is not homogeneous and thus show that there can be a magnetization current density even in the absence of a free current density. Similarly, find an expression for the volume density of magnetic poles for this situation.

20-21 A sphere of radius a and relative permeability κ_m is placed in a previously uniform magnetic field $\mathbf{H}_0 = H_0 \hat{\mathbf{z}}$. Find $\mathbf{H}$ everywhere.

20-22 A spherical shell of radii a and b has a relative permeability κ_m for $a \leq r \leq b$. There is vacuum everywhere else. It is placed in a previously uniform magnetic field $\mathbf{H}_0 = H_0 \hat{\mathbf{z}}$. Show that the "shielding factor," that is, the ratio H_i/H_0 where H_i is the magnetic field in the cavity, is given by $9\kappa_m[(\kappa_m + 2)(2\kappa_m + 1) - 2(\kappa_m - 1)^2(a/b)^3]^{-1}$.

20-23 A very long straight current I is parallel to the plane surface of a semiinfinite l.i.h. magnetic material and a distance d from it. Show that the magnetic field in the vacuum region can be obtained as the resultant of this current and an image current $I' = I(\kappa_m - 1)/(\kappa_m + 1)$ located a distance d inside the material. Find the current I'', which will give the correct value of $\mathbf{H}$ within the material if it is located at the same position as I. What is the force per unit length between I and the material? Is this force attractive or repulsive?

20-24 In our discussion of the coaxial line, we assumed that the inner conductor (region 1) was nonmagnetic. Suppose that it is actually diamagnetic with susceptibility $\chi_{md} < 0$ while everything else is unchanged. Find **H**, **B**, **M**, and $\mathbf{J}_m$ for all points of region 1 and plot your results. Verify that the appropriate boundary conditions are satisfied at $\rho = a$.

20-25 Consider the same coaxial line as discussed at the end of Section 20-5, except that region 2 between the conductors is filled with a nonhomogeneous material such that $\kappa_m = \kappa(\rho/a)$ where $\kappa = $ const. Find **H** and **B** within this region and the contribution L_2 of a length l of this region to the self-inductance.

20-26 Consider the same coaxial line as discussed at the end of Section 20-5, except that region 2 between the conductors is filled with two l.i.h. materials as follows. For $a < \rho < c$, the relative permeability is κ_{m1}, while it is κ_{m2} for $c < \rho < b$. Find the contribution L_2 of a length l of this region to the self-inductance.

20-27 A paramagnetic liquid of mass density d is contained in a U-tube of circular cross section of radius a. When a field $\mathbf{H}_0$ is applied to one arm of the U-tube, the liquid level is found to rise a distance h into the region where $\mathbf{H}_0 \neq 0$. Assume that $\chi_m \ll 1$, neglect all end effects, and find an expression from which the susceptibility can be found.

20-28 A very long straight wire carries a current I. Another long wire of relative permeability κ_m and circular cross section of radius a is parallel to the first and at a distance b away from it where $b \gg a$. Show that there will be an attractive force per unit length between them given approximately by $(\kappa_m - 1)\mu_0 a^2 I^2 / 2(\kappa_m + 1)\pi b^3$.

20-29 An electromagnet is made in the form of a torus with a rectangular cross section. The width of the cross section is 2 centimeters, while the inner and outer radii are 7 and 8 centimeters. The coil has 1000 turns and has a gap cut into the core of length 0.25 millimeters. When the flux produced in the ring is 2.60×10^{-4} webers, the permeability of the core is $4250\mu_0$. Find the induction, the reluctance, the current in the windings, the magnetic field in the gap and in the iron core, and the fraction of the total magnetomotive force that is "across" the gap.

20-30 Show that the energy stored in a toroidal magnetic circuit can be written in the form $\frac{1}{2}\mathcal{R}\Phi^2$.

20-31 Find the expression giving the force attracting one of the faces forming the gap in an electromagnet toward the other face. Evaluate this force for the case described numerically in Exercise 20-29, and then express the force per unit area in atmospheres.

20-32 We noted before the last example of Section 20-6 that forces on magnetic materials can be ascribed to magnetic forces acting on the bound magnetization currents. Therefore, it may seem puzzling that we found a force on the permeable rod of Figure 20-18 that is in the z direction since both the magnetic field and induced magnetization were assumed to be uniform and along this direction; this would make the magnetic forces perpendicular to the solenoid axis. The origin of the net force can be found in the very thing we neglected—the "edge effects" or curvature of the actual lines of induction as illustrated in Figure 18-3. Show qualitatively that the net force on the bound surface currents as produced by the curved field lines will really be in the z direction and that the sign of the force agrees with that deduced from (20-94) found from energy considerations.

21
_____

MAXWELL'S EQUATIONS
_____

Since our last summary at the beginning of Chapter 17, we have added Faraday's law and defined the vector $\mathbf{H}$ as a result of including the magnetic effects of matter. Thus, at this point, our overall results can be summarized by the equations

$$\nabla \cdot \mathbf{D} = \rho_f \qquad \nabla \cdot \mathbf{B} = 0$$

$$\nabla \times \mathbf{E} = -\frac{\partial \mathbf{B}}{\partial t} \qquad \nabla \times \mathbf{H} = \mathbf{J}_f \tag{21-1}$$

as given by (10-41), (16-3), (17-30), and (20-29). These field vectors are related to each other and to matter through (10-40) and (20-28): $\mathbf{D} = \epsilon_0 \mathbf{E} + \mathbf{P}$ and $\mathbf{H} = (\mathbf{B}/\mu_0) - \mathbf{M}$. In addition, we have the equation of continuity (12-19), which describes the conservation of free charge:

$$\nabla \cdot \mathbf{J}_f + \frac{\partial \rho_f}{\partial t} = 0 \tag{21-2}$$

As we pointed out previously, we assume that these equations are still correct when the fields can vary with time even though, with the exception of Faraday's law, the equations were obtained from action at a distance laws that describe time-independent situations.

A great contribution of Maxwell to electromagnetic theory was his first pointing out that two of the equations of (21-1) are incompatible with charge conservation as described by (21-2), and then showing how this situation could be resolved by the introduction of yet another "current."

21-1 THE DISPLACEMENT CURRENT

As stated in (1-49), the divergence of the curl of any vector is always zero. If we calculate the divergence of the curl of $\mathbf{H}$ as given in (21-1) and use (21-2), we obtain

$$\nabla \cdot (\nabla \times \mathbf{H}) = \nabla \cdot \mathbf{J}_f = -\frac{\partial \rho_f}{\partial t} \tag{21-3}$$

Since we can generally expect to have $(\partial \rho_f / \partial t) \neq 0$, we see that we have a fundamental contradiction between (21-3) and the requirement of (1-49).

In order to remedy this situation, Maxwell assumed, in effect, that the equation for $\nabla \times \mathbf{H}$ was not yet complete, but must have another "current density" in it, that is, we _assume_ that it should really be of the form

$$\nabla \times \mathbf{H} = \mathbf{J}_f + \mathbf{J}_d \tag{21-4}$$

where $\mathbf{J}_d$ is the additional necessary term. Our aim now is to find $\mathbf{J}_d$.

If we substitute (21-4) into (1-49), and use (21-2) and $\nabla \cdot \mathbf{D} = \rho_f$, we obtain

$$\nabla \cdot (\nabla \times \mathbf{H}) = 0 = \nabla \cdot \mathbf{J}_f + \nabla \cdot \mathbf{J}_d = -\frac{\partial \rho_f}{\partial t} + \nabla \cdot \mathbf{J}_d = -\frac{\partial}{\partial t} \nabla \cdot \mathbf{D} + \nabla \cdot \mathbf{J}_d$$

so that

$$\nabla \cdot \left(\mathbf{J}_d - \frac{\partial \mathbf{D}}{\partial t} \right) = 0 \qquad (21\text{-}5)$$

since we can interchange the order of partial differentiation. Therefore, whatever $\mathbf{J}_d$ is, it must satisfy (21-5). The simplest assumption, of course, is that the term in the parentheses is zero; accordingly, Maxwell assumed that

$$\mathbf{J}_d = \frac{\partial \mathbf{D}}{\partial t} \qquad (21\text{-}6)$$

so that (21-4) can also be written

$$\nabla \times \mathbf{H} = \mathbf{J}_f + \frac{\partial \mathbf{D}}{\partial t} \qquad (21\text{-}7)$$

and is now completely consistent with (21-2). We can also note that (21-5) could be satisfied in general by writing $\mathbf{J}_d = (\partial \mathbf{D}/\partial t) + \nabla \times \mathbf{G}$ where $\mathbf{G}$ is some arbitrary vector; however, it has never yet been found necessary to do this and it would be needlessly complicating to carry this extra term along.

This new current density $\mathbf{J}_d$ was called the *displacement current* by Maxwell. It has important consequences and is needed to make our later results agree with experiment, and, as we will see in Chapter 24, it is essential for the existence of electromagnetic waves.

Now that our form for $\nabla \times \mathbf{H}$ has been changed, we must investigate how this will affect some general results that depend on it. First, the integral form of Ampère's law for $\mathbf{H}$ is obtained by combining (21-7) with Stokes' theorem (1-67), and it becomes

$$\oint_C \mathbf{H} \cdot d\mathbf{s} = \int_S (\mathbf{J}_f + \mathbf{J}_d) \cdot d\mathbf{a} = \int_S \mathbf{J}_f \cdot d\mathbf{a} + \int_S \frac{\partial \mathbf{D}}{\partial t} \cdot d\mathbf{a}$$
$$= I_{f,\,\text{enc}} + I_{d,\,\text{enc}} \qquad (21\text{-}8)$$

where $I_{d,\,\text{enc}}$ is the total displacement current enclosed by the path of integration. When $\mathbf{D}$ is constant, $\partial \mathbf{D}/\partial t = 0$ and (21-8) properly reduces to our previous expression (20-32).

Another important result that depends on $\nabla \times \mathbf{H}$ is the boundary condition for the tangential components of $\mathbf{H}$ at a surface of discontinuity in properties. If we put (21-7) into our general result (9-13), we obtain

$$\hat{\mathbf{n}} \times (\mathbf{H}_2 - \mathbf{H}_1) = \lim_{h \to 0} \left[h \left(\mathbf{J}_f + \frac{\partial \mathbf{D}}{\partial t} \right) \right] \qquad (21\text{-}9)$$

As before, we can use the analogue of (15-14) for free currents to get $\lim_{h \to 0}(h\mathbf{J}_f) = \mathbf{K}_f$. Furthermore, just as we did preceding (17-13), we can expect that $\partial \mathbf{D}/\partial t$ will remain finite as the transition layer is shrunk to zero thickness so that, as $h \to 0$, $h(\partial \mathbf{D}/\partial t) \to 0$ and (21-9) will become $\hat{\mathbf{n}} \times (\mathbf{H}_2 - \mathbf{H}_1) = \mathbf{K}_f$, which is exactly (20-30) again. Therefore the introduction of the displacement current has not changed this boundary condition so that the tangential components of $\mathbf{H}$ will change only as a result of the existence of *free* surface currents.

■ **Example**

Charging capacitor. In Figure 21-1, we have a side view of a parallel plate capacitor with circular plates of radius a. For simplicity, we assume a vacuum between the plates, and, as usual, we also assume that the separation between the plates is so small

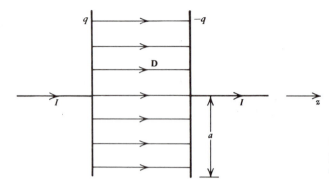

Figure 21-1. Side view of a parallel plate capacitor being charged.

compared with the radius that the electric field **E** can be taken to be uniform and confined entirely to the region between the plates. The electric field is given by (6-4) as **E** $= (\sigma_f/\epsilon_0)\hat{\mathbf{z}} = (q/\epsilon_0\pi a^2)\hat{\mathbf{z}}$ in terms of the surface density σ_f, the total charge q, and the area πa^2 of a plate. Therefore, in the region between the plates,

$$\mathbf{D} = \epsilon_0 \mathbf{E} = \frac{q}{\pi a^2}\hat{\mathbf{z}} \qquad (21\text{-}10)$$

These results are those corresponding to the electrostatic case in which the charge has a constant value and is uniformly distributed over the surfaces of the plates. Now let us assume that the capacitor is being charged at a constant rate dq/dt. We can imagine this being done by means of a current $I = dq/dt$ that is traveling in a long thin wire coinciding with the z axis, as illustrated. However, we want to continute to treat this case as if the charge is still uniformly distributed over the plates so that we can continue to use (21-10). In other words, we assume that the charge is essentially instantaneously distributed over the plate surface. What this means, in effect, is that the capacitor is being charged at such a low rate that any appreciable amount of charge transferred to the plate is added in a time large compared to the relaxation time of the conductor that we discussed in Section 12-6. We saw there that this time could well be as short as 10^{-14} second for a good conductor so that we will have no difficulty in assuming that I is small enough so that the charge distribution is nearly always the same as that of the equilibrium electrostatic case, that is, we are dealing with a quasistatic process.

Since q is not constant, the displacement current as obtained from (21-6) and (21-10) is

$$\mathbf{J}_d = \frac{\partial \mathbf{D}}{\partial t} = \frac{I}{\pi a^2}\hat{\mathbf{z}} \qquad (21\text{-}11)$$

Thus, we have the equivalent of a uniform current density between the plates as shown in Figure 21-2; we use the word "equivalent" because there is *no* transfer of real charge in the vacuum region between the plates, but, by (21-7), there is a source term for the magnetic field. There is a numerical equivalence, however, for we find from (21-11) and (12-6) that the total "current" between the plates is

$$I_d = \int_S \mathbf{J}_d \cdot d\mathbf{a} = \int_S \frac{I}{\pi a^2}\hat{\mathbf{z}} \cdot d\mathbf{a} = I \qquad (21\text{-}12)$$

and is thus equal to the total real current bringing charge up to the capacitor.

We can use (21-8) to calculate **H**. With circular plates, and the current I along the z axis, we will have axial symmetry about the z axis and we can conclude that **H** has the general form $\mathbf{H} = H_\varphi(\rho)\hat{\boldsymbol{\varphi}}$. Therefore, choosing a circular path of integration of radius ρ centered on the z axis and perpendicular to it, we will find that the line integral in

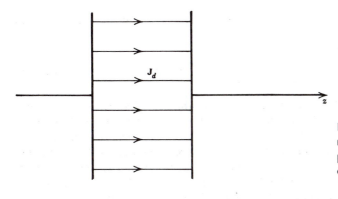

Figure 21-2. The displacement current between the plates of a parallel plate capacitor being charged.

(21-8) becomes $H_\varphi 2\pi\rho$ as usual so that

$$H_\varphi(\rho) = \frac{I_{f,\text{enc}} + I_{d,\text{enc}}}{2\pi\rho} \tag{21-13}$$

It will be helpful to divide space into the four regions shown in Figure 21-3. Region 1 is the volume between the capacitor plates, and 2 is the remainder of that enclosed by the two parallel planes which coincide in part with the plates. Regions 3 and 4 are the rest of space; I is different from zero within them.

In regions 3 and 4, $I_{f,\text{enc}} = I$ while $I_{d,\text{enc}} = 0$ and we get

$$H_{\varphi 3}(\rho) = H_{\varphi 4}(\rho) = \frac{I}{2\pi\rho} \tag{21-14}$$

For a path in region 2, $I_{f,\text{enc}} = 0$ while $I_{d,\text{enc}} = I_d = I$ by (21-12) so that

$$H_{\varphi 2}(\rho) = \frac{I}{2\pi\rho} \qquad (\rho > a) \tag{21-15}$$

Although (21-14) and (21-15) look alike, they were found by using *different sources*—a real current and a displacement current, respectively. Finally, for region 1 where $\rho \le a$,

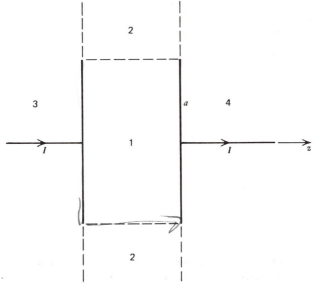

Figure 21-3. Regions used for calculating the magnetic field.

we have $I_{f,\text{enc}} = 0$ and $I_{d,\text{enc}}/I_d = I_{d,\text{enc}}/I = \pi\rho^2/\pi a^2 = \rho^2/a^2$ so that (21-13) gives us another familiar-looking expression:

$$H_{\varphi 1}(\rho) = \frac{I_d \rho}{2\pi a^2} = \frac{I\rho}{2\pi a^2} \qquad (\rho \le a) \qquad (21\text{-}16)$$

If we first consider values of $\rho > a$, we see from the above results that as we go from region 3 to 2 and then from 2 to 4, the tangential components of $\mathbf{H}$ are continuous as we expect from (20-31) since there are no free currents on these imaginary boundaries. On the other hand, if the displacement current had not been included in this particular calculation for this axially symmetric case, we would have concluded that $\mathbf{H} = 0$ in region 2 (and 1 as well); this would have led to a discontinuity in the tangential components of $\mathbf{H}$ across the boundary between 3 and 2 where there is *no* real current in direct contradiction to all of our experience and expectations. Maxwell also considered this problem and it was arguments such as these that helped persuade him of the necessity of the existence of the displacement current.

If we now consider values of $\rho < a$, there *is* a discontinuity in the tangential components of $\mathbf{H}$ as we go from region 3 to 1, say, and we find from (21-16) and (21-14) that the difference is

$$\mathbf{H}_1 - \mathbf{H}_3 = \frac{I}{2\pi\rho}\left(\frac{\rho^2}{a^2} - 1\right)\hat{\boldsymbol{\varphi}} \qquad (21\text{-}17)$$

and, according to (20-31), this must be equal to $\mathbf{K}_f \times \hat{\mathbf{n}}$ which in this case is $\mathbf{K}_f \times \hat{\mathbf{n}}_{3 \to 1} = \mathbf{K}_f \times \hat{\mathbf{z}}$. In order to verify this, we need to know $\mathbf{K}_f$, which we can find with the aid of Figure 21-4. In a time Δt, a charge $\Delta q = I\Delta t$ is brought to the plates. By our assumption, this is distributed "instantaneously" and uniformly over the plates. By symmetry, $\mathbf{K}_f$ must be radial and therefore has the form $\mathbf{K}_f = K_f(\rho)\hat{\boldsymbol{\rho}}$. As shown in the figure, the surface current through the circle of radius ρ must be distributing the charge that is to go to the remainder of the plate outside of the shaded circle of radius ρ. Therefore, in a time Δt, the total charge transferred across this circle as obtained from Figure 12-5a is

$$\int K_f \, ds \, \Delta t = K_f \Delta t \, 2\pi\rho = \Delta\sigma_f\left(\pi a^2 - \pi\rho^2\right)$$

$$= \frac{\Delta q}{\pi a^2}\left(\pi a^2 - \pi\rho^2\right) = I\Delta t\left(1 - \frac{\rho^2}{a^2}\right)$$

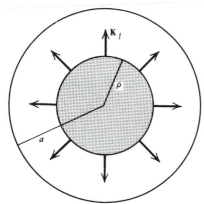

Figure 21-4.   Calculation of surface current density on the plate of a capacitor being charged.

and therefore $K_f = (I/2\pi\rho)[1 - (\rho^2/a^2)]$ so that

$$\mathbf{K}_f = K_f(\rho)\hat{\rho} = \frac{I}{2\pi\rho}\left(1 - \frac{\rho^2}{a^2}\right)\hat{\rho} \tag{21-18}$$

Thus, with the use of (1-76), we find that $\mathbf{K}_f \times \hat{\mathbf{z}} = K_f\hat{\rho} \times \hat{\mathbf{z}} = -K_f\hat{\varphi} = (I/2\pi\rho)[(\rho^2/a^2) - 1]\hat{\varphi}$, which is exactly the right-hand side of (21-17). Thus the boundary conditions are precisely satisfied, and we can point out again that this would not have been the case if we had failed to include the displacement current as a source of $\mathbf{H}$. ∎

If $\mathbf{J}_f = 0$, then $\nabla \times \mathbf{H} = \partial\mathbf{D}/\partial t$ and there is still a possible source of $\mathbf{H}$. In this form, there is a sort of analogy between it and Faraday's law given in (21-1), which one can state somewhat crudely by saying that Faraday's law describes an electric field as being produced by a changing magnetic field vector, while (21-7) describes a magnetic field as being produced by a changing electric field vector.

21-2 MAXWELL'S EQUATIONS IN GENERAL FORM

At this point, finally, our present knowledge of macroscopic descriptive electromagnetism is complete and can be summarized by the equations (21-1) as extended in (21-7):

$$\nabla \cdot \mathbf{D} = \rho_f \tag{21-19}$$

$$\nabla \times \mathbf{E} = -\frac{\partial\mathbf{B}}{\partial t} \tag{21-20}$$

$$\nabla \cdot \mathbf{B} = 0 \tag{21-21}$$

$$\nabla \times \mathbf{H} = \mathbf{J}_f + \frac{\partial\mathbf{D}}{\partial t} \tag{21-22}$$

These equations are known as *Maxwell's equations* and are assumed to be always true. [These actually correspond to a total of eight scalar equations, since (21-20) and (21-22) each have three components.]

It is worthwhile reminding ourselves of the physical content of these equations. Equation 21-19 summarizes Coulomb's law of force between point charges plus the electrical effects of matter, while (21-20) represents Faraday's law of induction, and is also compatible with Coulomb's law for static fields. The third member of the group (21-21), is a consequence of Ampère's law of force between currents and also reflects the fact that free magnetic charges are not known to exist. Finally, the last one, (21-22), includes Ampère's law of force between currents plus the magnetic effects of matter plus conservation of free charge; the last follows from the fact that the equation of continuity (21-2) can be derived from (21-22), (21-19), and (1-49), and thus need no longer be written separately.

These differential source equations must of course be supplemented by the defining equations (10-40) and (20-28), which relate pairs of the field vectors along with the description of matter in terms of the corresponding volume densities of dipole moments:

$$\mathbf{D} = \epsilon_0\mathbf{E} + \mathbf{P} \tag{21-23}$$

$$\mathbf{H} = \frac{\mathbf{B}}{\mu_0} - \mathbf{M} \tag{21-24}$$

Although the boundary conditions at a surface of discontinuity can always be obtained from Maxwell's equations and our general results from Chapter 9, it is usual and convenient to list them separately. They are given by (10-42), (9-16) and (17-13), (16-4), and (20-30) and (20-31):

$$\hat{\mathbf{n}} \cdot (\mathbf{D}_2 - \mathbf{D}_1) = \sigma_f \tag{21-25}$$

$$\hat{\mathbf{n}} \times (\mathbf{E}_2 - \mathbf{E}_1) = 0 \quad \text{or} \quad \mathbf{E}_{2t} = \mathbf{E}_{1t} \tag{21-26}$$

$$\hat{\mathbf{n}} \cdot (\mathbf{B}_2 - \mathbf{B}_1) = 0 \tag{21-27}$$

$$\hat{\mathbf{n}} \times (\mathbf{H}_2 - \mathbf{H}_1) = \mathbf{K}_f \quad \text{or} \quad \mathbf{H}_{2t} - \mathbf{H}_{1t} = \mathbf{K}_f \times \hat{\mathbf{n}} \tag{21-28}$$

where we remember that $\hat{\mathbf{n}}$ in these equations is always drawn from region 1 to region 2. (Note here that many books, particularly more advanced ones, drop the subscript f used to denote free charges and currents and require that the reader remember that this is what is meant; thus one commonly finds all of these equations written in terms of the symbols ρ, $\mathbf{J}$, σ, and $\mathbf{K}$. For convenience, however, we will not do this but continue using the subscript.)

These equations all describe the behavior of the field vectors. The basic connection between the fields and their effect on charged particles is described by the Lorentz force on a point charge q as given by (14-32):

$$\mathbf{F} = q(\mathbf{E} + \mathbf{v} \times \mathbf{B}) \tag{21-29}$$

where $\mathbf{v}$ is the velocity of the charge.

It is often convenient to have Maxwell's equations expressed in terms of only two vectors—one electric and one magnetic. For example, if we use (21-23) and (21-24) to eliminate $\mathbf{D}$ and $\mathbf{H}$ from (21-19) through (21-22), we obtain Maxwell's equations expressed only in terms of $\mathbf{E}$ and $\mathbf{B}$:

$$\nabla \cdot \mathbf{E} = \frac{1}{\epsilon_0}(\rho_f - \nabla \cdot \mathbf{P}) \tag{21-30}$$

$$\nabla \times \mathbf{E} = -\frac{\partial \mathbf{B}}{\partial t} \tag{21-31}$$

$$\nabla \cdot \mathbf{B} = 0 \tag{21-32}$$

$$\nabla \times \mathbf{B} = \mu_0 \left(\mathbf{J}_f + \nabla \times \mathbf{M} + \epsilon_0 \frac{\partial \mathbf{E}}{\partial t} + \frac{\partial \mathbf{P}}{\partial t} \right) \tag{21-33}$$

Two of these equations are unchanged from before and the other two are no longer quite as compact since the properties of the matter now appear explicitly. Nevertheless, their form is such as to be easily understood. The term in the parentheses of (21-30) is clearly the total charge density written as the sum of the free and bound charge densities as we found in (10-38). Similarly, the term in parentheses of (21-33) represents the total current density $\mathbf{J}_{tot}$. The first two parts are clearly the free current density and the magnetization current density of (20-10). The last two terms together are the displacement current density that now is seen to have two contributions: the first, $\epsilon_0(\partial \mathbf{E}/\partial t)$, is present even if there is no matter and is often called the *vacuum displacement current* density and the last term $\partial \mathbf{P}/\partial t$ we met before in (12-18) as the polarization current density associated with the motion of bound charges. Unless $\mathbf{P}$ and $\mathbf{M}$ represent permanent polarization and magnetization, there is an additional dependence on $\mathbf{E}$ and $\mathbf{B}$ in these equations in that we still have the possibility that $\mathbf{P} = \mathbf{P(E)}$ and $\mathbf{M} = \mathbf{M(B)}$ and until these functional relationships are known, these equations will be of limited utility.

The boundary conditions can also be expressed completely in terms of **E** and **B** and we find that the ones that do change in form become

$$\hat{\mathbf{n}} \cdot (\mathbf{E}_2 - \mathbf{E}_1) = \frac{1}{\epsilon_0} \left[\sigma_f - \hat{\mathbf{n}} \cdot (\mathbf{P}_2 - \mathbf{P}_1) \right] \tag{21-34}$$

$$\hat{\mathbf{n}} \times (\mathbf{B}_2 - \mathbf{B}_1) = \mu_0 \left[\mathbf{K}_f + \hat{\mathbf{n}} \times (\mathbf{M}_2 - \mathbf{M}_1) \right] \tag{21-35}$$

in agreement with our previous results (10-12) and (20-12).

In a similar manner, one can write the basic equations in terms of the pairs (**E, H**), (**D, B**), and (**D, H**) if it is desirable and the terms that arise can be interpreted in the same way.

The so-called *integral forms* of Maxwell's equations are obtained by combining the divergence theorem (1-59) and Stokes' theorem (1-67) with the appropriate members of (21-19) through (21-22) and the results are

$$\oint_S \mathbf{D} \cdot d\mathbf{a} = \int_V \rho_f \, d\tau \tag{21-36}$$

$$\oint_C \mathbf{E} \cdot d\mathbf{s} = -\int_S \frac{\partial \mathbf{B}}{\partial t} \cdot d\mathbf{a} \tag{21-37}$$

$$\oint_S \mathbf{B} \cdot d\mathbf{a} = 0 \tag{21-38}$$

$$\oint_C \mathbf{H} \cdot d\mathbf{s} = \int_S \mathbf{J}_f \cdot d\mathbf{a} + \int_S \frac{\partial \mathbf{D}}{\partial t} \cdot d\mathbf{a} \tag{21-39}$$

A little reflection will convince one that all of the general results that we have found in the previous chapters can be obtained from the equations that have been summarized in this section, often by working backward from the way in which we originally got them.

Finally, we can point out that Maxwell's equations are *linear* differential equations. Consequently, if two or more electromagnetic fields that are solutions of them are known, the *sum* of these fields will also be a solution. This is often referred to as the *superposition property* of the electromagnetic field or as the *superposition principle*.

21-3 MAXWELL'S EQUATIONS FOR LINEAR ISOTROPIC HOMOGENEOUS MEDIA

An important special and simplifying situation occurs when we are dealing with linear isotropic homogeneous media. For simplicity, we restrict ourselves to the case where the materials are l.i.h. in *all* of their properties. Then we can use the constitutive equations (10-51) and (20-53) to write

$$\mathbf{D} = \epsilon \mathbf{E} \qquad \mathbf{B} = \mu \mathbf{H} \tag{21-40}$$

In addition, the free current would be given by (12-25) as $\mathbf{J}_f = \sigma \mathbf{E}$. The quantities ϵ, μ, and σ are scalar constants, characteristic of the material. Now it may happen that there are free currents present other than those due to the conductivity of the material; for example, the material could have beams of charged particles incident on it from an external source. Thus, if we let $\mathbf{J}_f'$ be the free current density arising from sources other than conductivity, we can add this to $\sigma \mathbf{E}$ and write the total free current as

$$\mathbf{J}_f = \sigma \mathbf{E} + \mathbf{J}_f' \tag{21-41}$$

Although (21-41) represents a general possibility that should be kept in mind, we actually have little need of it in our applications, but we will continue nevertheless to include $\mathbf{J}_f'$ for completeness.

If we insert these expressions into (21-19) through (21-22), we can, for example, express them in terms of $\mathbf{E}$ and $\mathbf{B}$ only as

$$\nabla \cdot \mathbf{E} = \frac{\rho_f}{\epsilon} \tag{21-42}$$

$$\nabla \times \mathbf{E} = -\frac{\partial \mathbf{B}}{\partial t} \tag{21-43}$$

$$\nabla \cdot \mathbf{B} = 0 \tag{21-44}$$

$$\nabla \times \mathbf{B} = \mu\sigma\mathbf{E} + \mu\mathbf{J}_f' + \mu\epsilon\frac{\partial \mathbf{E}}{\partial t} \tag{21-45}$$

and the corresponding boundary conditions that are changed in form become

$$\hat{\mathbf{n}} \cdot (\epsilon_2 \mathbf{E}_2 - \epsilon_1 \mathbf{E}_1) = \sigma_f \tag{21-46}$$

$$\frac{\mathbf{B}_{2t}}{\mu_2} - \frac{\mathbf{B}_{1t}}{\mu_1} = \mathbf{K}_f \times \hat{\mathbf{n}} \tag{21-47}$$

Similarly, if the equations are expressed completely in terms of $\mathbf{E}$ and $\mathbf{H}$, we obtain

$$\nabla \cdot \mathbf{E} = \frac{\rho_f}{\epsilon} \tag{21-48}$$

$$\nabla \times \mathbf{E} = -\mu\frac{\partial \mathbf{H}}{\partial t} \tag{21-49}$$

$$\nabla \cdot \mathbf{H} = 0 \tag{21-50}$$

$$\nabla \times \mathbf{H} = \sigma\mathbf{E} + \mathbf{J}_f' + \epsilon\frac{\partial \mathbf{E}}{\partial t} \tag{21-51}$$

along with

$$\hat{\mathbf{n}} \cdot (\mu_2 \mathbf{H}_2 - \mu_1 \mathbf{H}_1) = 0 \tag{21-52}$$

In contrast to the equations of the previous section, these results are *not* always true, but they are *often* true.

The integral forms of Maxwell's equations can be obtained for this restricted case in the same manner by which one gets (21-36) through (21-39). The general superposition principle also applies to these materials, of course.

21-4 POYNTING'S THEOREM

We are now in a position to reconsider energy for these general fields described by Maxwell's equations. We recall our previous result of (12-35) that the rate at which electromagnetic energy is dissipated into heat per unit volume is given by $w = \mathbf{J}_f \cdot \mathbf{E}$. Therefore, if we integrate this expression over an arbitrary volume V, the total rate of loss of electromagnetic energy $\mathscr{W}$ will be

$$\mathscr{W} = \int_V w \, d\tau = \int_V \mathbf{J}_f \cdot \mathbf{E} \, d\tau \tag{21-53}$$

and we would now like to express this in terms of the field vectors. If we use $\mathbf{J}_f$ as given by (21-22), we find that we can also write

$$\mathscr{W} = \int_V \mathbf{E} \cdot (\nabla \times \mathbf{H}) \, d\tau - \int_V \mathbf{E} \cdot \frac{\partial \mathbf{D}}{\partial t} \, d\tau \qquad (21\text{-}54)$$

Combining (1-116) with (21-20), we can write the first integrand as

$$\mathbf{E} \cdot (\nabla \times \mathbf{H}) = \mathbf{H} \cdot (\nabla \times \mathbf{E}) - \nabla \cdot (\mathbf{E} \times \mathbf{H}) = -\mathbf{H} \cdot \frac{\partial \mathbf{B}}{\partial t} - \nabla \cdot (\mathbf{E} \times \mathbf{H})$$

If we now insert this into (21-54), use the divergence theorem (1-59) and transfer the result to the other side of the equation, we obtain

$$-\int_V \left(\mathbf{E} \cdot \frac{\partial \mathbf{D}}{\partial t} + \mathbf{H} \cdot \frac{\partial \mathbf{B}}{\partial t} \right) d\tau = \mathscr{W} + \oint_S (\mathbf{E} \times \mathbf{H}) \cdot d\mathbf{a} \qquad (21\text{-}55)$$

which is known as *Poynting's theorem*.

So far, our treatment has been perfectly general. However, it is easiest to interpret this result if we assume completely l.i.h. media, and since we will apply it only to these cases anyhow, we assume this to be the case. Using (21-40), we get $\mathbf{E} \cdot (\partial \mathbf{D}/\partial t) = \epsilon \mathbf{E} \cdot (\partial \mathbf{E}/\partial t) = \partial(\frac{1}{2}\epsilon \mathbf{E}^2)/\partial t$; similarly, we find that $\mathbf{H} \cdot (\partial \mathbf{B}/\partial t) = \partial[(\mathbf{B}^2/2\mu)]/\partial t$. Substituting these into (21-55), recalling that we are dealing with a volume V with fixed boundaries so that we can interchange the order of differentiation and integration, and reintroducing the integral expression for $\mathscr{W}$ from (21-53), we finally get

$$-\frac{\partial}{\partial t} \int_V \left(\frac{1}{2}\epsilon \mathbf{E}^2 + \frac{\mathbf{B}^2}{2\mu} \right) d\tau = \int_V \mathbf{J}_f \cdot \mathbf{E} \, d\tau + \oint_S (\mathbf{E} \times \mathbf{H}) \cdot d\mathbf{a} \qquad (21\text{-}56)$$

Now the integrand on the left-hand side is just the sum of the expressions (10-85) and (20-84) that we previously found to be the electric and magnetic energy densities, respectively. If we now make the reasonable assumption that they can be interpreted in exactly this same way when the fields vary with time, then the integrand will be the total electromagnetic energy density

$$u = u_e + u_m = \frac{1}{2}\epsilon \mathbf{E}^2 + \frac{\mathbf{B}^2}{2\mu} \qquad (21\text{-}57)$$

and the integral will be exactly the total electromagnetic energy contained in the volume V. Consequently, the left-hand side of (21-56) represents the rate of *decrease* of this total energy. Where does it go? The first term on the the right-hand side is the rate at which this energy is being converted into heat. If we now invoke conservation of energy, then whatever energy is not converted into heat *must* pass out from the volume V across its bounding surface S. Since the remaining term has just this form of an integral over the surface, we can interpret

$$\oint_S (\mathbf{E} \times \mathbf{H}) \cdot d\mathbf{a} = \left(\frac{dU}{dt} \right)_{\text{through } S} \qquad (21\text{-}58)$$

as being the rate of flow of energy across the bounding surface. The next thing we can do is to give a similar interpretation to the *integrand* of (21-58); thus

$$\mathbf{S} = \mathbf{E} \times \mathbf{H} \qquad (21\text{-}59)$$

will be taken to be the rate of flow of electromagnetic energy per unit area. In other words, it is an *energy current density* or *power flux* and will be measured in watts/(meter)^2. The quantity $\mathbf{S}$ is called the *Poynting vector*; its direction is the same as that of the instantaneous flow of energy, that is, it "points" in the direction of energy flow. Although we have abstracted this interpretation of the *integrand* from the physical meaning of the whole *integral*, it is certainly a plausible interpretation and it is one which has turned out to be consistent and useful, particularly in its application to time-dependent solutions of Maxwell's equations.

Even though most of the applications of this result are made to fields that change in time, we can consider an example to see that this interpretation of **S** gives reasonable and consistent results when applied to a simple case involving *steady-state* fields, that is, time-independent ones.

■ Example

Cylinder with constant current. Consider a portion of a long straight cylindrical conductor of length *l* and radius *a* as shown in Figure 21-5. We assume a constant current in the *z* direction, which is distributed uniformly across the cross section so that $\mathbf{J}_f = J_f\hat{\mathbf{z}} = \text{const.}$ We want to find **S** at the surface just outside of the conductor.

This is similar to the situation depicted in Figure 12-7 and, in the paragraph following (12-28), we found the electric field just outside the cylinder to be equal to the constant value inside and given by $\mathbf{E} = \mathbf{J}_f/\sigma$ making it parallel to the axis as shown. In (15-19) we have the value of **B** outside the cylinder so that, in this case, **H** just outside will be $\mathbf{H} = \mathbf{B}/\mu_0$, or

$$\mathbf{H} = \frac{I}{2\pi a}\hat{\boldsymbol{\varphi}} = \frac{1}{2}J_f a\hat{\boldsymbol{\varphi}} \tag{21-60}$$

since $J_f = I/\pi a^2$. Substituting these expressions into (21-59), and using (1-76), we find that

$$\mathbf{S} = \frac{J_f}{\sigma}\hat{\mathbf{z}} \times \frac{1}{2}J_f a\hat{\boldsymbol{\varphi}} = -\frac{J_f^2 a}{2\sigma}\hat{\boldsymbol{\rho}} \tag{21-61}$$

We see, as is also clear from the figure, that **S** is normal to the surface and directed radially inward so that there is a steady flow of energy into the conductor as *S* is constant. Now the element of area *d**a*** is directed outward from the surface; also, **S** is

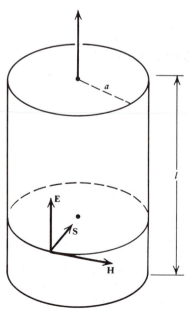

Figure 21-5.   Fields and Poynting vector for a cylinder carrying a constant current.

parallel to the ends of the cylinder. Therefore, the total rate at which energy is flowing *into* the volume is given by

$$-\oint \mathbf{S} \cdot d\mathbf{a} = S \int da = S(2\pi a l) = \frac{J_f^2}{\sigma}(\pi a^2 l) = \frac{J_f^2}{\sigma}(\text{volume}) \qquad (21\text{-}62)$$

with the use of (21-61) and where $\pi a^2 l$ is the volume of the conductor. If we compare this with (12-35), we see that (21-62) says that the total rate at which energy is flowing into the conductor is exactly equal to the rate at which energy is being dissipated into heat within the volume. This is exactly what is required for the steady-state situation we have assumed. The ultimate source of this energy is some device, like a battery, which maintains the potential difference between the ends of the conductor by continuously doing work on the charges as they pass through the complete circuit. Our description here involving **S** depicts the energy as being transferred from the source by means of the fields that it establishes throughout space as a result of the charge and current distributions within it. Finally, the energy flow passes normally through the surface of the conductor in just the right amount to be transformed into heat. Thus the interpretation given to the Poynting vector provides a complete and internally consistent description of the processes involved in this case.   ∎

21-5 ELECTROMAGNETIC MOMENTUM

We have just seen how energy density and energy flow can be ascribed to the electromagnetic field. It is also possible to associate momentum with it and it is, in fact, gratifying that this is so. We recall that the Coulomb forces between two point charges are equal and opposite, that is, they are in agreement with Newton's third law. On the other hand, we found in (13-19) that the forces between current elements as given by (13-17) are not generally equal and opposite so that these forces do not satisfy the third law. As we know from mechanics, it is the fact that the forces between the point masses of a composite system are equal and opposite that is the basis for the initial derivation of the *conservation of linear momentum* of an isolated system. Since this momentum conservation property is such a basic and far-reaching general principle, it would be extremely disturbing if our fundamental magnetic force law (13-17) somehow forced us to conclude that conservation of momentum was no longer a valid general principle in the presence of electromagnetic fields. Our aim now is to see how it can be retained. For simplicity, however, we do this only for the vacuum case.

We consider a distribution of free charges and currents occupying a volume V and described by the densities ρ_f and $\mathbf{J}_f$. If we let $d\mathbf{F}$ be the force on the charges and currents in a volume $d\tau$, we can combine (21-29) with (2-14) and (12-3) to get

$$d\mathbf{F} = (\rho_f \mathbf{E} + \mathbf{J}_f \times \mathbf{B}) \, d\tau \qquad (21\text{-}63)$$

Thus, we see that we can introduce a force per unit volume $\mathbf{f}_v$ by means of

$$\mathbf{f}_v = \frac{d\mathbf{F}}{d\tau} = \rho_f \mathbf{E} + \mathbf{J}_f \times \mathbf{B} \qquad (21\text{-}64)$$

Furthermore, the total electromagnetic force on our whole system can be obtained by summing (21-63) over the whole distribution to give

$$\mathbf{F} = \int_V \mathbf{f}_v \, d\tau = \int_V (\rho_f \mathbf{E} + \mathbf{J}_f \times \mathbf{B}) \, d\tau \qquad (21\text{-}65)$$

and now, in accordance with our usual custom, we want to express this completely in terms of the fields.

Setting **P** and **M** each equal to zero in (21-30) and (21-33), and using the results, we find that the integrand of (21-65) can be written as

$$\mathbf{f}_v = \epsilon_0 \mathbf{E}(\nabla \cdot \mathbf{E}) + \frac{1}{\mu_0}(\nabla \times \mathbf{B}) \times \mathbf{B} - \epsilon_0 \frac{\partial \mathbf{E}}{\partial t} \times \mathbf{B} \qquad (21\text{-}66)$$

We can improve the symmetry of this expression between **E** and **B** by adding zero to it as written in the form

$$-\epsilon_0 \mathbf{E} \times \frac{\partial \mathbf{B}}{\partial t} - \epsilon_0 \mathbf{E} \times (\nabla \times \mathbf{E}) + \frac{1}{\mu_0}\mathbf{B}(\nabla \cdot \mathbf{B}) = 0$$

which follows from (21-31) and (21-32). Doing this, and using (1-23), we get

$$\mathbf{f}_v + \epsilon_0 \frac{\partial}{\partial t}(\mathbf{E} \times \mathbf{B}) = \epsilon_0[\mathbf{E}(\nabla \cdot \mathbf{E}) - \mathbf{E} \times (\nabla \times \mathbf{E})]$$

$$+ \frac{1}{\mu_0}[\mathbf{B}(\nabla \cdot \mathbf{B}) - \mathbf{B} \times (\nabla \times \mathbf{B})] \qquad (21\text{-}67)$$

In order to see what we have, let us calculate the x component of the first bracketed term on the right; we find in a straightforward manner that it can be written as

$$[\mathbf{E}(\nabla \cdot \mathbf{E}) - \mathbf{E} \times (\nabla \times \mathbf{E})] \cdot \hat{\mathbf{x}} = \frac{\partial}{\partial x}\left(E_x^2 - \frac{1}{2}E^2\right) + \frac{\partial}{\partial y}(E_x E_y) + \frac{\partial}{\partial z}(E_x E_z)$$

and there will be a similar result for the term involving **B**. This can be written more compactly if we introduce a set of quantities T_{ij} defined as

$$T_{ij} = \epsilon_0\left(E_i E_j - \frac{1}{2}E^2\delta_{ij}\right) + \frac{1}{\mu_0}\left(B_i B_j - \frac{1}{2}B^2\delta_{ij}\right) \qquad (21\text{-}68)$$

with the use of the Kronecker delta symbol defined in (8-27) and where the indices i and j can independently take on the values x, y, and z. Using this definition, we see that the x component of (21-67) can be written in the form

$$\left[\mathbf{f}_v + \epsilon_0 \frac{\partial}{\partial t}(\mathbf{E} \times \mathbf{B})\right]_x = \frac{\partial T_{xx}}{\partial x} + \frac{\partial T_{xy}}{\partial y} + \frac{\partial T_{xz}}{\partial z} = \nabla \cdot \mathbf{X} \qquad (21\text{-}69)$$

where the vector $\mathbf{X} = T_{xx}\hat{\mathbf{x}} + T_{xy}\hat{\mathbf{y}} + T_{xz}\hat{\mathbf{z}}$. (The quantities T_{ij} are called the components of the *Maxwell stress tensor*; we will, however, not be using them again after this section.) If we now integrate (21-69) over the volume V and apply the divergence theorem (1-59), we get

$$\int_V \left[\mathbf{f}_v + \epsilon_0 \frac{\partial}{\partial t}(\mathbf{E} \times \mathbf{B})\right]_x d\tau = \int_V \nabla \cdot \mathbf{X}\, d\tau = \oint_S \mathbf{X} \cdot d\mathbf{a} \qquad (21\text{-}70)$$

Now let us extend the region of integration to include all space which is possible since ρ_f and $\mathbf{J}_f$ are zero outside of V. Since all of the sources ρ_f and $\mathbf{J}_f$ are within a finite region, at large distances R, we will, at worst, have $\mathbf{E} \sim 1/R^2$ and $\mathbf{B} \sim 1/R^3$ as we saw in (7-26) and (18-19), while the surface area of integration goes as R^2. Now the components of T_{ij} involve products of the field components, and we see from (21-69) that **X** will decrease much more rapidly with R than the surface area of integration will increase, so that its surface integral will vanish and (21-70) becomes

$$\int_{\text{all space}} \left[\mathbf{f}_v + \epsilon_0 \frac{\partial}{\partial t}(\mathbf{E} \times \mathbf{B})\right]_x d\tau = \oint_{\text{all space}} \mathbf{X} \cdot d\mathbf{a} = 0 \qquad (21\text{-}71)$$

We will get the same values for the y and z components of this integral so that the vector integral vanishes and thus

$$\int_{\text{all space}}\left[\mathbf{f}_v + \epsilon_0\frac{\partial}{\partial t}(\mathbf{E}\times\mathbf{B})\right]d\tau = \mathbf{F} + \frac{\partial}{\partial t}\int_{\text{all space}}\epsilon_0(\mathbf{E}\times\mathbf{B})\,d\tau = 0 \quad (21\text{-}72)$$

where we have used (21-65) to reintroduce the total force on the charges and currents. Now $\mathbf{F}$ acts on the charge and current carriers, that is, on the matter in the field; the matter has mechanical properties and we know from mechanics that $\mathbf{F}$ equals the rate of change of the momentum of the matter, that is, $\mathbf{F} = d\mathbf{p}_{\text{matter}}/dt$. Substituting this into (21-72), and noting that there is no longer any distinction between a partial and total time derivative in the second term since the integration is over all space, we find that (21-72) can be written as

$$\frac{d}{dt}\left[\mathbf{p}_{\text{matter}} + \int_{\text{all space}}\epsilon_0(\mathbf{E}\times\mathbf{B})\,d\tau\right] = 0 \quad (21\text{-}73)$$

so that the quantity in the brackets is a *constant*. But this is exactly the way in which we would expect a statement of the conservation of the *total momentum* of the whole system to be expressed. Therefore, we can interpret the integral as being the *momentum of the electromagnetic field* and write

$$\mathbf{p}_{\text{em field}} = \int_{\text{all space}}\epsilon_0\mathbf{E}\times\mathbf{B}\,d\tau \quad (21\text{-}74)$$

Then it is only natural to interpret the integrand as being a *momentum density* $\mathbf{g}$:

$$\mathbf{g} = \epsilon_0\mathbf{E}\times\mathbf{B} = \mu_0\epsilon_0\mathbf{S} \quad (21\text{-}75)$$

where $\mathbf{S}$ is the Poynting vector since $\mathbf{B} = \mu_0\mathbf{H}$ in this case.

If we recall from mechanics that the angular momentum $\mathbf{l}$ of a point mass with linear momentum $\mathbf{p}$ is defined by $\mathbf{l} = \mathbf{r}\times\mathbf{p}$ where $\mathbf{r}$ is its position vector, we see that by analogy we can define an *angular momentum density* $\mathscr{L}$ associated with the electromagnetic field by

$$\mathscr{L} = \mathbf{r}\times\mathbf{g} = \epsilon_0\mathbf{r}\times(\mathbf{E}\times\mathbf{B}) \quad (21\text{-}76)$$

EXERCISES

21-1 A parallel plate capacitor consists of two circular plates of area S with vacuum between them. It is connected to a battery of constant emf $\mathscr{E}$. The plates are then *slowly* oscillated so that they remain parallel but the separation d between them is varied as $d = d_0 + d_1\sin\omega t$. Find the magnetic field $\mathbf{H}$ between the plates produced by the displacement current. Similarly, find $\mathbf{H}$ if the capacitor is first disconnected from the battery and then the plates are oscillated in the same manner.

21-2 An infinitely long cylindrical capacitor of radii a and b ($b > a$) carries a free charge λ_f per unit length. The region between the plates is filled with a nonmagnetic dielectric of conductivity σ. Show that at every point in the dielectric, the conduction current is exactly compensated by the displacement current so that no magnetic field is produced in the interior. Find the rate of energy dissipation per unit volume at a point a distance ρ from the axis. Find the total rate of energy dissipation for a length l of the dielectric and show that this is equal to the rate of decrease of the electrostatic energy of the capacitor.

21-3 Express the general form of Maxwell's equations completely in terms of each of the following pairs: $(\mathbf{E},\mathbf{H})$, $(\mathbf{D},\mathbf{B})$, and $(\mathbf{D},\mathbf{H})$. In other words, find the analogues of (21-30) through (21-33) for each of these pairs.

21-4 Express the general boundary conditions completely in terms of each of the following pairs: $(\mathbf{E},\mathbf{H})$, $(\mathbf{D},\mathbf{B})$, and $(\mathbf{D},\mathbf{H})$. In other words, find the analogues of (21-34), (21-26), (21-27), and (21-35) for each of these pairs.

21-5   Starting with the Lorentz force (21-29) and Maxwell's equations for vacuum, derive Coulomb's law (2-3).

21-6   Assume that a certain distribution of free charges and currents $(\rho_1, \mathbf{J}_1)$ results in an electromagnetic field $\mathbf{E}_1$, $\mathbf{B}_1$, $\mathbf{H}_1$, and $\mathbf{D}_1$, that is, they are solutions of Maxwell's equations. Similarly, another distribution $(\rho_2, \mathbf{J}_2)$ produces $\mathbf{E}_2$, $\mathbf{B}_2$, $\mathbf{H}_2$, and $\mathbf{D}_2$. Verify the superposition property directly by showing that $\mathbf{E} = \mathbf{E}_1 + \mathbf{E}_2$, and so on, are also solutions of Maxwell's equations for the total source distributions $\rho_1 + \rho_2$ and $\mathbf{J}_1 + \mathbf{J}_2$.

21-7   Find the form of Maxwell's equations in terms of $\mathbf{E}$ and $\mathbf{B}$ for a linear isotropic but *non*-homogeneous medium.

21-8   Here are some genuine electromagnetic fields, that is, solutions of Maxwell's equations in a vacuum. (Later, we will see how they can be obtained.)

$$E_y = -H_0\mu_0\omega\left(\frac{a}{\pi}\right)\sin\left(\frac{\pi x}{a}\right)\sin(kz - \omega t)$$

$$H_x = H_0 k\left(\frac{a}{\pi}\right)\sin\left(\frac{\pi x}{a}\right)\sin(kz - \omega t)$$

$$H_z = H_0\cos\left(\frac{\pi x}{a}\right)\cos(kz - \omega t)$$

All other components are zero and $0 \le x \le a$. H_0, a, k, and ω are constants and $\mu_0\epsilon_0\omega^2 = k^2 + (\pi/a)^2$. (a) Verify by direct substitution that these satisfy Maxwell's equations for a vacuum containing no free charges or currents. (b) Find the displacement current. (c) Find the Poynting vector. (d) At $x = 0$, there is a perfectly conducting wall coincident with the yz plane. Assume that all fields are zero for negative x and find the free surface charge density and free surface current density as a function of position on this wall.

21-9   Find the Poynting vector on the bounding surface of region 1 of Figure 21-3. Find the total rate at which energy is entering region 1 and then show that it equals the rate at which the energy of the capacitor is changing.

21-10   Suppose that the very long coaxial line of Figure 20-14 is used as a transmission line between a battery and resistor. The battery of emf $\mathscr{E}$ has its two terminals connected to the two conductors at one end of the line, while, at the other end, the two conductors are connected by a resistor of resistance R. Find $\mathbf{S}$ in the region between the conductors. Show that the total power passing across a cross section of the line equals $\mathscr{E}^2/R$ and interpret. Will the direction of energy flow change if the connections to the battery are interchanged?

21-11   A conducting sphere of radius a is uniformly magnetized with a magnetization of magnitude M. It also has a net charge Q. Find the total angular momentum of the electromagnetic field of this system.

21-12   Consider a l.i.h. medium where ρ_f and $\mathbf{J}_f$ are both zero in the region of interest. Show that Maxwell's equations are invariant to the transformation

$$\mathbf{E}' = C\left[\mathbf{E}\cos\alpha + (\mu\epsilon)^{-1/2}\mathbf{B}\sin\alpha\right]$$
$$\mathbf{B}' = C\left[-(\mu\epsilon)^{1/2}\mathbf{E}\sin\alpha + \mathbf{B}\cos\alpha\right] \quad (21-77)$$

where C is a dimensionless constant and α is a constant but arbitrary angle. In other words, if $\mathbf{E}$ and $\mathbf{B}$ are solutions of Maxwell's equations, show that then $\mathbf{E}'$ and $\mathbf{B}'$ are too. Consider the special case $\alpha = \pi/2$, and thus show that, in this sense, the fields $\mathbf{E}$ and $\mathbf{B}$ can be interchanged. Illustrate this last result with a simple sketch. Is the relation we found after (16-40) compatible with these results? This property is often called the *duality* property of the electromagnetic field.

22

SCALAR AND VECTOR POTENTIALS

In the previous chapter, we summarized electromagnetism in terms of the field vectors themselves. We recall, however, that much of our previous work was simplified by the introduction of the scalar and vector potentials, and it is natural to wonder if they can also be usefully applied to these more general situations. It does turn out to be possible to formulate the subject in an analogous way. These more general potentials are found to be of the most use in discussing the production of fields from given sources and actually we will not make extensive use of them until Chapter 28. Nevertheless, it is convenient to discuss them in general terms at this point.

22-1 THE POTENTIALS IN GENERAL

We begin by noting that the equation $\nabla \cdot \mathbf{B} = 0$, which we originally used to introduce the vector potential $\mathbf{A}$ in (16-7), is still unchanged in form as shown in (21-21). Therefore, we can always satisfy (21-21) by writing

$$\mathbf{B}(\mathbf{r}, t) = \nabla \times \mathbf{A}(\mathbf{r}, t) \qquad (22\text{-}1)$$

The difference between (22-1) and (16-7) is that now both $\mathbf{B}$ and $\mathbf{A}$ are assumed to be possible functions of time t as well as of position $\mathbf{r}$.

If we substitute (22-1) into (21-20), we obtain $\nabla \times \mathbf{E} = -\partial(\nabla \times \mathbf{A})/\partial t = -\nabla \times (\partial \mathbf{A}/\partial t)$ so that

$$\nabla \times \left(\mathbf{E} + \frac{\partial \mathbf{A}}{\partial t} \right) = 0 \qquad (22\text{-}2)$$

Because of (1-48), the term in parentheses can always be written as the gradient of a scalar, that is, $\mathbf{E} + (\partial \mathbf{A}/\partial t) = -\nabla \phi$, or

$$\mathbf{E} = -\nabla \phi - \frac{\partial \mathbf{A}}{\partial t} \qquad (22\text{-}3)$$

This is, of course, of the same form as (17-12) since they both were derived in exactly the same way. Again, we must assume that the scalar potential ϕ is generally a function of both position and time.

Therefore, two of the set of Maxwell's equations will always be satisfied if $\mathbf{E}$ and $\mathbf{B}$ are written in the form (22-1) and (22-3). The remaining two members of the set should tell us how $\mathbf{A}$ and ϕ are to be found from the sources. If we substitute these expressions for $\mathbf{E}$ and $\mathbf{B}$ into (21-30) and (21-33), we obtain

$$\nabla^2 \phi + \frac{\partial}{\partial t} \nabla \cdot \mathbf{A} = -\frac{1}{\epsilon_0}(\rho_f - \nabla \cdot \mathbf{P}) \qquad (22\text{-}4)$$

$$\nabla^2 \mathbf{A} - \mu_0 \epsilon_0 \frac{\partial^2 \mathbf{A}}{\partial t^2} - \nabla \left(\nabla \cdot \mathbf{A} + \mu_0 \epsilon_0 \frac{\partial \phi}{\partial t} \right) = -\mu_0 \left(\mathbf{J}_f + \nabla \times \mathbf{M} + \frac{\partial \mathbf{P}}{\partial t} \right) \qquad (22\text{-}5)$$

with the use of (1-45) and (1-120). If we can now find solutions of these equations that satisfy appropriate boundary conditions, then we know from the way in which they were obtained that $\mathbf{E}$ and $\mathbf{B}$ as found from this $\mathbf{A}$ and ϕ will automatically be a solution of *all* of Maxwell's equations.

A difficulty with these two equations is that both $\mathbf{A}$ and ϕ appear in each equation, that is, the equations are not separated. In addition, if $\mathbf{P}$ and $\mathbf{M}$ do not represent distributions of permanent dipole moments that would be unaffected by the fields, then we have to assume that $\mathbf{P}$ and $\mathbf{M}$ are functions of $\mathbf{E}$ and $\mathbf{B}$ and thus of $\mathbf{A}$ and ϕ. Consequently, unless we have a specific dependence given to us, we cannot make any further progress with the general equations. Nonetheless, (22-1), (22-3), (22-4), and (22-5) are a set of equations that are assumed to be *always* true.

We can still draw some conclusions about boundary conditions at a surface of discontinuity in properties. The arguments that we used in connection with (16-21), which followed from $\mathbf{B} = \nabla \times \mathbf{A}$, are still valid when $\mathbf{B}$ is a function of the time, since we only required that $\mathbf{B}$ be finite. Since (22-1) still holds, we can conclude that the tangential components of $\mathbf{A}$ are always continuous:

$$\mathbf{A}_{2t} = \mathbf{A}_{1t} \tag{22-6}$$

Now we do not yet have a specific expression for $\nabla \cdot \mathbf{A}$; thus we cannot be sure of the behavior of the normal components of $\mathbf{A}$, but it is not unreasonable to expect that they will behave as in (16-20) and also be continuous.

In Section 9-5, we concluded that the electrostatic scalar potential was continuous. Now, however, (22-3) is applicable rather than simply $\mathbf{E} = -\nabla\phi$, but if we review our previous arguments, we see that if we add to them the usual requirement that $\partial\mathbf{A}/\partial t$ remain finite as the transition layer is shrunk to zero thickness we will again find that ϕ is continuous:

$$\phi_2 = \phi_1 \tag{22-7}$$

Finally, as a sort of check on our results, if we revert to the static case in which all time derivatives are zero, and use $\nabla \cdot \mathbf{A} = 0$ as given by (16-17), we find that (22-4) and (22-5) become $\nabla^2\phi = -\rho/\epsilon_0$ and $\nabla^2\mathbf{A} = -\mu_0\mathbf{J}$, which are exactly our previous results given by (11-1), (10-38), and (16-18). Thus our generalized scalar and vector potentials properly reduce to the previous ones that were applicable to static situations.

22-2 THE POTENTIALS FOR LINEAR ISOTROPIC HOMOGENEOUS MEDIA

As in Section 21-3, considerable simplification can be attained if we are dealing with completely l.i.h. materials, so that the constitutive equations (21-40) are applicable as is (21-41) for the free current density. It is somewhat more convenient to begin again with the forms of Maxwell's equations written for this case in (21-42) through (21-45). The middle two will again be satisfied if we use (22-1) and (22-3), and when these expressions for $\mathbf{E}$ and $\mathbf{B}$ are substituted into (21-42) and (21-45), we get

$$\nabla^2\phi + \nabla \cdot \frac{\partial\mathbf{A}}{\partial t} = -\frac{\rho_f}{\epsilon} \tag{22-8}$$

$$\nabla^2\mathbf{A} - \mu\sigma\frac{\partial\mathbf{A}}{\partial t} - \mu\epsilon\frac{\partial^2\mathbf{A}}{\partial t^2} - \nabla\left(\nabla \cdot \mathbf{A} + \mu\epsilon\frac{\partial\phi}{\partial t} + \mu\sigma\phi\right) = -\mu\mathbf{J}'_f \tag{22-9}$$

We can now give (22-8) a form more like (22-9) by adding and subtracting appropriate terms and we find that we can also write it as

$$\nabla^2\phi - \mu\sigma\frac{\partial\phi}{\partial t} - \mu\epsilon\frac{\partial^2\phi}{\partial t^2} + \frac{\partial}{\partial t}\left(\nabla \cdot \mathbf{A} + \mu\epsilon\frac{\partial\phi}{\partial t} + \mu\sigma\phi\right) = -\frac{\rho_f}{\epsilon} \tag{22-10}$$

These equations are the ones that we have to solve for **A** and ϕ. However, they each contain both **A** and ϕ and it would be desirable if we could somehow separate them. We recall from the Helmholtz theorem of Section 1-20 that the vector **A** will not be completely determined until we have specified both $\nabla \times \mathbf{A}$ and $\nabla \cdot \mathbf{A}$ everywhere. So far, we have only specified $\nabla \times \mathbf{A}$ by way of (22-1); consequently, we are free to choose $\nabla \cdot \mathbf{A}$ in any convenient way. Comparing (22-9) and (22-10), we see that a simple thing to do is to *require* that

$$\nabla \cdot \mathbf{A} + \mu\epsilon \frac{\partial \phi}{\partial t} + \mu\sigma\phi = 0 \qquad (22\text{-}11)$$

for then

$$\nabla^2 \mathbf{A} - \mu\sigma \frac{\partial \mathbf{A}}{\partial t} - \mu\epsilon \frac{\partial^2 \mathbf{A}}{\partial t^2} = -\mu \mathbf{J}_f' \qquad (22\text{-}12)$$

$$\nabla^2 \phi - \mu\sigma \frac{\partial \phi}{\partial t} - \mu\epsilon \frac{\partial^2 \phi}{\partial t^2} = -\frac{\rho_f}{\epsilon} \qquad (22\text{-}13)$$

The requirement (22-11) is known as the *Lorentz condition*. The advantage of this particular choice is that not only does it make the equations for **A** and ϕ independent of each other, but that the four scalar equations that result (one from each component of **A** and one for ϕ) are all differential equations of the *same general form* so that, in this sense, we are left with only *one* general problem to solve.

The formulation we have developed here thus anticipates that the general procedure to be used in solving an electromagnetic field problem by means of the potentials would be as follows. We assume that the external sources plus boundary conditions are given, that is, we assume that the functions $\rho_f(\mathbf{r}, t)$ and $\mathbf{J}_f'(\mathbf{r}, t)$ are known. Then (22-12) and (22-13) are to be solved for **A** and ϕ, making sure that they satisfy the Lorentz condition (22-11). When this has been done, **E** and **B** are found from (22-1) and (22-3) and the appropriate boundary conditions are satisfied as well. We know that **E** and **B** as found in this way will satisfy Maxwell's equations and thus will represent the complete solution. If we desire, we can then find $\mathbf{D} = \epsilon\mathbf{E}$ and $\mathbf{H} = \mathbf{B}/\mu$; we can also calculate the polarization and magnetization everywhere from $\mathbf{P} = (\kappa_e - 1)\epsilon_0\mathbf{E}$ and $\mathbf{M} = [(\kappa_m - 1)/\kappa_m\mu_0]\mathbf{B}$.

Actually the most important and convenient situations for which these potentials are used are for a nonconducting medium, and, since this will also be true for us, let us see to what our results reduce in this case. When $\sigma = 0$, $\mathbf{J}_f' = \mathbf{J}_f$ by (21-41), and (22-11) through (22-13) become

$$\nabla \cdot \mathbf{A} + \mu\epsilon \frac{\partial \phi}{\partial t} = 0 \qquad (22\text{-}14)$$

$$\nabla^2 \mathbf{A} - \mu\epsilon \frac{\partial^2 \mathbf{A}}{\partial t^2} = -\mu \mathbf{J}_f \qquad (22\text{-}15)$$

$$\nabla^2 \phi - \mu\epsilon \frac{\partial^2 \phi}{\partial t^2} = -\frac{\rho_f}{\epsilon} \qquad (22\text{-}16)$$

These are often written more compactly by defining the *D'Alembertian operator* as

$$\Box^2 = \nabla^2 - \mu\epsilon \frac{\partial^2}{\partial t^2} \qquad (22\text{-}17)$$

so that we obtain

$$\Box^2 \mathbf{A} = -\mu \mathbf{J}_f \qquad \Box^2 \phi = -\frac{\rho_f}{\epsilon} \qquad (22\text{-}18)$$

which are reminiscent of the Poisson equations (11-2) and (16-18).

If we assume that $\partial\phi/\partial t$ stays finite within the transition layer, we find from (22-14) and (9-7) that the normal components of **A** are continuous across a surface of discontinuity, that is,

$$A_{2n} - A_{1n} = 0 \tag{22-19}$$

which, when combined with (22-6), shows us that **A** itself is continuous,

$$\mathbf{A}_2 = \mathbf{A}_1 \tag{22-20}$$

just as it was in the static case as found in (16-22).

22-3 GAUGE TRANSFORMATIONS

Near the end of Section 16-2, we saw that the requirement $\mathbf{B} = \nabla \times \mathbf{A}$ still leaves some ambiguity in the vector potential, and we also saw, in (16-24) and (16-25), that a new potential $\mathbf{A}^\dagger$ defined by

$$\mathbf{A}^\dagger = \mathbf{A} + \nabla\chi \tag{22-21}$$

gives the induction **B** equally well. We see that this will still be the case here where we would now expect the scalar χ to be a function of time as well as of position, so that we write $\chi = \chi\,(\mathbf{r}, t)$. However, **E** now depends on the vector as well as the scalar potentials, and of course we want to obtain the same electric field. If we substitute (22-21) into (22-3), we get

$$\mathbf{E} = -\nabla\phi - \frac{\partial}{\partial t}(\mathbf{A}^\dagger - \nabla\chi) = -\nabla\left(\phi - \frac{\partial\chi}{\partial t}\right) - \frac{\partial\mathbf{A}^\dagger}{\partial t} = -\nabla\phi^\dagger - \frac{\partial\mathbf{A}^\dagger}{\partial t} \tag{22-22}$$

where we have set

$$\phi^\dagger = \phi - \frac{\partial\chi}{\partial t} \tag{22-23}$$

Thus, the *form* of (22-3) is preserved if we also change the scalar potential in this way. In other words, we get the same fields **E** and **B** whether we use the set of potentials $(\mathbf{A}, \phi)$ or $(\mathbf{A}^\dagger, \phi^\dagger)$. The kind of transformation defined by (22-21) and (22-23) is called a *gauge transformation*; thus, we have found that Maxwell's equations are invariant to a gauge transformation or to a "change in gauge." There are certain restrictions we want to put on χ, however, because it is desirable that both sets of potentials satisfy the Lorentz condition. If we substitute (22-21) and (22-23) into (22-14), we find that

$$\nabla \cdot \mathbf{A}^\dagger + \mu\epsilon\frac{\partial\phi^\dagger}{\partial t} - \nabla^2\chi + \mu\epsilon\frac{\partial^2\chi}{\partial t^2} = 0 \tag{22-24}$$

so that $\mathbf{A}^\dagger$ and $\phi^\dagger$ will also satisfy (22-14) provided that

$$\nabla^2\chi - \mu\epsilon\frac{\partial^2\chi}{\partial t^2} = \Box^2\chi = 0 \tag{22-25}$$

[In the case of static fields, (22-14) and (22-25) reduce to our previous results (16-26) and (16-27).]

It is also easily verified that under this assumption $\mathbf{A}^\dagger$ and $\phi^\dagger$ satisfy the same differential equations as do **A** and ϕ:

$$\Box^2\mathbf{A}^\dagger = -\mu\mathbf{J}_f \qquad \Box^2\phi^\dagger = -\frac{\rho_f}{\epsilon} \tag{22-26}$$

When one deals with potentials that fulfill all of these requirements, it is said that the *Lorentz gauge* is being used.

EXERCISES

22-1 Show that (22-8) and (22-9) can also be obtained by starting from (22-4) and (22-5) rather than by going back to Maxwell's equations first.

22-2 Show that, if the free charge and current distributions and the polarization and magnetization are all *given* functions of position and time, then the general equations satisfied by the potentials can be separated by using the form of the Lorentz condition appropriate for a *vacuum*. Find the differential equations satisfied by $\mathbf{A}$ and ϕ under these conditions.

22-3 Find the equations satisfied by the potentials for a linear isotropic but *non*homogeneous medium.

22-4 Find the boundary conditions satisfied by the normal components of $\mathbf{A}$ from the Lorentz condition (22-11) involving conductivity.

22-5 Consider a l.i.h. nonconducting medium. If one requires that $\nabla \cdot \mathbf{A} = 0$, one is said to be using the *Coulomb gauge*. Find the differential equations satisfied by $\mathbf{A}$ and ϕ for this case.

22-6 Show that, in a l.i.h. nonconducting region where $\rho_f = 0$ and $\mathbf{J}_f = 0$, the fields $\mathbf{E}$ and $\mathbf{B}$ can be found completely from a vector potential, that is, $\phi = $ const. (zero is usually chosen). What two equations determine $\mathbf{A}$?

22-7 Consider a region where $\rho_f = 0$, $\mathbf{J}_f = 0$, $\mathbf{M} = 0$, while $\mathbf{P}$ is a *given* function of position and time. Show that (22-4) and (22-5) can be satisfied and the electromagnetic field obtained from a *single* vector $\boldsymbol{\pi}_e$ such that $\mathbf{A} = \mu_0\epsilon_0(\partial\boldsymbol{\pi}_e/\partial t)$, $\phi = -\nabla \cdot \boldsymbol{\pi}_e$ and $\nabla^2\boldsymbol{\pi}_e - \mu_0\epsilon_0(\partial^2\boldsymbol{\pi}_e/\partial t^2) = -\mathbf{P}/\epsilon_0$. This vector $\boldsymbol{\pi}_e$ is called the *Hertz vector* or *polarization potential*. Express $\mathbf{E}$ and $\mathbf{B}$ completely in terms of $\boldsymbol{\pi}_e$, and $\mathbf{P}$ also if it leads to any simplification.

22-8 This exercise is similar in spirit to the preceding one. Assume that $\rho_f = 0$, $\mathbf{J}_f = 0$, $\mathbf{P} = 0$, while $\mathbf{M}$ is a given function of position and time. Show that the electromagnetic field can now be obtained from a single vector $\boldsymbol{\pi}_m$, which satisfies the equation $\nabla^2\boldsymbol{\pi}_m - \mu_0\epsilon_0(\partial^2\boldsymbol{\pi}_m/\partial t^2) = -\mu_0\mathbf{M}$. Find expressions for $\mathbf{A}$, ϕ, $\mathbf{E}$, and $\mathbf{B}$ in terms of $\boldsymbol{\pi}_m$.

23

SYSTEMS OF UNITS — A GUIDE FOR THE PERPLEXED

We have been using exclusively SI units which, for our purposes, are the same as those of the rationalized MKSA system, and we will continue to do so. This means that our unit system is based on *four* arbitrarily chosen and defined quantities—the meter, kilogram, second, and ampere. Primarily because of historical reasons based on the originally separate development of electricity and magnetism, other systems of units have been used and continue to be used. This is particularly the case in more advanced treatments of physics, especially in subjects like quantum mechanics and its application to the microscopic properties of matter. As a result, a student whose training has been conducted entirely in terms of MKSA units very often faces the problem of answering two questions: In what system of units is *this* equation written? What *numbers* do I put in it in order to work this problem? This chapter provides some guidance in how to answer these questions. Consequently, we do not give an exhaustive discussion of various unit systems but primarily try to indicate how they originated, the effects on the forms of our basic equations, and what to do about it. Thus, in a sense, this short chapter is a digression, but, since we have just finished the task of writing electromagnetism in its most general and fundamental form, this is a useful point at which to consider these questions.

23-1 ORIGIN OF OTHER SYSTEMS OF UNITS

In order to see how different systems can occur, it is sufficient for our purposes to consider two basic experimental results—one electrostatic and one magnetostatic. From Coulomb's law (2-3), we know that the magnitude of the force between two point charges has the form

$$F = C_e \frac{qq'}{R^2} \tag{23-1}$$

where C_e is a constant of proportionality whose numerical value will depend on the units that are used. We previously chose to write $C_e = 1/4\pi\epsilon_0$. Similarly, the magnitude of the force per unit length between two parallel currents as found from Ampère's law is given by (13-13) and (13-14) and can be written as

$$\frac{dF}{dz} = 2C_m \frac{II'}{\rho} \tag{23-2}$$

where C_m is another constant of proportionality that we previously wrote as $C_m = \mu_0/4\pi$. In addition, all systems of units use the definition of current given by (12-2): $I = dq/dt$.

If one always uses the same set of mechanical units in them, then the two forces involved will have the same dimensions and we see that the combinations $C_e qq'/R^2$

and $C_m II'$ also must have the same dimensions so that the ratio

$$\frac{C_e}{C_m} = c^2 \tag{23-3}$$

must have the dimensions of (distance/time)^2, that is, c has the dimensions of a *speed*. The value of this ratio has been measured many times and the *experimental result* is that

$$c = 3 \times 10^8 \text{ meters/second} \tag{23-4}$$

which is the same as the measured speed of light in a vacuum. The best value of c as presently known is 2.99792458×10^8 meters/second, but the value given in (23-4) is accurate enough for us here. (As we will see in the next chapter, the agreement between this ratio and the speed of light is not accidental.) We have already taken this numerical result into account in the values we gave for C_e and C_m and we find that $C_e/C_m = (4\pi\epsilon_0)^{-1}/(\mu_0/4\pi) = (\mu_0\epsilon_0)^{-1} = (9 \times 10^9)/(10^{-7}) = (3 \times 10^8 \text{ meters/second})^2$ with the use of (2-5) and (13-2); this is in agreement with (23-3) and (23-4). In other words, we have the fundamental result that

$$\mu_0\epsilon_0 = \frac{1}{c^2} \tag{23-5}$$

for the MKSA system that we are using.

The various systems of units used in electromagnetism essentially differ in the way in which these constants are chosen. It is clear that either C_e or C_m can be chosen arbitrarily, but then the value of the other is fixed by the requirement of (23-3).

All other systems of any interest are based on the use of the CGS system in which everything is expressed in terms of the arbitrary choice of *three* fundamental units for length, mass, and time; these are, respectively, the centimeter, gram, and second. The mechanical units are then found in the usual way from their definitions. Thus, the force unit is 1 gram $\times$ 1 centimeter/(second)^2 = 1 dyne. The unit of work or energy will be the product of a unit force and a unit displacement: 1 dyne-centimeter = 1 erg. The unit of power will be 1 erg/second, and so on.

Another distinction that is made between unit systems concerns whether they are "rationalized" or "unrationalized." What this means, in effect, is that for a rationalized system there are no numerical factors of 4π appearing in Maxwell's equations, while, as we will see, they do appear if an unrationalized system is used. Equations 21-19 through 21-22 show that we are using rationalized MKSA units. (The use of a rationalized system does not make 4π disappear, rather it simply means that 4π's are found elsewhere in results found from Maxwell's equations; thus, the choice of which type to use is somewhat a matter of taste.)

23-2 THE ELECTROSTATIC AND ELECTROMAGNETIC SYSTEMS

Suppose that you felt that Coulomb's law was a fundamental result that was the best place to start in defining a system of units for electromagnetism. Your natural inclination would be to give this equation as simple an appearance as possible; this can certainly be done by *choosing $C_e = 1$*. Then, from (23-3), C_m *must* be taken to be $1/c^2$ and (23-1) and (23-2) would become

$$F = \frac{qq'}{R^2} \qquad \frac{dF}{dz} = \frac{2II'}{c^2\rho} \qquad \text{(esu)} \tag{23-6}$$

This procedure leads one to the *electrostatic system of units* (esu). We see from the first expression in (23-6) that two equal unit charges a distance 1 centimeter apart will repel each other with a force of 1 dyne; the unit of charge defined in this way is called a *statcoulomb* (from electro*stat*ic). The unit of current will then be 1 statcoulomb/second = 1 statampere, and that of potential difference 1 erg/statcoulomb = 1 statvolt. One can continue this process and define a statfarad, statohm, and so on, and in this way develop a consistent and complete description. At some point, however, one has to decide how to define **B** and relate it to **E**; this is done in this system by writing Faraday's law as $\nabla \times \mathbf{E} = -\partial \mathbf{B}/\partial t$ or, equivalently, the Lorentz force as $q(\mathbf{E} + \mathbf{v} \times \mathbf{B})$. In this pure form, however, this system is seldom used anymore and we will not describe it further. Nevertheless, it should be pointed out that it is very common to find quantities measured in this system but *not* given in statamperes, statfarads, and so on, but merely stated as being so many "electrostatic units" or just so many "esu."

Now suppose that you were more interested in and had more experience with magnetostatics than in electrostatics; then you might feel that the expression (23-2) is a better starting point and you would like to simplify it as much as possible. This can be done by choosing $C_m = 1$ so that $C_e = c^2$ and (23-1) and (23-2) become

$$F = \frac{c^2 qq'}{R^2} \qquad \frac{dF}{dz} = \frac{2II'}{\rho} \qquad \text{(emu)} \qquad (23\text{-}7)$$

Such a procedure leads to the *electromagnetic system of units* (emu). We now see from the second expression in (23-7) that two very long equal parallel unit currents 1 centimeter apart will attract each other with a force of 2 dynes/centimeter; the unit of current defined in this way is called an *abampere* (from *ab*solute). The unit charge will be 1 abcoulomb = 1 abampere-second and one can continue this process to get abvolts, abfarads, and the like; it is also very common to use simply the terminology of "electromagnetic units" or "emu." Again the definitions of **E** and **B** are related by writing $\nabla \times \mathbf{E} = -\partial \mathbf{B}/\partial t$ or $\mathbf{F} = q(\mathbf{E} + \mathbf{v} \times \mathbf{B})$. In this pure form, the electromagnetic system is practically never used. What *is* still very much used, however, and the system which one needs to be able to deal with, is that which we consider next.

23-3 THE GAUSSIAN SYSTEM

This is an *unrationalized CGS* system that is *mixed* in the sense that electric quantities are measured in electrostatic units while magnetic quantities are measured in electromagnetic units. For our purposes, it will suffice simply to quote the form that the basic equations assume for this system. Maxwell's equations in general are

$$\nabla \cdot \mathbf{D} = 4\pi \rho_f \qquad \nabla \cdot \mathbf{B} = 0$$

$$\nabla \times \mathbf{E} = -\frac{1}{c}\frac{\partial \mathbf{B}}{\partial t} \qquad \nabla \times \mathbf{H} = \frac{4\pi}{c}\mathbf{J}_f + \frac{1}{c}\frac{\partial \mathbf{D}}{\partial t} \qquad (23\text{-}8)$$

where the various field vectors are related by

$$\mathbf{D} = \mathbf{E} + 4\pi \mathbf{P} \qquad \mathbf{H} = \mathbf{B} - 4\pi \mathbf{M} \qquad (23\text{-}9)$$

and the Lorentz force is

$$\mathbf{F} = q\left(\mathbf{E} + \frac{\mathbf{v}}{c} \times \mathbf{B}\right) \qquad (23\text{-}10)$$

[It follows from (23-8) that the equation of continuity still has the form $\nabla \cdot \mathbf{J}_f + (\partial \rho_f/\partial t) = 0$.]

Where they are applicable, the various constitutive equations are written

$$\mathbf{D} = \epsilon\mathbf{E} \qquad \mathbf{H} = \mathbf{B}/\mu \qquad \mathbf{J}_f = \sigma\mathbf{E}$$
$$\mathbf{P} = \chi_e\mathbf{E} \qquad \mathbf{M} = \chi_m\mathbf{H} \tag{23-11}$$

so that

$$\epsilon = 1 + 4\pi\chi_e \qquad \mu = 1 + 4\pi\chi_m \tag{23-12}$$

Expressions involving the potentials are easily seen to become

$$\mathbf{B} = \nabla \times \mathbf{A} \qquad \mathbf{E} = -\nabla\phi - \frac{1}{c}\frac{\partial\mathbf{A}}{\partial t} \tag{23-13}$$

while energy formulas of interest are

$$u = \frac{1}{8\pi}(\mathbf{E}\cdot\mathbf{D} + \mathbf{B}\cdot\mathbf{H}) \qquad \mathbf{S} = \frac{c}{4\pi}(\mathbf{E}\times\mathbf{H}) \tag{23-14}$$

where the first holds for linear media.

We see from the above that all of the field vectors **E**, **D**, **B**, **H**, **P**, and **M** have the *same dimensions*; this, of course, has not kept people from giving different *names* to the *units*. This is most prevalent with respect to magnetic quantities and the common usage is as follows: **B**, gauss; **H**, oersted; **M**, oersted (but see the next section); Φ, 1 gauss-(centimeter)^2 = 1 maxwell.

It also follows from (23-9) that in a vacuum, $\mathbf{D} = \mathbf{E}$ and $\mathbf{H} = \mathbf{B}$. The quantities ϵ, μ, χ_e, and χ_m are all dimensionless; we discuss their numerical values in the next section.

Furthermore, it is not uncommon to find a "modified" Gaussian system being used. It is just like the one we have described *except* that, while charge is still measured in statcoulombs (esu), current is measured in abamperes (emu). What this does, in effect, is to replace any symbol for current by c times that symbol (e.g., $\mathbf{J}_f \rightarrow c\mathbf{J}_f$). The only one of Maxwell's equations that is affected by this is Ampère's law, and then the equation of continuity, which become

$$\nabla \times \mathbf{H} = 4\pi\mathbf{J}_f + \frac{1}{c}\frac{\partial\mathbf{D}}{\partial t} \qquad \nabla \cdot \mathbf{J}_f + \frac{1}{c}\frac{\partial\rho_f}{\partial t} = 0 \tag{23-15}$$

Finally, the *Heaviside-Lorentz* system is simply a rationalized Gaussian system; when this is used, the effect is to replace every 4π in the equations (23-8) through (23-12) by unity; for example, one gets $\nabla \cdot \mathbf{D} = \rho_f$ and $\mathbf{D} = \mathbf{E} + \mathbf{P}$. The factors of c still remain, however.

If a particular author does not state the unit system that is being used, one can generally deduce what it is by looking at the form of some familiar results, preferably Maxwell's equations.

23-4 HOW TO COPE WITH THE GAUSSIAN SYSTEM

In principle, any desired result in the Gaussian system can be derived by starting with Maxwell's equations (23-8) and using the expressions (23-9) through (23-13) as required. This is not always convenient and it is desirable to have a method that will enable one to transform a given result written in the Gaussian system into the corresponding one in the MKSA system and *vice versa*. Table 23-1 provides a recipe for doing this. In order to use this table, one replaces a symbol in the column labeled by the system in which the formula is written by the symbol or combination listed for the other system. Symbols representing essentially mechanical quantities are unchanged.

Table 23-1. Conversion of symbols in equations
Symbols representing essentially mechanical quantities (length, mass, time, force, work, energy, power, etc.) are not changed (nor are derivatives). To convert an equation written in the MKSA system to the corresponding one in the Gaussian system, replace the symbol listed under the column labeled MKSA by that listed under Gaussian. The entries can also be used to convert a Gaussian equation to an MKSA one by going from right to left in the table.

Quantity	MKSA	Gaussian
Capacitance	C	$4\pi\epsilon_0 C$
Charge	q	$(4\pi\epsilon_0)^{1/2}q$
Charge density	$\rho, (\sigma, \lambda)$	$(4\pi\epsilon_0)^{1/2}\rho, (\sigma, \lambda)$
Conductivity	σ	$4\pi\epsilon_0\sigma$
Current	I	$(4\pi\epsilon_0)^{1/2}I$
Current density	$\mathbf{J}, (\mathbf{K})$	$(4\pi\epsilon_0)^{1/2}\mathbf{J}, (\mathbf{K})$
Dielectric constant	κ_e	ϵ
Dipole moment (electric)	$\mathbf{p}$	$(4\pi\epsilon_0)^{1/2}\mathbf{p}$
Dipole moment (magnetic)	$\mathbf{m}$	$(4\pi/\mu_0)^{1/2}\mathbf{m}$
Displacement	$\mathbf{D}$	$(\epsilon_0/4\pi)^{1/2}\mathbf{D}$
Electric field	$\mathbf{E}$	$(4\pi\epsilon_0)^{-1/2}\mathbf{E}$
Inductance	L	$(4\pi\epsilon_0)^{-1}L$
Magnetic field	$\mathbf{H}$	$(4\pi\mu_0)^{-1/2}\mathbf{H}$
Magnetic flux	Φ	$(\mu_0/4\pi)^{1/2}\Phi$
Magnetic induction	$\mathbf{B}$	$(\mu_0/4\pi)^{1/2}\mathbf{B}$
Magnetization	$\mathbf{M}$	$(4\pi/\mu_0)^{1/2}\mathbf{M}$
Permeability	μ	(1) $\kappa_m\mu_0$, then (2) $\kappa_m \to \mu$
Permeability (relative)	κ_m	μ
Permittivity	ϵ	(1) $\kappa_e\epsilon_0$, then (2) $\kappa_e \to \epsilon$
Polarization	$\mathbf{P}$	$(4\pi\epsilon_0)^{1/2}\mathbf{P}$
Resistance	R	$(4\pi\epsilon_0)^{-1}R$
Resistivity	ρ	$(4\pi\epsilon_0)^{-1}\rho$
Scalar potential	ϕ	$(4\pi\epsilon_0)^{-1/2}\phi$
Speed of light	$(\mu_0\epsilon_0)^{-1/2}$	c
Susceptibility	$\chi_e, (\chi_m)$	$4\pi\chi_e, (\chi_m)$
Vector potential	$\mathbf{A}$	$(\mu_0/4\pi)^{1/2}\mathbf{A}$

■ **Example**

Let us transform $\nabla \cdot \mathbf{D} = \rho_f$ as given by (21-19). Using the table, we get $\nabla \cdot [(\epsilon_0/4\pi)^{1/2}\mathbf{D}] = (4\pi\epsilon_0)^{1/2}\rho_f$, which reduces to $\nabla \cdot \mathbf{D} = 4\pi\rho_f$ as quoted in (23-8).

■

■ **Example**

Let us transform the expression for the Poynting vector given in (23-14) into MKSA form. Since $\mathbf{S}$ is an energy flow, it is not changed and, by going from the Gaussian column to the MKSA one, we get

$$\mathbf{S} = \frac{(\mu_0\epsilon_0)^{-1/2}}{4\pi}\left[(4\pi\epsilon_0)^{1/2}\mathbf{E} \times (4\pi\mu_0)^{1/2}\mathbf{H}\right] = \mathbf{E} \times \mathbf{H}$$

which is exactly (21-59).

■

The use of Table 23-1 may occasionally lead to a wrong result when applied to a Gaussian system equation that has been worked out for a *vacuum*. The reason is that, since $\mathbf{D} = \mathbf{E}$ and $\mathbf{B} = \mathbf{H}$ in this case, there is a tendency to use these symbols interchangeably which may lead to ambiguities since the conversion factors listed in Table 23-1 are different for the members of each of these pairs. For example, in such a situation it is quite common to find the equation connecting the field with the vector potential written as $\mathbf{H} = \nabla \times \mathbf{A}$, and we quickly find from the table that this will not transform directly back to the corresponding MKSA equation $\mathbf{B} = \nabla \times \mathbf{A}$, although it does lead to $\mu_0\mathbf{H} = \nabla \times \mathbf{A}$ which, *for a vacuum*, is all right.

The problem of converting the *form* of an equation is different from that of converting the *numerical values* of a given physical quantity from one unit system to another. For example, the data may be given numerically in Gaussian units and it is necessary to insert their equivalent values into an MKSA formula, or conversely. For this purpose, one requires a numerical conversion table and Table 23-2 is adequate for most purposes. The entry in each row gives the same amount of the given quantity expressed in different units; that is, the terms in each row are equal. The various factors of 3 arise from writing $c = 3 \times 10^8$ meters/second; this does not apply to powers of 10. Other needed conversions can be obtained in the usual manner by multiplying by unity as expressed by the appropriate ratio and canceling units; for example, one can write $1 = 10^3$ grams/1 kilogram.

Although $\mathbf{H}$ and $\mathbf{M}$ are both measured in ampere/meter in the MKSA system, we see that the conversion factors to oersted are different for each of them; this is a consequence of the 4π in (23-9), and similar remarks apply to $\mathbf{D}$ and $\mathbf{P}$. It is not unusual to find magnetization stated in gauss rather than in oersted; in the overwhelming majority of such cases, the author *really* means "oersted" and one can proceed by

Table 23-2. Conversion table for numerical values

Quantity	MKSA	Gaussian
Length	1 meter (m)	10^2 centimeters (cm)
Mass	1 kilogram	10^3 grams
Time	1 second	1 second
Force	1 newton	10^5 dynes
Work, energy	1 joule	10^7 ergs
Power	1 watt	10^7 ergs/second
Capacitance (C)	1 farad	9×10^{11} statfarads
Charge (q)	1 coulomb	3×10^9 statcoulombs
Charge density (ρ)	1 coulomb/m^3	3×10^3 statcoulomb/cm^3
Conductivity (σ)	1 (ohm-m)$^{-1}$	9×10^9 (statohm-cm)$^{-1}$
Current (I)	1 ampere	3×10^9 statamperes $= 10^{-1}$ abamperes
Current density ($\mathbf{J}$)	1 ampere/m^2	3×10^5 statampere/cm^2
Displacement ($\mathbf{D}$)	1 coulomb/m^2	$12\pi \times 10^5$ statvolt/cm
Electric field ($\mathbf{E}$)	1 volt/m	$\frac{1}{3} \times 10^{-4}$ statvolt/cm
Inductance (L)	1 henry	$\frac{1}{9} \times 10^{-11}$ stathenrys
Magnetic field ($\mathbf{H}$)	1 ampere/m	$4\pi \times 10^{-3}$ oersted
Magnetic flux (Φ)	1 weber	10^8 maxwells
Magnetic induction ($\mathbf{B}$)	1 weber/m^2 = 1 tesla	10^4 gauss
Magnetization ($\mathbf{M}$)	1 ampere/m	10^{-3} oersted
Polarization ($\mathbf{P}$)	1 coulomb/m^2	3×10^5 statvolt/cm
Potential (ϕ)	1 volt	$\frac{1}{300}$ statvolt
Resistance (R)	1 ohm	$\frac{1}{9} \times 10^{-11}$ statohms

changing the name and using the factor given in the table for **M**. Occasionally, the author *really* means "gauss"; this will normally signify that he or she has in mind an MKSA definition of magnetic dipole moment equal to μ_0 times the expression (19-20); this would make the relation among the magnetic vectors take the form $\mathbf{B} = \mu_0\mathbf{H} + \mathbf{M}$ rather than (21-24). In that case, it would be appropriate to measure **B** and **M** in the same units; this is a rare situation, however.

When one looks up numerical values of the quantities permeability, dielectric constant, and susceptibilities, one often finds them given in Gaussian terms. The numerical relations among these parameters in the two systems are:

$$\kappa_{e\,\text{MKSA}} = \left(\frac{\epsilon}{\epsilon_0}\right)_{\text{MKSA}} = \epsilon_{\text{Gaussian}} \tag{23-16}$$

$$\kappa_{m\,\text{MKSA}} = \left(\frac{\mu}{\mu_0}\right)_{\text{MKSA}} = \mu_{\text{Gaussian}} \tag{23-17}$$

$$\chi_{\text{MKSA}} = 4\pi\chi_{\text{Gaussian}} \tag{23-18}$$

where the last relation holds for both χ_e and χ_m. (Also see Exercise 20-17.)

EXERCISES

23-1 Using (23-6), express the dimensions of a statcoulomb in terms of centimeters, grams, and seconds. Similarly, use (23-7) to do the same for an abampere.

23-2 Show that 1 statcoulomb/(centimeter)^2 = 1 statvolt/centimeter. Also show that 1 statfarad = 1 centimeter, and that 1 statohm = 1 second/centimeter.

23-3 Show that all of the equations (23-8) through (23-13) can be obtained by applying Table 23-1 to the corresponding MKSA equations.

23-4 Beginning with the equations stated in Gaussian form, derive the differential equations satisfied by **A** and ϕ and the Lorentz condition for a l.i.h. medium. Verify that they are the same as those obtained with the use of Table 23-1.

23-5 Use (23-8) and (23-9) to obtain the capacitance of a parallel plate capacitor of plate area A and separation d with vacuum between the plates. Verify that your result is consistent with Table 23-1 and Exercise 23-2.

23-6 Use (23-8) to show that the induced emf will be written in Gaussian units as $\mathscr{E} = -c^{-1}(d\Phi/dt)$. If self-inductance is also defined in the usual way by $\mathscr{E} = -L(dI/dt)$, show that the analogue of (17-55) must be $L = \Phi/cI$. Then

show that 1 Gaussian unit of inductance = 1 stathenry = 1 (second)^2/centimeter. Now use (23-8) and (23-9) to find the self-inductance of a length l of an infinitely long ideal solenoid of cross-sectional area S, n turns per unit length, and vacuum inside. Verify that your result is consistent with Table 23-1 and the above result for its dimensions.

23-7 Use (23-8) and (23-11) to derive Poynting's theorem for a linear isotropic medium and thus show the suitability of the results quoted in (23-14).

23-8 As a simple numerical exercise in the use of Table 23-2, suppose that **H** and **M** are parallel and that they have the values α ampere/meter and β ampere/meter, respectively, where α and β are numbers. Find B in webers/(meter)^2, B in gauss, H in oersted, and M in oersted. Show that when the values just found for H and M are put into (23-9), the same value of B is obtained as found by direct use of the conversion factor for B itself.

23-9 Verify that (23-16) through (23-18) are correct. (*Hint*: as in the previous exercise, choose a specific numerical value for the appropriate quantity and carry out all of the conversions.)

24

PLANE WAVES

In principle, we regard any solution of Maxwell's equations that satisfies given boundary conditions as a possible electromagnetic field that would be produced by the appropriate distribution of charge and current. In practice, however, one does not go about trying to solve Maxwell's equations indiscriminately, but rather looks for solutions of a desired type, or for those thought to be suitable for the particular situation of interest at the moment. We first discuss several examples of this latter procedure for time-dependent fields, putting off for a while the question of how we would produce these fields if we so desired.

In the rest of this book, except for Section 25-7 and portions of Appendix B, we assume that we are dealing with media that are linear isotropic and homogeneous in *all* properties.

24-1 SEPARATE EQUATIONS FOR E AND B

Let us begin by assuming that there are no external free charges and currents in the region of interest, that is, we take $\rho_f = 0$ and $\mathbf{J}_f' = 0$. Then (21-42) through (21-45) become

$$\nabla \cdot \mathbf{E} = 0 \tag{24-1}$$

$$\nabla \times \mathbf{E} = -\frac{\partial \mathbf{B}}{\partial t} \tag{24-2}$$

$$\nabla \cdot \mathbf{B} = 0 \tag{24-3}$$

$$\nabla \times \mathbf{B} = \mu \sigma \mathbf{E} + \mu \epsilon \frac{\partial \mathbf{E}}{\partial t} \tag{24-4}$$

We can eliminate one of the fields in the following way. If we take the curl of (24-2), use (1-120), (24-1), and (24-4), we find that

$$\nabla \times (\nabla \times \mathbf{E}) = \nabla(\nabla \cdot \mathbf{E}) - \nabla^2 \mathbf{E} = -\nabla^2 \mathbf{E} = -\frac{\partial}{\partial t} \nabla \times \mathbf{B}$$

$$= -\mu\sigma \frac{\partial \mathbf{E}}{\partial t} - \mu\epsilon \frac{\partial^2 \mathbf{E}}{\partial t^2}$$

or

$$\nabla^2 \mathbf{E} - \mu\sigma \frac{\partial \mathbf{E}}{\partial t} - \mu\epsilon \frac{\partial^2 \mathbf{E}}{\partial t^2} = 0 \tag{24-5}$$

In exactly the same manner, $\nabla \times (\nabla \times \mathbf{B})$ will lead us to

$$\nabla^2 \mathbf{B} - \mu\sigma \frac{\partial \mathbf{B}}{\partial t} - \mu\epsilon \frac{\partial^2 \mathbf{B}}{\partial t^2} = 0 \tag{24-6}$$

We see that **E** and **B** separately satisfy the same equation. Thus, if $\psi(\mathbf{r}, t)$ is *any* of the six rectangular components of **E** and **B**, we find that

$$\nabla^2\psi - \mu\sigma\frac{\partial\psi}{\partial t} - \mu\epsilon\frac{\partial^2\psi}{\partial t^2} = 0 \tag{24-7}$$

so that, in effect, we have only one scalar equation to solve.

This last result, however, does *not* mean that we can take *any* arbitrary six solutions of (24-7), call them E_x, E_y,..., B_z in any manner and regard the result as a possible electromagnetic field. The reason is that the fields must still satisfy Maxwell's equations given above and this imposes restrictions on **E** and **B**. This has occurred because we differentiated Maxwell's equations to obtain (24-5) through (24-7) and the process of differentiation always results in some loss of information. Nevertheless, we can continue in this manner, but we will eventually have to satisfy (24-1) through (24-4) as well.

It will now be convenient to simplify things a bit more.

24-2 PLANE WAVES IN A NONCONDUCTING MEDIUM

First, let us assume that the medium is nonconducting so that $\sigma = 0$; then (24-7) takes the form

$$\nabla^2\psi - \mu\epsilon\frac{\partial^2\psi}{\partial t^2} = 0 \tag{24-8}$$

which is known as the *three-dimensional wave equation*. Although we will come back to (24-8), let us look for a function ψ of the special type $\psi = \psi(z, t)$. In other words, for any t, ψ is independent of x and y so that, for each value of z, ψ is constant on the corresponding infinite plane parallel to the xy plane; one such plane is indicated in Figure 24-1. In this case, (24-8) becomes

$$\frac{\partial^2\psi}{\partial z^2} - \mu\epsilon\frac{\partial^2\psi}{\partial t^2} = 0 \tag{24-9}$$

which is the *one-dimensional wave equation*.

In essence, the term *wave* is used to describe a pattern (or form) that propagates, that is, travels along. Then a function $f(z, t) = f(z - vt)$ with z and t appearing *only* in the combination $z - vt$ where v is a constant can be taken to be a wave traveling in the positive z direction with velocity v. (This is called a *plane wave* since f is constant on the plane $z = $ const.) We can verify the above with the aid of Figure 24-2. Consider the value of f at $t = 0$ and $z = 0$: $f(0) = f_0$. At a later time t_0, f will again equal the

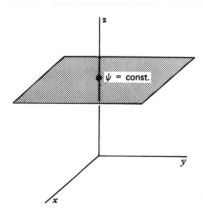

Figure 24-1. A plane of constant ψ, which is parallel to the xy plane.

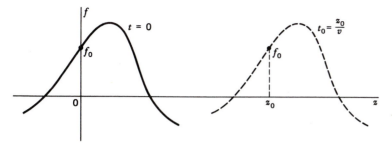

Figure 24-2. The shape at $t = 0$ has moved a distance z_0 in a time t_0.

same value f_0 at a position z_0 such that the argument is again zero, that is, $f(z_0 - vt_0) = f_0 = f(0)$, so that

$$z_0 = vt_0 \tag{24-10}$$

In other words, the particular feature $f = f_0$, which appeared at $z = 0$ and $t = 0$, now has moved to the new position z_0 given by (24-10), that is, to a point reached by traveling with constant velocity v. Similar remarks apply to the other values of f at $t = 0$, and the net result is that the whole picture has been shifted in the direction of positive z by the amount vt_0. Similarly, one can show that $g(z + vt)$ represents a wave of arbitrary shape moving in the direction of negative z, that is, to the left, with speed v and hence velocity $-v$.

We now want to show that a completely general solution of (24-9) can be written in the form

$$\psi(z, t) = f(z - vt) + g(z + vt) \tag{24-11}$$

where f and g are arbitrary functions. In order to do this, let $w = z - vt$; then we can write $f = f(w)$ and we find that

$$\frac{\partial f}{\partial z} = \frac{df}{dw}\frac{\partial w}{\partial z} = \frac{df}{dw} \quad \text{and} \quad \frac{\partial^2 f}{\partial z^2} = \frac{d^2 f}{dw^2}\frac{\partial w}{\partial z} = \frac{d^2 f}{dw^2}$$

while

$$\frac{\partial f}{\partial t} = \frac{df}{dw}\frac{\partial w}{\partial t} = -v\frac{df}{dw} \quad \text{and} \quad \frac{\partial^2 f}{\partial t^2} = -v\frac{d^2 f}{dw^2}\frac{\partial w}{\partial t} = v^2\frac{d^2 f}{dw^2}$$

Substituting these into (24-9), we get

$$\left(1 - v^2\mu\epsilon\right)\frac{d^2 f}{dw^2} = 0$$

which shows that $f(z - vt)$ is a solution provided that $v^2\mu\epsilon = 1$, or

$$v = \frac{1}{\sqrt{\mu\epsilon}} \tag{24-12}$$

Similarly, $g(z + vt)$ can be shown to be a solution of (24-9) with v given by the same expression (24-12). Thus we see already that our electromagnetic fields can have the form of plane waves propagating in the z direction and the speed of the wave is determined *solely* by the electromagnetic properties of the medium as given by the product $\mu\epsilon$.

It is usual to write (24-12) in a different manner; if we use (10-53), (20-55), and (23-5), we find that we can write

$$v = \frac{c}{n} \tag{24-13}$$

$$n = \sqrt{\kappa_e \kappa_m} \tag{24-14}$$

where n is called the *index of refraction* of the medium and is a dimensionless quantity. For a vacuum, $n = 1$ and the speed of the wave becomes simply c, which, as we remarked following (23-4), is the same as the measured speed of light in a vacuum. This result was first obtained by Maxwell and was taken by him, and others since then, as strong evidence for believing that light waves are actually electromagnetic waves. Since Maxwell's time, much more evidence has been accumulated in support of this idea—so much, in fact, that nowadays it is universally accepted. Furthermore, the concept of electromagnetic waves has been extended both experimentally and theoretically beyond merely that of describing visible light.

Substituting (24-12) into (24-9), we find that we can now write the wave equation in the form

$$\frac{\partial^2 \psi}{\partial z^2} = \frac{1}{v^2} \frac{\partial^2 \psi}{\partial t^2} \tag{24-15}$$

Although we know that (24-11) represents a general solution of this equation, it is too general for our purposes, and we want to consider solutions of a more specific form. In order to get an idea of the type we want to study, let us begin again with (24-15) and try the method of separation of variables that we found so helpful in Sections 11-4 and 11-5; in other words, let us try the form $\psi(z, t) = Z(z)T(t)$. Substituting this into (24-15), dividing by ZT, and proceeding in the usual manner, we get

$$\frac{1}{Z}\frac{d^2 Z}{dz^2} = \frac{1}{v^2 T}\frac{d^2 T}{dt^2} = \text{const.} = -k^2$$

so that we have the two differential equations

$$\frac{d^2 Z}{dz^2} + k^2 Z = 0 \quad \text{and} \quad \frac{d^2 T}{dt^2} + \omega^2 T = 0 \tag{24-16}$$

where

$$k^2 = \frac{\omega^2}{v^2} \tag{24-17}$$

The general solutions of (24-16) can be written as $Z_k(z) = \alpha_k e^{ikz} + \beta_k e^{-ikz}$ and $T_k(t) = \gamma_k e^{i\omega t} + \delta_k e^{-i\omega t}$ where α_k, β_k, γ_k, and δ_k are constants and $i = \sqrt{-1}$. If we multiply these two results together we get a solution to the wave equation for this particular value of k and the corresponding ω given by (24-17). Since there may well be many possible values of k, and since (24-15) is a linear differential equation, the sum of solutions each of this form will also be a solution. Thus, if we sum the products $Z_k T_k$ over all possible values of k, we get the general solution to the wave equation, and we find that it has the form

$$\psi(z, t) = \sum_k \left[\alpha_k \delta_k e^{i(kz - \omega t)} + \beta_k \gamma_k e^{-i(kz - \omega t)} \right]$$

$$+ \sum_k \left[\alpha_k \gamma_k e^{i(kz + \omega t)} + \beta_k \delta_k e^{-i(kz + \omega t)} \right] \tag{24-18}$$

The arguments of the exponential terms are seen, with the use of (24-17), to all be of the form $kz \mp \omega t = k[z \mp (\omega/k)t] = k(z \mp vt)$ so that ψ actually has turned out to look like (24-11), and, in fact, the first sum in (24-18) can be identified with f while the second sum represents g. Thus we see that we have found that the solution of the one-dimensional wave equation can be written as a *superposition of sinusoidal plane waves*.

Actually, we can reduce the complexity of (24-18) even more. Consider the term that involves $e^{i(kz-\omega t)}$; for reasons that will become clear shortly, we will always take ω to be a positive quantity. Now if k is also positive, this form represents a plane wave traveling in the direction of positive z with speed $v = \omega/k$. If k is negative, we can write $k = -|k|$, and this takes the form $e^{-i(|k|z+\omega t)}$, which is a plane wave in the direction of negative z with speed $v = \omega/|k|$. Therefore, as far as the first term in each sum of (24-18) is concerned, we need only concern ourselves with the single form $e^{i(kz-\omega t)}$ and the *sign* of k (positive or negative) will tell the sense of travel of the wave. The other exponential terms in (24-18) are simply the complex conjugates of those we have just discussed and still have the appropriate form. Accordingly, for our purposes it will be sufficient to study the behavior of a *single* prototype plane wave of a particular value of k (and corresponding ω) that we can take to be of the form

$$\psi(z, t) = \psi_0 e^{i(kz-\omega t)} \tag{24-19}$$

where ψ_0 is a constant; the general solution can be constructed as a sum of terms found from this as shown in (24-18). [The form (24-19) is the normal way of representing a plane wave in physics; however, in some other books, the exponential term is written as $e^{i(\omega t-kz)}$, that is, as the complex conjugate of (24-19). In addition, the notation j rather than i for $\sqrt{-1}$ is often used. Generally speaking, the results can usually be compared by making the replacement of i by $-i$ or $\pm j$ as the case may be.]

Since ψ is to be a component of **E** or **B**, it is a physical quantity and hence must be *real*. However, it is more convenient to deal with exponential functions than with sines and cosines, so that it is normal practice to continue to write ψ in the form (24-19) and to supplement this with a convention: If ψ (or **E** or **B**) is written or found in complex form, then the *real part* is taken to be the physically significant solution, that is,

$$\psi_{\text{physical}} = \text{Re } \psi = \text{Re}\left[\psi_0 e^{i(kz-\omega t)}\right] \tag{24-20}$$

(This convention is possible because the wave equation is linear so that the real part of ψ and its imaginary part are separately solutions.) Since it will usually turn out that the amplitude ψ_0 is itself complex, it will be useful to have explicit forms for the real part. If we write ψ_0 in terms of its real part ψ_{0R} and its imaginary part ψ_{0I} as

$$\psi_0 = \psi_{0R} + i\psi_{0I} \tag{24-21}$$

and use

$$e^{iu} = \cos u + i \sin u \qquad e^{-iu} = \cos u - i \sin u \tag{24-22}$$

then we find that (24-19) becomes $\psi = (\psi_{0R} + i\psi_{0I})[\cos(kz - \omega t) + i \sin(kz - \omega t)]$, so that

$$\text{Re } \psi = \psi_{0R} \cos(kz - \omega t) - \psi_{0I} \sin(kz - \omega t) \tag{24-23}$$

Another useful way of writing a complex number is in terms of a real amplitude ψ_{0a} and phase angle ϑ:

$$\psi_0 = \psi_{0a} e^{i\vartheta} \tag{24-24}$$

We see that these various quantities are related by

$$\psi_{0R} = \psi_{0a} \cos \vartheta \qquad \psi_{0I} = \psi_{0a} \sin \vartheta$$

$$\psi_{0a} = \left(\psi_{0R}^2 + \psi_{0I}^2\right)^{1/2} \qquad \tan \vartheta = \frac{\psi_{0I}}{\psi_{0R}} \tag{24-25}$$

When (24-24) is substituted into (24-19) it becomes $\psi = \psi_{0_a} e^{i(kz - \omega t + \vartheta)}$ so that

$$\text{Re } \psi = \psi_{0a} \cos(kz - \omega t + \vartheta) \tag{24-26}$$

which is often a very convenient form to use. A complex amplitude ψ_0 thus means that there exists a phase factor ϑ in the sinusoidal plane wave.

Equation 24-26, for example, shows us that a sinusoidal plane wave is periodic both in space and in time. The spatial period λ is called the *wavelength*, and the temporal period T is related to the frequency ν by $T = 1/\nu$. These quantities can be related to k and ω as follows. Since the period of a cosine is 2π, the function in (24-26) will repeat itself whenever its argument changes by 2π. Thus, for a fixed t, $\Delta|(kz - \omega t + \vartheta)| = 2\pi = |k \, \Delta z| = |k|\lambda$ so that $|k| = 2\pi/\lambda$. Similarly, at a fixed position, $\Delta|(kz - \omega t + \vartheta)| = 2\pi = \omega \Delta t = \omega T$ and $\omega = 2\pi/T = 2\pi\nu$. Hence the relations are

$$|k| = \frac{2\pi}{\lambda} \qquad \omega = 2\pi\nu = \frac{2\pi}{T} \tag{24-27}$$

From (24-17), we have $|k| = \omega/v$ and when (24-27) is combined with this, we get the familiar result for a sinusoidal plane wave

$$v = \nu\lambda \tag{24-28}$$

The quantity k is called the *propagation constant* and is measured in (meter)$^{-1}$; ω is the (angular) frequency and is in (radian)/second while ν is in hertz as we previously discussed before (17-14). As a result of all of these relations, there is a large number of different ways in which expressions involving a plane wave can be written; we will generally continue with (24-19) for economy in notation.

The argument of the exponential, $kz - \omega t$, is called the *phase* of the wave, so that the velocity v is the rate at which a definite value of this phase travels; accordingly, v is known as the *phase velocity*.

The three characteristics of the wave (k, ω, v) are related by our previous result (24-17). Now v is determined by the properties of the medium as given by (24-12); thus, the remaining two (k and ω) cannot be chosen independently. Since ω (or ν) is usually determined by the source which is producing the wave, it is customary to regard the frequency as the independent variable characterizing the wave, so that k becomes the dependent variable of this set.

Since we have decided to concentrate on ψ in the form (24-19), all of the components of **E** and **B** will be of this form as well, that is, for a sinusoidal plane wave we will have

$$\mathbf{E} = \mathbf{E}_0 e^{i(kz - \omega t)} \qquad \mathbf{B} = \mathbf{B}_0 e^{i(kz - \omega t)} \tag{24-29}$$

where $\mathbf{E}_0$ and $\mathbf{B}_0$ are appropriate *constant* vector amplitudes that must be related so as to satisfy Maxwell's equations. These forms enable us to simplify the writing of these equations. Since ψ is a function of z and t only, we find from (24-19) that

$$\frac{\partial \psi}{\partial x} = \frac{\partial \psi}{\partial y} = 0 \qquad \frac{\partial \psi}{\partial z} = ik\psi \qquad \frac{\partial \psi}{\partial t} = -i\omega\psi \tag{24-30}$$

and therefore $\nabla \cdot \mathbf{E} = \partial E_z / \partial z = ikE_z$ and $\nabla \cdot \mathbf{B} = ikB_z$. Similarly, we find that $\nabla \times \mathbf{E} = ik(-E_y\hat{\mathbf{x}} + E_x\hat{\mathbf{y}})$ with a corresponding expression for $\nabla \times \mathbf{B}$; also $\partial \mathbf{B}/\partial t = -i\omega \mathbf{B}$. Substituting these expressions into (24-1) through (24-4), setting $\sigma = 0$, and using (24-12), we find that Maxwell's equations for a plane wave propagating in the z direction in a nonconducting medium become

$$kE_z = 0 \qquad kB_z = 0$$
$$k(-E_y\hat{\mathbf{x}} + E_x\hat{\mathbf{y}}) = \omega\mathbf{B} \qquad k(-B_y\hat{\mathbf{x}} + B_x\hat{\mathbf{y}}) = -\frac{\omega}{v^2}\mathbf{E} \tag{24-31}$$

These can be written more compactly and usefully by noting that $E_z = \hat{\mathbf{z}} \cdot \mathbf{E}$ and $-E_y\hat{\mathbf{x}} + E_x\hat{\mathbf{y}} = \hat{\mathbf{z}} \times \mathbf{E}$ with similar expressions for $\mathbf{B}$; thus we can write (24-31) as

$$k\hat{\mathbf{z}} \cdot \mathbf{E} = 0 \qquad k\hat{\mathbf{z}} \cdot \mathbf{B} = 0$$

$$k\hat{\mathbf{z}} \times \mathbf{E} = \omega\mathbf{B} \qquad k\hat{\mathbf{z}} \times \mathbf{B} = -\frac{\omega}{v^2}\mathbf{E} \qquad (24\text{-}32)$$

and we remember that the direction of propagation is given by $\pm\hat{\mathbf{z}}$, corresponding to the sign of k.

Suppose that $k = 0$; then we see that it is possible to have $E_z \neq 0$ and $B_z \neq 0$. But we see from (24-17) that, in this case, $\omega = 0$ as well, so that we actually have a *static* situation. While this is certainly possible, it is not our primary interest here.

For a *nonstatic* case, $\omega \neq 0$ and therefore $k \neq 0$. Then we see from the first two equations of (24-32) that we must have $E_z = 0$ and $B_z = 0$. Thus, both $\mathbf{E}$ and $\mathbf{B}$ have no component in the direction of propagation, that is, both $\mathbf{E}$ and $\mathbf{B}$ are perpendicular to it; in other words, the electromagnetic plane wave is a *transverse* wave. We see from $k\hat{\mathbf{z}} \times \mathbf{E} = \omega\mathbf{B}$ that $\mathbf{B}$ is perpendicular to $\mathbf{E}$, which also follows from the last equation of (24-32). This all means that these vectors form a mutually perpendicular set; this is illustrated for positive k in Figure 24-3. As also shown in the figure, the Poynting vector $\mathbf{S} = \mathbf{E} \times \mathbf{H} = (\mathbf{E} \times \mathbf{B})/\mu$ is itself in the direction of propagation. If we solve the third equation (24-32) and use (24-17), we find that

$$\mathbf{B} = \frac{k}{\omega}\hat{\mathbf{z}} \times \mathbf{E} = \frac{1}{v}\hat{\mathbf{z}} \times \mathbf{E} \qquad (24\text{-}33)$$

showing the magnitudes of the fields to be related by

$$|\mathbf{B}| = \frac{|k|}{\omega}|\mathbf{E}| = \frac{1}{v}|\mathbf{E}| = \sqrt{\mu\epsilon}|\mathbf{E}| = \frac{n}{c}|\mathbf{E}| \qquad (24\text{-}34)$$

so that, for a vacuum, $|\mathbf{B}|/|\mathbf{E}| = 1/c$.

We see from (24-29) that the above relations also hold for the vector amplitudes $\mathbf{E}_0$ and $\mathbf{B}_0$ since the exponential factors will cancel from both sides. In addition, if $\mathbf{E}_0$ has the complex form, $\mathbf{E}_0 = \mathbf{E}_{0a}e^{i\vartheta}$, then these relations hold for the real amplitudes $\mathbf{E}_{0a}$ and $\mathbf{B}_{0a}$ as well. Consequently, if we were to take the real parts of our solutions, as we did to get (24-26), we will obtain

$$\mathbf{E}_{\text{real}} = \mathbf{E}_{0a}\cos(kz - \omega t + \vartheta) \qquad \mathbf{B}_{\text{real}} = \mathbf{B}_{0a}\cos(kz - \omega t + \vartheta) \qquad (24\text{-}35)$$

Since the arguments of the cosines are the same in both cases, the two fields are said to be *in phase*—they become zero together, reach their maximum together, and so on. This is illustrated for positive k and a definite time in Figure 24-4, that is the figure shows a "photograph" of the fields as a function of distance along the direction of

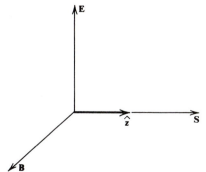

Figure 24-3. Relations among the field vectors, direction of propagation, and energy flow for a plane transverse wave.

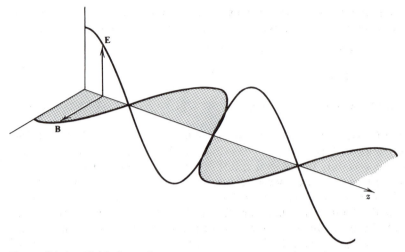

Figure 24-4.   Fields in a plane transverse wave at a given time.

travel. For an observer at a fixed point (z = const.), the time dependence of the fields described by (24-35) can be visualized by imagining this picture moving in the direction of the positive z axis with speed v. [For simplicity, we have assumed that $\mathbf{E}$ (and $\mathbf{B}$) always lies in the same plane, that is, it is *linearly polarized*; we return to this in Section 24-7.]

Although we haven't explicitly used the last relation of (24-32), that is, $k\hat{\mathbf{z}} \times \mathbf{B} = -(\omega/v^2)\mathbf{E}$, it is easily verified that it is consistent with (24-33).

24-3 PLANE WAVES IN A CONDUCTING MEDIUM

We now assume that $\sigma \neq 0$, but we will continue to assume that $\rho_f = 0$ and $\mathbf{J}_f' = 0$. The appropriate equation for any component of $\mathbf{E}$ and $\mathbf{B}$ is now (24-7), which becomes

$$\frac{\partial^2 \psi}{\partial z^2} - \mu\sigma \frac{\partial \psi}{\partial t} - \mu\epsilon \frac{\partial^2 \psi}{\partial t^2} = 0 \qquad (24\text{-}36)$$

when $\psi = \psi(z, t)$. We again try to solve this equation with a plane wave of the form (24-19), and when this is substituted into (24-36), we find that k and ω must now be related by the *dispersion relation*

$$k^2 = \omega^2 \mu\epsilon + i\omega\mu\sigma \qquad (24\text{-}37)$$

rather than (24-17). Since we are only interested in fields that vary harmonically in time, that is, are not damped in time, we require ω to be real as well as positive. Therefore, (24-37) can be satisfied only if k is a complex quantity; accordingly we assume k to have the form

$$k = \pm(\alpha + i\beta) \qquad (24\text{-}38)$$

where α and β are real and positive. Substituting this into (24-37), we obtain

$$\alpha^2 - \beta^2 + 2i\alpha\beta = \omega^2 \mu\epsilon + i\omega\mu\sigma \qquad (24\text{-}39)$$

Since the real and imaginary parts must be separately equal, we get

$$\alpha^2 - \beta^2 = \omega^2 \mu\epsilon \qquad (24\text{-}40)$$

$$2\alpha\beta = \omega\mu\sigma \qquad (24\text{-}41)$$

where we have assumed μ, ϵ, and σ to be real; we return to this point in Section 24-8. These can be solved for α and β and the results are

$$\alpha = \omega \sqrt{\frac{\mu\epsilon}{2}} \left[\sqrt{1 + \left(\frac{\sigma}{\omega\epsilon}\right)^2} + 1 \right]^{1/2} \tag{24-42}$$

$$\beta = \omega \sqrt{\frac{\mu\epsilon}{2}} \left[\sqrt{1 + \left(\frac{\sigma}{\omega\epsilon}\right)^2} - 1 \right]^{1/2} \tag{24-43}$$

(We note that when $\sigma = 0$, these reduce to $\alpha = \omega\sqrt{\mu\epsilon} = \omega/v$ and $\beta = 0$ as they should.)

These results are often written in terms of a dimensionless parameter Q, which, for reasons we will consider later, is defined by

$$Q = \frac{\omega\epsilon}{\sigma} \tag{24-44}$$

so that

$$\alpha = \omega \sqrt{\frac{\mu\epsilon}{2}} \left[\sqrt{1 + \frac{1}{Q^2}} + 1 \right]^{1/2} \tag{24-45}$$

$$\beta = \omega \sqrt{\frac{\mu\epsilon}{2}} \left[\sqrt{1 + \frac{1}{Q^2}} - 1 \right]^{1/2} \tag{24-46}$$

Furthermore, it will often be useful to write k in the form

$$k = |k|e^{i\Omega} \tag{24-47}$$

so that if we use (24-22) and (24-38), we get $k = \alpha + i\beta = |k|(\cos\Omega + i\sin\Omega)$, and therefore,

$$\alpha = |k| \cos\Omega \qquad \beta = |k| \sin\Omega \tag{24-48}$$

$$|k| = (\alpha^2 + \beta^2)^{1/2} = \omega\sqrt{\mu\epsilon} \left[1 + \frac{1}{Q^2} \right]^{1/4} \tag{24-49}$$

$$\tan\Omega = \frac{\beta}{\alpha} = Q\left[\sqrt{1 + \frac{1}{Q^2}} - 1 \right] = \sqrt{1 + Q^2} - Q \tag{24-50}$$

Now that we have all of these fairly complicated results, let us see what their significance is, and, in particular, what are the physical consequences of a complex propagation constant k.

Substituting $k = \alpha + i\beta$ into (24-19), we find that it becomes

$$\psi = \psi_0 e^{-\beta z} e^{i(\alpha z - \omega t)} \tag{24-51}$$

since $i^2 = -1$. A function of this form is called a *damped traveling wave* since its amplitude is not constant, but decreases with distance in the direction of propagation because of the factor $e^{-\beta z}$. The origin of this term arises from the existence of the conductivity since $\beta \neq 0$ only when $\sigma \neq 0$. Evidently it must be related to the loss of energy of the wave because of the resistive dissipation of energy into heat, and we will consider this in more detail shortly. Comparing the form of the propagation term with f of (24-11), we find that the velocity in this case is given by

$$v = \frac{\omega}{\alpha} = \frac{1}{\sqrt{\mu\epsilon}} \left\{ \frac{2}{[1 + (1/Q^2)]^{1/2} + 1} \right\}^{1/2} \tag{24-52}$$

and we see that in general $v < (\mu\epsilon)^{-1/2}$ or $v < v_{\text{nonconductor}}$ because of (24-12). Thus another effect of the conductivity of the medium is to make the wave travel more slowly. The wavelength can be found in the same manner by which we got (24-27), and we find that

$$\lambda = \frac{2\pi}{\alpha} = \frac{2\pi v}{\omega} = \frac{2\pi}{\omega\sqrt{\mu\epsilon}} \left\{ \frac{2}{[1 + (1/Q^2)]^{1/2} + 1} \right\}^{1/2} \tag{24-53}$$

which is, of course, in agreement with (24-28). Since v has decreased, we see that $\lambda < \lambda_{\text{nonconductor}}$ so that the same frequency wave has its wavelength shortened as compared to what it would be in a nonconducting medium with the same value of $\mu\epsilon$.

Since Q as given by (24-44) depends on the frequency ω, (24-52) shows that the wave velocity v is no longer a constant but is also a function of frequency, that is, we have $v = v(\omega)$. This phenomenon is called *dispersion* and hence a conducting medium is an example of what is known as a *dispersive medium*. In order to see the principal consequence of this, let us suppose that we are dealing with a more complicated field that is written as a superposition of sinusoidal traveling waves as is, say, the first sum of (24-18). Since k is no longer a constant but a function of ω, each component wave in the sum will travel at a different speed, so that, at a later time, the numerical value of the superposition at a given z will be different, that is, the superposition will have been changed in *form*. This is often described by saying that the superposition will alter its shape as it travels along.

From the damping factor $e^{-\beta z}$ in (24-51), we find that the distance of travel in which the amplitude decreases by a factor of $1/e$ is $\Delta z = 1/\beta$. This is generally written as δ and is called the *attenuation distance* or *skin depth* and is a convenient measure of the damping, since one can think of the wave as having essentially disappeared after traveling a few values of δ. It is possible to write this in a variety of ways with the use of (24-41), (24-52), (24-53), and (24-28):

$$\delta = \frac{1}{\beta} = \frac{2\alpha}{\mu\sigma\omega} = \frac{2}{\mu\sigma v} = \frac{4\pi}{\mu\sigma\omega\lambda} = \frac{2}{\mu\sigma v\lambda} \tag{24-54}$$

where λ is the wavelength *in the medium*.

For completeness, we should check to see that all of our conclusions are also appropriate for a wave traveling in the direction of negative z. Substituting $k = -(\alpha + i\beta)$ into (24-19), we find that it becomes

$$\psi = \psi_0 e^{\beta z} e^{-i(\alpha z + \omega t)} \tag{24-55}$$

and, on comparing this with (24-11), we see that it does represent a wave traveling in the correct sense with the same speed $v = \omega/\alpha$ as found in (24-52). Furthermore, since the values of z are now *decreasing*, the exponential factor $e^{\beta z}$ will decrease as the wave progresses, so that it still represents an attenuation.

There is another interesting and often very useful way of describing and thinking about our results. If we make a purely formal comparison between (24-19), (24-11), and (24-13), we see that we can always define a "wave velocity" V and an "index of refraction" N by

$$V = \frac{\omega}{k} = \frac{\omega}{\alpha + i\beta} = \frac{\omega}{|k|} e^{-i\Omega} \tag{24-56}$$

$$N = \frac{c}{V} = \frac{ck}{\omega} = \left(\frac{c\alpha}{\omega}\right) + i\left(\frac{c\beta}{\omega}\right) = \frac{c}{v} + i\left(\frac{c\beta}{\omega}\right)$$

$$= n' + i\left(\frac{c\beta}{\omega}\right) \tag{24-57}$$

where $n' = c/v$ is what one would call the usual index of refraction since it reduces to the value appropriate for a nonconductor as given by (24-14). Thus we can say that an absorbing medium ($\beta \neq 0$) can be described as having a *complex index of refraction*, with the appearance of the imaginary part being associated with attenuation.

So far, everything we have concluded about the effects of conductivity has been based on the behavior of a single component ψ. Now let us turn to the relation between the fields **E** and **B**. If we review our work from (24-29) on, we see that nothing depended on k being real; therefore, Maxwell's equations in the form (24-32) are still valid, except for the last one which came from (24-4) and that now takes the form

$$k\hat{\mathbf{z}} \times \mathbf{B} = -(\mu\epsilon\omega + i\mu\sigma)\mathbf{E} \tag{24-58}$$

Therefore, the fields are still transverse to the direction of propagation, while

$$\mathbf{B} = \frac{k}{\omega}\hat{\mathbf{z}} \times \mathbf{E} \tag{24-59}$$

again by (24-33). Thus all that is required is to see if (24-58) is satisfied. If we substitute (24-59) into it, use (1-30) and $\hat{\mathbf{z}} \cdot \mathbf{E} = 0$, we find that it becomes $(k^2 - \mu\epsilon\omega^2 - i\mu\sigma\omega)\mathbf{E} = 0$, which is satisfied with $\mathbf{E} \neq 0$ because of the dispersion relation (24-37). Now k is complex, and if we use the form (24-47) in (24-59), we find that the fields are now related by

$$\mathbf{B} = \frac{|k|}{\omega}e^{i\Omega}\hat{\mathbf{z}} \times \mathbf{E} \tag{24-60}$$

In order to see the significance of the phase factor $e^{i\Omega}$, we will find the real parts of **E** and **B**. If we again write $\mathbf{E}_0$ in (24-29) as $\mathbf{E}_{0a}e^{i\vartheta}$, where $\mathbf{E}_{0a}$ is real, and use (24-38) as well, we get

$$\mathbf{E} = \mathbf{E}_{0a}e^{-\beta z}e^{i(\alpha z - \omega t + \vartheta)}$$
$$\mathbf{B} = \frac{|k|}{\omega}\hat{\mathbf{z}} \times \mathbf{E}_{0a}e^{-\beta z}e^{i(\alpha z - \omega t + \vartheta + \Omega)} \tag{24-61}$$

so that

$$\mathbf{E}_{\text{real}} = \mathbf{E}_{0a}e^{-\beta z}\cos(\alpha z - \omega t + \vartheta)$$
$$\mathbf{B}_{\text{real}} = \frac{|k|}{\omega}\hat{\mathbf{z}} \times \mathbf{E}_{0a}e^{-\beta z}\cos(\alpha z - \omega t + \vartheta + \Omega) \tag{24-62}$$

and we can write the amplitude of $\mathbf{B}_{\text{real}}$ as $\mathbf{B}_{0a} = (|k|/\omega)\hat{\mathbf{z}} \times \mathbf{E}_{0a}$. Comparing these expressions with (24-35), we see that in a conducting medium **E** and **B** are *no longer in phase* and the phase difference Ω is positive according to (24-50). This means that **E** and **B** no longer reach their maxima and minima together, nor do they vanish together. We can compare them quantitatively. Suppose the phase, that is, the argument of the cosine, has the definite value P at a given position z_0 and a time t_E for the electric field, that is, $P = \alpha z_0 - \omega t_E + \vartheta$. Then, the induction will have reached the same *relative value* at this point at a time t_B when the phase has the same value, so that $P = \alpha z_0 - \omega t_B + \vartheta + \Omega$. Equating these expressions, we find that the times are related by

$$t_B = t_E + \frac{\Omega}{\omega} > t_E \tag{24-63}$$

which shows that **B** reaches its maximum value, for example, at this point at a *later* time than does **E**; in other words, **B** *lags* **E** as a function of *time*. On the other hand, at a given time t_0, the locations at which **B** and **E** have the same phase are seen to be

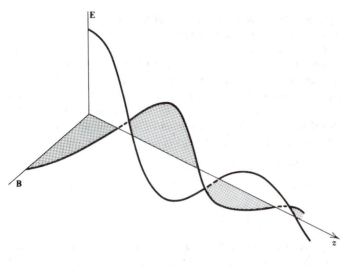

Figure 24-5. Fields in an attenuated plane wave in a conductor at a given time.

related by

$$z_B = z_E - \frac{\Omega}{\alpha} < z_E \tag{24-64}$$

so that we can speak of **B** *leading* **E** as a function of *position*. This effect, along with the attenuation, is illustrated schematically in Figure 24-5, which shows the situation at a given time and for positive k. (This figure should be compared with Figure 24-4.)

Another effect that follows from (24-62) is an alteration in the relative magnitudes, and, with the use of (24-49), we find that

$$\frac{|\mathbf{B}_{\text{real}}|}{|\mathbf{E}_{\text{real}}|} = \frac{|k|}{\omega} = \sqrt{\mu\epsilon} \left(1 + \frac{1}{Q^2}\right)^{1/4} \tag{24-65}$$

The value of this ratio is greater than the value $\sqrt{\mu\epsilon}$ applicable to a nonconductor, as is seen from (24-34). Thus the magnitude of **B** compared to that of **E** is relatively greater in a conductor than in a nonconductor.

[Actually, in a purely formal sense, one can justifiably assert that things have not changed at all. If we use (24-56) in (24-32), we can write $\mathbf{B} = (k/\omega)\hat{\mathbf{z}} \times \mathbf{E} = (\hat{\mathbf{z}} \times \mathbf{E})/V$ and when we compare this with (24-33), we see that in both cases the (complex) **B** is simply equal to $\hat{\mathbf{z}} \times \mathbf{E}$ divided by the appropriate "wave velocity."]

All of our results so far have been exact, although somewhat complicated. It is convenient and useful to obtain approximations to these results that are applicable to two extreme limiting cases. This process is most easily done in terms of the dimensionless parameter Q defined in (24-44). First of all, the definition is not frivolous as Q can be seen to have a simple physical interpretation. For an electric field varying harmonically in time, that is, proportional to $e^{-i\omega t}$, the magnitude of the displacement current will be $|\partial \mathbf{D}/\partial t| = |\epsilon \partial \mathbf{E}/\partial t| = \epsilon\omega|\mathbf{E}|$, while the magnitude of the conduction current will be $|\mathbf{J}_f| = \sigma|\mathbf{E}|$. If we now multiply the numerator and denominator of (24-44) by $|\mathbf{E}|$, we find that we can also write

$$Q = \frac{|\partial \mathbf{D}/\partial t|}{|\mathbf{J}_f|} \tag{24-66}$$

so that Q can be regarded as a measure of the relative importance of the displacement current and the conduction current. (Also see Exercise 24-29.) It is useful to use the value of Q to characterize a material as an "insulator" or as a "good conductor" for the given value of the frequency. We see from (24-44) that for small conductivity and/or high frequencies Q will be large, and we can say that we are dealing with an insulator. On the other hand, for large conductivity and/or low frequencies, Q will be very small and we are dealing with a good conductor; this is usually the situation applicable to metals. Accordingly, we can take our limiting cases as corresponding to $Q \gg 1$ and $Q \ll 1$, and we shall approximate our previous results to get formulas that are adequate and somewhat easier to use. We begin with the case more nearly like that of the previous section.

1. "Insulator" ($Q \gg 1$)

In this case, $1/Q^2 \ll 1$, and we can use the approximation $(1 + x)^n \simeq 1 + nx$ for $x \ll 1$. In this way, we find that, if we keep either the first-order correction term to something that stays finite for a nonconductor, or the first-order term for something that would vanish, our previous results become *approximately*

$$\alpha = \omega\sqrt{\mu\epsilon}\left(1 + \frac{1}{8Q^2}\right) \qquad \beta = \frac{1}{\delta} = \frac{\omega\sqrt{\mu\epsilon}}{2Q} = \frac{\sigma}{2}\left(\frac{\mu}{\epsilon}\right)^{1/2} \tag{24-67}$$

$$|k| = \omega\sqrt{\mu\epsilon}\left(1 + \frac{1}{4Q^2}\right) \qquad \tan\Omega = \frac{1}{2Q} = \frac{\sigma}{2\omega\epsilon} \tag{24-68}$$

$$v = \frac{1}{\sqrt{\mu\epsilon}}\left(1 - \frac{1}{8Q^2}\right) \qquad \lambda = \frac{2\pi}{\omega\sqrt{\mu\epsilon}}\left(1 - \frac{1}{8Q^2}\right) \tag{24-69}$$

$$\frac{|\mathbf{B}|}{|\mathbf{E}|} = \sqrt{\mu\epsilon}\left(1 + \frac{1}{4Q^2}\right) \tag{24-70}$$

and, of course, there are only small differences between the results and those of the last section.

2. "Good Conductor" ($Q \ll 1$)

Proceeding in a similar manner, we find the approximate expressions to be (with correction terms to order Q only):

$$\alpha = \left(\frac{1}{2}\mu\sigma\omega\right)^{1/2}\left(1 + \frac{1}{2}Q\right) \qquad \beta = \frac{1}{\delta} = \left(\frac{1}{2}\mu\sigma\omega\right)^{1/2}\left(1 - \frac{1}{2}Q\right) \tag{24-71}$$

$$|k| = (\mu\sigma\omega)^{1/2} \qquad \tan\Omega = 1 - Q \tag{24-72}$$

$$v = \left(\frac{2\omega}{\mu\sigma}\right)^{1/2}\left(1 - \frac{1}{2}Q\right) \qquad \lambda = 2\pi\left(\frac{2}{\mu\sigma\omega}\right)^{1/2}\left(1 - \frac{1}{2}Q\right) \tag{24-73}$$

$$\frac{|\mathbf{B}|}{|\mathbf{E}|} = \left(\frac{\mu\sigma}{\omega}\right)^{1/2} = \left(\frac{\mu\epsilon}{Q}\right)^{1/2} \tag{24-74}$$

It is actually common practice to approximate these even further by dropping the correction terms and we get

$$\alpha = \beta = \left(\frac{1}{2}\mu\sigma\omega\right)^{1/2} \tag{24-75}$$

$$\tan \Omega = 1, \text{ so that } \Omega = \frac{\pi}{4} \text{ or } 45° \tag{24-76}$$

$$v = \left(\frac{2\omega}{\mu\sigma}\right)^{1/2} = \left(\frac{2Q}{\mu\epsilon}\right)^{1/2} = \sqrt{2Q}\, v_{\text{nonconductor}} \tag{24-77}$$

$$\delta = \left(\frac{2}{\mu\sigma\omega}\right)^{1/2} = \frac{\lambda}{2\pi} \tag{24-78}$$

Comparing (24-74) and (24-70), we see that $|\mathbf{B}|$ is relatively much larger than $|\mathbf{E}|$ in a good conductor as compared to an insulator, while the phase difference is virtually always $\pi/4$ according to (24-76). Similarly, (24-77) shows that the wave speed is very much smaller in the case of a good conductor, while (24-78) shows that the penetration depth (attenuation distance) is of the order of a wavelength in *the conductor*, which in turn is very much smaller than in an insulator.

A numerical example is helpful here. A typical metal such as copper has $\sigma \simeq 6 \times 10^7$ (ohm-meter)$^{-1}$ and $\epsilon \simeq \epsilon_0$. Thus we find from (24-44) that

$$Q = \frac{2\pi\nu\epsilon}{\sigma} \simeq \left(\frac{2\pi\epsilon_0}{\sigma}\right)\nu = 9 \times 10^{-19}\nu \simeq 10^{-18}\nu$$

so that in order to have $Q \ll 1$, we must have $\nu \ll 10^{18}$ hertz. This means that this approximation is good all the way up in frequency until we get to the ultraviolet region, and there we may run into quantum-mechanical effects in any case. In other words, the case $Q \ll 1$ really fits ordinary metals very well.

If we use this same value of σ and take $\mu \simeq \mu_0$ in (24-78), we find that the penetration depth δ becomes $(6.5 \times 10^{-2})/\nu^{1/2}$ meter. At low frequencies, this is quite large so that there is very little attenuation. But by the time we get to the lower limit of the microwave region ($\nu \simeq 3 \times 10^9$ hertz), we find that $\delta \simeq 10^{-6}$ meter, so that the fields are essentially different from zero only within this small distance; this is the origin of the term "skin depth."

24-4 PLANE WAVES IN A CHARGED MEDIUM

Now that we have some experience, we can consider a more complicated situation. Let us assume that there is a free charge distribution ρ_f within our medium; we still assume that $\mathbf{J}_f' = 0$ since a nonzero value corresponds to such a rarely occurring case. Thus, instead of (24-1), we need to use $\nabla \cdot \mathbf{E} = \rho_f/\epsilon$ by (21-42), so that generally $\nabla \cdot \mathbf{E} \neq 0$. The remaining equations (24-2) through (24-4) are unchanged. Now, when we eliminate $\mathbf{B}$ by the same method we used before, we find that instead of (24-5), we obtain

$$\nabla^2\mathbf{E} - \mu\sigma\frac{\partial\mathbf{E}}{\partial t} - \mu\epsilon\frac{\partial^2\mathbf{E}}{\partial t^2} = \nabla(\nabla \cdot \mathbf{E}) \tag{24-79}$$

On the other hand, the equation satisfied by $\mathbf{B}$ alone is still (24-6):

$$\nabla^2\mathbf{B} - \mu\sigma\frac{\partial\mathbf{B}}{\partial t} - \mu\epsilon\frac{\partial^2\mathbf{B}}{\partial t^2} = 0 \tag{24-80}$$

since $\nabla \cdot \mathbf{B} = 0$ always.

We again consider the case in which the fields are functions only of z and t so that the only nonzero spatial derivatives are those with respect to z. Then (24-79) reduces to

$$\frac{\partial^2 \mathbf{E}}{\partial z^2} - \mu\sigma \frac{\partial \mathbf{E}}{\partial t} - \mu\epsilon \frac{\partial^2 \mathbf{E}}{\partial t^2} = \frac{\partial^2 E_z}{\partial z^2} \hat{\mathbf{z}} \qquad (24\text{-}81)$$

The x component of this equation is

$$\frac{\partial^2 E_x}{\partial z^2} - \mu\sigma \frac{\partial E_x}{\partial t} - \mu\epsilon \frac{\partial^2 E_x}{\partial t^2} = 0$$

and there will be a similar equation for E_y. Thus the equations for E_x, E_y, and the three components of $\mathbf{B}$ as obtained from (24-80) are still of the form (24-7). The only one which is different is the z component of (24-81):

$$\frac{\partial^2 E_z}{\partial z^2} - \mu\sigma \frac{\partial E_z}{\partial t} - \mu\epsilon \frac{\partial^2 E_z}{\partial t^2} = \frac{\partial^2 E_z}{\partial z^2}$$

which reduces to

$$\frac{\partial}{\partial t}\left(\frac{\sigma}{\epsilon} E_z + \frac{\partial E_z}{\partial t} \right) = 0 \qquad (24\text{-}82)$$

and shows that the term in parentheses is at most a function of z. If we let this function be $a(z)$, then the general solution of the resulting equation is $E_z = (\epsilon/\sigma)$ $[a(z) + b(z)e^{-\sigma t/\epsilon}]$. This component has one part that decays exponentially to zero with the same relaxation time ϵ/σ of (12-41) and is certainly not a wave. The other part represents a *static* field that can be a function of position; while this is a possibility, it is of no interest for a study of wave propagation and, rather than carry it along, we simply take $E_z = 0$. Thus, we conclude that a *traveling wave* in a charged l.i.h. medium is *still transverse* and all of the relevant components satisfy the equation (24-7). Consequently, all of our conclusions of the last two sections apply in this case also; hence we need consider it no further and can continue to use (24-1) through (24-4) for traveling waves.

24-5 PLANE WAVE IN AN ARBITRARY DIRECTION

For simplicity, we have considered only plane waves traveling in a specific direction that we chose to call the z axis. For further use, it is helpful to generalize our results in order that we can describe a plane wave traveling in an arbitrary direction with respect to a *given* set of coordinate axes.

Previously we wrote $\psi = \psi(z, t)$ so that ψ was constant on all points of an infinite plane perpendicular to the z axis as shown in Figure 24-1. What we want now is for ψ to depend only on the time t and the distance ζ of a given plane from the origin, so that $\psi = \psi(\zeta, t)$. This is illustrated by the edge-on view of the plane shown in Figure 24-6 where the orientation of the plane is described by its normal $\hat{\mathbf{n}}$ and where $\mathbf{r}$ is the position vector of an arbitrary point on the plane. We see from the figure that the constant distance ζ is given by

$$\zeta = \hat{\mathbf{n}} \cdot \mathbf{r} \qquad (24\text{-}83)$$

so that this is the equation of the plane. Thus, for a plane wave of propagation constant k, we have

$$\psi = \psi_0 e^{i(k\zeta - \omega t)} = \psi_0 e^{i(k\hat{\mathbf{n}} \cdot \mathbf{r} - \omega t)} \qquad (24\text{-}84)$$

Now if we always choose the normal $\hat{\mathbf{n}}$ to be in the direction of propagation of the

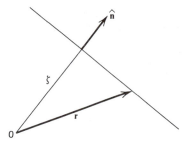

Figure 24-6. A plane wave traveling in the arbitrary direction n̂.

wave, we can take k to be positive and define a *propagation vector* $\mathbf{k}$ by

$$\mathbf{k} = k\hat{\mathbf{n}} = k\hat{\mathbf{k}} \qquad (24\text{-}85)$$

so that we can write

$$\psi = \psi_0 e^{i(\mathbf{k} \cdot \mathbf{r} - \omega t)} \qquad (24\text{-}86)$$

as our representation of a plane wave traveling in the direction $\mathbf{k}$ (with corresponding unit vector $\hat{\mathbf{k}} = \hat{\mathbf{n}}$). If, for example, we refer this to a specific set of rectangular axes, we can write

$$\mathbf{k} = k_x\hat{\mathbf{x}} + k_y\hat{\mathbf{y}} + k_z\hat{\mathbf{z}} \qquad (24\text{-}87)$$

and therefore

$$\psi = \psi_0 e^{i[(k_x x + k_y y + k_z z) - \omega t]} \qquad (24\text{-}88)$$

Since ψ can be a component of $\mathbf{E}$ or $\mathbf{B}$, the generalization of (24-29) will be

$$\mathbf{E} = \mathbf{E}_0 e^{i(\mathbf{k} \cdot \mathbf{r} - \omega t)} \qquad \mathbf{B} = \mathbf{B}_0 e^{i(\mathbf{k} \cdot \mathbf{r} - \omega t)} \qquad (24\text{-}89)$$

and now we need to see what Maxwell's equations become for this more general way of writing our fields.

With the assumed form (24-88), the analogues of (24-30) are now

$$\frac{\partial \psi}{\partial x} = ik_x\psi \qquad \frac{\partial \psi}{\partial y} = ik_y\psi \qquad \frac{\partial \psi}{\partial z} = ik_z\psi \qquad \frac{\partial \psi}{\partial t} = -i\omega\psi \qquad (24\text{-}90)$$

which means that the del operator is equivalent to the substitution $\nabla = i\mathbf{k} = ik\hat{\mathbf{k}}$ and Maxwell's equations (24-1) through (24-4) become

$$\begin{aligned} \mathbf{k} \cdot \mathbf{E} &= 0 & \mathbf{k} \cdot \mathbf{B} &= 0 \\ \mathbf{k} \times \mathbf{E} &= \omega\mathbf{B} & \mathbf{k} \times \mathbf{B} &= -(\mu\epsilon\omega + i\mu\sigma)\mathbf{E} \end{aligned} \qquad (24\text{-}91)$$

which are exactly the first three of (24-32) plus (24-58) with $k\hat{\mathbf{z}}$ replaced by $\mathbf{k} = k\hat{\mathbf{k}}$ as we would expect. Thus, we still have the fields related by

$$\mathbf{B} = \frac{k}{\omega}\hat{\mathbf{k}} \times \mathbf{E} \qquad (24\text{-}92)$$

where k and ω are connected by the dispersion relation (24-37) so that all of the results of the previous sections can be taken over to this general case simply by replacing $\hat{\mathbf{z}}$ by $\hat{\mathbf{k}}$ and kz by $\mathbf{k} \cdot \mathbf{r}$.

Later it will be helpful to have an explicit relation between $\mathbf{H}$ and $\mathbf{E}$, which can be found from (24-92) since $\mathbf{H} = \mathbf{B}/\mu$, and we get

$$\mathbf{H} = \frac{k}{\mu\omega}\hat{\mathbf{k}} \times \mathbf{E} = \frac{|k|}{\mu\omega}e^{i\Omega}\hat{\mathbf{k}} \times \mathbf{E} \qquad (24\text{-}93)$$

with the use of (24-60). In the important special case of a nonconducting medium, we can write $|k|/\omega = 1/v = \sqrt{\mu\epsilon}$ by (24-17) and (24-12) so that we also have

$$\mathbf{H} = \frac{\hat{\mathbf{k}} \times \mathbf{E}}{\mu v} = \left(\frac{\epsilon}{\mu}\right)^{1/2} \hat{\mathbf{k}} \times \mathbf{E} = \frac{\hat{\mathbf{k}} \times \mathbf{E}}{Z} \qquad (24\text{-}94)$$

where

$$Z = \left(\frac{\mu}{\epsilon}\right)^{1/2} = \left(\frac{\kappa_m \mu_0}{\kappa_e \epsilon_0}\right)^{1/2} = \left(\frac{\kappa_m}{\kappa_e}\right)^{1/2} Z_0 \qquad (24\text{-}95)$$

is called the *wave impedance*. The quantity $Z_0 = (\mu_0/\epsilon_0)^{1/2} = \mu_0 c = 377$ ohms is known as the *impedance of free space*.

24-6 COMPLEX SOLUTIONS AND TIME-AVERAGE ENERGY RELATIONS

As has already been amply illustrated, it is often convenient to obtain solutions in complex form. However, we must always remember to take the real part of the solution in order to get the physical quantity of interest as we did in (24-62), for example. Since it is sometimes inconvenient to find the real parts, it is desirable to rewrite some of our previous results so that the complex solutions can be substituted into them directly. We will, however, do this *only* for fields that have a sinusoidal time dependence, that is, are proportional to $e^{-i\omega t}$. It is important to remember that while this is applicable to plane waves, it is *not* restricted to them.

Let $\mathbf{E}_c$, $\mathbf{B}_c$, $\mathbf{H}_c$, and $\mathbf{D}_c$ represent solutions that have been obtained in complex form. Then the physically real electric and magnetic fields are

$$\mathbf{E} = \text{Re}\,(\mathbf{E}_c) = \text{Re}\,(\mathbf{E}_0 e^{-i\omega t}) \qquad (24\text{-}96)$$

$$\mathbf{H} = \text{Re}\,(\mathbf{H}_c) = \text{Re}\,(\mathbf{H}_0 e^{-i\omega t}) \qquad (24\text{-}97)$$

and that we previously wrote as $\mathbf{E}_{\text{real}}$ and so on. The quantities $\mathbf{E}_0$ and $\mathbf{H}_0$ are functions only of $\mathbf{r}$. If we now write them in terms of their real and imaginary parts, so that

$$\mathbf{E}_0 = \mathbf{E}_R + i\mathbf{E}_I \qquad \mathbf{H}_0 = \mathbf{H}_R + i\mathbf{H}_I$$

where $\mathbf{E}_R$, $\mathbf{E}_I$, $\mathbf{H}_R$, and $\mathbf{H}_I$ are all real, then (24-96) becomes

$$\mathbf{E} = \text{Re}\,[(\mathbf{E}_R + i\mathbf{E}_I)(\cos \omega t - i \sin \omega t)]$$
$$= \mathbf{E}_R \cos \omega t + \mathbf{E}_I \sin \omega t \qquad (24\text{-}98)$$

while (24-97) yields

$$\mathbf{H} = \mathbf{H}_R \cos \omega t + \mathbf{H}_I \sin \omega t \qquad (24\text{-}99)$$

Substituting (24-98) and (24-99) into (21-59), we find that the Poynting vector is given by

$$\mathbf{S} = \mathbf{E} \times \mathbf{H} = (\mathbf{E}_R \times \mathbf{H}_R) \cos^2 \omega t + (\mathbf{E}_I \times \mathbf{H}_I) \sin^2 \omega t$$
$$+ [(\mathbf{E}_R \times \mathbf{H}_I) + (\mathbf{E}_I \times \mathbf{H}_R)] \sin \omega t \cos \omega t \qquad (24\text{-}100)$$

which is generally a varying function of time.

In many situations, we are not primarily interested in the instantaneous value of the energy flow, often because it fluctuates too rapidly to be followed by our measuring instruments. The *time average* of the energy flow, $\langle \mathbf{S} \rangle$, is generally of much more significance. We see from (24-100) that this will be given by

$$\langle \mathbf{S} \rangle = \tfrac{1}{2}[(\mathbf{E}_R \times \mathbf{H}_R) + (\mathbf{E}_I \times \mathbf{H}_I)] \qquad (24\text{-}101)$$

since

$$\langle \cos^2 \omega t \rangle = \langle \sin^2 \omega t \rangle = \tfrac{1}{2} \qquad \langle \sin \omega t \cos \omega t \rangle = 0 \qquad (24\text{-}102)$$

These last results follow from (19-49) by dividing the appropriate integrals by 2π in order to average over a period of the trigonometric functions. Equation 24-101 is a fundamental result, but it can be written in a more convenient form.

Let $\mathbf{H}_c^* =$ complex conjugate of $\mathbf{H}_c = \mathbf{H}_c$ with i replaced by $-i = \mathbf{H}_0^* e^{i\omega t} = (\mathbf{H}_R - i\mathbf{H}_I)e^{i\omega t}$. Now consider

$$
\begin{aligned}
\mathbf{E}_c \times \mathbf{H}_c^* &= \left[(\mathbf{E}_R + i\mathbf{E}_I)e^{-i\omega t} \times (\mathbf{H}_R - i\mathbf{H}_I)e^{i\omega t} \right] \\
&= \left[(\mathbf{E}_R \times \mathbf{H}_R) + (\mathbf{E}_I \times \mathbf{H}_I) \right] + i\left[(\mathbf{E}_I \times \mathbf{H}_R) - (\mathbf{E}_R \times \mathbf{H}_I) \right] \quad (24\text{-}103)
\end{aligned}
$$

Comparing (24-101) and (24-103), we see that

$$\langle \mathbf{S} \rangle = \tfrac{1}{2}\operatorname{Re}(\mathbf{E}_c \times \mathbf{H}_c^*) \qquad (24\text{-}104)$$

Thus we see that we have succeeded in expressing the time average value of the Poynting vector entirely in terms of these two complex solutions of Maxwell's equations, so that we can calculate it directly without the necessity of first finding the real parts.

We can also carry through the same type of calculations for the average energy densities. The results are

$$\langle u_e \rangle = \langle \tfrac{1}{2}\epsilon E^2 \rangle = \tfrac{1}{4}\epsilon \mathbf{E}_c \cdot \mathbf{E}_c^* \qquad (24\text{-}105)$$

$$\langle u_m \rangle = \left\langle \frac{1}{2}\mu H^2 \right\rangle = \frac{1}{4}\mu \mathbf{H}_c \cdot \mathbf{H}_c^* = \frac{1}{4\mu}\mathbf{B}_c \cdot \mathbf{B}_c^* \qquad (24\text{-}106)$$

Now that we have obtained these important results, we no longer need to distinguish between real and complex fields for these purposes. Accordingly, in what follows, although we will continually be dealing with complex solutions of Maxwell's equations, we will not write them as $\mathbf{E}_c$ and $\mathbf{H}_c$, but simply as $\mathbf{E}$ and $\mathbf{H}$ and so on. If we actually need the real parts for specific purposes, we will point this out explicitly.

■ **Example**

Energy relations for a plane wave. We first consider the case $\sigma = 0$. Substituting (24-94) into (24-104), using (24-95), (1-30), and $\mathbf{k} \cdot \mathbf{E} = 0$ from (24-91), we obtain

$$\langle \mathbf{S} \rangle = \frac{1}{2}\left(\frac{\epsilon}{\mu}\right)^{1/2} \operatorname{Re}\left[\mathbf{E} \times (\hat{\mathbf{k}} \times \mathbf{E}^*)\right] = \frac{1}{2}\left(\frac{\epsilon}{\mu}\right)^{1/2}(\mathbf{E} \cdot \mathbf{E}^*)\hat{\mathbf{k}} \qquad (24\text{-}107)$$

since $\mathbf{E} \cdot \mathbf{E}^*$ is real. If we now use (24-89) and (24-94), we find that (24-107) can also be written as

$$\langle \mathbf{S} \rangle = \frac{1}{2}\left(\frac{\epsilon}{\mu}\right)^{1/2}|\mathbf{E}_0|^2\hat{\mathbf{k}} = \frac{1}{2}\left(\frac{\mu}{\epsilon}\right)^{1/2}|\mathbf{H}_0|^2\hat{\mathbf{k}} \qquad (24\text{-}108)$$

since $\mathbf{E} \cdot \mathbf{E}^* = \mathbf{E}_0 \cdot \mathbf{E}_0^* = |\mathbf{E}_0|^2$, and so on for $\mathbf{H}$. We see from this result that not only is the energy flow in the direction of propagation $\hat{\mathbf{k}}$, as we already knew, but also that it is proportional to the square of the amplitude of either $\mathbf{E}$ or $\mathbf{H}$.

In the same way, we find that the energy densities are

$$\langle u_e \rangle = \tfrac{1}{4}\epsilon \mathbf{E} \cdot \mathbf{E}^* = \tfrac{1}{4}\epsilon|\mathbf{E}_0|^2 = \tfrac{1}{4}\mu|\mathbf{H}_0|^2 = \langle u_m \rangle \qquad (24\text{-}109)$$

so that the average densities are equal. The total average energy density then becomes

$$\langle u \rangle = \langle u_e \rangle + \langle u_m \rangle = \tfrac{1}{2}\epsilon|\mathbf{E}_0|^2 = \tfrac{1}{2}\mu|\mathbf{H}_0|^2 \qquad (24\text{-}110)$$

which enables us to write (24-108) as

$$\langle \mathbf{S} \rangle = \frac{\langle u \rangle}{\sqrt{\mu \epsilon}} \hat{\mathbf{k}} = \langle u \rangle v \hat{\mathbf{k}} = \langle u \rangle \mathbf{v} \tag{24-111}$$

with the help of (24-12). Thus, the average energy current is the product of the average energy density and the velocity of the wave. This is in accord with the analogous result for current density $\mathbf{J} = \rho \mathbf{v}$ given by (12-3) and that of fluid kinematics where the mass flow per unit area is the product of the mass density and the fluid velocity.

In a conducting medium where $\sigma \neq 0$, we must use (24-93), and, when this is substituted into (24-104), we obtain

$$\langle \mathbf{S} \rangle = \frac{|k|}{2\mu \omega} \, \text{Re} \left[\mathbf{E} \times (\hat{\mathbf{k}} \times \mathbf{E}^*) e^{-i\Omega} \right]$$

$$= \frac{|k| \cos \Omega}{2\mu \omega} e^{-2\beta \zeta} |\mathbf{E}_0|^2 \hat{\mathbf{k}} = \frac{\alpha e^{-2\beta \zeta}}{2\mu \omega} |\mathbf{E}_0|^2 \hat{\mathbf{k}} = \frac{e^{-2\beta \zeta}}{2\mu v} |\mathbf{E}_0|^2 \hat{\mathbf{k}} \tag{24-112}$$

with the use of (24-22), (24-48), (24-52), (24-89), (24-85), (24-83), and (24-38), and where ζ now measures the distance in the direction of propagation. Similarly, the average total energy density becomes

$$\langle u \rangle = \frac{\alpha^2}{2\mu \omega^2} e^{-2\beta \zeta} |\mathbf{E}_0|^2 = \frac{e^{-2\beta \zeta}}{2\mu v^2} |\mathbf{E}_0|^2 \tag{24-113}$$

with the additional use of (24-49) and (24-40). We see that in this case $\langle \mathbf{S} \rangle = \langle u \rangle v \hat{\mathbf{k}}$ just as given by (24-111) for the nonconducting medium.

Both $\langle \mathbf{S} \rangle$ and $\langle u \rangle$ are seen to be proportional to $e^{-2\beta \zeta}$ so that they decrease with twice the attenuation factor of the fields; this is a result of the fact that they both are proportional to the square of the amplitudes. This energy is lost because of the resistive heating of the material arising from its conductivity. ∎

24-7 POLARIZATION

So far we have obtained many of our results from the assumed form (24-89) without being very specific about $\mathbf{E}_0$ and $\mathbf{B}_0$ other than that they are constants, that they will be related by $\mathbf{B}_0 = (k/\omega)\hat{\mathbf{k}} \times \mathbf{E}_0$ as found from (24-92), and that both lie in the plane perpendicular to the direction of propagation. The character of the wave depends on the nature of the amplitudes in this plane; we can concentrate on $\mathbf{E}_0$ since $\mathbf{B}_0$ can always be found from it.

In order to simplify our discussion somewhat, let us assume that our axes have been chosen so that the positive z axis is the direction of propagation; the transverse plane is then the xy plane, but we will not assume any specific orientation of the x and y axes with respect to the amplitude; for example, neither is chosen to be along the direction of $\mathbf{E}_0$. Since $\mathbf{E}_0$ lies in this plane, we can resolve it into its components and write

$$\mathbf{E}_0 = E_{0x} \hat{\mathbf{x}} + E_{0y} \hat{\mathbf{y}} \tag{24-114}$$

Since E_{0x} and E_{0y} are both generally complex numbers, we can write them in the form of (24-24) as

$$E_{0x} = E_1 e^{i\vartheta_1} \qquad E_{0y} = E_2 e^{i\vartheta_2} \tag{24-115}$$

so that $\mathbf{E}$ as given by (24-29) becomes

$$\mathbf{E} = \left(E_1 e^{i\vartheta_1} \hat{\mathbf{x}} + E_2 e^{i\vartheta_2} \hat{\mathbf{y}} \right) e^{i(kz - \omega t)} \tag{24-116}$$

For simplicity, we assume k to be real; we will see that this will not affect our basic conclusions. Then, taking the real parts of (24-116), we find the components of the electric field to be

$$E_x = E_1 \cos (kz - \omega t + \vartheta_1)$$
$$E_y = E_2 \cos (kz - \omega t + \vartheta_2)$$

(24-117)

The description of the electric field now depends on the relative values of the amplitudes (E_1, E_2) and phases $(\vartheta_1, \vartheta_2)$.

Since $-E_1 \le E_x \le E_1$ and $-E_2 \le E_y \le E_2$, according to (24-117), the tip of the electric field vector must always lie within the dashed rectangle shown in Figure 24-7; we note that *now* we are using **E** to represent the physical electric field, and *not* the complex expression of (24-116). The direction of propagation as given by the z axis is out of the page.

Suppose we imagine ourselves to be at a definite location. Then as time goes on, we see from (24-117) that the components of **E** will vary so that **E** itself changes and the tip of the **E** vector will trace out a path of some sort within the dashed rectangle. We can find this path or "orbit" by eliminating $(kz - \omega t)$ from (24-117). We find that

$$\frac{E_x}{E_1} = \cos (kz - \omega t) \cos \vartheta_1 - \sin (kz - \omega t) \sin \vartheta_1$$

$$\frac{E_y}{E_2} = \cos (kz - \omega t) \cos \vartheta_2 - \sin (kz - \omega t) \sin \vartheta_2$$

and therefore

$$\frac{E_x}{E_1} \sin \vartheta_2 - \frac{E_y}{E_2} \sin \vartheta_1 = -\cos (kz - \omega t) \sin (\vartheta_1 - \vartheta_2)$$

$$\frac{E_x}{E_1} \cos \vartheta_2 - \frac{E_y}{E_2} \cos \vartheta_1 = -\sin (kz - \omega t) \sin (\vartheta_1 - \vartheta_2)$$

Squaring each of these expressions, and then adding corresponding sides, we obtain

$$\left(\frac{E_x}{E_1} \right)^2 - 2 \left(\frac{E_x}{E_1} \right) \left(\frac{E_y}{E_2} \right) \cos (\vartheta_1 - \vartheta_2) + \left(\frac{E_y}{E_2} \right)^2 = \sin^2 (\vartheta_1 - \vartheta_2) \quad (24\text{-}118)$$

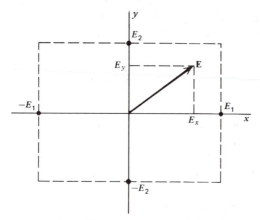

Figure 24-7. The components of the electric field in the plane transverse to the direction of propagation.

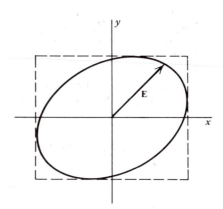

Figure 24-8. An elliptically polarized electric field.

This second-degree equation is generally that of an *ellipse* (since both E_x and E_y stay finite) and the electric field is said to be *elliptically polarized*. Thus the path traced out by its tip could be like that shown in Figure 24-8. The magnetic induction **B** will also be elliptically polarized since **B** is always perpendicular to **E**, and its ellipse will be rotated by 90° with respect to that for **E**.

The values of the principal axes of the ellipse and its orientation with respect to the axes clearly depend on the amplitudes E_1 and E_2 and the *relative phase* of the two components, since (24-118) depends only on the absolute value of the phase *difference* $|\vartheta_1 - \vartheta_2|$. It will be helpful to consider some special cases.

1. $\vartheta_1 - \vartheta_2 = 0$

In this case, (24-118) reduces to $[(E_x/E_1) - (E_y/E_2)]^2 = 0$ or

$$\frac{E_x}{E_1} = \frac{E_y}{E_2} \tag{24-119}$$

This is the equation of the straight line lying along the diagonal shown in Figure 24-9. The tip of **E** always lies on this line and the field is said to be *linearly polarized*. This can also be seen from (24-117) since, when $\vartheta_1 = \vartheta_2$, the two components are in phase, that is, they reach their maxima and minima together, and become zero together; a relation like this leads to the straight line shown in the figure.

2. $|\vartheta_1 - \vartheta_2| = \pi$

Now (24-118) reduces to $[(E_x/E_1) + (E_y/E_2)]^2 = 0$ or

$$\frac{E_x}{E_1} = -\frac{E_y}{E_2} \tag{24-120}$$

In this case, the field is again linearly polarized but, since E_x and E_y always have opposite signs, the line traced out by **E** lies along the other diagonal as shown in Figure 24-10.

3. $|\vartheta_1 - \vartheta_2| = \pi/2$

Here (24-118) becomes

$$\left(\frac{E_x}{E_1}\right)^2 + \left(\frac{E_y}{E_2}\right)^2 = 1 \tag{24-121}$$

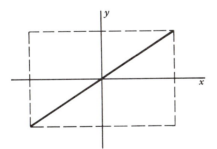

Figure 24-9. A linearly polarized electric field with zero phase difference between the *x* and *y* components.

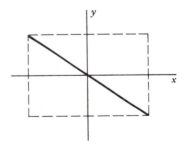

Figure 24-10. A linearly polarized electric field with a phase difference of π between the *x* and *y* components.

which is an ellipse with the major and minor axes along the coordinate axes as shown in Figure 24-11. A special case of this arises when $E_1 = E_2 = E_0$ and (24-121) becomes the equation of the circle $E_x^2 + E_y^2 = E_0^2$; the field is then said to be *circularly polarized*.

As we just saw, the *shape* of the ellipse is independent of the sign of the phase difference $\vartheta_1 - \vartheta_2$. However, the *sense* in which the path is traced out does depend on the sign, and this is what we now want to consider. It is convenient to introduce the phase difference Δ explicitly by writing

$$\vartheta_1 - \vartheta_2 = \Delta \tag{24-122}$$

and introducing the abbreviation for the phase associated with E_x

$$P = kz - \omega t + \vartheta_1 \tag{24-123}$$

so that (24-117) becomes

$$E_x = E_1 \cos P \qquad E_y = E_2 \cos (P - \Delta) \tag{24-124}$$

It will be helpful in analyzing these expressions to refer to Figure 24-12 in which we have plotted a portion of the cosine terms as a function of P. The solid line is $\cos P$ so that $\Delta = 0$ for it. The dashed line shows $\cos(P - \Delta)$ for $\Delta > 0$ and we see that, as a function of P, it lags the solid line. The dotted line shows $\cos(P - \Delta)$ for negative Δ,

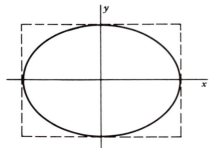

Figure 24-11.   An elliptically polarized electric field with a phase difference of $\frac{1}{2}\pi$ between the x and y components.

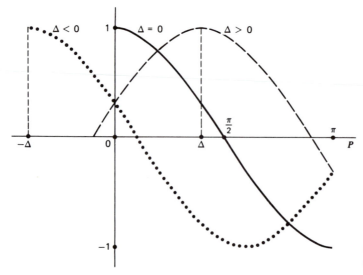

Figure 24-12. $\cos(P - \Delta)$ for various values of Δ.

and it is seen to lead the $\Delta = 0$ curve as a function of P. If we now consider (24-124) as a function of *increasing* P, then, if $\Delta > 0$, E_y lags E_x, that is, it reaches its maximum after E_x, becomes zero after E_x, and so on. This behavior is shown in Figure 24-13 and since the z axis is perpendicular to this plane and out of the page, we see that the ellipse is traced out in a counterclockwise sense when viewed opposite to the direction of propagation (and hence clockwise when viewed in the direction of propagation, that is, looking *out* from the page). If Δ is negative, then E_y leads E_x and the sense of rotation is opposite to the above, as shown in Figure 24-14.

Now, as seen from (24-123), P varies with both z and t, but in different ways. As a result, it is convenient to consider them separately.

Suppose t is a definite value, that is, we consider what one would observe as a function of position z, much like viewing a photograph of the wave. Then as z increases, P increases and the figures are directly applicable. Thus, if $\Delta > 0$, then looking *in* the direction of propagation (*out* of the page), the electric field will be seen to rotate *clockwise*, while it will be counterclockwise if Δ is negative. These are shown for a circularly polarized wave in Figure 24-15. By using the standard right-hand rule to define senses of rotations, we see that (a) corresponds to a positive sense of rotation, and (b) to a negative sense of rotation; a wave like (a) corresponding to $\Delta > 0$ would be

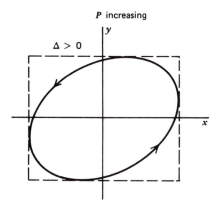

Figure 24-13. The ellipse is traced in a counterclockwise sense when viewed opposite to the direction of propagation.

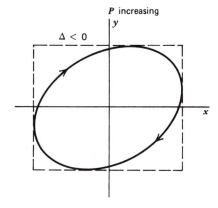

Figure 24-14. The sense of rotation is clockwise when viewed opposite to the direction of propagation.

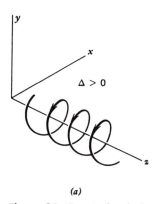

(a)

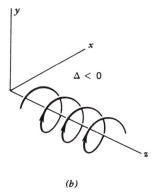

(b)

Figure 24-15. A circularly polarized wave of (a) positive helicity and (b) negative helicity.

described as a wave of *positive helicity*, while one for a negative phase difference would have *negative helicity*.

On the other hand, consider an observer at a definite position (z = const.). As t increases, P decreases according to (24-123). If we now review the way in which Figures 24-13 and 24-14 were constructed with the aid of Figure 24-12, we see that if we replace "P increasing" by "P decreasing" in these figures, we will also have to reverse the sense of rotation about the ellipses. Then the observer, looking into the page, and thus opposite to the direction of propagation, will see $\mathbf{E}$ trace out its ellipse in a clockwise sense for positive Δ and counterclockwise for $\Delta < 0$. (In other words, the observer is looking *toward the source*, which is the most reasonable thing to do.) The polarizations in these cases are called *right-handed* (positive) and *left-handed* (negative), respectively.

Thus, from either point of view, the correlation has turned out to be the same; a positive phase difference Δ corresponds to a positive sense of rotation, and a negative phase difference to a negative sense of rotation.

24-8 ARE THE ELECTROMAGNETIC PARAMETERS OF MATTER CONSTANTS?

We have found that the propagation properties of plane waves can be related to the parameters describing the medium, that is, μ, ϵ, and σ. As a specific example, we found in (24-24) that the index of refraction of a nonconducting medium is given very simply by $n = \sqrt{\kappa_e \kappa_m}$. This relation works very well for many materials, but as a contrary example, let us consider the case of water. If we look up these quantities for water in tables, we find that $\kappa_m \simeq 1$ and $\kappa_e \simeq 80$, so that $n \simeq 9$. But it is well known that the index of refraction of water for light is given very closely by $\frac{4}{3} = 1.33$. The solution to this apparent difficulty lies in the fact that our macroscopic formulation of electromagnetic theory gives no indication of the values to be expected for κ_e and κ_m but must rely on experiment to obtain them. In other words, what is actually done in practice is to take the results predicted by Maxwell's equations as *always* being correct, and then, by making them agree with experiment, *deducing* the values to be assigned to μ, ϵ, and σ. In this way, it is found that these parameters are not really constant for a given material but usually have a strong dependence on *frequency*. (We are not concerned here with a possible dependence on things like temperature and pressure.) In the above example, we used the value of the dielectric constant for water for the *static* case, whereas the appropriate one to use is that corresponding to the very high frequency of light waves.

The atomic nature of matter is the ultimate reason for this variation with frequency; the atomic charges that are polarized by the fields possess inertia that makes their response to the electromagnetic forces depend on frequency. Furthermore, a combination of the effects of inertia and damping forces can introduce a phase difference between the applied force and the response of the system and, as a result, it turns out that it is appropriate to write the parameters as complex numbers. These effects are considered from a detailed atomistic point of view in Appendix B. For purposes of illustration now, however, we can restrict ourselves to a simple case which we have already considered from a microscopic point of view, namely the conductivity of a material.

In Section 12-5, we discussed the microscopic origin of a finite conductivity in terms of the equation of motion (12-36) describing the net force on an electron of charge $-e$ as the sum of the electrical force and a mechanical "frictional" force described by an overall parameter ξ:

$$\mathbf{F}_{\text{net}} = m\mathbf{a} = -e\mathbf{E} - \xi\mathbf{v} \qquad (24\text{-}125)$$

We found the static value of the conductivity, which we now call σ_0, to be given by

(12-39) as

$$\sigma_0 = \frac{ne^2}{\xi} \tag{24-126}$$

where n is the number of electrons per unit volume. We now want to apply this same picture to the case in which the material is subject to the time varying electric field of a plane wave.

Substituting (24-89) into (24-125), we obtain

$$m\mathbf{a} = m\frac{d\mathbf{v}}{dt} = -e\mathbf{E}_0 e^{i(\mathbf{k}\cdot\mathbf{r}-\omega t)} - \xi\mathbf{v} \tag{24-127}$$

The equivalent of a steady-state solution here is one in which the response at a given location has the same time variation as the forcing function; accordingly we try to solve this equation by assuming a solution of the form $\mathbf{v} = \mathbf{v}_0 e^{-i\omega t}$ where $\mathbf{v}_0$ is independent of time. When this is substituted into (24-127), we obtain $m(d\mathbf{v}/dt) = -im\omega\mathbf{v} = -e\mathbf{E} - \xi\mathbf{v}$ so that

$$\mathbf{v} = \frac{-e\mathbf{E}}{\xi - im\omega} \tag{24-128}$$

The free current density can be found in the same way as was used in (12-38):

$$\mathbf{J}_f = \rho_f\mathbf{v} = n(-e)\mathbf{v} = \frac{ne^2\mathbf{E}}{\xi - im\omega} \tag{24-129}$$

Since, by definition of the conductivity, this would be written as $\mathbf{J}_f = \sigma\mathbf{E}$, we see that the conductivity is

$$\sigma = \sigma(\omega) = \frac{ne^2}{\xi - im\omega} = \frac{(ne^2/\xi)}{1 - i(m\omega/\xi)} = \frac{\sigma_0}{1 - i(\sigma_0 m\omega/ne^2)} \tag{24-130}$$

with the use of (24-126). First of all, we see that the conductivity has turned out to be a function of the frequency and the frequency dependent term involves both the inertia of the charge carriers (m) and the resistive forces (ξ); we also note that σ reduces properly to σ_0 for the static case $\omega = 0$. Second, we see that σ is complex, and by now we recognize that this means that the current density and the electric field are no longer in phase. [In fact, by analogy with mechanics, we recognize that (24-125) is simply the equation of motion of a damped "harmonic oscillator" with no restoring force, as is appropriate for unbound electrons, and we recall that then the displacement and velocity are no longer generally in phase with the applied oscillating force.]

We can find the real and imaginary parts of the conductivity by multiplying the numerator and denominator of (24-130) by the complex conjugate of the denominator and we get

$$\sigma = \sigma_R + i\sigma_I = \frac{\sigma_0}{\left[1 + \left(\sigma_0 m\omega/ne^2\right)^2\right]}\left[1 + i\left(\frac{\sigma_0 m\omega}{ne^2}\right)\right] \tag{24-131}$$

In order to investigate the physical significance of the complex conductivity a little more, let us assume that we are at a fixed location $\mathbf{r}$ so that $\mathbf{E}$ in (24-89) and hence $\mathbf{J}_f$ can be written in the form

$$\mathbf{E} = \mathbf{E}_0' e^{-i\omega t} \qquad \mathbf{J}_f = (\sigma_R + i\sigma_I)\mathbf{E}_0' e^{-i\omega t} \tag{24-132}$$

For simplicity, let us assume that $\mathbf{E}_0'$ is real; then the physical quantities of interest are given by the real parts of (24-132) and, with the aid of (24-22), they are found to be

$$\mathbf{E}_{\text{real}} = \mathbf{E}'_0 \cos \omega t$$
$$\mathbf{J}_{f\,\text{real}} = (\sigma_R \cos \omega t + \sigma_I \sin \omega t)\mathbf{E}'_0 \tag{24-133}$$

Thus the real part of a complex conductivity gives rise to a component of the current that is in phase with the applied electric field and it is the imaginary component σ_I that describes a component of the current that is completely *out of phase* with the field. [It is easily seen from (24-131) that both σ_R and σ_I are positive quantities.]

Now we can investigate the effect of a complex conductivity on the propagation characteristics of the medium. The dispersion relation (24-37) did not depend on the quantities being real, nor did the use of (24-38). However, when (24-39) is combined with (24-131), we find that now we have $\alpha^2 - \beta^2 + 2i\alpha\beta = \omega^2\mu\epsilon + i\omega\mu(\sigma_R + i\sigma_I)$, so that, instead of (24-40) and (24-41), α and β are related by

$$\alpha^2 - \beta^2 = \omega^2\mu\epsilon - \omega\mu\sigma_I, \qquad 2\alpha\beta = \omega\mu\sigma_R \tag{24-134}$$

We could go ahead and solve these for α and β as before, but there is an easier way. We can write the first expression as $\alpha^2 - \beta^2 = \omega^2\mu[\epsilon - (\sigma_I/\omega)]$ and upon comparing with (24-40) and (24-41), we see that (24-42) and (24-43) can still be used by simply making the replacements $\sigma \to \sigma_R$ and $\epsilon \to \epsilon - (\sigma_I/\omega)$.

This connection between the permittivity and the conductivity is sometimes emphasized by writing the dispersion relation (24-37) in the form

$$k^2 = \omega^2\mu\left(\epsilon + i\frac{\sigma}{\omega}\right) \tag{24-135}$$

so that, as far as wave propagation is concerned, the existence of a conductivity can be handled simply by adding the (generally complex) term $i\sigma/\omega$ to ϵ.

Rather than continuing with these general results, let us consider two extreme limiting cases that can be characterized by the relative size of the parameter ξ, which is a measure of the "friction" in the system, that is, of the overall effect of collisions.

1. "Large Friction" ($m\omega/\xi \ll 1$ or $\omega \ll ne^2/\sigma_0 m$)

We see from (24-130) that this means that $\sigma \simeq \sigma_0 \simeq$ const. so that the conductivity is essentially always real and equal to the static value. This is found to hold very well for metals. Using the numbers quoted for copper in Exercise 12-17, we find that $n \simeq 8.5 \times 10^{28}$ (meter)$^{-3}$. Using $\sigma_0 \simeq 6 \times 10^7$ (ohm-meter)$^{-1}$ and 1.60×10^{-19} coulomb and 9.11×10^{-31} kilogram for the electron's charge and mass, we find that this requirement on the frequency ν is $\nu \ll 6 \times 10^{12}$ hertz. This condition is satisfied until we are well into the microwave region, so that metals can justifiably be treated as having a constant and real conductivity as we implied in Section 24-3.

2. "Small Friction" ($m\omega/\xi \gg 1$ or $\omega \gg ne^2/\sigma_0 m$)

Here we can neglect the 1 in the denominator of (24-130), which yields

$$\sigma \simeq i\left(\frac{ne^2}{m\omega}\right) \tag{24-136}$$

and the conductivity is a pure imaginary, $\sigma_R = 0$, and the current density and the electric field are exactly 90° out of phase as shown by (24-133). Such a situation as this arises in an *ionized gas with low particle density* or *plasma* as it is known. Because of the tenuous nature of the medium, the effect of collisions is very small so that ξ is very small. Putting (24-136) directly into the dispersion relation (24-37), we find that

$$k^2 = \omega^2 \mu \epsilon \left(1 - \frac{ne^2}{m \epsilon \omega^2} \right) = \omega^2 \mu \epsilon \left(1 - \frac{\omega_P^2}{\omega^2} \right) \qquad (24\text{-}137)$$

where we have written

$$\omega_P^2 = \frac{ne^2}{m \epsilon} \qquad (24\text{-}138)$$

and where $\nu_P = \omega_P/2\pi$ is called the *plasma frequency*. Actually, for such a medium, $\mu \simeq \mu_0$ and $\epsilon \simeq \epsilon_0$ and we can approximate these expressions very well by

$$k^2 \simeq \left(\frac{\omega}{c} \right)^2 \left(1 - \frac{\omega_P^2}{\omega^2} \right) \qquad \omega_P^2 \simeq \frac{ne^2}{m \epsilon_0} \qquad (24\text{-}139)$$

Because of the minus sign in the expression for k^2, there are two separate cases to be considered.

If $\omega > \omega_P$, then $k^2 > 0$ so that k is real. Thus we have an undamped traveling wave of the form (24-84) with a phase velocity found from (24-139) to be

$$v = \frac{\omega}{k} = \frac{c}{\left[1 - (\omega_P/\omega)^2 \right]^{1/2}} \qquad (24\text{-}140)$$

which is greater than the value c, which would apply to a nondispersive medium with otherwise vacuum properties.

If $\omega < \omega_P$, then $k^2 < 0$ so that k is a pure imaginary. In this case, we can write k as found from (24-139) in the form

$$k = i\beta_P = i(\omega/c)\left[(\omega_P/\omega)^2 - 1 \right]^{1/2} \qquad (24\text{-}141)$$

and then a solution of the form (24-84) becomes

$$\psi = \psi_0 e^{-\beta_P \zeta} e^{-i\omega t} \qquad (24\text{-}142)$$

This is not a traveling wave, but a sinusoidally oscillating field component whose amplitude is exponentially decreasing with distance. If the plasma is large enough, then $\psi \to 0$, and the field will not penetrate the medium. Hence one can speak of a plasma as being an example of a "high-pass" filter in the sense that the frequency must be greater than the plasma frequency in order to get propagation rather than simply attenuation.

Although one would not describe the conduction electrons in the body of a neutral metal as a plasma of *low* particle density, they otherwise fit the requirements of this case for a high enough frequency to outweigh the effect of collisions, and it is of interest to evaluate the order of magnitude of ν_P for a metal. Using the same numbers that we have already used for copper, we find from (24-139) that $\nu_P = 2.6 \times 10^{16}$ hertz. What this extremely high value of ν_P tells us is that we cannot expect a typical metal to become "transparent" to electromagnetic waves until we are well into the ultraviolet region of the spectrum, a result that is completely consistent with our other conclusions about the behavior of metals. Other examples of plasmas will be considered in the exercises.

EXERCISES

24-1 The form (24-11) was verified to be the solution of (24-9) by direct substitution. Similarly, separation of variables in z and t was used to get the explicit form (24-18). Another approach to

(24-9) is the following. Define a new set of variables by $\xi = z + vt$ and $\eta = z - vt$ and express (24-9) in terms of ξ and η. If v is again given by (24-12), show that the resulting equation has a solution of the form $\psi = f(\eta) + g(\xi)$.

24-2 Phenomena that arise as a result of the superposition principle, that is, a sum of solutions is also a solution, are often characterized by the term *interference*. As an example, consider two plane waves ψ_1 and ψ_2 that are each of the form (24-19) with the same values of k and ω, that is, they are traveling in the same direction with the same velocity. Suppose, however, that they have different amplitudes and phases, so that $\psi_{01} = \psi_{1a}e^{i\vartheta_1}$ and $\psi_{02} = \psi_{2a}e^{i\vartheta_2}$. Find their sum $\psi = \psi_1 + \psi_2$. Find the real parts of ψ_1, ψ_2, and ψ. In the special case in which they have equal amplitudes ($\psi_{1a} = \psi_{2a} = \psi_a$), show that Re $\psi = 2\psi_a \cos\frac{1}{2}(\vartheta_1 - \vartheta_2)\cos[kz - \omega t + \frac{1}{2}(\vartheta_1 + \vartheta_2)]$ and interpret the result. If $|\vartheta_1 - \vartheta_2| = \pi$, what is Re ψ? Explain how this happened by considering the relation between the individual waves in the superposition.

24-3 As another example of interference, suppose that ψ_2 of the previous exercise is now propagating in the negative z direction. Show that if the waves have equal real positive amplitudes that Re $\psi = 2\psi_a \cos kz \cos \omega t$. This is an example of a *standing wave*. Plot Re ψ as a function of z for a convenient value of t, and as a function of t for a convenient value of z.

24-4 As was noted after (24-53), a superposition of plane waves traveling in a dispersive medium generally changes its form as it progresses. As an extreme example, consider two waves of equal real amplitudes and with almost the same propagation constants and frequencies that are traveling in the positive z direction, so that their sum has the form

$$\psi = \psi_0 e^{i[(k+dk)z - (\omega + d\omega)t]}$$
$$+ \psi_0 e^{i[(k-dk)z - (\omega - d\omega)t]}$$

Show that

$$\text{Re } \psi = 2\psi_0 \cos[(dk)z - (d\omega)t]\cos(kz - \omega t)$$
$$= \psi_m \cos(kz - \omega t) \qquad (24\text{-}143)$$

where ψ_m is called the "modulation." Thus (24-143) has the form of a wave with average propagation constant k and average frequency ω. Its amplitude is not constant, but itself is a wave and travels with the *group velocity* v_G

$$v_G = \frac{d\omega}{dk} \qquad (24\text{-}144)$$

What is the spatial period (wavelength) λ of the main wave? What is the wavelength λ_m of ψ_m? Find the ratio λ_m/λ. Sketch (24-143) at $t = 0$, and at a *slightly* later time, and thus verify that the form of Re ψ has changed. Identify which physical feature of your sketch travels with the phase velocity $v = \omega/k$ and which with the group velocity v_G. Why do you think the name "group" velocity was given to (24-144)? (This specific result is an example of the more general phenomenon known as *beats*.)

24-5 A plane electromagnetic wave traveling in a vacuum is given by $\mathbf{E} = \hat{\mathbf{y}}E_0 e^{i(kz - \omega t)}$ where E_0 is real. A circular loop of radius a, N turns, and resistance R is located with its center at the origin. The loop is oriented so that a diameter lies along the z axis and the plane of the loop makes an angle θ with the y axis. Find the emf induced in the loop as a function of time. Assume that $a \ll \lambda$. (Why?)

24-6 A particle of charge q and mass m is traveling with velocity $\mathbf{u}$ in the field of a plane electromagnetic wave in free space that itself is traveling in the z direction. Find the force on the particle. What does this become for the special case in which the particle is traveling in the same direction as the wave? What is the direction of the force? Under what conditions (if any) will the force vanish in this case?

24-7 A representative value for the conductivity of sea water is $\sigma = 4$ (ohm-meter)$^{-1}$. Take $\epsilon \simeq \epsilon_0$ and $\mu \simeq \mu_0$ and find Q, v, δ, the magnitude of the ratio E/cB, and Ω for each of these frequencies ν in hertz: 10^2, 10^7, 10^{10}, and 10^{15}. (They correspond roughly to "power," "radio," "microwave," and "light.") The result obtained in this way for δ for light is obviously ridiculous. Why? What is the probable reason for this erroneous value having been obtained?

24-8 Find the index of refraction n of a conductor as a function of Q. Show that, in the limit of a good conductor, your result reduces to $n^2 = \kappa_e \kappa_m/2Q$.

24-9 The solar constant, which is the average rate of incidence of radiation at the earth's atmosphere from the sun, is 1340 watts/(meter)^2. Assuming that the radiation is a linearly polarized plane wave, find the amplitudes of $\mathbf{E}$ and $\mathbf{B}$.

24-10 Verify (24-105) and (24-106).

24-11 Find the ratio $\langle u_m \rangle / \langle u_e \rangle$ for a plane wave in a conducting medium. Then find the approximate expressions for this ratio for the limiting cases of an insulator and a good conductor.

24-12 A plane wave travels in the positive z direction in a conductor with real conductivity. (a) Find the instantaneous and time average power loss per unit volume due to resistive heating for any z. (b) Find the total power loss per unit area between $z = 0$ and $z \to \infty$. (c) Find the time average Poynting vector at any z. (d) Compare the value of your result for (b) with the magnitude of your result in (c) evaluated at $z = 0$. Is your answer reasonable? Explain.

24-13 Consider a plane wave that is a superposition of two independent orthogonal plane waves and has the form $\mathbf{E} = \hat{\mathbf{x}} E_\alpha e^{i(kz - \omega t + \vartheta_\alpha)} + \hat{\mathbf{y}} E_\beta e^{i(kz - \omega t + \vartheta_\beta)}$ where k is real. Find $\langle \mathbf{S} \rangle$ and show that it equals the sum of the average Poynting vectors for the components.

24-14 Consider a plane wave that is a superposition of two independent plane waves with parallel electric fields and has the form $\mathbf{E} = \hat{\mathbf{x}} E_\alpha e^{i(kz - \omega t + \vartheta_\alpha)} + \hat{\mathbf{x}} E_\beta e^{i(kz - \omega t + \vartheta_\beta)}$ where k is real. Find $\langle \mathbf{S} \rangle$ and show that it is *not* equal to the sum of the average Poynting vectors for each component. (This is a result of "interference.")

24-15 Consider a superposition of plane waves all traveling in the same direction in a nonconducting medium. Thus, the real field will be a sum of terms like (24-35), that is, $\mathbf{E} = \Sigma_k \mathbf{E}_{0k} \cos(kz - \omega_k t + \vartheta_k)$. Show that the time-average electric energy density is the sum of the average energy densities associated with each component.

24-16 In Exercise 22-6, we found that in a l.i.h. nonconducting region where there are no free charges or free currents, $\mathbf{E}$ and $\mathbf{B}$ can be found from a vector potential $\mathbf{A}$ alone so that we can take $\phi = 0$. (a) What are the two equations that $\mathbf{A}$ must satisfy? (Actually, there are four scalar equations.) (b) Find the solution of these equations for which $\mathbf{A}$ is a plane wave traveling in the $\hat{\mathbf{k}}$ direction. (c) Find $\mathbf{E}$ and $\mathbf{B}$ from this $\mathbf{A}$, and verify that they are the same as those found in the text, that is, they satisfy the set of equations (24-91) with $\sigma = 0$.

24-17 A certain plane wave has a propagation vector $\mathbf{k} = 314\hat{\mathbf{x}} + 314\hat{\mathbf{y}} + 444\hat{\mathbf{z}}$ (meter)$^{-1}$. Assume it is traveling in a vacuum and find the wavelength, frequency, and angles made with the xyz axes.

24-18 Given the plane wave $\mathbf{E} = \gamma[(3 + 2i)\hat{\mathbf{x}} + (3 + 4i)\hat{\mathbf{y}}]e^{i(kz - \omega t)}$ where γ is a positive constant and k is positive and real. Find: E_1, E_2, ϑ_1, ϑ_2, Δ, and the angle φ between the major axis of the ellipse and the x axis. Does the wave have positive or negative helicity? Is its sense of polarization right-handed or left-handed?

24-19 Show that the electric field of a right-handed circularly polarized wave can be written in the form $\mathbf{E}_+ = E_0(\hat{\mathbf{x}} - i\hat{\mathbf{y}})e^{i(kz - \omega t + \vartheta)}$. What is the corresponding expression if it is left-handed?

24-20 Show that a linearly polarized wave can be written as a superposition of right- and left-handed circularly polarized waves of equal amplitudes. Show that $\langle \mathbf{S} \rangle$ for the linearly polarized wave equals the sum of the average Poynting vectors of each of its circularly polarized components.

24-21 Show that for a plasma for which $\epsilon \simeq \epsilon_0$ the plasma frequency in hertz can be found from $\nu_P = 8.97 n^{1/2}$. As examples of other important plasmas, find ν_P for (a) a typical gas discharge for which $n \simeq 10^{18}$ (meter)$^{-3}$; and (b) the ionosphere where $n \simeq 10^{10}$ (meter)$^{-3}$.

24-22 In an ionized gas, there are movable ions present as well as electrons. Why were we able to neglect their contribution to the conductivity and hence to the plasma frequency?

24-23 If a material has a conductivity that is a pure imaginary, such as is the case in (24-136), show that the average rate of energy dissipation per unit volume, $\langle w \rangle$, is zero. Is this result compatible with the assumption that collisions have been effectively neglected?

24-24 For a system with a conductivity given by (24-136), find the total current density $\mathbf{J}_t = \sigma \mathbf{E} + (\partial \mathbf{D}/\partial t)$ and show that it is *less* than it would be in the absence of conduction electrons. Explain how this occurs.

24-25 The plasma frequency was defined in a purely formal manner in (24-138), but it can be understood as a natural frequency of the system. This same expression can also be derived in these two ways. (a) Show that when an oscillatory solution for ρ_f proportional to $e^{-i\omega_P t}$ is combined with $\nabla \cdot \mathbf{J}_f + (\partial \rho_f / \partial t) = 0$, $\mathbf{J}_f = \sigma \mathbf{E}$, and $\nabla \cdot \mathbf{E} = \rho_f / \epsilon$, the result is again (24-138). (b) Suppose that the equal number of positive ions and electrons are displaced relative to each other by a small distance x. Find the force on one of the resulting surface charge densities that is produced by the other, and show that the corresponding equation of motion is that of a simple harmonic oscillator of frequency ω_P given by (24-138).

24-26 (a) Find the group velocity v_G defined in (24-144) for a system with the dispersion relation of (24-137). Plot v_G as a function of ω for $\omega > \omega_p$. Show that the product of the phase and group velocities is $vv_G = (\mu\epsilon)^{-1}$. (b) Show that if the dispersion relation has the form $k^2 = f(\omega^2)$, then the product vv_G will be constant provided that f has the form $f = A\omega^2 + B$ where A and B are constants.

24-27 Assume that a system is described by (24-137) and that $\omega > \omega_p$. Find $\langle u_e \rangle$, $\langle u_m \rangle$, $\langle u_m \rangle / \langle u_e \rangle$, $\langle u \rangle$, $\langle \mathbf{S} \rangle$, and show that $\langle \mathbf{S} \rangle = \langle u \rangle v_U \hat{\mathbf{k}}$ where v_U can be interpreted as the speed of energy flow. Show that $v_G < v_U < (\mu\epsilon)^{-1/2} < v$.

24-28 Consider the most general possibility in which a medium has all complex parameters, that is, $\mu = \mu_R + i\mu_I$, $\epsilon = \epsilon_R + i\epsilon_I$, and $\sigma = \sigma_R + i\sigma_I$. Show that k can still be written in the form (24-38) with α and β real. Find α and β. Now consider a strictly *nonconducting* material, and show that there can still be attenuation ($\beta \neq 0$) and that there will still be a phase difference between $\mathbf{E}$ and $\mathbf{B}$ ($\beta/\alpha \neq 0$).

24-29 The symbol Q was originally introduced as an abbreviation for "quality factor." Although we found a useful interpretation of Q in (24-66), there are two other ways of writing it that are helpful. (a) Show that Q measures the ratio of the relaxation time of a conducting medium as found in (12-41) and the temporal period T of the oscillation. (b) Express δ/λ as a function of Q only and show that when $Q \gg 1$, this reduces to $Q = \pi\delta/\lambda$. (c) Use this latter result to justify the following convenient way of remembering the physical meaning of Q: "The value of Q equals the number of cycles that a (nearly) free oscillation will persist."

24-30 A study of Figure 24-12 might lead one to wonder if our conclusions about leading or lagging might be changed if Δ turns out to have a value outside the range shown. Show that, for the discussion of elliptical polarization, one can restrict Δ to the range $-\pi$ to π so that, if Δ as found from (24-122) falls outside, one can add or subtract appropriate multiples of 2π to bring it within this range. Illustrate this with a simple sketch of a convenient example.

24-31 Consider two separate electromagnetic fields described by the pairs $(\mathbf{E}_1, \mathbf{H}_1)$ and $(\mathbf{E}_2, \mathbf{H}_2)$. Assume that the medium is l.i.h. in all properties, that $\mathbf{J}_f' = 0$, and that these two fields are sinusoidal functions of time with the same frequency, that is, are proportional to $e^{-i\omega t}$. Show that it follows from Maxwell's equations that $\nabla \cdot (\mathbf{E}_1 \times \mathbf{H}_2 - \mathbf{E}_2 \times \mathbf{H}_1) = 0$. This result is known as *Lorentz' lemma* and shows that these otherwise independent electromagnetic fields are related in this sense.

25

REFLECTION AND REFRACTION OF PLANE WAVES

In the last chapter, we considered plane waves of infinite extent, so that they could be thought of as traveling in an infinite medium. Eventually in any real situation, the wave will encounter another medium with different electromagnetic properties; for example, a light wave in air may impinge on a piece of glass. The incident wave will generally be able to enter the second medium to some extent and we want to find out what the overall situation will be when the resultant steady state has been attained. We will find that we can solve this problem by using the boundary conditions that the field vectors have to satisfy at a surface of discontinuity in properties, and we recall that these boundary conditions were obtained directly from Maxwell's equations.

25-1 THE LAWS OF REFLECTION AND REFRACTION

We assume that the boundary between the two media is an infinite plane surface. We further assume that there are no free charges or free surface currents on the boundary surface. The boundary conditions given by (21-25) through (21-28) then reduce simply to (1) the normal components of $\mathbf{D}$ and $\mathbf{B}$ are continuous, and (2) the tangential components of $\mathbf{E}$ and $\mathbf{H}$ are continuous. It will be seen soon that we need only the second set for our purposes; it will be left as an exercise to show that no additional information would be obtained by using the first set involving the normal components.

The most general situation we can visualize would therefore correspond to an incident wave traveling in a medium 1 of electromagnetic parameters $(\mu_1, \epsilon_1, \sigma_1)$ incident on a medium 2 described by $(\mu_2, \epsilon_2, \sigma_2)$. Although the general solution can be easily enough obtained when both media are conducting, the complexity of the results tends to obscure the understanding and interpretation of what is happening. Consequently, we assume for now that both media are nonconductors so that all propagation vectors are real; we briefly consider the effect of conductivity in Section 25-6. We can take advantage of the experience of others and begin at once with the knowledge that the boundary conditions can be satisfied only by assuming the existence of *three* waves: the *incident* wave in medium 1, a *reflected* wave also in 1, and a *transmitted* wave in medium 2; we use the subscripts i, r, and t, respectively, to label quantities referring to these waves.

Using (24-89), we can write the electric fields of these waves as

$$\mathbf{E}_i = \mathbf{E}_{0i} e^{i(\mathbf{k}_i \cdot \mathbf{r} - \omega_i t)} \qquad \mathbf{E}_r = \mathbf{E}_{0r} e^{i(\mathbf{k}_r \cdot \mathbf{r} - \omega_r t)}$$
$$\mathbf{E}_t = \mathbf{E}_{0t} e^{i(\mathbf{k}_t \cdot \mathbf{r} - \omega_t t)} \tag{25-1}$$

where

$$k_i^2 = \left(\frac{\omega_i}{v_1}\right)^2 = \left(\frac{n_1 \omega_i}{c}\right)^2 \qquad k_r^2 = \left(\frac{\omega_r}{v_1}\right)^2 = \left(\frac{n_1 \omega_r}{c}\right)^2 \qquad k_t^2 = \left(\frac{\omega_t}{v_2}\right)^2 = \left(\frac{n_2 \omega_t}{c}\right)^2 \tag{25-2}$$

and where we have introduced the phase velocity and index of refraction of the corresponding medium with the use of (24-17) and (24-13); we note that we have made no special assumptions about the frequencies. Similar equations can be written down for the magnetic fields.

In (25-1), the origin of time ($t = 0$) is arbitrary; $\mathbf{r}$ represents the position vector of a given point with respect to an arbitrary origin. For simplicity, however, we choose the origin to lie in the surface of separation so that the position vector $\mathbf{r}_B$ of a point in the bounding surface also lies in the surface. These quantities are all illustrated in Figure 25-1 where we also show the normal $\hat{\mathbf{n}}$ to the boundary drawn from medium 1 to medium 2 according to our standard convention.

The total electric field at a given point in medium 1 will be $\mathbf{E}_1 = \mathbf{E}_i + \mathbf{E}_r$, while that in 2 is $\mathbf{E}_2 = \mathbf{E}_t$. Now at any point *on the boundary* where $\mathbf{r} = \mathbf{r}_B$, we must have $\mathbf{E}_{1\,\text{tang}} = \mathbf{E}_{2\,\text{tang}}$ or

$$\left[\mathbf{E}_{0i}e^{i(\mathbf{k}_i \cdot \mathbf{r}_B - \omega_i t)} + \mathbf{E}_{0r}e^{i(\mathbf{k}_r \cdot \mathbf{r}_B - \omega_r t)}\right]_{\text{tang}} = \left[\mathbf{E}_{0t}e^{i(\mathbf{k}_t \cdot \mathbf{r}_B - \omega_t t)}\right]_{\text{tang}} \qquad (25\text{-}3)$$

and this is to be true for all values of t and all possible $\mathbf{r}_B$ where t and $\mathbf{r}_B$ can be independently varied. Regardless of the relative values of the amplitudes $\mathbf{E}_0$, it is clear that we cannot satisfy (25-3) unless the values of the exponents are all equal. Thus, we see first of all that $\omega_i = \omega_r = \omega_t = \omega$ so that all three waves must have the *same frequency* ω; then (25-2) simplifies to

$$k_i^2 = k_r^2 = \left(\frac{n_1 \omega}{c}\right)^2 = k_1^2 \qquad k_t^2 = \left(\frac{n_2 \omega}{c}\right)^2 = k_2^2 \qquad (25\text{-}4)$$

Similarly, we must have $\mathbf{k}_i \cdot \mathbf{r}_B = \mathbf{k}_r \cdot \mathbf{r}_B = \mathbf{k}_t \cdot \mathbf{r}_B$; it is helpful to subtract these in pairs to get the two independent equations

$$(\mathbf{k}_i - \mathbf{k}_r) \cdot \mathbf{r}_B = 0 \qquad (\mathbf{k}_i - \mathbf{k}_t) \cdot \mathbf{r}_B = 0 \qquad (25\text{-}5)$$

which shows that these *differences* in the propagation vectors must be perpendicular to $\mathbf{r}_B$, and hence to the bounding surface; this is also true for the pair $\mathbf{k}_r$ and $\mathbf{k}_t$ since we also have $(\mathbf{k}_r - \mathbf{k}_t) \cdot \mathbf{r}_B = 0$. The first equation of (25-5) relates the incident and reflected propagation vectors so we can call it the *law of reflection*; similarly, the second is the *law of refraction* since it relates the incident and transmitted (refracted) waves.

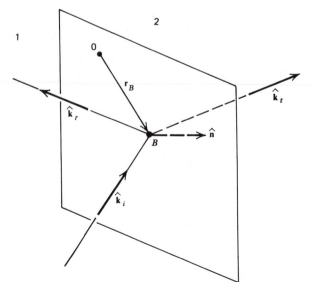

Figure 25-1.   Relations at a plane boundary between two media.

The plane defined by $\hat{n}$ and the incident vector $\mathbf{k}_i$ is called the *plane of incidence* and is shown in Figure 25-2. The angle θ_i made by $\mathbf{k}_i$ with the normal is the *angle of incidence*. We can resolve $\mathbf{k}_i$ into its components normal and parallel to the surface of discontinuity by defining a unit tangential vector $\hat{\tau}$, which is parallel to the surface and in the plane of incidence; thus we can write

$$\mathbf{k}_i = k_{in}\hat{n} + k_{i\tau}\hat{\tau} \tag{25-6}$$

Other vectors that do not necessarily lie in the plane of incidence can be written in component form by introducing a third unit vector $\hat{n} \times \hat{\tau}$, which is also in the surface between 1 and 2 and is perpendicular to both $\hat{n}$ and $\hat{\tau}$ as shown in Figure 25-3. Thus, we can write the other two propagation vectors in the form

$$\mathbf{k}_r = k_{rn}\hat{n} + k_{r\tau}\hat{\tau} + k_{rc}\hat{n} \times \hat{\tau} \qquad \mathbf{k}_t = k_{tn}\hat{n} + k_{t\tau}\hat{\tau} + k_{tc}\hat{n} \times \hat{\tau} \tag{25-7}$$

while

$$\mathbf{r}_B = r_{B\tau}\hat{\tau} + r_{Bc}\hat{n} \times \hat{\tau} \tag{25-8}$$

since it lies entirely in the surface and thus has no normal component.

Substituting (25-6) through (25-8) into (25-5), and remembering that the unit vectors are mutually perpendicular so that $\hat{n} \cdot \hat{\tau} = 0$, and so on, we obtain

$$(k_{i\tau} - k_{r\tau})r_{B\tau} - k_{rc}r_{Bc} = 0 \qquad (k_{i\tau} - k_{t\tau})r_{B\tau} - k_{tc}r_{Bc} = 0 \tag{25-9}$$

Since $\mathbf{r}_B$ is arbitrary, (25-9) must also be true for the special case for which $r_{B\tau} = 0$ while $r_{Bc} \neq 0$, and then (25-9) yields $k_{rc} = k_{tc} = 0$ so that $\mathbf{k}_r$ and $\mathbf{k}_t$ have no components perpendicular to the plane of Figure 25-2. In other words, all three propagation vectors *lie in the same plane*, that is, they are all in the plane of incidence. Now, since $r_{B\tau}$ can be different from zero, we see from (25-9) that

$$k_{r\tau} = k_{i\tau} \qquad k_{t\tau} = k_{i\tau} \tag{25-10}$$

so that all of the $\mathbf{k}$'s have the same tangential components. We can regard (25-10) as a second version of the laws of reflection and refraction. Now we can learn something about the normal components.

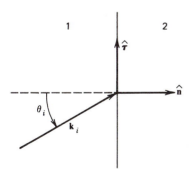

Figure 25-2. This is the plane of incidence.

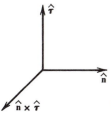

Figure 25-3. The three mutually perpendicular unit vectors.

Remembering that $k_{rc} = 0$, we find from (25-7), (25-4), (25-10), and (25-6) that $k_{rn}^2 = k_r^2 - k_{r\tau}^2 = k_i^2 - k_{r\tau}^2 = k_i^2 - k_{i\tau}^2 = k_{in}^2$ and therefore $k_{rn} = \pm k_{in}$. If the plus sign were chosen, then we see from Figure 25-2 that the reflected wave would be going toward the surface in the same way as the incident wave, whereas it should be tending away from it as in Figure 25-1. Thus we must have $k_{rn} = -k_{in}$. Similarly we find that $k_{tn}^2 = k_t^2 - k_{t\tau}^2 = k_2^2 - k_{i\tau}^2$. Therefore, the third versions of the laws of reflection and refraction are

$$k_{rn} = -k_{in} \qquad k_{tn}^2 = k_2^2 - k_{i\tau}^2 \tag{25-11}$$

Now $\mathbf{k}_i$ is a vector of magnitude k_1 by (25-4) and we see from Figure 25-2 that

$$k_{in} = k_1 \cos \theta_i \qquad k_{i\tau} = k_1 \sin \theta_i \tag{25-12}$$

and therefore $k_{rn} = -k_1 \cos \theta_i$, $k_{r\tau} = k_1 \sin \theta_i$ by (25-11) and (25-10). Since $\mathbf{k}_r$ also has the magnitude k_1, we can define an *angle of reflection* θ_r by

$$k_{rn} = -k_1 \cos \theta_r \qquad k_{r\tau} = k_1 \sin \theta_r \tag{25-13}$$

as shown in Figure 25-4a. By comparing these different expressions for the components of $\mathbf{k}_r$, we see that

$$\theta_r = \theta_i \tag{25-14}$$

so that the angle of reflection equals the angle of incidence. Thus, this ancient and well-known law of optics is seen to be a direct consequence of Maxwell's equations.

For the transmitted wave, we find from (25-10), (25-11), (25-12), and (25-4) that

$$k_{t\tau} = k_1 \sin \theta_i \tag{25-15}$$

$$k_{tn}^2 = k_2^2 - k_1^2 \sin^2 \theta_i = k_2^2 \left[1 - \left(\frac{n_1}{n_2} \right)^2 \sin^2 \theta_i \right] \tag{25-16}$$

These results are correct for any values of the ratio n_1/n_2 and of the angle of incidence.

Under many circumstances, it is useful to continue by *defining* an *angle of refraction* θ_t for $\mathbf{k}_t$ of magnitude k_2 by means of

$$k_{t\tau} = k_2 \sin \theta_t \qquad k_{tn} = k_2 \cos \theta_t \tag{25-17}$$

as is also illustrated in Figure 25-4b. Combining (25-17) with the law of refraction as given in (25-15) and (25-16), along with (25-4), we find that they lead to

$$n_1 \sin \theta_i = n_2 \sin \theta_t \tag{25-18}$$

$$\cos^2 \theta_t = 1 - \left(\frac{n_1}{n_2} \right)^2 \sin^2 \theta_i \tag{25-19}$$

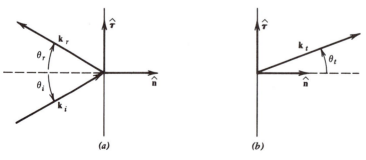

(a) (b)

Figure 25-4. (a) θ_r is the angle of reflection. (b) θ_t is the angle of refraction.

which are internally consistent since it is seen that $\cos^2 \theta_t = 1 - \sin^2 \theta_t$. The first equation (25-18) is known as *Snell's law* of refraction and was discovered experimentally by him; it is seen to relate the angles of incidence and refraction to the indices of refraction of the two media, and is also seen to be a consequence of Maxwell's equations *via* the boundary conditions obtained from them.

Although (25-17) defines two quantities $\sin \theta_t$ and $\cos \theta_t$, they do *not* always correspond to a *real* physical angle θ_t. Let us see how this comes about and what we should do about it. We find from (25-18) that $\sin \theta_t = (n_1/n_2) \sin \theta_i$. If $n_1 < n_2$, then $\sin \theta_t < \sin \theta_i \le 1$ and $\cos^2 \theta_t = 1 - \sin^2 \theta_t > 0$; thus θ_t is a real angle and $\theta_t < \theta_i$. This case is illustrated in Figure 25-5; note that $k_2/k_1 = n_2/n_1 > 1$. If $n_1 > n_2$, then we still have $\sin \theta_t = (n_1/n_2) \sin \theta_i > \sin \theta_i$ so that $\theta_t > \theta_i$ in general while $k_2/k_1 < 1$. However, we should also have $\sin \theta_t \le 1$ so that θ_t will be a real angle only as long as $\sin \theta_i \le n_2/n_1 < 1$. It is useful to define the *critical angle* θ_c by

$$\sin \theta_c = \frac{n_2}{n_1} \tag{25-20}$$

and we can write

$$\sin \theta_t = \frac{\sin \theta_i}{\sin \theta_c} \qquad (n_1 > n_2) \tag{25-21}$$

Therefore, for $\theta_i < \theta_c$, $\sin \theta_t < 1$ and $\theta_t > \theta_i$. This case is illustrated in Figure 25-6 (We note that the larger of the two angles θ_i and θ_t is always in the medium of smaller index of refraction.) If $\theta_i = \theta_c$, then $\sin \theta_t = 1$ and $\theta_t = 90°$, and the refracted wave is parallel to the interface as shown in Figure 25-7; we still have $k_2 < k_1$.

Since θ_i can be freely chosen by us, it is certainly possible to make $\theta_i > \theta_c$. Then (25-21) shows that $\sin \theta_t > 1$ while $\cos^2 \theta_t = 1 - \sin^2 \theta_t < 0$, which is not possible for a real angle and no figure like any of the last three can be drawn. (The quantity θ_t can still be interpreted as an angle as long as it is *complex* and of the form $\theta_t = \frac{1}{2}\pi + ix$. To see how this comes about, we use Equation 24-22 to evaluate the sine as $\sin \theta_t = (e^{i\theta_t} - e^{-i\theta_t})/2i = [e^{i(\pi/2)-x} - e^{-i(\pi/2)+x}]/2i = \frac{1}{2}(e^{-x} + e^x) = \cosh x \ge 1$ as required.)

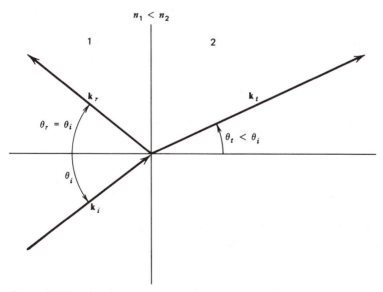

Figure 25-5. The three propagation vectors as related by the laws of reflection and refraction when $n_1 < n_2$.

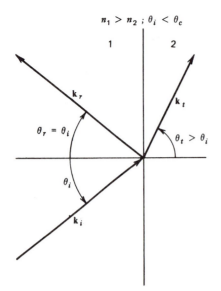

Figure 25-6. The three propagation vectors as related by the laws of reflection and refraction when $n_1 > n_2$ and the angle of incidence is less than the critical angle.

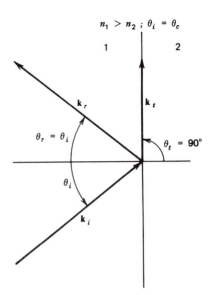

Figure 25-7. The three propagation vectors when the angle of incidence equals the critical angle.

Thus in this case it is not fruitful to continue with θ_t; we can, however, go back and evaluate the components of $\mathbf{k}_t$ in terms of θ_i by means of (25-15) and (25-16) since these are *always* valid. Consequently, there is no difficulty in handling this case in principle. It turns out to be convenient to discuss these two broad classes separately, and we put off the detailed consideration of the case ($n_1 > n_2$; $\theta_i > \theta_c$) until Section 25-4.

All of our results so far have been obtained solely from the exponential propagation terms in the single boundary condition (25-3), and we still have to find the necessary relations among the fields themselves. Now $\mathbf{E}_i$ must lie in a plane perpendicular to $\mathbf{k}_i$, but it need have no special orientation with respect to the plane of incidence of Figure 25-2. The general situation for an arbitrary orientation of $\mathbf{E}_i$ for an arbitrary angle of incidence is shown in Figure 25-8. The shaded plane is perpendicular to $\mathbf{k}_i$ and therefore contains $\mathbf{E}_i$; this plane is also perpendicular to the plane of incidence defined by $\mathbf{k}_i$ and $\hat{\mathbf{n}}$. We see that $\mathbf{E}_i$ can be written as the vector sum

$$\mathbf{E}_i = \mathbf{E}_{i\perp} + \mathbf{E}_{i\parallel} \qquad (25\text{-}22)$$

where $\mathbf{E}_{i\perp}$ is the component perpendicular to the plane of incidence and $\mathbf{E}_{i\parallel}$ is the component parallel to (in) the plane of incidence. It turns out that these two components behave differently at the plane of separation between the two media, so that it is desirable to consider the two cases separately. After we have found how each of these is affected, we can recombine the results to find the total reflected and transmitted fields, that is, we will be able to write

$$\mathbf{E}_r = \mathbf{E}_{r\perp} + \mathbf{E}_{r\parallel} \qquad \mathbf{E}_t = \mathbf{E}_{t\perp} + \mathbf{E}_{t\parallel} \qquad (25\text{-}23)$$

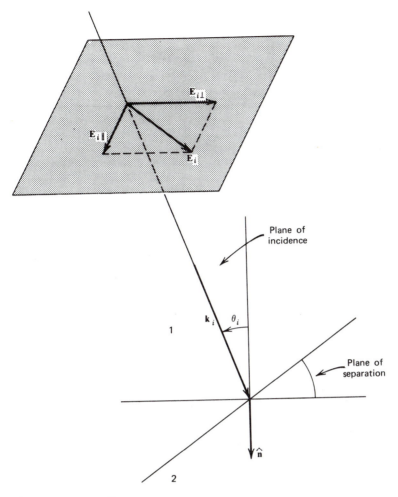

Figure 25-8. Definition of the components of E_i parallel to and perpendicular to the plane of incidence.

25-2 E PERPENDICULAR TO THE PLANE OF INCIDENCE

Since we have no *a priori* knowledge about the relations among the various electric fields, let us assume for definiteness that they are all directed *out* of the page at this particular instant and location on the bounding surface; if our assumption about the relative directions is incorrect, the final result will show us by means of its sign. The corresponding directions of **H** can be found from the condition that $\mathbf{E} \times \mathbf{H}$ is in the direction of propagation $\hat{\mathbf{k}}$. In this way, we get the situation shown in Figure 25-9; for clarity, the **E**'s and **H**'s are drawn at some distance from the interface between the two media, but we remember that they actually are to represent the situation *at* the surface where we apply the boundary conditions.

Since the electric fields are all parallel to the surface and to each other, they are all tangential components so that $\mathbf{E}_{1\,\text{tang}} = \mathbf{E}_{2\,\text{tang}}$ becomes simply

$$E_i + E_r = E_t \qquad (25\text{-}24)$$

All of the magnetic fields are in the plane of incidence for which $\hat{\tau}$ is the unit tangential

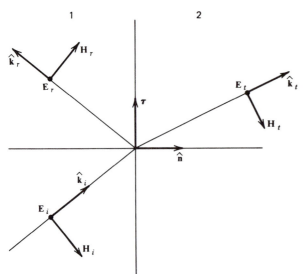

Figure 25-9. E_i is perpendicular to the plane of incidence.

vector and the other boundary condition $H_{1\,\text{tang}} = H_{2\,\text{tang}}$ becomes

$$(\mathbf{H}_i + \mathbf{H}_r) \cdot \hat{\boldsymbol{\tau}} = \mathbf{H}_t \cdot \hat{\boldsymbol{\tau}} \tag{25-25}$$

This last result can be expressed in terms of the electric fields by using

$$\mathbf{H} = \frac{\hat{\mathbf{k}} \times \mathbf{E}}{Z} \quad \text{and} \quad \hat{\boldsymbol{\tau}} = \frac{\mathbf{E}_i \times \hat{\mathbf{n}}}{E_i} \tag{25-26}$$

as obtained from (24-94) and an inspection of Figure 25-9, while recalling that $\mathbf{E}_i$ is out of the page. Therefore, the tangential component of any of the magnetic fields has the form

$$\mathbf{H} \cdot \hat{\boldsymbol{\tau}} = \frac{1}{ZE_i}(\hat{\mathbf{k}} \times \mathbf{E}) \cdot (\mathbf{E}_i \times \hat{\mathbf{n}}) = -\frac{E}{Z}(\hat{\mathbf{k}} \cdot \hat{\mathbf{n}}) \tag{25-27}$$

with the use of (1-106), $\hat{\mathbf{k}} \cdot \mathbf{E}_i = 0$ (they are perpendicular), and $\mathbf{E} \cdot \mathbf{E}_i = EE_i$ (they are parallel). The appropriate value of the impedance is $Z_1 = (\mu_1/\epsilon_1)^{1/2}$ for the incident and reflected waves in medium 1, while it is $Z_2 = (\mu_2/\epsilon_2)^{1/2}$ for $\mathbf{H}_t$. Taking this into account, and substituting (25-27) into (25-25), we get

$$\frac{1}{Z_1}\left[E_i(\hat{\mathbf{k}}_i \cdot \hat{\mathbf{n}}) + E_r(\hat{\mathbf{k}}_r \cdot \hat{\mathbf{n}})\right] = \frac{E_t}{Z_2}(\hat{\mathbf{k}}_t \cdot \hat{\mathbf{n}}) \tag{25-28}$$

Since E_i is arbitrary, the significant quantities are the two ratios E_r/E_i and E_t/E_i, and they can now be found from the two equations (25-24) and (25-28). The results are

$$\left(\frac{E_r}{E_i}\right)_\perp = \frac{Z_2(\hat{\mathbf{k}}_i \cdot \hat{\mathbf{n}}) - Z_1(\hat{\mathbf{k}}_t \cdot \hat{\mathbf{n}})}{Z_2(\hat{\mathbf{k}}_i \cdot \hat{\mathbf{n}}) + Z_1(\hat{\mathbf{k}}_t \cdot \hat{\mathbf{n}})} \tag{25-29}$$

$$\left(\frac{E_t}{E_i}\right)_\perp = \frac{2Z_2(\hat{\mathbf{k}}_i \cdot \hat{\mathbf{n}})}{Z_2(\hat{\mathbf{k}}_i \cdot \hat{\mathbf{n}}) + Z_1(\hat{\mathbf{k}}_t \cdot \hat{\mathbf{n}})} \tag{25-30}$$

where use has been made of $\hat{\mathbf{k}}_r \cdot \hat{\mathbf{n}} = -\hat{\mathbf{k}}_i \cdot \hat{\mathbf{n}}$ as found from (25-4) and the law of reflection given in (25-11). These ratios are the *general* solution for this case and are valid for any values of n_1/n_2 and θ_i. In order to emphasize this, we can express them completely in terms of θ_i by using $\hat{\mathbf{k}}_i \cdot \hat{\mathbf{n}} = k_{in}/k_1 = \cos\theta_i$ from (25-12) and $\hat{\mathbf{k}}_t \cdot \hat{\mathbf{n}} =$

k_{tn}/k_2 as found from (25-16) and we obtain

$$\left(\frac{E_r}{E_i}\right)_{\perp} = \frac{Z_2 \cos \theta_i - Z_1\left[1 - (n_1/n_2)^2 \sin^2 \theta_i\right]^{1/2}}{Z_2 \cos \theta_i + Z_1\left[1 - (n_1/n_2)^2 \sin^2 \theta_i\right]^{1/2}} \tag{25-31}$$

$$\left(\frac{E_t}{E_i}\right)_{\perp} = \frac{2 Z_2 \cos \theta_i}{Z_2 \cos \theta_i + Z_1\left[1 - (n_1/n_2)^2 \sin^2 \theta_i\right]^{1/2}} \tag{25-32}$$

In the special case of normal incidence ($\theta_i = 0$), these reduce to

$$\left(\frac{E_r}{E_i}\right)_{\perp} = \frac{Z_2 - Z_1}{Z_2 + Z_1} \qquad \left(\frac{E_t}{E_i}\right)_{\perp} = \frac{2 Z_2}{Z_2 + Z_1} \tag{25-33}$$

It is also very common to find these results written in terms of both the angle of incidence θ_i and the angle of refraction θ_t. Since θ_t is real only for $n_1 < n_2$ or $n_1 > n_2$ and $\theta_i \le \theta_c$, the following expressions are restricted to these cases. Then (25-29) and (25-30) become

$$\left(\frac{E_r}{E_i}\right)_{\perp} = \frac{Z_2 \cos \theta_i - Z_1 \cos \theta_t}{Z_2 \cos \theta_i + Z_1 \cos \theta_t} \qquad \left(\frac{E_t}{E_i}\right)_{\perp} = \frac{2 Z_2 \cos \theta_i}{Z_2 \cos \theta_i + Z_1 \cos \theta_t} \tag{25-34}$$

These are often put in other forms with the use of

$$\frac{Z_1}{Z_2} = \left(\frac{\mu_1 \epsilon_2}{\mu_2 \epsilon_1}\right)^{1/2} = \frac{\mu_1 n_2}{\mu_2 n_1} = \frac{\mu_1 \sin \theta_i}{\mu_2 \sin \theta_t} \tag{25-35}$$

which follows from (24-95), (24-12), (24-13), and (25-18). Dividing the numerator and denominator of each expression in (25-34) by Z_2 and using (25-35), we find that the ratios can also be written as

$$\left(\frac{E_r}{E_i}\right)_{\perp} = \frac{\mu_2 \tan \theta_t - \mu_1 \tan \theta_i}{\mu_2 \tan \theta_t + \mu_1 \tan \theta_i} \qquad \left(\frac{E_t}{E_i}\right)_{\perp} = \frac{2 \mu_2 \tan \theta_t}{\mu_2 \tan \theta_t + \mu_1 \tan \theta_i} \tag{25-36}$$

The results for normal incidence as found from (25-33) are

$$\left(\frac{E_r}{E_i}\right)_{\perp} = \frac{\mu_2 n_1 - \mu_1 n_2}{\mu_2 n_1 + \mu_1 n_2} \qquad \left(\frac{E_t}{E_i}\right)_{\perp} = \frac{2 \mu_2 n_1}{\mu_2 n_1 + \mu_1 n_2} \tag{25-37}$$

These are seen to be compatible with (25-36), since, as θ_i and θ_t both approach zero, Snell's law (25-18) becomes approximately

$$\frac{n_1}{n_2} = \frac{\sin \theta_t}{\sin \theta_i} \simeq \frac{\theta_t}{\theta_i} \simeq \frac{\tan \theta_t}{\tan \theta_i} \tag{25-38}$$

In the very important special case $\mu_1 = \mu_2$, which includes nonmagnetic media, these can also be written in general as

$$\left(\frac{E_r}{E_i}\right)_{\perp} = -\frac{\sin (\theta_i - \theta_t)}{\sin (\theta_i + \theta_t)} \qquad \left(\frac{E_t}{E_i}\right)_{\perp} = \frac{(n_1/n_2) \sin 2\theta_i}{\sin (\theta_i + \theta_t)} \tag{25-39}$$

while, for normal incidence, they reduce to

$$\left(\frac{E_r}{E_i}\right)_{\perp} = -\left(\frac{n_2 - n_1}{n_2 + n_1}\right) \qquad \left(\frac{E_t}{E_i}\right)_{\perp} = \frac{2 n_1}{n_2 + n_1} \tag{25-40}$$

The equations in (25-39) are usually called *Fresnel's equations*; he derived them from an elastic solid theory of light long before the development of Maxwell's equations and the electromagnetic theory of light. (It is perhaps not too surprising that he was able to do this, since, if an elastic wave in one medium is incident upon another medium of different properties, there will be boundary conditions that must also be satisfied. For example, if a wave is traveling on a stretched string, one would have to require that the displacement be continuous at the junction for otherwise the strings would separate.)

The general form of the results obtained from Fresnel's equations as a function of θ_i are shown in Figures 25-10 and 25-11 for the case $n_1 < n_2$. We note that the ratio $(E_r/E_i)_\perp$ is negative for all values of θ_i; this can also be seen from (25-39), since when $n_1 < n_2$, $\theta_i > \theta_t$, so that $\sin(\theta_i - \theta_t) > 0$. This means that $\mathbf{E}_r$ is actually *opposite* to the direction we assumed; in other words, the reflected wave has its phase changed by $180°$ upon reflection on a medium of greater index of refraction. On the other hand, $(E_t/E_i)_\perp$ is always positive, so that the transmitted wave undergoes no change in phase. (Thus, to represent the actual situation, Figure 25-9 should really have the direction of $\mathbf{H}_r$ reversed, while $\mathbf{E}_r$ is directed into the page; the rest of the figure will be unchanged.)

If $n_1 > n_2$, but $\theta_i \le \theta_c$, then $\theta_i < \theta_t$, as in Figure 25-6; thus $\sin(\theta_i - \theta_t)$ will be negative so that $(E_r/E_i)_\perp$ in (25-39) will always be positive, as will $(E_t/E_i)_\perp$ as before. Thus there is no phase change upon reflection against a medium of lower index, as long as the angle of incidence is less than the critical angle.

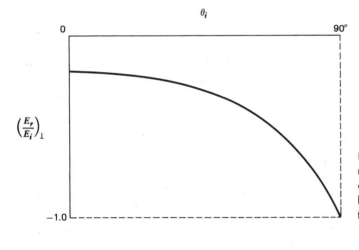

Figure 25-10. Ratio of reflected to incident electric field as given by Fresnel's equations for $n_1 < n_2$.

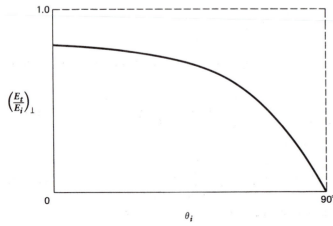

Figure 25-11. Ratio of transmitted to incident electric field as given by Fresnel's equations for $n_1 < n_2$.

It follows from (25-39), and is shown in Figure 25-11, that the transmitted wave is always zero for grazing incidence ($\theta_i = 90°$).

25-3 E PARALLEL TO THE PLANE OF INCIDENCE

We proceed in a manner similar to that of the last section. Let us assume that all of the magnetic fields are *into* the page; this leads to a situation as depicted in Figure 25-12. From $\mathbf{H} = \hat{\mathbf{k}} \times \mathbf{E}/Z$ and $\hat{\mathbf{k}} \cdot \mathbf{E} = 0$, as well as an inspection of the figure, we find the analogue to (25-26) to be

$$\mathbf{E} = Z\mathbf{H} \times \hat{\mathbf{k}} \quad \text{and} \quad \hat{\boldsymbol{\tau}} = \frac{\hat{\mathbf{n}} \times \mathbf{H}_i}{H_i} \tag{25-41}$$

so that

$$\mathbf{E} \cdot \hat{\boldsymbol{\tau}} = -ZH(\hat{\mathbf{k}} \cdot \hat{\mathbf{n}}) = -E(\hat{\mathbf{k}} \cdot \hat{\mathbf{n}}) \tag{25-42}$$

since $\hat{\mathbf{k}} \cdot \mathbf{H}_i = 0$ (they are perpendicular), and the values of E and H are related by the factor Z.

Consequently, the continuity of the tangential components of $\mathbf{E}$ ($\mathbf{E}_1 \cdot \hat{\boldsymbol{\tau}} = \mathbf{E}_2 \cdot \hat{\boldsymbol{\tau}}$) leads to

$$E_i(\hat{\mathbf{k}}_i \cdot \hat{\mathbf{n}}) + E_r(\hat{\mathbf{k}}_r \cdot \hat{\mathbf{n}}) = (E_i - E_r)(\hat{\mathbf{k}}_i \cdot \hat{\mathbf{n}}) = E_t(\hat{\mathbf{k}}_t \cdot \hat{\mathbf{n}}) \tag{25-43}$$

with the use of (25-11) and (25-4). Since the $\mathbf{H}$'s are always tangential components and parallel, $\mathbf{H}_{1\,\text{tang}} = \mathbf{H}_{2\,\text{tang}}$ yields

$$H_i + H_r = H_t = \frac{E_i + E_r}{Z_1} = \frac{E_t}{Z_2} \tag{25-44}$$

Equations 25-43 and 25-44 can now be solved for the two ratios E_r/E_i and E_t/E_i. Doing this, and using a fair amount of algebra and trigonometry, we can obtain all of the following results in the same way as used in the last section.

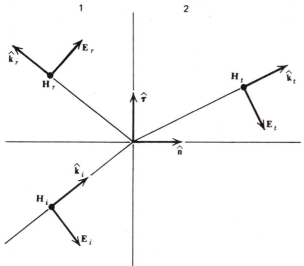

Figure 25-12. $\mathbf{E}_i$ is parallel to the plane of incidence.

The *general* solutions applicable to all values of n_1/n_2 and θ_i are:

$$\left(\frac{E_r}{E_i}\right)_{\parallel} = \frac{Z_1(\hat{\mathbf{k}}_i \cdot \hat{\mathbf{n}}) - Z_2(\hat{\mathbf{k}}_t \cdot \hat{\mathbf{n}})}{Z_1(\hat{\mathbf{k}}_i \cdot \hat{\mathbf{n}}) + Z_2(\hat{\mathbf{k}}_t \cdot \hat{\mathbf{n}})} = \frac{Z_1 \cos \theta_i - Z_2 \left[1 - (n_1/n_2)^2 \sin^2 \theta_i\right]^{1/2}}{Z_1 \cos \theta_i + Z_2 \left[1 - (n_1/n_2)^2 \sin^2 \theta_i\right]^{1/2}}$$

$$(25\text{-}45)$$

$$\left(\frac{E_t}{E_i}\right)_{\parallel} = \frac{2Z_2(\hat{\mathbf{k}}_i \cdot \hat{\mathbf{n}})}{Z_1(\hat{\mathbf{k}}_i \cdot \hat{\mathbf{n}}) + Z_2(\hat{\mathbf{k}}_t \cdot \hat{\mathbf{n}})} = \frac{2Z_2 \cos \theta_i}{Z_1 \cos \theta_i + Z_2 \left[1 - (n_1/n_2)^2 \sin^2 \theta_i\right]^{1/2}}$$

$$(25\text{-}46)$$

When θ_t is a real angle ($n_1 < n_2$ or $n_1 > n_2$ and $\theta_i \leq \theta_c$), these become

$$\left(\frac{E_r}{E_i}\right)_{\parallel} = \frac{Z_1 \cos \theta_i - Z_2 \cos \theta_t}{Z_1 \cos \theta_i + Z_2 \cos \theta_t} = \frac{\mu_1 \sin 2\theta_i - \mu_2 \sin 2\theta_t}{\mu_1 \sin 2\theta_i + \mu_2 \sin 2\theta_t}$$

$$(25\text{-}47)$$

$$\left(\frac{E_t}{E_i}\right)_{\parallel} = \frac{2Z_2 \cos \theta_i}{Z_1 \cos \theta_i + Z_2 \cos \theta_t} = \frac{4\mu_2 \cos \theta_i \sin \theta_t}{\mu_1 \sin 2\theta_i + \mu_2 \sin 2\theta_t}$$

$$(25\text{-}48)$$

and, in addition, when $\mu_1 = \mu_2$, we get the other two *Fresnel equations*

$$\left(\frac{E_r}{E_i}\right)_{\parallel} = \frac{\tan(\theta_i - \theta_t)}{\tan(\theta_i + \theta_t)} \qquad \left(\frac{E_t}{E_i}\right)_{\parallel} = \frac{2 \cos \theta_i \sin \theta_t}{\sin(\theta_i + \theta_t)\cos(\theta_i - \theta_t)}$$

$$(25\text{-}49)$$

The various expressions for normal incidence are

$$\left(\frac{E_r}{E_i}\right)_{\parallel} = \frac{Z_1 - Z_2}{Z_1 + Z_2} = \frac{\mu_1 n_2 - \mu_2 n_1}{\mu_1 n_2 + \mu_2 n_1} \quad \text{and} \quad \frac{n_2 - n_1}{n_2 + n_1} \quad (\mu_1 = \mu_2) \quad (25\text{-}50)$$

$$\left(\frac{E_t}{E_i}\right)_{\parallel} = \frac{2Z_2}{Z_1 + Z_2} = \frac{2\mu_2 n_1}{\mu_1 n_2 + \mu_2 n_1} \quad \text{and} \quad \frac{2n_1}{n_2 + n_1} \quad (\mu_1 = \mu_2) \quad (25\text{-}51)$$

With these last results, we have an apparent contradiction in sign. If we look again at Figures 25-2 and 25-8, we see that for normal incidence when $\hat{\mathbf{k}}_i$ and $\hat{\mathbf{n}}$ are in the same direction, there is no definite plane of incidence so that there is no distinction between $\mathbf{E}$ being perpendicular to or parallel to the plane of incidence and the results of both cases should reduce to the same thing. Now (25-51) does agree with the corresponding expressions in (25-33), (25-37), and (25-40), but the various forms for (E_r/E_i) all differ by a minus sign. This is a result of our original sign conventions for the two cases. In Figure 25-9, we assumed $\mathbf{E}_r$ and $\mathbf{E}_i$ to be parallel and thus treated E_r/E_i as a positive number; for $n_1 < n_2$, it actually turned out to be negative as we saw. On the other hand, the opposite directions of $\mathbf{E}_r$ and $\mathbf{E}_i$ were taken into account right from the start in Figure 25-12, so that the ratio E_r/E_i was also treated as positive. In other words, the sign difference reflects the difference in our treatment of the two, but each case is internally consistent and thus consistent with each other.

Plots of the ratios as given by the Fresnel equations (25-49) for $n_1 < n_2$ are shown in Figures 25-13 and 25-14. Looking at Figure 25-14 first, we see that it is similar to Figure 25-11 in that there is no phase change of the transmitted wave, and it again vanishes at grazing incidence. Since this is also what happened in the previous case, we conclude that electromagnetic radiation is *always* completely reflected from a surface of a dielectric at grazing incidence. (You can verify this for yourself right now by looking at a light source by reflection from a rough sheet of paper and then gradually increasing the angle of incidence to 90°.)

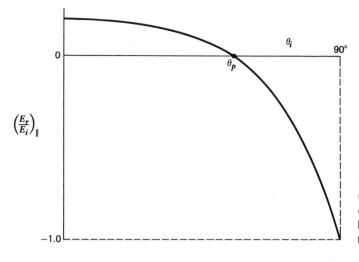

Figure 25-13. Ratio of reflected to incident electric field as given by Fresnel's equations for $n_1 < n_2$.

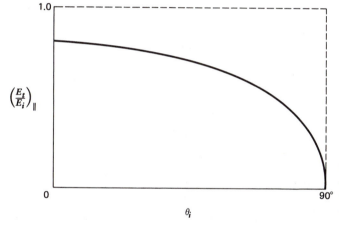

Figure 25-14. Ratio of transmitted to incident electric field as given by Fresnel's equations for $n_1 < n_2$.

We see, moreover, from Figure 25-13, that we have a case that did not occur before, that there is an angle θ_p for which the reflected wave is *zero*. The condition for this is seen from (25-49) to correspond to the denominator of $(E_r/E_i)_\parallel$ becoming infinite, or $\theta_p + \theta_{tp} = 90°$, where θ_{tp} is the corresponding angle of refraction. The law of refraction (25-18) then becomes $n_1 \sin\theta_p = n_2 \sin\theta_{tp} = n_2 \sin(90° - \theta_p) = n_2 \cos\theta_p$, so that

$$\tan\theta_p = \frac{n_2}{n_1} \tag{25-52}$$

a result known as *Brewster's law*. The angle θ_p is known as the *polarizing angle* for the following reason. Consider an incident wave that is unpolarized in the sense that its electric field has $\mathbf{E}_\perp$ and $\mathbf{E}_\parallel$ components that are both different from zero. If this wave is incident at the polarizing angle θ_p, then only the component $\mathbf{E}_\perp$ will be reflected, so that the reflected wave will have only such a component and hence will be *polarized*. The polarization of light by reflection was first discovered experimentally in just this way by Malus who was observing sunlight reflected from the windows of the Luxembourg Palace in Paris, and for us it is further evidence for the electromagnetic wave nature of light.

We also see from Figure 25-13, that, when $\theta_i > \theta_p$, the ratio $(E_r/E_i)_{\parallel}$ is negative. This means that the phase of the reflected wave is suddenly reversed, that is, changed by 180°, as the angle of incidence passes through the polarizing angle.

25-4 TOTAL REFLECTION ($n_1 > n_2$, $\theta_i > \theta_c$)

We will see the reason for this name shortly. In this case, θ_t is not a real angle, but we already have results expressed completely in terms of θ_i and they can be used directly. It will be helpful in interpreting our results if we be a little more specific about our coordinate system. Let us choose the xy plane to be the surface of separation and the xz plane to be the plane of incidence. This is shown in Figure 25-15 where the y axis is out of the page. We see that with this choice, $k_\tau = k_x$ and $k_n = k_z$, so that the incident electric field will have the form

$$\mathbf{E}_i = \mathbf{E}_{0i}e^{i(\mathbf{k}_i \cdot \mathbf{r} - \omega t)} = \mathbf{E}_{0i}e^{i[k_1(x \sin \theta_i + z \cos \theta_i) - \omega t]} \tag{25-53}$$

with the use of (25-12). The transmitted wave will be given by

$$\mathbf{E}_t = \mathbf{E}_{0t}e^{i(\mathbf{k}_t \cdot \mathbf{r} - \omega t)} = \mathbf{E}_{0t}e^{i(k_{tx}x + k_{tz}z - \omega t)} \tag{25-54}$$

and we have $k_{tx} = k_{t\tau} = k_1 \sin \theta_i$ from (25-15). The quantity in brackets in (25-16) is now negative so it is preferable to rewrite k_{tn}^2 in the form

$$k_{tn}^2 = k_{tz}^2 = -k_2^2\left[\left(\frac{n_1}{n_2}\right)^2 \sin^2 \theta_i - 1\right] = -k_2^2\left[\left(\frac{\sin \theta_i}{\sin \theta_c}\right)^2 - 1\right] = -K^2 \tag{25-55}$$

with the use of (25-20). Therefore, $k_{tz} = \pm iK$, making $e^{ik_{tz}z} = e^{\mp Kz}$. The plus sign in the exponent would lead to a wave that increases indefinitely as the wave travels in the positive z direction, so we reject it and write $k_{tz} = iK$ where

$$K = k_2\left[\left(\frac{n_1}{n_2}\right)^2 \sin^2 \theta_i - 1\right]^{1/2} = \left(\frac{n_2\omega}{c}\right)\left[\left(\frac{\sin \theta_i}{\sin \theta_c}\right)^2 - 1\right]^{1/2} \tag{25-56}$$

with the further use of (25-4). Substituting these various results into (25-54), we get the expression for the transmitted electric field in the medium of lower index of refraction:

$$\mathbf{E}_t = \mathbf{E}_{0t}e^{-Kz}e^{i(k_1 x \sin \theta_i - \omega t)} \tag{25-57}$$

This is a wave traveling in the x direction, that is, *parallel* to the surface of separation, but with its amplitude decreasing in the z direction—*perpendicular* to the direction of

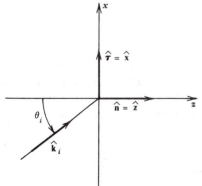

Figure 25-15. Coordinate system used in discussing total reflection.

propagation. The speed of the wave is

$$v_{2x} = \frac{\omega}{k_1 \sin \theta_i} = \frac{c}{n_1 \sin \theta_i} = \frac{(c/n_2)}{(n_1/n_2) \sin \theta_i} = \left(\frac{\sin \theta_c}{\sin \theta_i} \right) v_2 \qquad (25\text{-}58)$$

where v_2 is the normal speed of a wave in this medium. Since $\theta_i > \theta_c$, $v_{2x} < v_2$ and this wave travels more slowly than a usual plane wave; (25-57) is *not* a plane wave in the sense that its value is not constant on a plane perpendicular to the direction of propagation. Furthermore, the wave speed v_{2x} is not a constant that depends only on the properties of the medium, but it depends on the angle of incidence as well.

The amplitude decreases exponentially with increasing depth of penetration, and decreases by a factor of $1/e$ in a penetration depth δ_z given by

$$\delta_z = \frac{1}{K} = \frac{(\lambda_2/2\pi)}{\left[(n_1/n_2)^2 \sin^2 \theta_i - 1 \right]^{1/2}} \qquad (25\text{-}59)$$

where $\lambda_2 = 2\pi/k_2 = 2\pi c/n_2\omega$ is the wavelength of a normal plane wave in this medium. This wave that we have described is often called the *evanescent wave* and its properties are indicated very schematically in Figure 25-16. Although not indicated in the figure, this wave can also be characterized by saying that the planes of constant amplitude are perpendicular to the planes of constant phase. Now let us turn to the reflected electric field.

Using (25-56), we find that (25-31) and (25-45) can be written as

$$\left(\frac{E_r}{E_i} \right)_\perp = \frac{Z_2 \cos \theta_i - iZ_1(K/k_2)}{Z_2 \cos \theta_i + iZ_1(K/k_2)} \qquad \left(\frac{E_r}{E_i} \right)_\parallel = \frac{Z_1 \cos \theta_i - iZ_2(K/k_2)}{Z_1 \cos \theta_i + iZ_2(K/k_2)} \qquad (25\text{-}60)$$

We note that *both* of these have the same general form:

$$\frac{E_r}{E_i} = \frac{A - iB}{A + iB} = \frac{(A^2 + B^2)^{1/2} e^{-i\varphi}}{(A^2 + B^2)^{1/2} e^{i\varphi}} = e^{-2i\varphi} \qquad (25\text{-}61)$$

where $\tan \varphi = B/A$. Therefore, in both cases, $|E_r/E_i| = 1$ so that the reflected wave has the same amplitude as the incident wave. As we will see in the next section, the reflected energy also equals the incident energy; hence the name *total reflection* for this case.

The appearance of the factor $e^{-2i\varphi}$ in the ratio means, however, that there is a relative shift in phase of 2φ of the reflected wave with respect to the incident wave. The angles φ as found from (25-60), (25-56), and (25-35) are given by

$$\tan \varphi_\perp = \frac{Z_1(K/k_2)}{Z_2 \cos \theta_i} = \left(\frac{\mu_1 n_2}{\mu_2 n_1} \right) \frac{\left[(n_1/n_2)^2 \sin^2 \theta_i - 1 \right]^{1/2}}{\cos \theta_i} \qquad (25\text{-}62)$$

$$\tan \varphi_\parallel = \frac{Z_2(K/k_2)}{Z_1 \cos \theta_i} = \left(\frac{\mu_2 n_1}{\mu_1 n_2} \right) \frac{\left[(n_1/n_2)^2 \sin^2 \theta_i - 1 \right]^{1/2}}{\cos \theta_i} \qquad (25\text{-}63)$$

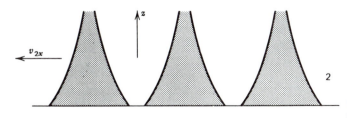

Figure 25-16. Schematic behavior of the evanescent wave.

We see that these angles are different and, in fact,

$$\tan \varphi_{\perp} = \left(\frac{\mu_1 n_2}{\mu_2 n_1} \right)^2 \tan \varphi_{\parallel} \tag{25-64}$$

Thus, for the case $\mu_1 = \mu_2$, $\tan \varphi_{\perp} < \tan \varphi_{\parallel}$ since $n_2 < n_1$ and therefore $\varphi_{\perp} < \varphi_{\parallel}$. Because of the similarity of the plane containing $\mathbf{E}_i$ in Figure 25-8 to the plane shown in Figure 24-7, we conclude that, if a linearly polarized wave is incident at an angle greater than the critical angle, its perpendicular and parallel components will be reflected with different phase shifts and the resultant reflected wave will be *elliptically polarized* and can be analyzed by the methods we used in Section 24-7.

25-5 ENERGY RELATIONS

Now we are in a position to see how the incoming energy flow is divided between the reflected and transmitted waves. The average incident power flow will be given by $\langle \mathbf{S}_i \rangle$ and it will have a component normal to the bounding surface and one parallel to the surface. What will be evidently of physical interest is what happens to the normal component since it gives what is reflected back into the original medium. Therefore, we define a *reflection coefficient* for power R as the magnitude of the ratio of the normal components:

$$R = \left| \frac{\langle \mathbf{S}_r \rangle \cdot \hat{\mathbf{n}}}{\langle \mathbf{S}_i \rangle \cdot \hat{\mathbf{n}}} \right| \tag{25-65}$$

Now we find from (24-107) that

$$\langle \mathbf{S} \rangle \cdot \hat{\mathbf{n}} = \tfrac{1}{2} (\epsilon / \mu)^{1/2} |\mathbf{E}|^2 (\hat{\mathbf{k}} \cdot \hat{\mathbf{n}}) \tag{25-66}$$

Since the incident and reflected waves are in the same medium, $(\epsilon / \mu)^{1/2}$ has the same value $(\epsilon_1 / \mu_1)^{1/2}$ for each of them; in addition, $|\hat{\mathbf{k}}_r \cdot \hat{\mathbf{n}}| = |\hat{\mathbf{k}}_i \cdot \hat{\mathbf{n}}| = \cos \theta_r = \cos \theta_i$ by the law of reflection (25-11) and (25-14). Taking all of this into account, we find from (25-65) and (25-66) that

$$R = \left| \frac{E_r}{E_i} \right|^2 \tag{25-67}$$

and since we have these ratios already, we can easily evaluate R.

In this way, we find from (25-34) and (25-47), for example, that

$$R_{\perp} = \left(\frac{Z_2 \cos \theta_i - Z_1 \cos \theta_t}{Z_2 \cos \theta_i + Z_1 \cos \theta_t} \right)^2 \tag{25-68}$$

$$R_{\parallel} = \left(\frac{Z_1 \cos \theta_i - Z_2 \cos \theta_t}{Z_1 \cos \theta_i + Z_2 \cos \theta_t} \right)^2 \tag{25-69}$$

and similar expressions can be found for all of the various ways in which we wrote the ratios of the fields. At normal incidence and $\mu_1 = \mu_2$, for example, both (25-40) and (25-50) give

$$R_{\text{normal}} = \left(\frac{n_2 - n_1}{n_2 + n_1} \right)^2 \quad (\mu_1 = \mu_2) \tag{25-70}$$

If we consider a specific case for which $n_2 / n_1 = 1.5$, which is typical for glass and air, we find that (25-70) gives $R_{\text{normal}} = 0.04$ so that about 4 percent of the incident energy flow is reflected at normal incidence.

Similarly, we can define a *transmission coefficient* for power T by

$$T = \left| \frac{\langle \mathbf{S}_t \rangle \cdot \hat{\mathbf{n}}}{\langle \mathbf{S}_i \rangle \cdot \hat{\mathbf{n}}} \right| = \frac{Z_1 \cos \theta_t}{Z_2 \cos \theta_i} \left| \frac{E_t}{E_i} \right|^2 \tag{25-71}$$

where we have used (25-66), (25-12), (25-17), and (24-95). For example, the cases corresponding to (25-68) and (25-69) can be found from (25-34) and (25-48) to be

$$T_\perp = \frac{4 Z_1 Z_2 \cos \theta_i \cos \theta_t}{\left(Z_2 \cos \theta_i + Z_1 \cos \theta_t \right)^2} \tag{25-72}$$

$$T_\parallel = \frac{4 Z_1 Z_2 \cos \theta_i \cos \theta_t}{\left(Z_1 \cos \theta_i + Z_2 \cos \theta_t \right)^2} \tag{25-73}$$

while, for normal incidence ($\theta_i = \theta_t = 0$), (25-51) and (25-71) give

$$T_{\text{normal}} = \frac{4 n_1 n_2}{\left(n_2 + n_1 \right)^2} \qquad (\mu_1 = \mu_2) \tag{25-74}$$

since $Z_1/Z_2 = n_2/n_1$ in this case by (25-35).

It is easily verified that all three of these pairs satisfy the relation

$$R + T = 1 \tag{25-75}$$

which can also be seen to hold for the most general cases described by (25-29), (25-30), (25-45), and (25-46). This result expresses the fact of energy conservation since it essentially says that all of the incident energy appears either in the reflected wave or in the transmitted wave so that none is lost during the process.

In the particular case of total reflection, the relation (25-61) shows us at once that $R = |E_r/E_i|^2 = 1$ so that all of the incident energy is reflected as we previously remarked. Therefore, we know that T must be zero for this case, and it will be left as an exercise to verify this in detail. In some respects, this seems to be a paradoxical situation. The transmitted fields are certainly different from zero, so that the average energy densities $\langle u_e \rangle$ and $\langle u_m \rangle$ are different from zero, and it seems strange that there can be energy in the second medium, when none can be transmitted into it. Our discussion of this case has actually been restricted to the *steady-state* in which all of the waves do not change their character with time so that quantities like R and T are constants. If we imagine how this system was started, say, by turning on a beam, then the initial transient behavior must have been different from the final steady state to which the system eventually settles down. It is these beginning processes that we must regard as the source of the original energy transmitted into the second medium, and we are not in a position to try to consider this in detail.

25-6 REFLECTION AT THE SURFACE OF A CONDUCTOR

An important situation arises when a wave in a dielectric is incident upon a conducting medium such as a metal. We know from Section 24-3 that the transmitted wave will be attenuated because of resistive heat losses, but we do not yet know how much of the incident energy will enter the conductor.

It will be sufficient for our purposes to consider only normal incidence; in this case, we do not need to distinguish between incident fields parallel to or perpendicular to the plane of incidence. Again, let us choose a specific coordinate system like that of Figure

25-15 for which the xy plane coincides with the surface of separation. We assume that all electric fields are along the x axis and thus are led to Figure 25-17. The unit propagation vectors are seen to be

$$\hat{\mathbf{k}}_i = -\hat{\mathbf{k}}_r = \hat{\mathbf{k}}_t = \hat{\mathbf{z}} \qquad (25\text{-}76)$$

and the electric fields will thus have the form

$$\mathbf{E}_i = \hat{\mathbf{x}} E_{0i} e^{i(k_1 z - \omega t)} \qquad \mathbf{E}_r = \hat{\mathbf{x}} E_{0r} e^{i(-k_1 z - \omega t)}$$

$$\mathbf{E}_t = \hat{\mathbf{x}} E_{0t} e^{i(k_2 z - \omega t)} \qquad (25\text{-}77)$$

where

$$k_1 = \frac{n_1 \omega}{c} \qquad k_2 = \alpha_2 + i\beta_2 = \frac{N_2 \omega}{c} \qquad (25\text{-}78)$$

by (25-4), (24-38), and (24-57); α_2 and β_2 can be found for the second medium from (24-42) and (24-43) or (24-45) and (24-46). The magnetic fields can be found from the general relation $\mathbf{H} = (k/\mu\omega)\hat{\mathbf{k}} \times \mathbf{E}$ given by (24-93), and, with the use of (25-76), we find that

$$\mathbf{H}_i = \hat{\mathbf{y}} \frac{k_1}{\mu_1 \omega} E_i \qquad \mathbf{H}_r = -\hat{\mathbf{y}} \frac{k_1}{\mu_1 \omega} E_r \qquad \mathbf{H}_t = \hat{\mathbf{y}} \frac{k_2}{\mu_2 \omega} E_t \qquad (25\text{-}79)$$

The boundary conditions giving the equality of the tangential components of $\mathbf{E}$ and $\mathbf{H}$ hold for $z = 0$ and all t. When the common factor $e^{-i\omega t}$ is canceled from both sides of the result obtained by using (25-77) and (25-79), we are led to the two equations

$$E_{0i} + E_{0r} = E_{0t} \qquad \frac{k_1}{\mu_1 \omega}(E_{0i} - E_{0r}) = \frac{k_2}{\mu_2 \omega} E_{0t} \qquad (25\text{-}80)$$

which, when solved for the two ratios, lead to

$$\frac{E_{0r}}{E_{0i}} = \frac{1 - (\mu_1/\mu_2)(k_2/k_1)}{1 + (\mu_1/\mu_2)(k_2/k_1)} = \frac{1 - (\mu_1/\mu_2)(c/n_1\omega)(\alpha_2 + i\beta_2)}{1 + (\mu_1/\mu_2)(c/n_1\omega)(\alpha_2 + i\beta_2)} \qquad (25\text{-}81)$$

$$\frac{E_{0t}}{E_{0i}} = \frac{2}{1 + (\mu_1/\mu_2)(k_2/k_1)} = \frac{2}{1 + (\mu_1/\mu_2)(c/n_1\omega)(\alpha_2 + i\beta_2)} \qquad (25\text{-}82)$$

Since both of these ratios are complex, the reflected and transmitted electric fields each have their phase shifted with respect to the incident field.

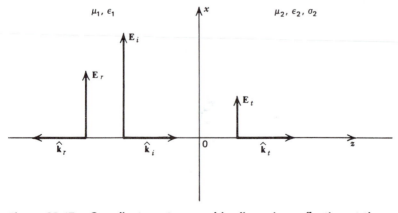

Figure 25-17.   Coordinate system used in discussing reflection at the surface of a conductor.

These can be written in an interesting manner by introducing the complex index of refraction N_2 by means of (25-78) and we find that

$$\frac{E_{0r}}{E_{0i}} = \frac{n_1 - (\mu_1/\mu_2)N_2}{n_1 + (\mu_1/\mu_2)N_2} \qquad \frac{E_{0t}}{E_{0i}} = \frac{2n_1}{n_1 + (\mu_1/\mu_2)N_2} \tag{25-83}$$

which are exactly of the form previously found in (25-37) and thus shows us how the complex index of refraction can really be used in a physically significant way. These ratios are often found written in terms of the real indices of refraction by writing $N_2 = n_2 + i(c\beta_2/\omega)$, as in (24-57), so that we get

$$\frac{E_{0r}}{E_{0i}} = \frac{[n_1 - (\mu_1/\mu_2)n_2] - i(\mu_1/\mu_2)(c\beta_2/\omega)}{[n_1 + (\mu_1/\mu_2)n_2] + i(\mu_1/\mu_2)(c\beta_2/\omega)} \tag{25-84}$$

$$\frac{E_{0t}}{E_{0i}} = \frac{2n_1}{[n_1 + (\mu_1/\mu_2)n_2] + i(\mu_1/\mu_2)(c\beta_2/\omega)} \tag{25-85}$$

In many cases of interest, these can be further approximated by taking $\mu_1 \simeq \mu_2$.

The reflection coefficient for power can now be found from (25-67) and (25-84) and is

$$R = \frac{[n_1 - (\mu_1/\mu_2)n_2]^2 + (\mu_1/\mu_2)^2(c\beta_2/\omega)^2}{[n_1 + (\mu_1/\mu_2)n_2]^2 + (\mu_1/\mu_2)^2(c\beta_2/\omega)^2} \tag{25-86}$$

As a special case, we see that as $\beta_2 \to \infty$, $R \to 1$. This limiting behavior tells us that waves that are very strongly absorbed are very strongly reflected. An excellent example of this behavior is provided by the optical properties of gold. This metal appears yellowish by reflection so that if a *sufficiently thin* sheet is illuminated by white light, the yellow to red portions will be practically all absorbed. As a result, the transmitted light will contain only the green to blue portions of the spectrum, so that the same sheet will appear greenish when viewed by transmission. A similar effect can be observed for copper.

All of our results so far in this section are exact, but it is useful to approximate R for an important limiting case.

■ **Example**

"Good conductors," (i.e., metals). We found in (24-75) that $\alpha_2 \simeq \beta_2 \simeq (\frac{1}{2}\mu_2\sigma_2\omega)^{1/2}$, and therefore $n_2 = c\alpha_2/\omega \simeq c\beta_2/\omega \simeq c(\mu_2\sigma_2/2\omega)^{1/2} = (\kappa_{m2}\kappa_{e2}\sigma_2/2\omega\epsilon_2)^{1/2} = (\kappa_{m2}\kappa_{e2}/2Q_2)^{1/2} \gg 1$ since $Q \ll 1$ for a good conductor. Therefore, we can write

$$R \simeq \frac{[n_1 - (\mu_1/\mu_2)n_2]^2 + (\mu_1/\mu_2)^2 n_2^2}{[n_1 + (\mu_1/\mu_2)n_2]^2 + (\mu_1/\mu_2)^2 n_2^2} = \frac{1 + [1 - (\mu_2 n_1/\mu_1 n_2)]^2}{1 + [1 + (\mu_2 n_1/\mu_1 n_2)]^2} \tag{25-87}$$

Since $n_1 \simeq 1$ for a typical dielectric, and if μ_2/μ_1 is also of order unity, we can take $\mu_2 n_1/\mu_1 n_2 = x \ll 1$ and expand (25-87) in a power series in x; keeping only the first nonvanishing term in x, we get $R = [1 + (1 - x)^2]/[1 + (1 + x)^2] \simeq (2 - 2x)/(2 + 2x) = (1 - x)/(1 + x) \simeq (1 - x)(1 - x) \simeq 1 - 2x$, so that for a good conductor

$$R \simeq 1 - \frac{2\mu_2 n_1}{\mu_1 n_2} = 1 - 2\left(\frac{\kappa_{e1}\kappa_{m2}}{\kappa_{m1}}\right)^{1/2}\left(\frac{2\omega\epsilon_0}{\sigma_2}\right)^{1/2} \tag{25-88}$$

where we also used (24-14) to simplify the result. This mildly complicated expression has the general form that $R = 1 - $ (small quantity) so that R will be close to unity for good conductors. Thus, essentially all of the energy will be reflected as is commonly observed to be the case for metals. What little energy does flow into the good conductor is quickly dissipated by the heat loss associated with the induced currents. ∎

One can find results for an arbitrary angle of incidence on a conducting medium in a manner similar to that we used for normal incidence. They turn out to be considerably more complicated, but for a *good conductor* one finds that the transmitted wave travels nearly normal to the bounding surface. In the limit $\sigma \to \infty$, its direction of travel *is* normal to the surface so that the effect is the same as if it arose from normal incidence.

25-7 CONTINUOUSLY VARYING INDEX OF REFRACTION

The two media that we have discussed were both assumed to be uniform so that their properties are independent of position. There are other possibilities that can occur. For example, the incident wave may be incident on a system composed of a series of parallel slabs, each slab with its own properties. Then one would have to satisfy the boundary conditions at each interface; a typical example of this will be illustrated by an exercise. Another possibility is that the second medium may have an index of refraction that varies continuously with position, and we want to look into a few of the consequences very briefly.

As a specific example, consider the case of a plasma for which the index of refraction can be found from (24-140), (24-139), and (24-13) to be

$$n_P = \left(1 - \frac{n_e e^2}{m\epsilon_0 \omega^2}\right)^{1/2} \tag{25-89}$$

where we have now written the number density of electrons as n_e. We see that when we have true propagation ($\omega > \omega_P$), then $n_P < 1$. Now suppose that n_e is not a constant but varies with position; then n_P will also be a function of position.

The ionosphere provides an example of this situation. The ionosphere is a portion of the earth's atmosphere where the air molecules have been ionized by radiation from the sun. Generally speaking, the density of electrons n_e increases with height so that n_P decreases with height. The ionosphere is not a stable feature of the earth's atmosphere in the sense that its thickness, amount of ionization, and location of its lower boundary can vary considerably with time, often within a matter of hours. Consequently, we consider only a simplified situation, which, however, is sufficient to illustrate the principal effects.

Let us assume that there is a definite boundary between the plasma and an unionized medium; again we choose the bounding surface as the xy plane with the z axis upward as shown in Figure 25-18. (The reflected wave is not shown.) We assume that n_e increases with increasing $z > 0$, thus making n_P decrease with z while remaining less than unity. A wave of direction $\hat{\mathbf{k}}_i$ is traveling in the air of index n_a and is incident on the boundary with an angle of incidence θ_i. Now Snell's law (25-18) says, in essence, that if we follow the transmitted wave, then the quantity $n \sin\theta = $ const., that is, it is "conserved." Therefore, if $\theta_P(z)$ is the angle made with the z axis by the transmitted wave at a distance z into the plasma, we will have

$$n_a \sin\theta_i = n_P(z)\sin\theta_P(z) = \text{const.} \tag{25-90}$$

Since $n_a > 1 > n_P(z)$, and n_P is decreasing, we see from this equation that $\theta_P(z)$ increases with distance of penetration. In other words, the direction of $\hat{\mathbf{k}}_t$ is turning continuously toward the horizontal, and the transmitted wave follows the curved path shown. Eventually, it will attain a maximum height z_m where $\hat{\mathbf{k}}_t$ is horizontal, and then

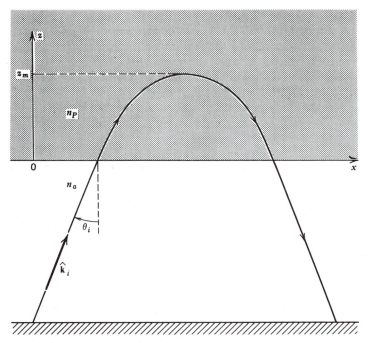

Figure 25-18. **Propagation of a wave in the ionosphere.**

it will turn back and follow the symmetrical path shown, finally returning to the air at the same angle θ_i and thence back to earth. (In radio transmission, the distance between the point of origin and return is called the "skip distance," and it can be many hundreds of kilometers.) In effect, the incident wave has been "reflected" from the plasma.

At the distance z_m, $\theta_P(z_m) = 90°$ so that (25-90) becomes $n_a \sin \theta_i = n_P(z_m)$. Combining this with (25-89), we find that

$$n_e(z_m) = \frac{m\epsilon_0\omega^2}{e^2}\left(1 - n_a^2 \sin^2 \theta_i\right) \tag{25-91}$$

which can be used to evaluate n_e at this particular value of z_m.

25-8 RADIATION PRESSURE

In Section 21-5, we saw that one can associate momentum with an electromagnetic field in a vacuum, and the momentum density as given by (21-75) is $\mathbf{g} = \mu_0\epsilon_0\mathbf{S}$. If we now want to apply this to plane waves, we will as usual be interested primarily in the time average of this quantity, which will be

$$\langle \mathbf{g} \rangle = \mu_0\epsilon_0\langle \mathbf{S} \rangle = \frac{\langle \mathbf{S} \rangle}{c^2} \tag{25-92}$$

We know from the example of Section 24-6 that for a plane wave $\langle \mathbf{S} \rangle$ is different from zero and is in the direction of propagation; therefore $\langle \mathbf{g} \rangle$ will also be different from zero and in the direction of propagation.

For situations involving reflection, the incident wave will be bringing momentum up to the bounding surface at a certain rate, while the reflected wave will be carrying it away at another rate. As a result, the momentum of the electromagnetic field in medium 1 will be changed, and we know from mechanics that this means that medium 2 has "exerted" a force on it. But by Newton's third law, the force *on* medium 2 will be equal and opposite to this; hence we conclude that an incident electromagnetic field exerts a force on the medium from which it is being reflected. This effect generally goes under the name of *radiation pressure* and we can easily calculate it from results we have already obtained. (This origin of radiation pressure is strongly reminiscent of that of the pressure exerted by a gas on the confining walls of its container in the kinetic theory of gases, and our calculation here will be recognized as very much like that used in kinetic theory.)

As shown in Figure 25-19a, let us consider the incident fields contained in an oblique cylinder of cross-sectional area ΔA and sides $c\,\Delta t$ where Δt is a small time interval. The volume of the cylinder is $\Delta \tau = c\,\Delta t\,\Delta A \cos \theta_i$. In this time interval, all of the momentum carried by the fields that is within this volume will be incident on the bounding surface. Since the normal component of momentum is of interest for finding the normal force, we see that, in the time interval, the initial normal momentum as found by using (25-92) is

$$G_{in} = \langle \mathbf{g}_i \rangle \cdot \hat{\mathbf{n}}\, \Delta \tau = \frac{\langle \mathbf{S}_i \rangle \cdot \hat{\mathbf{n}}\, \Delta t\, \Delta A \cos \theta_i}{c} \tag{25-93}$$

In Figure 25-19b, we have a similar picture for the reflected radiation, and since the distance of travel $c\,\Delta t$ is the same, the momentum of the fields in this volume is that associated with the same incident fields. Thus, the final normal momentum of the fields in medium 1 is

$$G_{fn} = \frac{\langle \mathbf{S}_r \rangle \cdot \hat{\mathbf{n}}\, \Delta t\, \Delta A \cos \theta_i}{c} \tag{25-94}$$

which is a negative quantity, since the angle between $\langle \mathbf{S}_r \rangle$ and $\hat{\mathbf{n}}$ is greater than 90°. The *change* in the normal momentum of the fields in medium 1 then is $\Delta G_{1n} = G_{fn} - G_{in} = (\Delta t\,\Delta A \cos \theta_i/c)[\langle \mathbf{S}_r \rangle \cdot \hat{\mathbf{n}} - \langle \mathbf{S}_i \rangle \cdot \hat{\mathbf{n}}]$ which, if we explicitly take the signs into account, becomes

$$\Delta G_{1n} = -\frac{\Delta t\,\Delta A \cos \theta_i}{c}\big[|\langle \mathbf{S}_r \rangle \cdot \hat{\mathbf{n}}| + |\langle \mathbf{S}_i \rangle \cdot \hat{\mathbf{n}}|\big] \tag{25-95}$$

If we let $\mathbf{F}_{em}$ be the average force on the fields in medium 1, and $\mathbf{F}_2$ be the force on medium 2, then $\mathbf{F}_2 = -\mathbf{F}_{em}$. Furthermore, from mechanics, the change in normal momentum during the time interval Δt is $\Delta G_{1n} = \mathbf{F}_{em} \cdot \hat{\mathbf{n}}\, \Delta t = -\mathbf{F}_2 \cdot \hat{\mathbf{n}}\, \Delta t = -F_{2n}\, \Delta t,$

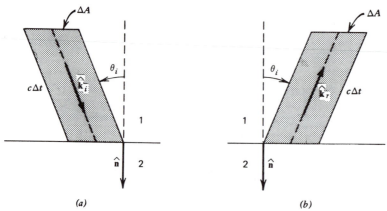

(a)   *(b)*

Figure 25-19. Calculation of radiation pressure. (*a*) Incident wave. (*b*) Reflected wave.

and therefore

$$F_{2n} = \frac{\Delta A \cos \theta_i}{c}[|\langle \mathbf{S}_r \rangle \cdot \hat{\mathbf{n}}| + |\langle \mathbf{S}_i \rangle \cdot \hat{\mathbf{n}}|] \tag{25-96}$$

This is a positive quantity and therefore is a force *on* medium 2, that is, in the direction of $\hat{\mathbf{n}}$. Since this is proportional to the area ΔA, we can define a *pressure* $P(\theta_i)$ as $F_{2n}/\Delta A$ and we get

$$P(\theta_i) = \frac{\cos \theta_i}{c}[|\langle \mathbf{S}_r \rangle \cdot \hat{\mathbf{n}}| + |\langle \mathbf{S}_i \rangle \cdot \hat{\mathbf{n}}|]$$

$$= \frac{(1 + R)\cos \theta_i}{c}|\langle \mathbf{S}_i \rangle \cdot \hat{\mathbf{n}}| = (1 + R)\cos^2 \theta_i \frac{|\langle \mathbf{S}_i \rangle|}{c} \tag{25-97}$$

with the use of (25-65) and (1-15). This is sometimes more conveniently written in terms of the average energy density since $|\langle \mathbf{S}_i \rangle| = \langle u_i \rangle c$ by (24-111) for an incident wave in a vacuum; thus (25-97) becomes

$$P(\theta_i) = (1 + R)\cos^2 \theta_i \langle u_i \rangle \tag{25-98}$$

The maximum value of P corresponds to $\theta_i = 0$ and $R = 1$, that is, normal incidence on a perfect conductor ($\sigma \to \infty$), and we see that

$$P_{\max} = 2\frac{|\langle \mathbf{S}_i \rangle|}{c} = 2\langle u_i \rangle \tag{25-99}$$

On the other hand, consider a complete absorber of radiation for which $R = 0$; this is *not* a conductor, but rather an idealized perfectly "black" object. Then, at normal incidence, $P_{\text{black}} = |\langle \mathbf{S}_i \rangle|/c = \langle u_i \rangle = \frac{1}{2}P_{\max}$.

The numerical values of the radiation pressure are really quite small for the usual situations encountered. For example, the value of the incident solar radiation at the earth, as quoted in Exercise 24-9, is 1340 watts/(meter)^2. Inserting this into (25-99), we find that the corresponding maximum radiation pressure is 8.9×10^{-6} newtons/(meter)$^2 = 8.8 \times 10^{-11}$ atmospheres. In spite of its very small value, the radiation pressure was measured for light and (25-99) thus verified as early as 1903; the fact that comets' tails always point away from the sun was once ascribed completely to the pressure of the solar radiation, but it is now thought to make only a partial contribution to this effect.

EXERCISES

25-1 Show that using the facts that the normal components of **D** and **B** are continuous gives no new information for the case in which **E** is perpendicular to the plane of incidence. Repeat when **E** is parallel to the plane of incidence.

25-2 Find the expressions analogous to (25-29), (25-30), (25-45), and (25-46) for the ratios H_r/H_i and H_t/H_i.

25-3 Show for the case $n_1 > n_2$ that the polarizing angle is less than the critical angle.

25-4 The expression $\tan \theta_p = n_2/n_1$ for the polarizing angle found in (25-52) involved the assumption that $\mu_1 = \mu_2$. Consider the general case in which media 1 and 2 are nonconductors but otherwise have arbitrary properties. (a) Show that $(E_r/E_i)_{\parallel} = 0$ when θ_i is such that $\tan^2 \theta_i = n_2^2(\mu_1^2 n_2^2 - \mu_2^2 n_1^2)[\mu_2^2 n_1^2(n_2^2 - n_1^2)]^{-1}$. (b) Find the corresponding expression for $\tan^2 \theta_i$ in order that $(E_r/E_i)_{\perp} = 0$. (c) Now assume that medium 1 is a vacuum and show that a polarizing angle is possible for the parallel case only when $\kappa_{e2} > \kappa_{m2}$.

(d) Find the corresponding requirement for the perpendicular case when medium 1 is a vacuum.

25-5 Find the ratios $(E_t/E_i)_\perp$ and $(E_t/E_i)_\parallel$ for the case of total reflection and express them in terms of the angle of incidence. Show that each can be written in the general form $A_t e^{-i\varphi_t}$ and find the ratio $(\tan\varphi_{t\perp}/\tan\varphi_{t\parallel})$. If the incident wave is linearly polarized, what kind of polarization will the transmitted electric field have in general?

25-6 Evaluate $\langle \mathbf{S}_t \rangle$ for the case of total reflection and then show that $\langle \mathbf{S}_t \rangle \cdot \hat{\mathbf{n}} = 0$ as required for T to be zero.

25-7 (a) For the case of total reflection, show that

$$\tan(\varphi_\parallel - \varphi_\perp) = \frac{(\mu_1 n_2/\mu_2 n_1)\left[(\mu_2 n_1/\mu_1 n_2)^2 - 1\right]\cos\theta_i\left[(n_1/n_2)^2\sin^2\theta_i - 1\right]^{1/2}}{\left[(n_1/n_2)^2 - 1\right]\sin^2\theta_i}$$

(b) An incident wave is traveling in a nonmagnetic medium of index of refraction 4.5. The second medium is a nonmagnetic glass of index 1.5. The incident wave is linearly polarized such that the components of $\mathbf{E}_{0i}$ perpendicular and parallel to the plane of incidence are equal. Find θ_i such that the totally reflected wave is circularly polarized. (c) Find the minimum value of the index of refraction of the incident medium for which it is possible to obtain a circularly polarized wave by one total reflection against this glass and find the corresponding angle of incidence.

25-8 Verify that the pairs (25-29), (25-30) and (25-45), (25-46) lead to expressions for R and T that satisfy (25-75).

25-9 In Figure 25-20, we show a beam of rectangular cross section $a \times b$ incident on a plane

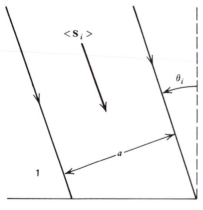

Figure 25-20. The incident beam of Exercise 25-9.

surface separating two media. (This is not a plane wave but, if the wavelength is very small compared to both a and b, it can be a good approximation to ascribe plane wave properties to it.) (a) What are the dimensions of the transmitted beam? (b) What is the total incident power P_i in terms of $\langle \mathbf{S}_i \rangle$? the transmitted power P_t in terms of $\langle \mathbf{S}_t \rangle$? (c) If we define a transmission coefficient $T' = P_t/P_i$, how does this one compare with the definition of T used in the text?

25-10 Show that for a good conductor, $E_{0r}/E_{0i} \simeq -1[1 - 2\pi(\mu_2/\mu_1)(\delta_2/\lambda_1)(1 - i)]$ where δ_2 is the skin depth in the conductor and λ_1 is the wavelength in the incident medium. For the remainder of this exercise, assume a perfect conductor ($\sigma_2 \to \infty$). Assume E_{0i} to be real for simplicity, and find $\mathbf{E}_{1\,\text{real}}$, that is, the total physical electric field in the incident medium and show that it is a standing wave with a node at the conducting surface. Similarly, show that $\mathbf{H}_{1\,\text{real}}$ is a standing wave with an antinode at the surface. Does $\mathbf{H}_{1\,\text{real}}$ lead or lag $\mathbf{E}_{1\,\text{real}}$ in time and by how much? Find $\langle u_e \rangle$ and $\langle u_m \rangle$ and show that the total energy density $\langle u \rangle$ is independent of position. Find $\mathbf{S}_1$ and show that it too is a standing wave with a node at the surface. Find $\langle \mathbf{S}_1 \rangle$ and interpret the result. Show that there must be a surface current density $\mathbf{K}_f$ and evaluate it.

25-11 Find the lowest-order approximation to the ratio E_{0t}/E_{0i} for a good conductor. If there is a phase difference between $\mathbf{E}_t$ and $\mathbf{E}_i$, evaluate it, and tell whether $\mathbf{E}_t$ leads or lags $\mathbf{E}_i$ in time.

25-12 Show that for a good conductor the reflection coefficient can also be written in the approximate form $R \simeq 1 - 4\pi(\mu_2/\mu_1)(\delta_2/\lambda_1)$ where δ_2 is the skin depth in the conductor and λ_1 is the wavelength in the incident medium.

25-13 Evaluate T for normal incidence on a good conductor by using the basic definition given in (25-71) and show that the result is in agreement with (25-75) and the result of the previous exercise.

25-14 See Figure 25-21. A plane wave traveling in a medium of impedance Z_1 is normally incident at $z = 0$ on a second medium of impedance Z_2. The second medium has thickness L and behind it is another medium of impedance Z_3, which occupies the rest of space. (a) Show that the ratio of the reflected and incident electric field

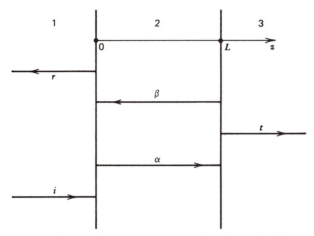

Figure 25-21. The waves and media of Exercise 25-14.

amplitudes in the incident medium is given by

$$\frac{E_{0r}}{E_{0i}} = \frac{Z_2(Z_3 - Z_1)\cos k_2 L - i(Z_2^2 - Z_1 Z_3)\sin k_2 L}{Z_2(Z_3 + Z_1)\cos k_2 L - i(Z_2^2 + Z_1 Z_3)\sin k_2 L}$$

(You will need the five *resultant* waves indicated, that is, the situation is in its final steady state. Remember: the boundary conditions must be satisfied at both boundaries simultaneously.) (b) Show that if $Z_1 \neq Z_3$, the reflected wave will be zero when L equals an odd multiple of a quarter wavelength in medium 2 and $Z_2 = (Z_1 Z_3)^{1/2}$. (c) Find the corresponding conditions for zero reflected wave when $Z_1 = Z_3 \neq Z_2$. (d) Light of wavelength 5×10^{-7} meters is normally incident in a vacuum upon a large slab of nonmagnetic glass of index of refraction 1.5. If the glass is to be coated with a layer of nonmagnetic material in order that the light not be reflected, find the required index of refraction and minimum thickness of the coating.

25-15 A curve that is tangent to the direction of $\hat{\mathbf{k}}$ at each point is called a *ray*; thus Figure 25-18 shows such a curve. (a) If the distance measured along the ray is s, show that it follows from (25-90) that $d\theta_P/ds = -(\tan\theta_P/n_P)(dn_P/ds) = -(\sin\theta_P/n_P)(dn_P/dz)$. (b) Show that this result is also consistent with the general appearance of the ray as shown in Figure 25-18. (c) Why, when θ_P becomes 90°, doesn't the ray continue horizontally, rather than bending back down as shown? (d) Find the dependence of n_P on z that will result in the path in the plasma being an arc of a circle of radius a.

25-16 Assume that the earth absorbs all of the solar radiation incident upon it, and find the ratio of the total force due to radiation pressure to the gravitational force of the sun. (The mass of the earth is 5.98×10^{24} kilograms, its radius is 6.37 $\times 10^6$ meters, and the average distance between the earth and the sun is 1.49×10^{11} meters.)

25-17 The radiation pressure on a boundary that is perfectly reflecting for any angle of incidence θ_i is found from (25-98) to be $P(\theta_i) = 2\cos^2\theta_i\langle u_i\rangle$. Now suppose that *one* side of the surface is isotropically irradiated by incident waves of the same frequency and equal values of $\langle u_i\rangle$. "Isotropically" means that all angles of incidence θ_i occur and with equal likelihood, that is, the probability that $\hat{\mathbf{k}}_i$ will be in an element of solid angle $d\Omega_i$ is proportional to $d\Omega_i$. Find the total radiation pressure on the surface and express it in terms of the total energy density $\langle u\rangle_{\text{tot}}$ at the surface, that is, $\langle u\rangle_{\text{tot}}$ includes both the incident and reflected waves.

25-18 In our derivation of the concept of radiation pressure, we used only very general concepts of momentum and the meaning of a change in momentum. It is also possible to approach this from a microscopic point of view by investigating the forces exerted by the transmitted electromagnetic field on the charged particles within the second medium. It is difficult to carry this out except for special cases, but when it is done it is concluded that radiation pressure arises from the *magnetic force* on the charges moving under the influence of the electric field. A general statement like this should be independent of the particular model that was used. Review our discussion of electromagnetic momentum in Section 21-5 and convince yourself that the conclusion quoted above is essentially correct.

26

FIELDS IN BOUNDED REGIONS

In the two preceding chapters, we have considered time-dependent solutions of Maxwell's equations in the form of plane waves of infinite extent so that they necessarily exist in unbounded regions. In more realistic cases, we can expect there to be boundaries (walls) of some sort about the region in which we wish to know the fields. In such situations, it is fairly clear that the solutions cannot generally be plane waves with definite values of the fields over an infinite plane since the fields will have to satisfy boundary conditions at the limits of the region as well as being solutions of Maxwell's equations.

As soon as we start thinking about bounded regions, it is evident that there can be many possibilities, both in the shape of the region and in the materials comprising the boundaries. Accordingly, it will be desirable to restrict ourselves to only a few cases. We assume that all bounding surfaces are surfaces of perfect conductors, as this will simplify the boundary conditions considerably. (This assumption is analogous to that used in mechanics in the study of vibrating strings and membranes, where one assumes the bounding regions to be perfectly rigid.) The first problems we will consider will concern the possible transfer of electromagnetic energy along a *wave guide*, that is, a tube with open ends. (From everyday experience, we already know that this is possible from the simple fact that we can see through long straight pipes.) We will also consider a *resonant cavity*, that is, a box of some shape enclosed by perfectly conducting walls.

26-1 BOUNDARY CONDITIONS AT THE SURFACE OF A PERFECT CONDUCTOR

By the term *perfect conductor*, we mean one for which $\sigma \to \infty$, or, more precisely, one for which the ratio $Q = \epsilon\omega/\sigma \to 0$. We have already seen in the paragraph following (24-78) that $Q \ll 1$ for common metals even at very high frequencies so that $Q = 0$ should be a good first approximation for metallic boundaries. Since we will be considering only fields that vary harmonically in time, that is, are proportional to $e^{-i\omega t}$, we see from the form of our general superposition given by (24-18) that we can study the fields in the conductor by considering the behavior of their plane wave components. In Section 25-6, we covered the case in which the field was normally incident on a conducting surface. At the end of that section, it was pointed out that, in the limit of infinite conductivity, the wave in the conductor travels normal to the surface regardless of the angle of incidence. Combining these results with (24-61), (24-54), and (24-84), we see that a component of the superposition for the electric field in the conductor will have the general form

$$\mathbf{E}_\tau = \mathbf{E}_{0\tau} e^{-\zeta/\delta} e^{i(\alpha\zeta - \omega t + \vartheta)} \tag{26-1}$$

where ζ is the distance of penetration into the conductor and δ is the skin depth. The subscript τ indicates that $\mathbf{E}_\tau$ is parallel to the surface of the conductor since it is transverse to the direction of propagation $\hat{n}$ of Figure 25-3. Thus, at the surface, $\mathbf{E}_\tau$ will be a tangential component. We found in (24-78) that $\delta = (2/\mu\sigma\omega)^{1/2}$ for a good conductor so that $\delta \to 0$ as $\sigma \to \infty$. Therefore, we see from (26-1) that $\mathbf{E}_r \to 0$ as $\sigma \to \infty$ for any value of $\zeta \neq 0$, that is, the electric field is zero at any point in a perfect conductor. Since the tangential components of $\mathbf{E}$ are always continuous, according to (21-26), we see that $\mathbf{E}_{\text{tang}} = 0$ just outside of the surface. In other words, $\mathbf{E}$ has no tangential component at the surface of a perfect conductor so that $\mathbf{E}$ must be normal to the surface.

The value of $\mathbf{B}$ inside the conductor is given by (24-92) to be $\mathbf{B} = (k/\omega)\hat{\mathbf{k}} \times \mathbf{E}_r$ so that $\mathbf{B}$ will also be transverse and $\sim e^{-\zeta/\delta}$ since $\hat{\mathbf{k}} = \hat{\mathbf{n}}$ here. Consequently, the transverse component of $\mathbf{B}$ inside will also vanish as $\sigma \to \infty$. Since $\mathbf{B}$ has no normal component, the continuity of the normal components as given by (21-27) shows that $\mathbf{B}_{\text{norm}} = 0$ just outside the conductor. Thus, at the surface of a perfect conductor and outside of it, $\mathbf{B}$ has no normal component, that is, it must be tangential to the surface. Similarly, all of the components of $\mathbf{D}$ and $\mathbf{H}$ inside the conductor will be $\sim e^{-\zeta/\delta}$ and will vanish as $\sigma \to \infty$.

Since these conclusions hold for each component of the superposition of (24-18), we see that *all* of the field vectors will be zero inside a perfect conductor. This enables us to simplify the general boundary conditions given by (21-25) through (21-28). If we choose medium 1 to be the conductor and medium 2 to be the region adjacent to it, we can set $\mathbf{D}_1$, $\mathbf{E}_1$, $\mathbf{B}_1$, and $\mathbf{H}_1$ all equal to zero, and, if we drop the subscript 2, the boundary conditions become

$$\hat{\mathbf{n}} \cdot \mathbf{D} = \sigma_f \qquad \hat{\mathbf{n}} \times \mathbf{E} = 0 \qquad \hat{\mathbf{n}} \cdot \mathbf{B} = 0 \qquad \hat{\mathbf{n}} \times \mathbf{H} = \mathbf{K}_f \qquad (26\text{-}2)$$

where $\hat{\mathbf{n}}$ is the unit normal vector pointing *outward* from the surface of the conductor (since our convention requires that $\hat{\mathbf{n}}$ always be drawn from region 1 to region 2). To repeat: at the surface of a perfect conductor, $\mathbf{E}$ is normal to the surface and $\mathbf{B}$ is tangential to the surface. To put it another way, $\mathbf{E}$ has no tangential component while $\mathbf{B}$ has no normal component. It is possible to have a finite surface current density $\mathbf{K}_f$ even though $\mathbf{E}_{\text{tang}} = 0$ since the conductivity is infinite. In fact it is precisely these surface currents that keep $\mathbf{H}_{\text{tang}}$ from being zero at the surface, even though it is zero inside. Similarly, the surface charge density σ_f arises because $\mathbf{D}_{\text{norm}}$ can be different from zero at the surface while it is zero inside the conductor.

26-2 PROPAGATION CHARACTERISTICS OF WAVE GUIDES

Figure 26-1 shows a wave guide that extends indefinitely in the z direction and of arbitrary and constant cross section in the xy plane. We assume that the bounding walls are perfect conductors and that the interior is filled with a l.i.h. nonconducting medium described by μ and ϵ. If ψ is any component of $\mathbf{E}$ or $\mathbf{B}$, we know from (24-7) and (24-12) that it satisfies the equation

$$\nabla^2 \psi - \frac{1}{v^2} \frac{\partial^2 \psi}{\partial t^2} = 0 \qquad (26\text{-}3)$$

where $v^2 = 1/\mu\epsilon$ and v would be the speed of a plane wave in the medium. We will try to find a solution of (26-3) in the form of a wave traveling in the z direction, that is, along the axis of the guide, by assuming that ψ has the form

$$\psi(x, y, z, t) = \psi_0(x, y)e^{i(k_g z - \omega t)} \qquad (26\text{-}4)$$

We note that this is *not* a plane wave since the amplitude ψ_0 is not a constant but can vary across the cross section. The quantity k_g is the *guide propagation constant* and can

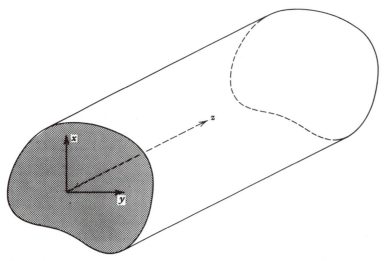

Figure 26-1. A wave guide with arbitrary and constant cross section.

be written as

$$k_g = \frac{2\pi}{\lambda_g} \tag{26-5}$$

where λ_g is the *guide wavelength*, that is, the spatial period along the guide.

If we substitute (26-4) into (26-3) and cancel the common exponent, we find that the equation satisfied by the amplitude is

$$\frac{\partial^2 \psi_0}{\partial x^2} + \frac{\partial^2 \psi_0}{\partial y^2} + k_c^2 \psi_0 = 0 \tag{26-6}$$

where

$$k_c^2 = k_0^2 - k_g^2 \tag{26-7}$$

$$k_0 = \frac{\omega}{v} = \frac{2\pi}{\lambda_0} \tag{26-8}$$

We see from (24-17) and (24-27) that, *if* ψ_0 were a plane wave of circular frequency ω, then k_0 and λ_0 would be the propagation constant and wavelength respectively; for this reason, λ_0 is often called the free space wavelength when there is a vacuum inside the guide. It is often useful to write k_c in the form

$$k_c = \frac{2\pi}{\lambda_c} \tag{26-9}$$

so that (26-7) can also be written as

$$\frac{1}{\lambda_c^2} = \frac{1}{\lambda_0^2} - \frac{1}{\lambda_g^2} \tag{26-10}$$

It is not enough that we be able to solve (26-6) because the corresponding component or components must also satisfy (26-2). It is not generally possible for these boundary conditions to be satisfied simultaneously for any arbitrary value of k_c but only for specific values of k_c and corresponding definite values of λ_c. Hence, even before we consider a specific problem, we can expect to find that only certain values of k_c are *allowed* or are the *eigenvalues* of the system. What is the physical significance of these allowed values that we are sure to encounter?

Suppose we consider a definite *mode* of the guide, that is, a particular value of k_c. If we also assume a given frequency ω, then k_g will be given by (26-7) as $k_g^2 = k_0^2 - k_c^2$. There are two possibilities of interest. If $k_0 > k_c$, so that $\lambda_0 < \lambda_c$ by (26-10), then $k_g^2 > 0$ and k_g will be real and we will have wave propagation along the guide. On the other hand, if $k_0 < k_c$, and hence $\lambda_0 > \lambda_c$, then $k_g^2 < 0$ and k_g will be a pure imaginary which we can write as $k_g = i|k_g|$. Substituting this into (26-4), we find that ψ will have the form $\psi = \psi_0 e^{-|k_g|z} e^{-i\omega t}$, which is not a wave but rather a harmonically oscillating "disturbance" whose amplitude is steadily decreasing as we go along the guide in the sense of increasing z. Therefore, we have found for a wave guide that we will get wave propagation *only* if $k_0 > k_c$, or $\lambda_0 < \lambda_c$. For this reason, λ_c is called the *cutoff wavelength* for this particular mode.

We can also state this result in terms of a *cutoff frequency* ω_c defined by

$$k_c = \frac{\omega_c}{v} \tag{26-11}$$

so that (26-7) can also be written as

$$k_g^2 = \frac{1}{v^2}\left(\omega^2 - \omega_c^2\right) \tag{26-12}$$

Then wave propagation is possible only if $\omega > \omega_c$, that is, if the applied frequency is greater than the cutoff frequency. Thus, a wave guide will also act as a high-pass filter in the same sense that we found for a plasma following (24-139). In fact, we see on comparing (26-12) with (24-137) that the dispersion relation for a wave guide is identical in form with that of a plasma and with the cutoff frequency for a definite mode corresponding to the plasma frequency ω_p.

26-3 FIELDS IN A WAVE GUIDE

In order to continue, we must use Maxwell's equations. We assume that there are no free charges or currents within the guide so that (21-48) through (21-51) become

$$\nabla \cdot \mathbf{E} = 0 \qquad \nabla \cdot \mathbf{H} = 0$$
$$\nabla \times \mathbf{E} = -\mu\frac{\partial \mathbf{H}}{\partial t} \qquad \nabla \times \mathbf{H} = \epsilon\frac{\partial \mathbf{E}}{\partial t} \tag{26-13}$$

Since each component of **E** and **H** is assumed to have the general form of (26-4), we will have

$$\mathbf{E} = \boldsymbol{\mathscr{E}}(x, y)e^{i(k_g z - \omega t)} \tag{26-14}$$

$$\mathbf{H} = \boldsymbol{\mathscr{H}}(x, y)e^{i(k_g z - \omega t)} \tag{26-15}$$

so that, if E_α is any component of **E**, we will have $\partial E_\alpha / \partial z = ik_g E_\alpha$ and $\partial E_\alpha / \partial t = -i\omega E_\alpha$, with similar expressions for the components of **H**. It is important to remember that, although the amplitudes $\mathscr{E}$ and $\mathscr{H}$ depend only on x and y, we are *not* assuming the fields to be transverse, that is, to have no z component. In other words, $\mathscr{E}$ has the general form

$$\boldsymbol{\mathscr{E}}(x, y) = \mathscr{E}_x(x, y)\hat{\mathbf{x}} + \mathscr{E}_y(x, y)\hat{\mathbf{y}} + \mathscr{E}_z(x, y)\hat{\mathbf{z}} \tag{26-16}$$

where the separate components $\mathscr{E}_x$, $\mathscr{E}_y$, and $\mathscr{E}_z$ can depend on x and y in different ways.

Substituting (26-14) and (26-15) into (26-13), and canceling the common exponential factor, we obtain the eight scalar equations:

$$\frac{\partial \mathscr{E}_x}{\partial x} + \frac{\partial \mathscr{E}_y}{\partial y} + ik_g \mathscr{E}_z = 0 \tag{26-17}$$

$$\frac{\partial \mathscr{H}_x}{\partial x} + \frac{\partial \mathscr{H}_y}{\partial y} + ik_g \mathscr{H}_z = 0 \tag{26-18}$$

$$\frac{\partial \mathscr{E}_z}{\partial y} - ik_g \mathscr{E}_y = i\omega\mu \mathscr{H}_x \tag{26-19}$$

$$ik_g \mathscr{E}_x - \frac{\partial \mathscr{E}_z}{\partial x} = i\omega\mu \mathscr{H}_y \tag{26-20}$$

$$\frac{\partial \mathscr{E}_y}{\partial x} - \frac{\partial \mathscr{E}_x}{\partial y} = i\omega\mu \mathscr{H}_z \tag{26-21}$$

$$\frac{\partial \mathscr{H}_z}{\partial y} - ik_g \mathscr{H}_y = -i\omega\epsilon \mathscr{E}_x \tag{26-22}$$

$$ik_g \mathscr{H}_x - \frac{\partial \mathscr{H}_z}{\partial x} = -i\omega\epsilon \mathscr{E}_y \tag{26-23}$$

$$\frac{\partial \mathscr{H}_y}{\partial x} - \frac{\partial \mathscr{H}_x}{\partial y} = -i\omega\epsilon \mathscr{E}_z \tag{26-24}$$

Solving for $\mathscr{E}_x$ by eliminating $\mathscr{H}_y$ between (26-20) and (26-22), and using (26-7) and (26-8), we find that

$$\mathscr{E}_x = \frac{i}{k_c^2} \left(k_g \frac{\partial \mathscr{E}_z}{\partial x} + \omega\mu \frac{\partial \mathscr{H}_z}{\partial y} \right) \tag{26-25}$$

Similarly, we find from (26-19) and (26-23) that

$$\mathscr{E}_y = \frac{i}{k_c^2} \left(k_g \frac{\partial \mathscr{E}_z}{\partial y} - \omega\mu \frac{\partial \mathscr{H}_z}{\partial x} \right) \tag{26-26}$$

and that

$$\mathscr{H}_x = \frac{i}{k_c^2} \left(-\omega\epsilon \frac{\partial \mathscr{E}_z}{\partial y} + k_g \frac{\partial \mathscr{H}_z}{\partial x} \right) \tag{26-27}$$

$$\mathscr{H}_y = \frac{i}{k_c^2} \left(\omega\epsilon \frac{\partial \mathscr{E}_z}{\partial x} + k_g \frac{\partial \mathscr{H}_z}{\partial y} \right) \tag{26-28}$$

also follow from the pairs (26-19)–(26-23) and (26-20)–(26-22), respectively. These results show that all four of the transverse components of $\mathscr{E}$ and $\mathscr{H}$ are independent of each other and can be found from the derivatives of the two longitudinal components; this should simplify our work.

However, in obtaining (26-25) through (26-28), we used only four of the eight Maxwell equations. If we substitute (26-25) and (26-26) into (26-17), we find that it will be satisfied provided that

$$\frac{\partial^2 \mathscr{E}_z}{\partial x^2} + \frac{\partial^2 \mathscr{E}_z}{\partial y^2} + k_c^2 \mathscr{E}_z = 0 \tag{26-29}$$

which must be the case as we already know from (26-6). This same result is easily seen to follow from (26-24). Similarly, (26-18) and (26-21) will be satisfied if

$$\frac{\partial^2 \mathscr{H}_z}{\partial x^2} + \frac{\partial^2 \mathscr{H}_z}{\partial y^2} + k_c^2 \mathscr{H}_z = 0 \tag{26-30}$$

which again agrees with (26-6).

Since each of the expressions (26-25) through (26-28) has the form of a sum, it is possible to regard the general transverse components as a superposition of two independent waves, one corresponding to $\mathscr{E}_z \neq 0$ and $\mathscr{H}_z = 0$, and the other to $\mathscr{E}_z = 0$ and $\mathscr{H}_z \neq 0$. Accordingly, it is convenient to separate the solutions into these two groups and consider them separately. If $\mathscr{E}_z = 0$, then $\mathscr{E}$ lies entirely in the xy plane and is perpendicular to the direction of propagation. Such a wave is called a *transverse electric* or TE mode. Similarly, the case $\mathscr{H}_z = 0$ corresponds to a *transverse magnetic* or TM mode. (If both $\mathscr{E}_z$ and $\mathscr{H}_z$ are zero, the corresponding mode is called *transverse electromagnetic* or TEM; this case requires a separate treatment that we defer to a later section.)

We can usefully summarize our results in the form of a step-by-step procedure for solving wave-guide problems.

For a TE mode: (1) set $\mathscr{E}_z = 0$ and find $\mathscr{H}_z$ as the general solution of (26-30) or its equivalent when written in another appropriate coordinate system such as cylindrical coordinates; (2) substitute this expression for $\mathscr{H}_z$ into (26-25) through (26-28) to find the transverse components of the amplitudes; (3) substitute these results into (26-14) and (26-15) to find the fields **E** and **H**; (4) make sure the expressions for **E** and **H** (or equivalently $\mathscr{E}$ and $\mathscr{H}$) satisfy the two boundary conditions of (26-2) that $\mathbf{E}_{\text{tang}} = 0$ and $\mathbf{H}_{\text{normal}} = 0$ (note that the tangential component E_z is already zero for this case); (5) take the real parts of the resulting expressions for **E** and **H** if it is desired to have the physical fields; and (6), the surface charge density and surface current density can now be found, if wanted, from the remaining boundary conditions $\hat{\mathbf{n}} \cdot \mathbf{D} = \sigma_f$ and $\hat{\mathbf{n}} \times \mathbf{H} = \mathbf{K}_f$ where they are of course evaluated *on* the bounding surface. [As we will see, it is sometimes more convenient to do step (4) before (2).]

For a TM mode, the same basic procedure is followed with $\mathscr{H}_z = 0$ and $\mathscr{E}_z$ found from (26-29). In this case, $\mathscr{E}_z$ is generally different from zero within the guide, but since it is a tangential component at the surface, it must be made to vanish there.

When these procedures are carried out, the fields obtained will automatically satisfy Maxwell's equations since they are used to obtain the basic results. Furthermore, the eigenvalues k_c, which characterize the various modes, will also be obtained during the process for the following reasons. When the general solution of (26-29) or (26-30) is obtained, it will involve constants of integration. It will then be found that the boundary conditions can be satisfied only for certain values of some of these constants, and they in turn will determine the allowed values of k_c. (As we will see in detail shortly, the ways in which the k_c are found are strongly reminiscent of what occurred when we solved Laplace's equation in Section 11-4.)

It will be left as exercises to show that the results of this section can be written in an even more compact and elegant form before it is necessary to specify the shape of the cross section of the guide. For our purposes, however, it is desirable to proceed at once to the consideration of a specific and important example.

26-4 RECTANGULAR GUIDE

This guide has a rectangular cross section of sides a and b, which we take to be located in the xy plane as shown in Figure 26-2. We see from (26-29) and (26-30) that for either type of mode we have to solve an equation of the form of (26-6). We use separation of

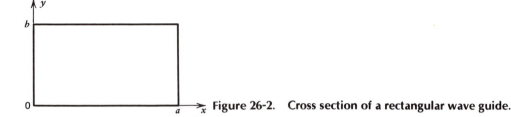

Figure 26-2. Cross section of a rectangular wave guide.

variables and write $\psi_0(x, y) = X(x)Y(y)$; then the same arguments as used in Section 11-4 lead us to the separated equations

$$\frac{1}{X}\frac{d^2X}{dx^2} = -\frac{1}{Y}\frac{d^2Y}{dy^2} - k_c^2 = \text{const.} = -k_1^2 \qquad (26\text{-}31)$$

so that $(d^2X/dx^2) + k_1^2 X = 0$ and $(d^2Y/dy^2) + k_2^2 Y = 0$ where

$$k_1^2 + k_2^2 = k_c^2 \qquad (26\text{-}32)$$

and k_2^2 is also a constant. Solving these in terms of trigonometric functions, we find that

$$\psi_0(x, y) = (C_1 \sin k_1 x + C_2 \cos k_1 x)(C_3 \sin k_2 y + C_4 \cos k_2 y) \qquad (26\text{-}33)$$

where the C's are constants of integration. This expression for ψ_0 contains a total of six constants, and, depending on the mode of interest, we identify (26-33) with either $\mathcal{H}_z$ or $\mathcal{E}_z$.

■ **Example**

TE modes. Here we set $\mathcal{E}_z = 0$ and write

$$\mathcal{H}_z = (C_1 \sin k_1 x + C_2 \cos k_1 x)(C_3 \sin k_2 y + C_4 \cos k_2 y) \qquad (26\text{-}34)$$

Since $\mathcal{H}_z$ is a tangential component, it need not vanish at the boundary so we will not learn much from (26-34) by itself. On the other hand, $\mathcal{E}_x$ and $\mathcal{E}_y$ will be tangential components at appropriate parts of the boundary and must vanish there, so it will be useful to find them next. (Of course, $\mathcal{H}_x$ and $\mathcal{H}_y$ would be normal components at parts of the boundary and must vanish there, so that we could deal with them instead. As we will see, however, we will end up with the same results.)

With $\mathcal{E}_z = 0$, we find that when (26-34) is substituted into (26-25) and (26-26) we obtain

$$\mathcal{E}_x = \frac{i\omega\mu k_2}{k_c^2}(C_1 \sin k_1 x + C_2 \cos k_1 x)(C_3 \cos k_2 y - C_4 \sin k_2 y) \qquad (26\text{-}35)$$

$$\mathcal{E}_y = -\frac{i\omega\mu k_1}{k_c^2}(C_1 \cos k_1 x - C_2 \sin k_1 x)(C_3 \sin k_2 y + C_4 \cos k_2 y) \qquad (26\text{-}36)$$

We see from Figure 26-2 that $\mathcal{E}_x$ will be a tangential component and must therefore vanish at $y = 0$ and $y = b$. Similarly, $\mathcal{E}_y$ must be zero at $x = 0$ and $x = a$. Evaluating for the zero values of x and y first, we get

$$\mathcal{E}_x(x,0) = 0 = \frac{i\omega\mu k_2 C_3}{k_c^2}(C_1 \sin k_1 x + C_2 \cos k_1 x) \qquad (26\text{-}37)$$

$$\mathcal{E}_y(0, y) = 0 = -\frac{i\omega\mu k_1 C_1}{k_c^2}(C_3 \sin k_2 y + C_4 \cos k_2 y) \qquad (26\text{-}38)$$

so that we must have $C_3 = 0$ and $C_1 = 0$. Therefore, at this stage, (26-34), (26-35), and (26-36) have simplified to

$$\mathcal{H}_z = C_2 C_4 \cos k_1 x \cos k_2 y \tag{26-39}$$

$$\mathcal{E}_x = -\frac{i\omega\mu k_2}{k_c^2} C_2 C_4 \cos k_1 x \sin k_2 y \tag{26-40}$$

$$\mathcal{E}_y = \frac{i\omega\mu k_1}{k_c^2} C_2 C_4 \sin k_1 x \cos k_2 y \tag{26-41}$$

We still have boundary conditions to satisfy at the two remaining faces. We see from (26-40) that the requirement $\mathcal{E}_x(x, b) = 0$ leads to $\sin k_2 b = 0$ so that $k_2 b = n\pi$ where n is an integer. Similarly, $\mathcal{E}_y(a, y) = 0$ gives the condition that $k_1 a = m\pi$ with m an integer. Thus we have found that

$$k_1 = \frac{m\pi}{a} \qquad k_2 = \frac{n\pi}{b} \tag{26-42}$$

so that (26-32) shows that the allowed values of k_c^2 are

$$k_c^2 = k_{c\ mn}^2 = \pi^2 \left[\left(\frac{m}{a} \right)^2 + \left(\frac{n}{b} \right)^2 \right] \tag{26-43}$$

The cutoff wavelengths and frequencies can now be found by using (26-43) in (26-9) and (26-11). The corresponding guide propagation constants and wavelengths can be found from (26-7) and (26-5):

$$k_g^2 = \left(\frac{2\pi}{\lambda_g} \right)^2 = k_0^2 - \pi^2 \left[\left(\frac{m}{a} \right)^2 + \left(\frac{n}{b} \right)^2 \right] \tag{26-44}$$

The only quantity left undetermined is the arbitrary amplitude $C_2 C_4$ of $\mathcal{H}_z$. If we set $C_2 C_4 = H_0$, then we can use (26-42) to write (26-39) through (26-41) more explicitly. Furthermore, we can use (26-39) in (26-27) and (26-28) to find the remaining amplitudes. When all this is done, we find that the amplitudes of a general TE mode in a rectangular guide become

$$\mathcal{E}_x = -\frac{i\omega\mu}{k_c^2} \left(\frac{n\pi}{b} \right) H_0 \cos \left(\frac{m\pi x}{a} \right) \sin \left(\frac{n\pi y}{b} \right) \tag{26-45}$$

$$\mathcal{E}_y = \frac{i\omega\mu}{k_c^2} \left(\frac{m\pi}{a} \right) H_0 \sin \left(\frac{m\pi x}{a} \right) \cos \left(\frac{n\pi y}{b} \right) \tag{26-46}$$

$$\mathcal{E}_z = 0 \tag{26-47}$$

$$\mathcal{H}_x = -\frac{i k_g}{k_c^2} \left(\frac{m\pi}{a} \right) H_0 \sin \left(\frac{m\pi x}{a} \right) \cos \left(\frac{n\pi y}{b} \right) \tag{26-48}$$

$$\mathcal{H}_y = -\frac{i k_g}{k_c^2} \left(\frac{n\pi}{b} \right) H_0 \cos \left(\frac{m\pi x}{a} \right) \sin \left(\frac{n\pi y}{b} \right) \tag{26-49}$$

$$\mathcal{H}_z = H_0 \cos \left(\frac{m\pi x}{a} \right) \cos \left(\frac{n\pi y}{b} \right) \tag{26-50}$$

where k_c and k_g are found from (26-43) and (26-44).

As a check on our results, we see from Figure 26-2 that $\mathcal{H}_x$ will be a normal component at $x = 0$ and $x = a$ and should vanish there; we see from (26-48) that this is indeed the case. Similarly, (26-49) shows that $\mathcal{H}_y = 0$ at $y = 0$ and $y = b$ as it should. [This is actually not such a surprise because the result of one of the exercises is that the transverse components in a TE mode are related by $\mathcal{H}_{\text{trans}} = (k_g/\omega\mu)\hat{z} \times \mathcal{E}_{\text{trans}}$

and therefore are perpendicular. Thus, if $\mathscr{E}_{\text{trans}}$ is made normal to the surface, $\mathscr{H}_{\text{trans}}$ will automatically be tangential as found above. A similar result holds for the TM modes.]

We recall that we still have to multiply each of these amplitude factors by the propagation term and take the real part to get the physical fields. For example, when (26-45) and (26-14) are combined, we get

$$E_x = -\frac{i\omega\mu}{k_c^2}\left(\frac{n\pi}{b}\right)H_0\cos\left(\frac{m\pi x}{a}\right)\sin\left(\frac{n\pi y}{b}\right)e^{i(k_g z - \omega t)} \tag{26-51}$$

Since $-i = e^{-i(1/2)\pi}$, the exponential factor can be written $\exp i[k_g z - (\omega t + \frac{1}{2}\pi)]$ which, when compared with (26-50) shows that E_x leads H_z in time by 90°. Similarly, (26-48) and (26-49) show us that H_x and H_y will lead H_z by 90° while E_y will lag H_z by this same amount according to (26-46).

So far we have not specified m and n other than to say that they are integers. First of all, we see from (26-45) through (26-50) that if m and/or n is negative, all of the field amplitudes are unchanged. Thus we can restrict m and n to be positive or zero. If m and n are *both* zero, (26-50) gives $\mathscr{H}_z = H_0 = \text{const.}$, while the other components are apparently zero, while (26-44) gives $k_g = k_0 = \omega/v$. Thus our field would seem to be of the form $H_z = H_0 e^{i(k_0 z - \omega t)}$, which would be a *longitudinal* field of only one component traveling along the axis of the guide with the speed v characteristic of a plane wave. However, $k_c^2 = 0$ by (26-43) and, since we divided by k_c^2 to get (26-25) through (26-28), we cannot actually use them in this case but must go back to Maxwell's equations (26-13). Then we would get $\nabla \cdot \mathbf{H} = 0 = \partial H_z/\partial z = ik_0 H_z$ so that $k_0 = 0$ and hence $\omega = 0$; this makes $H_z = \text{const.}$, which is also seen to be consistent with the rest of (26-13) with $\mathbf{E} = 0$. Thus, for $m = n = 0$, the solution is simply a constant magnetic field along the axis of the guide; while this is certainly possible, it is of no interest here.

In summary, m and n can be restricted to $m \geq 0$ and $n \geq 0$ but not $m = n = 0$. Thus a given TE mode of propagation can be characterized by the pair of independently assigned integers m and n. It is customary to describe it as a TE_{mn} mode.

Since the general TE case is quite complicated, let us consider only a particularly simple case in more detail. ∎

■ **Example**

TE_{10} mode. Let us assume that $a > b$ and look for the mode with the smallest k_c and hence the largest cutoff wavelength λ_c. We see from (26-43) that this will correspond to $m = 1$ and $n = 0$, that is, to the TE_{10} mode. Then $k_{c10} = \pi/a$ so that $\lambda_c = 2a$ and $\omega_c = \pi v/a$ by (26-9) and (26-11). The cutoff wavelength thus is twice the longest dimension of the cross section; in other words, it is not possible to "squeeze" a "bigger" wave into the guide. The guide propagation constant is found from (26-44) and (26-8) to be

$$k_g = \left[\left(\frac{\omega}{v}\right)^2 - \left(\frac{\pi}{a}\right)^2\right]^{1/2} \tag{26-52}$$

The field amplitudes as obtained from (26-45) through (26-50) are

$$\mathscr{E}_y = i\omega\mu\left(\frac{a}{\pi}\right)H_0\sin\left(\frac{\pi x}{a}\right) \tag{26-53}$$

$$\mathscr{H}_x = -ik_g\left(\frac{a}{\pi}\right)H_0\sin\left(\frac{\pi x}{a}\right) \tag{26-54}$$

$$\mathscr{H}_z = H_0\cos\left(\frac{\pi x}{a}\right) \tag{26-55}$$

while $\mathscr{E}_x = \mathscr{E}_z = 0$ and $\mathscr{H}_y = 0$. Inserting these into (26-14) and (26-15), assuming for simplicity that H_0 is real, and taking the real parts of the resulting expressions, we find the only nonzero field components to be

$$E_y = -H_0\omega\mu\left(\frac{a}{\pi}\right)\sin\left(\frac{\pi x}{a}\right)\sin\left(k_g z - \omega t\right) \qquad (26\text{-}56)$$

$$H_x = H_0 k_g\left(\frac{a}{\pi}\right)\sin\left(\frac{\pi x}{a}\right)\sin\left(k_g z - \omega t\right) \qquad (26\text{-}57)$$

$$H_z = H_0\cos\left(\frac{\pi x}{a}\right)\cos\left(k_g z - \omega t\right) \qquad (26\text{-}58)$$

We see that the values of E_y are independent of y; hence the electric fields are straight lines with constant magnitude at a given value of x but with a magnitude that does vary with x and is a maximum at the center where $x = \frac{1}{2}a$. The lines of H_x are straight with their maximum value at the center as well. The value of H_z, on the other hand, is zero in the center and has opposite signs on the two sides of the center. In Figure 26-3, we show E_y and H_x in the xy plane *at a given time* and where it is also assumed that H_0 and $\sin(k_g z - \omega t)$ are both positive. The solid lines are those of E_y while H_x is shown by the dashed lines; the arrows underneath indicate a variation of H_x with position. If we assume that we are at a position z where $\cos(k_g z - \omega t)$ is also positive, then, as also indicated in the figure, the lines of H_z as found from (26-58) will be out of the page in the left half of the figure and into the page in the right half since the z axis is out of the page. At a distance of half a guide wavelength further down the guide, $k_g z - \omega t$ will have changed by $k_g \Delta z = (2\pi/\lambda_g)(\lambda_g/2) = \pi$ so that the directions of all of the fields in Figure 26-3 will be reversed as is seen from (26-56) through (26-58).

Since E_y and H_x are each proportional to $\sin(k_g z - \omega t)$ while H_z varies as $\cos(k_g z - \omega t)$, E_y and H_x will both be zero when $|H_z|$ has a maximum. Similarly, $|E_y|$ and $|H_x|$ have maximum values when $H_z = 0$. The relationships of the fields at a given time are also illustrated in Figure 26-4, which shows a section of the guide parallel to the xz plane; the y axis is out of the page. In this case, the dashed lines show the resultant vector **H** in this plane; the situation depicted in Figure 26-3 could then correspond, for example, to a plane perpendicular to the z axis at the location indicated by the arrows labeled A.

Since our fields are actually waves, their behavior in time can be visualized by imagining that one is located at a given point and that these figures are traveling by in the positive z direction. They will be moving with the guide speed v_g found from (26-52) to be

$$v_g = \frac{\omega}{k_g} = \frac{v}{\left[1 - (\pi v/a\omega)^2\right]^{1/2}} \qquad (26\text{-}59)$$

from which it follows that $v_g > v$, the speed of a plane wave in the material within the guide; for a vacuum, we would have $v_g > c$.

Equations 26-57 and 26-58 show that H_x and H_z generally will have different amplitudes at a given location in the guide. If we recall our discussion in Section 24-7, we see that this means that the total magnetic field **H** will be elliptically polarized when regarded as a function of time.

Since there are nonzero normal components of **D** and tangential components of **H**, there must be free surface charges and currents on the walls of the guide according to (26-2), and these relations can be used to calculate them from our results. For example,

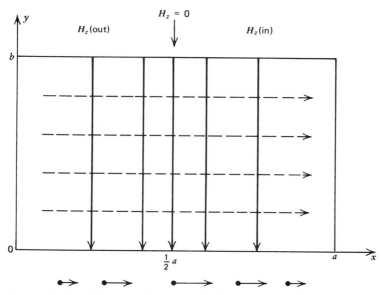

Figure 26-3. Fields of a TE_{10} mode in the cross section at a given time.

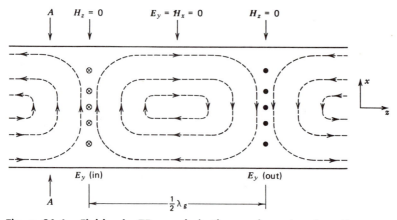

Figure 26-4. Fields of a TE_{10} mode in the xz plane at a given time.

on the face $y = 0$, the outward normal to the conductor is $\hat{\mathbf{n}} = \hat{\mathbf{y}}$ and we find that

$$\sigma_f = \hat{\mathbf{n}} \cdot \mathbf{D} = \epsilon E_y = -H_0 \omega \mu \epsilon \left(\frac{a}{\pi} \right) \sin \left(\frac{\pi x}{a} \right) \sin \left(k_g z - \omega t \right) \qquad (26\text{-}60)$$

while on the face $y = b$ where $\hat{\mathbf{n}} = -\hat{\mathbf{y}}$, σ_f has the same value but is of opposite sign. There are no surface charges on the faces perpendicular to the x axis since $E_x = 0$. We see from (26-60) that σ_f oscillates in time, and, in particular, its sign changes. Physically, this occurs because of the flow of free charge on the surfaces, that is, there must be surface currents. These can be calculated in a similar manner from $\mathbf{K}_f = \hat{\mathbf{n}} \times \mathbf{H}$ and (26-57) and (26-58). ∎

■ **Example**

TM modes. In this case, we set $\mathscr{H}_z = 0$ and set $\mathscr{E}_z$ equal to the expression for ψ_0 given in (26-33); hence (26-32) is again applicable. This case is actually simpler because $\mathscr{E}_z$

can be a tangential component and must vanish for $x = 0$ and a and $y = 0$ and b. Proceeding in the same manner as before, we see that we must now have $C_2 = C_4 = 0$ while k_1, k_2, and k_c are once again given by (26-42) and (26-43). Thus the TE and TM modes of a rectangular wave guide have the same set of cutoff wavelengths; the field configurations can be expected to be different however. Setting $C_1 C_3 = E_0$, we find that (26-33) gives the starting point for the TM calculation to be

$$\mathscr{E}_z = E_0 \sin\left(\frac{m\pi x}{a}\right) \sin\left(\frac{n\pi y}{b}\right) \tag{26-61}$$

which can now be used to calculate the rest of the field amplitudes by means of (26-25) through (26-28). We note that $m = n = 0$ makes $\mathscr{E}_z$, and then all of the other field components, zero; thus there is no TM_{00} mode. Furthermore, if $m = 0$ or $n = 0$, $\mathscr{E}_z = 0$ and all of the fields are zero. Thus, it is not possible to have a TM_{m0} or TM_{0n} mode, in contrast to the TE case. ∎

It is possible, of course, to have wave guides with cross-sectional shapes that are not rectangles; for example, ones that are circular. Such cases can be discussed in the same systematic manner that we have illustrated for the rectangular guide. The mathematical details are different (and generally more difficult) but the *principles* are exactly the same. We will not do this, however, but instead consider a different type of mode.

26-5 TEM WAVES

As we noted in the paragraph following (26-30), a TEM mode is one for which both $\mathscr{E}_z$ and $\mathscr{H}_z$ are zero. Looking at (26-25) through (26-28), we are tempted to conclude that all of the components are also zero so that such a mode is not possible. We did, however, divide through by k_c^2 to get these expressions, so it is best to go back to Maxwell's equations themselves as given by (26-17) through (26-24). Setting $\mathscr{E}_z = 0$ and $\mathscr{H}_z = 0$ in them, we find that they reduce to

$$\frac{\partial \mathscr{E}_x}{\partial x} + \frac{\partial \mathscr{E}_y}{\partial y} = 0 \qquad \frac{\partial \mathscr{E}_y}{\partial x} - \frac{\partial \mathscr{E}_x}{\partial y} = 0 \tag{26-62}$$

$$\frac{\partial \mathscr{H}_y}{\partial x} - \frac{\partial \mathscr{H}_x}{\partial y} = 0 \qquad \frac{\partial \mathscr{H}_x}{\partial x} + \frac{\partial \mathscr{H}_y}{\partial y} = 0 \tag{26-63}$$

$$\mathscr{H}_x = -\frac{k_g}{\omega\mu}\mathscr{E}_y = -\frac{\omega\epsilon}{k_g}\mathscr{E}_y \tag{26-64}$$

$$\mathscr{H}_y = \frac{k_g}{\omega\mu}\mathscr{E}_x = \frac{\omega\epsilon}{k_g}\mathscr{E}_x \tag{26-65}$$

From both (26-64) and (26-65), we find that $k_g/\omega\mu = \omega\epsilon/k_g$, so that

$$k_g^2 = \omega^2 \mu\epsilon = \frac{\omega^2}{v^2} = k_0^2 \tag{26-66}$$

by (26-8) and (24-12); then (26-7) show that $k_c^2 = 0$. Since $k_g = k_0$, the TEM wave will propagate along the guide with the same speed v as a plane wave would; since $k_c = 0$, the cutoff wavelength is infinite and a TEM wave can have any frequency or wavelength.

Because $\mathscr{E}$ only has components transverse to the direction of propagation, it has the form $\mathscr{E} = \mathscr{E}_x \hat{\mathbf{x}} + \mathscr{E}_y \hat{\mathbf{y}}$ and there will be a similar expression for $\mathscr{H}$; then (26-64) and

(26-65) can also be written simply as

$$\mathcal{H} = \frac{k_g}{\omega\mu}\hat{\imath} \times \mathcal{E} = \frac{\hat{\imath} \times \mathcal{E}}{Z} \tag{26-67}$$

where $Z = (\mu/\epsilon)^{1/2}$ is the plane wave impedance. Comparing (26-67) with (24-94), we see that a TEM wave has characteristics similar to those of an unbounded plane wave. For these reasons, TEM waves are also known as *principal modes*.

If we substitute (26-64) and (26-65) into the first equation of (26-63), we find that it gives the first equation of (26-62); the same is true for the second equation in each set. In other words, (26-63) will be satisfied if (26-62) is. Recalling that $\mathcal{E}_x$ and $\mathcal{E}_y$ are functions only of x and y by (26-14), we introduce a scalar function $\phi(x, y)$, and define $\mathcal{E}_x = -\partial\phi/\partial x$ and $\mathcal{E}_y = -\partial\phi/\partial y$, that is,

$$\mathcal{E} = -\nabla\phi \tag{26-68}$$

We then find that the second equation of (26-62) is satisfied because $\partial^2\phi/\partial y\,\partial x = \partial^2\phi/\partial x\,\partial y$ while the first one becomes

$$\frac{\partial^2\phi}{\partial x^2} + \frac{\partial^2\phi}{\partial y^2} = 0 \tag{26-69}$$

which is just the two-dimensional form of Laplace's equation. Therefore, if we take *any* solution of the appropriate two-dimensional *electrostatic* potential problem that gives an electric field normal to the perfectly conducting bounding surface, call this field $\mathcal{E}$, and then find $\mathcal{H}$ from (26-67), we can use the results in (26-14) and (26-15) to give a possible TEM mode for the system. The two-dimensional field pattern obtained in this way will then travel down the guide with the plane wave speed v characteristic of the medium.

Things are somewhat more complicated, however. If $\mathcal{E}$ is normal to the surface, then the gradient of the scalar potential ϕ is also normal to the surface by (26-68), and we saw in Section 1-9 that then ϕ must be constant on the surface as illustrated in Figure 1-18. In other words, the conducting surface must be an equipotential surface. But we found in Section 11-1 that a solution of Laplace's equation, which has the same value on *all* points of the boundary, has the same *constant* value everywhere within the region. If this is the case, $\nabla\phi = 0$ so that $\mathcal{E} = 0$ and $\mathcal{H} = 0$ by (26-68) and (26-67) and there will be no TEM wave after all. Thus a hollow pipe type of wave guide as shown in Figure 26-1 cannot have a TEM mode.

There can be a TEM mode, however, if the guide consists of at least *two* conductors such as the two parallel wires of Figure 11-10 or the coaxial cylinders of Figure 6-12. The reason is that, with there being more than one part to the bounding surface, the potential ϕ can have a constant value on one part of the surface and a *different* constant value on the other portion or portions. In such a case, ϕ need not be constant within the region between the conductors and $\mathcal{E} = -\nabla\phi$ can be different from zero, thus making a TEM mode possible.

To illustrate these ideas, let us consider the simplest possible TEM mode of a very important type of guide.

■ Example

Coaxial line. This consists of two coaxial cylinders of different radii a and b shown in Figure 26-5. We use plane polar coordinates ρ and φ in the plane perpendicular to the axis of the cylinders. The general solution of (26-69) for these coordinates is given by (11-141). We consider only the case in which ϕ is independent of angle, and we see that

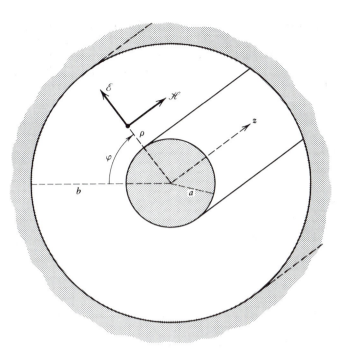

Figure 26-5. Fields in the principal mode of a coaxial line.

then

$$\phi = A + B \ln \rho \tag{26-70}$$

where A and B are constants. Inserting this into (26-68) and using (1-85), (26-67), and (1-76), we get

$$\mathscr{E} = -\frac{B}{\rho}\hat{\rho} \qquad \mathscr{H} = -\frac{B}{Z\rho}\hat{\varphi} \tag{26-71}$$

so that $\mathscr{E}$ has only a radial component while $\mathscr{H}$ has only a φ component as is shown in the figure. Inserting these results into (26-14) and (26-15), and using (26-66), we find the fields of the wave to be given by

$$\mathbf{E} = -\frac{B}{\rho}\hat{\rho}e^{i(k_0 z - \omega t)} \qquad \mathbf{H} = -\frac{B}{Z\rho}\hat{\varphi}e^{i(k_0 z - \omega t)} \tag{26-72}$$

The potential difference between the two conductors can be found from (5-11) to be

$$\Delta\phi = \int_a^b \mathbf{E} \cdot d\mathbf{s} = -B \ln\left(\frac{b}{a}\right)e^{i(k_0 z - \omega t)} = \Delta\phi_0 e^{i(k_0 z - \omega t)} \tag{26-73}$$

which has a maximum value $\Delta\phi_0 = -B \ln(b/a)$, so that (26-72) can also be written as

$$\mathbf{E} = \frac{\Delta\phi_0 e^{i(k_0 z - \omega t)}}{\rho \ln(b/a)}\hat{\rho} \tag{26-74}$$

The current on the inner conductor can be found from (21-39), if we note that when the path of integration is chosen to be in the plane of the cross section $(\partial\mathbf{D}/\partial t) \cdot d\mathbf{a} = 0$, so that

$$\oint_C \mathbf{H} \cdot d\mathbf{s} = -\frac{2\pi B}{Z}e^{i(k_0 z - \omega t)} = I_f = I_{f0}e^{i(k_0 z - \omega t)} \tag{26-75}$$

where the maximum value of the current is $I_{f0} = -2\pi B/Z$. (There will be an equal and opposite current on the surface at $\rho = b$, since, if we choose the path C to lie in the conductor for $\rho > b$, the line integral of $\mathbf{H}$ must vanish since $\mathbf{H} = 0$ within the perfect conductor, and then no net free current can be enclosed by the path.) Using (26-75), we can write the magnetic field in terms of the current as

$$\mathbf{H} = \frac{I_{f0}e^{i(k_0 z - \omega t)}}{2\pi\rho}\hat{\boldsymbol{\varphi}} \tag{26-76}$$

We see from (26-73) and (26-75) that the current and potential difference are in phase so that their ratio is constant and given by

$$\frac{\Delta\phi}{I_f} = \frac{\Delta\phi_0}{I_{f0}} = \frac{Z}{2\pi}\ln\left(\frac{b}{a}\right) = Z_c \tag{26-77}$$

where Z_c is called the characteristic impedance of the coaxial line. ∎

26-6 RESONANT CAVITIES

Suppose that we were to take a portion of length L of the wave guide of Figure 26-1 and cover the open ends with perfect conductors. In this way, we would obtain a volume enclosed by perfectly conducting walls. We can no longer expect to find the fields in the form of traveling waves because additional boundary conditions have been introduced and, at the very least, there will be reflections at the new boundary surfaces that will arise from the necessity of satisfying the boundary conditions. Recalling Exercises 25-10 and 24-3, we would now expect the fields to be in the form of standing waves corresponding to certain definite and characteristic frequencies of oscillation. Then the total fields can be regarded as a superposition of these normal modes. (This is very analogous to the mechanical problem of the vibrating string. The introduction of perfectly rigid supports at the ends of a finite length of string makes the normal modes take the form of standing waves arising from the superposition of incident and reflected traveling waves. An arbitrary displacement of the string can then be obtained as an appropriate superposition of these standing waves.)

Systems of this type are known as *resonant cavities* or *cavity resonators*. Evidently they can be of a wide variety of shapes and sizes, but the general principles for finding the fields are exactly the same and are most easily demonstrated by means of a specific example.

Let us consider a closed region with perfectly conducting rectangular walls of sides a, b, c and with the origin at one corner as shown in Figure 26-6. The cavity is filled with a l.i.h. nonconducting material described by μ and ϵ. Each field component still satisfies the wave equation (26-3) but we can no longer assume a solution of the form (26-4). It is evident from the treatment that led to (24-16), however, that the time variation can always be separated out in the form $e^{-i\omega t}$. Therefore, instead of (26-4), we assume that ψ can be written as

$$\psi(\mathbf{r}, t) = \psi_0(\mathbf{r})e^{-i\omega t} \tag{26-78}$$

which, when substituted into (26-3), leads to the equation that ψ_0 must satisfy:

$$\nabla^2\psi_0 + k_0^2\psi_0 = 0 \tag{26-79}$$

where $k_0^2 = (\omega/v)^2$ again. (This result is actually valid for any kind of coordinate system and is often called the wave equation without the time.)

We can solve (26-79) in rectangular coordinates by separation of variables. If we write $\psi_0 = X(x)Y(y)Z(z)$, and proceed in the same manner as we went from (26-6) to

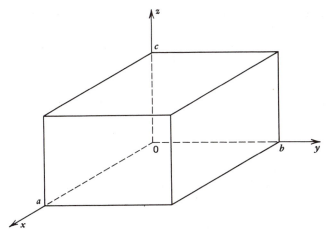

Figure 26-6. A rectangular resonant cavity.

(26-33), we find that

$$\psi_0(\mathbf{r}) = (C_1 \sin k_1 x + C_2 \cos k_1 x)(C_3 \sin k_2 y + C_4 \cos k_2 y)$$
$$\times (C_5 \sin k_3 z + C_6 \cos k_3 z) \tag{26-80}$$

where

$$k_1^2 + k_2^2 + k_3^2 = k_0^2 \tag{26-81}$$

and where the C's are constants. This can also be verified by direct substitution of (26-80) into (26-79). Combining this form with (26-78), we can write one of the components of the electric field as

$$E_x = (C_1 \sin k_1 x + C_2 \cos k_1 x)(C_3 \sin k_2 y + C_4 \cos k_2 y)$$
$$\times (C_5 \sin k_3 z + C_6 \cos k_3 z) e^{-i\omega t} \tag{26-82}$$

Now E_x will be a tangential component and must therefore vanish at the faces $y = 0$ and b and $z = 0$ and c. We see that this requires that $C_4 = C_6 = 0$ and that

$$k_2 = \frac{n\pi}{b} \qquad k_3 = \frac{p\pi}{c} \tag{26-83}$$

where n and p are integers. At this point, therefore, we have

$$E_x = (C_1' \sin k_1 x + C_2' \cos k_1 x) \sin k_2 y \sin k_3 z e^{-i\omega t} \tag{26-84}$$

where $C_1' = C_1 C_3 C_5$ and $C_2' = C_2 C_3 C_5$. Similarly, we find that

$$E_y = \sin k_1 x (C_3' \sin k_2 y + C_4' \cos k_2 y) \sin k_3 z e^{-i\omega t} \tag{26-85}$$

$$E_z = \sin k_1 x \sin k_2 y (C_5' \sin k_3 z + C_6' \cos k_3 z) e^{-i\omega t} \tag{26-86}$$

and

$$k_1 = \frac{m\pi}{a} \tag{26-87}$$

Substituting (26-83) and (26-87) into (26-81), and using (26-8), we find that the possible frequencies of oscillation in this cavity are given by

$$\left(\frac{\omega}{v} \right)^2 = \pi^2 \left[\left(\frac{m}{a} \right)^2 + \left(\frac{n}{b} \right)^2 + \left(\frac{p}{c} \right)^2 \right] \tag{26-88}$$

We note that if any two of the integers m, n, p are zero, then all of the components of $\mathbf{E}$ will be zero, as will all components of $\mathbf{H}$ as well.

The electric field must still satisfy Maxwell's equations and, in particular, we must have $\nabla \cdot \mathbf{E} = 0$. When (26-84) through (26-86) are substituted into this, we obtain

$$-(k_1 C_2' + k_2 C_4' + k_3 C_6') \sin k_1 x \sin k_2 y \sin k_3 z$$
$$+ \left[(k_1 C_1' \cos k_1 x \sin k_2 y \sin k_3 z) + (k_2 C_3' \sin k_1 x \cos k_2 y \sin k_3 z) \right.$$
$$\left. + (k_3 C_5' \sin k_1 x \sin k_2 y \cos k_3 z) \right] = 0 \quad (26\text{-}89)$$

A little thought will convince one that we cannot possibly have this expression always equal to zero for *all* values of x, y, and z (which can be chosen independently) unless $C_1' = C_3' = C_5' = 0$ and $k_1 C_2' + k_2 C_4' + k_3 C_6' = 0$. If we now let $C_2' = E_1$, $C_4' = E_2$, and $C_6' = E_3$, we find that the last condition can be written as

$$k_1 E_1 + k_2 E_2 + k_3 E_3 = 0 \quad (26\text{-}90)$$

while the expressions for the field components finally become

$$E_x = E_1 \cos k_1 x \sin k_2 y \sin k_3 z e^{-i\omega t} \quad (26\text{-}91)$$
$$E_y = E_2 \sin k_1 x \cos k_2 y \sin k_3 z e^{-i\omega t} \quad (26\text{-}92)$$
$$E_z = E_3 \sin k_1 x \sin k_2 y \cos k_3 z e^{-i\omega t} \quad (26\text{-}93)$$

so that E_1, E_2, and E_3 are seen to be the maximum values of the respective components.

If we define a vector $\mathbf{k}$ with components k_1, k_2, and k_3, and a vector $\mathbf{E}_0$ with components E_1, E_2, and E_3, (26-90) can be written as

$$\mathbf{k} \cdot \mathbf{E}_0 = 0 \quad (26\text{-}94)$$

which is very much like the result found for a plane wave in (24-91). Thus, for a given mode, $\mathbf{E}_0$ must be perpendicular to the vector $\mathbf{k} = (m\pi/a)\hat{\mathbf{x}} + (n\pi/b)\hat{\mathbf{y}} + (p\pi/c)\hat{\mathbf{z}}$.

The magnetic field can be found from $\nabla \times \mathbf{E} = -\mu(\partial \mathbf{H}/\partial t) = i\omega\mu\mathbf{H}$. For example, we get

$$i\omega\mu H_x = \frac{\partial E_z}{\partial y} - \frac{\partial E_y}{\partial z} = (k_2 E_3 - k_3 E_2) \sin k_1 x \cos k_2 y \cos k_3 z e^{-i\omega t} \quad (26\text{-}95)$$

Since $k_2 E_3 - k_3 E_2$ is the x component of $\mathbf{k} \times \mathbf{E}_0$, it is desirable to define a vector $\mathbf{H}_0$ by

$$\mathbf{H}_0 = \frac{1}{\omega\mu} \mathbf{k} \times \mathbf{E}_0 \quad (26\text{-}96)$$

for, if we let its rectangular components be H_1, H_2, and H_3, then we can write (26-95) as

$$H_x = -iH_1 \sin k_1 x \cos k_2 y \cos k_3 z e^{-i\omega t} \quad (26\text{-}97)$$

Again we note the similarity of (26-96) to the plane wave result given by the first expression of (24-93). Thus we can associate a set of mutually perpendicular vectors $\mathbf{k}$, $\mathbf{E}_0$, and $\mathbf{H}_0$ with each standing wave mode of the cavity.

Similarly, we find the other two components of $\mathbf{H}$ to be

$$H_y = -iH_2 \cos k_1 x \sin k_2 y \cos k_3 z e^{-i\omega t} \quad (26\text{-}98)$$
$$H_z = -iH_3 \cos k_1 x \cos k_2 y \sin k_3 z e^{-i\omega t} \quad (26\text{-}99)$$

We see that $H_x = 0$ at $x = 0$ and $x = a$, that is, at the walls for which it is a normal component; similarly, H_y and H_z vanish at $y = 0$ and b and $z = 0$ and c, respectively. Thus, the boundary conditions on $\mathbf{H}$ have been automatically satisfied once $\mathbf{E}$ was made to satisfy its own boundary conditions. Furthermore, it is easily verified that the two remaining Maxwell equations that we have not yet used are satisfied, that is, $\nabla \cdot \mathbf{H} = 0$ and $\nabla \times \mathbf{H} = \epsilon(\partial \mathbf{E}/\partial t)$. [One needs to use $\mathbf{k} \cdot \mathbf{H}_0 = 0$ as well as (26-96), (26-94), and (26-81), the last of which can also be written as $\mathbf{k} \cdot \mathbf{k} = k_0^2 = \omega^2/v^2 = \omega^2\mu\epsilon$.]

Each component of $\mathbf{E}$ varies as $e^{-i\omega t}$, while the components of $\mathbf{H}$ are proportional to $-ie^{-i\omega t} = e^{-i[\omega t + (1/2)\pi]}$. Thus, the electric and magnetic fields are not in phase in these standing waves but instead $\mathbf{H}$ leads $\mathbf{E}$ by $90°$.

A given $\mathbf{k}$ corresponds to a given mode, that is, a given set of integers m, n, p according to (26-83) and (26-87). Now (26-94) tells us that the vector $\mathbf{E}_0$ must be chosen to be perpendicular to $\mathbf{k}$. However, there are two *independent* mutually perpendicular directions along which $\mathbf{E}_0$ can be chosen and still be perpendicular to $\mathbf{k}$. Thus, for each possible value of $\mathbf{k}$, there are two possible independent directions of polarization of $\mathbf{E}_0$, so that there are two distinct modes for each allowed frequency given by (26-88). This property is known as *degeneracy* and is a fundamental and important feature of electromagnetic standing waves. If a, b, c are all different, then the various frequencies given by (26-88) will generally be different. However, if there are simple relations among the dimensions, it is possible that different choices of the integers will give the same frequency so that we will also have degeneracy, but arising in a different manner. As an extreme example, consider a cube for which $a = b = c$ so that (26-88) reduces to

$$\left(\frac{\omega}{v}\right)^2 = \left(\frac{\pi}{a}\right)^2 (m^2 + n^2 + p^2) \tag{26-100}$$

Thus, all combinations of integers that have the same value of $m^2 + n^2 + p^2$ will have the same frequency and the modes will be degenerate.

EXERCISES

26-1 Let $\boldsymbol{\mathscr{E}}_\tau = \mathscr{E}_x \hat{\mathbf{x}} + \mathscr{E}_y \hat{\mathbf{y}}$ be the transverse component of $\boldsymbol{\mathscr{E}}$, with a similar expression for $\boldsymbol{\mathscr{H}}_\tau$. For a TE mode, show that $\boldsymbol{\mathscr{H}}_\tau = (ik_g/k_c^2)\nabla\mathscr{H}_z$, $\boldsymbol{\mathscr{H}}_\tau = (\hat{\mathbf{z}} \times \boldsymbol{\mathscr{E}}_\tau)/Z_e$ where $Z_e = (k_0/k_g)Z = (\lambda_g/\lambda_0)Z$. Find the corresponding relations for a TM mode.

26-2 Using the results of the previous exercise, show that any solution of (26-6) that vanishes on the boundary of the guide will lead to a TM mode. Similarly, show that a solution of (26-6) for which $\hat{\mathbf{n}} \cdot \nabla\psi_0 = 0$ on the boundary will lead to a TE mode.

26-3 Suppose we define a group velocity v_G for propagation along a wave guide by $v_G = d\omega/dk_g$ as in (24-144). Find v_G as a function of ω and compare with v. Show that $v_G v_g = v^2 = 1/\mu\epsilon$ and compare with the results of Exercise 24-26.

26-4 A rectangular wave guide with vacuum inside has $a = 8$ centimeters and $b = 6$ centimeters. If a wave of frequency $\nu = 4 \times 10^9$ hertz is to be propagated down the guide, what modes are possible?

26-5 Consider a wave guide of square cross section. What conditions must the edge a satisfy in order that it be possible to propagate a TE_{10} mode but not TE_{11}, TM_{11}, or higher mode?

26-6 The interior of a hollow rectangular wave guide is filled with a l.i.h. nonmagnetic nonconducting dielectric. Show that the cutoff frequencies are smaller by a factor of $\kappa_e^{1/2}$ than if the interior were a vacuum. If one wanted to build the dielectric filled guide to operate in the same manner at a given frequency as the vacuum case, that is, to keep the cutoff frequencies the same, should it be made larger or smaller and by what factor should the dimensions a and b be changed?

26-7 For a TE_{10} mode in a rectangular guide: (a) find $\langle u \rangle$; (b) find $\langle \mathbf{S} \rangle$; (c) find the *spatial* average values $\langle\langle u \rangle\rangle$ and $\langle\langle \mathbf{S} \rangle\rangle$ from integrating your previous results for the time averages over the cross section of the guide; (d) define a speed of energy flow v_U by $\langle\langle \mathbf{S} \rangle\rangle = \langle\langle u \rangle\rangle v_U \hat{\mathbf{z}}$ (is this reasonable?) and compare v_U with the group velocity found in Exercise 26-3.

26-8 Find the surface current density $\mathbf{K}_f$ for a TE_{10} mode in a rectangular guide. Sketch the current at a given time over at least one spatial period along the guide.

26-9 Consider a rectangular guide with vacuum inside operating in the TE_{10} mode. Find the time average electrical force per unit area $\langle \mathbf{f}_e \rangle$ on the face $y = 0$.

26-10 For a rectangular guide, assume H_0 to be real, and find all of the real field components for a TE_{20} mode. Repeat for a TE_{11}.

26-11 Find the general expressions for the real field components in a TM mode of a rectangular guide (assume E_0 to be real).

26-12 Find the time average Poynting vector for a general TM mode in a rectangular guide. By integrating over the cross section of the guide, find the total power transmitted along the guide.

26-13 The rectangular wave guide satisfies the conditions of Exercise 22-6 for describing an electromagnetic field completely in terms of a vector potential $\mathbf{A}$. Find $\mathbf{A}$ for a TE_{mn} mode and verify that it satisfies $\nabla \cdot \mathbf{A} = 0$.

26-14 Find that superposition of a TE_{mn} and a TM_{mn} wave in a rectangular guide that will make $\mathbf{E}$ transverse to the y direction, that is, $E_y = 0$ while all of the other components of $\mathbf{E}$ and $\mathbf{H}$ are different from zero.

26-15 Consider a cylindrical wave guide, that is, one with a circular cross section of radius a and with the z axis coinciding with the axis of the cylinder. Write (26-3) in cylindrical coordinates, assume that the dependence of ψ on z and t is given by (26-4), and find the differential equation satisfied by ψ_0. Solve this equation by separation of variables and show that ψ_0 has the form $\psi_0 = C_m J_m(k_c \rho) e^{im\varphi}$ where C_m is a constant and m is an integer. [*Hint:* ψ_0 must be a single-valued function of φ, for example, $\psi_0(\varphi + 2\pi) = \psi_0(\varphi)$.] Furthermore, $J_m(\eta)$ satisfies *Bessel's differential equation*

$$\frac{d^2 J_m}{d\eta^2} + \frac{1}{\eta} \frac{dJ_m}{d\eta} + \left(1 - \frac{m^2}{\eta^2}\right) J_m = 0$$

so that J_m is a *Bessel function*. Find the condition which determines the k_c for a TM mode corresponding to the value m.

26-16 Find $\langle \mathbf{S} \rangle$ for the mode of the coaxial line discussed in the text. Find the total power transmitted along the line and express it in a form that involves Z_c.

26-17 Find the surface charge density on the inner conductor for the mode of the coaxial line discussed in the text. Show that the surface current density on the inner conductor equals the surface charge density times the velocity of the wave.

26-18 Consider a length l of the coaxial line with inductance L and capacitance C. Show that $Z_c = (L/C)^{1/2}$. (Note that the result is independent of l; accordingly, L and C are often described as the quantities per unit length.)

26-19 Show that there is no TEM mode for a coaxial line that can be obtained from any angular dependent term of (11-141), that is, for a ϕ given by $\phi = (A_m \rho^m + B_m \rho^{-m})(C_m \cos m\varphi + D_m \sin m\varphi)$.

26-20 Show that the integers m, n, p for the rectangular cavity can be restricted to be positive or zero. [Do not forget (26-94).]

26-21 Consider a cubical cavity of edge a. Show that the lowest frequency is threefold degenerate with respect to the integers m, n, p. Find all components of $\mathbf{E}$ and $\mathbf{H}$ for the mode $m = n = 1$, $p = 0$. Sketch the field lines at a given instant. Find $\langle \mathbf{S} \rangle$ and the total electromagnetic energy within the cavity.

26-22 Consider a cylindrical cavity of length l and circular cross section of radius a. The axis of the cylinder lies along the z axis and the two faces are at $z = 0$ and $z = l$. Find the possible frequencies of oscillation for this cavity. (You do not have to find all of the field components to do this. Also see Exercise 26-15.)

26-23 The Hertz vector $\boldsymbol{\pi}_e$ was introduced in Exercise 22-7. Show that a TM mode can be described by a Hertz vector of the form $\pi_e \hat{\mathbf{z}}$ and find π_e for a rectangular guide with a vacuum inside.

27

CIRCUITS AND TRANSMISSION LINES

Up to now, our primary interest and emphasis have been on describing electromagnetic phenomena in terms of fields. Our approach has often been one of obtaining results by considering macroscopic situations and then rewriting things in field terms, for example, in the way we went from (18-6) to (18-21). On the other hand, in many practical applications, such as in communications work, the emphasis is on the macroscopic arrangements of devices such as resistors, capacitors, and inductances known as *circuits*. Here the primary interest is in the currents and potential differences (often called *voltages*) within these systems, and the field aspects are generally not considered. Because of its practical importance, the study of circuits has been highly developed, and there are many books devoted to it. Consequently our brief discussion must be very limited in scope.

Our first aim is to find appropriate macroscopic equations for these systems. As we will see, they will generally turn out to be differential equations. There are a variety of possible approaches, but the simplest and most common one is to use a set of equations known as *Kirchhoff's laws*. These basic statements about circuits actually predate Maxwell's equations and are closely related to the conservation laws for charge and energy.

It is often convenient to divide these systems into two broad classes. In one, the various circuit components can be considered to be distinct physical entities connected together by wires whose electromagnetic properties are neglected. Such systems can be described as having *lumped parameters*. In the other class, the construction is such that the parameters cannot be assigned to particular circuit portions but are generally and collectively associated with every individual part—hence, they are said to be *distributed parameters*. We will consider examples of both classes.

27-1 KIRCHHOFF'S LAWS

We begin by considering steady currents as would be appropriate for direct current (DC) circuits. We know from (12-15) that this means that $\nabla \cdot \mathbf{J} = 0$ so that

$$\oint_S \mathbf{J} \cdot d\mathbf{a} = 0 \tag{27-1}$$

by the divergence theorem. Let us apply this to the situation shown in Figure 27-1 where we assume that all of the currents $I_1, I_2, \ldots, I_n$ are filamentary, as on wires, and the closed surface S surrounds a junction P at which the currents flow toward or away. We see that, in this case, (27-1) reduces to

$$\sum_{j=1}^{n} I_j = 0 \tag{27-2}$$

when the currents are assigned appropriate algebraic signs. This result is known as *Kirchhoff's first law* or the *junction theorem*. It says that the algebraic sum of the

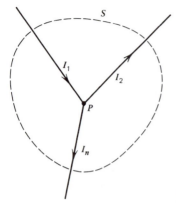

Figure 27-1. Currents at a junction.

currents at a junction is zero since charge cannot accumulate there in a steady-state situation and is clearly expressive of the conservation of charge.

If we take a charge around a complete loop in a circuit, then the net change in potential must be zero since the potential is a single-valued function and the initial and final points coincide. In other words, the sum of all of the potential changes encountered (or "voltages" V_j) must add to zero:

$$\sum_j V_j = 0 \qquad (27\text{-}3)$$

This is *Kirchhoff's second law* or *loop theorem* and is related to conservation of energy because potential differences are defined in terms of work done per unit charge.

If there is resistance in the circuit, we know from Sections 12-3 and 12-4 that there will be energy dissipated as heat and the circuit must include a nonconservative source of emf, such as a battery, to maintain the steady current.

The typical use of Kirchhoff's laws in DC circuits is to obtain enough independent equations to solve for the unknowns of the system. These are generally the currents if the emfs and resistances are given. One usually proceeds by assigning the directions of the currents in some arbitrary manner and using (27-2) and (27-3). The principal difficulties encountered are in assigning the correct signs to the voltages. This is also the main problem when the currents depend on the time and are just as easily illustrated for that.

Our discussion of time dependent currents (AC circuits) in this chapter will be restricted to "slowly" varying currents. To see what "slowly" means, consider the particular Maxwell equation (21-22):

$$\nabla \times \mathbf{H} = \mathbf{J} + \frac{\partial \mathbf{D}}{\partial t} \qquad (27\text{-}4)$$

(The subscript f has been dropped on $\mathbf{J}$ since we are only concerned with free currents anyhow.) A basic approximation in circuit theory is to neglect the displacement current $\partial \mathbf{D}/\partial t$. If we take the divergence of both sides of what remains we get $\nabla \cdot (\nabla \times \mathbf{H}) = 0 = \nabla \cdot \mathbf{J}$. Thus, in this approximation, (27-1) is still valid and we will be led again to (27-2), except that now it also holds for the instantaneous values of the time dependent currents:

$$\sum_j I_j(t) = 0 \qquad (27\text{-}5)$$

What is the physical significance of the neglect of the displacement current? As we will see in the next chapter, the inclusion of this term leads to radiation of electromagnetic waves which, from our present point of view, is an additional energy loss that we are neglecting. It will be shown that this means that we are assuming that $l \ll \lambda$ where l is the maximum linear dimension of the system and λ the wavelength. Consequently, any two equal and opposite current elements $\pm I\,d\mathbf{s}$ will have approximately the same phase since they are much less than a wavelength apart and, at large distances, their fields will cancel, that is there will be no radiation and associated energy loss. From another point of view, it means that the circuit is so small in physical size that the time needed for the propagation of electromagnetic signals can be neglected.

In order to evaluate these limitations on the theory, we can calculate the wavelength from $\lambda = c/\nu$. We find that for a "power" frequency of 60 hertz, and for a typical radio frequency of 10^6 hertz, the corresponding wavelengths are 5×10^6 meters and 300 meters. However, at a microwave frequency of 10^{10} hertz, the wavelength is only 3 centimeters and here we clearly have to be concerned. Thus, we conclude that standard AC circuit analysis cannot be slavishly used at very high frequencies but it certainly should be adequate for power and most communications systems.

The principal effect of time dependence on Kirchhoff's second law is the existence of induced emfs as described by Faraday's law (17-3), namely $\mathscr{E}_{\text{ind}} = -d\Phi/dt$. Thus we will be dealing with four sources of voltages: nonconservative emfs arising from generators, power supplies, or batteries; potential differences across resistors; potential differences across capacitors; and induced emfs associated with mutual or self-inductance. Since we are neglecting propagation times, all quantities can be evaluated throughout the circuit at the same time. The fact that there will be no net change of potential when a charge is taken around a complete loop will then still hold and we are led once more to (27-3), but now it applies to instantaneous values:

$$\sum_j V_j(t) = 0 \qquad (27\text{-}6)$$

The problem of giving the correct signs to the terms in (27-6) is best handled by assuming a direction for each current involved, a sign for each dI/dt, and a sense of traversal around the loop. If one does this, and is consistent with the assumptions made, then careful attention to the situation at each element will enable one to evaluate the voltages involved in an unambiguous way. Let us illustrate all this with an example.

■ **Example**

RL circuit. Figure 27-2 shows a battery of emf $\mathscr{E}$, a resistance R, and an inductance L connected in series. The circuit also contains a switch S. To discuss the general case, we assume that the switch is closed, there is a current I in the sense shown, and $dI/dt > 0$. Let us imagine starting at P and traversing the loop in the same sense as I. We know from (12-27) that the magnitude of the potential difference across R is IR, and from (12-25) that the direction of $\mathbf{E}$ is in the same sense as I. Therefore, the potential is

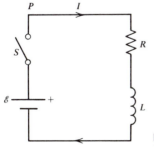

Figure 27-2.   An *RL* circuit.

decreasing as we traverse R, so that the change, found from the final value minus the initial value, will be negative; this gives the value $V_R = -IR$ for the voltage. We know from Lenz' law that the emf induced in L has a sense that tends to oppose the change. Because we have assumed that $dI/dt > 0$, the voltage V_L will tend to decrease I and hence will have a sense opposite to I; thus, with the use of (17-57), we conclude that $V_L = -L(dI/dt)$. The battery terminal marked $+$ is, by convention, the one of higher potential. This means that a charge passing through it has work done on it, thereby increasing its potential. Thus, the corresponding voltage is $V_B = \mathscr{E}$. Inserting these results into (27-6), we obtain $V_R + V_L + V_B = 0 = -IR - L(dI/dt) + \mathscr{E}$, or

$$L\frac{dI}{dt} + RI = \mathscr{E} \tag{27-7}$$

This first-order differential equation with constant coefficients can be used to find the current as a function of time, once the initial conditions are known.

The general solution of the homogeneous form of (27-7), that is, with zero on the right-hand side, is $I = Ke^{-Rt/L}$ where K is a constant of integration. A special solution of the inhomogeneous equation, that is, (27-7) as it stands, is clearly $I = \mathscr{E}/R$. Thus, the general solution, which is the sum of these two, is

$$I(t) = \frac{\mathscr{E}}{R} + Ke^{-Rt/L} \tag{27-8}$$

All that remains is the determination of K.

For example, let us assume that the switch S is initially open and, at $t = 0$, it is suddenly closed so that the emf $\mathscr{E}$ is instantaneously applied to the circuit. The initial condition for the current is then $I(0) = 0$. When this is substituted into (27-8), we find that $K = -\mathscr{E}/R$, so that the current at any subsequent time is

$$I(t) = \frac{\mathscr{E}}{R}(1 - e^{-Rt/L}) \tag{27-9}$$

Figure 27-3 shows the current as a function of time; the dimensionless variable t/τ is used where $\tau = L/R$ is called the *time constant* or *relaxation time* of this system. (We have already found this value of τ in Exercise 18-5 where a different initial condition was used.) We note that as $t \to \infty$, the current approaches the final steady-state value

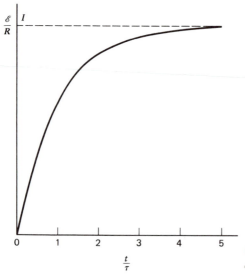

Figure 27-3. Increasing current in an *RL* circuit as a function of time.

of $\mathscr{E}/R$. Under these conditions, $dI/dt \to 0$ and hence $V_L \to 0$ so that there is no voltage across the inductance and the current is determined solely by the resistance of the circuit.   ∎

27-2 THE SERIES RLC CIRCUIT

We now consider the more elaborate circuit shown in Figure 27-4. A capacitance C has been added, and now the applied emf can be a function of the time so that $\mathscr{E} = \mathscr{E}(t)$. As before, we assume that I has the sense shown and that $dI/dt > 0$. We will still have $V_R = -IR$, $V_L = -L(dI/dt)$, while we can also write the "generator" voltage as $V_G = \mathscr{E}(t)$. Assuming the charges on the capacitor plates to be $\pm q$ as shown, and taking q to be positive, we find from (6-28) that $V_C = \Delta\phi = -q/C$. (We note that this will be correct in magnitude and sign even if q actually turns out to be negative.) Inserting these into (27-6), we get $V_R + V_L + V_C + V_G = 0 = -IR - L(dI/dt) - (q/C) + \mathscr{E}$, so that

$$L\frac{dI}{dt} + RI + \frac{q}{C} = \mathscr{E}(t) \tag{27-10}$$

Since we are primarily interested in the current, we can differentiate this with respect to the time, and use $I = dq/dt$ to obtain

$$L\frac{d^2I}{dt^2} + R\frac{dI}{dt} + \frac{I}{C} = \frac{d\mathscr{E}}{dt} \tag{27-11}$$

Thus, the application of Kirchhoff's second law has resulted in a second-order differential equation with constant coefficients.

We note that (27-11) is much like the mechanical equation of motion of a damped, forced, simple harmonic oscillator. Consequently, we can expect to get analogous behavior. For example, we recall that if we suddenly apply a periodic force to such an oscillator, the initial response (displacement) will be nonperiodic. However, after a long time, the displacement is found to be periodic in time with the same frequency as the applied force. The nonperiodic part of the solution that is eventually damped out is called a *transient*, while the periodic part that persists is called the *steady state*. It will be convenient to look at these two types of behavior separately.

1. Transient Response

A transient (damped) response can arise in general for any form of $\mathscr{E}(t)$, not only for a periodic one, since it corresponds to the solution of the homogeneous form of (27-11).

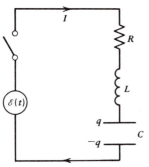

Figure 27-4.   A series *RLC* circuit.

As a simple illustration of such a case, let us assume that the external emf is provided by a battery so that $\mathcal{E} = \text{const.}$; then $d\mathcal{E}/dt = 0$ and (27-11) becomes

$$L\frac{d^2I}{dt^2} + R\frac{dI}{dt} + \frac{I}{C} = 0 \qquad (27\text{-}12)$$

The standard procedure is to look for a solution of exponential form; therefore we try a solution $I = ae^{\gamma t}$ where a and γ are constants. When this is substituted into (27-12), it gives $[L\gamma^2 + R\gamma + (1/C)]ae^{\gamma t} = 0$. Since we need $a \neq 0$ in order to have a current at all, the term in brackets must be zero. This requirement determines γ that is found to be

$$\gamma = -\frac{R}{2L} \pm \left[\left(\frac{R}{2L}\right)^2 - \frac{1}{LC}\right]^{1/2} = -\frac{R}{2L} \pm \delta \qquad (27\text{-}13)$$

so that there are two possible values of γ. We write them as γ_+ and γ_- corresponding, respectively, to the plus and minus signs in (27-13). For each possible γ, there will be an associated constant a, so that the general solution of (27-12) has the form

$$I(t) = a_+ e^{\gamma_+ t} + a_- e^{\gamma_- t}$$
$$= \left(a_+ e^{\delta t} + a_- e^{-\delta t}\right) e^{-Rt/2L} \qquad (27\text{-}14)$$

The only remaining unknowns are the constants a_+ and a_- that can be found from the initial conditions.

As an example, let us assume that, at $t = 0$, the capacitor is uncharged and the previously open switch is quickly closed. The corresponding initial conditions are that $I = 0$ and $q = 0$. First we see that $I(0) = 0 = a_+ + a_-$. We can get another independent equation from (27-10) and we find that $L(dI/dt)_{t=0} = \mathcal{E}$. When (27-14) is differentiated, evaluated at $t = 0$, and combined with the last result, we obtain $L(a_+\gamma_+ + a_-\gamma_-) = \mathcal{E}$. Solving the two equations for the a's, we find that $a_+ = -a_- = \mathcal{E}/[L(\gamma_+ - \gamma_-)] = \mathcal{E}/2L\delta$ and therefore

$$I(t) = \frac{\mathcal{E}}{2L\delta}\left(e^{\delta t} - e^{-\delta t}\right) e^{-Rt/2L} \qquad (27\text{-}15)$$

The nature of this solution now depends on δ that in turn depends on the relative values of R, L, and C.

For example, a common situation corresponds to R being very small or zero. In this case, the quantity under the square root in (27-13) will be negative, δ will be imaginary, and it is more convenient to set $\delta = i\omega_n$ where

$$\omega_n = \left[\frac{1}{LC} - \left(\frac{R}{2L}\right)^2\right]^{1/2} \qquad (27\text{-}16)$$

Then (27-15) becomes

$$I(t) = \frac{\mathcal{E}}{L\omega_n} e^{-Rt/2L} \sin\omega_n t \qquad (27\text{-}17)$$

This is a current that oscillates with the natural circular frequency ω_n but with an amplitude that decreases exponentially with time. Figure 27-5 shows this current as a function of time. This behavior clearly corresponds to the "underdamped" case for a mechanical system. Similarly, situations can be found corresponding to the overdamped and critically damped cases as will be shown in the exercises.

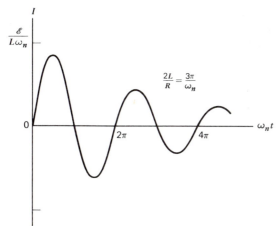

Figure 27-5. Underdamped current as a function of time.

2. Steady-state Response

Now let us assume that the applied emf is itself an oscillating function of the time and, in particular, has the form

$$\mathcal{E}(t) = \mathcal{E}_0 \cos \omega t \qquad (27\text{-}18)$$

where $\mathcal{E}_0 = $ const. We also assume that the switch in Figure 27-4 has been closed for a long time before $t = 0$ so that any transient currents will have long since been damped out. Now we want to find a special solution of the inhomogeneous form of (27-11).

We note that we can write $\mathcal{E} = \text{Re}[\mathcal{E}_0 e^{i\omega t}]$ and, since (27-11) is linear and has real coefficients, the real and imaginary parts of any solution will separately be solutions of the differential equation. Hence, as in Section 24-2, it will be convenient to deal with complex quantities and, by convention, take the real part to be the physical current. Accordingly, we write the emf as

$$\mathcal{E}(t) = \mathcal{E}_0 e^{i\omega t} \qquad (27\text{-}19)$$

and assume $\mathcal{E}_0$ to be real. (A sinusoidal time dependence written as $e^{i\omega t}$ is general usage in circuit theory in contrast to $e^{-i\omega t}$ that is common practice in the rest of physics. Therefore we will also use this notation, but only in this chapter.) We now try a solution of the form $I = I_0 e^{i\omega t}$ where I_0 is a constant that itself may turn out to be complex. Substituting this into (27-11) and using (27-19), we obtain

$$\left[-\omega^2 L + i\omega R + (1/C)\right] I_0 e^{i\omega t} = i\omega \mathcal{E}_0 e^{i\omega t}$$

so that, upon canceling the common factor $e^{i\omega t}$, we find that we can write I_0 in the form

$$I_0 = \frac{\mathcal{E}_0}{Z} \qquad (27\text{-}20)$$

where

$$Z = R + i\left(\omega L - \frac{1}{\omega C}\right) = R + iX \qquad (27\text{-}21)$$

Z is called the *impedance* and X the *reactance*. If we also write $X = X_L + X_C$, then $X_L = \omega L$ and $X_C = -1/\omega C$ are the inductive reactance and capacitive reactance, respectively.

The physical significance of a complex impedance is that the steady-state current is not in phase with the applied emf. We can show this explicitly by writing $Z = R + iX = |Z|e^{i\vartheta}$ and, by equating real and imaginary parts, we find that

$$|Z| = (R^2 + X^2)^{1/2} = \left[R^2 + \left(\omega L - \frac{1}{\omega C}\right)^2\right]^{1/2} \tag{27-22}$$

$$\tan \vartheta = \frac{X}{R} = \frac{\omega L - (1/\omega C)}{R} \tag{27-23}$$

and therefore

$$I(t) = I_0 e^{i\omega t} = \frac{\mathscr{E}_0}{|Z|} e^{i(\omega t - \vartheta)} \tag{27-24}$$

so that the real quantities are

$$\mathscr{E}(t) = \mathscr{E}_0 \cos \omega t \qquad I(t) = \frac{\mathscr{E}_0}{|Z|} \cos(\omega t - \vartheta) \tag{27-25}$$

We see that, if $\vartheta > 0$, then the current lags the emf in time, whereas it is in phase with or leads if $\vartheta = 0$ or $\vartheta < 0$, respectively. These cases correspond to X_L being greater than, equal to, or less than X_C. For given circuit parameters, we see from (27-23) that $\vartheta = 0$ for the frequency

$$\omega_0 = \frac{1}{\sqrt{LC}} \tag{27-26}$$

Although this is not the same as the natural frequency ω_n that we found in (27-16), the two will be very nearly the same if the resistance is small enough.

From the form of (27-20), we see that impedance is a generalization of resistance. However, the impedance is a function of the applied frequency and hence is not completely characterized by the electromagnetic parameters of the system. This means that the steady-state current will have different values depending on the frequency. The frequency for which the current has its maximum value is the *resonance frequency* and corresponds to the minimum value of $|Z|$. We see from (27-22) that this corresponds to $X = 0$, which shows that the resonance frequency is exactly ω_0 that we previously found to correspond to the current being in phase with the applied emf. At resonance, the current has the simple value of $\mathscr{E}/R$.

The resistance determines the "sharpness" of the resonance, that is, how quickly the current decreases from its value at resonance as the frequency is changed from ω_0. It is convenient to describe this by introducing the dimensionless quantity Q defined as the ratio of the inductive reactance at resonance to the resistance:

$$Q = \frac{\omega_0 L}{R} = \frac{1}{R}\left(\frac{L}{C}\right)^{1/2} \tag{27-27}$$

The impedance magnitude can then be written as

$$|Z| = R\left[1 + \left(\frac{Q\omega}{\omega_0}\right)^2\left(1 - \frac{\omega_0^2}{\omega^2}\right)^2\right]^{1/2} \tag{27-28}$$

Figure 27-6 shows the current amplitude as a function of ω/ω_0 for a few selected values of Q. One sees quite clearly how the sharpness of the resonance is decreased as the resistance is increased in order to make Q smaller. If Q is large enough, we can get an approximate relation between it and the broadness of the resonance curve. Because (27-28) is fairly complicated, it will suffice simply to ask at what frequency will the

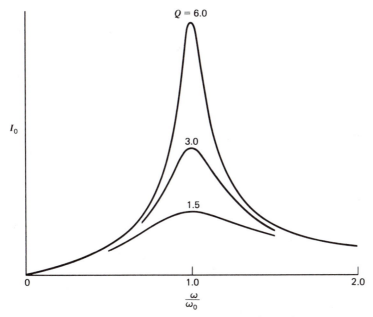

Figure 27-6. Resonance curves for various values of Q.

second term in the square root equal unity—then the current will be down by a factor $\sqrt{2}$ from its maximum. Setting $(Q\omega/\omega_0)^2[1 - (\omega_0^2/\omega^2)^2] = 1$, we can solve for $(\omega/\omega_0)^2$, and we find that $(\omega/\omega_0)^2 = 1 + (1/2Q^2) \pm (1/Q)[1 + (1/4Q^2)]^{1/2}$ which, for Q large enough, gives $(\omega/\omega_0)^2 \simeq 1 \pm (1/Q)$. If we introduce the magnitude of the corresponding frequency change $|\Delta\omega|$, assume that the resonance is fairly sharp, and use the plus sign, we get

$$\left(\frac{\omega}{\omega_0}\right)^2 = \left(\frac{\omega_0 + |\Delta\omega|}{\omega_0}\right)^2 \simeq 1 + \frac{2|\Delta\omega|}{\omega_0} = 1 + \frac{1}{Q}$$

so that

$$Q \simeq \frac{\omega_0}{2|\Delta\omega|} \qquad (27\text{-}29)$$

Now $2|\Delta\omega|$ can be taken as an approximation to the total "width" of the resonance curve so that we get this simple relation between Q and the width: the larger the Q of a circuit is the sharper will be the resonance and conversely. (We will leave to an exercise the question of how this Q is similar to the Q we introduced in Section 24-3.)

The most general solution for the current when the emf is oscillatory will be given by the sum of the solutions to the homogeneous and inhomogeneous forms of the basic differential equation; in other words, $I(t)$ will be given by the sum of (27-14) and (27-24). The constants of integration a_+ and a_- will have to be evaluated from the applicable initial conditions and, of course, will not necessarily have the values which led to (27-15).

27-3 MORE COMPLICATED SITUATIONS

Impedance as a generalization of resistance has proved to be a very useful concept. As might be expected, it is possible to combine impedances much as can be done with resistances and the results are quite analogous.

In Figure 27-7a, we have two impedances connected in series so that the same current I passes through them. By (27-20), the individual voltages will be $V_1 = Z_1 I$ and $V_2 = Z_2 I$ so that the total voltage across the combination is $V = V_1 + V_2 = (Z_1 + Z_2)I = Z_{\text{eff}} I$ where $Z_{\text{eff}} = Z_1 + Z_2$. Thus, the combination is equivalent to a single impedance of value Z_{eff}. This can clearly be generalized to more than two, so that the equivalent impedance of a group connected in series is

$$Z_s = Z_1 + Z_2 + Z_3 + \dots \tag{27-30}$$

We can analyze the parallel combination shown in Figure 27-7b in a similar way. The voltages V_1 and V_2 are equal to the same value V by (27-6) so that the individual currents are $I_1 = V_1/Z_1 = V/Z_1$ and $I_2 = V_2/Z_2 = V/Z_2$. The total current I as found from (27-5) is $I = I_1 + I_2 = [(1/Z_1) + (1/Z_2)]V = V/Z_{\text{eff}}$. Thus, for parallel connections, the impedances add in reciprocals, and the effective impedance is given by

$$\frac{1}{Z_p} = \frac{1}{Z_1} + \frac{1}{Z_2} + \frac{1}{Z_3} + \dots \tag{27-31}$$

(We note that these are exactly the same combining rules found for pure resistances in Exercise 12-6.)

The only real hazard in using these results is in forgetting that impedances are complex numbers so that the real and imaginary parts are added separately. For example, (27-30) would be written as

$$Z_s = (R_1 + R_2 + R_3 + \dots) + i(X_1 + X_2 + X_3 + \dots) \tag{27-32}$$

that has the magnitude

$$|Z_s| = \left[(R_1 + R_2 + R_3 + \dots)^2 + (X_1 + X_2 + X_3 + \dots)^2 \right]^{1/2} \tag{27-33}$$

Because of (27-31), it has been found to be convenient to define the admittance Y as the reciprocal of the impedance. Thus, we have

$$Y = \frac{1}{Z} = \frac{1}{R + iX} = \frac{R - iX}{R^2 + X^2} = G - iB \tag{27-34}$$

The real and imaginary parts of the admittance, G and B, are called the conductance and susceptance, respectively. When this is done, then (27-20) can be written as

$$I_0 = Y \mathscr{E}_0 \tag{27-35}$$

which is sometimes useful.

By using these results, it is often possible to analyze more complicated circuits by a step-by-step process of combining individual parts as either series or parallel combinations.

■ **Example**

Series-parallel circuit. We want to find the effective impedance of the circuit of Figure 27-8. The resistance R_L includes the resistance of the inductance as well as any other

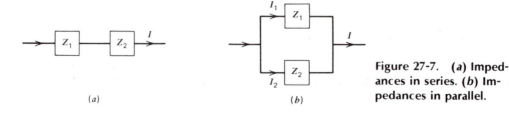

(a)

(b)

Figure 27-7. (a) Impedances in series. (b) Impedances in parallel.

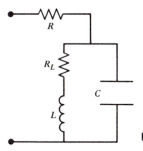

Figure 27-8. A series-parallel circuit.

resistance there might be in this branch. R_L and L are in series; they in turn are in parallel with C, and this whole combination is in series with R. By (27-21) and (27-30), the first pair have an impedance $R_L + i\omega L$. Since the impedance of C is $-i/\omega C$, we can use (27-31) and then (27-30) again to find the whole impedance of the circuit:

$$
\begin{aligned}
Z &= R + \cfrac{1}{\cfrac{1}{(R_L + i\omega L)} + \cfrac{1}{(-i/\omega C)}} \\[2mm]
&= R + \frac{(R_L + i\omega L)}{1 + i\omega C(R_L + i\omega L)} \\[2mm]
&= R + \frac{R_L + i\omega\left[L(1 - \omega^2 LC) - CR_L^2\right]}{(1 - \omega^2 LC)^2 + (\omega CR_L)^2}
\end{aligned}
\tag{27-36}
$$

When this circuit is connected to a generator, one must add the generator impedance to (27-36) to get the complete impedance of the system if one wants to find the current by (27-24).

An interesting special case occurs when R_L is small enough to be neglected. Then we find from (27-36) that

$$
|Z| = \left[R^2 + \frac{\omega^2 L^2}{(1 - \omega^2 LC)^2}\right]^{1/2}
\tag{27-37}
$$

Now, if $\omega^2 = \omega_0^2 = 1/LC$, we see that $|Z|$ becomes infinite so that the current will be *zero*. For frequencies different from this, the current will be different from zero and the curve of current as a function of frequency will be qualitatively like an inverted form of Figure 27-6. This is also a resonance phenomenon, but quite different from the previous example; this behavior is sometimes called *antiresonance*. ∎

Some circuits are deliberately constructed so that there is mutual inductance in the system, for example by winding one coil over another as in the second example of Section 17-4. Sometimes mutual inductance is present simply because of the proximity of two coils so that the flux produced by one can penetrate the turns of the other. In any case, one then has to take into account the induced voltage described by (17-50) that, in our present notation, we can write as $V_{k \to j} = -M_{jk}(dI_k/dt)$. When Kirchhoff's law (27-6) is used, these voltages must also be included. This is ordinarily no problem except for a sign ambiguity since we saw that mutual inductance can be a positive or negative quantity. What one generally needs is a knowledge of the physical arrangement of the windings so that Lenz' law can be used to find the sense of the induced voltages for the assumed current directions and signs of dI_k/dt. Once one has done all this in a consistent manner, the various loop equations can be written down and the currents calculated.

■ Example

Coupled circuits. We assume the two series circuits of Figure 27-9 to be coupled by means of the mutual inductance M of appropriate sign; for simplicity, we assume no resistance or applied emf in either circuit. (In order to have currents in such a system at all, we can assume that at one time there were emfs present but they have been switched out.) The application of Kirchhoff's law to each circuit is the same we used to get (27-10) with the addition of $V_{M1} = -M(dI_2/dt)$ and $V_{M2} = -M(dI_1/dt)$ where $M = M_{12} = M_{21}$ by (17-49). Including these in the equation that led to (27-10), and setting R and $\mathscr{E}$ equal to zero, we get two equations—one for each circuit:

$$L_1\frac{dI_1}{dt} + M\frac{dI_2}{dt} + \frac{q_1}{C_1} = 0$$

$$L_2\frac{dI_2}{dt} + M\frac{dI_1}{dt} + \frac{q_2}{C_2} = 0$$

In order to have only currents involved, we differentiate again to obtain

$$L_1\frac{d^2I_1}{dt^2} + M\frac{d^2I_2}{dt^2} + \frac{I_1}{C_1} = 0$$

$$L_2\frac{d^2I_2}{dt^2} + M\frac{d^2I_1}{dt^2} + \frac{I_2}{C_2} = 0 \qquad (27\text{-}38)$$

These equations are reminiscent of those applying to two coupled mechanical oscillators. This suggests that we try to solve them by looking for the normal modes and normal frequencies of the system, that is, solutions in which both currents oscillate with the same frequency. Accordingly, we try a solution of the form

$$I_1 = ae^{i\omega_n t} \qquad I_2 = be^{i\omega_n t} \qquad (27\text{-}39)$$

where a and b are constants. Substituting these into (27-38), we find that they give

$$\left(1 - \omega_n^2 L_1 C_1\right)a - \omega_n^2 M C_1 b = 0$$

$$-\omega_n^2 M C_2 a + \left(1 - \omega_n^2 L_2 C_2\right)b = 0 \qquad (27\text{-}40)$$

This set of homogeneous equations has a nontrivial solution only if the determinant of the coefficients is zero, that is, if

$$\left(1 - \omega_n^2 L_1 C_1\right)\left(1 - \omega_n^2 L_2 C_2\right) - \omega_n^4 M^2 C_1 C_2$$

$$= C_1 C_2\left(L_1 L_2 - M^2\right)\omega_n^4 - \left(L_1 C_1 + L_2 C_2\right)\omega_n^2 + 1 = 0 \qquad (27\text{-}41)$$

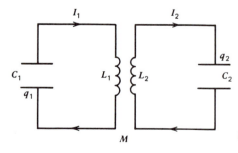

Figure 27-9. Two coupled circuits.

This is a quadratic equation in ω_n^2 and the solutions are

$$\omega_n^2 = \frac{(L_1C_1 + L_2C_2) \pm \left[(L_1C_1 + L_2C_2)^2 - 4C_1C_2(L_1L_2 - M^2)\right]^{1/2}}{2C_1C_2(L_1L_2 - M^2)} \tag{27-42}$$

The result of Exercise 17-26 shows us that $L_1L_2 - M^2 > 0$ so that there are two possible positive values of ω_n^2 that we call ω_+^2 and ω_-^2 according to which sign is used for the square root. [As a simple check on our work, we can let $M = 0$ in (27-42), and we find that $\omega_+^2 = 1/L_2C_2$ and $\omega_-^2 = 1/L_1C_1$. In other words, the normal frequencies are then just those of the individual uncoupled circuits.]

Since ω_+^2 and ω_-^2 are positive, upon taking the square roots, we find that there are four possible real values of ω_n we can use in (27-39), that is, $\pm\omega_+$ and $\pm\omega_-$. As each can have its own multiplicative constant, we see that each current can be written as the sum of four terms:

$$I_1 = a_1 e^{i\omega_+ t} + a_2 e^{-i\omega_+ t} + a_3 e^{i\omega_- t} + a_4 e^{-i\omega_- t}$$
$$I_2 = b_1 e^{i\omega_+ t} + b_2 e^{-i\omega_+ t} + b_3 e^{i\omega_- t} + b_4 e^{-i\omega_- t} \tag{27-43}$$

Thus each current generally has oscillatory components of both frequencies.

The eight constants of integration $(a_1, a_2, \ldots, b_4)$ that have appeared are not all independent as the equations (27-38) still have to be satisfied with the use of (27-43). Actually, the ratio of each corresponding pair can be found from either equation in (27-40) upon substitution of the appropriate ω_n^2. Thus, there are only four independent constants that can be evaluated from the initial values of the currents and charges. As can be expected from looking at (27-42), the algebra involved will be "straightforward but tedious," so we won't carry this further here but will leave the evaluation for a simpler case to an exercise. ∎

27-4 TRANSMISSION LINES

Up to this point, we have assumed that the circuits are small enough that the instantaneous values of the voltages and currents can be assumed to be independent of position. In many applications to power and communications systems, the circuit is so large that this is no longer a legitimate assumption and the relevant quantities do depend on position as well as on time. A typical example is shown in Figure 27-10. There are two long parallel conductors of length l. They have uniform cross section, although the cross sections need not be the same. The conductors need not be physically separated as shown but one may surround the other as in the coaxial line of Figure 18-1. At one end ($x = 0$), they are connected to an emf $\mathscr{E}(t) = \mathscr{E}_0 e^{i\omega t}$ and an impedance Z_0. At the other end ($x = l$), they are connected to a load impedance Z_l. We want to find the steady state voltage and current distributions along the line.

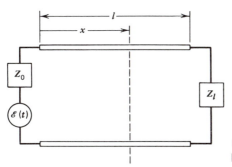

Figure 27-10. Typical example of a transmission line.

Since the voltage and current will vary with position, we cannot assume that parameters such as inductance and capacitance are localized at definite points but are generally distributed uniformly between the ends. If we let R' and L' be the resistance from all sources of loss and inductance *per unit length* of the pair, then the series impedance of a length dx will be $(R' + i\omega L')\,dx$ by (27-21). As there are two conductors involved, there will also be a capacitance C' per unit length. Furthermore, if there is imperfect insulation between the conductors, there can be a current leakage between them that we can describe by a conductance G' per unit length. In other words, we can describe the line schematically as a whole series of infinitesimal circuits as shown in Figure 27-11. Since the admittance of a capacitor is $Y_C = 1/Z_C = 1/X_C = i\omega C$ and admittances in parallel add, the (shunt) admittance between the conductors is $(G' + i\omega C')\,dx$.

Although we cannot apply Kirchhoff's laws to the line as a whole, we can for an infinitesimal portion of it. If $I(x)$ is the current at x, then the current at $x + dx$ is less than this by the current across the line that, by (27-35), is $YV(x)$ where $V(x)$ is the voltage between the conductors. Thus we get $I(x + dx) = I(x) - YV(x) = I(x) - (G' + i\omega C')V\,dx = I(x) + (\partial I/\partial x)\,dx$, so that

$$\frac{\partial I}{\partial x} = -(G' + i\omega C')V \tag{27-44}$$

Applying the loop theorem to the loop closed off by the dashed lines of the figure, we find that $-(R' + i\omega L')I\,dx - V(x + dx) + V(x) = 0$ so that

$$\frac{\partial V}{\partial x} = -(R' + i\omega L')I \tag{27-45}$$

By differentiating (27-45), and substituting from (27-44), we can get our equation involving V only:

$$\frac{\partial^2 V}{\partial x^2} - \gamma^2 V = 0 \tag{27-46}$$

where

$$\gamma = \alpha + i\beta = [(R' + i\omega L')(G' + i\omega C')]^{1/2} \tag{27-47}$$

Similarly, we find that

$$\frac{\partial^2 I}{\partial x^2} - \gamma^2 I = 0 \tag{27-48}$$

so that the voltage and current each satisfy differential equations of the same form.

Since we are looking only for steady-state solutions, we write these functions in the form

$$V(x, t) = V_0(x)e^{i\omega t} \qquad I(x, t) = I_0(x)e^{i\omega t} \tag{27-49}$$

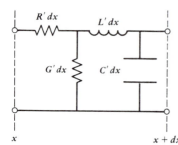

Figure 27-11. Schematic representation of an infinitesimal portion of a transmission line.

and we find that the spatially dependent amplitudes also satisfy equations of the same form:

$$\frac{d^2 V_0}{dx^2} - \gamma^2 V_0 = 0 \qquad \frac{d^2 I_0}{dx^2} - \gamma^2 I_0 = 0 \tag{27-50}$$

The general solution for I_0 can be written down at once as

$$I_0(x) = a_1 e^{\gamma x} + a_2 e^{-\gamma x} \tag{27-51}$$

where a_1 and a_2 are constants of integration. Substituting this into (27-44), while using (27-49), we find the general result for the voltage amplitude to be

$$V_0(x) = Z_i(-a_1 e^{\gamma x} + a_2 e^{-\gamma x}) \tag{27-52}$$

where

$$Z_i = \left[\frac{R' + i\omega L'}{G' + i\omega C'} \right]^{1/2} \tag{27-53}$$

is called the *characteristic impedance* of the line and will be measured in ohms.

We can easily see the physical significance of these solutions. If, for example, we combine the second part of (27-51) with (27-49), and use (27-47), we obtain $a_2 e^{i\omega t - \gamma x} = a_2 e^{-\alpha x} e^{i(\omega t - \beta x)}$. Upon comparing this with (24-51), and taking account of the difference in notation, we see that this is a damped wave traveling to the right (in the sense of increasing x) with a speed v and wavelength λ given by

$$v = \frac{\omega}{\beta} \qquad \lambda = \frac{2\pi}{\beta} \tag{27-54}$$

Similarly, the first term becomes $a_1 e^{\alpha x} e^{i(\omega t + \beta x)}$ that is a damped wave traveling to the left (in the sense of decreasing x). Both waves have the same damping factor α and speed v. We see from (27-47) that α and β will generally be functions of frequency as well as of the characteristics of the line. Thus the wave speed will also be a function of frequency so that a transmission line is an example of a dispersive medium.

One can obtain α and β from (27-47) in the same way that was used to obtain (24-42) and (24-43). The results are

$$\alpha = \frac{1}{\sqrt{2}} \left\{ (R'G' - \omega^2 L'C') + \left[(R'^2 + \omega^2 L'^2)(G'^2 + \omega^2 C'^2) \right]^{1/2} \right\}^{1/2} \tag{27-55}$$

$$\beta = \frac{1}{\sqrt{2}} \left\{ -(R'G' - \omega^2 L'C') + \left[(R'^2 + \omega^2 L'^2)(G'^2 + \omega^2 C'^2) \right]^{1/2} \right\}^{1/2} \tag{27-56}$$

We can find the constants of integration by considering conditions at the terminals. At the input end ($x = 0$), we can apply (27-6) to the amplitudes and use a loop going through $\mathscr{E}_0$, Z_0, and across the line. In this way we obtain $\mathscr{E}_0 - Z_0 I_0(0) - V_0(0) = 0 = \mathscr{E}_0 - Z_0(a_1 + a_2) - Z_i(-a_1 + a_2)$ so that

$$(Z_i - Z_0)a_1 - (Z_i + Z_0)a_2 = -\mathscr{E}_0 \tag{27-57}$$

Similarly, at the load end, we have $Z_l I_0(l) - V_0(l) = 0$ and thus

$$(Z_i + Z_l)e^{\gamma l}a_1 - (Z_i - Z_l)e^{-\gamma l}a_2 = 0 \tag{27-58}$$

Solving these two equations for a_1 and a_2, we obtain

$$a_1 = \frac{\mathscr{E}_0 \Gamma_l e^{-2\gamma l}}{(Z_i + Z_0)(1 - \Gamma_0 \Gamma_l e^{-2\gamma l})} \tag{27-59}$$

$$a_2 = \frac{\mathscr{E}_0}{(Z_i + Z_0)(1 - \Gamma_0 \Gamma_l e^{-2\gamma l})} \tag{27-60}$$

where

$$\Gamma_0 = \frac{Z_i - Z_0}{Z_i + Z_0} \qquad \Gamma_l = \frac{Z_i - Z_l}{Z_i + Z_l} \qquad (27\text{-}61)$$

are called the input and output *reflection coefficients*, respectively.

If we substitute these results into (27-51) and (27-52), we get the complete expressions for the current and voltage amplitudes:

$$I_0(x) = \frac{\mathscr{E}_0 e^{-\gamma x}\left[1 + \Gamma_l e^{-2\gamma(l-x)}\right]}{(Z_i + Z_0)(1 - \Gamma_0\Gamma_l e^{-2\gamma l})} \qquad (27\text{-}62)$$

$$V_0(x) = \frac{\mathscr{E}_0 Z_i e^{-\gamma x}\left[1 - \Gamma_l e^{-2\gamma(l-x)}\right]}{(Z_i + Z_0)(1 - \Gamma_0\Gamma_l e^{-2\gamma l})} \qquad (27\text{-}63)$$

As a first example, let us consider an infinitely long line; then $e^{-2\gamma l} \sim e^{-2\alpha l} \to 0$ as $l \to \infty$. Similarly, $e^{-2\gamma(l-x)} \to 0$ since $l - x$ will be positive; then (27-62) reduces to $I_0(x) = \mathscr{E}_0 e^{-\gamma x}/(Z_i + Z_0)$. The initial amplitude thus is $I_0(0) = \mathscr{E}_0/(Z_i + Z_0)$ and is determined solely by the input impedance Z_0 and the characteristic impedance Z_i of the line. This shows that the line impedance acts in this case as if it were simply in series with Z_0. The net result will be a damped wave traveling along the line with speed v given by (27-54).

Although (27-62) and (27-63) are exact, their physical significance is not too clear. We can get a useful interpretation of these results by expanding the denominator in a power series with the use of $1/(1 - y) = 1 + y + y^2 + y^3 + \dots$. In this way, we find that (27-62) can be written as

$$I_0(x) = \frac{\mathscr{E}_0}{(Z_i + Z_0)}\left[e^{-\gamma x} + \Gamma_l e^{-\gamma(2l-x)} + \Gamma_0\Gamma_l e^{-\gamma(2l+x)}\right.$$

$$+ \Gamma_0\Gamma_l^2 e^{-\gamma(4l-x)} + \Gamma_0^2\Gamma_l^2 e^{-\gamma(4l+x)}$$

$$\left.+ \Gamma_0^2\Gamma_l^3 e^{-\gamma(6l-x)} + \dots\right] \qquad (27\text{-}64)$$

If we now multiply through by $e^{i\omega t}$ to get the total current, we see that each term represents a traveling wave. Thus the whole current is a superposition of traveling waves, going in both directions; the amplitude of each component wave is proportional to various powers of the reflection coefficients. The Γ's are now seen to give the fraction of the incident amplitude that is reflected back into the line. The parenthetical portion of each exponent gives the distance the corresponding wave has traveled. For example, the sixth term in the expansion corresponds to a wave that has undergone a total of five reflections—three at the load end and two at the input end. It has traversed the line five times and is headed back to the input end and has gone a distance $l - x$ from the load end. This shows us how the steady state is finally attained after a sufficient number of reflections. If the load impedance equals the characteristic impedance so that $Z_l = Z_i$, then $\Gamma_l = 0$ and there are no reflections. In this case, (27-62) reduces to exactly the same result as for an infinite line.

For many lines, the resistance and conductance per unit length are very small and one can simplify our results by making some approximations. Let us assume that $\omega L' \gg R'$ and $\omega C' \gg G'$ and look for results with only first-order correction terms. Rather than dealing with (27-55) and (27-56), it is more convenient to go back to

(27-47) and we get

$$\gamma = \alpha + i\beta = i\omega\sqrt{L'C'}\left[\left(1 - \frac{iR'}{\omega L'}\right)\left(1 - \frac{iG'}{\omega C'}\right)\right]^{1/2}$$

$$\simeq i\omega\sqrt{L'C'}\left(1 - \frac{iR'}{2\omega L'}\right)\left(1 - \frac{iG'}{2\omega C'}\right)$$

$$\simeq i\omega\sqrt{L'C'}\left[1 - \frac{i}{2\omega}\left(\frac{R'}{L'} + \frac{G'}{C'}\right)\right]$$

so that

$$\alpha \simeq \frac{\sqrt{L'C'}}{2}\left(\frac{R'}{L'} + \frac{G'}{C'}\right) \qquad \beta \simeq \omega\sqrt{L'C'} \tag{27-65}$$

and the wave speed is

$$v \simeq \frac{1}{\sqrt{L'C'}} \tag{27-66}$$

by (27-54). We see that to this approximation both the wave speed and the attenuation constant are independent of the frequency and that the wave speed is the same as if the line had no resistance.

Similarly, the characteristic impedance obtained from (27-53) is

$$Z_i \simeq \sqrt{\frac{L'}{C'}}\left[1 + \frac{i}{2\omega}\left(\frac{G'}{C'} - \frac{R'}{L'}\right)\right] \simeq \sqrt{\frac{L'}{C'}} \tag{27-67}$$

since in many cases the line can be treated as resistanceless and then Z_i turns out to be approximately a pure resistance.

■ **Example**

Two-wire line. A very common example of a transmission line consists of two parallel wires as shown in cross section in Figure 11-10. From (11-53), (10-73), Exercise 17-23, and (20-66), we get

$$C' = \frac{\pi\epsilon}{\cosh^{-1}(D/2A)} \qquad L' = \frac{\mu}{\pi}\cosh^{-1}\left(\frac{D}{2A}\right) \tag{27-68}$$

and, if $A \ll D$, we can use

$$C' = \frac{\pi\epsilon}{\ln(D/A)} \qquad L' = \frac{\mu}{\pi}\ln\left(\frac{D}{A}\right) \tag{27-69}$$

where A is the wire radius and D the distance between centers.

In either case, we find from (27-66) that $v = 1/\sqrt{\mu\epsilon}$ and is exactly the same as the speed of a plane wave for a medium characterized by μ and ϵ as given by (24-12). The line impedance found from (27-67) is

$$Z_i = \frac{1}{\pi}\sqrt{\frac{\mu}{\epsilon}}\cosh^{-1}\left(\frac{D}{2A}\right) \simeq \frac{1}{\pi}\sqrt{\frac{\mu}{\epsilon}}\ln\left(\frac{D}{A}\right) \tag{27-70}$$

For simplicity, let us completely neglect the resistance of the line so that $\alpha = 0$ and assume that $Z_0 = Z_i$ so that $\Gamma_0 = 0$. Then (27-63) becomes

$$V_0 = \tfrac{1}{2}\mathscr{E}_0\left[e^{-i\beta x} - \Gamma_l e^{-2i\beta l + i\beta x}\right]$$

We will also assume that Z_l is real so that Γ_l is too. If we multiply our last result by $e^{i\omega t}$ and take the real part, we find the voltage to be given by

$$V = \tfrac{1}{2}\mathscr{E}_0\left[\cos\left(\omega t - \beta x\right) - \Gamma_l\cos\left(\omega t + \beta x - 2\beta l\right)\right] \qquad (27\text{-}71)$$

Suppose that the load end of the line is open, Z_l is infinite, $\Gamma_l = -1$ and (27-71) becomes

$$\begin{aligned}
V &= \tfrac{1}{2}\mathscr{E}_0\left[\cos\left(\omega t - \beta x\right) + \cos\left(\omega t + \beta x - 2\beta l\right)\right] \\
&= \mathscr{E}_0\cos\beta(x - l)\cos\left(\omega t - \beta l\right) \qquad (27\text{-}72)
\end{aligned}$$

This is a standing wave where the voltage oscillates sinusoidally at each point but with an amplitude that also varies sinusoidally along the line. In particular, the open end at $x = l$ has a maximum value so that it is a voltage antinode. There will be points at which the voltage is zero; these points are called nodes and their locations are given by the condition that $\beta(x - l) = n_o\pi/2$ where n_o is an odd integer. The distance Δx between adjacent nodes can be found since $\beta\Delta x = \Delta n_o\pi/2 = \pi$ or $\Delta x = \pi/\beta = \lambda/2$ showing that the nodes (and antinodes) are spaced a half wavelength apart.

Similarly, if the load end is short circuited so that $Z_l = 0$, then $\Gamma_l = 1$ and the voltage as obtained from (27-71) is

$$V = \mathscr{E}_0\sin\beta(x - l)\sin\left(\omega t - \beta l\right)$$

This is another standing wave with a voltage node at the short circuited end and with the same half wavelength spacing between nodes (and antinodes). Thus, in these respects, the two-wire line acts like an open or closed organ pipe.

Two parallel wires used in this way are often called *Lecher wires*.   ■

■ **Example**

Coaxial line. Another very common transmission line is the coaxial line as illustrated in Figures 6-12 and 18-1. From (6-45) and the principal portion of (18-34), we find that

$$C' = \frac{2\pi\epsilon}{\ln\left(b/a\right)} \qquad L' = \frac{\mu}{2\pi}\ln\left(\frac{b}{a}\right)$$

so that once again the waves travel with the speed of light since $v = 1/\sqrt{L'C'} = 1/\sqrt{\mu\epsilon}$. The characteristic impedance is

$$Z_i = \frac{1}{2\pi}\sqrt{\frac{\mu}{\epsilon}}\ln\left(\frac{b}{a}\right)$$

We note that this is exactly the same impedance (26-77) that we found for the coaxial line by treating it as a wave guide for the transmission of TEM waves.   ■

EXERCISES

27-1 In the RL circuit of Figure 27-2, $R = 4$ ohms, $L = 1.5$ henrys, and the battery emf is 12 volts. Find the current and the magnitudes of V_R and V_L for these times in seconds after the switch is closed: (a) 0.25, (b) 1.0, (c) 3.0.

27-2 In Figure 27-2, suppose that the inductance L is replaced by a capacitance C. Use Kirchhoff's laws to find a differential equation

from which the capacitor charge q can be found. Before the switch is closed at $t = 0$, there is a charge q_0 on C. Find q for all later times. What is the relaxation time of this system? Under what conditions will the corresponding current have a sense opposite to that shown in the figure?

27-3 Show that (27-10) can also be derived by using conservation of energy in the sense that the time rate of change of electric plus magnetic energy equals the power supplied by the emf less the rate of resistive loss.

27-4 If $(R/2L)^2 > (1/LC)$, then δ is real and the series RLC circuit is said to be *overdamped*. Find q as a function of time for the same initial conditions that led to (27-15). What does q become after a very long time?

27-5 If $(R/2L)^2 = 1/LC$, then the series *RLC* circuit is said to be *critically damped*. Since $\delta = 0$, (27-14) reduces to $I = (a_+ + a_-)e^{-Rt/2L} = ae^{-Rt/2L}$ so that there is only one constant of integration, whereas the general solution of a second-order differential equation requires two. Show that, in this case, the general solution of (27-12) has the form $I = (a + bt)e^{-Rt/2L}$ where a and b are constants. Find q as a function of time for the same initial conditions that led to (27-15). Show that $|V_R|$ has a maximum value at a time $2L/R$ after the switch is closed and find its maximum value.

27-6 Consider an underdamped series *RLC* circuit for which $\mathscr{E} = \mathscr{E}_0 \cos \omega t$ with $\omega \neq \omega_n$. At $t = 0$, the capacitor is uncharged and the previously open switch is suddenly closed. Find the current for all later times. Does your solution correctly reduce to (27-24) or (27-25) as $t \to \infty$?

27-7 For a system described by (27-25), show that the time-average power supplied by the emf is $(\mathscr{E}_0^2/2|Z|) \cos \vartheta$. In this context, $\cos \vartheta$ is called the *power factor*. What conditions will make the power factor zero? Physically, what is occurring to give rise to such a situation?

27-8 (a) Show that the equation determining the relative phase can be written as $\tan \vartheta = Q(\omega/\omega_0)[1 - (\omega_0/\omega)^2]$. Using this, show that, for a given ϑ, the larger the value of Q, the smaller the corresponding value of ω will be. (b) If the voltage across the capacitor is regarded as a function of frequency, show that its maximum amplitude occurs when $\omega = \omega_0[1 - (1/2Q^2)]^{1/2}$.

27-9 Show that Q also equals 2π times the time-average of the total energy of the circuit divided by the average energy dissipated per cycle when the system is at or very near resonance. How does this result compare with the interpretation found for the Q defined in Chapter 24 as given in the last sentence of Exercise 24-29?

27-10 A series *RLC* circuit has an additional inductance L'' in parallel with the whole combination. Find the resultant impedance of this new system.

27-11 There is a real current $I = I_0 \cos(\omega t + \epsilon)$ in a circuit consisting of a resistance R and inductance L connected in parallel. Find the voltage $V(t)$ across this circuit.

27-12 In the circuit of Figure 27-8, suppose that R_L is removed from the branch containing L and put in that containing C. Find Z. Verify that your result will correctly lead to (27-37) when R_L is negligible. Show that, when $\omega^2 LC = 1$, your answer and (27-36) both give the same value of $|Z|$.

27-13 Consider a special case of the coupled circuits of Figure 27-9 where $L_1 = L_2 = L$ and $C_1 = C_2 = C$. The initial conditions at $t = 0$ are: $I_1 = I_2 = 0$, $q_1 = q_0$, $q_2 = 0$. (How could you achieve this in practice?) Find the currents and charges for all later times.

27-14 A generator providing the emf $\mathscr{E}_0 e^{i\omega t}$ has an internal impedance Z_0. It is connected in series with a load whose impedance Z_l can be varied. Show that maximum time-average power will be transferred to the load when $Z_l = Z_0^*$.

27-15 Some lines are such that the resistance is large compared to its series inductive reactance and the capacitive reactance is large compared to the leakage conductance. For this case, find the lowest-order approximations for α, β, v, and Z_i.

27-16 By assuming that the line parameters themselves are independent of frequency, one can get a "distortionless" line, that is, the propagation quantities are independent of frequency, provided that $L'G' = R'C'$. Show this to be the case by finding α, v, and Z_i. (In practice, this generally requires unacceptably large values of L' and/or G'.)

27-17 It is often more convenient to express the voltage and current in terms of line input values, that is, values at $x = 0$. Show that (27-62) and (27-63) can be written in the form

$$V_0(x) = \frac{V_0(0) \cosh[\gamma(l - x) + \phi]}{\cosh(\gamma l + \phi)}$$

$$I_0(x) = \frac{V_0(0) \sinh[\gamma(l - x) + \phi]}{Z_i \cosh(\gamma l + \phi)}$$

where $\tanh \phi = Z_i/Z_l$. Also show that the effective input impedance can be usefully written in terms of the line constants and load impedance as

$$V_0(0)/I_0(0) = Z_i \coth(\gamma l + \phi)$$

$$= Z_i(Z_l + Z_i \tanh \gamma l)/$$

$$(Z_l \tanh \gamma l + Z_i).$$

27-18 If Z_{sc} and Z_{oc} are the input impedances when the line is terminated by a short circuit ($Z_l = 0$) and an open circuit ($Z_l \to \infty$), respectively, show that the characteristic impedance is given exactly by $Z_i = (Z_{sc} Z_{oc})^{1/2}$.

27-19 For the case of Lecher wires, show that voltage nodes are current antinodes and *vice versa*.

27-20 In order to get an idea of magnitudes, find the ratio b/a necessary to make Z_i of a coaxial line with vacuum inside equal to 60 ohms.

27-21 The time-average power transmitted past a given point on a transmission line is $\langle P \rangle = \mathrm{Re}\,(\frac{1}{2} I_0 V_0^*)$. (Do you agree?) If $\alpha = 0$ and Z_i is real, show that $\langle P \rangle \sim (1 - |\Gamma_l|^2)$. Thus, maximum power transfer will occur when $Z_l = Z_i$.

27-22 For simplicity, assume that all impedances are real and $\gamma = i\beta$. Show that the dependence of the voltage amplitude on position along the line can be generally described by $|V_0|^2 = A[1 + \Gamma_l^2 - 2\Gamma_l \cos 2\beta(l - x)]$ where $A = $ const. For this kind of standing wave, in which the minima are not zero, a useful and measurable quantity is the *voltage standing wave ratio* V_{SWR} defined by $V_{SWR} = |V_0|_{max} / |V_0|_{min}$. Show that

$$V_{SWR} = \frac{1 + |\Gamma_l|}{1 - |\Gamma_l|} \quad \text{and} \quad |\Gamma_l| = \frac{V_{SWR} - 1}{V_{SWR} + 1}$$

and that these are correct for both positive and negative values of Γ_l. (The second result provides a convenient way of measuring $|\Gamma_l|$ and hence of finding the load impedance.) Show that these results agree with what we found for the Lecher wire system. Explain the value found for V_{SWR} when $\Gamma_l = 0$. Show that the spacing between adjacent maxima (and adjacent minima) is $\lambda/2$.

28

RADIATION

In Chapters 24 through 26, we considered time-dependent electromagnetic fields in the forms of traveling and standing waves. We did not concern ourselves with the problem of how we could produce such fields if we so desired other than to note that it could be done in principle with an "appropriate" distribution of charges and currents. Now we want to consider some aspects of how electromagnetic fields in the form of waves can be generated; this part of electromagnetism is usually given the name *radiation*. The general subject of radiation covers a wide variety of problems and can become quite complex. Accordingly, we will have to confine ourselves to a few representative, reasonably simple, yet important examples.

If we recall our treatment of waves, we note that we managed very well by using Maxwell's equations and dealing with the fields **E** and **B** (or **H**) themselves. That is, we never found it necessary to use the general time-dependent vector and scalar potentials that were discussed in Chapter 22. The production of electromagnetic radiation is, however, much more easily handled in terms of these potentials **A** and ϕ.

28-1 RETARDED POTENTIALS

We restrict ourselves to regions with vacuum properties so that we can set $\mu\epsilon = \mu_0\epsilon_0 = 1/c^2$. If we simplify our notation somewhat by omitting the subscript f on the charge and current densities, the equations determining the potentials are then found from (22-14) through (22-16) to be

$$\nabla^2\mathbf{A} - \frac{1}{c^2}\frac{\partial^2\mathbf{A}}{\partial t^2} = -\mu_0\mathbf{J} \tag{28-1}$$

$$\nabla^2\phi - \frac{1}{c^2}\frac{\partial^2\phi}{\partial t^2} = -\frac{\rho}{\epsilon_0} \tag{28-2}$$

$$\nabla \cdot \mathbf{A} + \frac{1}{c^2}\frac{\partial\phi}{\partial t} = 0 \tag{28-3}$$

while the fields are obtained from

$$\mathbf{B} = \nabla \times \mathbf{A} \tag{28-4}$$

$$\mathbf{E} = -\nabla\phi - \frac{\partial\mathbf{A}}{\partial t} \tag{28-5}$$

according to (22-1) and (22-3). We recall that this set of equations is completely equivalent to Maxwell's equations.

It will also be helpful for us to remember that we have found the solutions for the potentials in the *static* case where all time derivatives are zero; these are given by

(16-12) and (5-7) and are

$$A(\mathbf{r}) = \frac{\mu_0}{4\pi} \int_{V'} \frac{\mathbf{J}(\mathbf{r}')\, d\tau'}{R} \tag{28-6}$$

$$\phi(\mathbf{r}) = \frac{1}{4\pi\epsilon_0} \int_{V'} \frac{\rho(\mathbf{r}')\, d\tau'}{R} \tag{28-7}$$

with $R = |\mathbf{r} - \mathbf{r}'|$. We know that the solutions for the potentials in the time-dependent case must reduce to (28-6) and (28-7) for a static situation. Unfortunately, we cannot, for example, simply replace $\mathbf{J}(\mathbf{r}')$ in (28-6) by $\mathbf{J}(\mathbf{r}', t)$ because of the time derivatives in the differential equation for $\mathbf{A}$.

It is possible to solve (28-1) and (28-2) by elegant and general methods, but they are quite complex. It is preferable for our purposes to use a simpler method that leads to the correct results and is much more valuable in aiding one's understanding. We will find the contribution to ϕ from a charge element $\Delta q = \rho \Delta \tau'$ in a volume $\Delta \tau'$ small enough so that Δq can be treated as a point charge. Then by summing the contributions of all the charge elements, we can get the total scalar potential. Now outside of $\Delta \tau'$ there is no charge and (28-2) becomes the homogeneous wave equation

$$\nabla^2 \phi - \frac{1}{c^2} \frac{\partial^2 \phi}{\partial t^2} = 0 \tag{28-8}$$

Because of the spherical symmetry associated with a point charge, $\phi(\mathbf{r}, t)$ can really depend only on the relative distance R from the charge and not on angles, that is, $\phi = \phi(R, t)$. Using (1-139), we find that $\nabla\phi = (\partial\phi/\partial R)\hat{\mathbf{R}} = (\partial\phi/\partial R)(\mathbf{R}/R)$, and then by successive use of (1-115), (1-139), and Exercise 1-21, we obtain

$$\nabla^2 \phi = \nabla \cdot \nabla\phi = \nabla \cdot \left[\frac{1}{R}\left(\frac{\partial\phi}{\partial R}\right)\mathbf{R} \right] = \nabla\left[\frac{1}{R}\left(\frac{\partial\phi}{\partial R}\right) \right] \cdot \mathbf{R} + \frac{1}{R}\left(\frac{\partial\phi}{\partial R}\right)\nabla \cdot \mathbf{R}$$

$$= \left\{ \frac{\partial}{\partial R}\left[\frac{1}{R}\left(\frac{\partial\phi}{\partial R}\right) \right]\frac{\mathbf{R}}{R} \right\} \cdot \mathbf{R} + \frac{3}{R}\left(\frac{\partial\phi}{\partial R}\right) = \frac{\partial^2\phi}{\partial R^2} + \frac{2}{R}\frac{\partial\phi}{\partial R}$$

so that (28-8) becomes

$$\frac{\partial^2\phi}{\partial R^2} + \frac{2}{R}\frac{\partial\phi}{\partial R} - \frac{1}{c^2}\frac{\partial^2\phi}{\partial t^2} = \frac{1}{R^2}\frac{\partial}{\partial R}\left(R^2\frac{\partial\phi}{\partial R} \right) - \frac{1}{c^2}\frac{\partial^2\phi}{\partial t^2} = 0 \tag{28-9}$$

[We note from (1-105) that this result is exactly what we would get by expressing ∇^2 in spherical coordinates with R as the radial distance and with an angle-independent ϕ.] If we now let

$$\phi(R, t) = \frac{\chi(R, t)}{R} \tag{28-10}$$

and substitute this into (28-9), we find that χ satisfies the one-dimensional wave equation

$$\frac{\partial^2\chi}{\partial R^2} - \frac{1}{c^2}\frac{\partial^2\chi}{\partial t^2} = 0 \tag{28-11}$$

In (24-9), (24-11), and (24-12), we found the solution of this equation to have the form $\chi = f(R - ct) + g(R + ct)$ where f and g are functions only of the combinations of variables $R \mp ct$ or, what amounts to the same thing, of the combinations $t \mp (R/c)$;

then we find from (28-10) that ϕ has the form

$$\phi(R, t) = \frac{f[t - (R/c)]}{R} + \frac{g[t + (R/c)]}{R} \tag{28-12}$$

Each term in (28-12) is a *spherical wave* traveling with speed c in the radial direction with respect to the location of Δq and with an amplitude that is constant on a sphere of radius R centered on the charge; f represents a wave traveling outward from Δq while g is one traveling inward. Since (28-12) was obtained as the solution of the source-free equation (28-8), f and g must be determined so that they correspond to the existence of the charge Δq located at $R \simeq 0$.

First of all, ϕ in (28-12) is to be the potential due to the changing source. If the function g were included as an ingoing wave, it would arrive at the source at a later time than t, specifically a time interval R/c later, and the value of g would have to be the correct value appropriate to a point very near to Δq. But this would mean that g was already correct at an earlier time t at a distance R away. In other words, the "effect" described by g would necessarily have to have occurred (or be known) *before* the "cause" due to the charge. Since this is contrary to the logical order of these events, we have to reject the term g on physical grounds and write $\phi = f[t - (R/c)]/R$. Now as we get in the immediate neighborhood of the nearly point charge Δq, that is, as $R \to 0$, the potential must reduce to that of a point charge at its instantaneous value and location, which as seen from (28-7), is $\Delta q(t)/4\pi\epsilon_0 R$. Thus $f[t - (R/c)]$ must have a form such that, for $R \to 0$, $f(t) = \Delta q(t)/4\pi\epsilon_0$ or, if γ is the general variable in f, such that

$$f(\gamma) = \frac{\Delta q(\gamma)}{4\pi\epsilon_0} \tag{28-13}$$

Thus, the form of f, and hence of ϕ, is finally determined to be

$$\phi(\mathbf{r}, t) = \frac{1}{4\pi\epsilon_0 R} \Delta q\left(t - \frac{R}{c}\right) = \frac{1}{4\pi\epsilon_0 R} \rho\left(\mathbf{r}', t - \frac{R}{c}\right) \Delta\tau' \tag{28-14}$$

This result tells us that the scalar potential at the field point $\mathbf{r}$ at the time t depends on the value of the charge at an *earlier* time $t - (R/c)$, and the time difference R/c is exactly the time required for the spherical wave to reach the field point. Thus the charge does not produce the potential instantaneously but a finite time, corresponding to the travel time of the wave, is required for the effect to be felt at the field point. The time $t' = t - (R/c)$ is usually called the *retarded time*.

Now that we have the contribution to the scalar potential from an element of charge, we can obtain the total scalar potential due to a charge distribution by summing (28-14) over all of the volume V' containing the charges; the result is

$$\phi(\mathbf{r}, t) = \frac{1}{4\pi\epsilon_0} \int_{V'} \frac{\rho[\mathbf{r}', t - (R/c)] \, d\tau'}{R} \tag{28-15}$$

We see that each charge element must be evaluated at its *own* retarded time; these times will generally vary from element to element since, for a given field point $\mathbf{r}$, the relative distance $R = |\mathbf{r} - \mathbf{r}'|$ will vary with the source point location $\mathbf{r}'$. Thus (28-15) depends on R explicitly because it appears in the denominator, and implicitly because it appears in the time variable of the charge density ρ.

Each rectangular component of (28-1) will have the same form as (28-2); for example, we have

$$\nabla^2 A_x - \frac{1}{c^2} \frac{\partial^2 A_x}{\partial t^2} = -\mu_0 J_x$$

Each such equation will lead to a result like (28-15), so that when they are added vectorially, we will find that **A** is given by

$$A(r, t) = \frac{\mu_0}{4\pi} \int_{V'} \frac{J[r', t - (R/c)] \, d\tau'}{R} \qquad (28\text{-}16)$$

The results (28-15) and (28-16) are known as *retarded potentials*; they are often written in shortened form with brackets about the source term in the integrands to remind one that they are to be evaluated at retarded times, that is, one often finds

$$A = \frac{\mu_0}{4\pi} \int_{V'} \frac{[J] \, d\tau'}{R} \qquad \phi = \frac{1}{4\pi\epsilon_0} \int_{V'} \frac{[\rho] \, d\tau'}{R} \qquad (28\text{-}17)$$

We note that (28-15) and (28-16) reduce correctly to (28-6) and (28-7) in the static case when the charge and current densities are independent of time. [Potentials that would arise from the g term of (28-12) are known as *advanced* potentials.]

28-2 MULTIPOLE EXPANSION FOR HARMONICALLY OSCILLATING SOURCES

We will restrict ourselves to finding the fields produced at a point in a vacuum outside of a finite region containing prescribed sources, that is, we assume that $J(r', t)$ and $\rho(r', t)$ are known functions of position and time. Furthermore, we consider only the important case in which the sources vary harmonically in time with circular frequency ω so that we can write

$$J(r', t) = J_0(r')e^{-i\omega t} \qquad \rho(r', t) = \rho_0(r')e^{-i\omega t} \qquad (28\text{-}18)$$

where J_0 and ρ_0 can be complex functions as discussed in connection with (24-19). Thus, if we write them in terms of a real amplitude and phase as

$$J_0(r') = \hat{j}|J_0(r')|e^{i\vartheta_c} \qquad \rho_0(r') = |\rho_0(r')|e^{i\vartheta_{ch}} \qquad (28\text{-}19)$$

where $\hat{j}$ gives the direction of J_0, then the real parts representing the physical currents and charges will be

$$J_{real} = \hat{j}|J_0| \cos(\omega t - \vartheta_c) \qquad \rho_{real} = |\rho_0| \cos(\omega t - \vartheta_{ch}) \qquad (28\text{-}20)$$

Following our previous results, if we replace t by $t - (R/c)$ in (28-18), and set

$$k = \frac{\omega}{c} = \frac{2\pi}{\lambda} \qquad (28\text{-}21)$$

we find that (28-15) and (28-16) become

$$\phi(r, t) = \frac{e^{-i\omega t}}{4\pi\epsilon_0} \int_{V'} \frac{e^{ikR}\rho_0(r') \, d\tau'}{R} \qquad (28\text{-}22)$$

$$A(r, t) = \frac{\mu_0 e^{-i\omega t}}{4\pi} \int_{V'} \frac{e^{ikR}J_0(r') \, d\tau'}{R} \qquad (28\text{-}23)$$

We see that the potentials will also vary harmonically in time (as will **E** and **B**) and with the same frequency as the sources. This will simplify our work somewhat as all time derivatives can then be replaced by $-i\omega$, that is

$$\frac{\partial \psi}{\partial t} = -i\omega\psi \qquad (28\text{-}24)$$

where ψ can be ϕ, as well as any component of **A**, **E**, or **B**.

We can, in fact, also express everything in terms of the vector potential in this case since the Lorentz condition (28-3) becomes $\nabla \cdot \mathbf{A} - (i\omega/c^2)\phi = 0$ so that

$$\phi = -i\frac{c^2}{\omega}\nabla \cdot \mathbf{A} \qquad (28\text{-}25)$$

If we now substitute (28-25) into (28-5), and use (28-4), (28-24), and (1-120), we obtain

$$\mathbf{E} = -\nabla\phi + i\omega\mathbf{A} = i(c^2/\omega)\left[\nabla(\nabla \cdot \mathbf{A}) + (\omega^2/c^2)\mathbf{A}\right]$$
$$= i(c^2/\omega)\left[\nabla \times \mathbf{B} + \nabla^2\mathbf{A} + (\omega^2/c^2)\mathbf{A}\right] \qquad (28\text{-}26)$$

At points where we want to find $\mathbf{E}$ and $\mathbf{B}$, the current density $\mathbf{J} = 0$ so that (28-1) becomes $\nabla^2\mathbf{A} + (\omega^2/c^2)\mathbf{A} = 0$; thus the last two terms of (28-26) vanish, and it reduces to

$$\mathbf{E} = i\frac{c^2}{\omega}\nabla \times \mathbf{B} = i\frac{c}{k}\nabla \times \mathbf{B} \qquad (28\text{-}27)$$

Actually this result is no surprise since it is just one of Maxwell's equations; thus, with $\sigma = 0$, (24-4) becomes $\nabla \times \mathbf{B} = \mu_0\epsilon_0(\partial\mathbf{E}/\partial t) = -i(\omega/c^2)\mathbf{E}$, which is exactly (28-27). In summary, then, we have found that we need only consider the vector potential $\mathbf{A}$ as given by (28-23); once $\mathbf{A}$ has been found, $\mathbf{B}$ is obtained from (28-4), and then $\mathbf{E}$ from (28-27).

The integral in (28-23) is difficult to deal with in general, so that it has been found useful to expand the integrand in a power series in much the same manner that we used in discussing electric and magnetic multipoles in Chapters 8 and 19. We choose our origin of coordinates at some arbitrary convenient location within the volume occupied by the sources, and we are led to the situation shown in Figure 28-1 where $\mathbf{r}$ is the position vector of the field point P. (Compare with Figures 19-1 and 8-1.) In addition, we can write

$$R = |\mathbf{r} - \mathbf{r}'| = (r^2 + r'^2 - 2rr'\cos\theta')^{1/2} = r(1 + \eta)^{1/2} \qquad (28\text{-}28)$$

by (19-2) and (8-4) where

$$\eta = -2\left(\frac{r'}{r}\right)\cos\theta' + \left(\frac{r'}{r}\right)^2 = -\frac{2(\hat{\mathbf{r}} \cdot \mathbf{r}')}{r} + \left(\frac{r'}{r}\right)^2 \qquad (28\text{-}29)$$

We will assume that r is large compared with the dimensions of the source volume V', and it will be sufficient for our purposes to keep only terms linear in r'/r. Then we can

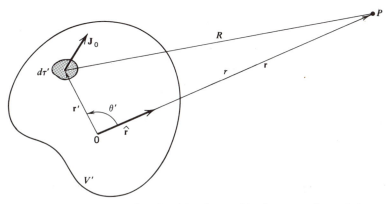

Figure 28-1. Relations involved in the multipole expansion of the vector potential.

write

$$\eta \simeq -\frac{2(\hat{\mathbf{r}} \cdot \mathbf{r}')}{r} \tag{28-30}$$

and, we can use the expansion (8-6) to write

$$\frac{1}{R} \simeq \frac{1}{r}\left(1 - \frac{1}{2}\eta\right) \tag{28-31}$$

Now let us consider the term e^{ikR}, which we also want to expand in an appropriate manner.

By (28-21), $kR = 2\pi R/\lambda$ so that this product essentially measures the distance R on a wavelength scale. If we are dealing with a compact source, $R \approx r$, and if we then were to expand e^{ikR} in powers of kR and keep only the first few terms, we would be unnecessarily restricting ourselves to field points nearby the source. However, if D is some typical dimension of the source volume V', the variations in $R - r$ will be of the order of magnitude of D, and it is not unreasonable to assume that $D \ll \lambda$; in this way, we do restrict ourselves to small sources on a wavelength scale, but we need not make any assumptions about r as compared to λ. In other words, what we really want to do is to expand the quantity $e^{ik(R-r)}$ rather than simply e^{ikR}. Before we proceed with this expansion, we want to consider another way of expressing our assumption that $D \ll \lambda$.

For harmonically oscillating systems with position vectors of the form $\mathbf{r}' = \mathbf{r}_0' e^{-i\omega t}$, the velocities will be $\mathbf{v}' = d\mathbf{r}'/dt = -i\omega \mathbf{r}'$ and hence the maximum values of the speeds will be $v \approx \omega D$. Assuming that $D \ll \lambda$, we then find that $v/c \approx \omega D/c \approx D/\lambda \ll 1$. In other words, our assumption, that all of the charges and currents can be thought of as being contained within a compact volume that remains small compared to the source-observer distance and to the wavelength during all of the times of observation, is equivalent to assuming that we are dealing with the "nonrelativistic" limit of "slowly moving" charges, that is, the charge speeds are small compared to c. These restrictions to slowly moving charges apply to all of the material in this chapter even though, in the last section, we will be able to find approximate means of dealing with certain sources with dimensions comparable to a wavelength.

In order to expand e^{ikR} appropriately, we need to express it in terms of η. We can do this by first writing $kR = kr + k(R - r)$ and then by using (28-28), we get $k(R - r) = kr[(1 + \eta)^{1/2} - 1] \simeq \frac{1}{2}kr\eta$. Inserting this into e^{ikR} and then using the expansion $e^u \simeq 1 + u$, and keeping only first-order terms, we get

$$e^{ikR} = e^{ikr}e^{ik(R-r)} \simeq e^{ikr}e^{i\frac{1}{2}kr\eta}$$
$$\simeq e^{ikr}\left(1 + \tfrac{1}{2}ikr\eta\right) \tag{28-32}$$

Combining this with (28-31), we get the approximation

$$\frac{e^{ikR}}{R} \simeq \frac{e^{ikr}}{r}\left\{1 - \frac{1}{2}(1 - ikr)\eta\right\} \tag{28-33}$$

Finally, inserting the expression for η given by (28-30) into (28-33), we obtain

$$\frac{e^{ikR}}{R} \simeq \frac{e^{ikr}}{r}\left\{1 + (1 - ikr)\left(\frac{\hat{\mathbf{r}} \cdot \mathbf{r}'}{r}\right)\right\} \tag{28-34}$$

which, when inserted into (28-23) gives us our desired expansion of the vector potential in the form

$$\mathbf{A} = \mathbf{A}_{\mathrm{I}} + \mathbf{A}_{\mathrm{II}} \tag{28-35}$$

where

$$\mathbf{A}_{\mathrm{I}} = \frac{\mu_0 e^{i(kr-\omega t)}}{4\pi r} \int_{V'} \mathbf{J}_0(\mathbf{r}') \, d\tau' \tag{28-36}$$

$$\mathbf{A}_{\mathrm{II}} = \frac{\mu_0(1 - ikr)e^{i(kr-\omega t)}}{4\pi r^2} \int_{V'} (\hat{\mathbf{r}} \cdot \mathbf{r}')\mathbf{J}_0(\mathbf{r}') \, d\tau' \tag{28-37}$$

since r is constant with respect to the primed variables of integration. (The subscripts on the $\mathbf{A}$'s are temporary and have merely been chosen to reflect the inverse powers of r that have naturally appeared.)

We note that all of the terms are proportional to $e^{i(kr-\omega t)}$ so that they represent spherical waves traveling outward from the origin with speed $\omega/k = c$. The amplitude of each wave has a dependence on r that becomes more complicated with each successive term in the expansion. Furthermore, each wave is proportional to an integral of the source current amplitude over the source volume, although the field point still remains in the integral of (28-37) since the integrand contains its direction $\hat{\mathbf{r}}$. Because of the use of expansions of R like those we used before, we suspect that these integrals are related to various multipole moments of the source and, in fact, we note that (28-36) and (28-37) contain exactly the first two integrals we found in (19-4) when we obtained the multipole expansion of the *magnetostatic* vector potential. Accordingly, we will investigate these integrals more closely with these ideas in mind.

The integral in (28-36) is what we called the magnetic monopole moment in Section 19-1 and we found in (19-5) that it was always zero. However, that was for steady currents for which the equation of continuity reduced to $\nabla' \cdot \mathbf{J} = 0$, which is not at all the case here. With the use of (28-24), the general equation of continuity (12-13) becomes $\nabla' \cdot \mathbf{J} + (\partial \rho/\partial t) = \nabla' \cdot \mathbf{J} - i\omega\rho = 0$ which, when combined with (28-18) yields

$$\nabla' \cdot \mathbf{J}_0 = i\omega\rho_0 \tag{28-38}$$

and shows explicitly that the charge and current distributions are not independent. Now we find from (1-121) that $(\mathbf{J}_0 \cdot \nabla')\mathbf{r}' = \mathbf{J}_0$, so that when (1-129) is applied to the source region it becomes

$$\oint_{S'} \mathbf{r}'(\mathbf{J}_0 \cdot d\mathbf{a}') = \int_{V'} [\mathbf{J}_0 + \mathbf{r}'(\nabla' \cdot \mathbf{J}_0)] \, d\tau' = 0 \tag{28-39}$$

since the currents $\mathbf{J}_0$ are zero on the bounding surface S'. Thus, if we also use (28-38), we see that

$$\int_{V'} \mathbf{J}_0 \, d\tau' = -\int_{V'} \mathbf{r}'(\nabla' \cdot \mathbf{J}_0) \, d\tau' = -i\omega \int_{V'} \rho_0(\mathbf{r}')\mathbf{r}' \, d\tau' = -i\omega\mathbf{p}_0 \tag{28-40}$$

where $\mathbf{p}_0$ is the (complex) electric dipole moment of the charge distribution amplitude according to (8-22). Therefore, (28-36) actually represents an electromagnetic field produced by a varying electric dipole moment and we can relabel $\mathbf{A}_{\mathrm{I}}$ and write it as

$$\mathbf{A}_{\mathrm{ed}} = -\frac{i\mu_0\omega\mathbf{p}_0}{4\pi r}e^{i(kr-\omega t)} \tag{28-41}$$

The integral appearing in (28-37) is exactly what we called $\mathscr{D}$ in (19-7). As before, it is useful to write the integrand as the sum of symmetric and antisymmetric parts by dividing the integrand into two equal parts and adding and subtracting a "suitable" quantity. Thus, we get

$$(\hat{\mathbf{r}} \cdot \mathbf{r}')\mathbf{J}_0 = \tfrac{1}{2}[(\hat{\mathbf{r}} \cdot \mathbf{r}')\mathbf{J}_0 - (\hat{\mathbf{r}} \cdot \mathbf{J}_0)\mathbf{r}'] + \tfrac{1}{2}[(\hat{\mathbf{r}} \cdot \mathbf{r}')\mathbf{J}_0 + (\hat{\mathbf{r}} \cdot \mathbf{J}_0)\mathbf{r}']$$

$$= \tfrac{1}{2}(\mathbf{r}' \times \mathbf{J}_0) \times \hat{\mathbf{r}} + \tfrac{1}{2}[(\hat{\mathbf{r}} \cdot \mathbf{r}')\mathbf{J}_0 + (\hat{\mathbf{r}} \cdot \mathbf{J}_0)\mathbf{r}'] \tag{28-42}$$

with the use of (1-30) and (1-23). If we now insert this into (28-37), we find that $\mathbf{A}_{\mathrm{II}}$ can be written as the sum

$$\mathbf{A}_{\mathrm{II}} = \frac{\mu_0(1 - ikr)e^{i(kr-\omega t)}}{4\pi r^2}\left[\frac{1}{2}\int_{V'}\mathbf{r}' \times \mathbf{J}_0\,d\tau'\right] \times \hat{\mathbf{r}}$$

$$+ \frac{\mu_0(1 - ikr)e^{i(kr-\omega t)}}{8\pi r^2}\left\{\int_{V'}[(\hat{\mathbf{r}} \cdot \mathbf{r}')\mathbf{J}_0 + (\hat{\mathbf{r}} \cdot \mathbf{J}_0)\mathbf{r}']\,d\tau'\right\} \quad (28\text{-}43)$$

By comparing the quantity in brackets in the first term of (28-43) with (19-20), we see that it is exactly the magnetic dipole moment $\mathbf{m}_0$ of the current amplitude distribution so that we can write this whole first term as

$$\mathbf{A}_{\mathrm{md}} = \frac{\mu_0(1 - ikr)(\mathbf{m}_0 \times \hat{\mathbf{r}})}{4\pi r^2}e^{i(kr-\omega t)} \quad (28\text{-}44)$$

The remaining integral in (28-43) will soon be seen to involve the electric quadrupole moment so that we will want to rewrite it in terms of the charge density by means of (28-38). Using $\mathbf{J}_0 = (\mathbf{J}_0 \cdot \nabla')\mathbf{r}'$, $\hat{\mathbf{r}} = \nabla'(\hat{\mathbf{r}} \cdot \mathbf{r}')$ by (19-11) and (1-115), we find that the integrand of the second term can be written as

$$\{[(\hat{\mathbf{r}} \cdot \mathbf{r}')\mathbf{J}_0] \cdot \nabla'\}\mathbf{r}' + \mathbf{r}'[\mathbf{J}_0 \cdot \nabla'(\hat{\mathbf{r}} \cdot \mathbf{r}')]$$

$$= \{[(\hat{\mathbf{r}} \cdot \mathbf{r}')\mathbf{J}_0] \cdot \nabla'\}\mathbf{r}' + \mathbf{r}'\{\nabla' \cdot [(\hat{\mathbf{r}} \cdot \mathbf{r}')\mathbf{J}_0]\}_- - \mathbf{r}'(\hat{\mathbf{r}} \cdot \mathbf{r}')(\nabla' \cdot \mathbf{J}_0)$$

and when this is integrated over V' and (1-129) and (28-38) are used, the second integral becomes

$$\oint_{S'}\mathbf{r}'(\hat{\mathbf{r}} \cdot \mathbf{r}')(\mathbf{J}_0 \cdot d\mathbf{a}') - \int_{V'}\mathbf{r}'(\hat{\mathbf{r}} \cdot \mathbf{r}')(\nabla' \cdot \mathbf{J}_0)\,d\tau' = -i\omega\int_{V'}\mathbf{r}'(\hat{\mathbf{r}} \cdot \mathbf{r}')\rho_0(\mathbf{r}')\,d\tau' \quad (28\text{-}45)$$

since the surface integral again vanishes as in (28-39). Putting this into the second part of (28-43), we find that it becomes

$$\mathbf{A}_{\mathrm{eq}} = -\frac{i\mu_0\omega(1 - ikr)e^{i(kr-\omega t)}}{8\pi r^2}\int_{V'}\mathbf{r}'(\hat{\mathbf{r}} \cdot \mathbf{r}')\rho_0(\mathbf{r}')\,d\tau' \quad (28\text{-}46)$$

so that we can also write $\mathbf{A}_{\mathrm{II}} = \mathbf{A}_{\mathrm{md}} + \mathbf{A}_{\mathrm{eq}}$. We see that the integral in (28-46) involves second moments of the charge density amplitude and hence is related to its electric quadrupole moment that we first discussed in detail following (8-22). We can now rewrite this result in a more easily recognizable form.

Since $\mathbf{B}$ will be obtained by differentiation *via* $\mathbf{B} = \nabla \times \mathbf{A}_{\mathrm{eq}}$, we can add a constant to the $\hat{\mathbf{r}}$ component of $\mathbf{A}_{\mathrm{eq}}$ without affecting the fields by (1-104). Accordingly, we add

$$-\int_{V'}\frac{1}{3}r'^2\hat{\mathbf{r}}\rho_0(\mathbf{r}')\,d\tau'$$

to the integral appearing in (28-46), which then becomes

$$\frac{1}{3}\int_{V'}[3\mathbf{r}'(\hat{\mathbf{r}} \cdot \mathbf{r}') - r'^2\hat{\mathbf{r}}]\rho_0(\mathbf{r}')\,d\tau' = \frac{1}{3}\mathbf{Q} \quad (28\text{-}47)$$

where $\mathbf{Q}$ is defined as the vector equal to the integral. We recall from (8-32) that the components of the quadrupole moment tensor $Q_{\alpha\beta}$ for a charge density $\rho_0(\mathbf{r}')$ are given by

$$Q_{\alpha\beta} = \int_{V'}(3\alpha'\beta' - r'^2\delta_{\alpha\beta})\rho_0(\mathbf{r}')\,d\tau' = \int_{V'}(3r'_\alpha r'_\beta - r'^2\delta_{\alpha\beta})\rho_0(\mathbf{r}')\,d\tau' \quad (28\text{-}48)$$

where α and β are independently taken to be the rectangular coordinates x, y, z. (We are now using α and β as indices as well as explicitly writing the components of $\mathbf{r}'$ as r'_α in order to avoid confusion in notation.) Let us also write $\hat{\mathbf{r}}$ in terms of its rectangular components (direction cosines) as $\hat{\mathbf{r}} = l_x\hat{\mathbf{x}} + l_y\hat{\mathbf{y}} + l_z\hat{\mathbf{z}}$. Now consider the sum

$$\sum_{\alpha=x,\,y,\,z} l_\alpha Q_{\alpha\beta} = \int_{V'} \sum_{\alpha=x,\,y,\,z} \left(3l_\alpha r'_\alpha r'_\beta - l_\alpha r'^2\delta_{\alpha\beta}\right)\rho_0\,d\tau'$$

$$= \int_{V'}\left[3r'_\beta(\hat{\mathbf{r}}\cdot\mathbf{r}') - r'^2 l_\beta\right]\rho_0\,d\tau' = Q_\beta$$

where (8-27) has been used. The components of the vector $\mathbf{Q}$ are thus given by

$$Q_\beta = \sum_{\alpha=x,\,y,\,z} l_\alpha Q_{\alpha\beta} \qquad (\beta = x,\,y,\,z) \tag{28-49}$$

and can be found once the components of the quadrupole moment tensor of ρ_0 are known and the direction of $\hat{\mathbf{r}}$ has been chosen. Now that we know how to find $\mathbf{Q}$, we can combine (28-47) and (28-46) to give

$$\mathbf{A}_{eq} = -\frac{i\mu_0\omega(1 - ikr)\mathbf{Q}}{24\pi r^2}e^{i(kr-\omega t)} \tag{28-50}$$

Finally, collecting all of our separate results, we can write our vector potential as the sum

$$\mathbf{A} = \mathbf{A}_{ed} + \mathbf{A}_{md} + \mathbf{A}_{eq} \tag{28-51}$$

as given by (28-41), (28-44), and (28-50). It is convenient to consider separately the electromagnetic fields arising from these terms. When we do this in the next sections, we will find that they will lead to fields corresponding to outward-traveling waves that have nonzero values of the time-average Poynting vector so that the source can be said to be "radiating." These fields are conventionally called electric dipole, magnetic dipole, and electric quadrupole radiation, respectively. In other fields of physics, especially in nuclear physics, these are often given the corresponding symbolic designations of E1, M1, and E2 radiations.

28-3 ELECTRIC DIPOLE RADIATION

We want to consider the field described by $\mathbf{A}_{ed}$. It is most convenient to use spherical coordinates and, to be specific, we assume that the dipole moment is along the polar axis (z axis). Thus, we can write $\mathbf{p}_0 = p_0\hat{\mathbf{z}} = p_0(\cos\theta\hat{\mathbf{r}} - \sin\theta\hat{\boldsymbol{\theta}})$ by (1-94), and then (28-41) becomes

$$\mathbf{A}_{ed} = -\frac{i\mu_0\omega p_0 e^{i(kr-\omega t)}}{4\pi r}(\cos\theta\hat{\mathbf{r}} - \sin\theta\hat{\boldsymbol{\theta}}) \tag{28-52}$$

which shows explicitly that, since $\mathbf{A}_{ed}$ is parallel to $\mathbf{p}_0$, it has only an r and a θ component and is also independent of φ.

The magnetic induction as found from (28-4) and (1-104) is

$$\mathbf{B} = -\frac{\mu_0 k^2\omega p_0}{4\pi}\left[\frac{1}{kr} + \frac{i}{(kr)^2}\right]\sin\theta\, e^{i(kr-\omega t)}\hat{\boldsymbol{\varphi}} \tag{28-53}$$

Combining this with (28-27), we find the electric field to be

$$\mathbf{E} = -\frac{k^3 p_0}{4\pi\epsilon_0}\left\{\left[\frac{2i}{(kr)^2} - \frac{2}{(kr)^3}\right]\cos\theta\,\hat{\mathbf{r}}\right.$$
$$\left. + \left[\frac{1}{kr} + \frac{i}{(kr)^2} - \frac{1}{(kr)^3}\right]\sin\theta\,\hat{\boldsymbol{\theta}}\right\} e^{i(kr - \omega t)} \qquad (28\text{-}54)$$

While these fields are quite complicated, they both have the form of waves traveling outward with speed c from the source at the origin. The amplitudes of the various components depend on r in different ways. The electric field has both an $\hat{\mathbf{r}}$ and a $\hat{\boldsymbol{\theta}}$ component, each of which has a different dependence on angle and on distance from the source. Thus, $\mathbf{E}$ is not proportional to $\hat{\mathbf{z}} = \cos\theta\,\hat{\mathbf{r}} - \sin\theta\,\hat{\boldsymbol{\theta}}$ and hence is not parallel to the dipole $\mathbf{p}_0$, although it does lie in the plane defined by $\mathbf{p}_0$ and $\hat{\mathbf{r}}$. Since $\mathbf{B}$ has only a $\hat{\boldsymbol{\varphi}}$ component, the lines of $\mathbf{B}$ are circles about the z axis, that is, about the axis of the dipole. In addition, $\mathbf{B}$ is perpendicular to the plane containing $\mathbf{p}_0$ and $\mathbf{E}$; hence $\mathbf{B}$ and $\mathbf{E}$ are perpendicular to each other. [Of course we still have to take the real parts of (28-53) and (28-54) in order to get the physical fields.]

Although these results are exact, it is customary and convenient to consider what they become in two extreme limiting cases corresponding to whether one is near to the source or far from it on a wavelength scale. If $r \ll \lambda$, so that $kr \ll 1$, one speaks of the "near zone," while if $r \gg \lambda$ ($kr \gg 1$), it is called the "far" or "radiation zone"; the case in which $kr \approx 1$ is called the "intermediate" or "induction zone" and the complete expressions for $\mathbf{E}$ and $\mathbf{B}$ should be used. We consider the two extreme cases separately.

1. Near Zone

Since $kr \ll 1$, we can set $e^{ikr} \simeq 1$. Furthermore, since the kr terms in the amplitudes all appear in the denominators, the highest power of kr will lead to the predominant term. In this way, (28-53) and (28-54) lead to

$$\mathbf{E}_N \simeq \frac{p_0 e^{-i\omega t}}{4\pi\epsilon_0 r^3}(2\cos\theta\,\hat{\mathbf{r}} + \sin\theta\,\hat{\boldsymbol{\theta}}) \qquad (28\text{-}55)$$

$$\mathbf{B}_N \simeq -\frac{i\mu_0\omega p_0 e^{-i\omega t}}{4\pi r^2}\sin\theta\,\hat{\boldsymbol{\varphi}} \qquad (28\text{-}56)$$

These can be put in a form that is even more understandable if we introduce the actual dipole moment $\mathbf{p}$ given by

$$\mathbf{p} = \mathbf{p}_0 e^{-i\omega t} = p_0 e^{-i\omega t}\hat{\mathbf{z}} \qquad (28\text{-}57)$$

for then we have

$$\mathbf{E}_N = \frac{p}{4\pi\epsilon_0 r^3}(2\cos\theta\,\hat{\mathbf{r}} + \sin\theta\,\hat{\boldsymbol{\theta}}) \qquad (28\text{-}58)$$

$$\mathbf{B}_N = \frac{\mu_0\sin\theta}{4\pi r^2}\frac{dp}{dt}\hat{\boldsymbol{\varphi}} = \frac{\mu_0}{4\pi r^2}\left(\frac{d\mathbf{p}}{dt}\right)\times\hat{\mathbf{r}} \qquad (28\text{-}59)$$

with the use of (28-24) and $\hat{\mathbf{z}}\times\hat{\mathbf{r}} = \sin\theta\,\hat{\boldsymbol{\varphi}}$ by (1-94) and (1-92). If we compare (28-58) with (8-50), we see that $\mathbf{E}_N$ has exactly the form of an electric dipole field arising from a dipole moment p located at the origin. In this case, of course, the dipole is oscillating with time, but the field point is so close to it that the retardation effects are negligible and the electric field follows the dipole moment in time as if the effects were propagated instantaneously.

If we now compare (28-59) with (14-6), we see that $\mathbf{B}_N$ has the same form as the static induction produced by a current element $I'd\mathbf{s}' = d\mathbf{p}/dt$. This is also consistent with the expression for the vector potential, since, with the use of (28-57) and $kr \ll 1$, (28-52) can be written as

$$\mathbf{A}_{ed\,N} = \frac{\mu_0}{4\pi r} \frac{d\mathbf{p}}{dt} \qquad (28\text{-}60)$$

which corresponds again to a current element $d\mathbf{p}/dt$ as is shown by (16-10).

This result that an oscillating dipole moment is equivalent to a current element can be understood quite easily if we refer to the specific case of our prototype of an electric dipole consisting of two equal and opposite charges q and $-q$ separated by a small distance $d\mathbf{s}'$. As we saw in (8-44), the dipole moment will be given by $\mathbf{p} = q\,d\mathbf{s}'$. Now if the charges vary periodically, there will be a current $I' = dq/dt$ between the ends of $d\mathbf{s}'$ and thus

$$\frac{d\mathbf{p}}{dt} = \frac{dq}{dt} d\mathbf{s}' = I' d\mathbf{s}' \qquad (28\text{-}61)$$

in agreement with the above. This is also consistent with our original assumption made in the paragraph following (28-31) that the source dimensions are small compared to a wavelength, that is, $ds' \ll \lambda$ in this case. Under these circumstances, it is appropriate to assume that the current I' is uniform over the length of the dipole.

2. Radiation Zone

Here we assume that $kr \gg 1$, and we can no longer set $e^{ikr} \simeq 1$. In this case, the predominant terms in (28-54) and (28-53) will be those with the lowest power of kr in the denominator, and we get the approximations

$$\mathbf{E}_R \simeq -\frac{k^2 p_0}{4\pi\epsilon_0 r} \sin\theta e^{i(kr-\omega t)}\hat{\boldsymbol{\theta}} \qquad (28\text{-}62)$$

$$\mathbf{B}_R \simeq -\frac{\mu_0 k \omega p_0}{4\pi r} \sin\theta e^{i(kr-\omega t)}\hat{\boldsymbol{\phi}} \qquad (28\text{-}63)$$

Both of these fields have amplitudes that vary only as $1/r$, and they have the same dependence upon $\sin\theta$. In addition, each is perpendicular to the direction of propagation to the field point as given by $\hat{\mathbf{r}}$ and they are perpendicular to each other. In other words, at large distances both of these fields become *transverse* waves. Since $\hat{\boldsymbol{\phi}} = \hat{\mathbf{r}} \times \hat{\boldsymbol{\theta}}$, we see that these fields are related as follows:

$$\mathbf{B}_R = \frac{1}{c}\hat{\mathbf{r}} \times \mathbf{E}_R \qquad (28\text{-}64)$$

where we used (28-21) and $\mu_0\epsilon_0 = 1/c^2$. Comparing this with (24-33) and remembering that we are in a vacuum and $\hat{\mathbf{r}}$ is the direction of propagation, we see that (28-64) is exactly the relation between the fields characteristic of a *plane wave in free space*. These relations are illustrated in Figure 28-2, which shows the fields at a given time.

Since both $\mathbf{E}_R$ and $\mathbf{B}_R$ are proportional to $\sin\theta$, they both become zero in the direction of the dipole $\mathbf{p}_0$ and have their maximum value for a given r in the direction perpendicular to the dipole axis.

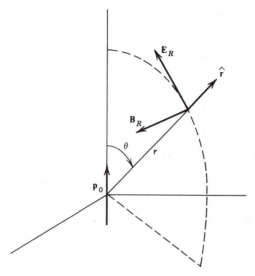

Figure 28-2. Fields at a given time in the radiation zone of an oscillating electric dipole.

Because of the relation between $\mathbf{E}_R$ and $\mathbf{B}_R$ given by (28-64) and shown in the figure, the Poynting vector $\mathbf{S}$ is seen to be radially outward. We can find its time average by using (24-104):

$$\langle \mathbf{S} \rangle = \frac{1}{2\mu_0} \operatorname{Re}(\mathbf{E}_R \times \mathbf{B}_R^*) = \mathscr{S}\hat{\mathbf{r}} = \frac{\mu_0 \omega^4 |p_0|^2}{32\pi^2 c r^2} \sin^2 \theta \hat{\mathbf{r}} \tag{28-65}$$

Since this is positive, there is a net outward flow of energy from the dipole; hence the appropriateness of the name "radiation zone." This energy must be supplied by whatever source is being used to keep the dipole oscillating with constant amplitude. The magnitude $\mathscr{S}$ is seen to be proportional to $1/r^2$ that is the familiar inverse square law of radiation intensity. (This has occurred because the individual fields vary as $1/r$ in this region.) We also see that $\mathscr{S}$ varies with angle as $\sin^2 \theta$; this is illustrated in Figure 28-3 that shows $\mathscr{S}$ as a function of θ for a definite value of r. (Since $\mathscr{S}$ is independent of φ, the figure can be imagined to be rotated about the dipole axis to give the surface representing radiation intensity as a function of direction of propagation.) Finally, we note that $\mathscr{S}$ is proportional to the absolute square of the dipole moment and to the fourth power of the frequency; thus, all other things being equal, it is easier to radiate at a higher frequency.

If we integrate (28-65) over a sphere of radius r, we can find the total rate at which the dipole is radiating energy, that is, the power $\mathscr{P}$. Using (28-65) along with (1-100), we get

$$\mathscr{P} = \int \langle \mathbf{S} \rangle \cdot d\mathbf{a} = \int_0^{2\pi} \int_0^{\pi} \mathscr{S} r^2 \sin \theta \, d\theta \, d\varphi = \frac{\mu_0 \omega^4 |p_0|^2}{12\pi c} \tag{28-66}$$

We have calculated these results by using the approximate values for the fields in the radiation zone. Physically, however, it is evident that there cannot be nonzero contributions to $\langle \mathbf{S} \rangle$ from the terms that drop off with a higher power of r, since if they existed in the near or intermediate zones, they also would have to be present far from the source because there is no matter in the intervening region to absorb these contributions. This can also be verified directly. If we use the complete expressions (28-53) and

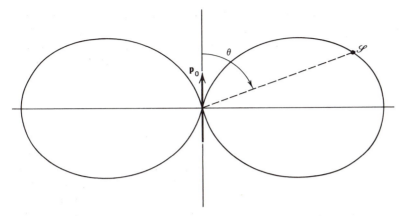

Figure 28-3. Magnitude of the Poynting vector as a function of angle for electric dipole radiation.

(28-54) to calculate $\mathbf{E} \times \mathbf{B}^*$, and omit the various constants we find that

$$\mathbf{E} \times \mathbf{B}^* \sim \left[\frac{1}{(kr)^2} + \frac{i}{(kr)^5} \right] \sin^2 \theta \hat{\mathbf{r}} - \left[\frac{i}{(kr)^3} + \frac{i}{(kr)^5} \right] \sin 2\theta \hat{\boldsymbol{\theta}} \quad (28\text{-}67)$$

Since it is only the real part of $\mathbf{E} \times \mathbf{B}^*$ that contributes to the time-average energy flow, we see that $\mathrm{Re}\,(\mathbf{E} \times \mathbf{B}^*) \sim \sin^2 \theta \hat{\mathbf{r}}/(kr)^2$, which involves only the $1/kr$ terms in the fields, that is, solely the contribution of the radiation zone. The rest of the terms in (28-67) are imaginary. The other components of the physical fields do contribute to the instantaneous value of $\mathbf{S}$, but they give rise to parts that flow alternately away from and toward the source as well as in the $\hat{\boldsymbol{\theta}}$ direction and, on the average, are zero.

As we previously noted, the energy that appears as radiation must be supplied by the external source producing the oscillating dipole. Thus, from the point of view of this external source, the dipole acts as a dissipative element in much the same way as a resistance dissipates electromagnetic energy into heat as we found in Section 12-4. It has been found convenient to describe a radiating system in terms of an equivalent or "radiation resistance" $\mathcal{R}$. In Exercise 12-13 we saw that the total rate at which heat is produced by a resistor $\mathcal{R}$ is $\mathcal{R}I^2$. If, now, the current is oscillating in time, this becomes $\mathcal{R}|I_0|^2 \cos^2(\omega t + \vartheta)$, and, when this is averaged over time, with the use of (24-102), we find the time-average rate of energy dissipation to be $\mathcal{P} = \frac{1}{2}\mathcal{R}|I_0|^2 = \mathcal{R}I_{\mathrm{rms}}^2$ in terms of the magnitude of the peak current $|I_0|$ or the "root-mean-square" current I_{rms}. Thus, the resistance can be written in terms of power loss as

$$\mathcal{R} = \frac{2\mathcal{P}}{|I_0|^2} = \frac{\mathcal{P}}{I_{\mathrm{rms}}^2} \quad (28\text{-}68)$$

In order to apply these ideas to an electric dipole radiator, it is conventional to use the prototype model of two charges separated by a distance ds'. Combining (28-57) and (28-61), we find that $I'\,ds' = I_0 e^{-i\omega t}\,ds' = -i\omega \mathbf{p}_0 e^{-i\omega t}$, so that $|\mathbf{p}_0|^2 = |I_0|^2 (ds')^2/\omega^2$. Inserting this equivalence into (28-66), and using (28-68) to define the radiation resistance, we find it to be given by

$$\mathcal{R}_{\mathrm{ed}} = \frac{\mu_0 \omega^2 (ds')^2}{6\pi c} = \frac{2}{3}\pi Z_0 \left(\frac{ds'}{\lambda} \right)^2 \quad (28\text{-}69)$$

with the use of (28-21) and (24-95). [The value of the numerical factor in (28-69) is 790 ohms.] The radiation resistance can be taken as a measure of the efficiency of a radiating system since we see from (28-68) that the larger the value of $\mathscr{R}$, the more power will be radiated for a given peak current. (We recall, however, that we have assumed that $ds' \ll \lambda$.)

■ **Example**

Accelerated point charge. We know from (8-19) and Exercise 8-2 that a single point charge can have a dipole moment with respect to an arbitrary origin and it is given by $\mathbf{p} = q\mathbf{r}'$. If the charge position is oscillating in time, we will have $\mathbf{r}' = \mathbf{r}_0 e^{-i\omega t}$. In order to use our previous results, let us assume that the charge is on the z axis so that $p = qz' = qz_0 e^{-i\omega t} = p_0 e^{-i\omega t}$. It is also convenient here to deal with instantaneous rather than time-average quantities. Now both (28-62) and (28-63) are proportional to $\omega^2 p_0$, which can be put in another form. The acceleration of the charge is $a = d^2 z'/dt^2 = -\omega^2 z'$ so that we can write $p = -(qa/\omega^2)$. Then the term $p_0 e^{i(kr-\omega t)} = p_0 e^{-i\omega[t-(r/c)]} = [p] = -(q/\omega^2)[a]$ where both $[p]$ and $[a]$ represent these quantities evaluated at the retarded time. Making these substitutions in (28-62) and (28-64), and using (28-21), we find that we can write

$$\mathbf{E}_R = \frac{q[a]\sin\theta\,\hat{\boldsymbol{\theta}}}{4\pi\epsilon_0 c^2 r} \qquad \mathbf{B}_R = \frac{q[a]\sin\theta\,\hat{\boldsymbol{\varphi}}}{4\pi\epsilon_0 c^3 r} \tag{28-70}$$

so that the instantaneous Poynting vector $\mathbf{S} = (\mathbf{E}_R \times \mathbf{B}_R)/\mu_0$ is

$$\mathbf{S} = \frac{q^2 [a]^2 \sin^2\theta\,\hat{\mathbf{r}}}{16\pi^2 \epsilon_0 c^3 r^2} \tag{28-71}$$

Integrating this over a sphere of radius r as in (28-66), we find the total instantaneous rate of radiation due to this accelerated charge to be

$$\mathscr{P}_{\text{instan}} = \frac{q^2 [a]^2}{6\pi\epsilon_0 c^3} \tag{28-72}$$

Although these results were obtained for this special case of a harmonically oscillating charge, their final form no longer contains this explicit assumption and, in fact, they can be shown to be correct for a slowly moving point charge whose motion otherwise is quite arbitrary. The commonly found expression (28-72) shows quite clearly that a moving charge will not radiate unless it is accelerating and (28-72) gives its total instantaneous rate of doing so. ■

28-4 MAGNETIC DIPOLE RADIATION

Let us assume the magnetic dipole moment to be along the z axis so that $\mathbf{m}_0 = m_0 \hat{\mathbf{z}}$. Then (28-44) can be written as

$$\mathbf{A}_{\text{md}} = \frac{\mu_0 m_0}{4\pi r^2} (1 - ikr) e^{i(kr-\omega t)} \sin\theta\,\hat{\boldsymbol{\varphi}} \tag{28-73}$$

with the use of (1-94) and (1-92). We can now proceed exactly as before to find the fields from (28-4) and (28-27); the results are

$$\mathbf{B} = -\frac{k^3 m_0}{4\pi\epsilon_0 c^2} \left\{ \left[\frac{2i}{(kr)^2} - \frac{2}{(kr)^3} \right] \cos\theta \hat{\mathbf{r}} \right.$$

$$\left. + \left[\frac{1}{kr} + \frac{i}{(kr)^2} - \frac{1}{(kr)^3} \right] \sin\theta \hat{\boldsymbol{\theta}} \right\} e^{i(kr-\omega t)} \qquad (28\text{-}74)$$

$$\mathbf{E} = \frac{\mu_0 k^2 \omega m_0}{4\pi} \left[\frac{1}{kr} + \frac{i}{(kr)^2} \right] \sin\theta e^{i(kr-\omega t)} \hat{\boldsymbol{\varphi}} \qquad (28\text{-}75)$$

If we now compare these with the electric dipole fields (28-53) and (28-54), we see that the dependence on r and on θ are the same in the two cases but that the electric field and magnetic induction have been "interchanged" in a sense. The actual relations among the field vectors are

$$\mathbf{B}_{md} = \frac{m_0}{c^2 p_0} \mathbf{E}_{ed} \qquad \mathbf{E}_{md} = -\frac{m_0}{p_0} \mathbf{B}_{ed} \qquad (28\text{-}76)$$

Put another way, the fields of one kind will be transformed into those of the other if we make the replacements

$$p_0 \rightarrow \frac{m_0}{c} \qquad \mathbf{E} \rightarrow c\mathbf{B} \qquad \mathbf{B} \rightarrow -\frac{\mathbf{E}}{c} \qquad (28\text{-}77)$$

[In the last two replacements, we recognize an example of the duality property of electromagnetic fields found in Exercise 21-12 for they correspond to (21-77) with $\alpha = -90°$ and $C = 1$. Since (21-77) was worked out for a source-free region, we were not able to predict the appropriate replacement for p_0.]

The lines of $\mathbf{E}$ given by (28-75) are circles about the polar axis that corresponds to the direction of $\mathbf{m}_0$ and thus $\mathbf{E}$ is perpendicular to the plane containing $\mathbf{B}$ and $\mathbf{m}_0$.

In the radiation zone, the fields are again transverse to the direction of propagation and are

$$\mathbf{B}_R = -\frac{k^2 m_0}{4\pi\epsilon_0 c^2 r} \sin\theta e^{i(kr-\omega t)} \hat{\boldsymbol{\theta}} \qquad (28\text{-}78)$$

$$\mathbf{E}_R = \frac{\mu_0 k \omega m_0}{4\pi r} \sin\theta e^{i(kr-\omega t)} \hat{\boldsymbol{\varphi}} \qquad (28\text{-}79)$$

They also satisfy (28-64) again which is compatible with (28-77) since it becomes $-\mathbf{E}_R/c = \hat{\mathbf{r}} \times \mathbf{B}_R$, which leads back to (28-64) with the use of (1-30) and $\hat{\mathbf{r}} \cdot \mathbf{B}_R = 0$. The relation between these fields is shown for a particular time in Figure 28-4. When we compare this figure with Figure 28-2 we see that it corresponds to a rotation of $-90°$ about the $\hat{\mathbf{r}}$ direction as implied by (28-77).

The time-average Poynting vector and total radiated power flow are easily calculated from (28-78) and (28-79) and it is found that

$$\langle \mathbf{S} \rangle = \mathscr{S} \hat{\mathbf{r}} = \frac{\mu_0 \omega^4 |m_0|^2}{32\pi^2 c^3 r^2} \sin^2\theta \hat{\mathbf{r}} \qquad (28\text{-}80)$$

$$\mathscr{P} = \frac{\mu_0 \omega^4 |m_0|^2}{12\pi c^3} \qquad (28\text{-}81)$$

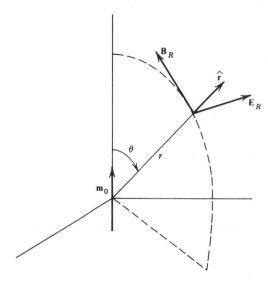

Figure 28-4. Fields at a given time in the radiation zone of an oscillating magnetic dipole.

which can equally well be obtained from (28-65) and (28-66) by making the replacement of p_0 by m_0/c. Thus magnetic dipole and electric dipole radiation are alike in having the same proportionality to the fourth power of the frequency. They also have the same $\sin^2 \theta$ angular dependence so that Figure 28-3 applies to this case also (with $\mathbf{p}_0$ replaced by $\mathbf{m}_0$).

■ **Example**

Current loop. In order to have a specific case, let us consider our prototype of a magnetic dipole, that is, the small circular current loop of radius a for which the dipole moment is $\pi a^2 I$ as given by (19-27). If $a \ll \lambda$, we can consider I to have the same value at all points on the loop, and if I_0 is its peak value, we have $m_0 = \pi a^2 I_0$. If we substitute this into (28-81) and use (28-68), we find the radiation resistance to be

$$\mathcal{R}_{md} = \frac{\pi \mu_0 \omega^4 a^4}{6c^3} = \frac{8}{3} \pi^5 Z_0 \left(\frac{a}{\lambda} \right)^4 \tag{28-82}$$

(The value of the numerical factor is 3.08×10^5 ohms.) Comparing this with (28-69), we see that there is a difference in the frequency dependence of the radiation resistance of our two model dipoles since $\mathcal{R}_{ed} \sim \omega^2$ while $\mathcal{R}_{md} \sim \omega^4$. ■

28-5 LINEAR ELECTRIC QUADRUPOLE RADIATION

The appropriate vector potential is given by (28-50). Since the situation in the radiation zone ($kr \gg 1$) is of most interest, we go directly to it and approximate (28-50) by

$$\mathbf{A}_{eq\,R} = -\frac{\mu_0 \omega^2 \mathbf{Q}}{24\pi cr} e^{i(kr - \omega t)} \tag{28-83}$$

where we remember that $\mathbf{Q}$ is a function of the direction of propagation given by $\hat{\mathbf{r}}$.

Since there are such a large number of possibilities for $Q_{\alpha\beta}$ and hence for $\mathbf{Q}$, we will consider only one simple but important example of electric quadrupole radiation—that arising from a linear quadrupole.

We previously saw that if a charge distribution has axial (rotational) symmetry about the z axis then there is only one independent component of the quadrupole moment tensor and the only nonvanishing values of $Q_{\alpha\beta}$ are given by (8-39) as

$$Q^a_{zz} = Q_0 \qquad Q^a_{xx} = Q^a_{yy} = -\tfrac{1}{2}Q_0 \qquad (28\text{-}84)$$

where we are using Q_0 rather than the previous Q^a for consistency of notation within this chapter. A conventional prototype of such a case is the point charge distribution of Figure 8-5b, which is repeated in Figure 28-5. The charge $2q$ is located at the origin while the two charges $-q$ are on the z axis, each a distance a from the origin. It is easily found from the point charge form of (28-48) given by (8-26) that if $q = q_0 e^{-i\omega t}$ then in this case we have

$$Q_0 = -4q_0 a^2 \qquad (28\text{-}85)$$

in agreement with Exercise 8-6. While the results of this section are applicable to the charge distribution of Figure 28-5, they are not restricted to it but can be used for any axially symmetric charge distribution described by (28-84). It is interesting to note, however, that the charges of Figure 28-5 have neither an electric dipole moment nor can they have a magnetic dipole moment. Thus the lowest-order radiation that they can produce will be electric quadrupole radiation.

Combining (28-84) and (28-49), we find the components of **Q** to be $Q_\beta = l_\beta Q_{\beta\beta}$ so that $Q_x = -\tfrac{1}{2}Q_0 l_x$, $Q_y = -\tfrac{1}{2}Q_0 l_y$, $Q_z = Q_0 l_z$ and therefore

$$\mathbf{Q} = Q_0\left(-\tfrac{1}{2}l_x\hat{\mathbf{x}} - \tfrac{1}{2}l_y\hat{\mathbf{y}} + l_z\hat{\mathbf{z}}\right)$$

$$= \tfrac{1}{2}Q_0\left[(3\cos^2\theta - 1)\hat{\mathbf{r}} - \tfrac{3}{2}\sin 2\theta\hat{\boldsymbol{\theta}}\right] \qquad (28\text{-}86)$$

where we have used (1-93) and (1-94) to express **Q** in spherical coordinates. Substitution of this **Q** into (28-83) yields the radiation zone vector potential for the linear quadrupole:

$$\mathbf{A}^a_{\text{eq } R} = -\frac{\mu_0\omega^2 Q_0 e^{i(kr-\omega t)}}{48\pi cr}\left[(3\cos^2\theta - 1)\hat{\mathbf{r}} - \frac{3}{2}\sin 2\theta\hat{\boldsymbol{\theta}}\right] \qquad (28\text{-}87)$$

The fields can now be found in the usual way from (28-4) and (28-27) while keeping only the largest terms which are $\sim 1/r$ since we are in the radiation zone with $kr \gg 1$.

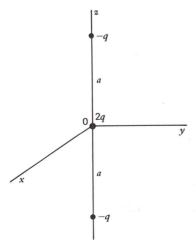

Figure 28-5. Prototype charge distribution of a linear electric quadrupole.

The results turn out to be

$$\mathbf{B}_R = \frac{i\mu_0\omega^3 Q_0}{32\pi c^2 r}\sin 2\theta\, e^{i(kr-\omega t)}\hat{\boldsymbol{\varphi}} \tag{28-88}$$

$$\mathbf{E}_R = \frac{i\mu_0\omega^3 Q_0}{32\pi c r}\sin 2\theta\, e^{i(kr-\omega t)}\hat{\boldsymbol{\theta}} \tag{28-89}$$

which again are transverse to the direction of propagation, are related by (28-64), and will be described by a figure similar to Figure 28-2.

The time-average Poynting vector is found to be

$$\langle\mathbf{S}\rangle = \frac{1}{2\mu_0}\operatorname{Re}\left(\mathbf{E}_R \times \mathbf{B}_R^*\right) = \mathcal{S}\hat{\mathbf{r}} = \frac{\mu_0\omega^6|Q_0|^2}{2048\pi^2 c^3 r^2}\sin^2 2\theta\,\hat{\mathbf{r}} \tag{28-90}$$

which is proportional to the sixth power of the frequency. The angular dependence of $\mathcal{S}$ is given by $\sin^2 2\theta = 4\sin^2\theta\cos^2\theta$ and is shown in Figure 28-6. The value of $\mathcal{S}$ is zero at $\theta = 0°$, $90°$, and $180°$, and has maxima at $\theta = 45°$ and $135°$.

Integrating (28-90) over a sphere of radius r, we find the total radiated power to be

$$\mathcal{P} = \frac{\mu_0\omega^6|Q_0|^2}{960\pi c^3} \tag{28-91}$$

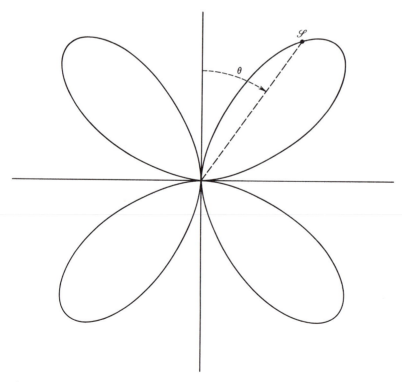

Figure 28-6. Magnitude of the Poynting vector as a function of angle for linear electric quadrupole radiation.

28-6 ANTENNAS

Up to now we have assumed that the dimensions of our radiating system are small compared to a free space wavelength, so that we could take the current in our simple models to be uniform over the system. In practical cases, it is convenient to use radiators whose dimensions are comparable to a wavelength; such a system is usually called an *antenna*. In this case, the current distribution is not uniform over the length and this must be taken into account. While antenna design is an art by itself, we can illustrate the general principles involved by means of a few examples. However, we will consider only examples involving electric dipole radiation.

The basic situation is illustrated in Figure 28-7. We have a straight conductor of length $2l$ along the z axis. An oscillating current is produced and maintained in this conductor usually by means of connections at its center to an external power supply. We want to find the fields at a point P with coordinates r and θ; it is assumed that P is far enough away so that we need deal only with the radiation zone. Since we found in (28-61) that an oscillating dipole is equivalent to a current element, a natural approach is to visualize the antenna as a series of current elements $I'(z')\,dz'$. Each element produces an electric field $d\mathbf{E}$ and the total electric field at P will be the sum of these because of the superposition property of electromagnetic fields. In other words, we are dealing with an interference phenomenon.

Immediately after (28-68), we found the equivalence $-i\omega p_0 = I_0\,ds'$, so that in this situation we make the replacement $p_0 = (i/\omega)I'(z')\,dz'$ in (28-62) and use (28-21) to obtain

$$d\mathbf{E} = -\frac{i\mu_0\omega I'(z')\,dz'}{4\pi r'}\sin\theta' e^{i(kr'-\omega t)}\hat{\boldsymbol{\theta}}' \qquad (28\text{-}92)$$

where

$$r' = \left(r^2 + z'^2 - 2rz'\cos\theta\right)^{1/2} \qquad (28\text{-}93)$$

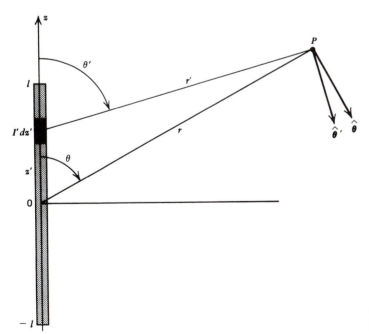

Figure 28-7. Calculation of the radiation field of a linear antenna.

and $\hat{\boldsymbol{\theta}}'$ is the unit vector perpendicular to $\mathbf{r}'$. The total electric field at the field point will then be obtained by integrating (28-92) over the antenna.

We have already assumed that $r \gg \lambda$. We shall now also assume that the total length $2l \ll r$ so that $|z'| \ll r$. Under these conditions, we can approximate r' by

$$r' \simeq r - z' \cos\theta \qquad (28\text{-}94)$$

The amplitude of $d\mathbf{E} \sim 1/r'$ and we will not make much error by simply taking $r' \simeq r$ here; we cannot do this in $e^{ikr'}$, however, because even small changes in r' can make large changes in the exponent. Finally, we neglect the differences between all of the $\hat{\boldsymbol{\theta}}'$ and $\hat{\boldsymbol{\theta}}$ as well as between θ' and θ so that, in effect, we are assuming that all of the contributions to the total $\mathbf{E}$ are parallel to each other. Doing all of this, we finally get our expression for the total electric field to be

$$\mathbf{E} = -\frac{i\mu_0\omega \sin\theta\, e^{i(kr-\omega t)}\hat{\boldsymbol{\theta}}}{4\pi r} \int_{-l}^{l} e^{-ikz'\cos\theta} I'(z')\, dz' \qquad (28\text{-}95)$$

Once $\mathbf{E}$ has been found, we can evaluate $\mathbf{B} = (\hat{\mathbf{r}} \times \mathbf{E})/c$ according to (28-64) for these fields in the radiation zone.

We can go no further until we have a definite current amplitude distribution $I'(z')$ along the antenna.

Example

Half-wave antenna. Clearly the current I' must be zero at the ends $z' = \pm l$ of the antenna, that is, the ends are current nodes. Stated in these terms, the problem is reminiscent of the boundary conditions for a string rigidly fixed at both ends. One would think then, that by proper excitation, one could obtain current distributions analogous to the modes of oscillation of a vibrating string. We recall that the simplest such mode is one with a displacement antinode at the center; in that case, the total length of the string equals one half wavelength of the oscillation. By analogy, we assume a similar situation here. Thus, $2l = \frac{1}{2}\lambda$ so that $l = \frac{1}{4}\lambda$ and we write the current

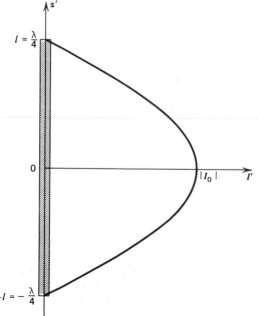

Figure 28-8. Assumed current amplitude distribution for a half-wave antenna.

amplitude distribution as

$$I'(z') = I_0 \cos\left(\frac{\pi z'}{2l}\right) = I_0 \cos\left(\frac{2\pi z'}{\lambda}\right) = I_0 \cos kz' \qquad (28\text{-}96)$$

This distribution is illustrated in Figure 28-8. The limits of integration in (28-95) then become $\pm \frac{1}{4}\lambda = \pm \pi/2k$. Such an antenna is called a *half-wave antenna*. It can be obtained by using a conductor that is actually a half-wavelength long and excited at the center or by bottom excitation of one that is one quarter-wavelength long and that is perpendicular to a conducting plane such as the earth or a large metal sheet. In the latter case, the image currents in the plane provide the other half of the antenna. It turns out in practice that the existence of the radiation results in a current distribution that is different from the simple one assumed in (28-96). The effect is small, however, and we will not consider it further.

If we let $\mathscr{I}$ be the integral in (28-95), substitute (28-96) into it, change to the dimensionless variable $u = kz'$, and use (24-22), we find that it becomes

$$\mathscr{I} = I_0 \int_{-\pi/2k}^{\pi/2k} e^{-ikz'\cos\theta} \cos kz'\, dz'$$

$$= \frac{I_0}{k} \int_{-\pi/2}^{\pi/2} e^{-iu\cos\theta} \cos u\, du = \frac{I_0}{2k} \int_{-\pi/2}^{\pi/2} \left[e^{iu(1-\cos\theta)} + e^{-iu(1+\cos\theta)} \right] du$$

$$= \frac{I_0}{k} \left\{ \frac{\sin\left[\frac{1}{2}\pi(1-\cos\theta)\right]}{1-\cos\theta} + \frac{\sin\left[\frac{1}{2}\pi(1+\cos\theta)\right]}{1+\cos\theta} \right\}$$

$$= \frac{2cI_0}{\omega \sin^2\theta} \cos\left(\tfrac{1}{2}\pi\cos\theta\right) \qquad (28\text{-}97)$$

Substituting this into (28-95), we obtain

$$\mathbf{E} = -\frac{i\mu_0 c I_0}{2\pi r \sin\theta} \cos\left(\frac{1}{2}\pi\cos\theta\right) e^{i(kr-\omega t)} \hat{\boldsymbol{\theta}} \qquad (28\text{-}98)$$

which, interestingly, has an amplitude that is independent of the frequency. We see from (28-97) that this occurred because the integration introduced a factor $1/\omega$, which canceled the ω in (28-95). Since each element of the antenna does not radiate along its axis, that is, for $\theta = 0$ or π, we expect $\mathbf{E}$ to vanish at these angles also. The angular dependence found in (28-98) becomes indeterminate at these values of θ, but it is easily verified by the use of L'Hôpital's Rule that $\mathbf{E} \to 0$ as $\theta \to 0$ or π.

The power flow is given by

$$\langle \mathbf{S} \rangle = \frac{1}{2\mu_0} \operatorname{Re}\left(\mathbf{E} \times \mathbf{B}^*\right) = \frac{|\mathbf{E}|^2}{2\mu_0 c} \hat{\mathbf{r}} = \frac{\mu_0 c |I_0|^2}{8\pi^2 r^2 \sin^2\theta} \cos^2\left(\frac{1}{2}\pi\cos\theta\right)\hat{\mathbf{r}} \quad (28\text{-}99)$$

The angular distribution given by $\cos^2\left(\frac{1}{2}\pi\cos\theta\right)/\sin^2\theta$ is shown as the solid curve of Figure 28-9. The dashed curve shows the $\sin^2\theta$ distribution characteristic of a single dipole. The two are qualitatively similar, although the antenna is seen to radiate even more predominantly in the horizontal plane.

The total radiated power is

$$\mathscr{P} = \int \langle \mathbf{S} \rangle \cdot d\mathbf{a} = \frac{\mu_0 c |I_0|^2}{8\pi^2} \int_0^{2\pi} \int_0^{\pi} \frac{\cos^2\left(\frac{1}{2}\pi\cos\theta\right)}{\sin^2\theta} \sin\theta\, d\theta\, d\varphi \quad (28\text{-}100)$$

The integration over φ gives 2π. The integral over θ cannot be expressed in terms of elementary functions, but it can be put into the form of tabulated functions by defining a new variable v by means of $\frac{1}{2}\pi\cos\theta = \frac{1}{2}(\pi - v)$ so that $v = \pi(1 - \cos\theta)$. The

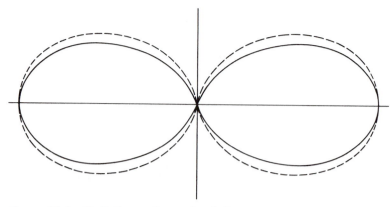

Figure 28-9. Radiation pattern of a half-wave antenna (solid) as compared to that of a single dipole (dashed).

integral then becomes

$$\int_0^\pi \frac{\cos^2\left(\frac{1}{2}\pi\cos\theta\right)d\theta}{\sin\theta} = \frac{1}{\pi}\int_0^{2\pi}\frac{\sin^2\left(\frac{1}{2}v\right)dv}{\sin^2\theta} = \frac{\pi}{2}\int_0^{2\pi}\frac{(1-\cos v)\,dv}{v(2\pi-v)}$$

$$= \frac{1}{4}\int_0^{2\pi}(1-\cos v)\,dv\left(\frac{1}{v}+\frac{1}{2\pi-v}\right)$$

$$= \frac{1}{4}\left[\int_0^{2\pi}\frac{(1-\cos v)\,dv}{v}+\int_0^{2\pi}\frac{(1-\cos v)\,dv}{2\pi-v}\right] \quad (28\text{-}101)$$

If we make the change of variable $w = 2\pi - v$ in the second integral we see that it equals the first so that (28-101) becomes

$$\frac{1}{2}\int_0^{2\pi}\frac{(1-\cos v)\,dv}{v} = \frac{1}{2}(2.438)$$

as found from tables.[1] Combining all of this, we find that (28-100) is

$$\mathscr{P} = \frac{2.438\mu_0 c|I_0|^2}{8\pi} \quad (28\text{-}102)$$

The radiation resistance as found from (28-68) is $\mathscr{R} = 2.438\mu_0 c/4\pi = 0.194 Z_0 = 73.1$ ohms. ∎

The angular dependence of the radiated power shown in Figure 28-9 is independent of the azimuthal angle φ; thus, for example, the antenna radiates equally in all horizontal directions. It is often possible to enhance the radiation in a given direction by utilizing the interference effects obtained by using more than one antenna. Such combinations of antennas are called *antenna arrays*. The general principles used are simple to describe but may be difficult to apply in detail.

Suppose we have a number n of half-wave antennas with their axes parallel to the z direction. The electric field produced by any one of them at a field point P far away from the array can be taken to be given by (28-98). Then the total electric field

[1] M. Abramowitz and I. A. Stegun, eds., *Handbook of Mathematical Functions*, Dover, New York, 1965, pp. 231 ff.

produced by the array will be a sum of terms of this form, that is, we will have

$$\mathbf{E} = -\frac{i\mu_0 c}{2\pi} \sum_{j=1}^{n} \frac{I_{0j}}{r_j \sin\theta_j} \cos\left(\frac{1}{2}\pi\cos\theta_j\right) e^{i(kr_j - \omega t)}\hat{\boldsymbol{\theta}}_j \qquad (28\text{-}103)$$

where, as in Figure 28-7, r_j is the distance from the jth antenna to P, θ_j and $\hat{\boldsymbol{\theta}}_j$ the corresponding angle and unit vector, and I_{0j} the peak current of the (complex) current distribution. One then chooses a convenient origin, usually near the center of the array, so that the spherical coordinates of P are r, θ, and φ with respect to it. It is then possible to express r_j and θ_j in terms of r, θ, φ, and the location of the jth antenna with respect to the origin, much as was done to obtain (28-93). If the array is compact enough so that its dimensions are small compared to r, then one can replace r_j by r in the amplitude denominator, and take each $\hat{\boldsymbol{\theta}}_j$ to be the same as the $\hat{\boldsymbol{\theta}}$ for P, so that (28-103) can be written as

$$\mathbf{E} \simeq -\frac{i\mu_0 c}{2\pi r} e^{i(kr - \omega t)}\hat{\boldsymbol{\theta}} \sum_{j=1}^{n} \frac{I_{0j}}{\sin\theta_j} \cos\left(\frac{1}{2}\pi\cos\theta_j\right) e^{ik(r_j - r)} \qquad (28\text{-}104)$$

We note the appearance of the term $e^{ik(r_j - r)}$ in the sum; this factor is very important in the final result because even though $r_j - r$ may be small in absolute value, $k(r_j - r)$ can have large variations so that each exponential term can be quite different in value. The relative phases of the current distributions of the antennas are contained in the amplitude terms I_{0j} because they can be written as $|I_{0j}|e^{i\vartheta_j}$, as was similarly done in (28-19). We see from (28-104) that there are a large number of possibilities even for such a special case as this. One can freely choose the number n of antennas in the array, the relative magnitudes and phases of their currents ($|I_{0j}|$ and ϑ_j), and the relative locations and spacings of the antennas (which will appear in the expressions for r_j and θ_j). Because of the great variety of specific possibilities, we will not consider antenna arrays in any more detail, but a few simple examples are considered in the exercises.

So far we have only considered antennas as radiating sources. However, as every user of a radio or television set knows, antennas are also used to detect electromagnetic waves. (We considered this aspect briefly in Exercise 24-5 where the emf induced by a plane wave in a small circular loop was calculated.) The question then naturally arises as to how the receiving properties of an antenna are related to its radiative properties if, in fact, there is any relation at all. The answer to this question is indicated by our final topic in this chapter.

■ **Example**

A reciprocity theorem. Consider two length $d\mathbf{s}_1$ and $d\mathbf{s}_2$ of two different conducting wires as shown in Figure 28-10. Suppose the first has an oscillating current I in it; we want to find the emf $\mathscr{E}$ produced in the other. If we assume that they are far apart, the electric field produced by the element $I\,d\mathbf{s}_1$ can be obtained from (28-92), but it will be helpful to write that expression in a more general form. First of all, we drop the primes and replace $I'(z')\,dz'$ by $I\,ds_1$; second, (28-92) involves the assumption that the dipole was oriented along the z axis so that we have $I\,d\mathbf{s}_1 = I\,ds_1\hat{\mathbf{z}}$. Now we write $\hat{\boldsymbol{\theta}} = \hat{\boldsymbol{\varphi}} \times \hat{\mathbf{r}}$ $= [(\hat{\mathbf{z}} \times \hat{\mathbf{r}})/\sin\theta] \times \hat{\mathbf{r}}$ with the use of (1-92) and (1-94), so that $I\,ds_1\sin\theta\hat{\boldsymbol{\theta}} = I(d\mathbf{s}_1 \times \hat{\mathbf{r}})$ $\times \hat{\mathbf{r}}$. The electric field $d\mathbf{E}$ produced by the current element $I\,d\mathbf{s}_1$ at $d\mathbf{s}_2$ can thus be written as

$$d\mathbf{E} = -\frac{i\mu_0\omega[I]}{4\pi r}(d\mathbf{s}_1 \times \hat{\mathbf{r}}) \times \hat{\mathbf{r}} \qquad (28\text{-}105)$$

where $[I]$ is the current evaluated at the retarded time. Now the tangential components of this electric field are continuous across the surface of the wire by (21-26), so that the

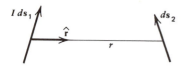

Figure 28-10.   Calculation of a reciprocity theorem.

emf induced in $d\mathbf{s}_2$ as given by (17-7) is found to be

$$d\mathscr{E} = d\mathbf{E} \cdot d\mathbf{s}_2 = -\frac{i\mu_0\omega[I]}{4\pi r}\left[(d\mathbf{s}_1 \times \hat{\mathbf{r}}) \times \hat{\mathbf{r}}\right] \cdot d\mathbf{s}_2$$

$$= \frac{i\mu_0\omega[I]}{4\pi r}\left[(\hat{\mathbf{r}} \times d\mathbf{s}_1) \cdot (\hat{\mathbf{r}} \times d\mathbf{s}_2)\right] \qquad (28\text{-}106)$$

with the use of (1-29) and (1-23). This result is completely symmetric in $d\mathbf{s}_1$ and $d\mathbf{s}_2$ and is unaltered if the sign of $\hat{\mathbf{r}}$ is changed. This means that if the element $d\mathbf{s}_2$ were carrying the current I, exactly this same emf would be induced in the element $d\mathbf{s}_1$. A result of this type is called a *reciprocity theorem*.

What (28-106) tells us, then, is that the value of the induced emf or "received signal" does not depend on whether the element $d\mathbf{s}$ is being employed as a radiator or as a receiver. When extended to the whole antenna, this reciprocity theorem shows that the antenna characteristics, such as angular dependence, are the same whether the antenna radiates or absorbs radiation. For example, since a half-wave antenna does not radiate in the direction of its axis ($\theta = 0$) as we found from (28-98), an electromagnetic wave traveling parallel to its axis will not induce an emf in it either. In this special case, it is clear that this will be so since the incident electric field is transverse to the direction of propagation and will be normal to the conductor forming the antenna. Thus, the electric field will have no tangential component along the antenna length and cannot produce an induced emf. ∎

EXERCISES

28-1 An infinitely long thin straight wire coincides with the z axis. It carries no current for $t < 0$. At $t = 0$, a current I is applied over the whole length of the wire, and it remains at this constant value for all later times. At a point a distance ρ from the wire find $\mathbf{A}$, $\mathbf{B}$, $\mathbf{E}$, and $\mathbf{S}$ for all times. Show that your results reduce to sensible values as $t \to \infty$. (*Hint*: recall Section 16-4.)

28-2 A system of charges and currents occupies a fixed finite volume V'. Show that in general

$$\int_{V'} \mathbf{J}(\mathbf{r}')\, d\tau' = \frac{d\mathbf{p}}{dt}$$

where $\mathbf{p}$ is the total electric dipole moment of the distribution.

28-3 Find the amplitude of the electric field at a point in the equatorial plane that is 10 kilometers from an electric dipole radiating one kilowatt.

28-4 Find the fraction of the energy radiated by an electric dipole within $\pm 10°$ of the equatorial plane.

28-5 Assume that p_0 is real and find the general physical fields $\mathbf{E}$ and $\mathbf{B}$ for an electric dipole from

(28-53) and (28-54). Then find $\mathbf{S}$ and show that it has oscillatory components along $\hat{\mathbf{r}}$ and $\hat{\boldsymbol{\theta}}$. Finally show that your expression for $\mathbf{S}$ reduces correctly to (28-65).

28-6 Show that the fields in the radiation zone due to an electric dipole can be written in the form

$$\mathbf{E}_R = \frac{\mu_0}{4\pi r}\left\{\left[\frac{d^2\mathbf{p}}{dt^2}\right] \times \hat{\mathbf{r}}\right\} \times \hat{\mathbf{r}}$$

$$\mathbf{B}_R = \frac{\mu_0}{4\pi cr}\left[\frac{d^2\mathbf{p}}{dt^2}\right] \times \hat{\mathbf{r}}$$

where the quantity in brackets is evaluated at the retarded time.

28-7 Show that the electric dipole fields in the radiation zone can be written in the form

$$\mathbf{E}_R = \frac{\mu_0}{4\pi r}\left[\frac{dI'}{dt}\right](d\mathbf{s}' \times \hat{\mathbf{r}}) \times \hat{\mathbf{r}}$$

$$\mathbf{B}_R = \frac{\mu_0}{4\pi cr}\left[\frac{dI'}{dt}\right]d\mathbf{s}' \times \hat{\mathbf{r}}$$

where the derivative of the current is evaluated at the retarded time.

28-8 Show that the magnetic dipole fields in the radiation zone can be written in the form

$$\mathbf{E}_R = -\frac{\mu_0}{4\pi c r}\left[\frac{d^2\mathbf{m}}{dt^2}\right] \times \hat{\mathbf{r}}$$

$$\mathbf{B}_R = \frac{\mu_0}{4\pi c^2 r}\left\{\left[\frac{d^2\mathbf{m}}{dt^2}\right] \times \hat{\mathbf{r}}\right\} \times \hat{\mathbf{r}}$$

where $[d^2\mathbf{m}/dt^2]$ is evaluated at the retarded time.

28-9 Show that all of the fields in the radiation zone whether due to an electric dipole, magnetic dipole, or linear electric quadrupole can be expressed in terms of the corresponding vector potential as follows:

$$\mathbf{E}_R = \left\{\left[\frac{d\mathbf{A}}{dt}\right] \times \hat{\mathbf{r}}\right\} \times \hat{\mathbf{r}} \qquad \mathbf{B}_R = \frac{1}{c}\left[\frac{d\mathbf{A}}{dt}\right] \times \hat{\mathbf{r}}$$

where $[d\mathbf{A}/dt]$ is evaluated at the retarded time.

28-10 Find $\mathbf{E}$ and $\mathbf{B}$ in the near zone for the linear electric quadrupole and compare your result with (8-55).

28-11 Show that the radiation fields of a slowly moving point charge can be written in terms of the retarded value [a] of the acceleration as:

$$\mathbf{E}_R = \frac{q\{[\mathbf{a}] \times \hat{\mathbf{r}}\} \times \hat{\mathbf{r}}}{4\pi\epsilon_0 c^2 r} \qquad \mathbf{B}_R = \frac{q[\mathbf{a}] \times \hat{\mathbf{r}}}{4\pi\epsilon_0 c^3 r}$$

28-12 A commonly used approximate method of deriving the fields in the *radiation zone* of a linear electric quadrupole is based on the model illustrated in Figure 28-5. This can be interpreted as two oppositely directed dipoles of amplitude $\mp q_0 a$ located at $z = \pm \frac{1}{2}a$, respectively, and the total field $\mathbf{E}_R$ can be written as a superposition of fields obtained from (28-62). Do this, and then by using the approximation $a \ll r$, show that the first-order result gives (28-89).

28-13 A simple model for calculating the fields in the radiation zone from a magnetic quadrupole can be devised by analogy with the previous exercise and Exercise 19-16. Consider two magnetic dipoles of equal dipole moments $\pm m_0\hat{\mathbf{z}}$ located at $z = \pm b$; the total magnetic dipole moment of this system is zero. Such a situation could be obtained with two small current loops, each parallel to the xy plane, and with oppositely directed currents. The total induction $\mathbf{B}_R$ can then be written as a superposition of fields obtained

from (28-78). Do this, and then by using the approximation $b \ll r$, find the fields $\mathbf{B}_R$ (and hence $\mathbf{E}_R$) as the first-order result.

28-14 An antenna for which the total length $2l = \lambda$ is called a full-wave antenna. Show that two possible current amplitude distributions consistent with this are $I' = I_0 \sin kz'$ and $I' = I_0 \sin k|z'|$. Show that these are consistent with the model of a full-wave antenna as equivalent to two colinear half-wave antennas spaced a half-wavelength apart and excited with relative phases of π and 0, respectively. Find $\mathbf{E}$ and $\langle\mathbf{S}\rangle$ for the first case and thus show that the angular distribution is proportional to $[\sin(\pi\cos\theta)/\sin\theta]^2$. Make a rough sketch of this angular distribution.

28-15 Two half-wave antennas lie in the yz plane. They are parallel to the z axis and their centers lie on the y axis at $y = a$ and $y = -a$. Assume that $a \ll r$, and find the resultant electric field in the xy plane. Assume the currents to have equal amplitude. Now assume that they are a half-wavelength apart, and consider the two cases in which their currents (1) are in phase and (2) have a phase difference of π. Sketch the magnitude of the electric field as a function of φ for these last two cases.

28-16 If the source has both an electric dipole moment and a magnetic dipole moment, the total electric field in the radiation zone will be the sum of (28-62) and (28-79) and we have the possibility of interference. Show, however, that $\langle\mathbf{S}\rangle = \langle\mathbf{S}\rangle_{ed} + \langle\mathbf{S}\rangle_{md}$, that is, the total radiated power does not contain any interference terms. Will this same result be true if it has both an electric dipole moment and a linear electric quadrupole moment?

28-17 The Hertz vector $\boldsymbol{\pi}_e$ was introduced in Exercise 22-7. The Hertz vector corresponding to a point electric dipole is asserted to be $\boldsymbol{\pi}_e = \mathbf{p}_0 e^{i(kr-\omega t)}/4\pi\epsilon_0 r$. Verify that this gives the correct vector potential for an electric dipole.

28-18 Consider a system of charged particles for which the ratio of charge to mass, q_i/m_i, is the same for each particle. Show that the total dipole moment can be written as $\mathbf{p} = (q_i/m_i)M\mathbf{R}_{cm}$ where M is the total mass of the system and $\mathbf{R}_{cm}$ is the position vector of the center of mass. Show that, if there is no net external force on the system, it will not produce electric dipole radiation. Similarly, show that, if there is no net external torque on this system, it will not produce magnetic dipole radiation either.

29

SPECIAL RELATIVITY

The basic concern of the theory of relativity is the comparison of results obtained by observers of physical phenomena who are moving with respect to each other. The special theory, which was propounded by Einstein in 1905 and that is the only form we will consider, deals with observers who have a constant velocity with respect to each other. The term "relativity" refers to the assumed *equivalence* of these observers and their respective coordinate systems for the description of phenomena. As we will see, the origins of special relativity can be found in the structure of electromagnetic theory so that we can usefully begin by considering how moving observers describe some electromagnetic effects.

29-1 HISTORICAL ORIGINS OF SPECIAL RELATIVITY

When we consider, for example, the electric field vector $\mathbf{E}(\mathbf{r}, t)$, we mean the value of this quantity at a point $\mathbf{r}$ and a time t that are specified with respect to some particular set of coordinate axes. The observer who is determining $\mathbf{E}$ at $\mathbf{r}$ and t is assumed to be at rest with respect to these axes. In the paragraph following (1-14), we briefly alluded to the question of what coordinate system was to be used and we concluded that it was one of the inertial systems of mechanics. As we also know from mechanics, a coordinate frame fixed to the surface of the earth is, for most purposes, a satisfactory approximation to an inertial system.

In the discussion of Faraday's law for moving media in Section 17-3, however, we did encounter the question of how observers moving with respect to each other would describe the same phenomena and we saw that in general their descriptions would be different. In particular, we found (17-29), which stated that

$$\mathbf{E}' = \mathbf{E} + \mathbf{v} \times \mathbf{B} \tag{29-1}$$

was the way in which the fields in the two systems were related; then Faraday's law $\nabla \times \mathbf{E} = -\partial \mathbf{B}/\partial t$ had a form that was independent of the motion of the medium. We saw that in some cases it was easier to discuss a problem by going into the moving system and dealing with the "motional" electric field $\mathbf{E}'_m = \mathbf{v} \times \mathbf{B}$. What is of more interest to us here, however, is that (29-1) shows that these two relatively moving observers describe the same phenomena in *different ways*. As an extreme example, if $\mathbf{E} = 0$, then what appears to one observer as a purely magnetic induction $\mathbf{B}$ is seen by the other as an electric field $\mathbf{E}'$. Thus an electric field is not an absolute property of a given point and time but also depends on the motion of the observer.

In order to obtain (29-1), we used the fact that $\mathbf{F}' = \mathbf{F}$ where the forces being compared are measured in two inertial systems, that is, they have no acceleration with respect to each other and with respect to the primary inertial system. Let us remind ourselves of how this situation comes about.

In Figure 29-1, we show two coordinate systems S and S' with rectangular axes chosen to be parallel to each other for simplicity. The relative position of their origins is given by $\mathbf{R}_0 = \mathbf{V}t$ so that their origins coincide at $t = 0$ and $\mathbf{V}$ is the *constant* velocity of S' with respect to S. The position vectors of a point P are $\mathbf{r}$ and $\mathbf{r}'$, so that

$$\mathbf{r}' = \mathbf{r} - \mathbf{R}_0 = \mathbf{r} - \mathbf{V}t \tag{29-2}$$

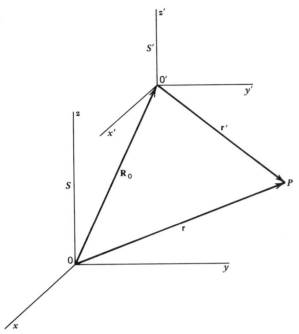

Figure 29-1. Two rectangular coordinate systems with parallel axes of coordinates.

The velocities of the point P as measured in the two systems are therefore related by the familiar result

$$\mathbf{v}' = \frac{d\mathbf{r}'}{dt} = \frac{d\mathbf{r}}{dt} - \frac{d\mathbf{R}_0}{dt} = \mathbf{v} - \mathbf{V} \tag{29-3}$$

Since $\mathbf{V}$ is constant, the accelerations are equal, that is, $\mathbf{a}' = d\mathbf{v}'/dt = d\mathbf{v}/dt = \mathbf{a}$. Therefore, the forces on a point mass m located at P are also equal: $\mathbf{F}' = m\mathbf{a}' = m\mathbf{a} = \mathbf{F}$. Thus, the basic law of mechanics has the *same form* in these two coordinate systems moving with constant velocity with respect to each other. In other words, they are *equally suitable* for use as a reference system. This result is known as the *classical* (or *Galilean*) *principle of relativity*. The formula (29-2) that relates the two sets of coordinates and time ($t' = t$) for a given "event" is known as the *classical* (or *Galilean*) *transformation*. While (29-2) is perfectly general, it is more convenient to specialize it to the case in which the relative motion is along the common xx' direction so that $\mathbf{V} = V\hat{\mathbf{x}}$; then (29-2) can be written in component form as

$$x' = x - Vt \qquad y' = y \qquad z' = z \qquad t' = t \tag{29-4}$$

where, for completeness, we have included the usual assumption that $t' = t$. In this case, (29-3) becomes simply

$$v'_x = v_x - V \qquad v'_y = v_y \qquad v'_z = v_z \tag{29-5}$$

Since (29-1) indicates that the expression for the field changes depending on the coordinate system one is using, it is natural to wonder how a transformation of coordinates like (29-4) will affect the equations determining the fields, that is, Maxwell's equations. For our purposes, it will be sufficient to consider a consequence of Maxwell's equations, that is, the wave equation

$$\nabla^2\psi - \frac{1}{c^2}\frac{\partial^2\psi}{\partial t^2} = 0 \tag{29-6}$$

which is satisfied in a vacuum for any component of **E** or **B**. We know that (29-6) has plane wave solutions with speed c. We will show that the *form* of this equation is not preserved by the Galilean transformation (29-4) but that plane wave solutions are still possible.

For simplicity, let us restrict ourselves to the case in which the direction of propagation of the wave is along the x axis so that (29-6) reduces to

$$\frac{\partial^2 \psi}{\partial x^2} = \frac{1}{c^2} \frac{\partial^2 \psi}{\partial t^2} \tag{29-7}$$

Now let us assume that ψ is written as a function of x' and t' by means of (29-4) so that

$$\frac{\partial \psi}{\partial x} = \frac{\partial \psi}{\partial x'} \frac{\partial x'}{\partial x} + \frac{\partial \psi}{\partial t'} \frac{\partial t'}{\partial x} = \frac{\partial \psi}{\partial x'}$$

and therefore $\partial^2 \psi / \partial x^2 = \partial^2 \psi / \partial x'^2$. Similarly, we find that

$$\frac{\partial \psi}{\partial t} = -V \frac{\partial \psi}{\partial x'} + \frac{\partial \psi}{\partial t'}$$

$$\frac{\partial^2 \psi}{\partial t^2} = V^2 \frac{\partial^2 \psi}{\partial x'^2} - 2V \frac{\partial^2 \psi}{\partial x' \partial t'} + \frac{\partial^2 \psi}{\partial t'^2}$$

When these results are substituted into (29-7), we obtain

$$\left(c^2 - V^2 \right) \frac{\partial^2 \psi}{\partial x'^2} = \frac{\partial^2 \psi}{\partial t'^2} - 2V \frac{\partial^2 \psi}{\partial x' \partial t'} \tag{29-8}$$

which is clearly different in form from the equation (29-7) satisfied in the unprimed axes. Let us nevertheless try to find a plane wave solution of (29-8) of the form

$$\psi(x', t') = \psi_0 e^{ik'(x' - u't')} \tag{29-9}$$

where u' is the phase velocity. Substituting (29-9) into (29-8), we find that it is a solution provided that $u' = \pm c - V = \pm(c \mp V)$. In other words, if the wave is traveling in the positive x' direction, it has the speed $c - V$, while its speed is $c + V$ for propagation in the negative x' direction. These results are therefore consistent with the particular case given by (29-5), and for a general direction of propagation, the wave velocity in the primed system can be found from (29-3).

This calculation has shown us that *if* (29-4) and (29-7) are simultaneously valid then electromagnetic effects will generally be different when observed from different coordinate systems moving with constant velocity with respect to each other and, in particular, the speed of a plane wave in vacuum would not always have the value c. Since (29-6) is a consequence of Maxwell's equations, these results imply that there can be *only one* frame of reference in which Maxwell's equations have the form in which we have been writing them and in which electromagnetic waves have the speed c. This preferred reference system was generally identified with the primary inertial system of mechanics, that is, one fixed with respect to the "fixed stars." This frame also had to be endowed with electromagnetic properties so that it could propagate waves; this augmented system was then called the *ether*.

Electromagnetism and the Galilean transformation taken together thus imply the existence of only one system of reference in which Maxwell's equations are valid. On the other hand, we have also seen that the whole idea of a preferred system is foreign to mechanics because of the existence of the Galilean principle of relativity. However, because the concept of equivalence of different coordinate systems has such a simplicity and attractiveness, we are led to consider yet another possibility, namely that there exists a *common* principle of relativity valid for both mechanics and electromagnetism. If this were to be the case though, then the Galilean principle of relativity based on the Galilean transformation and Newtonian mechanics can not actually be the appropriate one. (We are not considering the possibility that Maxwell's equations may need alteration.)

The proper choice among these possibilities can only be made with the aid of the results of experiment. Historically, one approach was to assume the validity of both Maxwell's equations and the Galilean transformation, that is, the existence of a preferred reference frame. Then a laboratory frame attached to the earth would be moving with respect to the primary inertial system (ether), so that one could look for effects associated with this motion and that could be calculated by transforming Maxwell's equations into this system much as we did to obtain (29-8). We will consider two of these experiments—one that does not involve wave propagation and one that does.

■ **Example**

The Trouton-Noble experiment. In this experiment, a charged parallel plate capacitor was suspended so that its plates were vertical and it could rotate about an axis parallel to and between the plates. Because of the assumed motion of the earth with respect to the ether, there will be a torque on the capacitor that tends to align the plane of the plates parallel to their velocity $\mathbf{V}$. We will simplify things somewhat by replacing the charges on the plates by two point charges whose separation is the same as that of the plates as shown in Figure 29-2. The magnetic force $\mathbf{F}_m$ on the positive charge due to the negative charge is found from (14-31) to be

$$\mathbf{F}_m = -\frac{\mu_0 q^2 \mathbf{V} \times (\mathbf{V} \times \mathbf{R})}{4\pi R^3} \tag{29-10}$$

and there will be an equal and opposite force on the negative charge. These two forces will produce a torque $\boldsymbol{\tau}$ on the system given by

$$\boldsymbol{\tau} = \mathbf{R} \times \mathbf{F}_m = \frac{\mu_0 q^2 (\mathbf{V} \cdot \mathbf{R})(\mathbf{V} \times \mathbf{R})}{4\pi R^3} \tag{29-11}$$

which has the direction shown in the figure for $\theta > 90°$. (The sign of $\boldsymbol{\tau}$ is reversed for $\theta < 90°$.) We see that $\boldsymbol{\tau}$ is parallel to the axis and tends to rotate the charges until the

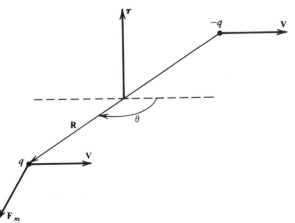

Figure 29-2.   The Trouton-Noble experiment.

line **R** connecting them is perpendicular to **V**. The magnitude of the torque is

$$\tau = \frac{\mu_0 q^2 V^2 \sin\theta \cos\theta}{4\pi R} = \frac{1}{2} U_e \left(\frac{V}{c}\right)^2 \sin 2\theta \tag{29-12}$$

where $U_e = q^2/4\pi\epsilon_0 R$ is the magnitude of the electrostatic energy of the system by (5-49). We note that the effect measured by τ is proportional to $(V/c)^2$, which turns out to be characteristic of these experiments. (The result of taking the actual distribution of the charges on the plates into account is to double the value given above for τ. Furthermore, the attractive electrostatic force is always along the direction of **R** and hence does not contribute to the torque.)

Taking V to be the speed of the earth in its orbit, which is about 3×10^4 meters/second, one finds that the magnitude of the torque is large enough to be observable. However, the most careful experiments failed to find such a torque, so that this experiment provides no evidence in favor of a preferred reference frame for electromagnetism. ∎

■ **Example**

The Michelson-Morley experiment. This experiment was based on the result following (29-9), which showed, in effect, that the velocity combination formula (29-3) applied to electromagnetic waves as well as to point masses. If **V** is interpreted as the velocity of S' with respect to the ether, then (29-3) says that the velocity of the wave will depend on the motion of the medium and will be c only with respect to the ether. Because the orbital speed of the earth is so small compared to c, it has not been feasible to make a direct measurement of the wave velocity in various directions with respect to the earth's surface in order to check (29-3). It is possible, however, to compare these directions and look for this small effect by using the wave itself in a particular manner; this idea had been suggested by Maxwell, and the experiment was first performed by Michelson and Morley in 1887.

The experimental arrangement is illustrated in Figure 29-3. Light from the source L is incident on the partially silvered glass plate M and divided into two beams that travel in directions that are perpendicular to each other. Each beam travels the distance l to the mirror M_1 or M_2 and is reflected back over its original path. Part of the light

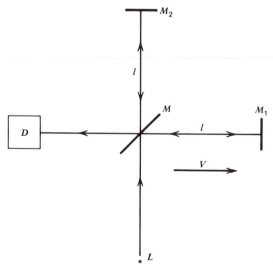

Figure 29-3.   The Michelson-Morley experiment.

from each beam then passes along the fourth arm of the device toward a detector D. The beams have thus been recombined and any phase difference between them resulting from their trips back and forth will produce interference effects observable as a variable amplitude in the resultant beam. In order to calculate the phase difference, we need to find the times taken by the beams to travel along their respective paths.

If we let the arm 1 be in the direction of motion of the apparatus, then, from before, the speed of the light will be $c - V$ on the outward path and $c + V$ on the return path. Thus the time taken to pass to and fro along this arm is

$$t_1 = \frac{l}{c - V} + \frac{l}{c + V} = \frac{2l}{c}\left(1 - \frac{V^2}{c^2}\right)^{-1} \simeq \frac{2l}{c}\left(1 + \frac{V^2}{c^2}\right) \qquad (29\text{-}13)$$

In the perpendicular direction along the second arm, the resultant relative light velocity as found from (29-3) is $(c^2 - V^2)^{1/2}$, and the corresponding time for the round trip is

$$t_2 = \frac{2l}{(c^2 - V^2)^{1/2}} \simeq \frac{2l}{c}\left(1 + \frac{V^2}{2c^2}\right) \qquad (29\text{-}14)$$

Thus the two times are not equal and their difference is

$$\Delta t = t_1 - t_2 \simeq \frac{l}{c}\left(\frac{V}{c}\right)^2 \qquad (29\text{-}15)$$

Taking $l = 30$ meters and $V = 3 \times 10^4$ meters/second as before, we find that $\Delta t = 10^{-15}$ second. This small time difference corresponds to a phase difference of $\omega \Delta t = 2\pi c \Delta t/\lambda = 0.6(2\pi) = 3.8$ radians for visible light with a wavelength of 0.5×10^{-6} meters. This is a sizable relative phase difference and could result in detectable changes as the apparatus is rotated to interchange the role of the two arms. However, no such effects were observed by Michelson and Morley. Since then this experiment and variations on it have been frequently repeated, particularly as more sensitive and accurate equipment has become available. The results have always been the same—that for electromagnetic waves there is no evidence to support the use of (29-3), which results from combining the concepts of a preferred reference frame for electromagnetism with the Galilean relativity principle for mechanics. ∎

While we have discussed only two experiments of this general type, there have not as yet been any results of any kind that support a preferred frame for electromagnetism. Accordingly, we are led to assume that there is a common principle of relativity for both electromagnetism and mechanics. It cannot, however, be described by the Galilean transformation given by (29-2) and (29-4). The necessary new concepts are provided by the postulates of special relativity.

One should not get the idea, however, that the whole of special relativity stands or falls on the results of a *single* "crucial" experiment, in particular, the Michelson-Morley experiment. By now, special relativity has been extended and tested by means of such a large number of interwoven experiments that it provides us with a single coherent and useful description of a great variety of phenomena.

29-2 THE POSTULATES AND THE LORENTZ TRANSFORMATION

There is some doubt as to whether Einstein was strongly influenced by, or was even aware of, the results of the experiments we have just described. There is no doubt, however, that he thought deeply and hard about the questions we outlined in the last section. The result was that in 1905 he proposed the equivalent of the two following postulates:

- **I** All systems of coordinates are equally suitable for the description of physical phenomena.

- **II** The speed of light in vacuum is the same for all observers and is independent of the motion of the source.

The first postulate literally refers to *all* systems of coordinates, although we restrict ourselves to those systems that have a constant velocity with respect to each other. When we do this, we are discussing the *special (restricted) theory* of relativity. An unrestricted discussion would correspond to the *general theory* of relativity, which we do not consider.

The first postulate also asserts that there is a common principle of relativity for *all* of physics, although we consider it only in terms of particle mechanics and electromagnetism. Another way of stating the first postulate is to say that no theory can contain any reference whatsoever to an *absolute velocity* of translational motion of the coordinate system being used. We saw in the last section that Maxwell's equations combined with the Galilean transformation did indicate a possibility of effects due to an absolute velocity, but such predictions were not borne out by experiment. This could mean that Maxwell's equations are themselves not completely correct, or it could mean that the Galilean transformation is not correct. Einstein preferred the second possibility but rather than postulating that Maxwell's equations were covariant (unchanged in form) to the proper transformation of coordinates, he found it to be sufficient (and equivalent) to state his second postulate in terms of the speed of light.

It is this second postulate that leads to the more unfamiliar results of special relativity, including those that apparently violate "common sense." It also implies that Newtonian mechanics is not the correct form for the exact representation of mechanical phenomena, since the Galilean transformation representing the relativity principle natural to mechanics is incompatible with the second postulate as we found after (29-9). Thus we first have to find the transformation rules for the coordinates that will satisfy the second postulate. Then we will have to find such laws of mechanics as will satisfy the first postulate when these transformation rules are used. Finally, we will then have to see how electromagnetism as described by Maxwell's equations fits with these postulates. Whether these postulates are actually applicable to the description of physical phenomena can only then be appraised by comparing predictions of experimental results with the actual results themselves.

Before we begin these tasks, however, it will be well to see that these postulates are compatible with the results of the two experiments discussed in the last section. As far as the Trouton-Noble experiment is concerned, we can invoke the first postulate to allow us to calculate the effect in a coordinate system in which the charges are at rest. Here the only force is the attractive electrostatic force and, as previously pointed out, this will lead to no torque and hence no rotation of the capacitor in agreement with observation. In the Michelson-Morley experiment, the speed of the waves will be c for both arms, making $t_1 = t_2$ and thence $\Delta t = 0$, again in agreement with observation.

The transformation of coordinates, which is quantitatively in accord with the second postulate, is called the *Lorentz transformation*. We consider two coordinate systems S and S' in relative motion with velocity V along their common xx' direction. For a given point event, an observer in S will assign the spatial coordinates and time (x, y, z, t), while those obtained by an observer in S' will be (x', y', z', t'). Thus our transformation equations must enable us to calculate the position (x', y', z') and time t' of the event, if it is known that the first observer assigns the values (x, y, z) and t to it, and conversely. Our derivation will be somewhat simplified, but it will lead to the correct result and will include all of the principal features of a more elaborate treatment.

Suppose that, at the instant the two origins of the coordinate systems coincide, a small pulse of light is produced at the origin. The second postulate requires that each observer see the light propagating with the same speed c in all directions. In other words, with respect to his or her own coordinate system, each observer must see the wave front as a sphere centered on the corresponding origin of radius c times the time. The equations of the wavefronts are thus

$$x^2 + y^2 + z^2 - c^2t^2 = 0 \qquad x'^2 + y'^2 + z'^2 - c^2t'^2 = 0 \qquad (29\text{-}16)$$

so that we must satisfy the identity

$$x'^2 + y'^2 + z'^2 - c^2t'^2 = x^2 + y^2 + z^2 - c^2t^2 \qquad (29\text{-}17)$$

An equivalent way of saying this is that the quantity on either side of (29-17) must be *invariant* with respect to the transformation from one system to the other.

Since we have taken the x and x' axes in the direction of the relative velocity of S and S', it is reasonable to assume that the coordinates transverse to the motion are unchanged, that is, $y' = y$ and $z' = z$; then (29-17) becomes

$$x'^2 - c^2t'^2 = x^2 - c^2t^2 \qquad (29\text{-}18)$$

Now there is only *one* relative velocity V. From the point of view of S, the origin $0'$ must have the x coordinate Vt as shown in Figure 29-4a. Similarly, the x' coordinate of 0 as seen in S' must be $-Vt'$ as shown in (b) of the figure. Therefore, the transformation equations must satisfy the conditions that

$$\begin{aligned} x &= Vt & \text{when } x' = 0 \\ x' &= -Vt' & \text{when } x = 0 \end{aligned} \qquad (29\text{-}19)$$

We will try the simplest relations that will satisfy (29-19), that is, the linear equations

$$x' = \gamma(x - Vt) \qquad x = \gamma'(x' + Vt') \qquad (29\text{-}20)$$

where γ and γ' are constants to be determined. We can now express t' in terms of the unprimed quantities by eliminating x' from (29-20) and we find that

$$t' = \gamma\left[t - \frac{x}{V}\left(1 - \frac{1}{\gamma\gamma'}\right)\right] \qquad (29\text{-}21)$$

Putting these expressions for x' and t' into (29-18) and grouping terms, we get

$$x^2\left[1 - \gamma^2 + \frac{\gamma^2c^2}{V^2}\left(1 - \frac{1}{\gamma\gamma'}\right)^2\right] + 2xt\left[\gamma^2V - \frac{\gamma^2c^2}{V}\left(1 - \frac{1}{\gamma\gamma'}\right)\right]$$
$$+ t^2\left[\gamma^2(c^2 - V^2) - c^2\right] = 0 \qquad (29\text{-}22)$$

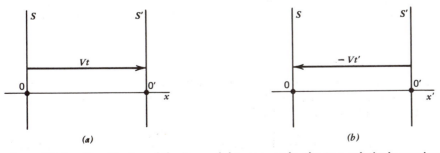

(a) (b)

Figure 29-4. Coordinates of the two origins as seen by the two relatively moving observers.

In order that this will always be zero for all possible values of x and t, it is necessary that the coefficients of x^2, xt, and t^2 individually vanish. Equating each of these coefficients to zero, we find that

$$\gamma' = \gamma = \frac{1}{\left[1 - (V^2/c^2)\right]^{1/2}} \tag{29-23}$$

will satisfy (29-22).

Combining all of these results, we get the Lorentz transformation formulas

$$x' = \gamma(x - Vt) \qquad y' = y \qquad z' = z \qquad t' = \gamma\left(t - \frac{V}{c^2}x\right) \tag{29-24}$$

$$x = \gamma(x' + Vt') \qquad y = y' \qquad z = z' \qquad t = \gamma\left(t' + \frac{V}{c^2}x'\right) \tag{29-25}$$

where γ is given by (29-23). The equations (29-25) can be obtained from (29-24) by solving for the unprimed quantities in terms of the primed ones, or, equivalently, by interchanging primed and unprimed symbols and changing the sign of V as would be expected from the overall symmetry of the situation.

If we let $V/c \to 0$, so that $\gamma \to 1$, we see that (29-24) reduces to the Galilean transformation (29-4) that we now see as a first approximation to the Lorentz transformation for finite values of V such that $V/c \ll 1$.

It will be convenient to introduce a special symbol for the ratio of V to c by

$$\beta = \frac{V}{c} \tag{29-26}$$

so that

$$\gamma = \frac{1}{\left(1 - \beta^2\right)^{1/2}} \tag{29-27}$$

while (29-24) becomes

$$x' = \gamma(x - \beta ct) \qquad y' = y \qquad z' = z \qquad t' = \gamma\left(t - \frac{\beta x}{c}\right) \tag{29-28}$$

Actually, any linear transformation of coordinates and time that satisfies (29-17) is called a Lorentz transformation. The one given in (29-28) is that particular one that applies only to the case in which the relative motion of S and S' is along their common x direction. In the next section, we consider more general Lorentz transformations but for now we restrict ourselves to (29-28).

The first physical results implied by the Lorentz transformation that we will investigate are all basically kinematic ones. The existence of the first three effects we consider can be deduced qualitatively from the postulates alone, but for our purposes it will be more efficient to calculate them by direct use of the Lorentz transformation.

■   **Example**

Relativity of simultaneity. Suppose that two events occur at the points x_1 and x_2 in the S system and are simultaneous there so that $t_1 = t_2$. The times at which these events are observed in S' as obtained from (29-28) are

$$t_1' = \gamma\left(t_1 - \frac{\beta x_1}{c}\right) \qquad t_2' = \gamma\left(t_2 - \frac{\beta x_2}{c}\right) \tag{29-29}$$

so that the time interval between these events is

$$\Delta t' = t'_2 - t'_1 = -\frac{\gamma\beta}{c}(x_2 - x_1) \neq 0 \tag{29-30}$$

This result shows that two events that occur at *different points* x_1 and x_2, and that are simultaneous for an observer at rest in S, will no longer appear to be simultaneous to an observer who is at rest in S'. In other words, simultaneity is not an absolute property of a pair of events but is also dependent on the relative state of motion of the observer. ■

■ **Example**

Time dilation. Suppose that we have a clock located at the point x_1 and that it is emitting signals of some sort at regular intervals Δt, where $\Delta t = t_2 - t_1 = t_3 - t_2$, and so on. The corresponding times in S' are given by (29-29), except that now $x_2 = x_1$ since the clock is fixed in S. Therefore, when seen from the system that is moving relative to the clock, we find that these signals will be separated by the time intervals $\Delta t' = t'_2 - t'_1 = t'_3 - t'_2$, and so on, given by

$$\Delta t' = \gamma\Delta t > \Delta t \tag{29-31}$$

since $\gamma > 1$. Thus the time interval appears to be longer to the moving observer than it does to the one at rest with respect to the clock. ■

■ **Example**

The Lorentz contraction. In principle, we measure a length by placing a measuring rod along the distance to be measured and then finding the difference between the scale marks that *simultaneously coincide* with the ends of the length of interest. Although such a detailed specification appears to be trivial when the measuring rod and the length are relatively at rest, it is essential when they are moving with respect to each other.

Let $L = x_2 - x_1$ be the distance between two points as found with a scale that is at rest with respect to them. The length as found by the moving observer who assigns coordinates x'_2 and x'_1 to the points is found from (29-28) to be

$$L' = x'_2 - x'_1 = \gamma[x_2 - x_1 - \beta c(t_2 - t_1)] \tag{29-32}$$

However, the values x'_2 and x'_1 must have been determined simultaneously in S' so that $t'_2 = t'_1$. This means that the time interval $t_2 - t_1$ in (29-32) must be given by $t_2 - t_1 = (\beta/c)(x_2 - x_1) = (\beta/c)L$ according to (29-29). When this is substituted into (29-32) and (29-27) is used, we find that

$$L' = \gamma(1 - \beta^2)L = \frac{L}{\gamma} = (1 - \beta^2)^{1/2}L < L \tag{29-33}$$

Thus, as found by the moving observer, the length in the direction of motion will be contracted by the factor $1/\gamma$. ■

■ **Example**

Velocity addition formulas. The rectangular components of the velocity $\mathbf{v}'$ of a point as observed in S' are given by

$$v'_x = \frac{dx'}{dt'} \qquad v'_y = \frac{dy'}{dt'} \qquad v'_z = \frac{dz'}{dt'} \tag{29-34}$$

We now want to find the components v_x, v_y, v_z of the velocity of this same point when its motion is referred to the system S.

Differentiating the first expression in (29-25) and the last in (29-28), and using (29-26), we obtain

$$v_x = \frac{dx}{dt} = \gamma\left(\frac{dx'}{dt'}\frac{dt'}{dt} + \beta c\frac{dt'}{dt}\right) \tag{29-35}$$

$$\frac{dt'}{dt} = \gamma\left(1 - \frac{\beta v_x}{c}\right) \tag{29-36}$$

When these are combined so as to eliminate dt'/dt, we find that

$$v_x = \gamma^2(v_x' + V)\left(1 - \frac{\beta v_x}{c}\right)$$

which can be solved for v_x; the result is that

$$v_x = \frac{v_x' + V}{1 + (Vv_x'/c^2)} \tag{29-37}$$

We can now express (29-36) completely in terms of primed quantities by substituting (29-37) into it and we get

$$\frac{dt'}{dt} = \frac{1}{\gamma\left[1 + (Vv_x'/c^2)\right]} \tag{29-38}$$

Similarly, we find that

$$v_y = \frac{dy}{dt} = \frac{dy'}{dt'}\frac{dt'}{dt} = \frac{v_y'}{\gamma\left[1 + (Vv_x'/c^2)\right]} \tag{29-39}$$

$$v_z = \frac{v_z'}{\gamma\left[1 + (Vv_x'/c^2)\right]} \tag{29-40}$$

These last two relations, together with (29-37), enable us to convert velocity components observed in one system to those observed in another, and are the relativistic analogues to the classical formulas (29-5) that are based on the Galilean transformation.

We see that these formulas are more complicated than the classical ones and, in particular, v_x' is involved in the formulas (29-39) and (29-40) used to convert y and z components of the velocities. On the other hand, if $\beta = V/c \ll 1$, we see that (29-37), (29-39), and (29-40) become approximately $v_x \simeq v_x' + V$, $v_y \simeq v_y'$, and $v_z \simeq v_z'$, which are just (29-5) again. Thus, the simple addition of components that follows from the Galilean transformation is an approximation appropriate to small relative velocities of the two coordinate systems.

The relativistic velocity transformation formulas have the interesting property that the sum of two velocities can never exceed c. This is most easily illustrated with an extreme example. Suppose that $V = c$, while $v_x' = c$ also. The Galilean transformation would yield $v_x = v_x' + V = c + c = 2c$, while the Einstein formula (29-37) becomes $v_x = (c + c)/[1 + (c^2/c^2)] = c$ as was asserted.   ∎

◼ **Example**

The Doppler effect and aberration. The question of interest here is how the frequencies and directions of propagation of a wave are related for the two relatively moving observers. Suppose that a light source Q', at rest in S', emits a spherical wave. We know from the results of the last chapter that this wave will be represented in S' by

$$\psi' = \frac{\psi'_0}{r'} e^{i(k'r' - \omega't')} \tag{29-41}$$

where r' is the distance from Q', ψ'_0 is a constant amplitude, and the circular frequency ω' and propagation constant k' must be related by

$$\omega' = k'c \tag{29-42}$$

since the wave has speed c.

Now suppose that there is an observer at rest in S at a point P in the xy plane. The coordinates of P are (x, y) in S, and (x', y') in S'. The S observer will interpret the wave as a spherical wave originating from a point source Q in that system, so that it will be described by

$$\psi = \frac{\psi_0}{r} e^{i(kr - \omega t)} \tag{29-43}$$

where

$$\omega = kc \tag{29-44}$$

since the wave also has speed c in S because of the second postulate.

As shown in Figure 29-5, the direction of propagation from Q' to P can be described by the angle θ' made with the x' axis, while θ is the corresponding angle in S. We also see from the figure that

$$r' = x' \cos \theta' + y' \sin \theta' \qquad r = x \cos \theta + y \sin \theta \tag{29-45}$$

Now the *numerical* value of the phase of the wave (zero, for example) must be independent of whatever system is being used to express it: thus the exponents in (29-41) and (29-43) must be always equal. Furthermore, the coordinates and times must be related by the Lorentz transformation (29-28). Equating these exponents, substituting from (29-42), (29-44), (29-45), and (29-28), we find that we must have

$$\omega\left(\frac{x \cos \theta + y \sin \theta}{c} - t\right) = \omega'\left[\frac{\gamma(x - \beta ct) \cos \theta' + y \sin \theta'}{c} - \gamma\left(t - \frac{\beta x}{c}\right)\right] \tag{29-46}$$

This equality must hold for all possible values of x, y, and t; this is possible only if their coefficients are separately equal.

Equating the coefficients of t on both sides of (29-46), we find that

$$\omega = \omega'\gamma(1 + \beta \cos \theta') \tag{29-47}$$

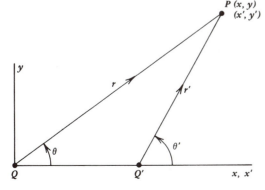

Figure 29-5. Coordinates for the calculation of the Doppler effect and aberration.

which relates the frequencies and is the relativistic Doppler effect formula. We see that if $\beta \ll 1$, then $\gamma \simeq 1$, and (29-47) reduces to the classical formula

$$\omega \simeq \omega'(1 + \beta \cos \theta') \tag{29-48}$$

Similarly, when the coefficients of x are equated, we find that $\omega \cos \theta = \omega'\gamma(\cos \theta' + \beta)$, which can be used to eliminate the frequencies from (29-47) with the result that

$$\cos \theta = \frac{\cos \theta' + \beta}{1 + \beta \cos \theta'} \tag{29-49}$$

relates the directions of propagation and is called the aberration formula.

Finally, by equating the coefficients of y in (29-46), we obtain $\omega \sin \theta = \omega' \sin \theta'$, which is easily shown to be consistent with (29-47) and (29-49).

The formula (29-47) differs from the classical acoustical result (29-48) in that it predicts a *transverse* effect. Suppose that the frequency is observed in S at right angles to the motion, that is, $\theta = 90°$. Then $\cos \theta = 0$ and (29-49) gives $\cos \theta' = -\beta$. If this is substituted into (29-47) and (29-27) is used, we find that $\omega = \omega'(1 - \beta^2)^{1/2} < \omega'$. This predicted effect has been observed by studying the radiation from an electric discharge in hydrogen and provides some experimental evidence for the basic correctness of the relativity postulates.   ∎

In order to see the motivation for what we will do in the next section, it is useful to look at some of our results from another point of view. The Lorentz transformation applies to coordinate differentials too, and we find from (29-28) that

$$dx' = \gamma(dx - \beta c\, dt) \quad dy' = dy \quad dz' = dz \quad dt' = \gamma\left(dt - \frac{\beta}{c} dx\right) \tag{29-50}$$

In addition, (29-17) shows us that the Lorentz transformation corresponds to the equality

$$(dx)^2 + (dy)^2 + (dz)^2 - c^2(dt)^2 = (dx')^2 + (dy')^2 + (dz')^2 - c^2(dt')^2 \tag{29-51}$$

Dividing out $-(dt)^2$ and $-(dt')^2$, taking the square root, and using (29-34) and (1-6), we find that the quantity $d\tau$ given by

$$d\tau = dt\left(1 - \frac{v^2}{c^2}\right)^{1/2} = dt'\left(1 - \frac{v'^2}{c^2}\right)^{1/2} \tag{29-52}$$

has the same value for all reference frames related by the Lorentz transformation. This invariant quantity $d\tau$ is called the *proper time interval* and we see from (29-52) that $d\tau = dt_0 = $ the time interval measured in the system in which the point is at rest ($v = 0$).

If we now divide both sides of each term in (29-50) by $d\tau$, we obtain

$$\frac{dx'}{d\tau} = \gamma\left(\frac{dx}{d\tau} - \beta c \frac{dt}{d\tau}\right) \qquad \frac{dy'}{d\tau} = \frac{dy}{d\tau}$$

$$\frac{dz'}{d\tau} = \frac{dz}{d\tau} \qquad \frac{dt'}{d\tau} = \gamma\left(\frac{dt}{d\tau} - \frac{\beta}{c} \frac{dx}{d\tau}\right) \tag{29-53}$$

Using (29-52), (29-34), and comparing (29-53) with (29-28), we see that the four quantities

$$\frac{dx}{d\tau} = \frac{v_x}{[1 - (v^2/c^2)]^{1/2}} \qquad \frac{dy}{d\tau} = \frac{v_y}{[1 - (v^2/c^2)]^{1/2}}$$

$$\frac{dz}{d\tau} = \frac{v_z}{[1 - (v^2/c^2)]^{1/2}} \qquad \frac{dt}{d\tau} = \frac{1}{[1 - (v^2/c^2)]^{1/2}} \tag{29-54}$$

transform in exactly the same way as do x, y, z, t. Therefore, since we know from (29-25) that

$$x = \gamma(x' + Vt') \qquad t = \gamma\left[t' + (V/c^2)x'\right]$$

we can immediately write the transformation equations for the *analogous* quantities $dx/d\tau$ and $dt/d\tau$ as

$$\frac{v_x}{\left[1 - (v^2/c^2)\right]^{1/2}} = \gamma\left\{ \frac{v'_x}{\left[1 - (v'^2/c^2)\right]^{1/2}} + V\frac{1}{\left[1 - (v'^2/c^2)\right]^{1/2}} \right\} \quad (29\text{-}55)$$

$$\frac{1}{\left[1 - (v^2/c^2)\right]^{1/2}} = \gamma\left\{ \frac{1}{\left[1 - (v'^2/c^2)\right]^{1/2}} + \frac{V}{c^2}\frac{v'_x}{\left[1 - (v'^2/c^2)\right]^{1/2}} \right\} \quad (29\text{-}56)$$

If we now divide (29-55) by (29-56), we obtain

$$v_x = \frac{v'_x + V}{1 + (Vv'_x/c^2)}$$

which is exactly the velocity transformation formula (29-37). You should be sure to verify that (29-39) and (29-40) can be obtained in a similar way.

Our use of (29-55) and (29-56) is quite different from the previous method of obtaining (29-37), which involved direct differentiations of the transformation equations and subsequent elimination of terms. We were able to obtain (29-37) in this latter manner because we knew how all of the quantities involved were affected by a Lorentz transformation, that is, *we knew their transformation properties.*

The whole point of this example, then, is to show us that we should be able to write the transformation formulas for certain quantities quite easily, provided that we know something about their general properties. This turns out to be an extremely valuable and efficient way of looking at things and is what we consider in the next section.

29-3 GENERAL LORENTZ TRANSFORMATIONS, 4-VECTORS, AND TENSORS

Let us begin by considering the two rectangular coordinate systems shown in Figure 29-6. They have a common origin but differ by a rotation; that is, if we rotate one set of axes as a rigid body, it can be made to coincide with the other. The exact rotation could be described, for example, by specifying the direction cosines of the primed axes with respect to the unprimed axes, or *vice versa.*

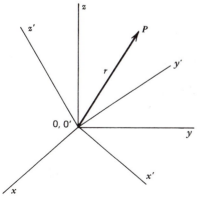

Figure 29-6.   The two sets of axes are rotated with respect to each other.

Consider a point P that has the coordinates x, y, z in one set of axes while its coordinates are x', y', z' in the other set. Both of these different sets of numbers locate the same point P. The distance r of P from the origin has the same numerical value regardless of which coordinate system we are using. In other words, r is an *invariant* and we must have

$$r^2 = x^2 + y^2 + z^2 = x'^2 + y'^2 + z'^2 \qquad (29\text{-}57)$$

It will be convenient to introduce the notation

$$x_1 = x \qquad x_2 = y \qquad x_3 = z \qquad (29\text{-}58)$$

for then (29-57) can be written somewhat more compactly as

$$r^2 = \sum_{j=1}^{3} x_j^2 = \sum_{j=1}^{3} x_j'^2 \qquad (29\text{-}59)$$

The equations that relate the two sets of coordinates will be linear so that we can write the three equations

$$x_j' = \sum_{k=1}^{3} a_{jk} x_k \qquad (j = 1, 2, 3) \qquad (29\text{-}60)$$

that is, the three equations

$$\begin{aligned} x_1' &= a_{11}x_1 + a_{12}x_2 + a_{13}x_3 \\ x_2' &= a_{21}x_1 + a_{22}x_2 + a_{23}x_3 \\ x_3' &= a_{31}x_1 + a_{32}x_2 + a_{33}x_3 \end{aligned} \qquad (29\text{-}61)$$

The set of nine numbers a_{jk} characterize the rotation relating the primed and unprimed axes. It is also evident that these a_{jk} cannot all be independent because the transformation equations (29-60) have not yet been made to satisfy the fundamental physical requirement that the axes are related by a *rotation*, that is, that r^2 be an invariant as described by (29-59).

In this particular three-dimensional case, it is easy to identify the a_{jk}. The position vector of P in the primed system can be written as $\mathbf{r}' = x'\hat{\mathbf{x}}' + y'\hat{\mathbf{y}}' + z'\hat{\mathbf{z}}'$ by (1-11), while for the unprimed system it is $\mathbf{r} = x\hat{\mathbf{x}} + y\hat{\mathbf{y}} + z\hat{\mathbf{z}}$. Since they represent the same physical vector (that is, displacement from the origin), we must also have $\mathbf{r}' = \mathbf{r}$. Then we can use (1-21) to find x' as

$$x' = \mathbf{r}' \cdot \hat{\mathbf{x}}' = \mathbf{r} \cdot \hat{\mathbf{x}}' = x\hat{\mathbf{x}}' \cdot \hat{\mathbf{x}} + y\hat{\mathbf{x}}' \cdot \hat{\mathbf{y}} + z\hat{\mathbf{x}}' \cdot \hat{\mathbf{z}} = l_{xx}x + l_{xy}y + l_{xz}z \quad (29\text{-}62)$$

where l_{xx}, l_{xy}, l_{xz} are the direction cosines of $\hat{\mathbf{x}}'$ with respect to the x, y, z axes, respectively, according to (1-8). [These direction cosines satisfy $l_{xx}^2 + l_{xy}^2 + l_{xz}^2 = 1$ by (1-9).] Hence by comparing (29-62) with (29-61), we see that $a_{11} = l_{xx}$, $a_{12} = l_{xy}$, and $a_{13} = l_{xz}$. In the same way, we will find that the a_{2k} are the direction cosines of $\hat{\mathbf{y}}'$, and the a_{3k} are those of $\hat{\mathbf{z}}'$.

In Section 1-1, we defined a vector as any quantity with the same mathematical properties as the displacement of a point. Now the position vector $\mathbf{r}$ certainly is such a quantity and its components are exactly the coordinates of a point. Thus (29-60) describes the effect of a rotation on the *components* of this particular vector. Therefore, we can immediately say that if A_1, A_2, and A_3 are the rectangular components of any vector $\mathbf{A}$, then they will satisfy a relation like (29-60), that is,

$$A_j' = \sum_{k=1}^{3} a_{jk} A_k \qquad (j = 1, 2, 3) \qquad (29\text{-}63)$$

where the a_{jk} are the *same set* of coefficients that describe the effect of a rigid rotation of axes on the coordinates of a point.

While we could continue with this three-dimensional case, it is just as easy to carry out a further discussion for a more general situation to which we now turn.

If we look back at (29-17) we see that this basic condition for the Lorentz transformation can be expressed as requiring the invariance of the quantity s^2 given by

$$s^2 = x^2 + y^2 + z^2 - c^2 t^2 = x'^2 + y'^2 + z'^2 - c^2 t'^2 \tag{29-64}$$

If we now introduce the notation

$$x_1 = x \qquad x_2 = y \qquad x_3 = z \qquad x_4 = ict \tag{29-65}$$

we see that (29-64) can be written

$$s^2 = \sum_{\mu=1}^{4} x_\mu^2 = \sum_{\mu=1}^{4} x_\mu'^2 \tag{29-66}$$

On comparing (29-66) and (29-59), we see that the *most general* Lorentz transformation can be interpreted as a *rigid rotation of axes in a four dimensional space* with axes $x_1, x_2, x_3, x_4 = ict$. This interesting, although somewhat formal, result has many far-reaching and useful consequences, as we will see.

We can write the transformation equations relating the two sets of coordinates as the four general linear equations

$$x_\mu' = \sum_{\nu=1}^{4} a_{\mu\nu} x_\nu \qquad (\mu = 1, 2, 3, 4) \tag{29-67}$$

If the transformation were not linear, it would give a preferred status to whatever origin we happened to choose, or to some other point; but this would violate the first postulate since it would provide us with an objective means of distinguishing one coordinate system from another.

In order that (29-67) represent a Lorentz transformation, (29-66) must also be satisfied. This will impose some requirements on the 16 coefficients $a_{\mu\nu}$ and we want to find out what they are. Substituting (29-67) into (29-66), we get

$$\sum_\lambda x_\lambda'^2 = \sum_\lambda \left(\sum_\mu a_{\lambda\mu} x_\mu \right) \left(\sum_\nu a_{\lambda\nu} x_\nu \right) = \sum_{\mu\nu} \left(\sum_\lambda a_{\lambda\mu} a_{\lambda\nu} \right) x_\mu x_\nu \tag{29-68}$$

Since (29-68) must also equal $\sum_\mu x_\mu^2$, we see that we must have

$$\sum_\lambda a_{\lambda\mu} a_{\lambda\nu} = \delta_{\mu\nu} \tag{29-69}$$

with the use of (8-27) since the coefficient of $x_\mu x_\nu$ must equal 1 if $\nu = \mu$ and 0 if $\nu \neq \mu$.

Now we could equally well have written our transformation in the form

$$x_\nu = \sum_{\lambda=1}^{4} b_{\nu\lambda} x_\lambda' \qquad (\nu = 1, 2, 3, 4) \tag{29-70}$$

where the $b_{\nu\lambda}$ are an appropriate set of coefficients. It is evident that the $b_{\nu\lambda}$ must be related to the $a_{\mu\nu}$ and we can determine this by substituting (29-70) into (29-67) to give

$$x_\mu' = \sum_\nu a_{\mu\nu} \left(\sum_\lambda b_{\nu\lambda} x_\lambda' \right) = \sum_\lambda x_\lambda' \left(\sum_\nu a_{\mu\nu} b_{\nu\lambda} \right) = \sum_\lambda \delta_{\mu\lambda} x_\lambda' \tag{29-71}$$

with the use of (8-27) again. Comparing the last two terms, we see that

$$\sum_\nu a_{\mu\nu} b_{\nu\lambda} = \delta_{\mu\lambda} \tag{29-72}$$

We can solve (29-72) for the b's by multiplying both sides of it by $a_{\mu\rho}$, summing over μ, and using (29-69):

$$\sum_{\mu\nu} a_{\mu\rho} a_{\mu\nu} b_{\nu\lambda} = \sum_{\mu} a_{\mu\rho} \delta_{\mu\lambda} = a_{\lambda\rho} = \sum_{\nu} b_{\nu\lambda} \left(\sum_{\mu} a_{\mu\rho} a_{\mu\nu} \right) = \sum_{\nu} b_{\nu\lambda} \delta_{\rho\nu} = b_{\rho\lambda} \quad (29\text{-}73)$$

Thus, $b_{\rho\lambda} = a_{\lambda\rho}$, so that we can say that

$$\text{if } x'_{\mu} = \sum_{\nu} a_{\mu\nu} x_{\nu}, \quad \text{then} \quad x_{\mu} = \sum_{\nu} a_{\nu\mu} x'_{\nu} \quad (29\text{-}74)$$

Furthermore, (29-72) can then be written entirely in terms of the a's by substituting the result of (29-73) into it, and interchanging the indices λ and ν to give

$$\sum_{\lambda} a_{\mu\lambda} a_{\nu\lambda} = \delta_{\mu\nu} \quad (29\text{-}75)$$

Thus, there is only one set of coefficients describing the transformation and (29-69) and (29-75) represent the conditions required by (29-66).

When the particular Lorentz transformation (29-28) that we have been using is written in the notation of (29-65), it becomes

$$x'_1 = \gamma x_1 + i\beta\gamma x_4 \qquad x'_2 = x_2$$
$$x'_3 = x_3 \qquad x'_4 = \gamma x_4 - i\beta\gamma x_1 \quad (29\text{-}76)$$

Upon comparing (29-76) with (29-67), we see that the coefficients $a_{\mu\nu}$ representing this particular transformation can be written as the matrix

$$a_{\mu\nu} = \begin{vmatrix} a_{11} & a_{12} & a_{13} & a_{14} \\ a_{21} & a_{22} & a_{23} & a_{24} \\ a_{31} & a_{32} & a_{33} & a_{34} \\ a_{41} & a_{42} & a_{43} & a_{44} \end{vmatrix} = \begin{vmatrix} \gamma & 0 & 0 & i\beta\gamma \\ 0 & 1 & 0 & 0 \\ 0 & 0 & 1 & 0 \\ -i\beta\gamma & 0 & 0 & \gamma \end{vmatrix} \quad (29\text{-}77)$$

It will be left as exercises to verify that the $a_{\mu\nu}$ given in (29-77) satisfy the requirements (29-69) and (29-75).

It is evident that a rigid rotation of axes in three-dimensional space will also keep the expression $x^2 + y^2 + z^2 - c^2 t^2$ invariant because of (29-57) and its independence of the time. Thus such a physical rotation of axes should also be included in the group of general Lorentz transformations described by (29-67).

At this stage, it will be useful to introduce some new quantities, mostly by analogy with the three-dimensional case. We remind ourselves that an *invariant* is a quantity whose numerical value does not change as the result of a Lorentz transformation. Examples are s^2 of (29-66) and the proper time $d\tau$ of (29-52).

As a generalization of what we found to hold for a vector $\mathbf{A}$ in (29-63), we define a *4-vector* A_{μ} as a set of four quantities (A_1, A_2, A_3, A_4) whose *transformation properties* are the same as those of the coordinates. That is, if the Lorentz transformation is described by (29-74), then the components A_{μ} and A'_{μ} are related by

$$A'_{\mu} = \sum_{\nu=1}^{4} a_{\mu\nu} A_{\nu} \quad \text{and} \quad A_{\mu} = \sum_{\nu=1}^{4} a_{\nu\mu} A'_{\nu} \quad (29\text{-}78)$$

where the coefficients $a_{\mu\nu}$ are exactly those appearing in (29-67) and $\mu = 1, 2, 3, 4$. Since we saw that a rigid rotation of axes is also a Lorentz transformation, it follows from this definition and (29-63) that the first three components (A_1, A_2, A_3) of A_{μ} must be the components of an ordinary three-dimensional vector $\mathbf{A}$.

The sum of two 4-vectors must be a 4-vector so that

$$C_{\mu} = (A + B)_{\mu} = A_{\mu} + B_{\mu} \quad (29\text{-}79)$$

in analogy with (1-10).

We can define a scalar product of two 4-vectors by generalizing the result obtained for $\mathbf{A} \cdot \mathbf{B}$ in (1-20):

$$\text{Scalar product} = \sum_\mu A_\mu B_\mu \tag{29-80}$$

After reviewing the three-dimensional definition (1-15), we suspect that this scalar product is an invariant. This is easily proven by using (29-80), (29-78), and (29-69):

$$\sum_\mu A'_\mu B'_\mu = \sum_\mu \left(\sum_\nu a_{\mu\nu} A_\nu \right) \left(\sum_\lambda a_{\mu\lambda} B_\lambda \right)$$

$$= \sum_{\nu\lambda} A_\nu B_\lambda \left(\sum_\mu a_{\mu\nu} a_{\mu\lambda} \right) = \sum_{\nu\lambda} A_\nu B_\lambda \delta_{\nu\lambda} = \sum_\nu A_\nu B_\nu$$

From (1-17), we are led to define the square of the "length" of a 4-vector as the scalar product of the 4-vector with itself; thus

$$(\text{"length"})^2 = \sum_\mu A_\mu^2 \tag{29-81}$$

In addition to the coordinates, we have already encountered a 4-vector; it is the *4-velocity* U_μ whose components as found from (29-54) and (29-65) are

$$U_1 = \frac{dx_1}{d\tau} = \frac{v_x}{\left[1 - (v^2/c^2) \right]^{1/2}} \qquad U_2 = \frac{dx_2}{d\tau} = \frac{v_y}{\left[1 - (v^2/c^2) \right]^{1/2}}$$

$$U_3 = \frac{dx_3}{d\tau} = \frac{v_z}{\left[1 - (v^2/c^2) \right]^{1/2}} \qquad U_4 = \frac{dx_4}{d\tau} = \frac{ic}{\left[1 - (v^2/c^2) \right]^{1/2}} \tag{29-82}$$

where $v^2 = v_x^2 + v_y^2 + v_z^2$. These four quantities form a 4-vector because we showed in (29-55) and (29-56) that their transformation properties are the same as those of the coordinates.

Inserting (29-82) into (29-81), we obtain the interesting result that the square of the "length" of the 4-velocity is negative:

$$(\text{"length"})^2 = \sum_\mu U_\mu^2 = \frac{v_x^2 + v_y^2 + v_z^2 - c^2}{1 - (v^2/c^2)} = -c^2 \tag{29-83}$$

A *second-rank tensor* $T_{\mu\nu}$ is defined as a set of 16 quantities such that each index transforms in the same way as do the coordinate indices, that is, if (29-67) holds, then

$$T'_{\mu\nu} = \sum_\lambda \sum_\rho a_{\mu\lambda} a_{\nu\rho} T_{\lambda\rho} \tag{29-84}$$

Two indices are needed to specify each component; hence the term "second rank." Similarly, a 4-vector with its single index is seen to be a tensor of the first rank, while an invariant scalar is a tensor of zero rank. It is possible to define tensors of higher rank by a straightforward generalization of (29-84) but we will have no need of them.

A tensor is said to be *symmetric* if $T_{\mu\nu} = T_{\nu\mu}$ and therefore it has only 10 independent components. A tensor is *antisymmetric* if $T_{\nu\mu} = -T_{\mu\nu}$ and has only 6 independent components since the diagonal elements must be zero, that is $T_{\mu\mu} = 0$. A tensor does *not* have to be symmetric or antisymmetric, but if it is, this symmetry property is not changed by a Lorentz transformation; that is, if $T_{\mu\nu}$ is symmetric (or antisymmetric), $T'_{\mu\nu}$ is also symmetric (or antisymmetric). We can prove this for both cases at once by writing $T_{\mu\nu} = \pm T_{\nu\mu}$ and using the upper sign for the symmetric case and the lower for the antisymmetric case. Using this in (29-84), we find that

$$T'_{\mu\nu} = \sum_{\lambda\rho} a_{\mu\lambda} a_{\nu\rho} T_{\lambda\rho} = \pm \sum_{\lambda\rho} a_{\mu\lambda} a_{\nu\rho} T_{\rho\lambda} \qquad (29\text{-}85)$$

If we now interchange μ and ν in (29-84) and then interchange λ and ρ, we obtain

$$T'_{\nu\mu} = \sum_{\lambda\rho} a_{\nu\lambda} a_{\mu\rho} T_{\lambda\rho} = \sum_{\lambda\rho} a_{\nu\rho} a_{\mu\lambda} T_{\rho\lambda} \qquad (29\text{-}86)$$

Comparing (29-86) with (29-85), we see that $T'_{\mu\nu} = \pm T'_{\nu\mu}$ with the choice of signs being the same as originally used, thus showing that the symmetry property has not been changed.

An important example is

$$F_{\mu\nu} = A_\mu B_\nu - A_\nu B_\mu \qquad (29\text{-}87)$$

While we have written these 16 numbers as $F_{\mu\nu}$, this does *not* by itself prove that $F_{\mu\nu}$ is actually a tensor, although if it is, it will be antisymmetric. In order to show that $F_{\mu\nu}$ does indeed have the correct transformation properties, we substitute (29-78) into (29-87) and we find that

$$F'_{\mu\nu} = A'_\mu B'_\nu - A'_\nu B'_\mu = \sum_{\lambda\rho} a_{\mu\lambda} a_{\nu\rho} A_\lambda B_\rho - \sum_{\lambda\rho} a_{\nu\rho} a_{\mu\lambda} A_\rho B_\lambda$$

$$= \sum_{\lambda\rho} a_{\mu\lambda} a_{\nu\rho} (A_\lambda B_\rho - A_\rho B_\lambda) = \sum_{\lambda\rho} a_{\mu\lambda} a_{\nu\rho} F_{\lambda\rho} \qquad (29\text{-}88)$$

which is exactly what is required by the definition (29-84). We also see that if $\mu, \nu = 1, 2, 3$ then the components of $F_{\mu\nu}$ are exactly those of the usual three-dimensional cross product $\mathbf{A} \times \mathbf{B}$ as shown by (1-27). Accordingly, the cross product is seen to be actually a tensor rather than a vector.

In a similar manner, one can show that the following quantities are 4-vectors:

$$B_\mu = \sum_\nu A_\nu T_{\nu\mu} \qquad C_\mu = \sum_\nu T_{\mu\nu} A_\nu \qquad (29\text{-}89)$$

According to the chain rule for differentiating a function of given unprimed variables with respect to different primed variables, one has

$$\frac{\partial}{\partial x'_\mu} = \sum_\nu \frac{\partial x_\nu}{\partial x'_\mu} \frac{\partial}{\partial x_\nu} \qquad (29\text{-}90)$$

as we have already used to obtain (29-8) from (29-7). We see from (29-74) that $\partial x_\nu / \partial x'_\mu = a_{\mu\nu}$, so that (29-90) becomes

$$\frac{\partial}{\partial x'_\mu} = \sum_\nu a_{\mu\nu} \frac{\partial}{\partial x_\nu} \qquad (29\text{-}91)$$

On comparing (29-91) with (29-78), we see that the four operators $\partial/\partial x_\mu$ transform exactly like a 4-vector. This fact enables us to define a four-dimensional del operator $\square$ with the components

$$\square = \left[\frac{\partial}{\partial x_1}, \frac{\partial}{\partial x_2}, \frac{\partial}{\partial x_3}, \frac{\partial}{\partial x_4} \right] \qquad (29\text{-}92)$$

and use it in a fashion analogous to the three-dimensional operator ∇ as written in (1-41). [We note that this result justifies the statement following (1-41) that ∇ has the same properties as the displacement of a point, that is, it has vector characteristics.]

If ψ is an invariant, we can define a Gradient, $\Box\psi$, as the 4-vector whose components are $\partial\psi/\partial x_\mu$ much as in (1-37). Similarly, we can define

Divergence: $$\sum_\mu \frac{\partial A_\mu}{\partial x_\mu} \qquad \text{(an invariant)} \qquad (29\text{-}93)$$

Four-dimensional Laplacian:

$$\Box^2 = \sum_\mu \frac{\partial^2}{\partial x_\mu^2} = \nabla^2 - \frac{1}{c^2}\frac{\partial^2}{\partial t^2} \qquad \text{(an invariant)} \qquad (29\text{-}94)$$

Curl: $$C_{\mu\nu} = \frac{\partial A_\nu}{\partial x_\mu} - \frac{\partial A_\mu}{\partial x_\nu} \qquad \text{(an antisymmetric tensor)} \qquad (29\text{-}95)$$

Divergence of a tensor:

$$\sum_\nu \frac{\partial T_{\nu\mu}}{\partial x_\nu} = D_\mu \qquad \text{(a 4-vector)} \qquad (29\text{-}96)$$

We note that the four-dimensional Laplacian is actually the vacuum d'Alembertian operator of (22-17). Furthermore, if $\mu, \nu \neq 4$ in (29-95), then we see from (1-43) that the corresponding components of $C_{\mu\nu}$ are the components of the three-dimensional curl, $\nabla \times \mathbf{A}$.

While all of this material in this section is very interesting, it is only natural to wonder what is the point of discussing it. If we ask ourselves what the first postulate means here, we see that it says, in effect, that there should be no way of making an absolute distinction between two systems moving with constant velocity with respect to each other. In addition, we found from the second postulate that observations made in the two systems must be correlated by means of the Lorentz transformation. Combining these two results, we will see immediately that we can say that the two postulates taken together require that the laws of physics when properly formulated must have their *form unchanged* when subjected to a Lorentz transformation, that is, they must be *covariant* with respect to the transformation. In order to show that this must be so, let us see what the consequences would be if this were not so in a very simple case. Suppose that a particular law under consideration had the general form $\mathscr{F} = \mathscr{G} + \mathscr{H}$. Let us suppose also that when we referred everything to the primed system by means of a Lorentz transformation, it became $\mathscr{F}' = \mathscr{G}' + \mathscr{H}' + \epsilon'$ where ϵ' depended on the particular primed system involved. This equation is clearly not covariant and, in fact, the very existence of the ϵ' term would enable us to distinguish among the various systems in an absolute way. Since this would violate the first postulate, we can see the reason for requiring covariance with respect to a Lorentz transformation.

We have just seen that 4-vectors and second-rank tensors are covariant by their very definition. Thus it is evident that, if we were to express all of our physical laws in 4-vector or tensor form, they would *automatically* be covariant with respect to a Lorentz transformation and would thereby satisfy both the postulates. With this in mind, we can see, in essence, what we will do next. We will look at some physical laws and see if they can be written in terms of 4-vectors or tensors. If they already are or easily can be, we need do nothing more because we know that they are already valid in special relativity. If the laws are not covariant, yet are correct in the nonrelativistic case, then our task is to try to *generalize* these results so that they can be expressed in 4-vector or tensor form and thus be compatible with special relativity. In this latter situation, however, we will have to bear two points in mind. First of all, our generalizations will always have to reduce to the known valid results in their nonrelativistic limit. Second, our generalizations will still need to be tested by *experiment* because the process of generalizing to 4-vector or tensor form does not necessarily lead to a unique result.

29-4 PARTICLE MECHANICS

While electromagnetism is our primary interest, the mechanics of mass points is so important in physics and provides such a good illustration of what we just discussed at the end of the last section, that we will briefly consider it in this section.

The basic equation of the nonrelativistic mechanics of a mass point subject to the net force $\mathbf{f}$ is

$$\mathbf{f} = \frac{d\mathbf{p}}{dt} \tag{29-97}$$

where

$$\mathbf{p} = m_0 \mathbf{v} \tag{29-98}$$

is the linear momentum of the particle in terms of its velocity $\mathbf{v}$ and its inertial (or rest) mass m_0. Equation 29-97 is clearly not in 4-vector or tensor form because it has only three components and the differentiation is with respect to the time which we know is not an invariant scalar. Therefore these equations must be generalized in order to make them satisfy the requirements of special relativity.

We begin with the momentum. While $\mathbf{v}$ is not a 4-vector, it is closely related to the 4-velocity $U_\mu = dx_\mu/d\tau$. Hence a plausible generalization of (29-98) is to use the invariant scalar m_0 and define the *4-momentum* P_μ as

$$P_\mu = m_0 U_\mu \tag{29-99}$$

We can write this in component form by using (29-82) and we get

$$P_j = \frac{m_0 v_j}{\left[1 - \left(v^2/c^2\right)\right]^{1/2}} \quad (j = 1,2,3) \qquad P_4 = \frac{im_0 c}{\left[1 - \left(v^2/c^2\right)\right]^{1/2}} \tag{29-100}$$

We see that as $v/c \to 0$, $P_j \to m_0 v_j$, which is (29-98) so that (29-99) appears to be a reasonable choice. For the moment, we disregard the extra component P_4 that we have introduced in this way.

In order to get an equation of motion analogous to (29-97), we differentiate (29-99) with respect to the invariant $d\tau$ and define the *4-force* F_μ by

$$F_\mu = \frac{dP_\mu}{d\tau} = \frac{d}{d\tau}\left(m_0 U_\mu\right) = m_0 \frac{d^2 x_\mu}{d\tau^2} \tag{29-101}$$

to obtain the desired generalization of the Newtonian equation; the 4-vector F_μ is also called the *Minkowski force*.

In order to relate (29-101) to the ordinary force components, we can substitute for $d\tau$ from (29-52) and we get

$$F_\mu\left(1 - \frac{v^2}{c^2}\right)^{1/2} = \frac{d}{dt}\left(m_0 U_\mu\right) \tag{29-102}$$

and the first three components of this as found with the use of (29-82) are

$$F_j\left(1 - \frac{v^2}{c^2}\right)^{1/2} = \frac{d}{dt}\left\{\frac{m_0 v_j}{\left[1 - \left(v^2/c^2\right)\right]^{1/2}}\right\} = f_j \tag{29-103}$$

where the f_j must be the x, y, z components of the ordinary force $\mathbf{f}$ since (29-103) has to reduce to (29-97) and (29-98) combined when $v/c \ll 1$. Therefore the first three components of the Minkowski force are related to the ordinary force by

$$F_j = \frac{f_j}{\left[1 - \left(v^2/c^2\right)\right]^{1/2}} \tag{29-104}$$

In practical calculations, one usually does not want to deal with the 4-force but prefers to use just the three equations (29-103); thus it is common practice to write

$$\mathbf{f} = \frac{d}{dt}\left\{\frac{m_0\mathbf{v}}{\left[1 - \left(v^2/c^2\right)\right]^{1/2}}\right\} \tag{29-105}$$

and to call this the *relativistic equation of motion*. Furthermore, if one still wishes to regard force as the rate of change of momentum, (29-105) can be written as $\mathbf{f} = d\mathbf{p}/dt$ where

$$\mathbf{p} = m\mathbf{v} \quad \text{and} \quad m = \frac{m_0}{\left[1 - \left(v^2/c^2\right)\right]^{1/2}} \tag{29-106}$$

The quantity m introduced in this way is then called the *mass* of the particle because of the *analogy* with (29-98). If this procedure is followed, (29-106) provides the basis for the common statement that the mass of a particle is no longer a constant but increases as the speed increases.

On the other hand, it is not at all necessary to interpret these results in this way, and doing so follows only from a natural desire to write momentum as always being the product of the mass and the *ordinary velocity* $\mathbf{v}$ rather than as the rest mass m_0 times a *new function* of the velocity. In fact, such an approach actually contradicts the basic philosophy of the covariant approach of relativity because (29-106) is not in 4-vector form. It is much more consonant with relativistic concepts to ascribe an invariant scalar property—the rest mass m_0—to the particle and then define the 4-vector momentum as the product of this scalar invariant with the 4-vector velocity, exactly as we did in (29-99).

We have seen that our process of generalization led from the three component equation (29-98) to (29-99), which has *four* components. We saw that the first three components of P_μ can be given an adequate interpretation and now we must look at the "extra" one, which turns out to be not so unfamiliar after all.

Let us define the quantity W by

$$P_4 = \frac{iW}{c} \tag{29-107}$$

so that (29-100) gives

$$W = \frac{m_0c^2}{\left[1 - \left(v^2/c^2\right)\right]^{1/2}} \tag{29-108}$$

In order to interpret this, we see what it becomes when $v/c \ll 1$. Expanding the denominator with the use of (8-6), we find that

$$W = m_0c^2\left(1 + \frac{v^2}{2c^2} + \frac{3v^4}{8c^4} + \dots\right) = m_0c^2 + \frac{1}{2}m_0v^2 + \dots \tag{29-109}$$

The second term can be recognized immediately for it is simply the *kinetic energy* of the particle in Newtonian mechanics. Accordingly, it seems quite reasonable in this more general case to call W the *total energy* of the particle. We see that, if the particle is at rest so that $v = 0$, the value of W is m_0c^2, which is called the *rest energy*. It is customary then to regard the total energy of the particle as being composed of two parts—an intrinsic part due to its rest mass (the rest energy) and the additional part, which appears when the particle is moving. Thus, if we let T be the kinetic energy, we can write

$$W = m_0 c^2 + T \tag{29-110}$$

so that

$$T = m_0 c^2 \left\{ \frac{1}{[1 - (v^2/c^2)]^{1/2}} - 1 \right\} \tag{29-111}$$

according to (29-108).

The fourth component of the 4-momentum thus has turned out to be simply proportional to the particle energy. This shows that the linear momentum and energy of a particle should not be regarded as different entities, but simply as two aspects of the same attributes of the particle since they appear as separate components of the same 4-vector.

This last remark can be illustrated quantitatively. Since P_μ is a 4-vector it transforms according to (29-78), and if we use the second form of this for the particular Lorentz transformation of (29-77), we obtain

$$P_x = \gamma \left(P_x' + \frac{\beta}{c} W' \right) \qquad P_y = P_y'$$
$$P_z = P_z' \qquad W = \gamma(W' + \beta c P_x') \tag{29-112}$$

clearly showing that what appears as energy in one system appears as momentum in another and, conversely, what appears as momentum in one system is energy in another.

■ Example

Particle at rest in S'. Here $\mathbf{v}' = 0$ so that $P_x' = P_y' = P_z' = 0$ and $W' = m_0 c^2$ by (29-100) and (29-108). Inserting these into (29-112) and using (29-26) and (29-27), we find that

$$P_x = \frac{m_0 V}{[1 - (V^2/c^2)]^{1/2}} \qquad W = \frac{m_0 c^2}{[1 - (V^2/c^2)]^{1/2}} \tag{29-113}$$

and $P_y = P_z = 0$. These results are exactly what we should expect by (29-100) and (29-108) since, from the point of view of the observer in S, the particle is moving along the x axis with speed V. Thus the particle has only (rest) energy in S', but has both energy and momentum with respect to S. ■

We can now see the significance of the fourth component of the Minkowski force. We find from (29-101), (29-107), and (29-52) that

$$F_4 = \frac{dP_4}{d\tau} = \frac{i}{c} \frac{dW}{d\tau} = \frac{i}{c[1 - (v^2/c^2)]^{1/2}} \frac{dW}{dt} \tag{29-114}$$

and is proportional to the time rate of change of the energy, or to the rate at which the force is doing work on the particle.

This can be seen in quite another way that provides further justification for the interpretation of W as the energy. We find from (29-99) and (29-83) that

$$\sum_\mu P_\mu^2 = m_0^2 \sum_\mu U_\mu^2 = -m_0^2 c^2 \tag{29-115}$$

and, on differentiating this with respect to τ and using (29-101), we obtain

$$\sum_{\mu} P_{\mu} \frac{dP_{\mu}}{d\tau} = \sum_{\mu} P_{\mu} F_{\mu} = 0 \tag{29-116}$$

[Since the scalar product of P_{μ} and F_{μ} is zero, we can say that the 4-momentum and 4-force are always "perpendicular" by analogy with (1-15).] If we write out (29-116) in detail and use (29-100), (29-104), and (29-114), we obtain

$$P_4 F_4 = -\sum_{j=1}^{3} P_j F_j = -\mathbf{P} \cdot \mathbf{F} = -\frac{m_0 \mathbf{v} \cdot \mathbf{f}}{1 - (v^2/c^2)}$$

$$= \left\{ \frac{i m_0 c}{[1 - (v^2/c^2)]^{1/2}} \right\} \left\{ \frac{i}{c[1 - (v^2/c^2)]^{1/2}} \frac{dW}{dt} \right\}$$

so that

$$\frac{dW}{dt} = \mathbf{v} \cdot \mathbf{f} \tag{29-117}$$

This result shows explicitly that the time rate of change of W equals the rate at which the force does work on the particle and, since this is the definition of rate of increase of energy in mechanics, it is just what we should expect to find from our interpretation of W.

The energy can also be expressed in terms of the linear momentum by writing out (29-115) and using (29-107). The result is that

$$W^2 = c^2 \sum_{j=1}^{3} P_j^2 + \left(m_0 c^2 \right)^2 = (\mathbf{P}c)^2 + \left(m_0 c^2 \right)^2 \tag{29-118}$$

which is a convenient starting point for the development of the Hamiltonian formulation of relativistic mechanics.

Other important aspects of mechanics that we have not yet discussed explicitly are the conservation laws of momentum and energy. Since we have seen that it is no longer adequate to consider momentum and energy as separate entities, it would seem that the natural relativistic generalization would simply be the conservation of the 4-momentum. In fact, this is exactly what has been found to be correct experimentally, and, in addition, this generalized conservation law holds for a system of particles, even when the number of particles and their rest masses are different in the initial and final states. The concept of 4-momentum conservation is particularly useful in the study of collisions. In quantitative form this conservation law can be written as the four equations

$$\sum_{j=1}^{N} P_{\mu}^{b}(j) = \sum_{k=1}^{N'} P_{\mu}^{a}(k) \qquad (\mu = 1, 2, 3, 4)$$

where $P_{\mu}^{b}(j)$ is the μth component of the 4-momentum of the jth particle before the collision (or general interaction); similarly the superscript a on the right labels the values after the collision. This equation also provides for the number of particles to change from N to N'. The fact that the sum of P_4 is also conserved shows that the rest energies and kinetic energies need *not* be individually conserved, although their sum must be. In other words, rest mass and kinetic energy can be converted into each other. This conservation law has been well verified experimentally, particularly in reactions involving atomic nuclei and collisions of high-energy particles of various kinds. The excellent agreement with experiment provides additional evidence for the basic correctness of special relativity.

29-5 ELECTROMAGNETISM IN VACUUM

In contrast to mechanics, we will see that electromagnetism as described by Maxwell's equations for a vacuum is already covariant with respect to Lorentz transformations. We did not require this directly, although the second postulate did deal with the invariance of one of the consequences of Maxwell's equations.

In principle we know from (29-91) how the differential operators in Maxwell's equations transform, but we would also like the transformation properties of ρ, **J**, **E**, **B**, **A**, and ϕ.

We begin with the equation of continuity (12-13), which expresses the fundamental property of conservation of charge. We certainly want this to be covariant, that is, we want

$$\nabla \cdot \mathbf{J} + \frac{\partial \rho}{\partial t} = 0 \quad \text{and} \quad \nabla' \cdot \mathbf{J}' + \frac{\partial \rho'}{\partial t'} = 0 \qquad (29\text{-}119)$$

If we introduce four quantities J_μ by means of the respective equalities given by

$$(J_1, J_2, J_3, J_4) = (J_x, J_y, J_z, ic\rho) \qquad (29\text{-}120)$$

and use (29-65), we find that (29-119) can be written as

$$\sum_\mu \frac{\partial J_\mu}{\partial x_\mu} = 0 \quad \text{and} \quad \sum_\mu \frac{\partial J_\mu'}{\partial x_\mu'} = 0 \qquad (29\text{-}121)$$

Comparison of this with (29-93) shows that it has the form of the divergence of a 4-vector, making us suspect that J_μ is itself actually a 4-vector.

Let us consider a volume element $d\mathcal{V}$ in a coordinate system S in which the charges have a velocity **v**; the total charge in $d\mathcal{V}$ is $\rho \, d\mathcal{V}$. Now let us consider a coordinate system S_0 in which the charges are at rest so that $\mathbf{v}_0 = 0$; this system is called the *rest system*. In the volume element $d\mathcal{V}_0$ of S_0, which corresponds to $d\mathcal{V}$ of S, the total charge is $\rho_0 \, d\mathcal{V}_0$ where ρ_0 is the charge density in the rest system. It seems quite unreasonable to expect that the basic elementary unit of charge—that on an electron or a proton—will be changed merely because we observe it in another coordinate system. Then since all charge is an integral multiple of this unit, as we mentioned following (12-1), the determination of the total charge is essentially one of *counting* to find a definite integer. Since an integer is an invariant, we conclude that it is reasonable to assume that *total charge is an invariant*. This gives us

$$\rho_0 \, d\mathcal{V}_0 = \rho \, d\mathcal{V} \qquad (29\text{-}122)$$

In this case, the relative velocity of the systems S and S_0 is **v**. Since the dimensions along the relative velocity are contracted according to (29-33) while the dimensions transverse to the relative motion are unaffected, the two volumes are related by

$$d\mathcal{V} = \left(1 - \frac{v^2}{c^2}\right)^{1/2} d\mathcal{V}_0 \qquad (29\text{-}123)$$

Combining this with (29-122), we obtain the transformation formula for charge densities:

$$\rho = \frac{\rho_0}{\left[1 - \left(v^2/c^2\right)\right]^{1/2}} \qquad (29\text{-}124)$$

In (12-3) we found the current and charge densities to be related by $\mathbf{J} = \rho\mathbf{v}$. Using this, along with (29-124) and (29-82), we find that

$$J_x = \rho v_x = \frac{\rho_0 v_x}{\left[1 - \left(v^2/c^2\right)\right]^{1/2}} = \rho_0 U_1$$

Similarly, we find that $J_y = \rho_0 U_2$, $J_z = \rho_0 U_3$, $ic\rho = \rho_0 U_4$, so that (29-120) can be written as

$$J_\mu = \rho_0 U_\mu \qquad (29\text{-}125)$$

This shows that J_μ is actually a 4-vector since it is the product of another 4-vector (the 4-velocity) and the scalar invariant ρ_0 (rest system charge density). This 4-vector is called the *4-current*. Thus the equation of continuity (29-121) has been correctly written in covariant form.

The transformation properties of J_μ are given by (29-78), or, to put it another way, we see that J_x, J_y, J_z, ρ transform like x, y, z, t, respectively. Therefore, for the particular Lorentz transformation (29-24), we can immediately say that

$$J_x' = \gamma(J_x - V\rho) \qquad J_y' = J_y$$

$$J_z' = J_z \qquad \rho' = \gamma\left(\rho - \frac{V}{c^2}J_x\right) \qquad (29\text{-}126)$$

while the inverse equations

$$J_x = \gamma(J_x' + V\rho') \qquad \rho = \gamma\left(\rho' + \frac{V}{c^2}J_x'\right) \qquad (29\text{-}127)$$

follow from (29-25).

■ **Example**

Convection current. Suppose that S' is fixed within a material body that moves with constant speed v relative to S. Then we can use (29-127) if we replace V by v. If we now go to the nonrelativistic limit where $v/c \ll 1$ so that $\gamma \simeq 1$, then (29-127) gives

$$J_x \simeq J_x' + v\rho' \qquad \rho \simeq \rho' \qquad (29\text{-}128)$$

While the charge density remains constant, the current density observed in S is the sum of J_x' observed in S' and the *convection current* $v\rho'$ due to the motion of the charge density ρ' with respect to S; we have already mentioned convection currents arising in just this way at the end of Section 12-2.   ■

If we define the four quantities A_μ by

$$(A_1, A_2, A_3, A_4) = \left(A_x, A_y, A_z, \frac{i\phi}{c}\right) \qquad (29\text{-}129)$$

then, with the use of (29-94) and (29-120), the four equations (28-1) and (28-2) can be combined into

$$\Box^2 A_\mu = -\mu_0 J_\mu \qquad (29\text{-}130)$$

since

$$\Box^2(i\phi/c) = -i\rho/c\epsilon_0 = -J_4/c^2\epsilon_0 = -\mu_0 J_4$$

The Lorentz condition (28-3) then becomes simply

$$\sum_\mu \frac{\partial A_\mu}{\partial x_\mu} = 0 \qquad (29\text{-}131)$$

Since (29-130) shows that operating upon A_μ with an invariant yields a 4-vector, we see that A_μ is a 4-vector; it is called the *4-potential*.

We see from (29-129) that A_x, A_y, A_z, ϕ transform like x, y, z, $c^2 t$, respectively. Therefore, for the particular Lorentz transformation (29-24), we can immediately say that

$$A'_x = \gamma\left(A_x - \frac{V}{c^2}\phi\right) \qquad A'_y = A_y$$

$$A'_z = A_z \qquad \phi' = \gamma(\phi - VA_x) \tag{29-132}$$

Equations 29-130 and 29-131 are the covariant forms of Maxwell's equations as written in terms of the potentials. Now we can turn to the fields **E** and **B**.

Since we have been obtaining **B** from $\mathbf{B} = \nabla \times \mathbf{A}$, it seems that it would be useful to investigate the four-dimensional version of this as given by (29-95). Accordingly, we define the antisymmetric *electromagnetic field tensor* $f_{\mu\nu}$ by

$$f_{\mu\nu} = \frac{\partial A_\nu}{\partial x_\mu} - \frac{\partial A_\mu}{\partial x_\nu} \tag{29-133}$$

Then we find, for example, with the use of (29-129) that

$$f_{12} = \frac{\partial A_2}{\partial x_1} - \frac{\partial A_1}{\partial x_2} = \frac{\partial A_y}{\partial x} - \frac{\partial A_x}{\partial y} = B_z$$

Similarly,

$$f_{14} = \frac{\partial A_4}{\partial x_1} - \frac{\partial A_1}{\partial x_4} = \frac{i}{c}\frac{\partial \phi}{\partial x} - \frac{1}{ic}\frac{\partial A_x}{\partial t} = -\frac{iE_x}{c}$$

since $\mathbf{E} = -\nabla\phi - (\partial\mathbf{A}/\partial t)$. Continuing in this way, we find that **E** and **B** appear in the components of $f_{\mu\nu}$ as follows:

$$f_{\mu\nu} = \begin{pmatrix} 0 & B_z & -B_y & -\dfrac{iE_x}{c} \\[2mm] -B_z & 0 & B_x & -\dfrac{iE_y}{c} \\[2mm] B_y & -B_x & 0 & -\dfrac{iE_z}{c} \\[2mm] \dfrac{iE_x}{c} & \dfrac{iE_y}{c} & \dfrac{iE_z}{c} & 0 \end{pmatrix} \tag{29-134}$$

Since we know how the components of a tensor transform, we can find the transformation properties of the field components; we will return to this shortly.

It can then be shown that all of Maxwell's equations as written in terms of the fields are contained in the following covariant system of equations:

$$\frac{\partial f_{\lambda\rho}}{\partial x_\nu} + \frac{\partial f_{\rho\nu}}{\partial x_\lambda} + \frac{\partial f_{\nu\lambda}}{\partial x_\rho} = 0 \tag{29-135}$$

$$\sum_\nu \frac{\partial f_{\mu\nu}}{\partial x_\nu} = \mu_0 J_\mu \tag{29-136}$$

In order to see how this goes, let us consider the first component of (29-135), that is, the form in which the index *one* does *not* appear; with the use of (29-134), we find that it is

$$\frac{\partial f_{34}}{\partial x_2} + \frac{\partial f_{42}}{\partial x_3} + \frac{\partial f_{23}}{\partial x_4} = 0 = -\frac{i}{c}\frac{\partial E_z}{\partial y} + \frac{i}{c}\frac{\partial E_y}{\partial z} + \frac{1}{ic}\frac{\partial B_x}{\partial t}$$

which can be rearranged as

$$\frac{\partial E_z}{\partial y} - \frac{\partial E_y}{\partial z} = -\frac{\partial B_x}{\partial t}$$

which is the x component of $\nabla \times \mathbf{E} = -\partial \mathbf{B}/\partial t$. Similarly, one can show that the remaining three components of (29-135) give the y and z components of Faraday's law as well as $\nabla \cdot \mathbf{B} = 0$.

If we set $\mu = 1$ in (29-136), and use (29-134) and (29-120), we get

$$\frac{\partial f_{11}}{\partial x_1} + \frac{\partial f_{12}}{\partial x_2} + \frac{\partial f_{13}}{\partial x_3} + \frac{\partial f_{14}}{\partial x_4} = \mu_0 J_1 = \frac{\partial B_z}{\partial y} - \frac{\partial B_y}{\partial z} + \frac{1}{ic}\frac{\partial}{\partial t}\left(-\frac{iE_x}{c}\right)$$

which can also be written as

$$\frac{\partial B_z}{\partial y} - \frac{\partial B_y}{\partial z} = \mu_0 J_x + \frac{1}{c^2}\frac{\partial E_x}{\partial t}$$

which is the x component of $\nabla \times \mathbf{B} = \mu_0 \mathbf{J} + c^{-2}(\partial \mathbf{E}/\partial t)$, that is, the vacuum form of (21-33). The remaining two components of this equation are obtained for $\mu = 2$ and $\mu = 3$, while when $\mu = 4$ is used in (29-136) we find as needed that the result is the remaining Maxwell equation of the vacuum version of (21-30) through (21-33), that is, $\nabla \cdot \mathbf{E} = \rho/\epsilon_0$.

According to (29-84), the components of the electromagnetic field tensor will transform as

$$f_{\mu\nu}' = \sum_{\lambda\rho} a_{\mu\lambda} a_{\nu\rho} f_{\lambda\rho} \tag{29-137}$$

and we can use this in conjunction with (29-134) to obtain the transformation formulas for $\mathbf{E}$ and $\mathbf{B}$. We will do this only for the particular Lorentz transformation described by (29-77).

As an example, we consider the 14 component of (29-137). Remembering that $f_{\rho\lambda} = -f_{\lambda\rho}$, we obtain

$$f_{14}' = -\frac{iE_x'}{c} = \sum_{\lambda\rho} a_{1\lambda} a_{4\rho} f_{\lambda\rho} = \sum_{\lambda} a_{1\lambda}(a_{41} f_{\lambda 1} + a_{44} f_{\lambda 4})$$

$$= a_{11}(a_{41} f_{11} + a_{44} f_{14}) + a_{14}(a_{41} f_{41} + a_{44} f_{44})$$

$$= (a_{11} a_{44} - a_{14} a_{41}) f_{14} = (\gamma^2 - \beta^2\gamma^2) f_{14}$$

$$= f_{14} = -\frac{iE_x}{c}$$

and therefore $E_x' = E_x$. Similarly, we find that the 42 component of (29-137) leads to

$$f_{42}' = \frac{iE_y'}{c} = \sum_{\lambda\rho} a_{4\lambda} a_{2\rho} f_{\lambda\rho} = \sum_{\lambda} a_{4\lambda} a_{22} f_{\lambda 2}$$

$$= a_{41} f_{12} + a_{44} f_{42} = (-i\beta\gamma) B_z + \gamma\left(\frac{iE_y}{c}\right)$$

so that $E_y' = \gamma(E_y - \beta c B_z) = \gamma(E_y - V B_z)$. By continuing this process, we find the complete set of transformation formulas to be:

$$E'_x = E_x \qquad\qquad B'_x = B_x$$

$$E'_y = \gamma(E_y - VB_z) \qquad B'_y = \gamma\left(B_y + \frac{VE_z}{c^2}\right) \qquad (29\text{-}138)$$

$$E'_z = \gamma(E_z + VB_y) \qquad B'_z = \gamma\left(B_z - \frac{VE_y}{c^2}\right)$$

The inverse transformations are

$$E_x = E'_x \qquad E_y = \gamma(E'_y + VB'_z), \text{ etc.}$$

and are obtained by interchanging primed and unprimed quantities and changing the sign of V.

Now that we have these results, we can see that we need no longer be restricted to the case in which the relative velocity $\mathbf{V}$ of S' with respect to S is along the x axis. Since the orientation of the x axis is completely arbitrary, we can introduce the field components parallel ($\parallel$) and perpendicular ($\perp$) to the direction of relative translation, and we can write (29-138) in the general form:

$$\mathbf{E}'_{\parallel} = \mathbf{E}_{\parallel} \qquad\qquad \mathbf{B}'_{\parallel} = \mathbf{B}_{\parallel}$$

$$\mathbf{E}'_{\perp} = \gamma(\mathbf{E}_{\perp} + \mathbf{V} \times \mathbf{B}_{\perp}) \qquad \mathbf{B}'_{\perp} = \gamma\left(\mathbf{B}_{\perp} - \frac{\mathbf{V}}{c^2} \times \mathbf{E}_{\perp}\right) \qquad (29\text{-}139)$$

$$= \gamma(\mathbf{E} + \mathbf{V} \times \mathbf{B})_{\perp} \qquad = \gamma\left(\mathbf{B} - \frac{\mathbf{V}}{c^2} \times \mathbf{E}\right)_{\perp}$$

where we have $\mathbf{E} = \mathbf{E}_{\parallel} + \mathbf{E}_{\perp}$ and $\mathbf{B} = \mathbf{B}_{\parallel} + \mathbf{B}_{\perp}$.

We note that in the nonrelativistic limit, $V/c \ll 1$, $\gamma \simeq 1$ and (29-139) becomes $\mathbf{E}' \simeq \mathbf{E} + \mathbf{V} \times \mathbf{B}$ and $\mathbf{B}' \simeq \mathbf{B}$. The first result is exactly (29-1) with $\mathbf{v}$ now appropriately replaced by $\mathbf{V}$ and that we previously deduced by considering Faraday's law with the help of the Lorentz force law.

These results show us clearly that the electric and magnetic field vectors $\mathbf{E}$ and $\mathbf{B}$ actually are not independent and separate entities. The fundamental quantity is the field tensor $f_{\mu\nu}$ and the way in which it is resolved into electric and magnetic components is determined by the relative motion of the observer. This can be illustrated very well by considering two extreme examples.

■ **Example**

Pure electric case in S. Suppose that $\mathbf{E} \neq 0$, but that $\mathbf{B} = 0$. Then we find from (29-139) that in S' one will have

$$\mathbf{E}'_{\parallel} = \mathbf{E}_{\parallel} \qquad \mathbf{E}'_{\perp} = \gamma\mathbf{E}_{\perp}$$

$$\mathbf{B}'_{\parallel} = 0 \qquad \mathbf{B}'_{\perp} = -\frac{\gamma}{c^2}\mathbf{V} \times \mathbf{E}_{\perp}$$

so that

$$\mathbf{B}' = \mathbf{B}'_{\perp} = -\frac{\mathbf{V} \times \mathbf{E}'_{\perp}}{c^2} = -\frac{\mathbf{V} \times \mathbf{E}'}{c^2} \qquad (29\text{-}140)$$

since $\mathbf{V} \times \mathbf{E}'_{\parallel} = 0$. Thus what appears to be purely an electric field to one observer is seen as both an electric and a magnetic field to a second observer moving with respect to the first. ■

■ **Example**

Pure magnetic case in S. Now suppose that $\mathbf{E} = 0$ while $\mathbf{B} \neq 0$. Then we find from (29-139) that in S' one will have

$$\mathbf{B}'_{\parallel} = \mathbf{B}_{\parallel} \qquad \mathbf{B}'_{\perp} = \gamma \mathbf{B}_{\perp}$$

$$\mathbf{E}'_{\parallel} = 0 \qquad \mathbf{E}'_{\perp} = \gamma \mathbf{V} \times \mathbf{B}_{\perp}$$

so that

$$\mathbf{E}' = \mathbf{E}'_{\perp} = \mathbf{V} \times \mathbf{B}'_{\perp} = \mathbf{V} \times \mathbf{B}' \tag{29-141}$$

and what appears to be a purely magnetic field for one observer will appear to be both an electric and a magnetic field to a relatively moving observer.   ■

The transformation equations (29-138) or (29-139) sometimes make it easier to solve certain problems because one can choose a coordinate system in which the answer can be very easily found and then the desired results can be obtained by transforming back to the actual system of interest. One will not get any results this way that could not be obtained by solving Maxwell's equations, but when such a method is applicable it is generally easier and faster. In the next section, we illustrate this approach with an instructive and important example.

29-6 FIELDS OF A UNIFORMLY MOVING POINT CHARGE

We consider a point charge q moving with constant velocity $\mathbf{v}$ with respect to the system S as shown in Figure 29-7. We are interested in the fields associated with this steady-state situation, that is, we will not consider the effects of the initial acceleration of the charge during which we know from (28-70) that it will radiate.

A good choice for S' should be the system in which q is at rest at the origin for then the field in S' is just the electrostatic Coulomb field of a point charge and we have

$$\mathbf{E}' = \frac{q\mathbf{r}'}{4\pi\epsilon_0 r'^3} \qquad \mathbf{B}' = 0 \tag{29-142}$$

where $\mathbf{r}'$ is the position vector of the point where the field is to be evaluated and

$$r'^2 = x'^2 + y'^2 + z'^2 \tag{29-143}$$

Inserting (29-142) into (29-138), and using (29-143), (29-24), and (29-23), with v replacing V, we find one of the field components in S to be

$$E_x = E'_x = \frac{qx'}{4\pi\epsilon_0 r'^3} = \frac{q\gamma(x - vt)}{4\pi\epsilon_0 \left[\gamma^2(x - vt)^2 + y^2 + z^2\right]^{3/2}} \tag{29-144}$$

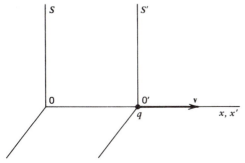

Figure 29-7.   The point charge q is moving with constant velocity with respect to S.

where now

$$\gamma = \frac{1}{\left[1 - (v^2/c^2)\right]^{1/2}} \tag{29-145}$$

It will be convenient to express this in terms of the position of q with respect to S. The S coordinates of q are $(vt, 0, 0)$, so that the relative position vector of the field point with respect to the charge location is $\mathbf{R} = (x - vt)\hat{\mathbf{x}} + y\hat{\mathbf{y}} + z\hat{\mathbf{z}}$. Then we can also write (29-144) as

$$E_x = \frac{q\gamma R_x}{4\pi\epsilon_0 \left(\gamma^2 R_x^2 + R_y^2 + R_z^2\right)^{3/2}} \tag{29-146}$$

The other two components of $\mathbf{E}$ can be found from the appropriate inverse forms of (29-138) and the results are

$$E_y = \frac{q\gamma R_y}{4\pi\epsilon_0 \left(\gamma^2 R_x^2 + R_y^2 + R_z^2\right)^{3/2}} \tag{29-147}$$

$$E_z = \frac{q\gamma R_z}{4\pi\epsilon_0 \left(\gamma^2 R_x^2 + R_y^2 + R_z^2\right)^{3/2}} \tag{29-148}$$

We see that these three results are the components of the single vector equation

$$\mathbf{E} = \frac{q\gamma \mathbf{R}}{4\pi\epsilon_0 \left(\gamma^2 R_x^2 + R_y^2 + R_z^2\right)^{3/2}} \tag{29-149}$$

The value of $\mathbf{B}$ could also be found from (29-138), but it is simpler just to use the result (29-140) obtained for the pure electric case. Thus, when we take the interchange of S and S' into account, we obtain

$$\mathbf{B} = \frac{\mathbf{v} \times \mathbf{E}}{c^2} \tag{29-150}$$

from which we could find the explicit components of $\mathbf{B}$ if we so desired. We see that the lines of $\mathbf{B}$ are circles whose centers lie on the line of motion of the charge; this is reasonable because of our previous recognition of the equivalence of a moving charge to a current element.

Thus, we have found the exact solution to this problem in a comparatively simple manner. We see that $\mathbf{E}$ is directed radially outward from the charge, while (29-150) shows that $\mathbf{B}$ is perpendicular to the plane of $\mathbf{v}$ and $\mathbf{E}$. This situation is illustrated in Figure 29-8. The basic structure of the field will be the same for all times, and can be obtained by simply translating this figure along the x axis with speed v.

It is helpful to express our results in terms of the distance R from the charge and the angle θ made by $\mathbf{R}$ with the direction of the velocity. We see from the figure that $R_x = R \cos\theta$ while $R_y^2 + R_z^2 = R^2 - R_x^2 = R^2 \sin^2\theta$ so that

$$\gamma^2 R_x^2 + R_y^2 + R_z^2 = R^2\gamma^2\left(1 - \beta^2 \sin^2\theta\right) \tag{29-151}$$

where now $\beta = v/c$. Substituting (29-151) into (29-149), we obtain

$$\mathbf{E} = \frac{q\left(1 - \beta^2\right)\hat{\mathbf{R}}}{4\pi\epsilon_0 R^2\left(1 - \beta^2 \sin^2\theta\right)^{3/2}} \tag{29-152}$$

which shows that the field is inverse square in its dependence on radial distance from

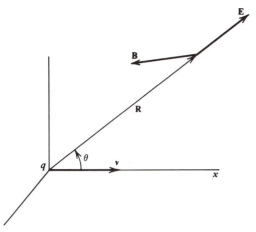

Figure 29-8. The fields of a uniformly moving point charge.

the charge but that, at a given distance, its magnitude depends strongly on direction, in contrast to the simple Coulomb field.

In order to illustrate the dependence of the magnitude upon direction, let us first consider two extreme situations. Directly in front or behind ($\theta = 0$ or $180°$), we have

$$E_{\parallel} = \frac{q}{4\pi\epsilon_0 R^2}(1 - \beta^2)$$

while to one side ($\theta = 90°$), we find that

$$E_{\perp} = \frac{q}{4\pi\epsilon_0 R^2}\frac{1}{(1 - \beta^2)^{1/2}}$$

Therefore, for a very rapidly moving charge for which $\beta \simeq 1$, we see that, at a given distance, $E_{\parallel}$ is very small, while $E_{\perp}$ is very large. (As $\beta \to 0$, both of these components become equal to each other and to the static Coulomb field.)

These effects are also illustrated in Figure 29-9, which shows the magnitude of $\mathbf{E}$ at a given distance plotted as a function of angle θ for $\beta = 0.0$, 0.5, and 0.9, that is, $E/[q/4\pi\epsilon_0 R^2]$ is shown. We see again that for a rapidly moving charge the field is concentrated in a small angle around the equatorial plane.

It is also desirable to look at this problem in terms of the potentials. The electrostatic field in S' described by (29-142) corresponds to the potentials

$$\phi' = \frac{q}{4\pi\epsilon_0 r'} \qquad \mathbf{A}' = 0 \tag{29-153}$$

and therefore $A_1' = A_2' = A_3' = 0$ and $A_4' = i\phi'/c$ by (29-129). Inserting these values into the inverse form of (29-132), we find that the corresponding potentials in S are

$$\phi = \gamma\phi' = \frac{\gamma q}{4\pi\epsilon_0\left[\gamma^2(x - vt)^2 + y^2 + z^2\right]^{1/2}} \tag{29-154}$$

$$\mathbf{A} = A_x\hat{\mathbf{x}} = \frac{v\gamma\phi'}{c^2}\hat{\mathbf{x}} = \frac{\mu_0\gamma q v}{4\pi\left[\gamma^2(x - vt)^2 + y^2 + z^2\right]^{1/2}} \tag{29-155}$$

since $A_y = A_z = 0$.

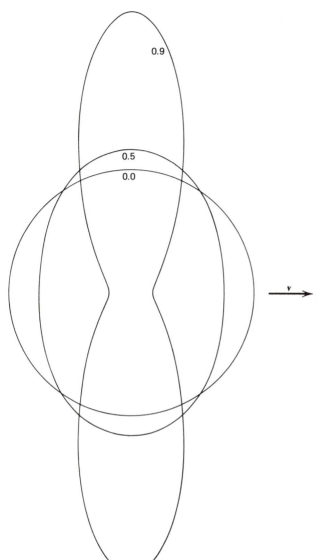

Figure 29-9. The magnitude of the electric field as a function of angle for three different values of β.

If we also use (29-151), we find that they can be written as

$$\phi = \frac{q}{4\pi\epsilon_0 R \left(1 - \beta^2 \sin^2 \theta\right)^{1/2}} \qquad (29\text{-}156)$$

$$\mathbf{A} = \frac{\mu_0 q \mathbf{v}}{4\pi R \left(1 - \beta^2 \sin^2 \theta\right)^{1/2}} \qquad (29\text{-}157)$$

where R is the distance from the charge to the field point.

We note that the vector potential has the direction of the charge velocity, and that, when $\beta \ll 1$, $\mathbf{A} \simeq (\mu_0/4\pi)(q\mathbf{v}/R)$ in agreement with (16-14) and thus justifying the remarks in Section 14-5 about the applicability of our previous results to slowly moving charges.

Both of these potentials depend on θ in the same way. Directly forward (or back) of the charge where $\theta = 0°$ (or 180°), $\phi = q/4\pi\epsilon_0 R$ and equals the static value. However,

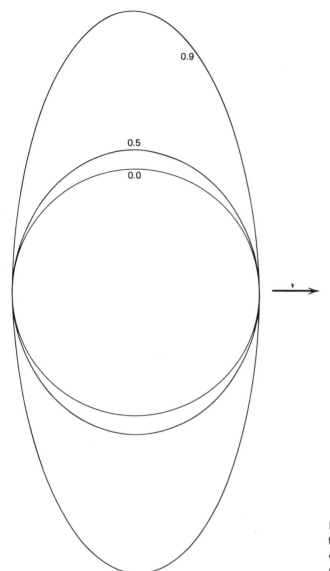

0.9

0.5

0.0

v

Figure 29-10. The magnitude of the scalar potential as a function of angle for three different values of β.

when $\theta = 90°$, $\phi = q/[4\pi\epsilon_0 R(1 - \beta^2)^{1/2}] = \gamma\phi_{\text{static}}$ and is larger by the factor γ. In Figure 29-10, we illustrate the general angular dependence by showing $\phi/[q/4\pi\epsilon_0 R]$ as a function of θ for $\beta = 0.0$, 0.5, and 0.9 corresponding to the values of Figure 29-9.

EXERCISES

29-1 Show by direct substitution of (29-24) into (29-7) that the Lorentz transformation preserves the form of the wave equation, that is, if $\partial^2\psi/\partial x^2 = \partial^2\psi/c^2\partial t^2$, then $\partial^2\psi/\partial x'^2 = \partial^2\psi/c^2\partial t'^2$.

29-2 If the Lorentz contraction (29-33) is assumed to represent an actual physical contraction of the moving rod, show that this will account for the result of the Michelson-Morley experiment.

29-3 The average lifetime of a μ-meson before radioactive decay as measured in its "rest" system is 2.22×10^{-6} second. What will be its average lifetime for an observer with respect to whom the meson has a speed of $0.99c$? How far will the meson travel in this time?

29-4 A rigid rod of length L makes an angle θ with the x axis of the system in which it is at rest.

Show that, for an observer moving with respect to the rod, the apparent length L' and angle θ' are given by

$$L' = L\left[(\cos\theta/\gamma)^2 + \sin^2\theta\right]^{1/2} \qquad \tan\theta' = \gamma\tan\theta$$

29-5 Show that two successive Lorentz transformations corresponding to speeds V_1 and V_2 in the same direction are equivalent to a single Lorentz transformation with a speed $V = (V_1 + V_2)/[1 + (V_1V_2/c^2)]$. Is this result compatible with (29-37)?

29-6 Show that the velocity transformation formulas can be written in vector form as $\mathbf{v} = [\mathbf{v}'_{\parallel} + \mathbf{V} + (\mathbf{v}'_{\perp}/\gamma)]/[1 + (\mathbf{V}\cdot\mathbf{v}')/c^2]$ where $\mathbf{v}'_{\parallel}$ is the velocity component parallel to the relative velocity $\mathbf{V}$ of S and S' while $\mathbf{v}'_{\perp}$ is the component perpendicular to $\mathbf{V}$.

29-7 Find the transformation laws for the components of the acceleration, $a_x = dv_x/dt$, and so on.

29-8 Fizeau measured the index of refraction of light in flowing water. His result can be expressed in terms of the phase velocity v in the moving medium of index of refraction n and flow velocity V as $v = v_0 + V[1 - (1/n^2)]$ where $v_0 = c/n$ is the phase velocity in the stationary medium. Show that this follows from (29-37) for the case $V/c \ll 1$.

29-9 If the angles θ and θ' of Figure 29-5 are both small and if $\beta \ll 1$, show that the angles are approximately related by $\theta \simeq (1 - \beta)\theta'$ so that their fractional difference is $(\theta' - \theta)/\theta' \simeq \beta = V/c$.

29-10 Show that

$$\left[1 - (v'^2/c^2)\right]^{1/2}/\left[1 - (v^2/c^2)\right]^{1/2}$$
$$= \gamma[1 + (\beta v'_x/c)] = \left\{\gamma[1 - (\beta v_x/c)]\right\}^{-1}$$

29-11 Verify that the $a_{\mu\nu}$ given in (29-77) satisfy the requirements (29-69) and (29-75).

29-12 Show that, if $\sum_\mu A_\mu B_\mu$ is an invariant for any arbitrary 4-vector A_μ, then B_μ is also a 4-vector.

29-13 Show that the quantities defined in (29-89) are 4-vectors.

29-14 The *4-acceleration* is defined by $a_\mu = dU_\mu/d\tau = d^2x_\mu/d\tau^2$. Show that its components are related to the ordinary acceleration $\mathbf{a}$ and velocity $\mathbf{v}$ by

$$a_1 = \frac{1}{[1 - (v^2/c^2)]}\left\{a_x + \frac{(v_x/c^2)(\mathbf{v}\cdot\mathbf{a})}{[1 - (v^2/c^2)]}\right\}, \text{etc.}$$

$$a_4 = \frac{(i/c)(\mathbf{v}\cdot\mathbf{a})}{[1 - (v^2/c^2)]^2}$$

29-15 If $\mathbf{k}$ and ω are the propagation vector and circular frequency of a plane wave in free space, show that $k_\mu = (\mathbf{k}, i\omega/c)$ is a 4-vector. Use the transformation properties of k_μ to derive the Doppler and aberration formulas (29-47) and (29-49).

29-16 Derive the set of coefficients $a_{\mu\nu}$ that describe a general Lorentz transformation consisting of a 30° rotation about the y axis plus a translation along the rotated x' axis with a constant speed $V = \frac{1}{2}c$.

29-17 Using the transformation properties of the Minkowski force F_μ, find the transformation laws for the ordinary force $\mathbf{f}$ and power dW/dt. (The results of Exercise 29-10 should prove to be helpful.)

29-18 For many purposes outside the scope of this book, electromagnetic radiation can be treated as composed of small, localized "clumps" of radiation called *photons*. Show that, if we regard a photon as a particle of zero rest mass and total energy $W = h\nu$ where h is Planck's constant, the Doppler and aberration formulas (29-47) and (29-49) can be obtained from the transformation laws for the 4-momentum P_μ.

29-19 Two particles, each of rest mass $\frac{1}{2}M_0$, are connected by a compressed spring of negligible rest mass. The particles are tied together with a massless string and the whole system is at rest in a coordinate frame S_0. The string is then cut, and the two particles fly off in opposite directions, each with speed v_0. What is the initial potential energy of the system in S_0? By explicitly transforming the velocities to another frame S, find the final energy and momentum, and thus the initial energy and momentum, in S. Then find the rest mass of the initial system in the frame S, and interpret the result.

29-20 (a) Show that the quantities $\mathbf{E}\cdot\mathbf{B}$ and $E^2 - c^2B^2$ are invariants. (b) Evaluate these quantities for a plane wave. (c) Show that the statement that $\mathbf{E}$ and $\mathbf{B}$ are perpendicular has absolute significance, that is, if they are perpendicular for one observer they will be perpendicular for all observers. (d) Show that a field that is purely magnetic in one frame cannot be transformed into one that is purely electric in a different reference frame, and conversely.

29-21 A charge q is instantaneously at rest in S' and subject to an electric field $\mathbf{E}'$ so that the force on it is $\mathbf{f}' = q\mathbf{E}'$. Use the transformation laws for the force found in Exercise 29-17 and for the fields found in (29-138) to show that the force in S is exactly the Lorentz force $\mathbf{f} = q(\mathbf{E} + \mathbf{V} \times \mathbf{B})$.

29-22 Show that the relativistic equation of motion (29-105) can be written in the form

$$\frac{m_0}{\left[1 - \left(v^2/c^2\right)\right]^{1/2}} \frac{d\mathbf{v}}{dt} = \mathbf{f} - \frac{(\mathbf{v} \cdot \mathbf{f})}{c^2} \mathbf{v}$$

and that if $\mathbf{f}$ is the Lorentz force on a point charge, $q(\mathbf{E} + \mathbf{v} \times \mathbf{B})$, the right-hand side becomes $q\{\mathbf{E} + \mathbf{v} \times \mathbf{B} - [(\mathbf{v} \cdot \mathbf{E})/c^2]\mathbf{v}\}$.

29-23 Show that the equation of motion of a particle of charge q in an electromagnetic field where the force is given by $\mathbf{f} = q(\mathbf{E} + \mathbf{v} \times \mathbf{B})$ can be written as

$$m_0 \frac{d^2 x_\mu}{d\tau^2} = q\sum_\nu f_{\mu\nu} U_\nu$$

29-24 Find the explicit expression for $\mathbf{B}$ due to a uniformly moving point charge in terms of $\mathbf{R}$, the relative position vector of the field point and the location of the charge. Show that your result reduces to (14-28) when $\beta \ll 1$.

29-25 Show that the potentials ϕ and $\mathbf{A}$ found for a uniformly moving charge in (29-154) and (29-155) give the correct values of $\mathbf{E}$ and $\mathbf{B}$.

29-26 Find the Poynting vector $\mathbf{S}$ for the uniformly moving point charge. [Use (29-152) for simplicity.] Show that the net power radiated by this charge is zero.

29-27 A point charge q_1 at $\mathbf{r}_1$ has a constant velocity $\mathbf{v}_1$. A second point charge q_2 is located at $\mathbf{r}_2$ and has an instantaneous velocity $\mathbf{v}_2$. Find the total electromagnetic force on q_2 due to q_1. Show that when your result is applied to the system of Figure 29-2 and if $V/c \ll 1$, then the magnetic force which you found reduces correctly to (29-10).

29-28 An infinitely long straight line charge with linear charge density λ coincides with the x axis of system S. Find the fields $\mathbf{E}'$ and $\mathbf{B}'$ in a system S' with speed V in the direction of positive x. Compare your expression for $\mathbf{B}'$ with that of a current-carrying wire as given by (14-17) and show that they are in agreement. Note that $\mathbf{E}'$ is not zero in the moving system. What is the physical difference between a current-carrying wire and a line charge moving in the direction of its length that accounts for the appearance of $\mathbf{E}'$?

29-29 An infinitely long ideal solenoid is at rest in the frame S' with its axis parallel to the y' axis. It has n' turns per unit length and carries a steady current I'. Find $\mathbf{E}$ and $\mathbf{B}$ inside and outside the solenoid for an observer in S for whom the solenoid is traveling with constant velocity $\mathbf{V} = V\hat{\mathbf{y}}$. Sketch the lines of $\mathbf{E}$ and indicate the nature of the charge distribution that must be associated with such an electric field. Show that this charge distribution is in qualitative agreement with (29-127). Will an electric field be observed by someone who sees the solenoid moving with constant speed along the direction of the solenoid axis?

29-30 A point electric dipole of moment $\mathbf{p}'$ is at rest at the origin in S'. Find the scalar and vector potentials produced by this dipole as observed by someone in S. How do the components of the dipole moment transform? Qualitatively account for the fact that $\mathbf{A} \neq 0$.

29-31 In (28-72), we found an expression for the total rate of radiation from an accelerated charge. It was proportional to the square of the retarded value of the acceleration, that is, to $[\mathbf{a}]^2 = [\mathbf{a}] \cdot [\mathbf{a}]$. A plausible way of generalizing this would be to replace the square of the actual acceleration by the "square" of the 4-acceleration a_μ. In this way we obtain

$$\mathscr{P} = \frac{q^2}{6\pi\epsilon_0 c^3} \sum_\mu [a_\mu]^2$$

Use the results of Exercise 29-14 to express this in terms of the ordinary acceleration and velocity and verify that the result reduces to (28-72) in the appropriate limit. (This general result is known as Liénard's formula.)

A

MOTION OF CHARGED PARTICLES

Our principal interest so far has been in the properties of the macroscopic electromagnetic field. We have found how it is produced by its sources and have learned how to describe the overall effects of the presence of matter. An important use of $\mathbf{E}$ and $\mathbf{B}$ is to affect the motion of charged particles. Such effects have applications in the design of apparatus for producing particles with large kinetic energies that are then used in the study of reactions in nuclear physics and in "high-energy" physics. Other applications arise in astrophysics and geophysics, plasma physics in general, magnetohydrodynamics, the study of thermonuclear reactions, and the construction of devices using beams of charged particles. We will, however, confine ourselves to only a comparatively few simple and basic situations.

We assume that the net force $\mathbf{f}$ on a particle of charge q and position $\mathbf{r}$ is the Lorentz force $q(\mathbf{E} + \mathbf{v} \times \mathbf{B})$ where $\mathbf{v} = d\mathbf{r}/dt$ is the particle velocity. Equating this to the mass m_0 times the acceleration, we get the equation of motion

$$m_0 \frac{d\mathbf{v}}{dt} = q(\mathbf{E} + \mathbf{v} \times \mathbf{B}) \tag{A-1}$$

Our use of m_0 shows that we will consider only nonrelativistic cases. In the most general situation, $\mathbf{E}$ and $\mathbf{B}$ can each be a function of position and time. In many applications, the particles used can interact with each other and with other particles, primarily by collisions. The effect of collisions then must be somehow included in (A-1), but we will not try to do so.

As we know from mechanics, we will generally also need to know the initial conditions of the motion, that is, the initial position $\mathbf{r}_0$ and velocity $\mathbf{v}_0$ at $t = 0$ in order to get a complete solution of (A-1).

A-1 STATIC ELECTRIC FIELD

We assume that $\mathbf{E}$ is independent of time. If $\mathbf{B} = 0$, then (A-1) reduces to

$$m_0 \frac{d\mathbf{v}}{dt} = q\mathbf{E} \tag{A-2}$$

If we now also assume that $\mathbf{E}$ is independent of position so that $\mathbf{E}$ is a uniform static field, then $d\mathbf{v}/dt$ is constant. We can immediately integrate (A-2) twice to obtain the complete solution

$$\mathbf{v} = \mathbf{v}(t) = \frac{q}{m_0} \mathbf{E}t + \mathbf{v}_0 \tag{A-3}$$

$$\mathbf{r} = \mathbf{r}(t) = \frac{q}{2m_0} \mathbf{E}t^2 + \mathbf{v}_0 t + \mathbf{r}_0 \tag{A-4}$$

530

In order to apply these results to specific problems, we can, for example, write the components of (A-3) and (A-4) in rectangular coordinates. Thus, we find $v_x(t) = (q/m_0)E_x t + v_{0x}$, $x(t) = (q/2m_0)E_x t^2 + v_{0x}t + x_0$, with similar expressions for the y and z components.

For other applications, it is often useful to write some quantities in terms of their components parallel ($\parallel$) and perpendicular ($\perp$) to the field **E**; thus we could write

$$\mathbf{v} = \mathbf{v}_\parallel + \mathbf{v}_\perp \qquad \mathbf{r} = \mathbf{r}_\parallel + \mathbf{r}_\perp \tag{A-5}$$

Applying this to (A-3) and (A-4), we obtain

$$\mathbf{v}_\parallel = \frac{q}{m_0}\mathbf{E}t + \mathbf{v}_{0\parallel} \qquad \mathbf{v}_\perp = \mathbf{v}_{0\perp}$$

$$\mathbf{r}_\parallel = \frac{q}{2m_0}\mathbf{E}t^2 + \mathbf{v}_{0\parallel}t + \mathbf{r}_{0\parallel} \qquad \mathbf{r}_\perp = \mathbf{v}_{0\perp}t + \mathbf{r}_{0\perp} \tag{A-6}$$

which show explicitly that only the component of **v** parallel to **E** is affected while its component perpendicular to **E** remains constant. This problem is similar to that of a mass point in a uniform gravitational field and, just as in that case, the path described by (A-4) is a parabola.

If **E** is not uniform, the particle trajectories can be more or less complicated depending upon the precise dependence of **E** on position. We will consider only an integral consequence of (A-2) which involves energy changes.

If the particle moves from $\mathbf{r}_1$ to $\mathbf{r}_2$, the work done on the particle *by* the field is

$$W_{1 \to 2} = \int_{\mathbf{r}_1}^{\mathbf{r}_2} \mathbf{f} \cdot d\mathbf{s} = q\int_{\mathbf{r}_1}^{\mathbf{r}_2} \mathbf{E} \cdot d\mathbf{s} = -q\,\Delta\phi$$

$$= -q[\phi(\mathbf{r}_2) - \phi(\mathbf{r}_1)] \tag{A-7}$$

by (5-11) where $\Delta\phi$ is the change in the scalar potential. Now we also know from mechanics that the work done by the net force equals the change in the kinetic energy

$$T = \tfrac{1}{2}m_0 v^2 \tag{A-8}$$

so that $W_{1 \to 2} = \Delta T = T_2 - T_1$. Equating this to (A-7), we find that $T_2 + q\phi(\mathbf{r}_2) = T_1 + q\phi(\mathbf{r}_1)$ showing that

$$T + q\phi(\mathbf{r}) = \tfrac{1}{2}m_0 v^2 + q\phi(\mathbf{r}) = \text{const.} \tag{A-9}$$

which is just the statement of conservation of energy where $q\phi(\mathbf{r})$ is the potential energy of the particle in the field. This last result is of course in agreement with (5-48) where we came to the same conclusion.

A-2 STATIC MAGNETIC FIELD

We assume $\mathbf{B} \neq 0$ and independent of time. If $\mathbf{E} = 0$, then (A-1) becomes

$$m_0 \frac{d\mathbf{v}}{dt} = \mathbf{f} = q\mathbf{v} \times \mathbf{B} \tag{A-10}$$

Since **f** and **v** are always perpendicular, $\mathbf{f} \cdot \mathbf{v} = 0$ and the magnetic induction does no work on the particle; consequently the kinetic energy (A-8) will be constant.

If we use the resolution of **v** into parallel and perpendicular components given by (A-5), we find that

$$\frac{d\mathbf{v}}{dt} = \frac{d\mathbf{v}_\parallel}{dt} + \frac{d\mathbf{v}_\perp}{dt} = \frac{q}{m_0}\mathbf{v} \times \mathbf{B} = \frac{q}{m_0}\mathbf{v}_\perp \times \mathbf{B}$$

since $\mathbf{v}_\parallel \times \mathbf{B} = 0$ by (1-22). Thus, we obtain

$$\frac{d\mathbf{v}_\parallel}{dt} = 0 \qquad \frac{d\mathbf{v}_\perp}{dt} = \frac{q}{m_0}\mathbf{v}_\perp \times \mathbf{B} \tag{A-11}$$

since $\mathbf{v}_\perp \times \mathbf{B}$ is also perpendicular to $\mathbf{B}$.

Now let us assume that $\mathbf{B}$ is independent of position so that we have a uniform magnetic induction. We see then that $v_\parallel = \text{const.}$, so that the particle moves with uniform velocity along the direction of $\mathbf{B}$. We see that $d\mathbf{v}_\perp/dt$ is always perpendicular to both $\mathbf{v}_\perp$ and $\mathbf{B}$, and so $\mathbf{v}_\perp$ itself has a constant magnitude $v_\perp$. Consequently, the magnitude of $d\mathbf{v}_\perp/dt$ will be constant and equal to $(q/m_0)v_\perp B$. This situation is illustrated in Figure A-1, which is drawn for positive q and $\mathbf{B}$ out of the paper. We recall that an acceleration of constant magnitude that is always perpendicular to the velocity results in motion in a circular path. We can find the radius r_C of the circle by recognizing that $d\mathbf{v}_\perp/dt$ must have the magnitude of the necessary centripetal acceleration $v_\perp^2/r_C$. Equating this to $(q/m_0)v_\perp B$, we get

$$r_C = \frac{m_0 v_\perp}{qB} \tag{A-12}$$

which involves the particle properties only through the ratio of the charge to the mass. We can also obtain the period of revolution τ_C and the angular frequency ω_C by noting that in one period the particle travels a total distance $2\pi r_C$ so that $\tau_C = (2\pi r_C/v_\perp)$; then ω_C is given by $\omega_C = 2\pi/\tau_C$. Combining these expressions with (A-12), we find that

$$\tau_C = \frac{2\pi m_0}{qB} \qquad \omega_C = \left(\frac{q}{m_0}\right)B \tag{A-13}$$

The name *cyclotron frequency* is given to ω_C. We see that both these quantities in (A-13) are independent of both r_C and $v_\perp$. The faster particles move in circles of greater radius by (A-12), but all particles of the same value of q/m_0 take the same time to go around their path.

We can also write these results in another way. We let $\mathbf{r}_C$ be the position vector of the particle with respect to the center of the circular orbit G (for "guiding center"). Then we will have

$$\mathbf{r} = \mathbf{r}_G + \mathbf{r}_C \tag{A-14}$$

where $\mathbf{r}_G$ is the position vector of G with respect to the origin of our fixed coordinate system. We now see from the figure that the motion of the particle can be described as a rigid rotation of $\mathbf{r}_C$ about G with angular velocity ω_C. We know from kinematics that,

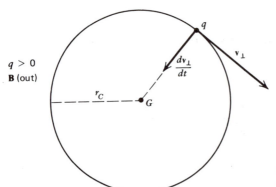

Figure A-1.   Circular motion of a point charge in a uniform induction.

in such a case, we can write the rate of change of the vector $\mathbf{v}_\perp$ in the form

$$\frac{d\mathbf{v}_\perp}{dt} = \boldsymbol{\omega}_C \times \mathbf{v}_\perp \tag{A-15}$$

Equating (A-15) with (A-11), we find the vector angular velocity to be

$$\boldsymbol{\omega}_C = -\left(\frac{q}{m_0}\right)\mathbf{B} \tag{A-16}$$

and whose sign agrees with Figure A-1 since $\boldsymbol{\omega}_C$ is into the page.

If $\mathbf{v}_\parallel \neq 0$, the total motion of the particle can be described as a constant velocity $\mathbf{v}_\parallel$ of the guiding center along the direction of $\mathbf{B}$ with the circular motion arising from $\mathbf{v}_\perp$ superimposed on it. In other words, the motion will be a *helix* as shown in Figure A-2 for positive q and $\mathbf{v}_\parallel$ in the same direction as $\mathbf{B}$.

If we combine (A-8) and (A-5), we find that we can write

$$T = T_\parallel + T_\perp = \tfrac{1}{2}m_0 v_\parallel^2 + \tfrac{1}{2}m_0 v_\perp^2 \tag{A-17}$$

where $T_\parallel$ and $T_\perp$ can be identified with those contributions to the total kinetic energy associated, respectively, with the motion parallel to and perpendicular to $\mathbf{B}$. We have just seen that $v_\parallel$ and $v_\perp$ are both individually constant for a uniform induction so that the individual contributions are themselves constants:

$$T_\parallel = \tfrac{1}{2}m_0 v_\parallel^2 = \text{const.} \qquad T_\perp = \tfrac{1}{2}m_0 v_\perp^2 = \text{const.} \tag{A-18}$$

A point charge traveling around the circle of Figure A-1 is equivalent to a current element $I\,d\mathbf{s} = q\mathbf{v}_\perp$ by (14-29) and thus will have a magnetic dipole moment $\mathbf{m}$. We see from Figure A-1 that $\mathbf{m}$ will be directed into the page and hence in the direction of $\boldsymbol{\omega}_C$. According to (19-27), $m = IS = I\pi r_C^2$ where I is the equivalent current. Since the total charge q passes a given point on the perimeter each time around, and since each circuit takes a time τ_C, the equivalent current is $I = q/\tau_C = \omega_C q/2\pi$. In this way we find that

$$m = I\pi r_C^2 = \frac{1}{2}\left(\frac{m_0}{q}\right)\frac{(m_0 v_\perp^2)}{B^2}\omega_c = \frac{m_0 v_\perp^2}{2B} = \frac{T_\perp}{B} \tag{A-19}$$

with the use of (A-12), (A-13), and (A-18). If we take the direction of $\mathbf{m}$ into account we can write

$$\mathbf{m} = \left(\frac{m_0}{q}\right)\frac{T_\perp}{B^2}\boldsymbol{\omega}_C = -\frac{T_\perp}{B^2}\mathbf{B} \tag{A-20}$$

with the use of (A-16). We see that $\mathbf{m}$ is opposite to the direction of $\mathbf{B}$ and hence is diamagnetic in the sense discussed following (20-52).

The magnitude of the magnetic flux enclosed by the circular orbit is

$$\Phi = B\pi r_C^2 = 2\pi\left(\frac{m_0}{q}\right)\frac{T_\perp}{B} = 2\pi\left(\frac{m_0}{q}\right)m \tag{A-21}$$

and is, of course, a constant.

B Figure A-2. The general motion in a uniform induction is a helix.

■ **Example**

Rectangular coordinates. Although we know that the general motion in a uniform induction will be a helix as in Figure A-2, it is instructive to see how this turns out in detail in terms of a specific coordinate system. Let us use rectangular coordinates and let $\mathbf{B} = B\hat{\mathbf{z}}$ where $B = $ const. Then $\mathbf{v}_\| = v_z\hat{\mathbf{z}}$ while $\mathbf{v}_\perp = v_x\hat{\mathbf{x}} + v_y\hat{\mathbf{y}}$ and we find that (A-11) gives

$$\frac{dv_z}{dt} = 0 \qquad \frac{dv_x}{dt} = \frac{qB}{m_0}v_y \qquad \frac{dv_y}{dt} = -\frac{qB}{m_0}v_x \qquad (A\text{-}22)$$

From the first of these we obtain $v_z = v_{0z} = $ const. so that $z = z_0 + v_{0z}t$.

If we differentiate the second equation of (A-22) and substitute from the third, we get an equation for v_x alone:

$$\frac{d^2v_x}{dt^2} = \frac{qB}{m_0}\frac{dv_y}{dt} = -\left(\frac{qB}{m_0}\right)^2 v_x \qquad (A\text{-}23)$$

This has a solution

$$v_x = v_{0x}\cos\left(\frac{qB}{m_0}\right)t \qquad (A\text{-}24)$$

corresponding, as we will see, to choosing $v_{0y} = 0$. Since $v_x = dx/dt$, we can integrate (A-24) once more to obtain

$$x = x_0 + v_{0x}\left(\frac{m_0}{qB}\right)\sin\left(\frac{qB}{m_0}\right)t \qquad (A\text{-}25)$$

Now we can find v_y directly from the second equation of (A-22) and (A-24):

$$v_y = \frac{m_0}{qB}\frac{dv_x}{dt} = -v_{0x}\sin\left(\frac{qB}{m_0}\right)t \qquad (A\text{-}26)$$

and therefore since $v_y = dy/dt$, we get

$$y = y_0 + v_{0x}\left(\frac{m_0}{qB}\right)\left[\cos\left(\frac{qB}{m_0}\right)t - 1\right] \qquad (A\text{-}27)$$

First of all, we see from (A-24) and (A-26) that $v_\perp^2 = v_x^2 + v_y^2 = v_{0x}^2 = $ const. in agreement with (A-18). If we define $y_0' = y_0 - v_{0x}(m_0/qB)$, then (A-27) can be written somewhat more simply as

$$y = y_0' + v_{0x}\left(\frac{m_0}{qB}\right)\cos\left(\frac{qB}{m_0}\right)t \qquad (A\text{-}28)$$

We now find from (A-25) and (A-28) that

$$(x - x_0)^2 + (y - y_0')^2 = \left(\frac{m_0 v_{0x}}{qB}\right)^2 \qquad (A\text{-}29)$$

which is the equation of a circle with center at (x_0, y_0') and a radius $r_C = m_0 v_{0x}/qB = m_0 v_\perp/qB$ in agreement with (A-12) and Figure A-1. We note that the center of this circle is *not* at the initial position in the xy plane, that is, at (x_0, y_0) but that the y coordinate of the center is y_0' as given by (A-28) and (A-27).

We also see that all of the xy coordinates and velocity components oscillate with the circular frequency qB/m_0 in agreement with (A-13).

Finally, we see that if $x - x_0$ is increasing from zero, say, then $y - y_0'$ will be decreasing from r_C. Thus the position vector in the xy plane, $\mathbf{r}_C = (x - x_0)\hat{\mathbf{x}} + (y - y_0')\hat{\mathbf{y}}$, will be rotating clockwise as seen looking toward the xy plane from the direction of the positive z axis, that is, opposite to the direction of $\mathbf{B}$. But this is exactly the rotation shown in Figure A-1 so that we have verified in detail all of the previous results by means of these specific calculations.   ∎

■ Example

Cyclotron. We have seen that the time τ_C required to go around the circular path is independent of the radius of the path and the speed of the particle. This fact has been used as the basis for the device, used to accelerate charged particles to large kinetic energies, called the *cyclotron*. The principal features are shown in Figure A-3. A thin circular hollow conducting cylinder is cut along a diameter and the two portions separated by a small gap. A uniform $\mathbf{B}$ is normal to the plane of these "dees," while an alternating potential difference $\Delta\phi$ of frequency ω_C is applied between them. A charged particle that is produced within the gap will be accelerated toward one of the dees and enter its interior. Inside it will be shielded from $\mathbf{E}$ but will travel in a semicircular path whose radius is given by (A-12). It will reenter the gap at a time $\frac{1}{2}\tau_C$ later where it will find that $\mathbf{E}$ in the gap has been reversed. It will be accelerated across the gap, gaining kinetic energy $q|\Delta\phi|$, according to (A-7), and will enter the other dee. Since $v_\perp$ is now larger, its radius will be larger by (A-12), but it will travel its new semicircle in the same time $\frac{1}{2}\tau_C$ and will reenter the gap to find $\mathbf{E}$ again reversed. It will again be accelerated across the gap and will travel in a new semicircle of larger radius. The whole process will continue in this manner, the particle gaining kinetic energy $q|\Delta\phi|$ each time it crosses the gap, and thus traveling a path of ever increasing radius. Finally, when the radius is nearly equal to the radius R of the dees, the beam of particles can be extracted with a deflecting field near an opening in the wall of one of the dees. Since $r_{C\,\max} = R$, we find from (A-12) that the maximum speed $v_{\perp\,\max}$ which the particle can be given is

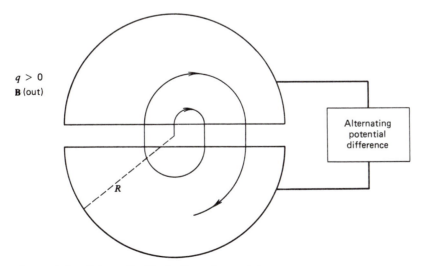

Figure A-3. Schematic illustration of a cyclotron.

qBR/m_0, so that the final kinetic energy of a particle in the emerging beam is

$$T_{max} = \frac{1}{2}m_0 v_{\perp\,max}^2 = \frac{q^2 B^2 R^2}{2m_0} \tag{A-30}$$

which is proportional to B^2 and to R^2.

When the particle speed gets very large, the relativistic equation of motion (29-105) has to be used. If we interpret this as corresponding to an increase in the particle mass, we see from (A-13) that the cyclotron frequency will no longer be a constant but will vary with the particle speed. Consequently, the frequency of the applied potential difference must be correspondingly altered to keep "in step" with the particles in order to prevent them from finding an **E** of the wrong sign when they enter the gap, as this would decelerate them and decrease their kinetic energy. ∎

■ **Example**

Magnetic focusing. In Figure A-4 we show a particle that is projected at $t = 0$ from a point P with an initial velocity $\mathbf{v}_0$, which makes the angle θ with the direction of a uniform **B**. It will travel in a helical path like that of Figure A-2. After a time τ_C, it will have made one turn of the helix and will be at the point Q, which has the same coordinates in the plane perpendicular to **B** as does P. We can also see this from (A-25) and (A-27) since, when $t = \tau_C = 2\pi m_0/qB$, the argument of the sine and cosine is 2π and $x(\tau_C) = x_0$ and $y(\tau_C) = y_0$. During this time, the particle has traveled the horizontal distance

$$l = v_\parallel \tau_C = 2\pi \left(\frac{m_0}{q}\right)\frac{v_0 \cos\theta}{B} \tag{A-31}$$

Now consider a group of particles for which P, v_0 and θ are the same but whose azimuthal angles φ about the axis of **B** are different. Each of them will travel its own helix, but they will all strike the line PQ at the same point Q. In other words, they have been brought to a *focus* at Q, and one can call the distance l the *focal length*. We see from (A-31) that a measurement of l could then be used as a way of measuring the charge to mass ratio q/m_0. It turns out that this arrangement does not lead to great accuracy for such a measurement; the focusing property described does, however, prove to be of use in other applications. ∎

Each case in which **B** is not uniform generally has to be considered separately. However, there is one situation of considerable practical interest that we can discuss approximately. This is the one for which **B** has axial symmetry and is slowly varying with position. Thus if we take the z axis of cylindrical coordinates to be the axis of symmetry, we can write $\mathbf{B}(z, \rho)$ and assume that the variation of **B** with both z and ρ is

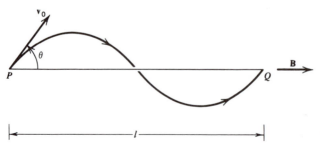

Figure A-4. Magnetic focusing.

not great. This would correspond to lines of **B** like those of Figure A-5. We can expect that the motion of the particle will still be approximately helical as indicated in the figure.

There will now be a radial component B_ρ in addition to B_z. We can find its approximate value by using $\nabla \cdot \mathbf{B} = 0$, which, with the use of (1-87), becomes

$$\frac{1}{\rho} \frac{\partial}{\partial \rho} (\rho B_\rho) = -\frac{\partial B_z}{\partial z} \tag{A-32}$$

For a slowly varying B_z, we can set $\partial B_z / \partial z \simeq$ const., and integrate (A-32) to give

$$B_\rho = -\frac{1}{2} \rho \frac{\partial B_z}{\partial z} \tag{A-33}$$

since $B_\rho(0) = 0$ from Figure A-5, that is, from the way in which the z axis was chosen.

Using (1-76), we find that the z component of (A-10) becomes

$$m_0 \frac{dv_z}{dt} = -q v_\varphi B_\rho \tag{A-34}$$

In a slowly varying **B**, the radius of the helix will not change very much for each turn so that we can take $v_\rho \ll v_\varphi$ and then $v_\varphi \simeq -v_\perp$ since $\mathbf{v}_\perp$ and $\hat{\varphi}$ are oppositely directed for positive q. Making this substitution into (A-34), along with (A-33), and using (A-12) to express ρ as the radius of the orbit, we find that

$$m_0 \frac{dv_z}{dt} = -\frac{1}{2} q v_\perp \rho \frac{\partial B_z}{\partial z} = -\frac{1}{2} \frac{m_0 v_\perp^2}{B} \frac{\partial B_z}{\partial z} = -m \frac{\partial B_z}{\partial z} \tag{A-35}$$

with the use of (A-19). If we now multiply both sides of this by v_z, we obtain

$$m_0 v_z \frac{dv_z}{dt} = \frac{d}{dt} \left(\frac{1}{2} m_0 v_z^2 \right) = -m \frac{\partial B_z}{\partial z} \frac{dz}{dt} = -m \frac{dB_z}{dt} \tag{A-36}$$

where dB_z / dt is the total rate of change of B_z as seen by the particle as it moves along.

Since the magnetic force does no work on the particle, the total kinetic energy T is constant and we find from (A-17) and (A-19) that

$$\frac{dT_\parallel}{dt} = \frac{d}{dt} \left(\frac{1}{2} m_0 v_z^2 \right) = -\frac{dT_\perp}{dt} = -\frac{d}{dt} (m B_z) \tag{A-37}$$

to the same order of approximation that we have been using. Equating the right-hand sides of (A-36) and (A-37), we find that $d(mB_z)/dt = m(dB_z/dt) + B_z(dm/dt) = m(dB_z/dt)$ and therefore

$$\frac{dm}{dt} = 0 \quad \text{and} \quad m = \text{const.} \tag{A-38}$$

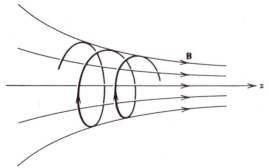

Figure A-5. General motion in a nonuniform B with axial symmetry.

In other words, the particle moves in such a way that its magnetic dipole moment is approximately a constant. If we look again at (A-21), we see that an equivalent way of saying this is that during the motion the particle orbit encloses the same amount of flux Φ. Thus as it moves into regions of larger field magnitude, its radius will decrease. Since Φ is constant, we can also say that the orbit will enclose the same lines of $\mathbf{B}$ as is indicated in Figure A-5.

We can integrate (A-37) and we find that

$$\tfrac{1}{2}m_0 v_z^2 + mB_z = T_\parallel - \mathbf{m} \cdot \mathbf{B} = \text{const.} = \mathscr{E} \tag{A-39}$$

if we take into account the opposite directions of $\mathbf{m}$ and $\mathbf{B}$ shown in (A-20). We see from this that as the particle moves into a region of larger B_z, its longitudinal kinetic energy will decrease, and if mB_z becomes equal to $\mathscr{E}$, the particle will reverse its direction and move back into the region of weaker field; this is in agreement with (A-35) which indicates a force on the particle opposite to the direction of increasing B_z. In other words, the particle will have been *reflected*. This fact is the basis of *magnetic mirrors*, which are used in attempts to confine the ionized particles produced at the high temperatures of fusion reactors.

During the process described above in which $T_\parallel$ will decrease, $T_\perp$ will increase in order to keep T constant. Thus the particle will go about its helix of decreasing radius with greater and greater speed until the point of reflection is reached. This same conclusion also follows from (A-38) and (A-19), since if B increases, $T_\perp$ must also increase in order to keep the ratio m constant.

A-3 STATIC ELECTRIC AND MAGNETIC FIELDS

We now assume that both $\mathbf{E}$ and $\mathbf{B}$ are different from zero. We also assume that they are independent of position and time. The complete equation of motion (A-1) must then be used. It is helpful to consider it in stages.

First we introduce the components of both $\mathbf{v}$ and $\mathbf{E}$ that are parallel and perpendicular *to the direction of the magnetic induction*. Then (A-1) becomes

$$m_0 \frac{d\mathbf{v}_\parallel}{dt} + m_0 \frac{d\mathbf{v}_\perp}{dt} = q\left(\mathbf{E}_\parallel + \mathbf{E}_\perp + \mathbf{v}_\perp \times \mathbf{B}\right) \tag{A-40}$$

since $\mathbf{v}_\parallel \times \mathbf{B} = 0$. This yields the two equations

$$m_0 \frac{d\mathbf{v}_\parallel}{dt} = q\mathbf{E}_\parallel \tag{A-41}$$

$$m_0 \frac{d\mathbf{v}_\perp}{dt} = q(\mathbf{E}_\perp + \mathbf{v}_\perp \times \mathbf{B}) \tag{A-42}$$

The solution to (A-41) is given by the parallel parts of (A-6) with $\mathbf{E}_\parallel$ replacing $\mathbf{E}$; thus $\mathbf{v}_\parallel$ is unaffected by $\mathbf{B}$.

The effect of the remaining portion of $\mathbf{E}$ can be accounted for by writing

$$\mathbf{v}_\perp = \mathbf{v}_D + \mathbf{v}' \tag{A-43}$$

where $\mathbf{v}_D$ will be chosen in a way that we will soon determine. We note that $\mathbf{v}_D$ and $\mathbf{v}'$ are each perpendicular to $\mathbf{B}$. Using (A-43), we can write (A-42) as

$$m_0 \frac{d\mathbf{v}_\perp}{dt} = q(\mathbf{E}_\perp + \mathbf{v}_D \times \mathbf{B} + \mathbf{v}' \times \mathbf{B}) \tag{A-44}$$

and we would like to find $\mathbf{v}_D$ such that

$$\mathbf{E}_\perp + \mathbf{v}_D \times \mathbf{B} = 0 \tag{A-45}$$

We can solve this for $\mathbf{v}_D$ by crossing $\mathbf{B}$ with it and using (1-30); the result is $\mathbf{B} \times \mathbf{E}_\perp + \mathbf{B} \times (\mathbf{v}_D \times \mathbf{B}) = 0 = \mathbf{B} \times \mathbf{E}_\perp + B^2 \mathbf{v}_D$ since $\mathbf{v}_D \cdot \mathbf{B} = 0$. Thus we find that

$$\mathbf{v}_D = -\frac{\mathbf{B} \times \mathbf{E}_\perp}{B^2} = \frac{\mathbf{E}_\perp \times \mathbf{B}}{B^2} = \text{const.} \tag{A-46}$$

Since $\mathbf{v}_D$ is constant, we find that (A-44) reduces to

$$m_0 \frac{d\mathbf{v}'}{dt} = q\mathbf{v}' \times \mathbf{B} \tag{A-47}$$

which is exactly of the form of the second equation of (A-11) and for which we know the complete solution from the last section, that is, a circular motion like that of Figure A-1.

The general motion in the simultaneous presence of uniform $\mathbf{E}$ and $\mathbf{B}$ thus is the superposition of three parts. There is a constant acceleration along the direction of $\mathbf{B}$ produced by $\mathbf{E}_\parallel$. There is a constant velocity in the direction of $\mathbf{E}_\perp \times \mathbf{B}$ and hence perpendicular to $\mathbf{B}$. Finally, there is the circular motion about the axis of $\mathbf{B}$ with radius and frequency given by (A-12) and (A-13) with the constant speed v' replacing $v_\perp$. The velocity $\mathbf{v}_D$ is known as the *drift velocity*; its magnitude is $E_\perp/B$ and is independent of both the charge and the mass of the particle.

For simplicity, we will concentrate on the combined drift and circular motion, that is, on the projection of the motion on the plane perpendicular to $\mathbf{B}$; $\mathbf{E}_\perp$ and $\mathbf{v}_D$ are both in this plane. This combined motion can be described in convenient geometrical terms. The circular motion associated with $\mathbf{v}'$ has a radius $r_C' = m_0 v'/qB$ and will make one complete rotation in the time $\tau_C = 2\pi m_0/qB$. During this time, the guiding center G will have traveled a distance $d_G = v_D \tau_C = 2\pi m_0 E_\perp/qB^2$. This is just the circumference of a circle of radius $r_G = m_0 E_\perp/qB^2 = v_D/\omega_C$. Thus the motion of G corresponds to that of a circle of this radius that is rolling without slipping on a horizontal line and with angular velocity ω_C as illustrated in Figure A-6. Since the circular motion described by $\mathbf{v}'$ is about G, we see that the net motion of the charge will be the same as if it were rigidly fastened to the rolling circle at the point P a distance r_C' from the center G. This is exactly the situation that produces the curve known as a *cycloid*. If $r_C' = r_G$, then P is on the periphery of the circle and the curve that P traces out is the ordinary cycloid shown in Figure A-7a. If $r_C' > r_G$, then P is outside of the circle and produces the *prolate cycloid* of (b) of the figure, while if

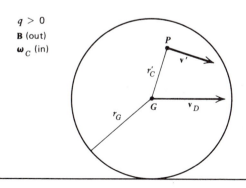

Figure A-6. Calculation of the motion in a uniform E and B.

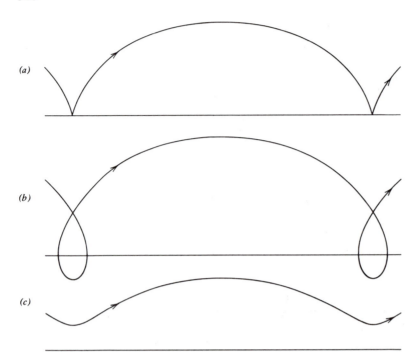

(a)

(b)

(c)

Figure A-7. Possible types of cycloidal paths.

$r'_C < r_G$, the curve is the *curtate cycloid* shown in (c). From our previous results, we have

$$\frac{r'_C}{r_G} = \frac{(m_0 v'/qB)}{(v_D \tau_C/2\pi)} = \frac{v'}{v_D} = \frac{B}{E_\perp} v' \tag{A-48}$$

■ **Example**

Rectangular coordinates. As we did in the last section, let us verify these results in terms of a specific coordinate system. We assume that $\mathbf{E} = E\hat{y}$ and $\mathbf{B} = B\hat{z}$. Thus $\mathbf{E}_\parallel = 0$ and $v_\parallel = v_z$ will be constant; we choose $v_z = 0$ so that the motion is confined to the xy plane. We also see from (A-46) that $\mathbf{v}_D = v_D\hat{x} = (E/B)\hat{x}$ in this case. However, let us begin anew with (A-1), which, for these fields, yields the two equations of interest:

$$\frac{dv_x}{dt} = \frac{qB}{m_0} v_y = \omega_C v_y \tag{A-49}$$

$$\frac{dv_y}{dt} = \frac{q}{m_0}(E - Bv_x) = -\omega_C(v_x - v_D) \tag{A-50}$$

If we differentiate (A-49) and substitute from (A-50), we obtain

$$\frac{d^2 v_x}{dt^2} = \omega_C \frac{dv_y}{dt} = -\omega_C^2(v_x - v_D)$$

which has a solution

$$v_x = v_D + (v_{0x} - v_D)\cos\omega_C t \tag{A-51}$$

corresponding, as we will see, to choosing $v_{0y} = 0$. We can now find v_y from (A-49) and (A-51):

$$v_y = \frac{1}{\omega_C} \frac{dv_x}{dt} = -(v_{0x} - v_D) \sin \omega_C t \tag{A-52}$$

Before we go on to find the coordinates, we can verify that these results for the velocity components are consistent with our previous ones. The components of **v′** as obtained from (A-43) are

$$v'_x = v_x - v_D \qquad v'_y = v_y \tag{A-53}$$

so that (A-51) and (A-52) can also be written as

$$v'_x = v'_{0x} \cos \omega_C t \qquad v'_y = -v'_{0x} \sin \omega_C t \qquad v'_{0x} = v_{0x} - v_D \tag{A-54}$$

The expressions for v'_x and v'_y are just those expected for a velocity **v′** of magnitude $|\mathbf{v}'| = v' = |v'_{0x}|$ rotating in a clockwise sense with angular speed ω_C as in Figure A-6.

We can find the particle coordinates by integrating (A-51) and (A-52), and evaluating the constants of integration in terms of the initial position. The results are

$$x = x_0 + v_D t + \frac{(v_{0x} - v_D)}{\omega_C} \sin \omega_C t \tag{A-55}$$

$$y = y_0 + \frac{(v_{0x} - v_D)}{\omega_C} (\cos \omega_C t - 1) \tag{A-56}$$

which can be rearranged as

$$x = x_0 + \frac{v_D}{\omega_C}(\omega_C t) - \frac{(v_D - v_{0x})}{\omega_C} \sin \omega_C t$$

$$y = y_0 + \frac{(v_D - v_{0x})}{\omega_C}(1 - \cos \omega_C t)$$

These are in the usual parametric form of the equation of a cycloid with associated radii $r_G = v_D/\omega_C$, $r'_C = |v_D - v_{0x}|/\omega_C = v'/\omega_C$ in agreement with previous results.

An interesting special case arises when $v_{0x} = v_D$ for then (A-55) and (A-56) become simply $x = x_0 + v_D t$, $y = y_0$ showing that the particle travels in a straight line with constant velocity equal to the drift velocity and hence is otherwise unaffected by the fields. We see from (A-45) that this occurs because, with this initial velocity, the net force is zero and remains so since there is no acceleration to change **v** from its initial value of $\mathbf{v}_D$. We note that this straight line motion is independent of either q or m_0. In this sense, this field combination acts as a velocity selector since all particles traveling horizontally must have the same definite velocity $\mathbf{v}_D$. ∎

■ **Example**

Magnetron. This is an example of a case in which **B** is uniform but **E** is not. (See Figure A-8.) The system consists of two coaxial conducting cylinders of radii a and b with a uniform induction **B** parallel to the axis. There is a potential difference $\Delta\phi$ between the cylinders so that a radial field **E** is produced. Suppose a charge q is introduced at the surface of the inner cylinder. The field **E** will accelerate it toward the outer cylinder. Because of the **B**, however, the path will be curved and the particle may be turned back before it can reach the outer cylinder. In this case there will be no current between the cylinders. For a smaller **B**, the charge q can reach the plate and there will be a current.

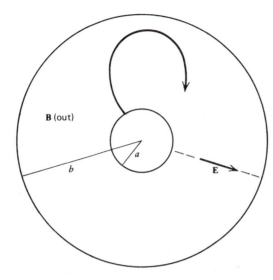

Figure A-8. Magnetron.

Thus a knowledge of the conditions for the transition between no current and current will enable one to use this device to measure $\mathbf{B}$. (In an actual magnetron, the inner cylinder is a heated filament and the particles that are produced are electrons of charge $-e$. It is just as easy to analyze this problem for an arbitrary charge q, however.)

The variation of both $\mathbf{E}$ and ϕ with position can easily be found by using Gauss' law (4-1) or the general expression (11-141). As we will see, we actually do not need to know these details. Since $\mathbf{B}$ does no work on the charge, the energy conservation expression (A-9) is applicable. If we use cylindrical coordinates and note that $\phi = \phi(\rho)$, by symmetry, (A-9) can be written as

$$\frac{1}{2}m_0\left(v_\rho^2 + v_\varphi^2\right) + q\phi(\rho) = \frac{1}{2}m_0\left[\left(\frac{d\rho}{dt}\right)^2 + \rho^2\left(\frac{d\varphi}{dt}\right)^2\right] + q\phi(\rho) = \text{const.} \quad \text{(A-57)}$$

where the velocity components can be found from (1-82). The constant can be evaluated by setting $\rho = a$; if we assume that the particle is produced at the inner cylinder with negligible velocity, its kinetic energy is zero and the constant in (A-57) is just $q\phi(a)$ and we have

$$\frac{1}{2}m_0\left[\left(\frac{d\rho}{dt}\right)^2 + \rho^2\left(\frac{d\varphi}{dt}\right)^2\right] + q\phi(\rho) = q\phi(a) \quad \text{(A-58)}$$

We want the condition that the particle just misses the outer cylinder at $\rho = b$. If this is the case, its path will be tangent to the surface of the cylinder so that $v_\rho(b) = (d\rho/dt)_b = 0$ and (A-58) gives

$$b^2\left(\frac{d\varphi}{dt}\right)_b^2 = \frac{2q}{m_0}[\phi(a) - \phi(b)] = \frac{2q\,\Delta\phi}{m_0} \quad \text{(A-59)}$$

We can get another expression for the quantity on the left by turning to the equation of motion.

With $\mathbf{v} = v_\rho\hat{\boldsymbol{\rho}} + v_\varphi\hat{\boldsymbol{\varphi}}$, $\mathbf{E} = E\hat{\boldsymbol{\rho}}$, and $\mathbf{B} = B\hat{\mathbf{z}}$, we find with the use of (1-76) and a result from kinematics that the φ component of (A-1) is

$$m_0 a_\varphi = m_0\left[\rho\frac{d^2\varphi}{dt^2} + 2\left(\frac{d\rho}{dt}\right)\left(\frac{d\varphi}{dt}\right)\right] = -qBv_\rho = -qB\frac{d\rho}{dt}$$

If we multiply both sides of this by ρ, we find that it can be written as

$$\frac{d}{dt}\left[m_0\rho^2\left(\frac{d\varphi}{dt}\right)\right] = -qB\rho v_\rho = -qB\frac{d}{dt}\left(\frac{1}{2}\rho^2\right) \tag{A-60}$$

In this form, it is perhaps more familiar. The term in brackets on the left-hand side is the angular momentum of the particle about the z axis, while the middle expression is the z component of the torque as found from $\mathbf{r} \times \mathbf{f} = \rho\hat{\boldsymbol{\rho}} \times q[E\hat{\boldsymbol{\rho}} + (v_\rho\hat{\boldsymbol{\rho}} + v_\varphi\hat{\boldsymbol{\varphi}}) \times B\hat{\mathbf{z}}]$ using (1-76) again. Equating the first and third expressions of (A-60), we can integrate the result to obtain

$$\rho^2\frac{d\varphi}{dt} + \frac{qB}{2m_0}\rho^2 = \text{const.} \tag{A-61}$$

We can evaluate the constant by considering the initial situation at the inner cylinder $\rho = a$. Since we have assumed the particle to originate with negligible velocity, we have $(d\varphi/dt)_a = 0$ and the constant equals $(qB/2m_0)a^2$ so that (A-61) becomes

$$\rho^2\frac{d\varphi}{dt} = -\frac{qB}{2m_0}(\rho^2 - a^2) \tag{A-62}$$

We note that since $\rho \geq a$, $d\varphi/dt$ is negative for $q > 0$, which agrees with the path shown in Figure A-8.

If we now set $\rho = b$ in (A-62), we obtain

$$b^2\left(\frac{d\varphi}{dt}\right)_b = -\frac{qB}{2m_0}(b^2 - a^2)$$

If we solve this for $(d\varphi/dt)_b$ and substitute the result in (A-59), we find that

$$B^2 = \frac{8m_0b^2\Delta\phi}{q(b^2 - a^2)^2} \tag{A-63}$$

which relates the value of B to the value of the potential difference $\Delta\phi$ for which the current between the cylinders just vanishes. ∎

A-4 A TIME-DEPENDENT MAGNETIC FIELD

Although a magnetic induction that is constant in time does no work on a charged particle, a time-varying one can do so by means of the induced emf described by Faraday's law. This fact has been used as the basis for the charged particle accelerator known as the *betatron*; the name arises because it was originally used to produce electrons of large kinetic energy and thus they were similar to the beta particles of radioactivity.

The essential features of this device are shown in Figure A-9. A particle is constrained to move in a circle of radius R that encloses a cylindrically symmetric induction $\mathbf{B} = B_z\hat{\mathbf{z}}$, which is normal to the plane of the orbit. The induction $\mathbf{B}$ is varied in time so that the work done on the particle in each complete circuit equals its charge times the induced emf $\mathscr{E}_{\text{ind}}$ given by (17-6) as

$$\mathscr{E}_{\text{ind}} = \oint\mathbf{E}_{\text{ind}} \cdot d\mathbf{s} = -\frac{d\Phi}{dt} = -\pi R^2\frac{d\langle B_z\rangle}{dt} \tag{A-64}$$

where $\langle B_z\rangle$ is the average value of the induction over the area enclosed by the orbit. Because of the cylindrical symmetry we can write $\mathbf{E} = E_\varphi\hat{\boldsymbol{\varphi}}$ and the integral becomes

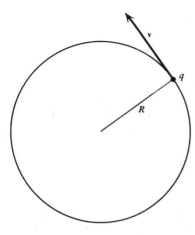

Figure A-9. Essential aspects for describing the action of a betatron.

$2\pi R E_\varphi$. Thus the magnitude of the induced electric field is

$$E_\varphi = \frac{1}{2} R \frac{d\langle B_z \rangle}{dt} \qquad (A\text{-}65)$$

and it will produce a tangential force of magnitude

$$f_\varphi = q E_\varphi = \frac{1}{2} q R \frac{d\langle B_z \rangle}{dt} = m_0 \frac{dv_\varphi}{dt} \qquad (A\text{-}66)$$

We require that the particle be constrained to move in a circle of fixed R. This can be done with an induction normal to the plane of the orbit and with a value B_R; this is known as the "guiding field." The required value of B_R can be obtained from (A-12) and is $B_R = (m_0 v_\varphi / q R)$ since $v_\varphi = v_\perp$ in this case. Since v_φ is changing by (A-66), B_R must be changing as well in order to keep R constant, and we find that it is necessary that

$$\frac{dB_R}{dt} = \frac{m_0}{qR} \frac{dv_\varphi}{dt} \qquad (A\text{-}67)$$

When we eliminate $m_0(dv_\varphi/dt)$ between (A-66) and (A-67), we find that we must have $(dB_R/dt) = \frac{1}{2}(d\langle B_z \rangle / dt)$, which leads to

$$B_R = \tfrac{1}{2}\langle B_z \rangle + \text{const.} \qquad (A\text{-}68)$$

which is known as the betatron condition and must be satisfied at all times if the betatron is to work. Normally, the two fields B_R and $\langle B_z \rangle$ are changed simultaneously with both starting from zero. In that case, the constant in (A-68) must be zero and the condition simplifies to $B_R = \tfrac{1}{2}\langle B_z \rangle$. In other words, the induction at the particle position must always equal one half the average induction enclosed by the circular orbit.

EXERCISES

A-1 At $t = 0$, a particle is at the origin with initial velocity $\mathbf{v}_0 = v_0 \hat{\mathbf{x}}$ and enters a region of uniform electric field $\mathbf{E} = E\hat{\mathbf{y}}$. Find the equation of its path and its deflection d from the x axis after it has gone a distance D along x. Find the angle α made by $\mathbf{v}$ with the x axis at this point.

A-2 A particle moves in two dimensions in a region where the scalar potential is $\phi = \frac{1}{2}A(x^2 - y^2)$ where A is a constant with the same sign as q. At $t = 0$, the particle is at rest at (x_0, y_0). Find x and y for subsequent times and sketch the resultant path of the particle.

A-3 Suppose that $\mathbf{B} = 0$ for the system of coaxial cylinders of Figure A-8. Show that circular orbits are possible when $\Delta\phi \neq 0$. Evaluate the kinetic energy and show that it is independent of the radius of the circular orbit.

A-4 A charged particle is traveling horizontally with speed v_0 when it enters a region of uniform $\mathbf{B}$ directed vertically downward. Find the deflection d from its original direction of travel when the projection of its position on its original direction is D where $D \leq r_C$.

A-5 In one type of mass spectrometer a particle of charge q is produced at rest and then accelerated through a potential difference $\Delta\phi$. It then enters a region of uniform $\mathbf{B}$, which is perpendicular to its velocity. After traveling a semicircle, the particle strikes a detector at a distance D from its point of entrance. Find an expression from which the mass m_0 can be found in terms of the quantities given.

A-6 Suppose that Figure A-4 shows the xz plane with $\mathbf{B}$ along the positive z axis and P is the origin. Find the position vector $\mathbf{r}_G$ of the guiding center of this motion as a function of t.

A-7 Suppose particles of the same q, m_0, and v_0 are projected from P in Figure A-4 over a range of angles $\theta_0 - \Delta\theta \leq \theta \leq \theta_0 + \Delta\theta$. Their focal points Q will vary over a corresponding range of position Δl. Find Δl and the ratio $\Delta l/l_0$ where l_0 is the focal length for the central angle θ_0. What is the largest allowable value of $\Delta\theta$ if $\theta_0 = 10°$ and the spread Δl is required to be less than 5 percent of l_0?

A-8 After being accelerated through a potential difference of 100 volts, a beam of electrons spreads out into a cone of very small angle. A long ideal solenoid of 10 turns per centimeter has its axis aligned with the axis of the cone. What current will be required in the solenoid in order to refocus the electrons within a distance of 10 centimeters?

A-9 Suppose that charged particles are produced with negligible velocity at one plate of a parallel plate capacitor of plate separation d. There is a potential difference $\Delta\phi$ between the plates and a uniform induction $\mathbf{B}$ is parallel to the plates. Show that there will be no current between the plates if $\Delta\phi$ is less than $\frac{1}{2}(q/m_0)B^2d^2$.

A-10 The curves of Figure A-7 were drawn on the assumption that q was positive. What changes, if any, will there be in this figure if $\mathbf{E}_\perp$ and $\mathbf{B}$ are unchanged but q is negative?

A-11 At $t = 0$, a charged particle at the origin has a velocity $v_0\hat{\mathbf{z}}$. It is in the presence of parallel uniform fields $\mathbf{E} = E\hat{\mathbf{x}}$ and $\mathbf{B} = B\hat{\mathbf{x}}$. Find the coordinates x and y when the particle has reached one half of its maximum value of z for the first time.

A-12 A charged particle is in a region of uniform $\mathbf{B}$ and is also subject to a uniform gravitational field so that the gravitational force on it is $m_0\mathbf{g}$. Show that it has a drift velocity equal to $m_0(\mathbf{g}_\perp \times \mathbf{B})/qB^2$. (Note that this drift velocity depends on the charge to mass ratio of the particle.)

A-13 At $t = 0$, a charged particle at the origin has a velocity $\mathbf{v}_0 = v_0\hat{\mathbf{x}}$. There is a uniform electric field $\mathbf{E} = E\hat{\mathbf{y}}$ present. Solve the relativistic equation of motion (29-105) to find the coordinates for later times. Verify that for small values of t, the path of the particle is approximately a parabola.

B

ELECTROMAGNETIC PROPERTIES OF MATTER

We have described the overall effects of the presence of matter on electromagnetic fields in terms of parameters such as the electric susceptibility χ_e. From a purely macroscopic point of view such parameters are assumed to be characteristic of a given material and their numerical values are to be determined from experiment. In other words, the formulation of the electromagnetic field that we have obtained does not attempt to predict their values. The observed fact that different materials have different values of these quantities indicates that they depend on the details of the microscopic structure of the material, that is, on the atomic and molecular properties of its constituents as well as on its state of aggregation, whether solid, liquid, or gas.

In this appendix we consider some of the principal features that are involved in a microscopic approach; we have already seen an example of this in our brief discussion of conductivity in Sections 12-5 and 24-8. Whole books can be, and have been, written on these general subjects so that we will necessarily be very limited in what we can do. In particular, while quantum mechanics is the appropriate way of dealing with atoms and molecules, we can get sufficiently good results by using classical mechanics to investigate the mechanical response of a collection of interacting charged mass points to an applied electromagnetic field. The results we will obtain in this way will not be completely accurate in all details, but are surprisingly good and will give us an adequate understanding of what is going on.

B-1 STATIC ELECTRIC PROPERTIES

We want to evaluate the electric susceptibility χ_e defined in (10-50) for linear isotropic materials by

$$\mathbf{P} = \chi_e \epsilon_0 \mathbf{E} \qquad (\text{B-1})$$

This shows that we have to calculate the electric dipole moment per unit volume $\mathbf{P}$. In turn, $\mathbf{P}$ will be related to the average dipole moment of the molecules within the volume.

At the beginning of Section 10-1, we discussed the possibility that a material could become polarized by the production of an induced dipole moment. This induced moment arises when an electric field applied to a neutral molecule with no permanent dipole moment shifts the positions of the positive and negative charge distributions with respect to each other. The relative shift is finite because the electrostatic forces on the charges will be opposed by internal forces within the molecule and a new state of mechanical equilibrium will be attained.

To be specific, let us consider a neutral molecule constructed from n atoms. The ith atom has a positive nucleus of charge $Z_i e$ together with Z_i electrons each of charge $-e$ where Z_i is the atomic number. The total number of electrons will therefore be $Z_t = Z_1 + Z_2 + \ldots + Z_n$. If we let $\mathbf{r}_i$ be the position vector of the ith nucleus and $\mathbf{r}_j$

that of the jth electron, the total dipole moment as found from (8-19) is

$$\mathbf{p} = \sum_{i=1}^{n} Z_i e\mathbf{r}_i - e\sum_{j=1}^{Z_t} \mathbf{r}_j \qquad \text{(B-2)}$$

Since the molecule has no monopole moment, we know from (8-43) that we can use any convenient origin for our coordinate system, for example, one at the center of mass; for a monatomic molecule, we could choose the origin at the location of the nucleus so that the first term of (B-2) would be zero. Since the charges can be assumed to be moving, the electrons in particular traversing some "orbits," $\mathbf{p}$ can be a complicated function of time. We are, however, concerned here only with time average values, so that our interest is really in

$$\langle\mathbf{p}\rangle = \sum_{i} Z_i e\langle\mathbf{r}_i\rangle - e\sum_{j} \langle\mathbf{r}_j\rangle \qquad \text{(B-3)}$$

so that

$$\mathbf{P} = N\langle\mathbf{p}\rangle \qquad \text{(B-4)}$$

where N is the number of molecules per unit volume.

If we let values in the absence of an applied field be designated by the subscript zero, we can write (B-3) as

$$\mathbf{p}_0 = \langle\mathbf{p}\rangle_0 = \sum_{i} Z_i e\langle\mathbf{r}_i\rangle_0 - e\sum_{j} \langle\mathbf{r}_j\rangle_0 \qquad \text{(B-5)}$$

If $\mathbf{p}_0 \neq 0$, the molecule has a permanent moment, while if $\mathbf{p}_0 = 0$ it does not and $\mathbf{P}$ as given by (B-4) can arise only from induced moments. For example, in a monatomic molecule, we can assume the electron charge distribution to be spherically symmetric about the nucleus. Thus, if we choose the origin at the nucleus, we can take $\langle\mathbf{r}_i\rangle_0 = 0$ and then $\langle\mathbf{r}_j\rangle_0$ will be zero also because positive and negative values of x_j, y_j, z_j will be equally likely so that $\langle x_j\rangle_0 = \langle y_j\rangle_0 = \langle z_j\rangle_0 = 0$. The result will be that $\mathbf{p}_0 = 0$ as we expect.

In the presence of a polarizing field we can expect each $\langle\mathbf{r}\rangle$ to be different from $\langle\mathbf{r}\rangle_0$. Since the shift arises from the balance between an electrostatic force and an internal mechanical "restoring" force $\mathbf{F}_m$, we can write

$$\mathbf{F}_{\text{net}} = 0 = \mathbf{F}_m + q\mathbf{E}_p \qquad \text{(B-6)}$$

for the new equilibrium situation for any charge q as we did in (2-9). In (B-6), $\mathbf{E}_p$ is the "effective" or "polarizing" electric field that is producing the displacement of the charges. [It is sometimes called the local field, but we have already used this after (10-70) to refer to the field produced by the bound charges.] It is not necessarily equal to the macroscopic average field $\mathbf{E}$ in the material since the molecule in question is itself contributing to $\mathbf{E}$; we return to this point shortly.

Since $\mathbf{F}_m$ is to describe some sort of restoring force, we expect it to be a function of the displacement $(\langle\mathbf{r}\rangle - \langle\mathbf{r}\rangle_0)$ and also that $\mathbf{F}_m(0) = 0$. Furthermore, the displacement is not expected to be large so that it is appropriate to expand $\mathbf{F}_m$ in a power series in its argument and keep only the first term; this gives us

$$\mathbf{F}_m(\langle\mathbf{r}\rangle - \langle\mathbf{r}\rangle_0) = -K(\langle\mathbf{r}\rangle - \langle\mathbf{r}\rangle_0) \qquad \text{(B-7)}$$

where K is an effective "spring constant" for this charge q. Substituting (B-7) into (B-6), we obtain

$$\langle\mathbf{r}\rangle - \langle\mathbf{r}\rangle_0 = \frac{q}{K}\mathbf{E}_p \qquad \text{(B-8)}$$

If we now substitute this into (B-3), using the appropriate charge in each case, and then use (B-5) along with our assumption that $\mathbf{p}_0 = 0$, we find that the average induced dipole moment is

$$\langle \mathbf{p} \rangle = e^2 \left(\sum_i \frac{Z_i^2}{K_i} + \sum_j \frac{1}{K_j} \right) \mathbf{E}_p = \alpha \mathbf{E}_p \qquad (B\text{-}9)$$

and is linearly proportional to the polarizing field $\mathbf{E}_p$. The constant of proportionality α is called the *polarizability*. [We recall that in (11-112) we found a similar result for the grounded conducting sphere when in a previously uniform field.] We see that when (B-9) is substituted into (B-4), the polarization will be proportional to the inducing field $\mathbf{E}_p$ as expected for a linear isotropic material.

The polarizability in (B-9) is expressed in terms of sums involving the various "spring constants" of the molecule. We can also get a crude estimate of its value in a somewhat different manner as follows. Let us consider a single atom and assume that the electron charge distribution is not only spherically symmetric but *uniform* throughout a sphere of radius a. We further assume that the effect of the field $\mathbf{E}_p$ is to shift the negative charge distribution *rigidly* with respect to the nucleus so that the final result is a relative displacement $\mathbf{l}$ between their centers as shown in Figure B-1. Then the induced dipole moment will be $\langle \mathbf{p} \rangle = (Ze)\mathbf{l}$ by (8-44). The nuclear charge Ze can then be considered to be in equilibrium under the influence of $\mathbf{E}_p$ and the electric field at the surface of the sphere of uniform negative charge of radius l. Equating E_p to the magnitude of this field as given by (4-21), we get $E = (Ze)l/4\pi\epsilon_0 a^3 = E_p$ and therefore $\langle p \rangle = Zel = 4\pi\epsilon_0 a^3 E_p$. Thus the polarizability is

$$\alpha = 4\pi\epsilon_0 a^3 \qquad (B\text{-}10)$$

and is proportional to the volume of the atom. [We note that (B-10) is the same as was found for the conducting sphere in (11-113).]

Now we turn to the problem of evaluating the polarizing field $\mathbf{E}_p$. We assume the material to be homogeneous as well as linear and isotropic. We want the total electric field at the location of the molecule except for that arising from the molecule itself. If we think of the usual case of a material in a capacitor, the field $\mathbf{E}_p$ will have contributions from the free charges on the plates, the bound surface charges on the dielectric boundaries, and from *all of the rest* of the dipoles in the material. Thus we can write

$$\mathbf{E}_p = \mathbf{E}_f + \mathbf{E}_b + \mathbf{E}_D \qquad (B\text{-}11)$$

[If we assume no free charge within the dielectric, there is no bound volume charge density either by (10-58); the charges contributing to $\mathbf{E}_f$ and $\mathbf{E}_b$ are illustrated for a

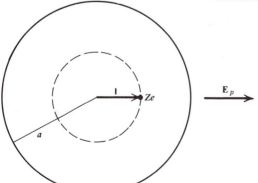

Figure B-1. The nucleus and the electron charge distribution are rigidly displaced by **l**.

parallel plate capacitor in Figure 10-6.] Now we know that the macroscopic average field is produced by the free and bound surface charges only so that $\mathbf{E}_f + \mathbf{E}_b$ is actually $\mathbf{E}$ so that we can write

$$\mathbf{E}_p = \mathbf{E} + \mathbf{E}_D \qquad \text{(B-12)}$$

The dipole moment of each of the remaining dipoles contributes to $\mathbf{E}_D$. If we are very far from the molecule in question, the dipoles will appear to it to be continuously distributed so that it would be appropriate to describe them in terms of the uniform dipole moment per unit volume $\mathbf{P}$. On the other hand, the other dipoles that are very close on a microscopic scale will appear as discrete dipoles located at specific points; in other words, they cannot be treated as a continuous distribution. This suggests that if we draw a sphere of some appropriate radius R about the point in question, we can divide space into two regions—outside the sphere, the dipoles will be described by $\mathbf{P}$ and will contribute $\mathbf{E}_O$ to $\mathbf{E}_D$, while inside the sphere, we treat them as individuals and their contribution to $\mathbf{E}_D$ will be written as $\mathbf{E}_I$. Thus (B-12) becomes

$$\mathbf{E}_p = \mathbf{E} + \mathbf{E}_O + \mathbf{E}_I \qquad \text{(B-13)}$$

Figure B-2 shows this hypothetical sphere. It is evident that the proper value for R is not unique since the distinction between "continuous" and "discrete" distributions is somewhat fuzzy; on the other hand, a reasonable value for R can be decided upon. For the material outside the sphere, $\rho_b = 0$ by (10-58), and we see that it is completely equivalent only to the bound surface charges on the surface of the sphere arising from the discontinuity of $\mathbf{P}$ there. The surface density of these bound charges can be found from (10-8) as $\sigma_b = \mathbf{P} \cdot \hat{\mathbf{n}}' = P \cos \theta'$. This is exactly the surface charge density of (10-27) that is illustrated in Figure 10-9. The field produced by it was found to have the magnitude $P/3\epsilon_0$ in (10-37) and we see from the signs of the charge distribution of Figure B-2 that it will have the same direction as $\mathbf{P}$. Thus $\mathbf{E}_O = \mathbf{P}/3\epsilon_0$ and (B-13) becomes

$$\mathbf{E}_p = \mathbf{E} + \frac{\mathbf{P}}{3\epsilon_0} + \mathbf{E}_I \qquad \text{(B-14)}$$

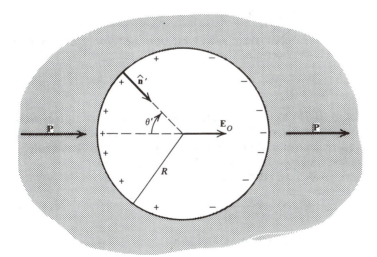

Figure B-2. Hypothetical spherical cavity used to calculate the electric field due to the polarized material outside of it.

In the most general situation, E_I can be different from zero, but we will confine ourselves to the large number of cases for which $E_I = 0$. Let us see how this can occur. Equation 8-84 gives the electric field produced by a point dipole located at the origin. If, instead, the dipole p_k is located at a point r_k with respect to the origin chosen at the center of our sphere, we can replace r in (8-84) by $-r_k$ so that the contribution of p_k to the field at the center will be

$$E_k = \frac{1}{4\pi\epsilon_0}\left[-\frac{p_k}{r_k^3} + \frac{3(p_k \cdot r_k)r_k}{r_k^5}\right] \tag{B-15}$$

When we sum (B-15) over all the dipoles within the sphere, we get E_I. Doing this, and noting that in a uniformly polarized material all molecules have the same average dipole moment, we can let $p_k = \langle p \rangle$ and we get

$$E_I = \frac{1}{4\pi\epsilon_0}\left[-\langle p \rangle \sum_k \frac{1}{r_k^3} + 3\sum_k \frac{(\langle p \rangle \cdot r_k)r_k}{r_k^5}\right] \tag{B-16}$$

The x component of this is

$$E_{Ix} = \frac{1}{4\pi\epsilon_0}\left(-\langle p_x \rangle \sum_k \frac{1}{r_k^3} + 3\langle p_x \rangle \sum_k \frac{x_k^2}{r_k^5} + 3\langle p_y \rangle \sum_k \frac{y_k x_k}{r_k^5} + 3\langle p_z \rangle \sum_k \frac{z_k x_k}{r_k^5}\right) \tag{B-17}$$

If the dipoles are distributed isotropically, then for every molecule with coordinates (x, y, z) there will be one with coordinates $(-x, -y, -z)$ or $(-x, y, z)$ and so on. Since there will actually be a large number of molecules within the sphere, the terms in the last two sums will be positive and negative with equal likelihood for a given magnitude of r_k so that

$$\sum_k \frac{y_k x_k}{r_k^5} = \sum_k \frac{z_k x_k}{r_k^5} = 0$$

Similarly, for a given r_k, the squares of the coordinates will appear with equal likelihood so that on the average $\langle x_k^2 \rangle = \langle y_k^2 \rangle = \langle z_k^2 \rangle$ and $r_k^2 = \langle x_k^2 \rangle + \langle y_k^2 \rangle + \langle z_k^2 \rangle = 3\langle x_k^2 \rangle$. Since this will be true for each r_k, when we sum up over the large number of molecules, we get

$$\sum_k \frac{x_k^2}{r_k^5} = \frac{1}{3}\sum_k \frac{1}{r_k^3}$$

so that (B-17) becomes $E_{Ix} = 0$. Similar results will be obtained for the y and z components. Thus $E_I = 0$ and (B-14) becomes simply

$$E_p = E + \frac{P}{3\epsilon_0} \tag{B-18}$$

The reasoning that led us from (B-16) to (B-18) would be applicable if the dipoles were distributed on a simple cubic lattice. Similarly, E_I would be zero if the positions of the dipoles were completely random as would be the case in a liquid or a gas. Hence there are a great many situations for which we can expect (B-18) to be suitable.

If we now combine (B-18), (B-9), and (B-4), we find that

$$P = N\alpha\left(E + \frac{P}{3\epsilon_0}\right) \tag{B-19}$$

so that when we solve this for P and use (B-1) we find that

$$\chi_e = \frac{(N\alpha/\epsilon_0)}{1 - (N\alpha/3\epsilon_0)} \tag{B-20}$$

which relates the macroscopic susceptibility to the microscopic polarizability α and to the number density N. Conversely, if we write $\chi_e = \kappa_e - 1$ by (10-52) and solve (B-20) for α, we get

$$\alpha = \frac{3\epsilon_0(\kappa_e - 1)}{N(\kappa_e + 2)} \tag{B-21}$$

which is known as the *Clausius-Mossotti relation*. Since α is a constant, (B-21) tells us that if we vary the density of the material the dielectric constant κ_e will also vary in such a way that the quantity on the right-hand side will remain constant. On the other hand, one can use measured values of κ_e to evaluate α; if this is then combined with (B-10), one can find the atomic radius a. Values of a obtained in this manner agree reasonably well with those obtained by other means.

Now let us suppose that the molecule has a permanent dipole moment, that is, $\mathbf{p}_0$ of (B-5) is different from zero. We also considered this possibility in Section 10-1 and concluded that, in the presence of a field, the torque on $\mathbf{p}_0$ described by (8-75) will tend to align the permanent dipoles in the direction of the field. This tendency will be opposed by randomizing processes like collisions associated with temperature agitation of the molecules. In the final equilibrium state, however, we still expect there to be a net dipole moment in the direction of $\mathbf{E}_p$ so that $\langle \mathbf{p} \rangle \neq 0$, but we also expect that $\langle \mathbf{p} \rangle < \mathbf{p}_0$. In such a case, $\langle \mathbf{p} \rangle$ can also be expected to have a dependence on the absolute temperature T, and this, in turn, will make the susceptibility depend on temperature in addition to any variation associated with a change in the number density N with T.

In order to find the average dipole moment for equilibrium conditions at a given temperature, we must involve a result from statistical mechanics that says that the probability of finding a system in a given quantum state of energy W_m is given by

$$\mathscr{P}_m(W_m) = \frac{e^{-\beta W_m}}{\sum_m e^{-\beta W_m}} \tag{B-22}$$

where $\beta = 1/kT$ and k is Boltzmann's constant equal to 1.38×10^{-23} joule/K. The denominator given in (B-22) is needed so that the total probability of finding the system in some state is unity, that is, so that $\sum_m \mathscr{P}_m = 1$.

If ψ is a dynamical variable of interest whose value in the mth state is ψ_m, then its average value will be given by the usual definition of an average as

$$\langle \psi \rangle = \sum_m \psi_m \mathscr{P}_m = \frac{\sum_m \psi_m e^{-\beta W_m}}{\sum_m e^{-\beta W_m}} \tag{B-23}$$

Strictly speaking, (B-22) and (B-23) apply to the quantum states of a single system of interest. It can be shown that if the system consists of a collection of independent, or nearly independent, molecules, then these results can also be applied to a single molecule where W_m is now the energy of a molecule and $\langle \psi \rangle$ is the average value for a given molecule. Furthermore, it can also be shown that in the limit where classical mechanics is applicable, the sum in (B-23) can be replaced by integrals over the momentum components and position coordinates. Thus, in rectangular coordinates, (B-23) would become

$$\langle \psi \rangle = \frac{\int \psi e^{-\beta W} \, dv_x \, dv_y \, dv_z \, dx \, dy \, dz}{\int e^{-\beta W} \, dv_x \, dv_y \, dv_z \, dx \, dy \, dz} \tag{B-24}$$

since $\dot{p}_x = d(m_0 v_x) = m_0\, dv_x$, and so on, and the factors m_0^3 would cancel from the numerator and denominator. The integrals will generally be sixfold integrals with appropriate limits of integration, and, in order to use (B-24), we need to know both W and ψ as functions of (v_x, v_y, v_z, x, y, z). [For example, if we were dealing with an ideal gas in which interactions between the molecules could be neglected, then W would be simply the kinetic energy given by (A-8) as $\frac{1}{2}m_0 v^2$ so that (B-22) would give $\mathscr{P} \sim e^{-m_0 v^2/2kT}\, dv_x\, dv_y\, dv_z\, dx\, dy\, dz$, which is the familiar Maxwell velocity distribution function.]

In general, W will include potential energy as well as kinetic energy. Of particular interest to us here is the interaction or orientational energy of a dipole in an applied electric field; this is obtained from (8-73) as $U_D = -p_0 E_p \cos\theta$ since now we are dealing with a permanent dipole $\mathbf{p}_0$ in a polarizing field $\mathbf{E}_p$ and the angle between them is now being written as θ. Then if we let W' be all of the rest of the energy, we can write

$$W = W' - p_0 E_p \cos\theta \tag{B-25}$$

and assume W' to be independent of θ. The quantity we are interested in here is the component of $\mathbf{p}_0$ in the direction of $\mathbf{E}_p$ so we let $\psi = p_0 \cos\theta$. If we write the volume element as $dx\, dy\, dz = r^2 \sin\theta\, dr\, d\theta\, d\varphi$ and insert $p_0 \cos\theta$ into (B-24), we see that the integrals involving W', r, φ, and each component of $\mathbf{v}$ will cancel from the numerator and denominator and we will be left with

$$\langle p_0 \cos\theta \rangle = \frac{\displaystyle\int_0^\pi p_0 \cos\theta\, e^{\beta p_0 E_p \cos\theta} \sin\theta\, d\theta}{\displaystyle\int_0^\pi e^{\beta p_0 E_p \cos\theta} \sin\theta\, d\theta} = \frac{\displaystyle\int_{-1}^1 p_0 \mu e^{\beta p_0 E_p \mu}\, d\mu}{\displaystyle\int_{-1}^1 e^{\beta p_0 E_p \mu}\, d\mu} \tag{B-26}$$

with the use of (2-22). Finally, if we let

$$y = \beta p_0 E_p = \frac{p_0 E_p}{kT} \tag{B-27}$$

we can write (B-26) as

$$\langle p_0 \cos\theta \rangle = p_0 \frac{d}{dy} \ln \int_{-1}^1 e^{y\mu}\, d\mu = p_0 \frac{d}{dy} \ln\left(\frac{2\sinh y}{y}\right) = p_0 L\left(\frac{p_0 E_p}{kT}\right) \tag{B-28}$$

where

$$L(y) = \coth y - \frac{1}{y} \tag{B-29}$$

is called the *Langevin function* and measures the ratio of the actual average component of $\mathbf{p}_0$ to its maximum value p_0. The function $L(y)$ is shown as a function of y in Figure B-3. We see that $L(y) \to 1$ as y becomes very large, that is, for large polarizing fields and/or very low temperatures. In this case $\langle p_0 \cos\theta \rangle \simeq p_0$, which corresponds to almost complete alignment of the permanent dipoles with the polarizing field. The material is then said to be saturated.

The general behavior of $\langle p_0 \cos\theta \rangle$ is therefore not linear in E_p as we might have expected. For very small values of y, corresponding to small E_p and/or high temperature, we can expand $L(y)$ in a power series and we find that

$$L(y) \simeq \left(\frac{1}{y} + \frac{y}{3} - \frac{y^3}{45} + \dots\right) - \frac{1}{y} \simeq \frac{y}{3} \qquad (y \ll 1) \tag{B-30}$$

Combining this with (B-28), (B-27), and (B-4), we find the magnitude of the polariza-

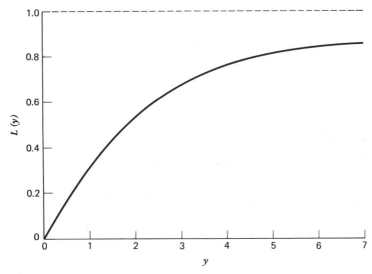

Figure B-3. The Langevin function.

tion to be

$$P = N\langle p_0 \cos\theta \rangle = \frac{Np_0^2}{3kT}E_p \tag{B-31}$$

which is linear in E_p. The corresponding polarizability is

$$\alpha_C = \frac{p_0^2}{3kT} \tag{B-32}$$

and is inversely proportional to the temperature. The result (B-31) is known as *Curie's law* for dielectrics.

We can make some numerical estimates to see whether (B-30) is a reasonable approximation. The value of p_0 will be typically of the order of magnitude of the electronic charge times some fraction of the radius a of a molecule. Since a is of the order of 10^{-10} meter, we get $p_0 \approx 10^{-30}$ coulomb-meter. At temperatures around room temperature ($T \simeq 290$ K), we find that $y \ll 1$ provided that $E_p \ll kT/p_0 \approx 10^9$ volts/meter. Since this is a very large field, we see that (B-31) is generally a very good approximation.

Under these same conditions, the polarizability as found from (B-32) is $\alpha_C \approx 10^{-40}$ farad-(meter)^2. The polarizability associated with induced polarization given by (B-10) is found to be $\alpha \approx 10^{-40}$ farad-(meter)^2 so that these quantities are of comparable magnitude. Consequently, for polar molecules, it is appropriate to speak of a total polarizability as the sum of these:

$$\alpha_t = \alpha + \alpha_C = \alpha + \frac{p_0^2}{3kT} \tag{B-33}$$

The expression comparable to the Clausius-Mossotti relation (B-21) would then be

$$\frac{3\epsilon_0(\kappa_e - 1)}{N(\kappa_e + 2)} = \alpha + \frac{p_0^2}{3kT} \tag{B-34}$$

which is known as the *Debye equation*. If the quantity on the left-hand side is evaluated from measured values and then plotted as a function of $1/T$, we see that the result should be a straight line whose intercept will give the induced polarizability α. Similarly the slope of the line will give $p_0^2/3k$ from which p_0 can be found.

Ferroelectric materials that can have a polarization in the absence of a field are most conveniently discussed after the somewhat analogous magnetic materials are considered, and we do this at the end of the next section.

B-2 STATIC MAGNETIC PROPERTIES

Initially we consider only linear isotropic homogeneous magnetic materials so that we are interested in finding the magnetic susceptibility χ_m defined in (20-52) as $\mathbf{M} = \chi_m \mathbf{H}$. For practically all materials that can be adequately described by this relation, it is found that $|\chi_m| \ll 1$ so that from (20-56) it will be almost equally accurate to use $\mathbf{M} = \chi_m \mathbf{B}/\mu_0$. Hence we write the defining equation for our purposes as

$$\mathbf{M} = \chi_m \mathbf{H} = \chi_m \left(\frac{\mathbf{B}}{\mu_0} \right) \tag{B-35}$$

and within the material we generally just use $\mathbf{H} = \mathbf{B}/\mu_0$ to relate the two fields.

In the last section, we introduced the polarizing field $\mathbf{E}_p$ as the field at the location of the molecule due to all of the rest of the charges. Similarly we want to define a magnetizing field $\mathbf{H}_m$ (and hence $\mathbf{B}_m$), which is the field affecting the molecule and arises from all currents except those that we can ascribe to the molecule itself. The arguments that led to (B-13) can be repeated almost verbatim in this case, so that we can write down at once that

$$\mathbf{H}_m = \mathbf{H} + \mathbf{H}_O + \mathbf{H}_I \tag{B-36}$$

where $\mathbf{H}_O$ is the magnetic field due to the continuous distribution of magnetization $\mathbf{M}$ outside the sphere of radius R shown in Figure B-4. Since there is a discontinuity in the normal component of $\mathbf{M}$ at the surface of this sphere and no free currents within the material, we can find $\mathbf{H}_O$ at the center either *via* the bound surface currents given by (20-8) as $\mathbf{K}_m = \mathbf{M} \times \hat{\mathbf{n}}'$ or from the surface density of poles $\sigma_m = \mathbf{M} \cdot \hat{\mathbf{n}}'$ according to

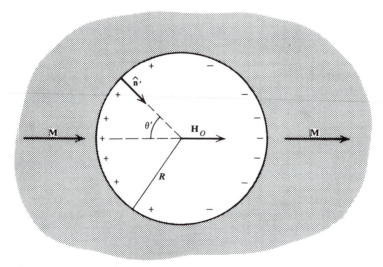

Figure B-4. Hypothetical spherical cavity used to calculate the magnetic field due to the magnetized material outside of it.

(20-42); the result will be the same either way. In this case, we see from the figure that $\sigma_m = M\cos\theta'$ and has the signs indicated; the direction of $\mathbf{H}_O$ is thus the same as that of $\mathbf{M}$. This pole distribution is exactly that of (20-48) for which the magnitude of $\mathbf{H}_O$ as found from (20-49) is $M/3$; substituting this into (B-36) we get

$$\mathbf{H}_m = \mathbf{H} + \frac{\mathbf{M}}{3} + \mathbf{H}_I \tag{B-37}$$

and the corresponding induction would be $\mathbf{B}_O = \mu_0 \mathbf{M}/3$. [The apparent discrepancy in sign between (B-37) and (20-49), as well as the difference in value of $\mathbf{B}$ inside the spheres, arises from the fact that (20-49) was obtained for a sphere of magnetization $\mathbf{M}$ for which $\mathbf{M}$ is different from zero inside, while in this case, we are interested in the field inside an empty sphere within a uniformly magnetized material. This can be seen by comparing Figures 20-8 and B-4; note that the pole distributions have the same angular dependence in the two cases but *opposite signs* with respect to the direction of $\mathbf{M}$—the angles θ' are defined differently in the two figures.]

Similarly, the arguments that showed that there are a great many cases for which $\mathbf{E}_I = 0$ in the electric case can be equally well used in the magnetic case because of the complete analogy between (8-84) and (19-55), which give the fields of the corresponding dipoles. Consequently we restrict ourselves to cases in which $\mathbf{H}_I$ can be taken to be zero and, combining (B-37) with our approximation $\mathbf{H} = \mathbf{B}/\mu_0$, we can write

$$\mathbf{B}_m = \mu_0\mathbf{H}_m = \mu_0\left(\mathbf{H} + \frac{\mathbf{M}}{3}\right) = \mu_0\left(1 + \frac{1}{3}\chi_m\right)\mathbf{H} \tag{B-38}$$

as the induction acting on the molecule in question.

It was pointed out after (20-52) that, in contrast to the electric case, the susceptibility can be negative. Such a material is called *diamagnetic* and we now show that it is due to induced dipoles that arise from the altered orbital motion of the electrons when a field is applied; as we will see, the induced moments are opposite in direction to the inducing field.

Suppose an electron in an atom is circulating in a circular orbit of radius a and angular speed ω_0 in the absence of a field. There will necessarily be a centripetal force $\mathbf{F}_c$ on it arising from internal interactions within the atom and whose magnitude is given by

$$F_c = \frac{m_e v_0^2}{a} = m_e\omega_0^2 a \tag{B-39}$$

where m_e is the electron rest mass. In the presence of a field $\mathbf{B}_m$ there will be an additional force on it given by $\mathbf{F}_m = -e\mathbf{v} \times \mathbf{B}_m$ according to (14-30). If we assume that $\mathbf{v}$ and $\mathbf{B}_m$ are perpendicular, for simplicity, the magnitude of this force will be evB_m and will be directed either radially inward or outward depending on the sense of $\mathbf{v}$ relative to $\mathbf{B}_m$. This is shown in Figure B-5 for the case in which $\mathbf{B}_m$ is out of the page; the vector angular velocity $\boldsymbol{\omega}$ is seen to be out of the page in (a) and hence parallel to $\mathbf{B}_m$, while it is into the page and antiparallel to $\mathbf{B}_m$ in (b). If we assume for now that the radius of the orbit remains constant, and thus F_c will as well, the new net centripetal force can be written as

$$F_c \pm evB_m = F_c \pm e\omega aB_m = m_e\omega^2 a \tag{B-40}$$

where the choice of signs corresponds to the sign of ω as compared to $\mathbf{B}_m$, that is, upper sign for (a) and lower sign for (b) of the figure. We see that ω must be different in the presence of $\mathbf{B}_m$. Substituting for F_c from (B-39), we find that (B-40) leads to

$$\omega^2 - \omega_0^2 = (\omega - \omega_0)(\omega + \omega_0) = \pm\frac{e}{m_e}\omega B_m \tag{B-41}$$

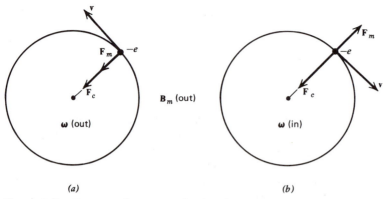

Figure B-5.   Forces acting on an electron in an atom in the presence of a magnetizing field.

It turns out that the frequency change $\Delta\omega = \omega - \omega_0$ is very small even for very large values of $\mathbf{B}_m$. As a result we can approximate $\omega + \omega_0$ by $2\omega_0$ and replace ω by ω_0 in the right-hand side; we therefore find that

$$\Delta\omega = \omega - \omega_0 = \pm \frac{e}{2m_e} B_m \qquad (B\text{-}42)$$

so that ω has been increased for ω parallel to $\mathbf{B}_m$ and decreased if ω is antiparallel. The numerical value of $eB_m/2m_e$ is called the *Larmor frequency*.

Since we assumed a to be constant, a new value of ω means that the electron speed $v = \omega a$ has been changed. This means that the kinetic energy was changed so that work must have been done on the electron. We now have to evaluate this work and see if it is compatible with the above results. During the process of changing the induction from zero to $\mathbf{B}_m$, there is an induced emf given by (17-6) to be

$$\mathscr{E}_{\text{ind}} = \oint \mathbf{E}_{\text{ind}} \cdot d\mathbf{s} = -\frac{d\Phi}{dt} = -\pi a^2 \frac{dB_m}{dt} \qquad (B\text{-}43)$$

and corresponds to an induced electric field $\mathbf{E}_{\text{ind}}$, which is clockwise in Figure B-5. However, because of the negative charge of the electron, the corresponding force $-e\mathbf{E}_{\text{ind}}$ will be opposite to the induced field. Consequently, the work done *on* the electron for each trip around its orbit will be $\pm e|\mathscr{E}_{\text{ind}}|$ with the same choice of signs as before. Since the number of times that it circulates in one second is the frequency $\omega'/2\pi$, where ω' is an intermediate value of the circular frequency, the rate at which the changing induction is doing work on the charge is

$$\frac{dW}{dt} = \pm \frac{\omega' e}{2\pi} |\mathscr{E}_{\text{ind}}| = \pm \frac{1}{2} ea^2 \omega' \frac{dB_m}{dt}$$

Again assuming that the radius a is constant, we can set this equal to the rate at which the kinetic energy $\frac{1}{2} m_e v'^2 = \frac{1}{2} m_e a^2 \omega'^2$ is changing so that

$$\frac{d}{dt}\left(\frac{1}{2} m_e a^2 \omega'^2\right) = m_e a^2 \omega' \frac{d\omega'}{dt} = \pm \frac{1}{2} ea^2 \omega' \frac{dB_m}{dt}$$

This shows that $d\omega' = \pm(e/2m_e)\, dB_m$ and thus the total change in ω will be

$$\Delta\omega = \int_0^{B_m} \pm \left(\frac{e}{2m_e}\right) dB_m = \pm \frac{e}{2m_e} B_m$$

in agreement with (B-42).

It is desirable to put these results in vector form. In Figure B-6 we show what is essentially a side view of Figure B-5; parts (a) and (b) of the two figures correspond, while (c) shows the direction of $\mathbf{B}_m$. Since (a) corresponds to the upper sign of (B-42), the magnitude of $\boldsymbol{\omega}$ is increased and we see that $\Delta\boldsymbol{\omega}$ is in the same direction as $\mathbf{B}_m$; the lower sign of (B-42) indicates a decrease in the magnitude of $\boldsymbol{\omega}$ and (b) of Figure B-6 shows that $\Delta\boldsymbol{\omega}$ is again in the same direction as $\mathbf{B}_m$. Therefore we have the general result that

$$\Delta\boldsymbol{\omega} = \frac{e}{2m_e}\mathbf{B}_m \qquad \text{(B-44)}$$

for electrons in definite orbits. Now we need to relate these results to induced dipole moments.

An electron traveling around either of the circles of Figure B-5 is equivalent to a current element $I\,d\mathbf{s} = q\mathbf{v} = -e\mathbf{v}$ by (14-29) and will have a magnetic dipole moment $\mathbf{m}$. We see from Figure B-5a that $\mathbf{m}$ will be directed into the page and thus opposite to $\boldsymbol{\omega}$; $\mathbf{m}$ is also opposite to $\boldsymbol{\omega}$ for (b) since $\mathbf{m}$ is out of the page in this case. According to (19-27), $m = IS = I\pi a^2$ where I is the equivalent current magnitude. Since the total charge e passes a given point on the perimeter each time around, and since each circuit takes a time equal to the period $\tau = 2\pi/\omega$, the equivalent current is $I = e/\tau = e\omega/2\pi$ and therefore $m = \frac{1}{2}ea^2\omega$. Combining this with the fact that $\mathbf{m}$ and $\boldsymbol{\omega}$ are oppositely directed we get

$$\mathbf{m} = -\tfrac{1}{2}ea^2\boldsymbol{\omega} \qquad \text{(B-45)}$$

for an electron in a definite orbit.

We are actually interested in the *induced moment* produced by $\mathbf{B}_m$, that is, the change $\Delta\mathbf{m}$. Calculating $\Delta\mathbf{m}$ from (B-45) and using (B-44), we get

$$\Delta\mathbf{m} = -\tfrac{1}{2}ea^2\,\Delta\boldsymbol{\omega} = -\frac{e^2}{4m_e}a^2\mathbf{B}_m \qquad \text{(B-46)}$$

which we note is actually independent of the sign of the charge since it is $\sim e^2$. Each electron in the molecule will give a contribution like this where a^2 has the appropriate value. Now we obtained (B-46) by assuming that $\mathbf{B}_m$ was perpendicular to the plane of the orbit; if $\mathbf{B}_m$ were along the z axis, we would have $a^2 = x_j^2 + y_j^2$ for the jth electron. Hence if we sum up over all the electrons and average over all orientations, we find that the average induced dipole moment for a molecule is

$$\langle\Delta\mathbf{m}\rangle = -\frac{e^2}{4m_e}\left\langle\sum_j\left(x_j^2 + y_j^2\right)\right\rangle\mathbf{B}_m = -\frac{e^2}{4m_e}\left[\sum_j\left(\langle x_j^2\rangle + \langle y_j^2\rangle\right)\right]\mathbf{B}_m$$

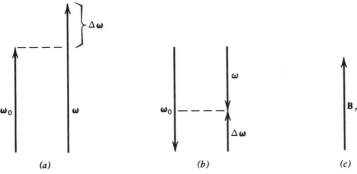

Figure B-6. Changes in the angular velocity of an electron in an atom resulting from a magnetizing field.

and if we multiply this by the number of molecules per unit volume N and use (B-38), we find the magnetization to be

$$\mathbf{M} = -\frac{\mu_0 N e^2}{4m_e}\left[\sum_j \left(\langle x_j^2\rangle + \langle y_j^2\rangle\right)\right]\mathbf{H}_m \tag{B-47}$$

We will soon see that $|\chi_m| \ll 1$ in this case so that $\mathbf{H}_m \simeq \mathbf{H}$ by (B-38) and then a comparison of (B-47) with (B-35) shows that the susceptibility is

$$\chi_m = -\frac{\mu_0 N e^2}{4m_e}\sum_j \left(\langle x_j^2\rangle + \langle y_j^2\rangle\right) \tag{B-48}$$

and is negative since all terms in this expression are positive. Thus the mechanism we have been discussing actually does result in diamagnetism.

For an atom with isotropic charge distribution, we can use the result obtained after (B-17) that $\langle x_j^2\rangle = \langle y_j^2\rangle = \frac{1}{3}r_j^2$ for a given distance r_j from the nucleus and when this is put into (B-48) we finally obtain

$$\chi_m = -\frac{\mu_0 N e^2}{6m_e}\sum_j r_j^2 = -\frac{\mu_0 N Z e^2}{6m_e}\langle r^2\rangle \tag{B-49}$$

where Z is the number of electrons.

As a numerical example, suppose we consider helium at standard temperature and pressure ($0°$ C and 1 atmosphere) for which $N \simeq 3 \times 10^{25}$ (meter)$^{-3}$. Also taking $Z = 2$ and $\langle r^2\rangle \approx 10^{-20}$ (meter)^2, we find that $\chi_m \approx -3 \times 10^{-9}$. This is about equal to what is observed, and also justifies our assertion that $|\chi_m| \ll 1$.

The paramagnetic case for which χ_m is positive arises from the existence of permanent magnetic dipoles. In (B-45) we found a magnetic dipole moment for an electron in a definite orbit; if we multiply and divide this by the electron mass m_e we get

$$\mathbf{m}_{\text{orb}} = -\left(\frac{e}{2m_e}\right)(m_e a^2)\boldsymbol{\omega} = -\frac{e}{2m_e}\mathbf{l} \tag{B-50}$$

where $\mathbf{l}$ is its orbital angular momentum since $m_e a^2$ is the moment of inertia of the particle. The proportionality factor $(-e/2m_e)$ is called the *gyromagnetic ratio*. Thus (B-50) provides a connection between the magnetic and mechanical properties of an electron and we see that $\mathbf{m}$ is always oppositely directed to $\mathbf{l}$. [If we were to repeat the derivation that led to (B-45) for an arbitrary charge q, we see that we would get $\mathbf{m}_{\text{orb}} = (q/2m_q)\mathbf{l}$ where m_q is the particle mass; thus $\mathbf{m}$ and $\mathbf{l}$ are in the same direction for a positive charge.] If we sum (B-50) over all the electrons in the molecule, we find the total moment to be

$$\mathbf{m}_{\text{orb}} = -\frac{e}{2m_e}\sum_j \mathbf{l}_j = -\frac{e}{2m_e}\mathbf{L} \tag{B-51}$$

where $\mathbf{L}$ is the total orbital angular momentum of the molecule.

Things are more complicated than this, however. It has been shown by quantum-mechanical methods that in addition to its orbital angular momentum, an electron has an *intrinsic angular momentum* (or *spin*) $\mathbf{s}$ and associated magnetic moment $\mathbf{m}_s$; the relation between them is twice that of (B-50):

$$\mathbf{m}_s = -2\left(\frac{e}{2m_e}\right)\mathbf{s} \tag{B-52}$$

If we sum (B-52) over all the electrons in a molecule, and add the result to (B-51), the total magnetic dipole moment is

$$\mathbf{m} = -\frac{e}{2m_e}(\mathbf{L} + 2\mathbf{S}) \tag{B-53}$$

where $\mathbf{S} = \sum_j \mathbf{s}_j$. In order for the molecule to have a permanent dipole moment, this quantity must be different from zero. Furthermore, it can be shown that under many circumstances (B-53) can be written in terms of a total angular momentum $\mathbf{J} = \mathbf{L} + \mathbf{S}$ as

$$\mathbf{m} = -g\frac{e}{2m_e}\mathbf{J} \tag{B-54}$$

where g is a dimensionless quantity known as the *Landé g-factor*; g has a numerical value of the order of unity and must be calculated by specifically quantum-mechanical means. Since the numerical values of $\mathbf{J}$ can be shown to be multiples of $h/2\pi$, where h is Planck's constant, it is common practice to measure atomic dipole moments in units of the *Bohr magneton* $\mu_e = eh/4\pi m_e = 9.27 \times 10^{-24}$ ampere-(meter)^2.

Our primary interest in (B-54) is that it shows the possibility of a permanent magnetic dipole moment $\mathbf{m}_0$. In the presence of the induction $\mathbf{B}_m$, the moment will have an interaction or orientational energy given by (19-40) as $U'_D = -\mathbf{m}_0 \cdot \mathbf{B}_m = -\mu_0\mathbf{m}_0 \cdot \mathbf{H}_m$ with the use of (B-38). If we compare this with (B-25), which led to (B-28), we see that we could go through exactly the same calculation as before and we would find the average component of $\mathbf{m}_0$ in the direction of the magnetizing field to be

$$\langle m_0 \cos\theta \rangle = m_0 L\left(\frac{\mu_0 m_0 H_m}{kT}\right) \tag{B-55}$$

where $L(y)$ is again the Langevin function (B-29). Similarly, for conditions for which $L(y) \simeq \frac{1}{3}y$, the magnetization will be

$$M = N\langle m_0 \cos\theta \rangle = \frac{N\mu_0 m_0^2}{3kT}H_m \tag{B-56}$$

with a resultant susceptibility

$$\chi_m = \frac{N\mu_0 m_0^2}{3kT} \tag{B-57}$$

which is positive and thus corresponds to paramagnetism. This result is *Curie's law* for magnetic materials.

As a numerical example, let us take m_0 to be a Bohr magneton and consider a gas at standard conditions so that $N \simeq 3 \times 10^{25}$ (meter)$^{-3}$. We find that $\chi_m \simeq 3 \times 10^{-7}$, which is about what is observed. We note that this has a magnitude that is about 100 times greater than the comparable diamagnetic case. This is generally what occurs. While we expect all materials to have a diamagnetic contribution to the susceptibility, when they have a permanent moment as well, the corresponding paramagnetic susceptibility is so much larger that it is almost the whole susceptibility. We also note that $\chi_m \ll 1$ as we have asserted; consequently, H_m in (B-56) can usually be simply replaced by H as we see from (B-38).

One interesting property of ferromagnetic materials is that the permanent dipoles can be aligned to some extent even in the absence of an external **B** field, and then they are said to have a *permanent magnetization*. It is possible to see how this can occur by means of a simple phenomenological theory introduced by Weiss. We assume that the magnetizing field can be written in analogy with (B-38) as

$$H_m = H + \lambda M \tag{B-58}$$

where we consider only the case where **M** and **H** are parallel. The coefficient λ is to be found by experiment; we will see shortly why we do not take $\lambda = \frac{1}{3}$ as we might have expected from (B-38). If we further assume that we can still use (B-55), then we can write

$$M = Nm_0 L(y) \tag{B-59}$$

$$y = \frac{\mu_0 m_0}{kT}(H + \lambda M) \tag{B-60}$$

If we solve (B-60) for M, it becomes

$$M = \left(\frac{kT}{\lambda\mu_0 m_0}\right)y - \frac{H}{\lambda} \tag{B-61}$$

Equations B-59 and B-61 are simultaneous equations that must be satisfied by M. If both of these are plotted as functions of y, then the value of M is given by the point of intersection of the two curves since M must lie on both of them. This is illustrated in Figure B-7. As T is increased, the slope of the straight line of (B-61) is increased, and we see from the figure that the point of intersection will correspond to a smaller value of M.

If $H = 0$, it is still possible for the curves to intersect and give $M \neq 0$ as shown in Figure B-8; the magnetization is then usually called the *spontaneous magnetization* M_s. As T increases, the point of intersection moves down the curve $L(y)$ so that $M_s(T)$ decreases. We also see that $M_s = 0$ when the slope of the straight line is greater than or equal to the initial slope of (B-59) that is $\frac{1}{3}Nm_0$ according to (B-30). If T_c is this limiting temperature, then $(kT_c/\lambda\mu_0 m_0) = \frac{1}{3}Nm_0$ from (B-61) so that

$$T_c = \frac{\lambda\mu_0 Nm_0^2}{3k} \tag{B-62}$$

and

$$M_s = 0 \quad \text{for} \quad T \geq T_c \tag{B-63}$$

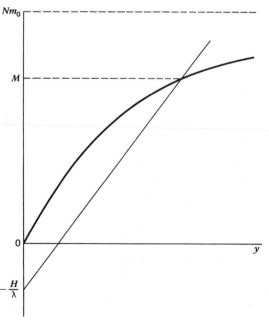

Figure B-7. Calculation of the magnetization according to the Weiss theory.

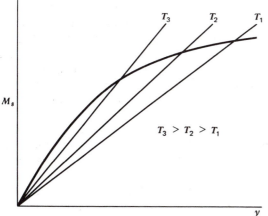

Figure B-8. Calculation of spontaneous magnetization.

T_c is called the *Curie temperature*. The values of M_s as found by the method of Figure B-8 are shown in Figure B-9. The overall behavior agrees surprisingly well with what is observed when one considers that we are using a classical calculation rather than a more exact quantum-mechanical one.

If $y \ll 1$, we can use (B-30) in (B-59), and when we substitute (B-60) into the result and solve for M we find that

$$M = \frac{N\mu_0 m_0^2 H}{3k(T - T_c)} \tag{B-64}$$

with the use of (B-62) as well. Thus, well above the Curie temperature, the magnetization is proportional to H; in other words, the material has a "paramagnetic" susceptibility given by

$$\chi_m = \frac{N\mu_0 m_0^2}{3k(T - T_c)} \tag{B-65}$$

which is known as the *Curie-Weiss law* and also agrees quite well with experiment.

Now we are in a position to evaluate the parameter λ. For iron, $m_0 \simeq 2\mu_e$ and $T_c = 1043$ K. The value of N as found from the molecular weight A, mass density d,

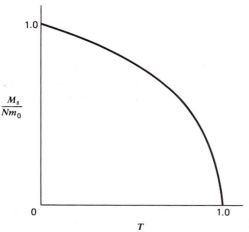

Figure B-9. Spontaneous magnetization as a function of temperature.

and Avogadro's number N_A is $N = N_A d/A \simeq 8.5 \times 10^{28}$ (meter)$^{-3}$. Inserting these into (B-62), we find that $\lambda \simeq 4700$. This is an enormous number and cannot be understood on any classical calculation since we saw in the considerations that led to (B-38) that we would expect a value of $\lambda \approx 1$. Heisenberg showed that these large values of λ are specifically quantum-mechanical and electrostatic in origin; when the electron spins are aligned parallel or antiparallel, the charge distribution is altered in such a way as to favor parallel orientations that will lead to a spontaneous magnetization.

Hysteresis effects that were discussed in Section 20-7 have been found to be associated with the existence of *domains*. In each domain, the permanent dipoles are aligned, but the resultant magnetization of individual domains are themselves not all parallel. When an external field is applied, there is a combination of rotation of domains into alignment with the field and an increase in size of domains that are generally parallel to the field and a decrease in those that are generally oriented in other directions. The overall result is a magnetization curve of the type shown in Figure 20-19.

Many ferroelectric materials that can have a permanent electric polarization can be discussed in terms of a Weiss-type theory with an appropriate change of notation. Thus, instead of (B-18), one could write $\mathbf{E}_p = \mathbf{E} + (\lambda_e/\epsilon_0)\mathbf{P}$ and the analogues of (B-59), (B-61), (B-62), and (B-65) are found to be

$$P = N p_0 L(y) \tag{B-66}$$

$$P = \left(\frac{\epsilon_0 k T}{\lambda_e p_0} \right) y - \frac{\epsilon_0 E}{\lambda_e} \tag{B-67}$$

$$T_c = \frac{\lambda_e N p_0^2}{3 \epsilon_0 k} \tag{B-68}$$

$$\chi_e = \frac{N p_0^2}{3 \epsilon_0 k (T - T_c)} \tag{B-69}$$

Actually, the Weiss theory does not work quite so well for ferroelectric materials as it does for ferromagnetic ones; when it does seem to be reasonably applicable, the values of the parameter λ_e usually turn out to be much *smaller* than unity. Ferroelectric materials also exhibit hysteresis properties and there is evidence for the existence of domains in them.

B-3   RESPONSE TO TIME-VARYING FIELDS

In Section 24-8 we pointed out that, when detailed comparisons are made between the predictions of Maxwell's equations and the results of experiment, it has been found necessary to treat the electromagnetic parameters μ, ϵ, and σ as functions of the circular frequency ω rather than as simply constants. We also discussed the conductivity from a microscopic point of view and found in (24-130) that σ was a complex function of frequency. The conductivity expression involved both the inertia of the charge carriers and a mechanical resistive force proportional to the velocity. This general dependence of the parameters upon frequency is called *dispersion* and we now want to consider its origins in more detail. We restrict ourselves to the case of the response of a medium to a plane wave that is propagating within it. While this is not the most general possibility, it will be sufficient to illustrate the basic aspects involved.

We recall from (24-38) and (24-29) that the most general case of a plane wave propagating in the z direction involves a complex propagation constant

$$k = \alpha + i\beta \tag{B-70}$$

so that

$$\mathbf{E} = \mathbf{E}_0 e^{i(kz - \omega t)} = \mathbf{E}_0 e^{-\beta z} e^{i(\alpha z - \omega t)} \tag{B-71}$$

Thus, the real part of k determines the phase velocity while the imaginary part describes an attenuation of the wave. The basic relation between k and the properties of the medium was given in (24-135) as

$$k^2 = \omega^2 \mu \left(\epsilon + i\frac{\sigma}{\omega} \right) = \frac{\omega^2 \kappa_m}{c^2} \left(\kappa_e + \frac{i\sigma}{\omega \epsilon_0} \right) \tag{B-72}$$

where we have used (10-53), (20-55), and (23-5). We also saw in (24-56) and (24-57) that it is also possible to describe this situation in terms of a complex index of refraction that we write here as

$$\mathcal{N} = \left(\frac{c}{\omega} \right) k = n + i\eta \tag{B-73}$$

where n is the ordinary index of refraction [written in (24-57) as n'] and where η will describe attenuation through (B-70) and (B-71). Comparing (B-73) and (B-72), we see that

$$n + i\eta = \left[\kappa_m \left(\kappa_e + \frac{i\sigma}{\omega \epsilon_0} \right) \right]^{1/2} \tag{B-74}$$

For simplicity, we assume the material to be nonmagnetic and set $\kappa_m = 1$. If, in addition, it is nonconducting, $\sigma = 0$ as well and (B-74) reduces to

$$n + i\eta = \sqrt{\kappa_e} \tag{B-75}$$

We are continuing to write $\mathcal{N}$ as complex even though the only source of attenuation that we have considered so far has been conductivity; as we will see shortly, a nonconducting medium can also lead to attenuation because of the existence of damping of the motion of the charges. We will return to the more general case of $\sigma \neq 0$ later in this section.

The form of (B-75) indicates what our approach will be. As in Section B-1, we expect the incident fields to induce electric dipole moments because the charges will be displaced from their equilibrium positions. From that we can find the polarization and then the susceptibility χ_e from (B-1); finally κ_e will be given by $1 + \chi_e$ and this can be substituted into (B-75) to find the overall response of the medium.

Because of the very large mass of the nucleus as compared to that of an electron, it should be a good approximation to assume the nuclei to remain at rest; hence we need only consider the behavior of the electrons. As a result this treatment is often referred to as the *electron theory of matter*. If we let $\mathbf{r}$ be the *displacement* of the electron from its equilibrium position, we can assume a mechanical restoring force $\mathbf{F}_m$ as we did in (B-7); we write it as $\mathbf{F}_m = -K\mathbf{r} = -m_e \omega_0^2 \mathbf{r}$ where ω_0 is the natural frequency of oscillation of the charge. As we did in (12-36) and (24-125), we also assume a frictional damping force of some sort proportional to the velocity, that is, $\mathbf{F}_d = -\xi \mathbf{v} = -m_e \gamma \mathbf{v}$ where γ is now used as a measure of the damping. Finally, the fields arising from the wave produce a Lorentz force that is given by (21-29) as $\mathbf{F}_l = -e(\mathbf{E}_p + \mathbf{v} \times \mathbf{B}_m)$ where $\mathbf{E}_p$ and $\mathbf{B}_m$ are the resultant inducing fields of (B-18) and (B-38). Setting the mass times the acceleration equal to the net force, we obtain the equation of motion

$$\mathbf{F}_{\text{net}} = m_e \mathbf{a} = -m_e \omega_0^2 \mathbf{r} - m_e \gamma \mathbf{v} - e(\mathbf{E}_p + \mathbf{v} \times \mathbf{B}_m) \tag{B-76}$$

We will be considering only media for which n is of the order of unity; in that case, we know from (24-34) that $|\mathbf{B}_m| \approx |\mathbf{E}_p|/c$ since the fields will not differ too much from those in the incident wave. The ratio of the magnetic force to the electric force will then be approximately

$$\frac{F_{\text{mag}}}{F_{\text{elec}}} \approx \frac{evB_m}{eE_p} \approx \frac{v}{c} \ll 1$$

and we can simplify (B-76) even further by neglecting the term $\mathbf{v} \times \mathbf{B}_m$. The steady-state displacement of the electron will then be parallel to $\mathbf{E}_p$. The problem then reduces to a one-dimensional one and if we let x be the displacement, we can write (B-76) as

$$m_e\left(\frac{d^2x}{dt^2} + \gamma\frac{dx}{dt} + \omega_0^2 x\right) = -eE_p = -eE_{p0}e^{i(kz-\omega t)} \tag{B-77}$$

for a wave propagating in the z direction. Now the variation in kz over the molecule will be of the order of $2\pi a/\lambda$ where a is the radius of the molecule. We now assume that $a \ll \lambda$; since $a \approx 10^{-10}$ meter, this is satisfied even into the ultraviolet where one begins to need a quantum-mechanical description anyhow. In other words, the electric field will be approximately constant over the volume of the molecule located at z_0. If we absorb the constant factor e^{ikz_0} into the amplitude E_{p0}, all that remains is the sinusoidal time variation and (B-77) becomes

$$m_e\left(\frac{d^2x}{dt^2} + \gamma\frac{dx}{dt} + \omega_0^2 x\right) = -eE_{p0}e^{-i\omega t} = -eE_p \tag{B-78}$$

which is recognizable as the equation of motion of a forced damped harmonic oscillator.

The only solution of (B-78) of interest to us is the steady-state displacement that we can find by assuming a solution of the form $x = x_0 e^{-i\omega t}$ where x_0 is a constant. When this is substituted into (B-78), we obtain

$$x = x_0 e^{-i\omega t} = \frac{-(e/m_e)E_{p0}e^{-i\omega t}}{\omega_0^2 - \omega^2 - i\gamma\omega} \tag{B-79}$$

The corresponding induced dipole moment in the direction of the field can be found from the one-dimensional form of (B-3), remembering that here x is the displacement, and is

$$p = -ex = p_0 e^{-i\omega t} = \frac{(e^2/m_e)E_{p0}e^{-i\omega t}}{\omega_0^2 - \omega^2 - i\gamma\omega} \tag{B-80}$$

since we are assuming the nuclei to remain stationary. If there are n_0 electrons of this type in the molecule, their total contribution to the dipole moment will be

$$n_0 p = \frac{n_0(e^2/m_e)E_p}{\omega_0^2 - \omega^2 - i\gamma\omega} \tag{B-81}$$

since $E_p = E_{p0}e^{-i\omega t}$. This result can be easily generalized to the situation in which there is more than one type of electron since we do not expect them all to be in identical situations. Thus if we let n_k be the number of electrons in the molecule that are characterized by the natural frequency ω_k and damping constant γ_k, we can sum (B-81) over all types to get the total induced dipole moment:

$$\langle p \rangle = E_p \sum_k \frac{n_k(e^2/m_e)}{\omega_k^2 - \omega^2 - i\gamma_k\omega} = \alpha E_p \qquad (B\text{-}82)$$

where α is the polarizability according to the definition (B-9). We see that now α is both complex and frequency dependent. Before we go on, we can compare this with the static case ($\omega = 0$) and we find from (B-82) that

$$\alpha(0) = \sum_k \frac{n_k e^2}{m_e \omega_k^2} = \sum_k \frac{n_k e^2}{K_k} = \sum_j \frac{e^2}{K_j}$$

where, in the last step, we went from a sum over *types* k to a sum over *individual electrons j*. (The total number of electrons is $Z_t = \sum_k n_k$.) Comparing this result with (B-9), we see that they are the same if we recall that we are assuming the nuclear force constants to be so very large that $1/K_i \simeq 0$.

The derivation of (B-21) from (B-20) still applies, of course, and if we combine (B-21), (B-82), and (B-75), we get our basic result that

$$\frac{(n + i\eta)^2 - 1}{(n + i\eta)^2 + 2} = \frac{N\alpha}{3\epsilon_0} = \frac{N}{3\epsilon_0} \sum_k \frac{n_k(e^2/m_e)}{\omega_k^2 - \omega^2 - i\gamma_k\omega} \qquad (B\text{-}83)$$

where N is the number of molecules per unit volume. (We remember that the sum is over a single molecule and hence is independent of N.) We could now find n and η by separately equating the real and imaginary parts of each side of (B-83). As a qualitative check on our results, we see that if all the $\gamma_k = 0$, the right-hand side is real and $\eta = 0$; thus, as expected, it is the damping terms in the equations of motion that lead to attenuation and energy absorption from the wave. Similarly, $\eta = 0$ when $\omega = 0$; hence there is no attenuation in the static case.

As a special case, let us consider a perfectly transparent material for which $\eta = 0$. Then, for a *fixed frequency* ω, (B-83) becomes

$$\left(\frac{n^2 - 1}{n^2 + 2} \right) \frac{1}{N} = \frac{\alpha(\omega)}{3\epsilon_0} = \text{const.} \qquad (B\text{-}84)$$

This result is known as the *Lorenz-Lorentz law* and describes how the index of refraction varies with the density of the material. It is found to be extremely accurate for many materials, even to being approximately correct for the case in which a liquid changes to its vapor phase. We note that in the static case ($\omega = 0$) where we can replace n^2 by κ_e, (B-84) reduces to the Clausius-Mossotti relation (B-21). Instead of continuing with the general result (B-83), we turn to a somewhat simpler situation.

For many cases, particularly gases, α turns out to be so small that $n \simeq 1$ and $\eta \simeq 0$. Consequently, $|n + i\eta| \simeq 1$ and we can approximate the denominator on the left of (B-83) by 3. Then $(n + i\eta)^2 \simeq 1 + (N\alpha/\epsilon_0)$ and $n + i\eta \simeq [1 + (N\alpha/\epsilon_0)]^{1/2} \simeq 1 + (N\alpha/2\epsilon_0)$ so that

$$n + i\eta = 1 + \frac{N}{2\epsilon_0} \sum_k \frac{n_k(e^2/m_e)}{\omega_k^2 - \omega^2 - i\gamma_k\omega} \qquad (B\text{-}85)$$

We can now find n and η separately by multiplying the numerator and denominator in the sum by $(\omega_k^2 - \omega^2 + i\gamma_k\omega)$ and equating real and imaginary parts on each side; the results are

$$n = 1 + \frac{N}{2\epsilon_0} \sum_k \frac{n_k\left(e^2/m_e\right)\left(\omega_k^2 - \omega^2\right)}{\left(\omega_k^2 - \omega^2\right)^2 + \left(\gamma_k\omega\right)^2} \tag{B-86}$$

$$\eta = \frac{N}{2\epsilon_0} \sum_k \frac{n_k\left(e^2/m_e\right)\gamma_k\omega}{\left(\omega_k^2 - \omega^2\right)^2 + \left(\gamma_k\omega\right)^2} \tag{B-87}$$

We can see the general nature of the frequency dependence more easily if we assume only one type of electron with one set of constants n_0, ω_0^2, and γ; then these results reduce to

$$n - 1 = \frac{Nn_0\left(e^2/2m_e\epsilon_0\right)\left(\omega_0^2 - \omega^2\right)}{\left(\omega_0^2 - \omega^2\right)^2 + \left(\gamma\omega\right)^2} \tag{B-88}$$

$$\eta = \frac{Nn_0\left(e^2/2m_e\epsilon_0\right)\gamma\omega}{\left(\omega_0^2 - \omega^2\right)^2 + \left(\gamma\omega\right)^2} \tag{B-89}$$

which are shown as functions of ω in Figure B-10. The frequency region around ω_0 where n and η are both changing rapidly is called the region of *anomalous dispersion*. In the general cases (B-86) and (B-87), the curves of n and η will consist of a superposition of ones like those shown with anomalous dispersion near each of the ω_k, and this is the general type of dispersion curve found for gases. By fitting curves like these to experimental results, one can evaluate the quantities n_k, ω_k, and γ_k.

If we look again at (B-83) and assume $\eta \simeq 0$ for simplicity, we see that in the more general situation we will observe anomalous dispersion about the same frequencies, ω_k, except for the quantity $(n^2 - 1)/(n^2 + 2)$ rather than for $(n - 1)$.

Now let us suppose that our medium also contains some electrons that are free to move so that it will have conducting properties as well. We see from (B-74) that we can handle this by simply adding $(i\sigma/\omega\epsilon_0)$ to κ_e. Thus, if $(n + i\eta)_{nc}$ refers to the values found in the nonconducting case, we find from (B-74), (B-75), and (B-83) that

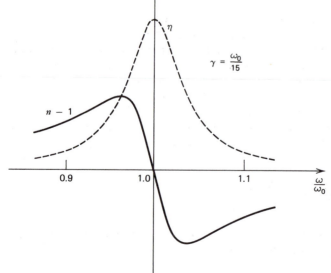

Figure B-10.   Behavior of the real and imaginary parts of the index of refraction near a resonant frequency.

when $\kappa_m \simeq 1$

$$(n + i\eta)^2 = (n + i\eta)^2_{nc} + \frac{i\sigma}{\omega\epsilon_0} = \frac{1 + (2N\alpha/3\epsilon_0)}{1 - (N\alpha/3\epsilon_0)} + \frac{i\sigma}{\omega\epsilon_0} \qquad (B-90)$$

An electron that is free to move about corresponds to one not subject to a restoring force so that $\omega_0^2 = 0$ for it. Thus its equation of motion as found from (B-78) will be of the form

$$m_e\left(\frac{d^2x}{dt^2} + \gamma_0\frac{dx}{dt}\right) = -eE_{p0}e^{-i\omega t}$$

This is just the equation (24-127) that we previously used to find the conductivity where $\xi = m_e\gamma_0$. Consequently we can use the value of σ given by (24-130):

$$\sigma = \sigma(\omega) = \frac{\sigma_0}{1 - i(\sigma_0 m_e\omega/N_0 e^2)} \qquad (B-91)$$

where N_0 is now used for the number of free electrons per unit volume and the static value of the conductivity σ_0 is

$$\sigma_0 = \frac{N_0 e^2}{\xi} = \frac{N_0 e^2}{\gamma_0 m_e} \qquad (B-92)$$

Substituting (B-91) into (B-90), we obtain

$$(n + i\eta)^2 = \frac{1 + (2N\alpha/3\epsilon_0)}{1 - (N\alpha/3\epsilon_0)} + \frac{i(\sigma_0/\omega\epsilon_0)}{1 - i(\sigma_0 m_e\omega/N_0 e^2)} \qquad (B-93)$$

as the complete expression from which to find n and η separately. [Remember that α as given by the sum in (B-83) is complex.]

For our further considerations, it will be sufficient to restrict ourselves to the case in which $|u| = |N\alpha/\epsilon_0| \ll 1$ so that the first term on the right-hand side of (B-93) can be approximated by $[1 + (2u/3)]/[1 - (u/3)] \simeq [1 + (2u/3)][1 + (u/3)] \simeq 1 + u = 1 + (N\alpha/\epsilon_0)$. Then if we multiply the numerator and denominator of each term in the sum of (B-83) by the complex conjugate of its denominator, do a similar thing to the last term on the right of (B-93), multiply out the left side of (B-93), and then equate real and imaginary parts on each side, we find that

$$n^2 - \eta^2 = 1 + \frac{N}{\epsilon_0}\sum_k \frac{n_k(e^2/m_e)(\omega_k^2 - \omega^2)}{(\omega_k^2 - \omega^2)^2 + (\gamma_k\omega)^2} - \frac{(\sigma_0^2 m_e/N_0\epsilon_0 e^2)}{1 + (\sigma_0 m_e\omega/N_0 e^2)^2} \qquad (B-94)$$

$$2n\eta = \frac{N}{\epsilon_0}\sum_k \frac{n_k(e^2/m_e)\gamma_k\omega}{(\omega_k^2 - \omega^2)^2 + (\gamma_k\omega)^2} + \frac{(\sigma_0/\omega\epsilon_0)}{1 + (\sigma_0 m_e\omega/N_0 e^2)^2} \qquad (B-95)$$

which now can be solved for n and η individually if desired. We will not do this, but mostly content ourselves with some general remarks about these results.

Let us first consider the low frequency case, which we take as corresponding to $\sigma_0/\omega\epsilon_0 \gg 1$, that is, the "good conductor" ($Q \ll 1$), which we discussed in detail immediately following (24-70). Here the frequency is much lower than any natural resonant frequencies of the bound electrons and their contribution to (B-94) and (B-95) is negligible. The contribution of the conductivity to (B-94) is also small for small ω because the numerator is constant and small; on the other hand, the conductivity term in (B-95) is very large. Thus these expressions take the form $n^2 - \eta^2 = 1 + \delta$ where δ is a very small quantity while $2n\eta \simeq (\sigma_0/\omega\epsilon_0) \gg 1$. This can only be the case if n and

η are both large and approximately equal; thus $n \simeq \eta \simeq (\sigma_0/2\omega\epsilon_0)^{1/2}$. Then we see from (B-73) and (B-70) that the real and imaginary parts of the propagation constant are approximately equal and given by $\alpha \simeq \beta \simeq (\omega/c)n = (\frac{1}{2}\mu_0\sigma_0\omega)^{1/2}$, which is exactly what we found before in (24-75) for the corresponding case.

As the frequency increases into the infrared and visible portions of the spectrum, it becomes equal to at least some of the natural frequencies of the bound electrons. Then their resonant contributions become important and comparable to or greater than that of the free electrons and one has to deal with the very complicated general expressions (B-94) and (B-95).

EXERCISES

B-1 The susceptibility of helium at one atmosphere and $140°C$ is $\chi_e = 6.84 \times 10^{-5}$. Find the polarizability α and use the result to estimate the radius of a helium atom.

B-2 For air at $0°C$ and one atmosphere, $\kappa_e = 1.000590$. Use the Clausius-Mossotti relation to find κ_e at 100 atmospheres and the same temperature. Find the percentage difference between your result and the measured value of 1.0548.

B-3 Integrate the equation of motion $d\mathbf{l}/dt = $ torque, where the torque is due to the induced electric field given in (B-43), and thus show that the induced diamagnetic moment is again given by (B-46).

B-4 It is found that $\chi_m = 2.2 \times 10^{-5}$ for aluminum at $20°C$. Assume that this is completely paramagnetic in origin and find the corresponding permanent dipole moment m_0. How many Bohr magnetons is this? [Aluminum has a density of 2.7 grams/(centimeter)^3 and an atomic weight of 27 grams/mole.]

B-5 Verify (B-66) through (B-69).

B-6 Show that the Weiss theory of ferromagnetism violates the third law of thermodynamics by showing that $(dM_s/dT)_{T=0} \neq 0$.

B-7 The susceptibility of barium titanate (BaTiO$_3$) is $\chi_e = (1.7 \times 10^5)/(T - 393)$ for T sufficiently large. Use this to estimate λ_e for this material.

B-8 Beginning with (B-54), show that the equation of motion of the magnetic dipole can be written in the form $(d\mathbf{m}/dt) = -(ge/2m_e)\mathbf{m} \times \mathbf{B}$. Now assume that $\mathbf{B} = B_0\hat{\mathbf{z}}$ where B_0 is a constant. Solve the equation of motion and show that the solution describes a precession of $\mathbf{m}$ about z, that is, $\mathbf{m}$ sweeps out a cone of constant angle with an axis coinciding with the z axis. What is the angular frequency of the motion? What is the sense of rotation of the projection of $\mathbf{m}$ on the xy plane when viewed from a point on the positive z axis?

B-9 The index of refraction of water at $15°C$ for a particular optical frequency is 1.3337; its density is 0.9991 grams/(centimeter)^3. Use the Lorenz-Lorentz law to find n for water vapor at $0°C$ and one atmosphere by treating the vapor as an ideal gas. Find the percentage deviation of your result from the measured value of 1.000250.

B-10 Assume $\gamma \ll \omega_0$ and use (B-88) and (B-89) to show that in a region of anomalous dispersion, the maximum and minimum values of n occur at frequencies where η is half its maximum value. Find the relation between γ and the half-width $\Delta\omega$ of η, that is, the frequency difference between ω_0 and the frequency where η has half its maximum value.

B-11 Since the induced dipole in (B-80) is oscillating in time, it will radiate energy; this is called scattered energy. Find the total rate at which p is radiating by assuming a gas so that E_{p0} can be taken to be approximately the same as the incident electric field. Show that when $\omega \ll \omega_0$, the radiated power is proportional to $1/\lambda^4$ where λ is the incident wavelength. This is called *Rayleigh scattering* and is used to account for the blue color of the sky.

ANSWERS TO ODD-NUMBERED EXERCISES

1-3 $5\hat{x} - 3\hat{y} - \hat{z}$; $32.3°$, $120°$, $99.7°$

1-5 -14.4

1-9 $\left(\dfrac{x}{a^2}\hat{x} + \dfrac{y}{b^2}\hat{y} + \dfrac{z}{c^2}\hat{z}\right)\left(\dfrac{x^2}{a^4} + \dfrac{y^2}{b^4} + \dfrac{z^2}{c^4}\right)^{-(1/2)}$

1-13 $\frac{1}{2}abc(a + b + c)$

1-15 $(2k)^{1/2}[(8/21) - (4k/5)]$

1-19 No; a/ρ; $(b/\rho)\hat{z}$; $(x^2 + y^2)^{-(1/2)}[(ax - by)\hat{x} + (ay + bx)\hat{y}] + c\hat{z}$; $(a \sin\theta + c \cos\theta)\hat{r}$ $+ (a \cos\theta - c \sin\theta)\hat{\theta} + b\hat{\phi}$

1-21 3

1-23 $-\pi r_0$

1-25 No

2-1 $\dfrac{qq'}{4\pi\epsilon_0}\left\{\dfrac{[(x - a)\hat{x} + y\hat{y}]}{[(x - a)^2 + y^2]^{3/2}} - \dfrac{[(x + a)\hat{x} + y\hat{y}]}{[(x + a)^2 + y^2]^{3/2}}\right\}$

2-3 $-1.90(q^2/4\pi\epsilon_0 a^2)(\hat{x} + \hat{y} + \hat{z})$

2-7 $\hat{z}(\lambda\rho a^3 L)/[3\epsilon_0 z_0(z_0 + L)]$

2-9 $\hat{x}\dfrac{\lambda^2}{2\pi\epsilon_0}\left[\left(1 + \dfrac{L^2}{a^2}\right)^{1/2} - 1\right]$

2-11 coulomb/(meter)^4; $\frac{1}{2}\pi a^4 A$; $\hat{z}\dfrac{qAz}{2\epsilon_0}\left[\dfrac{(a^2 + 2z^2)}{(a^2 + z^2)^{1/2}} - 2z\right]$

3-1 $(q/4\pi\epsilon_0)\{[x\hat{x} + (y - a)\hat{y}][x^2 + (y - a)^2]^{(-3/2)} - [x\hat{x} + (y + a)\hat{y}]$ $[x^2 + (y + a)^2]^{-(3/2)}$; $E_x = 0$ when $x = 0$ or $y = 0$

3-3 $1.90(q/4\pi\epsilon_0 a^2)(\hat{x} + \hat{y} - \hat{z})$

3-7 $(\lambda/2\pi\epsilon_0)[-a\hat{x} + (c - b)\hat{y}][a^2 + (c - b)^2]^{-1}$

3-9 $-(\sigma/\epsilon_0)\hat{z}$ for $-a < z < a$; 0 for $|z| > a$

3-11 $(\lambda/4\pi\epsilon_0\rho)[(\sin\alpha_2 + \sin\alpha_1)\hat{\rho} + (\cos\alpha_2 - \cos\alpha_1)\hat{z}]$; $(\lambda L\hat{\rho})/[2\pi\epsilon_0\rho(L^2 + \rho^2)^{1/2}]$

3-13 $\rho_{ch}a^2\hat{x}/2\epsilon_0 x$ for $x > a$; $\rho_{ch}x\hat{x}/2\epsilon_0$ for $x < a$

4-1 $\rho abc/\epsilon_0$ for both

4-3 $\lambda\hat{\rho}/2\pi\epsilon_0\rho$, $(\rho < \rho_0)$; $(\lambda + 2\pi\rho_0\sigma)\hat{\rho}/2\pi\epsilon_0\rho$, $(\rho_0 < \rho)$; $-\lambda/2\pi\rho_0$; yes, since $2\pi\rho_0\sigma$ is the charge per unit length on the cylinder

4-5 $2Ar^{3/2}\hat{r}/7\epsilon_0$, $(r < a)$; $2Aa^{7/2}\hat{r}/7\epsilon_0 r^2$, $(a < r)$

4-7 $\rho_{ch}\rho\hat{\rho}/2\epsilon_0$, $(\rho < a)$; $\rho_{ch}a^2\hat{\rho}/2\epsilon_0\rho$, $(a < \rho)$; yes, since $\lambda = \pi a^2\rho_{ch}$

4-9 0, $(\rho < a)$; $Af(\rho)\hat{\rho}/\epsilon_0\alpha^2\rho$, $(a < \rho < b)$ where $f(\rho) = (e^{-\alpha a} - e^{-\alpha\rho}) + \alpha(ae^{-\alpha a} - \rho e^{-\alpha\rho})$; $Af(b)\hat{\rho}/\epsilon_0\alpha^2\rho$, $(b < \rho)$; $\alpha = 0$ and $n = 0$, yes

4-11 $4\epsilon_0 E_0\rho^2/a^3$, $(0 < \rho < a)$; 0, $(a < \rho)$

5-1 Yes, since $\nabla \times \mathbf{E} = 0$; $\phi = -xyz + x^2 + \text{const.}$

5-3 $(q/4\pi\epsilon_0)\{[x^2 + y^2 + (z - a)^2]^{-(1/2)} - [x^2 + y^2 + (z + a)^2]^{-(1/2)}\}$; $\phi(z = 0) = 0$; each point is equidistant from the equal and opposite charges

5-5 $0.710(q/\epsilon_0 a)$

5-9 $\phi_o = Q/4\pi\epsilon_0 r$; $\phi_i = [Q/4(n+2)\pi\epsilon_0 a][(n+3) - (r/a)^{n+2}]$

5-11 $\rho(b^3 - a^3)/3\epsilon_0 r$, $(r > b)$; $(\rho/3\epsilon_0)[(3b^2/2) - \frac{1}{2}r^2 - (a^3/r)]$, $(a < r < b)$; $\rho(b^2 - a^2)/2\epsilon_0$, $(0 < r < a)$

5-13 $(\lambda a \alpha/2\pi\epsilon_0)(a^2 + z^2)^{-(1/2)}$; because ϕ found only for the *single* value of $x = 0$

5-15 $-(\sigma/2\epsilon_0)|z| + C = -(\sigma/2\epsilon_0)\sqrt{z^2} + C$; planes parallel to the charged sheet

5-17 $-(\sigma/2\epsilon_0)(|z| - a)$; $\simeq 5a$

5-21 $(\lambda/\pi\epsilon_0)\ln[(b-a)/(b+a)]$

5-23 $-q^2/8\pi\epsilon_0 a$

6-1 Q on surface of radius c, $-Q$ for b, and Q for a; $Q/4\pi\epsilon_0 r$, $(c \le r)$; $Q/4\pi\epsilon_0 c$, $(b \le r \le c)$; $(Q/4\pi\epsilon_0 rbc)[bc - r(c - b)]$, $(a \le r \le b)$; $(Q/4\pi\epsilon_0 abc)[bc - a(c - b)]$, $(0 \le r \le a)$

6-3 7.08×10^{-4} farad

6-5 $c_{11} = -c_{12} = -c_{21} = 4\pi\epsilon_0 ab/(b - a)$; $c_{22} = 4\pi\epsilon_0(ab + bc - ca)/(b - a)$; $C = c_{11}$

6-9 $4\Delta\phi/b$

6-15 $2\pi\epsilon_0 L[\ln(c^2/ab)]^{-1}$

7-1 $-0.918(q^2/\epsilon_0 a)$

7-5 $\pi\sigma^2 a^3/4\epsilon_0$

7-7 2.25×10^{32} joule; 6.14×10^{-61} meter

7-9 $5/6$

7-11 $(q_l^2 L/4\pi\epsilon_0)\ln(b/a)$

7-15 1.81×10^{-4} kilogram

7-17 $\frac{1}{2}\epsilon_0\{\Delta\phi/[a\ln(b/a)]\}^2$; $\hat{\boldsymbol{\rho}}$; 0

8-3 $(p/4\pi\epsilon_0)(z^2 - \frac{1}{4}l^2)^{-1}$; $z > 5l$; $z > 7.06l$

8-5 $Q = 9q$; $\mathbf{p} = qa(4\hat{\mathbf{x}} + 5\hat{\mathbf{y}} + 14\hat{\mathbf{z}})$; $Q_{xx} = -11qa^2$, $Q_{yy} = -8qa^2$, $Q_{zz} = 19qa^2$, $Q_{xy} = 6qa^2$, $Q_{yz} = 15qa^2$, $Q_{zx} = 21qa^2$; at $(a/9)(4\hat{\mathbf{x}} + 5\hat{\mathbf{y}} + 14\hat{\mathbf{z}})$

8-7 $Q = \lambda L$; $\mathbf{p} = \frac{1}{2}\lambda L^2(\cos\alpha\hat{\mathbf{x}} + \sin\alpha\hat{\mathbf{y}})$; $Q_{xx} = (\frac{1}{3})\lambda L^3(3\cos^2\alpha - 1)$, $Q_{yy} = (\frac{1}{3})\lambda L^3(3\sin^2\alpha - 1)$, $Q_{zz} = -(\frac{1}{3})\lambda L^3$, $Q_{xy} = \lambda L^3\cos\alpha\sin\alpha$, $Q_{yz} = Q_{zx} = 0$; $\phi_Q = (\lambda L^3/24\pi\epsilon_0 r^5)[3(x\cos\alpha + y\sin\alpha)^2 - r^2]$

8-9 $Q = \rho abc$; $\mathbf{p} = \frac{1}{2}Q(a\hat{\mathbf{x}} + b\hat{\mathbf{y}} + c\hat{\mathbf{z}}) = Q$(position vector of center); $Q_{xx} = (\frac{1}{3})Q(2a^2 - b^2 - c^2)$, $Q_{yy} = (\frac{1}{3})Q(2b^2 - c^2 - a^2)$, $Q_{zz} = (\frac{1}{3})Q(2c^2 - a^2 - b^2)$, $Q_{xy} = (\frac{3}{4})Qab$, $Q_{yz} = (\frac{3}{4})Qbc$, $Q_{zx} = (\frac{3}{4})Qca$; $(3Qa^2\sin\theta/16\pi\epsilon_0 r^3)[\sin\theta\sin\varphi\cos\varphi + \cos\theta(\sin\varphi + \cos\varphi)]$

8-11 $Q_{xx} - 2(2a_x p_x - a_y p_y - a_z p_z) + (2a_x^2 - a_y^2 - a_z^2)Q$

8-13 No; replace $\mathbf{r}$ by $\mathbf{R} = \mathbf{r} - \mathbf{r}'$ in (8-84)

8-15 With $\alpha = 3qa^2/8\pi\epsilon_0$, $\phi_Q = \alpha(\rho^2\sin 2\varphi)(\rho^2 + z^2)^{-5/2}$; $E_\rho = \alpha\rho(3\rho^2 - 2z^2)(\sin 2\varphi)(\rho^2 + z^2)^{-7/2}$, $E_\varphi = -2\alpha\rho(\cos 2\varphi)(\rho^2 + z^2)^{-5/2}$, $E_z = 5\alpha z\rho^2(\sin 2\varphi)(\rho^2 + z^2)^{-7/2}$; $\rho^3 = (\alpha/\phi_Q)\sin 2\varphi$; $\rho^4 = K(\cos 2\varphi)^3$

8-17 $-(q\mathbf{p}\cdot\hat{\mathbf{r}})/4\pi\epsilon_0 r^2$; $(q\mathbf{p}\times\hat{\mathbf{r}})/4\pi\epsilon_0 r^2$; $(q/4\pi\epsilon_0 r^3)[\mathbf{p} - 3(\mathbf{p}\cdot\hat{\mathbf{r}})\hat{\mathbf{r}}]$

8-19 No; $\mathbf{F}_2 = (3/4\pi\epsilon_0 R^4)\{[(\mathbf{p}_1\cdot\mathbf{p}_2) - 5(\mathbf{p}_1\cdot\hat{\mathbf{R}})(\mathbf{p}_2\cdot\hat{\mathbf{R}})]\hat{\mathbf{R}} + (\mathbf{p}_2\cdot\hat{\mathbf{R}})\mathbf{p}_1 + (\mathbf{p}_1\cdot\hat{\mathbf{R}})\mathbf{p}_2\}$; $3p_1 p_2\hat{\mathbf{R}}/4\pi\epsilon_0 R^4$; $-3p_1 p_2\hat{\mathbf{R}}/2\pi\epsilon_0 R^4$

8-21 $(3Q_1^a Q_2^a/64\pi\epsilon_0 r^9)(3r^4 - 30z^2 r^2 + 35z^4)$

9-1 $\mathbf{E}_{1n} = 2\hat{\mathbf{x}} + \hat{\mathbf{y}} + \hat{\mathbf{z}}$; $\mathbf{E}_{1t} = 2\hat{\mathbf{x}} - 4\hat{\mathbf{z}}$

9-3 With $A = \sigma_0 z/\epsilon_0 a^2$, $\mathbf{E}_2 = (\alpha + Ax)\hat{\mathbf{x}} + (\beta + Ay)\hat{\mathbf{y}} + (\gamma + Az)\hat{\mathbf{z}}$

9-5 Potential difference independent of path; follows from (9-22)

10-1 1.2×10^{-4} coulomb/(meter)^2

10-3 $-\alpha P$, $P(1 + \alpha t)$, $-P, 0$

10-5 $\phi_o = -Pa^3/3\epsilon_0 z^2$, $E_{zo} = -2Pa^3/3\epsilon_0 z^3$, $\phi_i = Pz/3\epsilon_0$, $E_{zi} = -P/3\epsilon_0$

10-7 $\rho_b = -\alpha(n + 2)r^{n-1}$, $\sigma_b = \alpha a^n$; $\mathbf{E}_o = 0$, $\mathbf{E}_i = -\mathbf{P}/\epsilon_0$; $\phi_o = 0$,
$\phi_i = [\alpha/\epsilon_0(n + 1)](r^{n+1} - a^{n+1})$

10-9 $-P/3\epsilon_0$

10-11   (a) $\rho_b = 0$, $\sigma_{b \text{ top}} = -\sigma_{b \text{ bottom}} = P$; (b) $\mathbf{E} = E\hat{\mathbf{z}}$; $E = -(P/2\epsilon_0)\{2 + (z - L)$
$[(z - L)^2 + a^2]^{-(1/2)} - (z + L)[(z + L)^2 + a^2]^{-(1/2)}\}$,$(0 \le z < L)$; $E = (P/2\epsilon_0)$
$\{-(z - L)[(z - L)^2 + a^2]^{-(1/2)} + (z + L)[(z + L)^2 + a^2]^{-(1/2)}\}$, $(z > L)$;
(c) $E_{2n} - E_{1n} = P/\epsilon_0$; (d) $E(0) = -(P/\epsilon_0)\{1 - [1 + (a/L)^2]^{-(1/2)}\}$; (e) $|E|_{\max} = P/\epsilon_0$ for
$a \gg L$; yes, since like two infinite plane sheets

10-13   (10-56) and (10-57)

10-15 $\rho_b = -[(\kappa_e - 1)\rho_f/\kappa_e] - (\mathbf{P} \cdot \nabla \ln \kappa_e)/(\kappa_e - 1)$;
$\rho = (\rho_f/\kappa_e) - (\mathbf{P} \cdot \nabla \ln \kappa_e)/(\kappa_e - 1)$; if $\mathbf{P}$ and $\nabla\kappa_e$ are perpendicular

10-17 $\mathbf{D} = q\hat{\mathbf{r}}/4\pi r^2$, everywhere; inside: $\mathbf{E} = q\hat{\mathbf{r}}/4\pi\kappa_e\epsilon_0 r^2$, $\mathbf{P} = (\kappa_e - 1)q\hat{\mathbf{r}}/4\pi\kappa_e r^2$; outside:
$\mathbf{E} = q\hat{\mathbf{r}}/4\pi\epsilon_0 r^2$, $\mathbf{P} = 0$; $(\kappa_e - 1)q/\kappa_e$

10-19 $\mathbf{D} = \lambda_f\hat{\boldsymbol{\rho}}/2\pi\rho$, $\mathbf{E} = \lambda_f\hat{\boldsymbol{\rho}}/(2\pi\alpha\epsilon_0\rho^{n+1})$, $\rho_b = -n\lambda_f/(2\pi\alpha\rho^{n+2})$; $n = -1$; $\mathbf{D}$ is unchanged, $\rho_b = \lambda_f/2\pi\alpha\rho$

10-21 $\mathbf{D} = \lambda_f\hat{\boldsymbol{\rho}}/2\pi\rho$, $\mathbf{E} = \lambda_f\hat{\boldsymbol{\rho}}[2\pi\epsilon_0\rho(\alpha + \beta z)]^{-1}$, $\mathbf{P} = (\alpha + \beta z - 1)\lambda_f\hat{\boldsymbol{\rho}}[2\pi\rho(\alpha + \beta z)]^{-1}$, $\rho_b = 0$, yes

10-23 $D = Q/A$, everywhere; in vacuum, $E = Q/\epsilon_0 A$, $P = 0$; in dielectric, $E = Q/\kappa_e\epsilon_0 A$,
$P = (\kappa_e - 1)Q/\kappa_e A$; $C = \kappa_e\epsilon_0 A[\kappa_e(d - t) + t]^{-1}$

10-25 $[(\kappa_{e2} - \kappa_{e1})\epsilon_0 A/d][\ln(\kappa_{e2}/\kappa_{e1})]^{-1}$

10-27
$$C = 2\pi\epsilon_0 L\left[\frac{1}{\kappa_{e1}}\ln\left(\frac{\rho_0}{a}\right) + \frac{1}{\kappa_{e2}}\ln\left(\frac{b}{\rho_0}\right)\right]^{-1}$$

10-29
$$4\pi\epsilon_0\left[\frac{1}{\kappa_{e1}a} - \frac{1}{\kappa_{e2}b} + \frac{1}{r_0}\left(\frac{1}{\kappa_{e2}} - \frac{1}{\kappa_{e1}}\right)\right]^{-1}$$

10-31 $\mathbf{D} = \rho_0\mathbf{r}/3$; $\mathbf{E} = \rho_0\mathbf{r}/3\kappa_e\epsilon_0$; $2\pi\rho_0^2 a^5/45\kappa_e\epsilon_0$; $-4\pi(\kappa_e - 1)a^3\rho_0/3\kappa_e$; $\phi_o = \rho_0 a^3/3\epsilon_0 r$,
$$\phi_i = \left(\frac{\rho_0 a^2}{3\epsilon_0}\right)\left[1 + \frac{1}{2\kappa_e}\left(1 - \frac{r^2}{a^2}\right)\right]; (2\kappa_e + 1)\rho_0 a^2/6\kappa_e\epsilon_0$$

10-33 $\Delta\phi = (\kappa_e + 1)\Delta\phi_0/2\kappa_e$; $Q = \epsilon_0(\kappa_e + 1)L^2\Delta\phi_0/2d$, $E = (\kappa_e + 1)\Delta\phi_0/2\kappa_e d$

10-35 $(\epsilon_0 L/2d)(\kappa_e - 1)t(\Delta\phi)^2[\kappa_e(d - t) + t]^{-1}$

11-3 q at $(-a, -b, 0)$, $-q$ at $(-a, b, 0)$ and $(a, -b, 0)$; $(q/4\pi\epsilon_0)$
$\{[(x - a)^2 + (y - b)^2 + z^2]^{-(1/2)} - [(x + a)^2 + (y - b)^2 + z^2]^{-(1/2)} +$
$[(x + a)^2 + (y + b)^2 + z^2]^{-(1/2)} - [(x - a)^2 + (y + b)^2 + z^2]^{-(1/2)}\}$;
$(q/4\pi\epsilon_0)[(y - b)\{[(x - a)^2 + (y - b)^2 + z^2]^{-3/2}$
$-[(x + a)^2 + (y - b)^2 + z^2]^{-3/2}\} + (y + b)\{[(x + a)^2 + (y + b)^2 + z^2]^{-3/2}$
$-[(x - a)^2 + (y + b)^2 + z^2]^{-3/2}\}]$; $E_y(0, y, z) = 0$;
$\sigma_f(x, 0, z) = -(qb/2\pi)\{[(x - a)^2 + b^2 + z^2]^{-3/2} - [(x + a)^2 + b^2 + z^2]^{-3/2}\}$; opposite to q

11-5 $-3p^2\hat{\mathbf{x}}/64\pi\epsilon_0 d^4$

11-9 $(4\pi\epsilon_0)^{-1}[(q/d) + (Q/a)]$; $(q\hat{\mathbf{z}}/4\pi\epsilon_0)[(Q/d^2) - q a(a/d)^3(2d^2 - a^2)(d^2 - a^2)^{-2}]$

11-11 $-\lambda a/\pi(a^2 + \rho^2)$; $-\lambda$; $-\lambda^2\hat{\mathbf{x}}/4\pi\epsilon_0 a$

11-13 $2\pi\epsilon_0/[\cosh^{-1}(h/A)]$; $-\pi\epsilon_0(\Delta\phi)^2[\cosh^{-1}(h/A)]^{-2}(h^2 - A^2)^{-(1/2)}$

11-15 $-(2\epsilon_0\phi_0/L)[\sinh(\pi y/L)]^{-1}$

11-17 $\phi = 2\phi_0 \sum_{n \text{ odd}} \left(\frac{2}{n\pi}\right) \sin\left(\frac{n\pi y}{a}\right) \sinh\left[\frac{n\pi}{2}\left(\frac{2x}{a} - 1\right)\right]\left[\sinh\left(\frac{n\pi}{2}\right)\right]^{-1}$; $E_y = 0$,

$E_x = -4\left(\frac{\phi_0}{a}\right) \sum_{n \text{ odd}} (-1)^{(n-1)/2}\left[\sinh\left(\frac{n\pi}{2}\right)\right]^{-1} = -1.669\left(\frac{\phi_0}{a}\right)$

11-19 $X'' + Y'' + Z'' = \alpha + \beta + \gamma = 0$; $X(x) = \frac{1}{2}\alpha x^2 + Cx + C'$; constant field plus one varying linearly with x

11-21 $ra^2/(r^3 + 2a^3) = K \sin^2 \theta$

11-23 With $R_1 = (r^2 + b^2 - 2rb\cos\theta)^{1/2}$, $R_2 = [(br/a)^2 + a^2 - 2rb\cos\theta]^{1/2}$;

$\phi = \frac{q}{4\pi\epsilon_0}\left(\frac{1}{R_1} - \frac{1}{R_2}\right)$; $E_r = \frac{q}{4\pi\epsilon_0}\left\{\frac{(r - b\cos\theta)}{R_1^3} - \frac{[(b^2r/a^2) - b\cos\theta]}{R_2^3}\right\}$,

$E_\theta = \frac{qb\sin\theta}{4\pi\epsilon_0}\left(\frac{1}{R_1^3} - \frac{1}{R_2^3}\right)$; $\mathbf{E}(r = 0) = -\hat{\mathbf{z}}q(a^3 - b^3)/4\pi\epsilon_0 b^2 a^3$;

$\sigma_f(\theta) = -(qa/4\pi)[1 - (b^2/a^2)][a^2 + b^2 - 2ab\cos\theta]^{-3/2}$; $-q$

11-25 $\phi = -E_0[1 - (a^2/\rho^2)]\rho\cos\varphi$; $\sigma_f = 2\epsilon_0 E_0 \cos\varphi$

11-27 $\frac{4\phi_0}{\pi} \sum_{m \text{ odd}} \frac{(-1)^{(m-1)/2}}{m}\left(\frac{\rho}{a}\right)^m \cos m\varphi$; $-(4\phi_0/\pi a)\hat{\mathbf{x}}$

11-29

$(r > a)$: $\left(\frac{\lambda}{2\epsilon_0}\right)\left[\left(\frac{a}{r}\right) - \frac{1}{2}\left(\frac{a}{r}\right)^3 P_2(\cos\theta) + \frac{3}{8}\left(\frac{a}{r}\right)^5 P_4(\cos\theta) - \ldots\right]$; $(r < a)$:

$\left(\frac{\lambda}{2\epsilon_0}\right)\left[1 - \frac{1}{2}\left(\frac{r}{a}\right)^2 P_2(\cos\theta) + \frac{3}{8}\left(\frac{r}{a}\right)^4 P_4(\cos\theta) - \ldots\right]$

11-31 $\phi = \phi_0 + (z/d)\{\phi_d - \phi_0 + (\rho_0 d^2/12\epsilon_0)[1 - (z/d)^3]\}$; $\sigma(0) = -(\epsilon_0/d)(\phi_d - \phi_0) - (\rho_0 d/12)$; $\sigma(d) = (\epsilon_0/d)(\phi_d - \phi_0) - \frac{1}{4}\rho_0 d$

12-1 Ampere/(meter)^5; $-63A$; $-12\pi Aa^5/5$, decreasing

12-3 $(3Q\omega r\sin\theta/4\pi a^3)\hat{\boldsymbol{\varphi}}$; $Q\omega/2\pi$

12-5 $-(\sigma\phi_0/a)\hat{\mathbf{z}}$

12-7 IR/l

12-9 $[\sigma_1(d - x)\phi_1 + \sigma_2 x\phi_2]/[\sigma_1(d - x) + \sigma_2 x]$; $(\sigma_1\epsilon_2 - \sigma_2\epsilon_1)(\phi_1 - \phi_2)/[\sigma_1(d - x) + \sigma_2 x]$

12-11 $\pi\sigma d\Delta\phi/\cosh^{-1}(D/2A)$

12-17 1.51×10^{-4} meter/second

13-5 $-\frac{1}{2}\mu_0 IK'\hat{\mathbf{z}}$

13-7 $-\hat{\mathbf{y}}\frac{\mu_0 II'a^2}{\pi} \int_0^\pi \frac{\sin^2\varphi \, d\varphi}{d^2 + a^2\sin^2\varphi} = -\hat{\mathbf{y}}\mu_0 II'\left[1 - \frac{d}{(d^2 + a^2)^{1/2}}\right]$

13-9 $\mu_0 I^2(\hat{\mathbf{x}} + \hat{\mathbf{y}})/4\pi a$

14-1 $(\mu_0/\pi\rho)[-I\sin\alpha\hat{\mathbf{y}} + (I\cos\alpha - I')\hat{\mathbf{z}}]$

14-3 $(\mu_0/2\pi)\{I_1[-(y - y_1)\hat{\mathbf{x}} + (x - x_1)\hat{\mathbf{y}}][(x - x_1)^2 + (y - y_1)^2]^{-1}$

$+ I_2[-(y - y_2)\hat{\mathbf{x}} + (x - x_2)\hat{\mathbf{y}}][(x - x_2)^2 + (y - y_2)^2]^{-1}\}$

14-5 $\mu_0 nI'\{1 - (2/\pi)\arcsin[a^2/(L^2 + a^2)]\}$; $\mu_0 nI'$

14-7 $(\mu_0 I'a/2\pi)(a^2 + z^2)^{-3/2}(z\sin\alpha\hat{\mathbf{x}} + a\alpha\hat{\mathbf{z}})$

14-9 $\mu_0 K'\hat{\mathbf{x}}$, $(0 < z < d)$; 0, elsewhere

14-11 $\mu_0\sigma\omega\hat{\mathbf{z}}[(z^2 + \frac{1}{2}a^2)(z^2 + a^2)^{-(1/2)} - |z|]$; $\frac{1}{2}\mu_0\sigma\omega a\hat{\mathbf{z}}$

14-13 $\frac{2}{5}\mu_0 P\omega a\hat{\mathbf{z}}$; 0

14-15 $qv\mu_0 I'(b - a)/4ab$, horizontal and to the right

15-1 $2\pi[1 - z(z^2 + a^2)^{-(1/2)}]$

15-5 An infinitely long straight current NI perpendicular to the plane of the torus and passing through the center 0.

15-7 $B_\varphi = \mu_0 I\rho/2\pi a^2$, $(0 < \rho < a)$; $\mu_0 I/2\pi\rho$, $(a < \rho < b)$; $(\mu_0 I/2\pi\rho)[(c^2 - \rho^2)/(c^2 - b^2)]$, $(b < \rho < c)$; 0, $(c < \rho)$

15-9 $\frac{1}{2}\mu_0\rho_{ch}\mathbf{v} \times \boldsymbol{\rho}$, $(0 < \rho < a)$; $\frac{1}{2}\mu_0\rho_{ch}(a/\rho)^2\mathbf{v} \times \boldsymbol{\rho}$, $(a < \rho)$

16-3 $f(x, y, z) = -(\alpha z/y^2) - (\beta z/x^2) + g(x, y)$; $\mu_0\mathbf{J} = \hat{\mathbf{x}}[(2\alpha z/y^3) + (\partial g/\partial y)] - \hat{\mathbf{y}}[(2\beta z/x^3) + (\partial g/\partial x)] + \hat{\mathbf{z}}[(2\alpha x/y^3) - (2\beta y/x^3)]$; $\nabla \cdot \mathbf{J} = 0$

16-5 $\chi(a \to b) = -xyB$; $\chi(a \to c) = -\frac{1}{2}xyB$; $\chi(b \to c) = \frac{1}{2}xyB$

16-7 $B_\rho(\rho, z) \simeq (3\mu_0 Ia^2 z\rho/4)(a^2 + z^2)^{-5/2}$; $B_z(\rho, z) \simeq (\mu_0 Ia^2/2)(a^2 + z^2)^{-3/2}\{1 + [\frac{3}{4}\rho^2(a^2 - 4z^2)(a^2 + z^2)^{-2}]\}$

16-9 $4\pi\mathbf{A}/\mu_0 I = \hat{\mathbf{x}}\ln[(\alpha + a + x)(\beta - a - x)/(\delta + a - x)(\gamma - a + x)] + \hat{\mathbf{y}}\ln[(\gamma + a + y)(\alpha - a - y)/(\beta + a - y)(\delta - a + y)]$ where $\alpha^2 = (a + x)^2 + (a + y)^2$, $\beta^2 = (a + x)^2 + (a - y)^2$, $\gamma^2 = (a - x)^2 + (a + y)^2$, $\delta^2 = (a - x)^2 + (a - y)^2$; 0

16-11 $A_z = (\mu_0 I/4\pi)[1 - (\rho^2/a^2) + 2\ln(\rho_0/a)]$, $(\rho < a)$; $(\mu_0 I/2\pi)\ln(\rho_0/\rho)$, $(\rho > a)$; $A_z(0) = (\mu_0 I/4\pi)[1 + 2\ln(\rho_0/a)]$

16-13 $A_z = A_0 - (\mu_0 I\rho^2/4\pi a^2)$, $(0 < \rho < a)$; $A_0 - (\mu_0 I/4\pi)[1 + 2\ln(\rho/a)]$, $(a < \rho < b)$; $A_0 - (\mu_0 I/4\pi)[1 + 2\ln(b/a)] + [\mu_0 I/4\pi(c^2 - b^2)][\rho^2 - b^2 - 2c^2\ln(\rho/b)]$, $(b < \rho < c)$; $A_0 - (\mu_0 I/2\pi)[\ln(b/a) + c^2(c^2 - b^2)^{-1}\ln(c/b)]$, $(c < \rho)$; $A_0 = (\mu_0 I/2\pi)\{\ln(b/a) + [1 - (b/c)^2]^{-1}\ln(c/b)\}$

16-15 Zero

17-3 $\lambda I_0 e^{-\lambda t}(\mu_0 b/2\pi)\ln[1 + (a/d)]$; clockwise

17-5 $\omega B_0 ab \sin(2\omega t + \varphi_0 + \alpha)$; no

17-7 $\mathbf{P} = (\kappa_e - 1)\epsilon_0\omega B\rho\hat{\boldsymbol{\rho}}$; $Q_b = 2\pi(\kappa_e - 1)\epsilon_0\omega Ba^2 l$

17-9 4.71 volts

17-11 (a) $l/\sigma ab$; (b) $vB_d l$; (c) $v\sigma abB_d$; (d) 5 ohms, 8.25×10^{-4} volts, 1.65×10^{-4} amperes

17-13 $3\epsilon_0[(\kappa_e - 1)VvB]^2/2\pi R^4$; attractive

17-15 $\frac{1}{2}\sigma\omega B_0\rho \sin(\omega t + \alpha)\hat{\boldsymbol{\varphi}}$

17-17 1.42×10^{-4} henrys

17-19 $(\mu_0 b/2\pi)\ln[(a + d)(d + D)/d(a + d + D)]$

17-21 $(\mu_0 N^2 a/2\pi)\ln[(2b + a)/(2b - a)]$

17-23 $LC = \mu_0\epsilon_0 l^2$; $L = (\mu_0 l/\pi)\cosh^{-1}(D/2A)$

18-5 $i = i_0 e^{-t/\tau}$ where $\tau = L/R$

18-7 $I^2(\mu_0 l/16\pi)(b^2 - a^2)^{-2}[(b^2 - a^2)(b^2 - 3a^2) + 4a^4\ln(b/a)]$

18-9 $b < 1.28a$

18-11 $\mathbf{F}_{mj} = \sum_{k \neq j} I_j I_k \nabla_j M_{jk}$

18-13 $\hat{\mathbf{x}}\mu_0 II'[1 - b(b^2 - a^2)^{-(1/2)}]$

18-15 $-\hat{\mathbf{z}}(3\pi\mu_0 a^2 b^2 I_j I_k/2c^4)$

18-17 $-\frac{1}{2}\mu_0 II'nSa^2\{[a^2 + (\frac{1}{2}l - \delta)^2]^{-3/2} - [a^2 + (\frac{1}{2}l + \delta)^2]^{-3/2}\}$

18-19 0.504 tesla

19-1 $I\pi a^2\hat{\mathbf{z}}$

19-3 $\frac{1}{4}Qa^2\omega$

19-7 $\pi\mu_0 a^2 b^2/4c^3$

19-11 $\tan\alpha_2 = -\frac{1}{2}\tan\alpha_1$

19-13 $IabB\sin\varphi\hat{\mathbf{z}}$

20-1 182 amperes/meter

20-3 $\mathbf{J}_m = (2M/a)\hat{\mathbf{z}}$; $\mathbf{K}_m$: 0, $(x = 0)$; $-M\hat{\mathbf{z}}$, $(x = a)$; 0,$(y = 0)$; $-M\hat{\mathbf{z}}$, $(y = a)$; $-(M/a)(x\hat{\mathbf{x}} + y\hat{\mathbf{y}})$, $(z = 0)$; $(M/a)(x\hat{\mathbf{x}} + y\hat{\mathbf{y}})$, $(z = a)$

20-5 α in amperes/(meter)^3, β in amperes/meter; $\mathbf{J}_m = 0$; $\mathbf{K}_m = (\alpha a^2 \cos^2\theta + \beta)\sin\theta\,\hat{\boldsymbol{\varphi}}$

20-7 $B_z = \frac{1}{2}\mu_0 M\{(z + \frac{1}{2}l)[a^2 + (z + \frac{1}{2}l)^2]^{-1/2} - (z - \frac{1}{2}l)[a^2 + (z - \frac{1}{2}l)^2]^{-1/2}\}$, $H_z = (B_z/\mu_0) - M$; for $l \ll a$: $B_{z0} \simeq \frac{1}{2}\mu_0 Mla^2(a^2 + z^2)^{-3/2}$, $B_{zi} \simeq \mu_0 Ml/2a \simeq 0$, $H_{zi} \simeq -M[1 - (l/2a)] \simeq -M$

20-9 $\mathbf{J}_m = 0$, $\mathbf{K}_m = 0$; $\mathbf{B}_i = 0$, $\mathbf{H}_i = -\mathbf{M}$, $\mathbf{H}_0(a) = 0$

20-13 $\rho_m = 0$, $\sigma_{m\,\text{top}} = M$, $\sigma_{m\,\text{bottom}} = -M$; $\mathbf{H}_i = -\mathbf{M}$, $\mathbf{H}_0 = 0$; 1; yes

20-15 $\phi_m = 0$, $H_z = 0$, $(0 < z < a)$; $\phi_m = (M/3z^2)(z^3 - a^3)$, $H_z = -(M/3z^3)(z^3 + 2a^3)$, $(a < z < b)$; $\phi_m = (M/3z^2)(b^3 - a^3)$, $H_z = (2M/3z^3)(b^3 - a^3)$, $(b < z)$

20-17 $\chi_m = \chi_{m,\text{mass}}d$; $\chi_m = \chi_{m,\text{molar}}(d/A)$

20-19 (b) $\delta \simeq \frac{1}{2}(\chi_{m2} - \chi_{m1})\sin 2\alpha_1$, 6.3×10^{-4} degrees; (c): (i) $-42.1°$, (ii) $44.9°$

20-21 Outside: $H_r = H_0 \cos\theta\{1 + [(\kappa_m - 1)/(\kappa_m + 2)](2a^3/r^3)\}$, $H_\theta = -H_0 \sin\theta\{1 - [(\kappa_m - 1)/(\kappa_m + 2)](a^3/r^3)\}$, $H_\varphi = 0$; inside: $\mathbf{H} = [3/(\kappa_m + 2)]\mathbf{H}_0$

20-23 $2I/(\kappa_m + 1)$; $[(\kappa_m - 1)/(\kappa_m + 1)](\mu_0 I^2/4\pi d)$; attractive

20-25 $H_{\varphi 2} = I/2\pi\rho$, $B_{\varphi 2} = \kappa\mu_0 I/2\pi a$, $L_2 = \kappa\mu_0(b - a)l/2\pi a$

20-27 $\chi_m = 2dgh/\mu_0 H_0^2$

20-29 1.30 tesla; 1.44×10^6 amperes/weber; 0.374 ampere; $H_0 = 1.03 \times 10^6$ amperes/meter, $H_i = 247$ amperes/meter; 0.689

20-31 $(\mu - \mu_0)B_0^2 S/2\mu\mu_0$; 134 newtons; 6.63 atmospheres

21-1 $-\frac{1}{2}\epsilon_0 \mathcal{E}\omega d_1\rho \cos\omega t(d_0 + d_1 \sin\omega t)^{-2}\hat{\boldsymbol{\varphi}}$; 0

21-3 $\nabla \cdot \mathbf{E} = (\rho_f - \nabla \cdot \mathbf{P})/\epsilon_0$, $\nabla \times \mathbf{E} = -\mu_0[(\partial\mathbf{H}/\partial t) + (\partial\mathbf{M}/\partial t)]$, $\nabla \cdot \mathbf{H} = -\nabla \cdot \mathbf{M}$, $\nabla \times \mathbf{H} = \mathbf{J}_f + \epsilon_0(\partial\mathbf{E}/\partial t) + (\partial\mathbf{P}/\partial t)$; $\nabla \cdot \mathbf{D} = \rho_f$, $\nabla \times \mathbf{D} = -\epsilon_0(\partial\mathbf{B}/\partial t) + \nabla \times \mathbf{P}$, $\nabla \cdot \mathbf{B} = 0$, $\nabla \times \mathbf{B} = \mu_0[\mathbf{J}_f + (\partial\mathbf{D}/\partial t) + \nabla \times \mathbf{M}]$; $\nabla \cdot \mathbf{D} = \rho_f$, $\nabla \times \mathbf{D} = -\mu_0\epsilon_0[(\partial\mathbf{H}/\partial t) + (\partial\mathbf{M}/\partial t)] + \nabla \times \mathbf{P}$, $\nabla \cdot \mathbf{H} = -\nabla \cdot \mathbf{M}$, $\nabla \times \mathbf{H} = \mathbf{J}_f + (\partial\mathbf{D}/\partial t)$

21-7 $\nabla \cdot \mathbf{E} = (\rho_f - \mathbf{E} \cdot \nabla\epsilon)/\epsilon$, $\nabla \times \mathbf{E} = -\partial\mathbf{B}/\partial t$, $\nabla \cdot \mathbf{B} = 0$, $\nabla \times \mathbf{B} = \mu[\mathbf{J}_f' + \sigma\mathbf{E} + \epsilon(\partial\mathbf{E}/\partial t)] + (\nabla\mu \times \mathbf{B})/\mu$

21-9 $\mathbf{S} = -(qI/2\epsilon_0\pi^2 a^3)\hat{\boldsymbol{\rho}}$; $dU/dt = qId/\epsilon_0\pi a^2 = d(q^2/2C)/dt$

21-11 $\hat{\mathbf{z}}(2/9)\mu_0 a^2 QM$

22-3 $\nabla^2\phi + \dfrac{\partial}{\partial t}\left(\nabla \cdot \mathbf{A} + \dfrac{1}{\epsilon}\mathbf{A} \cdot \nabla\epsilon\right) + \dfrac{1}{\epsilon}(\nabla\phi) \cdot (\nabla\epsilon) = -\dfrac{\rho_f}{\epsilon}$; $\nabla^2\mathbf{A} - \mu\sigma\dfrac{\partial\mathbf{A}}{\partial t} - \mu\epsilon\dfrac{\partial^2\mathbf{A}}{\partial t^2} - $

$\nabla(\nabla \cdot \mathbf{A}) - \mu\sigma\nabla\phi - \mu\epsilon\nabla\left(\dfrac{\partial\phi}{\partial t}\right) + \dfrac{1}{\mu}(\nabla\mu) \times (\nabla \times \mathbf{A}) = -\mu\mathbf{J}_f'$

22-5 $\nabla^2\phi = -\rho_f(\mathbf{r}, t)/\epsilon$; $\nabla^2\mathbf{A} - \mu\epsilon(\partial^2\mathbf{A}/\partial t^2) = -\mu\mathbf{J}_f + \mu\epsilon\nabla(\partial\phi/\partial t)$

22-7 $\mathbf{E} = \nabla(\nabla \cdot \boldsymbol{\pi}_e) - \mu_0\epsilon_0(\partial^2\boldsymbol{\pi}_e/\partial t^2) = \nabla \times (\nabla \times \boldsymbol{\pi}_e) - (\mathbf{P}/\epsilon_0)$, $\mathbf{B} = \mu_0\epsilon_0\nabla \times (\partial\boldsymbol{\pi}_e/\partial t)$

23-1 [statcoulomb] = (gram)$^{1/2}$(centimeter)$^{3/2}$(second)$^{-1}$; [abampere] = (gram)$^{1/2}$(centimeter)$^{1/2}$(second)$^{-1}$

23-5 $C = A/4\pi d$

24-1 $\partial^2\psi/\partial\xi\,\partial\eta = 0$

24-5 $\mathcal{E} = \pi a^2 NkE_0 \cos\theta \sin\omega t$; so that $\mathbf{B} \simeq$ const. $= \mathbf{B}(z = 0)$

24-7 Listed in the order Q, v(meters/second), δ(meter), E/cB, Ω(radian). For $\nu = 10^2$ hertz: 1.39×10^{-9}, 1.58×10^4, 25.2, 3.73×10^{-5}, $\pi/4 = 0.785$; for 10^7: 1.39×10^{-4}, 5.00×10^6, 7.96×10^{-2}, 1.18×10^{-2}, 0.785; for 10^{10}: 0.139, 1.48×10^8, 2.70×10^{-3}, 0.371, 0.716; for 10^{15}: 1.39×10^4, 3×10^8, 1.33×10^{-3}, $1.00 - 1.29 \times 10^{-9}$, 3.60×10^{-5}; everyday observation shows that light penetrates many meters into seawater; the static value for σ was used

24-9 $E_0 = 1005$ volts/meter, $B_0 = 3.35 \times 10^{-6}$ tesla

24-11 $[1 + (1/Q^2)]^{1/2}$; $1 + (1/2Q^2)$, $(Q \gg 1)$; $(1/Q)(1 + \frac{1}{2}Q^2)$, $(Q \ll 1)$

24-13 $\hat{\mathbf{z}}(E_\alpha^2 + E_\beta^2)/2Z$

24-15 $\langle u_e \rangle = \sum_k \frac{1}{4}\epsilon \mathbf{E}_{0k}^2$

24-17 $\lambda = 10^{-2}$ meters, $\nu = 3 \times 10^{10}$ hertz, $\alpha = \beta = 60°$, $\gamma = 45°$

24-19 $\mathbf{E}_- = E_0(\hat{\mathbf{x}} + i\hat{\mathbf{y}})e^{i(kz - \omega t + \vartheta)}$

24-21   (a) 8.97×10^9 hertz, (b) 8.97×10^5 hertz

24-23   Yes

24-27   In the order asked for: $\frac{1}{4}\epsilon|\mathbf{E}_0|^2$, $\frac{1}{4}(k^2/\mu\omega^2)|\mathbf{E}_0|^2$, $1 - (\omega_P/\omega)^2$, $\frac{1}{2}\epsilon|\mathbf{E}_0|^2[1 - (\omega_P^2/2\omega^2)]$, $\frac{1}{2}\{(\epsilon/\mu)[1 - (\omega_P/\omega)^2]\}^{1/2}|\mathbf{E}_0|^2\mathbf{k}$, $v_U = (\mu\epsilon)^{-1/2}[1 - (\omega_P/\omega)^2]^{1/2}[1 - (\omega_P^2/2\omega^2)]^{-1}$

24-29   (a) $Q = 2\pi(\tau/T)$; (b) $2\pi\delta/\lambda = (1 + Q^2)^{1/2} + Q$

25-5 $(E_t/E_i)_\perp = 2Z_2 \cos\theta_i[Z_2 \cos\theta_i + i(Z_1 K/k_2)]^{-1}$; $(E_t/E_i)_\parallel = 2Z_2 \cos\theta_i[Z_1 \cos\theta_i + i(Z_2 K/k_2)]^{-1}$; $(Z_1/Z_2)^2$; elliptical

25-7   (b) $\theta_i = 21.2°$ or $40.7°$; (c) $n_1 = 2.41n_2 = 3.62$, $\theta_i = 32.8°$

25-9   (a) $a_t = a(\cos\theta_t/\cos\theta_i)$, $b_t = b$; (b) $P_i = |\langle \mathbf{S}_i \rangle|ab$, $P_t = |\langle \mathbf{S}_t \rangle|a_t b_t$; (c) $T' = T$

25-11 $E_{0t}/E_{0i} \simeq 2\pi(\mu_2/\mu_1)(\delta_2/\lambda_1)(1 - i)$; E_t leads by $45°$

25-13 $T = \mathrm{Re}\,(E_{0t}H_{0t}^*)/\mathrm{Re}\,(E_{0i}H_{0i}^*) = 4\pi(\mu_2/\mu_1)(\delta_2/\lambda_1)$

25-15   (c) $d\theta_P/ds = n_P^{-1}|dn_P/dz| \neq 0$; (d) $n_P(z)/n_P(0) = an_a \sin\theta_i[n_P(0)z + an_a \sin\theta_i]^{-1}$

25-17 $P_{\text{tot}} = \frac{1}{3}\langle u \rangle_{\text{tot}}$

26-1 $\mathscr{E}_\tau = (ik_g/k_c^2)\nabla\mathscr{E}_z$, $\mathscr{H}_\tau = (\hat{\mathbf{z}} \times \mathscr{E}_\tau)/Z_m$, $Z_m = (k_g/k_0)Z = (\lambda_0/\lambda_g)Z$

26-3 $v_G = v[1 - (\omega_c/\omega)^2]^{1/2} < v$. Like Exercise 24-26 with $k \to k_g$, $v \to v_g$, and $(\mu\epsilon)^{-1} \to v^2$

26-5 $(\lambda/2) < a < (\lambda/\sqrt{2})$

26-7   (a) $\langle u \rangle = \frac{1}{4}\mu|H_0|^2[2(\omega a/\pi v)^2 \sin^2(\pi x/a) + \cos(2\pi x/a)]$; (b) $\langle \mathbf{S} \rangle = \hat{\mathbf{z}}\frac{1}{2}\mu\omega k_g (a/\pi)^2|H_0|^2 \sin^2(\pi x/a)$; (c) $\langle\langle u \rangle\rangle = \frac{1}{4}\mu(a\omega/\pi v)^2|H_0|^2$; $\langle\langle \mathbf{S} \rangle\rangle = \hat{\mathbf{z}}\frac{1}{4}\omega\mu k_g(a/\pi)^2|H_0|^2$; (d) $v_U = v^2/v_g = v_G$

26-9 $\frac{1}{4}\mu_0 H_0^2(\omega a/\pi c)^2 \sin^2(\pi x/a)\hat{\mathbf{y}}$

26-11   With $P = k_g z - \omega t$: $H_z = 0$, $E_x = -E_0(k_g/k_c^2)(m\pi/a)\cos(m\pi x/a)\sin(n\pi y/b)\sin P$, $E_y = -E_0(k_g/k_c^2)(n\pi/b)\sin(m\pi x/a)\cos(n\pi y/b)\sin P$, $E_z = E_0 \sin(m\pi x/a)\sin(n\pi y/b)\cos P$, $H_x = E_0(\omega\epsilon/k_c^2)(n\pi/b)\sin(m\pi x/a)\cos(n\pi y/b)\sin P$, $H_y = -E_0(\omega\epsilon/k_c^2)(m\pi/a)\cos(m\pi x/a)\sin(n\pi y/b)\sin P$

26-13 $\mathbf{A} = -(i/\omega)\mathbf{E}$, so that with $P = k_g z - \omega t$: $A_x = -(\mu/k_c^2)(n\pi/b)H_0 \cos(m\pi x/a)\sin(n\pi y/b)e^{iP}$, $A_y = (\mu/k_c^2)(m\pi/a)H_0 \sin(m\pi x/a)\cos(n\pi y/b)e^{iP}$, $A_z = 0$

26-15 $\dfrac{1}{\rho}\dfrac{\partial}{\partial\rho}\left(\rho\dfrac{\partial\psi_0}{\partial\rho}\right) + \dfrac{1}{\rho^2}\dfrac{\partial^2\psi_0}{\partial\varphi^2} + k_c^2\psi_0 = 0$; $J_m(k_c a) = 0$

26-17 $\sigma_f = \epsilon\Delta\phi_0[a \ln(b/a)]^{-1}\cos(k_0 z - \omega t)$

26-21 $E_x = E_y = 0$, $E_z = E_3 \sin(\pi x/a)\sin(\pi y/a)e^{-i\omega t}$, $H_x = -i(\pi E_3/\omega\mu a)\sin(\pi x/a)\cos(\pi y/a)e^{-i\omega t}$, $H_y = i(\pi E_3/\omega\mu a)\cos(\pi x/a)\sin(\pi y/a)e^{-i\omega t}$, $H_z = 0$; $\langle \mathbf{S} \rangle = 0$; $U = \frac{1}{8}a^3\epsilon|E_3|^2$

26-23 $\pi_e = E_z/k_c^2$ with E_z given by (26-61) and (26-14)

27-1 $I = 1.46, 2.21, 2.79$ amperes; $|V_R| = 5.84, 8.84, 11.17$ volts; $|V_L| = 6.16, 3.16, 0.83$ volts

27-5 $q = C\mathscr{E}\{1 - [1 + (Rt/2L)]e^{-Rt/2L}\}$; $(2/e)\mathscr{E}$

27-7 $R = 0$

27-11 $\omega RLI_0 \cos(\omega t + \epsilon + \phi)/(R^2 + \omega^2 L^2)^{1/2}$ where $\tan\phi = R/\omega L$

27-13 $I_1(t) = -\frac{1}{2}q_0(\omega_+ \sin \omega_+ t + \omega_- \sin \omega_- t)$; $I_2(t) = \frac{1}{2}q_0(\omega_+ \sin \omega_+ t - \omega_- \sin \omega_- t)$;
$q_1(t) = \frac{1}{2}q_0(\cos \omega_+ t + \cos \omega_- t)$; $q_2(t) = \frac{1}{2}q_0(-\cos \omega_+ t + \cos \omega_- t)$ where $\omega_+^2 = 1/[C(L - M)]$ and $\omega_-^2 = 1/[C(L + M)]$

27-15 $\alpha = \beta = (\omega R'C'/2)^{1/2}$, $v = (2\omega/R'C')^{1/2}$, $Z_i = (R'/2\omega C')(1 - i)$

28-1   All fields are zero for $t < (\rho/c)$. For $t \geq (\rho/c)$, and with $F = [1 - (\rho/ct)^2]^{1/2}$:
$\mathbf{A} = \hat{\mathbf{z}}(\mu_0 I/4\pi)\ln[(1 + F)/(1 - F)]$, $\mathbf{B} = \hat{\boldsymbol{\varphi}}(\mu_0 I/2\pi\rho F)$, $\mathbf{E} = -\hat{\mathbf{z}}(\mu_0 I/2\pi tF)$,
$\mathbf{S} = \hat{\boldsymbol{\rho}}(\mu_0 I^2/4\pi^2\rho tF^2)$

28-3   0.03 volts/meter

28-5   With $P = kr - \omega t$:

$$\mathbf{E} = \frac{k^3 p_0}{4\pi\epsilon_0}\left\{\hat{\mathbf{r}}\cos\theta\left[\frac{2\sin P}{(kr)^2} + \frac{2\cos P}{(kr)^3}\right] + \hat{\boldsymbol{\theta}}\sin\theta\left[\left(-\frac{1}{kr} + \frac{1}{k^3 r^3}\right)\cos P + \frac{\sin P}{(kr)^2}\right]\right\},$$

$$\mathbf{B} = -\frac{\mu_0 k^2 \omega p_0}{4\pi}\hat{\boldsymbol{\varphi}}\sin\theta\left[\frac{\cos P}{kr} - \frac{\sin P}{(kr)^2}\right],$$

$$\mathbf{S} = \frac{k^5 \omega p_0^2}{16\pi^2\epsilon_0}\left\{\hat{\mathbf{r}}\sin^2\theta\left[\frac{\cos^2 P}{(kr)^2} - \frac{\cos 2P}{(kr)^4} + \left(-\frac{1}{k^3 r^3} + \frac{1}{2k^5 r^5}\right)\sin 2P\right] + \right.$$

$$\left. \hat{\boldsymbol{\theta}}\sin\theta\cos\theta\left[\left(\frac{1}{k^3 r^3} - \frac{1}{k^5 r^5}\right)\sin 2P + \frac{2\cos 2P}{(kr)^4}\right]\right\}$$

28-13 $\mathbf{B}_R = (i\mu_0 \omega^3 m_0 b/4\pi c^3 r)\sin 2\theta\, e^{i(kr - \omega t)}\hat{\boldsymbol{\theta}}$, $\mathbf{E}_R = -(i\mu_0 \omega^3 m_0 b/4\pi c^2 r)\sin 2\theta\, e^{i(kr - \omega t)}\hat{\boldsymbol{\varphi}}$

28-15   With $\mathbf{E}_0 = (i\mu_0 c|I_0|/\pi r)e^{i[kr - \omega t + \frac{1}{2}(\vartheta_2 + \vartheta_1)]}\hat{\mathbf{z}}$: $\mathbf{E} = \mathbf{E}_0\cos[ka\sin\varphi + \frac{1}{2}(\vartheta_2 - \vartheta_1)]$,
$\mathbf{E}_1 = \mathbf{E}_0\cos(\frac{1}{2}\pi\sin\varphi)$, $\mathbf{E}_2 = -\mathbf{E}_0\sin(\frac{1}{2}\pi\sin\varphi)$

29-3 15.7×10^{-6} seconds; 4.66×10^3 meters

29-5   Yes

29-7   With $\eta = 1 + (Vv_x'/c^2)$: $a_x = a_x'/\gamma^3\eta^3$, $a_y = [a_y' - (V/c^2\eta)v_y'a_x']/\gamma^2\eta^2$,
$a_z = [a_z' - (V/c^2\eta)v_z'a_x']/\gamma^2\eta^2$

29-17   With $\epsilon = 1 - (\beta v_x/c)$: $f_x' = f_x - (V/c^2\epsilon)(v_y f_y + v_z f_z)$, $f_y' = f_y/\gamma\epsilon$, $f_z' = f_z/\gamma\epsilon$,
$(dW'/dt') = [(dW/dt) - Vf_x]/\epsilon$

29-19 $(PE)_0 = M_0 c^2\{[1 - (v_0^2/c^2)]^{-(1/2)} - 1\}$, $W_S = M_{0S}c^2[1 - (V^2/c^2)]^{-(1/2)}$,
$P_S = M_{0S}V[1 - (V^2/c^2)]^{-(1/2)}$, where $M_{0S} = M_0/[1 - (v_0^2/c^2)]^{1/2} = M_0 + [(PE)_0/c^2]$
$= $ (inertial rest mass) $+$ (mass equivalent of initial potential energy)

29-27 $\mathbf{F}_{\text{on }2} = (\gamma_1 q_1 q_2/4\pi)[\gamma_1^2 R_x^2 + R_y^2 + R_z^2]^{-3/2} \times \{(\mathbf{R}/\epsilon_0) + \mu_0\mathbf{v}_2 \times (\mathbf{v}_1 \times \mathbf{R})\}$ where $\mathbf{R} = \mathbf{r}_2 - \mathbf{r}_1$ and $\gamma_1 = [1 - (v_1^2/c^2)]^{-(1/2)}$

29-29   Outside: $\mathbf{B} = 0$, $\mathbf{E} = 0$; inside: $\mathbf{B} = \gamma\mu_0 n'I'\hat{\mathbf{y}}$, $\mathbf{E} = -\gamma V\mu_0 n'I'\hat{\mathbf{z}}$; no

29-31 $(q^2/6\pi\epsilon_0 c^3)[1 - (v^2/c^2)]^{-3}\{[\mathbf{a}]^2 - ([\mathbf{v}] \times [\mathbf{a}])^2/c^2\}$

A-1 $y = (qE/2m_0 v_0^2)x^2$; $d = qED^2/2m_0 v_0^2$; $\tan\alpha = qED/m_0 v_0^2$

A-3 $T = \frac{1}{2}q\Delta\phi[\ln(b/a)]^{-1}$

A-5 $m_0 = qB^2 D^2/8\Delta\phi$

A-7 $\Delta l = 4\pi(m_0 v_0/qB)\sin\theta_0\sin\Delta\theta$; $\Delta l/l_0 = 2\tan\theta_0\sin\Delta\theta$; $\Delta\theta \leq 8.15°$

A-11 $x = (\pi^2/72)(m_0/q)(E/B^2)$, $y = 0.134(m_0 v_0/qB)$

A-13   With $A = (qE/m_0 c)[1 - (v_0^2/c^2)]^{(1/2)}$: $z = 0$, $y = (c/A)\{[1 + (At)^2]^{1/2} - 1\}$,
$x = (v_0/A)\ln\{At + [1 + (At)^2]^{1/2}\} = (v_0/A)\sinh^{-1}(At)$

B-1 $\alpha = 3.42 \times 10^{-41}$ farad-(meter)^2; $a = 6.75 \times 10^{-11}$ meter

B-7 2.3×10^{-3}

B-9   1.000240; 0.001 percent

B-11 $(\mu_0/12\pi c)(e^2/m_e)^2\omega^4|E_{p0}|^2[(\omega_0^2 - \omega^2)^2 + (\gamma\omega)^2]^{-1}$

INDEX

CONVERSION OF SYMBOLS IN EQUATIONS

Symbols representing essentially mechanical quantities (length, mass, time, force, work, energy, power, etc.) are not changed (nor are derivatives). To convert an equation written in the MKSA system to the corresponding one in the Gaussian system, replace the symbol listed under the column labeled MKSA by that listed under Gaussian. The entries can also be used to convert a Gaussian equation to an MKSA one by going from right to left in the table.

Quantity	MKSA	Gaussian
Capacitance	C	$4\pi\epsilon_0 C$
Charge	q	$(4\pi\epsilon_0)^{1/2} q$
Charge density	$\rho, (\sigma, \lambda)$	$(4\pi\epsilon_0)^{1/2}\rho, (\sigma, \lambda)$
Conductivity	σ	$4\pi\epsilon_0\sigma$
Current	I	$(4\pi\epsilon_0)^{1/2} I$
Current density	$\mathbf{J}, (\mathbf{K})$	$(4\pi\epsilon_0)^{1/2}\mathbf{J}, (\mathbf{K})$
Dielectric constant	κ_e	ϵ
Dipole moment (electric)	$\mathbf{p}$	$(4\pi\epsilon_0)^{1/2}\mathbf{p}$
Dipole moment (magnetic)	$\mathbf{m}$	$(4\pi/\mu_0)^{1/2}\mathbf{m}$
Displacement	$\mathbf{D}$	$(\epsilon_0/4\pi)^{1/2}\mathbf{D}$
Electric field	$\mathbf{E}$	$(4\pi\epsilon_0)^{-1/2}\mathbf{E}$
Inductance	L	$(4\pi\epsilon_0)^{-1} L$
Magnetic field	$\mathbf{H}$	$(4\pi\mu_0)^{-1/2}\mathbf{H}$
Magnetic flux	Φ	$(\mu_0/4\pi)^{1/2}\Phi$
Magnetic induction	$\mathbf{B}$	$(\mu_0/4\pi)^{1/2}\mathbf{B}$
Magnetization	$\mathbf{M}$	$(4\pi/\mu_0)^{1/2}\mathbf{M}$
Permeability	μ	(1) $\kappa_m\mu_0$, then
		(2) $\kappa_m \to \mu$
Permeability (relative)	κ_m	μ
Permittivity	ϵ	(1) $\kappa_e\epsilon_0$, then
		(2) $\kappa_e \to \epsilon$
Polarization	$\mathbf{P}$	$(4\pi\epsilon_0)^{1/2}\mathbf{P}$
Resistance	R	$(4\pi\epsilon_0)^{-1} R$
Resistivity	ρ	$(4\pi\epsilon_0)^{-1}\rho$
Scalar potential	ϕ	$(4\pi\epsilon_0)^{-1/2}\phi$
Speed of light	$(\mu_0\epsilon_0)^{-1/2}$	c
Susceptibility	$\chi_e, (\chi_m)$	$4\pi\chi_e, (\chi_m)$
Vector potential	$\mathbf{A}$	$(\mu_0/4\pi)^{1/2}\mathbf{A}$